The Initiation of DNA Replication in Eukaryotes

Daniel L. Kaplan
Editor

The Initiation of DNA Replication in Eukaryotes

Springer

Editor
Daniel L. Kaplan
Florida State Univ. College of Medi
Tallahassee, FL, USA

ISBN 978-3-319-24694-9 ISBN 978-3-319-24696-3 (eBook)
DOI 10.1007/978-3-319-24696-3

Library of Congress Control Number: 2015954360

Springer Cham Heidelberg New York Dordrecht London

Printed on acid-free paper

Springer International Publishing AG Switzerland is part of Springer Science+Business Media (www.springer.com)

Contents

Contributors

Hiroyuki Araki Department of Genetics, SOKENDAI, Mishima, Shizuoka, Japan

Division of Microbial Genetics, National Institute of Genetics, Mishima, Shizuoka, Japan

Ryan M. Baxley Biochemistry, Molecular Biology and Biophysics, University of Minnesota, Minneapolis, MN, USA

Anja-Katrin Bielinsky Biochemistry, Molecular Biology and Biophysics, University of Minnesota, Minneapolis, MN, USA

Dominik Boos Vertebrate DNA Replication, Centre for Medical Biotechnology (ZMB), University of Duisburg-Essen, Essen, Germany

Arturo Calzada Department of Microbial Biotechnology, Centro Nacional de Biotecnologia (CNB), Consejo Superior de Investigaciones Cientificas (CSIC), Madrid, Spain

Patrizia Di Crescenzio Structural biology laboratory, Elettra-Sincrotrone Trieste S.C.p.A., Trieste, Italy

Dipartimento di Scienze della Vita, Università degli Studi di Trieste, Trieste, Italy

Nalini Dhingra Department of Biomedical Sciences, Florida State University College of Medicine, Tallahassee, FL, USA

Anne D. Donaldson Institute of Medical Sciences, University of Aberdeen, Aberdeen, UK

Roberto A. Donnianni Department of Microbiology & Immunology, Columbia University Medical Center, New York, NY, USA

Pedro Ferreira Molecular Genetics II, Vertebrate DNA Replication, Centre for Medical Biotechnology (ZMB), University of Duisburg-Essen, Essen, Germany

Catherine A. Fox Department of Biomolecular Chemistry, School of Medicine and Public Health, University of Wisconsin, Madison, WI, USA

Agnieszka Gambus School of Cancer Sciences, University of Birmingham, Birmingham, UK

David M. Gilbert Department of Biological Science, Florida State University, Tallahassee, FL, USA

Serena Giovinazzi Department of Biomedical Sciences, College of Medicine, Florida State University, Tallahassee, FL, USA

Akash Gunjan Department of Biomedical Sciences, College of Medicine, Florida State University, Tallahassee, FL, USA

Monica P. Gutiérrez University Program in Genetics and Genomics, Duke University Medical Center, Durham, NC, USA

Shin-ichiro Hiraga Institute of Medical Sciences, University of Aberdeen, Aberdeen, UK

Timothy Hoggard Department of Biomolecular Chemistry, School of Medicine and Public Health, University of Wisconsin, Madison, WI, USA

Manzar Hossain Cold Spring Harbor Laboratory, Cold Spring Harbor, NY, USA

Olivier Hyrien Institut de Biologie de l'Ecole Normale Superieure (IBENS), CNRS UMR8197, Paris, France

Kerstin Köhler Molecular Genetics II, Vertebrate DNA Replication, Centre for Medical Biotechnology (ZMB), University of Duisburg-Essen, Essen, Germany

Daniel L. Kaplan Department of Biomedical Sciences, Florida State University College of Medicine, Tallahassee, FL, USA

Kyle N. Klein Department of Biological Science, Florida State University, Tallahassee, FL, USA

Huilin Li Biosciences Department, Brookhaven National Laboratory, Upton, NY, USA

Department of Biochemistry and Cell Biology, Stony Brook University, Stony Brook, NY, USA

Ivan Liachko Department of Genome Sciences, University of Washington, Seattle, WA, USA

David M. MacAlpine Department of Pharmacology and Cancer Biology, Duke University Medical Center, Durham, NC, USA

Matteo De March Structural biology laboratory, Elettra-Sincrotrone Trieste S.C.p.A., Trieste, Italy

Hisao Masai Department of Genome Medicine, Tokyo Metropolitan Institute of Medical Science, Tokyo, Japan

Barbara Medagli Structural biology laboratory, Elettra-Sincrotrone Trieste S.C.p.A., Trieste, Italy

Conrad A. Nieduszynski Sir William Dunn School of Pathology, University of Oxford, Oxford, UK

Silvia Onesti Structural biology laboratory, Elettra-Sincrotrone Trieste S.C.p.A., Trieste, Italy

Johanna Paik Department of Biomedical Sciences, College of Medicine, Florida State University, Tallahassee, FL, USA

Boris Pfander DNA Replication and Genome Integrity, Max Planck Institute of Biochemistry, Martinsried, Munich, Germany

Giacomo De Piccoli Warwick Medical School, University of Warwick, Coventry, UK

Helmut Pospiech Research Group Biochemistry, Leibniz Institute on Aging-Fritz Lipmann Institute, Jena, Germany

Faculty of Biochemistry and Molecular Medicine, University of Oulu, Oulu, Finland

M.K. Raghuraman Department of Genome Sciences, University of Washington, Seattle, WA, USA

Dirk Remus Molecular Biology Program, Memorial Sloan-Kettering Cancer Center, New York, NY, USA

Karl-Uwe Reußwig DNA Replication and Genome Integrity, Max Planck Institute of Biochemistry, Martinsried, Munich, Germany

Alberto Riera DNA Replication Group, Institute of Clinical Science, Faculty of Medicine, Imperial College London, London, UK

Daniel Rossbach Biochemistry and Molecular Genetics, University of Colorado School of Medicine, Aurora, CO, USA

Nanda Kumar Sasi Laboratory of Genome Integrity and Tumorigenesis, Van Andel Research Institute, Grand Rapids, MI, USA

Graduate Program in Genetics, Michigan State University, East Lansing, MI, USA

Robert A. Sclafani Biochemistry and Molecular Genetics, University of Colorado School of Medicine, Aurora, CO, USA

Christian Speck DNA Replication Group, Institute of Clinical Sciences, Faculty of Medicine, Imperial College London, London, UK

Michael J.R. Stark Centre for Gene Regulation & Expression, College of Life Sciences, University of Dundee, Dundee, UK

Bruce Stillman Cold Spring Harbor Laboratory, Cold Spring Harbor, NY, USA

Jingchuan Sun Biosciences Department, Brookhaven National Laboratory, Upton, NY, USA

Lorraine S. Symington Department of Microbiology & Immunology, Columbia University Medical Center, New York, NY, USA

Anna Szambowska Research Group Biochemistry, Leibniz Institute on Aging-Fritz Lipmann Institute, Jena, Germany

Seiji Tanaka Department of Genetics, SOKENDAI, Mishima, Shizuoka, Japan

Division of Microbial Genetics, National Institute of Genetics, Mishima, Shizuoka, Japan

Yee Mon Thu Biochemistry, Molecular Biology and Biophysics, University of Minnesota, Minneapolis, MN, USA

Lei Wei Gerstner Sloan Kettering Graduate School of Biomedical Sciences, New York, NY, USA

Molecular Biology Program, Memorial Sloan Kettering Cancer Center, New York, NY, USA

Michael Weinreich Laboratory of Genome Integrity and Tumorigenesis, Van Andel Research Institute, Grand Rapids, MI, USA

Satoshi Yamazaki Department of Genome Medicine, Tokyo Metropolitan Institute of Medical Science, Tokyo, Japan

Naoko Yoshizawa-Sugata Department of Genome Medicine, Tokyo Metropolitan Institute of Medical Science, Tokyo, Japan

Zuanning Yuan Biosciences Department, Brookhaven National Laboratory, Upton, NY, USA

Department of Biochemistry and Cell Biology, Stony Brook University, Stony Brook, NY, USA

Xiaolan Zhao Gerstner Sloan Kettering Graduate School of Biomedical Sciences, New York, NY, USA

Molecular Biology Program, Memorial Sloan Kettering Cancer Center, New York, NY, USA

Chapter 1
Introduction to Eukaryotic DNA Replication Initiation

Nalini Dhingra and Daniel L. Kaplan

Abstract Every time a cell divides, a copy of its genomic DNA has to be faithfully copied to generate new genomic DNA for the daughter cells. The process of DNA replication needs to be precisely regulated to ensure that replication of the genome is complete and accurate, but that re-replication does not occur. Errors in DNA replication can lead to genome instability and cancer. The process of replication initiation is of paramount importance, because once the cell is committed to replicate DNA, it is optimal to complete replication with minimal errors. Furthermore, agents that inhibit DNA replication initiation are now being targeted for cancer therapy. A great deal of progress has been made in understanding how DNA replication is initiated in eukaryotic cells in the past 10 years. This chapter introduces how the position of replication initiation, called the replication origin, is chosen. This chapter also introduces how replication initiation is integrated with the phases of the cell cycle, and how replication initiation is regulated in the case of damage to DNA. It is the cellular protein machinery that enables replication initiation to be activated and regulated. We now have an in-depth understanding of how cellular proteins work together to start DNA replication. A mechanistic description of DNA replication initiation is introduced in this chapter as well.

Keywords Origin • Initiation • DNA replication • Helicase • Polymerase • Kinase • DNA damage • Checkpoint • Cell cycle • Replication fork

Introduction

Eukaryotic replication is a highly controlled process and is tightly regulated to ensure that chromosomes duplicate only once per cell cycle and that the genomic stability of a cell is maintained. Replication occurs in three distinct steps: initiation,

N. Dhingra • D.L. Kaplan (✉)
Department of Biomedical Sciences, Florida State University College of Medicine, 1115 W. Call St., Tallahassee, FL 32306, USA
e-mail: nalini.dhingra@med.fsu.edu; Daniel.kaplan@med.fsu.edu

D.L. Kaplan (ed.), *The Initiation of DNA Replication in Eukaryotes*,
DOI 10.1007/978-3-319-24696-3_1

elongation, and termination. To ensure that DNA replication occurs only once per cell cycle and that the entire DNA is faithfully replicated, a cell employs various control mechanisms at various steps that lead to the initiation of DNA replication. In general, the initiation process is divided into two distinct phases:

1. Origin licensing: which occurs in the late mitosis or early G_1 phase where the Mcm2-7 complex (which forms the core of the replication fork helicase) is assembled onto replication origins as a double hexamer in the inactive form.
2. Conversion of the Mcm2-7 complex to an active replication fork helicase, which occurs in the S phase.

The ORC protein present at the replication origins facilitates the loading of Mcm2-7 double hexamer onto origin. The loaded Mcm2-7 double hexamer encircles double-stranded DNA, which then dissociates to form bidirectional replication forks. In this chapter, we briefly discuss how origins are defined and activated in eukaryotes. We also discuss the mechanism of origin licensing and helicase formation. Finally, we also mention the various mechanisms employed by a living cell in situations of DNA damage and replication stress (Fig. 1.1).

Origin Selection

Replication initiates at distinct DNA regions called the origins of replication initiation (*ori*). Replication initiator proteins bind at these replication origins. Origin sequences in budding yeast are characterized by autonomously replicating sequences (ARS), present at an interval of 30 kbps throughout the chromosome [1]. There are estimated to be approximately 400 ARSs in the yeast genome. Each ARS is 100–200 bps long and is characterized by the presence of A, B1, B2, and B3 elements. The A and B1 elements are highly conserved and form the binding site for the initiator protein, the origin recognition complex (ORC). The B element consists of a region of helical instability that helps in the unwinding of DNA. The A element also contains the ARS consensus sequence (ACS), which is an 11 bp region rich in adenines and thymines and is required for the ARS function [2]. However, a match to the ACS is not sufficient for origin function, as there are more than 12,000 potential matches for ACS in the yeast genome, pointing out a need for additional sequence or chromatin requirements for defining replication origin [3]. In contrast, origin sequences in higher eukaryotes including fission yeast are not defined by consensus sequences but are rather defined by chromatin structure and epigenetic modifications [4]. Due to the lack of a consensus sequence, the ORC protein complex in metazoans may be targeted to specific sites (mainly the transcription start sites of actively transcribed genes) by various protein factors. In mammalian cells, many potential genetic and epigenetic determinants for replication origins have been reported using genome-wide mapping techniques [5]. These studies also show the enrichment of origins near active promoter elements at CpG islands [6]. Nucleosome positioning on chromatin is also a defining feature of replication origins.

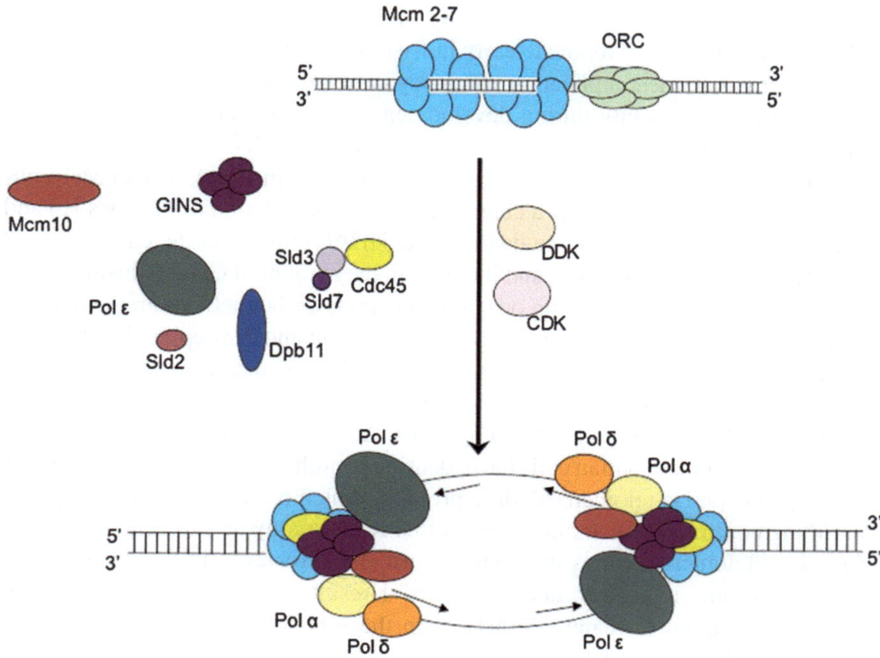

Fig. 1.1 A schematic of replication initiation in *Saccharomyces cerevisiae*. At replication origins, Mcm2-7 is present as a double hexamer encircling double-stranded DNA in G_1 phase. Along with CDK and DDK, the cell employs a complex machinery of protein factors that facilitate the formation of bidirectional replication forks in S phase. Some of these factors become a part of the replication fork while others do not travel with the replication fork. Cdc45-Mcm2-7-GINS (CMG) forms the active replication fork helicase, unwinding double-stranded DNA to generate single-stranded templates for the replicative polymerases. Pol δ is devoted to replication of the lagging strain, and Pol ε is devoted to synthesis of the leading strand. Sld3, Sld2, and Dpb11 are required for initiation, but these three proteins do not travel with the replication apparatus. CDK phosphorylates Sld2 and Sld3 to form a ternary complex with Dpb11, while DDK phosphorylates subunits of the Mcm2-7 complex

Various studies in yeast and higher eukaryotes show a decrease in nucleosome occupancy at origins. This nucleosome-free region may be a determinant of ORC binding and may also facilitate the loading of Mcm2-7 complexes [7].

Origin Activation

The replication origins are further classified as early- or late-replicating origins based on their timing of replication. Various studies show strong correlation between replication timing and chromatin structure [8]. Studies on replication timing in budding yeast have revealed early replication of origins in the centromeric region and late replication of origins in the subtelomeric regions. This repression of origin activation at telomeres is attributed mainly to the local chromatin structure [7, 9]. Centromeric regions also replicate early in fission yeast [9].

In higher eukaryotes, replication efficiency correlates with histone modification and transcriptional activity during development such that replication of gene-rich regions occurs earlier in S phase. Studies have shown that local changes in histone acetylation can also alter the replication program. For instance, sequences replicated in early S phase from HeLa cells exhibit hyperacetylation of histones H3 and H4, depletion of Rpd3 (a histone deacetylase, HDAC) results in the early firing of normally late-activating origins, and increase in the amount of Gcn5 (a histone acetyltransferase) around a late-firing origin results in its earlier activation. Local chromatin environment may also regulate ORC recruitment and Pre-replication complex (pre-RC) assembly [6, 7]. Replication timing is also regulated by various protein factors. Forkhead box (Fox) transcription factors, Fkh1 and Fkh2, have been shown to regulate replication timing in budding yeast. Advanced studies in budding yeast, fission yeast, and mammalian cells have identified another protein factor Rif1 (Rap1 interacting factor 1-a telomere-binding protein), which has a broad role in replication timing control. These studies suggest that Rif1 directly acts to delay the origin firing of subtelomeric origins in budding yeast [10]. However, in fission yeast and higher eukaryotic organisms, deletion of Rif 1 resulted in an advanced timing of replication initiation of many late origins in the subtelomeric as well as internal chromosomal loci while delaying the activation of many early origins. Rif1 along with Taz1 also regulates the timing of Cdc45-Sld3 loading in G1 phase. Cdc45-Sld3 loading to replication origins has been suggested to be a limiting step controlling replication timing. Studies in several replication systems have shown that Cdc45 along with other initiation factors (Sld2, Sld3, Dpb11, DDK) are rate limiting for replication initiation [9, 11]. Finally, the DNA replication checkpoint also regulates replication timing, as it is known to suppress the activation of late origins in response to replication stress.

In general, early-replicating domains are euchromatin DNA regions localized in the interior of the nucleus, and characterized with a high gene density and high GC content. Whereas late-replicating domains characterized by fewer genes are packaged into heterochromatin and localized to the nuclear periphery. This temporal organization of genome replication, which is cell type specific, allows the cell to balance replication with limiting resources such as initiation factors and nucleotide pool and is conserved from yeast to humans, suggesting that timing of origin firing is regulated independently of origin selection [5, 12, 13].

Replication and Chromatin

The basic unit of chromatin is the nucleosome that comprises of a histone octamer (consisting of two molecules each of histones H2A, H2B, H3, and H4) and 147 bps of DNA wrapped around the octamer 1.7 times. The epigenetic state of chromatin is defined by DNA methylation and posttranslational modifications (PTMs) of histones like acetylation, ribosylation, ubiquitination, and SUMOylation. These histones and various histone variants contribute to the diverse chromatin structure and are deposited in a replication-dependent or -independent manner. It is well established that the diversity in chromatin structure and modifications influence the selection and activity of replication origins. However, replication fork progression leads to the disruption of existing chromatin structure by removal of nucleosomes from the DNA. Thus, it is essential to reestablish the epigenetic information on the newly synthesized chromatin. Chromatin reassembly on nascent strand occurs via two pathways: in the first, parental histones generated by the disruption of nucleosomes are recycled behind the fork and in the second pathway newly synthesized histones are deposited onto nascent DNA [14]. Various biochemical and genetic studies have identified chromatin remodelers that contribute to the disruption and assembly of the chromatin structure during replication and also maintain its epigenetic states. Some of these remodelers include the ATP-utilizing chromatin assembly and remodeling factor (ACF) and the INO80 complex and its catalytic subunit SNF2. Biochemical studies have also identified histone chaperones that are responsible for the deposition of histones onto replicating DNA. Some of these chaperones are the human chromatin assembly factor-1 (CAF-1), antisilencing function 1 (ASF1), the nucleosome assembly protein 1 (NAP1), and the nuclear autoantigenic sperm protein (NASP) [7].

Role of ORC in Replication Initiation

ORC is a hetero-hexameric protein complex, consisting of the Orc1, Orc2, Orc3, Orc4, Orc5, and Orc6 subunits, that is conserved throughout species. It was first identified and purified from budding yeast as a factor that remained bound to the ACS upon DNaseI digestion [15]. ORC is present on the origin sequences throughout the cell cycle and binds DNA in an ATP-dependent manner. In contrast to budding yeast, the fission yeast and metazoan ORC complex binds replication origins periodically during the cell cycle. Orc1, Orc4, and Orc5 have AAA+ATPase domains [16]. The ORC protein complex functions as a scaffold for the recruitment of Cdc6, Cdt1, and Mcm2-7 complex in G_1 phase, which together with the ORC complex form the pre-RC. In addition to its role in recruiting the Mcm2-7 complex to origins, Orc1 protein has also been shown to interact with histone H4 via its conserved bromo-adjacent homology (BAH) domain in both yeast and human cells. This interaction may be involved in the local chromatin organization at some replication origins, affecting their activity [7]. The Orc1 subunit of human ORC complex also associates with centrosomes and is involved in duplication of centrioles. It controls the cyclin E-CDK-dependent reduplication of centrioles [17]. Genome-wide studies with budding yeast have classified ORC binding to the replication origins as either DNA dependent or chromatin dependent with chromatin-dependent ORC-binding origins being associated with early activation [18]. Studies in *Drosophila* and humans have shown that some of the ORC subunits interact with the heterochromatin protein (HP1) and maintain the heterochromatic environment [6].

ORC along with various chromatin remodelers also facilitates the positioning of nucleosomes at the origins during their activation, thus remodeling the chromatin, which may be critical for the assembly of pre-RC [6].

Mutations in the *ORC1*, *ORC4*, and *ORC6* subunits of human ORC complex have been associated with a rare autosomal recessive disorder called the Meier-Gorlin syndrome (MGS). This disorder is characterized by postnatal growth retardation, dwarfism, microcephaly, and developmental abnormalities in the ear and patella. Cells from these patients have a delayed cell cycle progression resulting in reduced cell number [19]. Studies have identified mutations R105Q and E127G in Orc1, which affect the centriole copy number and cause centrosome reduplication in human cells. This may contribute to dwarfism and microcephaly [20]. Studies in zebrafish show that mutations in the H4K20me2 binding pocket of Orc1 also influence the recruitment of ORC onto replication origins, resulting in a diminished pre-RC assembly [21]. Interestingly, depletion of Orc1 in zebrafish embryos resulted in abnormal body curvature and reduced viability. This defect might be a direct consequence of impaired origin licensing [22]. A missense mutation in Tyr174 of human Orc4 was also found in patients with MGS. This residue is present in the highly conserved region of AAA+ATPase domain of Orc4. Mutation studies of the orthologous residue of Tyr174 present in budding yeast (Tyr232) demonstrate that the strain exhibits a reduced growth rate with a defect in S-phase progression [23]. In Drosophila, a mutation in the C-terminal region of Orc6, which is implicated in MGS, has been shown to impair binding of Orc6 to the rest of the Orc complex, thus preventing the loading of ORC onto replication origins [24]. Additional mutations in MGS patients were also identified in *CDC6* and *CDT1* genes.

Role of Mcm2-7 Complex in Replication Initiation

Mcm2-7 is a hexameric protein complex that consists of six distinct but evolutionarily related Mcm (minichromosome maintenance) proteins having an ATPase domain at their C terminal end. These Mcm proteins were first isolated in a screen for yeast mutants that were defective in the maintenance of circular plasmids containing an ARS sequence [25]. The six subunits of Mcm2-7 complex are assembled as a ring in the order Mcm3-Mcm5-Mcm2-Mcm6-Mcm4-Mcm7 forming the core of the eukaryotic replication fork helicase [26]. The ATPase active sites in Mcm2-7 complex are formed at the dimer interfaces with one subunit contributing the Walker A motif and the adjoining subunit contributing an essential arginine [27]. A study with budding yeast proteins shows that Mcm2-7 complex by itself has a weak helicase activity in vitro, which depends on the specific buffer conditions. It was also demonstrated that the Mcm2-7 complex has an ATP-regulated gate at the Mcm2/Mcm5 interface that might facilitate the loading of Mcm2-7 at replication origins or extrusion of single-stranded DNA during replication initiation [27]. In addition, ATP binding also plays a role in the stabilization of the Mcm complex. ATP hydrolysis by Mcm further facilitates the assembly of Mcm2-7 double hexamers on DNA [28].

The Mcm2-7 is loaded as a double hexamer in an ATP-dependent manner to surround double-stranded DNA, in a process called the licensing of replication origins. However, the number of loaded Mcm2-7 complexes is way more than actually required to establish replication forks. These additional copies of Mcm2-7 may be used to establish new forks in case of replication fork stalling due to DNA damage [29, 30].

Licensing of Replication Origins

Origin licensing is best studied in the budding yeast system. In the late M and G_1 phase, Cdc6 (also a AAA + ATPase protein) binds to the ORC protein complex and together they function with Cdt1 to load Mcm2-7 double hexamer onto origin sequences. These four factors together form the pre-RC. Earlier studies showed a concerted loading of the Mcm2-7 double hexamer onto double-stranded DNA [31]; however recent studies support a step-by-step loading of the two hexamers. ORC binds to Cdc6 in an ATP-dependent manner and together they recruit Cdt1-Mcm2-7 to form an OCCM (Orc-Cdc6-Cdt1-Mcm2-7) complex in the absence of ATP hydrolysis. This recruitment of Mcm2-7 is facilitated by Mcm3 and Cdt1. The Cdt1-Mcm2-7 interacts with ORC-Cdc6 via the C-terminal region of the Mcm2-7 hexamer leaving its N-terminal region free to bind the second Mcm2-7 hexamer. Once the OCCM is formed, ATP hydrolysis of Orc1 and Cdc6 causes the release of Cdt1 from the OCCM complex to produce an OCM (Orc-Cdc6-Mcm2-7) complex. This OCM complex, which is a transient and salt-sensitive intermediate, is capable of recruiting a second Mcm2-7 hexamer via the N-terminal domain (NTD) region of the first loaded Mcm2-7 hexamer. This results in the formation of another intermediate complex, the OCMM (Orc-Cdc6-Mcm2-7-Mcm2-7) complex in which the two hexamers are associated to each other head to head via their N-terminal region. This loaded double hexamer appears to have a twisted structure which is relaxed in the active helicase [29, 32]. Recent studies have also revealed that during licensing, the Mcm2-7 hexamer is loaded onto the double-stranded DNA via the interface between Mcm2 and Mcm5 at the stage of OCCM formation prior to ATP hydrolysis, thus separating the two events of helicase loading and double hexamer formation [33]. The loaded Mcm2-7 double hexamer is then transformed into an active replication fork helicase during S phase by the action of various protein factors.

Helicase Activation

The Mcm2-7 complex is activated in S phase by the action of two kinases, the cyclin-dependent kinase (S-CDK) and the Dbf4-dependent kinase (DDK) along with a number of other protein factors, some of which also travel along the replication fork. Studies in budding yeast have identified a number of initiation factors, namely Cdc45, GINS [Go-Ichi-Ni-San, Japanese for 5-1-2-3, for Sld5, Psf1 (partner with Sld5), Psf2, Psf3], Sld7, Sld3, Sld2, Dpb11, Pol-ε, RPA, and Mcm10. All these proteins together form the pre-initiation complex (Pre-IC) that is finally converted to an active helicase composed of Cdc45, GINS, and Mcm2-7 (CMG complex), through a series of highly regulated molecular events. Formation of the CMG complex constitutes two complex reactions. First, the Mcm2-7 double hexamer dissociates to form two single Mcm2-7 hexamers. Second, the Mcm2-7 ring opens for single-stranded DNA extrusion and then closes such that a single hexamer of Mcm2-7 is present around single-stranded DNA along with Cdc45 and GINS. Once the CMG complex is formed, the different polymerases are recruited to single-stranded DNA at the origins to start DNA synthesis and DNA replication proceeds bidirectionally. Pol-α synthesizes short DNA strands, while Pol-ε and Pol-δ elongate the leading and lagging strand, respectively.

The Dbf4-Dependent Kinase (DDK)

DDK consists of a catalytic subunit (Cdc7) and a regulatory subunit (Dbf4). The Cdc7 subunit is stable throughout the cell cycle, whereas the level of Dbf4 is regulated such that it remains high during S phase and then decreases in the late M and G_1 phase [34]. Dbf4 becomes ubiquitinated and is subjected to proteasomal degradation by the anaphase-promoting complex (APC/C) in the late M and G_1 phase. However, in the S phase, APC/C remains inactive allowing the accumulation of Dbf4 [35]. Studies in budding yeast identified a mutant in MCM5 called the *mcm5-bob1* that can bypass the requirement of DDK for replication initiation [36]. The *mcm5-bob1* mutation causes a conformational change in the Mcm2-7 ring such that it allows for the binding of Cdc45 protein in the early G_1 phase, suggesting that DDK may be required for Cdc45 binding at the origins [37]. It has also been shown that DDK phosphorylates Mcm2-7 complex at its Mcm2, Mcm4, and Mcm6 subunits at their amino terminals. Phosphorylation at the amino terminal serine/threonine domain (NSD) of Mcm4 has been shown to alleviate an inhibitory activity at Mcm4 [38, 39]. These studies suggest that even though Mcm5 is not directly phosphorylated by DDK, the phosphorylation of other Mcm subunits is sufficient to cause a conformational change in Mcm5, such that Cdc45 binds Mcm2-7 [40, 41]. A recent study in budding yeast demonstrates how the *mcm5-bob1* mutation bypasses the requirement for DDK phosphorylation of Mcm subunits. This report shows that Mcm2 phosphorylation by DDK is essential for cell growth and DNA replication. Absence of Mcm2 phosphorylation by DDK also results in a decreased amount of origin single-stranded DNA in S phase in contrast to the cells with the *mcm5-bob1* mutation. Dbf4-Cdc7 phosphorylation of Mcm2 weakens its interaction with Mcm5 and helps in the opening of Mcm2-7 ring at the Mcm2/Mcm5 gate, to allow for the extrusion of single-stranded DNA from the central channel of Mcm2-7. Similarly, cells with the *mcm5-bob1* mutation also exhibit a weak interaction between Mcm2 and Mcm5, suggesting that the *mcm5-bob1* mutation bypasses the requirement of DDK phosphorylation of Mcm2 by an alternate mechanism that leads to the Mcm2-7 ring opening [42]. Other independent studies have also shown that DDK facilitates the association of Sld3, Sld7, and Cdc45 with the Mcm2-7 complex [34]. It has also been suggested that DDK might facilitate Mcm2-7 double hexamer dissociation to form Mcm2-7 single hexamer, prior to single-stranded DNA extrusion [35].

As mentioned above, DDK-dependent phosphorylation of Mcm4 is one of the key events for pre-RC formation. However, recent studies have shown that this event is under the control of Rif1-mediated phosphatase action and loss of Rif1 partially compensates for impaired DDK function [43, 44]. Rif1 was mentioned earlier as a factor that regulates replication timing. Budding yeast Rif1 has also been shown to have protein phosphatase 1 (PP1) docking motif. Rif1 binds Glc7 (the budding yeast PP1) and recruits it to telomeres and possibly to late origins of DNA replication. This is important to maintain the replication timing at telomeres, as the replication timing at budding yeast telomeres advanced when the ability of Rif1 to recruit Glc7 was compromised [44]. It has been suggested that in G_1 phase Rif1 recruits Glc7 onto chromatin and directs it to dephosphorylate Mcm4, thus preventing early initiation of replication. However, in S phase, when DDK levels are high, Rif1 gets phosphorylated by DDK and releases Glc7, thus favoring DDK-dependent phosphorylation of Mcm4 and consequent origin activation. PP1 interaction motifs of Rif1 are conserved from yeast through higher eukaryotes. Studies in *Xenopus* and human cells also support the model, where Rif1 may counteract DDK-dependent phosphorylation of Mcm4 by targeting PP1 to dephosphorylate Mcm4 [43].

S Phase-Cyclin-Dependent Kinase

S phase-cyclin-dependent kinase (S-CDK) inhibits origin licensing and promotes DNA replication during S phase. Its concentration is regulated by APC/C-mediated proteasomal degradation [35]. S-CDK inhibits origin licensing by phosphorylating Cdc6, which results in its SCF-dependent degradation. It phosphorylates Mcm3, which causes the nuclear export of Mcm2-7/Cdt1. S-CDK sterically inhibits ORC function by binding to the Orc6 RXL motif (a cyclin-binding motif). Finally, it also phosphorylates Orc2 and Orc6, thus inhibiting the interaction of ORC with Cdt1 [45].

During replication initiation, budding yeast S-CDK (Clb5-Cdc28 and Clb6-Cdc28) phosphorylates Sld2 and Sld3 to facilitate their interaction with Dpb11. Dpb11 consists of two pairs of BRCT (BRCA1 C-terminus) domains that bind phosphorylated proteins. The N-terminal pair of BRCT domains binds phosphorylated Sld3, while the C-terminal pair binds phosphorylated Sld2 [46]. Phosphorylation of Thr84 of Sld2 is essential for its association with Dpb11. Phosphorylation of Thr84 also stimulates Sld2 association with ssDNA [47]. However, Thr84 phosphorylation requires prior phosphorylation of other Sld2 sites. Sld2 has a cluster of 11 CDK phosphorylation motifs. This pre-phosphorylation of Sld2 causes a conformational change in Sld2 protein in order to expose Thr84 to CDK activity, thereby facilitating its phosphorylation. The multisite phosphorylation of Sld2 also creates a high threshold for CDK activity that prevents premature replication. Sld3 has 12

CDK phosphorylation sites. However, binding of Sld3 to Dpb11 requires the simultaneous phosphorylation of Thr600 and Ser622, which may require high CDK activity [34]. The S-CDK-dependent formation of Dpb11-Sld3-Sld2 complex is an essential step during replication initiation. In vivo studies in budding yeast have shown that the fusion of Sld3-Dpb11 when combined with Sld2T84D (a phosphomimetic mutant of Sld2) bypasses the requirement for S-CDK [48]. The phosphorylation-dependent interaction of Sld2-Dpb11 is also important for the formation of a pre-loading complex (pre-LC), which consists of Sld2, Dpb11, GINS, and Pol-ε. The pre-LC is an important intermediate that may facilitate GINS loading to the origin [49].

Roles of Sld2, Sld3, and Dpb11 in Replication Initiation

The budding yeast proteins Sld2, Sld3, and Dpb11 are essential proteins required for the initiation of DNA replication. These proteins however do not travel along the replication fork. *DPB11* (DNA polymerase B-binding protein subunit 11) was first isolated as a multicopy suppressor of mutations in the *DPB2* subunit of Pol-ε and was shown to have a dual role in chromosomal replication and at the cell cycle checkpoint [50]. *SLD2* (synthetically lethal with *dpb11-1*) and *SLD3* were isolated in screens for identifying factors that interact with *DPB11*.

In vitro studies with purified budding yeast proteins have shown that Sld3 and Dpb11 independently interact with Cdc45 and help in its recruitment to the Mcm2-7 complex [51–53]. Sld2, Sld3, and Dpb11 also associate with Mcm2-7 independently before the activation of S-CDK and this association prevents premature binding of GINS to Mcm2-7. This ensures that GINS does not associate with Mcm2-7-Cdc45 complex prior to the dissociation of Mcm2-7 double hexamer and extrusion of single-stranded DNA. However, once DDK and S-CDK are activated in S phase and single-stranded DNA is extruded from the central channel of Mcm2-7 complex, Sld2, Sld3, and Dpb11 dissociate from Mcm2-7 and bind origin single-stranded DNA. S-CDK phosphorylation of Sld2 and Sld3 results in the formation of Dpb11-Sld3-Sld2 complex that interacts tightly with origin single-stranded DNA via its three different binding sites. This may allow for the subsequent association of GINS with Mcm2-7-Cdc45 complex. Binding of GINS completes the formation of the CMG complex [54–56].

The orthologs of Dpb11, Sld3, and Sld2 in fission yeast are Cut5, Sld3, and Drc1, respectively. In fission yeast, Sld3 associates with the origins in a DDK-dependent manner; however, this association is independent of Cdc45 association. In fact, association of Sld3 with origins is essential for the subsequent recruitment of Cut5, Drc1, GINS, and Cdc45. Similar to budding yeast, fission yeast Drc1 and Sld3 interact with Cut5 in a CDK-dependent manner [34]. In vertebrates, the functional homologs of Dpb11, Sld3, and Sld2 are reported to be TopBP1, Treslin/Ticrr, and RecQL4, respectively, even though they show very limited sequence similarity. *Xenopus* TopBP1 (Xmus101) has been shown to directly interact with Cdc45, thereby facilitating the loading of Cdc45 onto replication origins [57]. *Xenopus*

Treslin/Ticrr also interacts with Cdc45 and is required for its association with chromatin. In addition, both human and *Xenopus* Treslin/Ticrr associate with the N-terminal BRCT domain of TopBP1 and this interaction is S-CDK dependent [58]. Finally, RecQL4 has a very weak similarity to Sld2 and this similarity is restricted to the first 400 amino acids of the N-terminal region of RecQL4, which is essential for cell growth and DNA replication. The N-terminal region of *Xenopus* RecQL4 associates with TopBP1; however unlike budding yeast this interaction is CDK independent. The N-terminal region of human RecQL4 also binds TopBP1 and shows an interaction with ssDNA, dsDNA, and Y-shaped DNA [59]. *Xenopus* and human RecQL4 also binds to Mcm10 and associates with the CMG complex in Mcm10-dependent manner [35]. Two additional metazoan protein factors GEMC1 and DUE-B are also required for the recruitment of Cdc45 to chromatin and show binding to both Cdc45 and TopBP1. These two factors however have no identified homologs in lower eukaryotes suggesting that helicase activation is a more complicated process in vertebrates than in yeast.

These studies demonstrate that Dpb11, Sld3, and Sld2 play a critical role in replication initiation and their levels are significant for normal cell proliferation. Overexpression of these limiting factors leads to increased origin firing while their low levels result in low levels of replication initiation. Therefore, regulation of these essential protein factors is important for genome stability [60].

Role of Mcm10 in Replication

Minichromosome maintenance protein 10 (Mcm10) is an essential replication protein present in eukaryotes and it has been shown to genetically interact with a wide array of proteins. These interacting proteins include replication initiation proteins, polymerases, replication checkpoint proteins, double-strand break (DSB) repair proteins, and proteins involved in the SUMO pathway. Structural studies on Mcm10 show the presence of a coiled coil (CC) motif in its N-terminal domain (NTD), an oligosaccharide/oligonucleotide binding (OB) fold in its internal domain (ID), and a variable C-terminal domain (CTD) which is absent in unicellular eukaryotes. The NTD of *S. pombe*, *Xenopus laevis*, and humans has been implicated in self-interaction to form Mcm10 oligomers. A recent study has shown the interaction of NTD of Mcm10 with the Mec3 subunit of 9-1-1 checkpoint clamp [61]. The OB fold present in the ID forms a DNA-binding site and is also involved in interactions with Mcm2-7 complex, Pol-α, and proliferating cell nuclear antigen (PCNA). Interaction with Pol-α is mediated through a conserved hydrophobic patch, known as the Hsp10-like domain, and interaction with PCNA occurs via the PIP (PCNA interacting peptide) box. The variable CTD provides an additional surface for interaction with proteins and DNA.

Mcm10 has been shown to be indispensible for CMG helicase activation [62, 63]. It has also been shown to be involved in DNA unwinding since it has affinity for both single-stranded (ss) and double-stranded (ds) DNA [64]. In addition to its role during replication initiation, Mcm10 is also required for polymerase loading and replication elongation. It has been identified as a component of replication forks and shown to recruit Pol-α to chromatin. It also interacts with PCNA (processivity factor for DNA polymerases) and this interaction in budding yeast is regulated by ubiquitylation of Mcm10 [65]. However, the presence of Mcm10 at moving replication forks has recently been questioned [62, 65].

Mcm10 depletion in cells creates a requirement for checkpoint signaling and double-strand break repairs. Due to its essential role in genome maintenance, misregulation of Mcm10 expression correlates with cancer development. In addition, mutations within the conserved regions of Mcm10 have been identified during sequencing of various cancer genomes [66].

The Cdc45-Mcm2-7-GINS (CMG) Complex

The CMG complex also called the active replication fork helicase is formed in the S phase around single-stranded DNA and translocates in a 3′–5′ direction. The CMG complex catalyzes DNA unwinding during replication. It is composed of three essential proteins: Cdc45, Mcm2-7, and GINS, which are conserved throughout eukaryotes. The Cdc45 protein is conserved among eukaryotes and shows sequence similarity to archaeal proteins of the DHH family of phosphoesterases [67]. Cdc45 has been predicted to have a strong structural similarity to the bacterial RecJ proteins [68]. In addition, yeast and human Cdc45 also binds single-stranded DNA [69, 70]. GINS was identified as a heterotetramer protein complex required for DNA replication in budding yeast, comprising of four subunits Psf1, Psf2, Psf3, and Sld5, which are highly conserved among eukaryotes [71]. GINS complex was also purified from *Xenopus* egg extracts and was showed to have a ringlike structure [72]. Several independent studies have isolated the human GINS complex and described its crystal structure [73–75].

In vitro studies with *Drosophila* proteins show that *Drosophila* Mcm2-7 has a very minimal helicase activity. However, the helicase activity of Mcm2-7 increases by approximately 300-fold when it associates with Cdc45 and GINS to form the CMG complex. This complex also has a higher affinity for both single-stranded DNA and forked DNA substrate than does the Mcm2-7 complex and this DNA binding is ATP dependent [76]. Single-particle EM studies using *Drosophila* proteins have shown that the Mcm2-7 by itself exists in two conformations, the planar notched-ring conformation and the spiral lock-washer conformation, with an opening present at the Mcm2/Mcm5 interface. The Mcm2-7 present within the CMG complex adopts a planar notched ring conformation with a gap between Mcm2 and Mcm5 subunits. However, the Mcm2-Mcm5 gate closes upon nucleotide binding. In addition, Cdc45 and GINS were seen to form a handle-like structure that also helps to bridge the gap between Mcm2 and Mcm5 gate. This study using *Drosophila* proteins also demonstrated that in the CMG complex, Cdc45 associates with the N-terminal of Mcm2 and the four subunits of GINS (Psf1, Psf2, Psf3, and Sld5) form extensive interactions with the N- and C-termini of Mcm3 and Mcm5. GINS and Cdc45 also make extensive contacts with each other [77]. The association of Mcm2-7 proteins with Cdc45 and GINS provides stability to the Mcm2-7 ring and aids in its efficient functioning. The CMG complex was also isolated from human cells and was shown to have properties similar to those of the *Drosophila* CMG complex [78]. However, how this CMG complex aids in the unwinding of double-stranded DNA still remains unclear.

DNA Damage Response

Cells are constantly exposed to various endogenous and exogenous DNA-damaging agents, such as reactive oxygen species (ROS) generated within a cell, ionizing, or UV light-mediated irradiation or enzymes involved in DNA compaction like DNA topoisomerases. As a result, DNA damage caused by these agents challenges the maintenance of cellular genome integrity. In order to maintain genomic integrity, eukaryotic cells activate the DNA damage response (DDR), which detects DNA lesions and coordinates various cellular processes important for recovery. Depending on the extent of DNA damage, DDR can either lead the cell to senescence or apoptosis, or activate specific mechanisms that repair the DNA damage or help the cell to tolerate DNA damage. Various repair mechanisms used by a cell in response to DNA lesions include base excision repair (BER) to repair single-strand breaks (SSBs) or subtle changes to DNA, nucleotide excision repair (NER) for bulkier single-strand lesions that distort the DNA helical structure, homologous recombination, and non-homologous end joining (NHEJ) to cope with double-strand breaks (DSBs), mismatch repair, and finally translesion synthesis and template switching [79].

DDR causes a cell to arrest either in G_1 phase or in G_2 phase. In addition, replication fork-associated DDR delays progression through S phase and controls initiation events [80]. The signal transduction pathways of DDR that regulate cell cycle progression and activate the effector kinases in order to repair DNA lesions constitute the checkpoint machinery. Thus, DNA damage checkpoint is activated as a result of the initial processing of DNA damage.

The checkpoint signaling is mediated through two main kinases that belong to the phosphatidylinositol 3-kinase-related protein kinase (PIKK) family:

1. Mec1 (mitosis entry checkpoint 1) also called ATR (ATM and Rad3-related) in mammals, which is activated in response to ssDNA coated with RPA (replication protein A, an ssDNA-binding protein).
2. Tel1 (telomere maintenance 1) also called ATM (ataxia telangiectasia mutated) in mammals, which is activated in response to DSBs. At DSBs, Tel1/ATM is first recruited and activated by the MRN complex (Mre11-Rad50-Xrs2 in budding yeast or Mre11-Rad50-Nbs1 in mammals). This promotes resection at DSBs, generating ssDNA, which then activates Mec1/ATR kinase.

Mec1/ATR is recruited onto chromatin via its regulatory subunit called Ddc2 (or ATRIP, ATR interacting protein, in mammals). Mec1/ATR activation requires activator proteins, which are the 9-1-1 checkpoint clamp (Rad9-Hus1-Rad1 in mammals or Ddc1-Rad17-Mec3 in budding yeast) and Dpb11 (or TopBP1 I mammals). The 9-1-1 checkpoint clamp is loaded onto chromatin via the clamp loader

Rad24-RFC (replication factor C, Rad17-RFC in humans). In budding yeast, Mec1 also phosphorylates Ddc1 (component of the 9-1-1 complex), which then recruits Dpb11 to stimulate Mec1 kinase activity. This Dpb11-Ddc1 interaction is conserved in higher eukaryotes. Activation of Mec1 results in the phosphorylation of various proteins in the cell including effector kinases Chk1 and Rad53 (Chk2 in humans). These effector kinases undergo trans-autophosphorylation with the aid of mediator proteins like Rad9 (53BP1/MDC1/BRCA1 in mammals) or Mrc1 (claspin in mammals). The hyper-phosphorylated effector kinases finally regulate various downstream processes by transmitting the checkpoint response to a range of effector proteins. Mec1 and Tel1 also phosphorylate chromatin-bound proteins like the histone variant H2A (H2AX in mammals) to cause local chromatin changes [81].

Studies in budding yeast have demonstrated that during G_1 phase, the DNA damage checkpoint is mediated through the Ddc1 subunit of the 9-1-1 complex which directly activates Mec1, while Dpb11 was shown to be dispensable during the G_1 phase. However, the G_2/M DNA damage checkpoint requires both the 9-1-1- and Dpb11-dependent activation of Mec1 [82, 83].

In addition to DNA damage checkpoint signaling, DNA damage-induced sumoylation (DDIS) of several protein factors also forms an integral part of DDR and enhances the cell's ability to replicate and repair damaged DNA. Sumoylation involves covalent addition of small ubiquitin-like modifier (SUMO) to one or more lysines of the target protein. SUMO targets proteins involved in DNA replication and in DNA repair pathways like recombination, base excision repair, nucleotide excision repair, and nonhomologous end joining. Some of these proteins identified in budding yeast are Dpb11, Mcm2, Mcm4, Orc2, Orc6, Pol1, Rad 52, Rad59, Apn1, Rad1, and Rad2. In addition, sumoylation of some DNA lesion sensor proteins may also contribute to achieve checkpoint activation [84].

Replication Checkpoint Signaling

The progressing replication fork can encounter obstacles mainly DNA breaks that partially block the progression of replication fork and disturb its stability. Replication fork stalling can also occur in situations of replication stress like nucleotide depletion. Under such conditions that threaten DNA replication and cause replication fork stalling, the cell activates a replication or S-phase checkpoint signaling pathway. The replication checkpoint signaling regulates cell cycle progression through S phase in response to DNA damage or replication stress. This is important to maintain genome integrity and to ensure error-free duplication of the entire genome. The replication checkpoint promotes DNA repair and stabilizes the stalled replication fork by the activation of a signal transduction cascade involving various protein factors. It also inhibits origin firing and slows down DNA synthesis to facilitate DNA repair [85].

During conditions of replicative stress, the helicase uncouples itself from DNA polymerases. As a result, the helicase keeps unwinding DNA, while DNA synthesis halts, creating an excess of ssDNA bound with RPA (replication protein A, an ssDNA-binding protein). This ssDNA generated at stalled replication forks activates

Mec1/ATR-mediated replication checkpoint signaling [81]. Studies in budding yeast show that Mec1 activation in S-phase checkpoint is regulated by three activator proteins, which act in a redundant manner. These three proteins are the 9-1-1 checkpoint clamp, Dpb11 and Dna2 (a conserved nuclease, essential for Okazaki fragment maturation), which activate Mec1 by a similar mechanism that ultimately phosphorylates Rad53. In addition to the Mec1-mediated checkpoint signaling, the replication checkpoint also has a secondary pathway for Rad53 phosphorylation that involves Tel1. Thus, the complete elimination of S-phase checkpoint signaling can be achieved only by the elimination of Mec1 activation function of all the three activators (9-1-1 complex, Dpb11, and Dna2) and the elimination of Tel1-mediated Rad53 phosphorylation [85, 86].

An activated replication checkpoint regulates cell cycle progression and blocks the G_2/M transition. It also phosphorylates various components of the replication machinery to facilitate stabilization of stalled replication forks and also blocks further origin firing [87]. Studies in budding yeast identified Dbf4 (regulatory subunit of DDK) and Sld3 as Rad53 substrates, demonstrating the direct regulation of DNA replication machinery by checkpoints. Rad53-dependent phosphorylation of Sld3 prevents its interaction with Cdc45 and Dpb11, which is essential for activation of replication origins. Sld3 and Dbf4 phosphorylation thus interferes with the CDK- and DDK-dependent activation of origins [88]. Another key feature of replication checkpoint response is the regulation of ribonucleotide reductase (RNR) in order to maintain the optimum level of dNTPs, since too little or too much dNTP can be mutagenic [87]. Mec1/ATR activation also prevents chromosome breakage at fragile sites, which experience slow movement of replication fork [85]. However, some studies argue that replisome stability might not be a key feature of checkpoint response. Studies using *Xenopus* egg extracts indicated that replication could resume even in the absence of checkpoint kinases under certain circumstances [87]. Another study in budding yeast showed the stable association of replisome with replication forks during replication stress even in the absence of Mec1 or Rad53. This suggests that checkpoint kinases might regulate the function of replisome proteins rather than its stability during conditions of replication fork stalling [89].

Break-Induced DNA Replication

Break-induced replication (BIR) is a DSB repair pathway that is used by a cell in situations where only one end of the DSB shares homology with a template. BIR contributes to replication restart at stalled or collapsed replication forks. It also plays an important role in telomere maintenance in the absence of telomerase. In eukaryotes, BIR is best studied in the budding yeast model system. BIR initiates when a single strand invades into the homologous DNA template and forms a displacement loop (D-loop). This process is mediated by Rad51 and is followed by the assembly of a unidirectional replication fork and extensive DNA synthesis. However, formation of a replication fork from the D-loop is not very well understood [90]. Studies in budding yeast have shown that BIR requires almost all the components of normal DNA replication including Mcm2-7, Cdt1, Cdc45, GINS, DDK, Dpb11, Sld3, Pol α-primase, Mcm10, and Ctf4. Whereas Cdc6 and ORC are not necessary for BIR [91], BIR initiation requires Pol δ and Pol ε is required to continue DNA synthesis later. However, how the replication fork is established outside S phase, in the G_2 phase, still remains an important question in the field. Replication during BIR has a much higher mutation rate than normal replication. BIR may also result in various chromosomal rearrangements like template switching, copy number variation, or nonreciprocal translocations. In humans, BIR is mainly involved in alternative lengthening of telomeres (ALT) or chromosomal rearrangements, which cause genetic instability and are particularly associated with several human cancers [92].

The following chapters will discuss these various aspects of DNA replication initiation in eukaryotes in greater detail. This textbook will provide an excellent introduction in DNA replication initiation in eukaryotes for those who are new to the field, and will also provide detailed information in DNA replication to those who are more advanced. Many of the great advances in DNA replication initiation have been discovered in the past several years, and we know that the timely publication of this volume will encompass the important, recent developments in replication initiation.

Abbreviations

AAA+	ATPases associated with diverse cellular activities
ACS	ARS consensus sequence
ALT	Alternative lengthening of telomeres
APC	Anaphase-promoting complex
ARS	Autonomously replicating sequence
ASF1	Anti-silencing function 1
ATM	Ataxia telangiectasia mutated
ATP	Adenosine triphosphate
ATR	ATM and Rad3 related
BAH	Bromo adjacent homology
BER	Base excision repair
BIR	Break-induced replication
BRCT	BRCA1 C-terminus
CAF-1	Chromatin assembly factor-1
CC	Coiled coil
Cdc	Cell division cycle
CDK	Cyclin-dependent kinase
Cdt 1	Cdc10-dependent transcript 1
CMG	Cdc45-Mcm-GINS
CTD	C-terminal domain
Dbf4	Dumb bell forming 4
DDK	Dbf4-dependent kinase

DDR	DNA damage response
DNA	Deoxyribonucleic acid
dNTPs	Deoxynucleotide triphosphates
Dpb11	DNA polymerase B-binding subunit 11
ds	Double stranded
DSB	Double-strand break
EM	Electron microscopy
Fox	Forkhead box
GINS	Go Ichi Ni San (5-1-2-3)
HDAC	Histone deacetylase
ID	Internal domain
Mcm	Minichromosome maintenance
Mec1	Mitosis entry checkpoint 1
MGS	Meier-Gorlin syndrome
NAP1	Nucleosome assembly protein 1
NASP	Nuclear autoantigenic sperm protein
NER	Nucleotide excision repair
NHEJ	Nonhomologous end joining
NTD	N-terminal domain
OB	Oligosaccharide/oligonucleotide binding
ORC	Origin recognition complex
ori	Origin of replication initiation
PCNA	Proliferating cell nuclear antigen
PIKK	Phosphatidylinositol 3-kinase-related protein kinase
Pol-α	DNA polymerase-α primase
Pol-δ	DNA polymerase-δ
Pol-ε	DNA polymerase-ε
PP1	Protein phosphatase 1
Pre-IC	Pre-initiation complex
Pre-LC	Pre-loading complex
Pre-RC	Pre-replication complex
PTM	Posttranslational modification
RFC	Replication factor C
Rif1	Rap1 interacting factor 1
RNR	Ribonucleotide reductase
ROS	Reactive oxygen species
RPA	Replication protein A
SCF	Skp, Cullin, F-box containing
Sld	Synthetically lethal with *dpb11-1*
ss	Single-stranded
SSBs	Single-strand breaks
SUMO	Small ubiquitin-like modifier
Tel1	Telomere maintenance 1
TopBP1	Topoisomerase II-binding protein I
UV	Ultraviolet

References

1. Newlon CS, Theis JF. The structure and function of yeast ARS elements. Curr Opin Gen Dev. 1993;3:752–8.
2. Huang R, Kowalski D. A DNA unwinding element and an ARS consensus comprise a replication origin within a yeast chromosome. EMBO J. 1993;12:4521–31.
3. Nieduszynski C, Knox Y, Donaldson A. Genome-wide identification of replication origins in yeast by comparative genomics. Genes Dev. 2006;20:1874–9.
4. Hyrien O. Peaks cloaked in the mist: the landscape of mammalian replication origins. J Cell Biol. 2015;208:147–60.
5. Rhind N, Gilbert D. DNA replication timing. Cold Spring Harb Perspect Biol. 2013;5:a010132.
6. Ding Q, MacAlpine D. Defining the replication program through the chromatin landscape. Crit Rev Biochem Mol Biol. 2011;46:165–79.
7. MacAlpine D, Almouzni G. Chromatin and DNA replication. Cold Spring Harb Perspect Biol. 2013;5:a010207.
8. Ma E, Hyrien O, Goldar A. Do replication forks control late origin firing in Saccharomyces cerevisiae? Nucleic Acids Res. 2012;40:2010–9.
9. Aparicio O. Location, location, location: it's all in the timing for replication origins. Genes Dev. 2013;27:117–28.
10. Peace J, Ter-Zakarian A, Aparicio O. Rif1 regulates initiation timing of late replication origins throughout the S. cerevisiae genome. PLoS One. 2014;9:e98501.
11. Mantiero D, Mackenzie A, Donaldson A, Zegerman P. Limiting replication initiation factors execute the temporal programme of origin firing in budding yeast. EMBO J. 2011; 30:4805–14.
12. Pope B, Gilbert D. The replication domain model: regulating replicon firing in the context of large-scale chromosome architecture. J Mol Biol. 2013;425:4690–5.
13. Pope B, Hiratani I, Gilbert D. Domain-wide regulation of DNA replication timing during mammalian development. Chromosome Res. 2010;18:127–36.
14. Whitehouse I, Smith DJ. Chromatin dynamics at the replication fork: there's more to life than histones. Curr Opin Genet Dev. 2013;23(2):140–6.
15. Bell S, Stillman B. ATP-dependent recognition of eukaryotic origins of DNA replication by a multiprotein complex. Nature. 1992;357:128–34.
16. Costa A, Hood I, Berger J. Mechanisms for initiating cellular DNA replication. Annu Rev Biochem. 2013;82:25–54.
17. Hemerly A, Prasanth S, Siddiqui K, Stillman B. Orc1 controls centriole and centrosome copy number in human cells. Science. 2009;323:789–93.
18. Hoggard T, Shor E, Müller C, Nieduszynski C, Fox C. A Link between ORC-origin binding mechanisms and origin activation time revealed in budding yeast. PLoS Genet. 2013;9: e1003798.
19. McIntosh D, Blow J. Dormant origins, the licensing checkpoint, and the response to replicative stresses. Cold Spring Harb Perspect Biol. 2012;4:a012955.
20. Hossain M, Stillman B. Meier-Gorlin syndrome mutations disrupt an Orc1 CDK inhibitory domain and cause centrosome reduplication. Genes Dev. 2012;26:1797–810.
21. Shen Z. The origin recognition complex in human diseases. Biosci Rep. 2013;33:e00044.
22. Bicknell L, Bongers E, Leitch A, Brown S, Schoots J, Harley M, et al. Mutations in the pre-replication complex cause Meier-Gorlin syndrome. Nat Genet. 2011;43:356–9.
23. Guernsey D, Matsuoka M, Jiang H, Evans S, Macgillivray C, Nightingale M, et al. Mutations in origin recognition complex gene ORC4 cause Meier-Gorlin syndrome. Nat Genet. 2011;43:360–4.
24. Bleichert F, Balasov M, Chesnokov I, Nogales E, Botchan M, Berger J. A Meier-Gorlin syndrome mutation in a conserved C-terminal helix of Orc6 impedes origin recognition complex formation. Elife. 2013;2:e00882.
25. Maine G, Sinha P, Tye B. Mutants of S. cerevisiae defective in the maintenance of minichromosomes. Genetics. 1984;106:365–85.

26. Davey MJ, Indiani C, O'Donnell M. Reconstitution of the Mcm2-7p heterohexamer, subunit arrangement, and ATP site architecture. J Biol Chem. 2003;278:4491–9.
27. Bochman M, Schwacha A. The Mcm2-7 complex has in vitro helicase activity. Mol Cell. 2008;31:287–93.
28. Coster G, Frigola J, Beuron F, Morris E, Diffley J. Origin licensing requires ATP binding and hydrolysis by the MCM replicative helicase. Mol Cell. 2014;55:666–77.
29. Riera A, Tognetti S, Speck C. Helicase loading: how to build a MCM2-7 double-hexamer. Semin Cell Dev Biol. 2014;30:104–9.
30. Woodward AM, Göhler T, Luciani MG, Oehlmann M, Ge X, Gartner A, et al. Excess Mcm2-7 license dormant origins of replication that can be used under conditions of replicative stress. J Cell Biol. 2006;173(5):673–83.
31. Remus D, Beuron F, Tolun G, Griffith J, Morris E, Diffley J. Concerted loading of Mcm2-7 double hexamers around DNA during DNA replication origin licensing. Cell. 2009; 139:719–30.
32. Sun J, Fernandez-Cid A, Riera A, Tognetti S, Yuan Z, Stillman B, et al. Structural and mechanistic insights into Mcm2-7 double-hexamer assembly and function. Genes Dev. 2014;28:2291.
33. Samel S, Fernández-Cid A, Sun J, Riera A, Tognetti S, Herrera M, et al. A unique DNA entry gate serves for regulated loading of the eukaryotic replicative helicase MCM2-7 onto DNA. Genes Dev. 2014;28:1653–66.
34. Tanaka S, Araki H. Helicase activation and establishment of replication forks at chromosomal origins of replication. Cold Spring Harb Perspect Biol. 2013;5:a010371.
35. Tognetti S, Riera A, Speck C. Switch on the engine: how the eukaryotic replicative helicase MCM2-7 becomes activated. Chromosoma. 2015;124:13.
36. Hardy CFJ, Dryga O, Seematter S, Pahl PMB, Sclafani RA. *mcm5/cdc46-bob1* bypasses the requirement for the S phase activatorCdc7p. Proc Natl Acad Sci U S A. 1997;94:3151–5.
37. Sclafani R, Tecklenburg M, Pierce A. The mcm5-bob1 bypass of Cdc7p/Dbf4p in DNA replication depends on both Cdk-1 independent and Cdk-1-dependent steps in Saccharomyces cerevisiae. Genetics. 2002;161:47–57.
38. Sheu Y, Stillman B. The Dbf4-Cdc7 kinase promotes S phase by alleviating an inhibitory activity in Mcm4. Nature. 2010;463:113–7.
39. Sheu Y-J, Stillman B. Cdc7-Dbf4 phosphorylates MCM proteins via a docing site-mediated mechanism to promote S phase progression. Mol Cell. 2006;24:101–13.
40. Sclafani R, Holzen T. Cell cycle regulation of DNA replication. Annu Rev Genet. 2007;41:237–80.
41. Hoang ML, Leon RP, Pessoa-Brandao L, Hunt S, Raghuraman MK, Fangman WL, et al. Structural changes in Mcm5 protein bypass Cdc7-Dbf4 function and reduce replication origin efficiency in S. cerevisiae. Mol Cell Biol. 2007;27:7594–602.
42. Bruck I, Kaplan DL. The Dbf4-Cdc7 kinase promotes Mcm2-7 ring opening to allow for single-stranded DNA extrusion and helicase assembly. J Biol Chem. 2015;290:1210–21.
43. Hiraga S, Alvino G, Chang F, Lian H, Sridhar A, Kubota T, et al. Rif1 controls DNA replication by directing Protein Phosphatase 1 to reverse Cdc7-mediated phosphorylation of the MCM complex. Genes Dev. 2014;28:372–83.
44. Davé A, Cooley C, Garg M, Bianchi A. Protein phosphatase 1 recruitment by Rif1 regulates DNA replication origin firing by counteracting DDK activity. Cell Rep. 2014;7:53–61.
45. Chen S, Bell S. CDK prevents Mcm2-7 helicase loading by inhibiting Cdt1 interaction with Orc6. Genes Dev. 2011;25:363–72.
46. Tanaka S, Umemori T, Hirai K, Muramatsu S, Kamimura Y, Araki H. CDK-dependent phosphorylation of Sld2 and Sld3 initiates DNA replication in budding yeast. Nature. 2007;445(7125):328–32.
47. Kanter D, Kaplan D. Sld2 binds to origin single-stranded DNA and stimulates DNA annealing. Nucleic Acids Res. 2011;39:2580–92.
48. Zegerman P, Diffley JF. Phosphorylation of Sld2 and Sld3 by cyclin-dependent kinases promotes DNA replication in budding yeast. Nature. 2007;445(7125):281–5.
49. Muramatsu S, Hirai K, Tak Y, Kamimura Y, Araki H. CDK-dependent complex formation between replication proteins Dpb11, Sld2, Pol (epsilon}, and GINS in budding yeast. Genes Dev. 2010;24:602–12.

50. Araki H, Leem S, Phongdara A, Sugino A. Dpb11, which interacts with DNA polymerase II(epsilon) in Saccharomyces cerevisiae, has a dual role in S-phase progression and at a cell cycle checkpoint. Proc Natl Acad Sci U S A. 1995;92:11791–5.
51. Kamimura Y, Tak YS, Sugino A, Araki H. Sld3, which interacts with Cdc45 (Sld4), functions for chromosomal DNA replication in Saccharomyces cerevisiae. EMBO J. 2001;20(8): 2097–107.
52. Bruck I, Kaplan D. GINS and Sld3 compete with one another for Mcm2-7 and Cdc45 binding. J Biol Chem. 2011;286:14157–67.
53. Dhingra N, Bruck I, Smith S, Ning B, Kaplan D. Dpb11 helps control assembly of the Cdc45-Mcm2-7-GINS replication fork helicase. J Biol Chem. 2015;290:7586.
54. Bruck I, Kanter DM, Kaplan DL. Enabling association of the GINS tetramer with the Mcm2-7 complex by phosphorylated Sld2 protein and single-stranded origin DNA. J Biol Chem. 2011;286:36414–26.
55. Bruck I, Kaplan D. The replication initiation protein sld2 regulates helicase assembly. J Biol Chem. 2014;289:1948–59.
56. Bruck I, Kaplan D. Origin single-stranded DNA releases Sld3 protein from the Mcm2-7 complex, allowing the GINS tetramer to bind the Mcm2-7 complex. J Biol Chem. 2011; 286:18602–13.
57. Van Hatten RA, Tutter AV, Holway AH, Khederian AM, Walter JC, Michael WM. The Xenopus Xmus101 protein is required for the recruitment of Cdc45 to origins of DNA replication. J Cell Biol. 2002;159:541–7.
58. Boos D, Sanchez-Pulido L, Rappas M, Pearl LH, Oliver AW, Ponting CP, et al. Regulation of DNA replication through Sld3-Dpb11 interaction is conserved from yeast to humans. Curr Biol. 2011;21(13):1152–7.
59. Ohlenschläger O, Kuhnert A, Schneider A, Haumann S, Bellstedt P, Keller H, et al. The N-terminus of the human RecQL4 helicase is a homeodomain-like DNA interaction motif. Nucleic Acids Res. 2012;40:8309–24.
60. Zegerman P. Evolutionary conservation of the CDK targets in eukaryotic DNA replication initiation. Chromosoma. 2015;124:309.
61. Alver R, Zhang T, Josephrajan A, Fultz B, Hendrix C, Das-Bradoo S, et al. The N-terminus of Mcm10 is important for interaction with the 9-1-1 clamp and in resistance to DNA damage. Nucleic Acids Res. 2014;42:8389–404.
62. van Deursen F, Sengupta S, De Piccoli G, Sanchez-Diaz A, Labib K. Mcm10 associates with the loaded DNA helicase at replication origins and defines a novel step in its activation. EMBO J. 2012;31(9):2195–206.
63. Watase G, Takisawa H, Kanemaki M. Mcm10 plays a role in functioning of the eukaryotic replicative DNA helicase, Cdc45-Mcm-GINS. Curr Biol. 2012;22:343–9.
64. Kanke M, Kodama Y, Takahashi TS, Nakagawa T, Masukata H. Mcm10 plays an essential role in origin DNA unwinding after loading of the CMG components. EMBO J. 2012;31(9): 2182–94.
65. Thu YM, Bielinsky AK. Enigmatic roles of Mcm10 in DNA replication. Trends Biochem Sci. 2013;38(4):184–94.
66. Thu YM, Bielinsky AK. MCM10: one tool for all-Integrity, maintenance and damage control. Semin Cell Dev Biol. 2014;30:121–30.
67. Onesti S, MacNeill SA. Structure and evolutionary origins of the CMG complex. Chromosoma. 2013;122(1-2):47–53.
68. Krastanova I, Sannino V, Amenitsch H, Gileadi O, Pisani FM, Onesti S. Structural and functional insights into the DNA replication factor Cdc45 reveal an evolutionary relationship to the DHH family of phosphoesterases. J Biol Chem. 2012;287(6):4121–8.
69. Bruck I, Kaplan DL. Cdc45 protein-single-stranded DNA interaction is important for stalling the helicase during replication stress. J Biol Chem. 2013;288(11):7550–63.
70. Szambowska A, Tessmer I, Kursula P, Usskilat C, Prus P, Pospiech H, et al. DNA binding properties of human Cdc45 suggest a function as molecular wedge for DNA unwinding. Nucleic Acids Res. 2014;42:2308–19.

71. Takayama Y, Kamimura Y, Okawa M, Muramatsu S, Sugino A, Araki H. GINS, a novel multiprotein complex required for chromosomal DNA replication in budding yeast. Genes Dev. 2003;17(9):1153–65.
72. Kubota Y, Takase Y, Komori Y, Hashimoto Y, Arata T, Kamimura Y, et al. A novel ring-like complex of Xenopus proteins essential for the initiation of DNA replication. Genes Dev. 2003;17(9):1141–52.
73. Boskovic J, Coloma J, Aparicio T, Zhou M, Robinson CV, Méndez J, et al. Molecular architecture of the human GINS complex. EMBO Rep. 2007;8(7):678–84.
74. Kamada K, Kubota Y, Arata T, Shindo Y, Hanaoka F. Structure of the human GINS complex and its assembly and functional interface in replication initiation. Nat Struct Mol Biol. 2007;14(5):388–96.
75. Chang YP, Wang G, Bermudez V, Hurwitz J, Chen XS. Crystal structure of the GINS complex and functional insights into its role in DNA replication. Proc Natl Acad Sci U S A. 2007;104(31):12685–90.
76. Ilves I, Petojevic T, Pesavento JJ, Botchan MR. Activation of the MCM2-7 helicase by association with Cdc45 and GINS proteins. Mol Cell. 2010;37(2):247–58.
77. Costa A, Ilves I, Tamberg N, Petojevic T, Nogales E, Botchan MR, et al. The structural basis for MCM2-7 helicase activation by GINS and Cdc45. Nat Struct Mol Biol. 2011;18(4):471–7.
78. Kang Y, Galal W, Farina A, Tappin I. J. H. Properties of the human Cdc45/Mcm2-7/GINS helicase complex and its action with DNA polymerase {varepsilon} in rolling circle DNA synthesis. Proc Natl Acad Sci U S A. 2012;109:6042–7.
79. Lord CJ, Ashworth A. The DNA damage response and cancer therapy. Nature. 2012; 481(7381):287–94.
80. Hustedt N, Gasser SM, Shimada K. Replication checkpoint: tuning and coordination of replication forks in s phase. Genes (Basel). 2013;4(3):388–434.
81. Recolin B, van der Laan S, Tsanov N, Maiorano D. Molecular mechanisms of DNA replication checkpoint activation. Genes (Basel). 2014;5(1):147–75.
82. Navadgi-Patil VM, Burgers PM. Cell-cycle-specific activators of the Mec1/ATR checkpoint kinase. Biochem Soc Trans. 2011;39(2):600–5.
83. Navadgi-Patil VM, Kumar S, Burgers PM. The unstructured C-terminal tail of yeast Dpb11 (human TopBP1) protein is dispensable for DNA replication and the S phase checkpoint but required for the G2/M checkpoint. J Biol Chem. 2011;286(47):40999–1007.
84. Cremona CA, Sarangi P, Yang Y, Hang LE, Rahman S, Zhao X. Extensive DNA damage-induced sumoylation contributes to replication and repair and acts in addition to the mec1 checkpoint. Mol Cell. 2012;45(3):422–32.
85. Segurado M, Tercero JA. The S-phase checkpoint: targeting the replication fork. Biol Cell. 2009;101:617–27.
86. Kumar S, Burgers PM. Lagging strand maturation factor Dna2 is a component of the replication checkpoint initiation machinery. Genes Dev. 2013;27(3):313–21.
87. Labib K, Piccoli GD. Surviving chromosome replication: the many roles of the S-phase checkpoint pathway. Phil Trans R Soc B. 2011;366:3554–61.
88. Zegerman P, Diffley JF. Checkpoint-dependent inhibition of DNA replication initiation by Sld3 and Dbf4 phosphorylation. Nature. 2010;467(7314):474–8.
89. De Piccoli G, Katou Y, Itoh T, Nakato R, Shirahige K, Labib K. Replisome stability at defective DNA replication forks is independent of S phase checkpoint kinases. Mol Cell. 2012; 45:699–704.
90. Anand RP, Lovett ST, Haber JE. Break-induced DNA replication. Cold Spring Harb Perspect Biol. 2013;5(12):a010397.
91. Lydeard JR, Lipkin-Moore Z, Sheu YJ, Stillman B, Burgers PM, Haber JE. Break-induced replication requires all essential DNA replication factors except those specific for pre-RC assembly. Genes Dev. 2010;24(11):1133–44.
92. Malkova A, Ira G. Break-induced replication: functions and molecular mechanism. Curr Opin Genet Dev. 2013;23(3):271–9.

Chapter 2
Choice of Origins and Replication Timing Control in Budding Yeast

Arturo Calzada

Abstract A complete and exact replication of every eukaryotic chromosome within each cell division cycle is essential to maintain stable genomes during cell proliferation. Abundant origins of DNA replication where the replication machinery assembles into replisomes to initiate DNA synthesis are widespread along chromosomes. DNA replication shows characteristic spatio-temporal patterns of origin usage and replication timing during S phase, which are conserved through evolution and are cell type specific, indicating an active process of regulation. Important advances have recently been made to elucidate the determinants and molecular mechanisms that regulate the patterns of origin activation. Among these, *cis*-acting elements, chromatin determinants, the timing of origin licensing and factors regulating the choice of origins and the firing timing during S phase have been described in *Saccharomyces cerevisiae*. Much less understood is the biological significance of this replication programme, but it could be significant in providing both robustness and plasticity to the DNA replication process in terms of replication completion and the maintenance of genome integrity.

Keywords Budding yeast • DNA replication origins • Cell cycle regulation • Origin specification • Origin activation • Firing timing • Replication completion • Genome stability

Introduction

Life perpetuates through the continued generation of daughter cells and requires the complete and exact replication of an accurate genome in every cell division cycle. The transmission of inexact genetic contents threatens the stability of progeny with potentially harmful consequences for viability or health. Reaching and maintaining cell populations in unicellular and multicellular organisms require vast numbers of cell divisions, providing ample opportunities for errors to occur. Successful DNA

A. Calzada (✉)
Department of Microbial Biotechnology, Centro Nacional de Biotecnologia (CNB), Consejo Superior de Investigaciones Cientificas (CSIC), Darwin 3, Madrid 28049, Spain
e-mail: arturo.calzada@csic.es

D.L. Kaplan (ed.), *The Initiation of DNA Replication in Eukaryotes*,
DOI 10.1007/978-3-319-24696-3_2

replication is thus a significant process that is further complicated in eukaryotic cells by the large size and fragmentation of eukaryotic genomes into chromosomes, the complex structure of chromatin and the pressure to complete replication in the relatively short duration of the S phase before segregation of sister chromatids starts in anaphase. To initiate DNA synthesis all eukaryotes display very abundant origins of DNA replication [1] that collectively expedite DNA synthesis, but that by being so numerous complicate their individual regulation to block re-replication while ensuring that no regions are left incompletely replicated.

Validating the 'replicon model' proposed by Jacob, Brenner and Cuzin in 1963, the initiation of eukaryotic DNA replication relies on the bipartite system of origins and initiation factors that are both necessary and together sufficient to initiate DNA synthesis [2]. The factors involved in the two-step mechanism of origin activation are now well known (for recent detailed reviews see [3, 4]). Briefly, in the first step, known as origin licensing, pre-replicative complexes (pre-RC) [5] form at origins by the sequential binding of the origin recognition complex (ORC), Cdc6, Cdt1 and two head-to-head Mcm2-7 hexamers. In the second step, known as origin firing, licensed origins are selected to initiate DNA synthesis, by the attraction of additional factors including Sld3, Sld7 and Cdc45 to the pre-RCs and by the phosphorylation by the Dbf4-dependent kinase (DDK) of at least some subunits of Mcm2-7, to form the pre-initiation complex (pre-IC) [6]. In parallel, a pre-loading complex (pre-LC) [7] containing GINS, Sld2, Polε and Dpb11 forms outside origins. Phosphorylation of Sld2 and Sld3 by the S-phase cyclin-dependent kinases (S-CDKs) [8, 9] is essential for the pre-LC to be recruited to pre-IC origins. The active Cdc45/Mcm2-7/GINS (CMG) helicase assembles [10, 11], and upon the attraction of additional replication factors two replisomes form which depart from each origin in opposite directions after DNA unwinding. This reaction leaves the origin in an inactive post-replication state in which it is bound only by ORC and with which it forms a post-replicative complex (post-RC) [5]. In synchrony with the cell cycle, licensing only occurs from late mitosis and during the G1 phase up to START, depending on the expression or recycling of the licensing factors and the inactivity of the S-, mitotic- and G1-CDKs. In late G1 phase the activity of G1-CDKs precludes licensing, while the lack of S-CDK impedes firing [12]. Firing initiates as soon as S-CDKs activate at the beginning of S phase. The persistence of active licensing-inhibitory CDKs up to the metaphase-to-anaphase transition prohibits new licensing. This dependency of licensing on the absence of CDK activity, and of firing on the presence of S-CDK, ensures that the activation of any origin is unique to each cell cycle.

However, in spite of this common machinery of origin activation, only a subset of origins is selected for firing during S phase, and origins display characteristic origin efficiencies (the percentage of firing in a cell population) and firing timing, leading to characteristic spatio-temporal patterns of replication initiation [13–18], both evidencing the active regulation of origin choice. In contrast to these conclusions obtained from cell populations, a stochastic choice of origins among single cells has been found when individual cells have been studied, revealing randomness in origin selection [19, 20]. The combination of global control in the order of firing and local stochastic competition among origins for firing has led to the suggestion of a 'controlled stochastic' model of origin choice [21, 22]. The factors and molecular mechanism that control the choice of origins and the firing timing, and the significance of having defined replication patterns, are as yet incompletely understood and are the focus of intense research.

Here I focus on the current understanding of the determinants of the choice of origins, and their effect on the timing of replication in budding yeast, and also compile evidence supporting the biological significance of this replication programme. In brief, knowledge of the precise map of origin location and the temporal replication profile in budding yeast has facilitated the discovery of determinants of origin usage. Multiple factors are found to influence origin selection and firing timing in budding yeast, including *cis*-acting sequences at origins, local chromatin structure and epigenetic marks, origin positioning within chromosomes, timing of pre-RC formation and maintenance, recruitment of firing timing factors and competition for limiting origin firing factors (reviewed in [23–26]). All these regulators commonly display differential influences among origins; indeed, they provide diversity to the population of origins. The competition among origins for limiting firing factors is a source of plasticity in origin selection. The integration of these multiple controls at each origin could explain the differential activation probability and timing of choice among origins observed in cell populations, and the stochastic origin selection observed in single cells [22, 24–28]. Together with the non-random distribution of exceeding origin numbers, this organisation presumably adds redundancy and robustness to replication, for example against incomplete termination in agreement with the 'origin redundancy' model [27, 29]. Importantly, altering this programme is found to have negative consequences for chromosome integrity and genome stability.

Cis-Acting Elements and Chromatin Determinants at Origins for Origin Selection

Origins were first found in budding yeast and defined as autonomous replicating sequences (ARS) because they confer autonomous replication and maintenance to plasmids and are the sites where bidirectional replication starts [30–33]. The study of some ARSs by scanning mutagenesis showed that origins are a modular combination of distinct *cis*-acting elements including an essential A element which is constant to origins, and a variable composition of a few individually non-essential B elements that provide diversity among origins [34]. The sequence conservation of A elements allowed the definition of an extended ARS consensus sequence (EACS) of 17 AT-rich base pairs [35, 36] that further extends up to 33 base pairs if the ACS–ORC binding consensus [37] is considered. The ACS is insufficient to define an origin; thousands of sequences match the ACS on the yeast genome but data from genome-wide studies reveal that only around 800 are confirmed or likely ARSs [38]. ACS-B1 provides a bipartite sequence for ORC recognition [39, 40] and the B2 element of ARS1 facilitates pre-RC formation or maintenance [41]. In spite of the sequence specificity of pre-RC formation in *S. cerevisiae* in vivo, ORC can bind and load Mcm2-7 complexes to non-origin sequences [42] and support plasmid replication in vitro [43], which is similar to forced ORC binding to DNA in *Drosophila* [44].

Origin sequences and chromatin structure can modulate the probability of origin specification by influencing ORC recruitment and pre-RC assembly during licensing, and contribute to the timely selection and efficiency of origin activation during S phase. In support of origin sequence features influencing origin specification and activity in *S. cerevisiae*, it is found that mutations in the *cis*-acting sequences modify origin efficiency [45, 46], presumably by distinct abilities of differing sequences to attract or retain pre-RC factors (Fig. 2.1a). In further support, direct ORC–DNA chromatin-independent interactions also contribute to ORC recruitment to origins and to replication timing, because origins with this interaction are enriched for late firing [47] (Fig. 2.1a). Consistently, distinct sequence elements within origins influence that origins are differentially tolerant to the mutation of licensing factors or to CDK deregulation in the G1 phase, supporting a hierarchy of replication origins [48, 49]. Similarly, certain origin sequences predispose origins to re-replication, presumably by increasing the competency of origins to recruit or maintain pre-RC factors [50].

Chromatin determinants also regulate the dynamics of origin specification by licensing, and influence the activation timing. The positioning of origins close to those with earlier or more efficient activation can favour passive replication, so that in proximal origins the choice of one inactivates the others [51]. Pioneering experiments in budding yeast showed that the chromatin environment and origin position determine origin efficiency and firing timing independently of origin sequences, as evidenced by moving an early-firing origin to a subtelomeric late-replicating region or a late-firing origin to a plasmid [52]. This work also predicted the existence of *cis*-acting elements that determine the firing timing of proximal origins. Indeed, surrounding sequences and not the origin itself advance the firing timing of nearby origins [53]. Centromeres are normally early replicating, and they influence the replication timing of close regions as shown by the relocation of a functional centromere to a late-replicating region advancing the replication timing of surrounding origins even at long distances [54, 55] (Fig. 2.1b). Mechanistically, the effect of centromeres on the early firing timing of nearby origins can be contributed by kinetochores attracting DDK to recruit Sld3 and Sld7 to promote early replication [54, 55] (Fig. 2.1b), in a similar manner to the finding that the HP1 protein stimulates Sld3 loading and binding of Dfp1/Dbf4 for early replication of pericentromeric chromatin in the fission yeast [56]. Telomeres cause replication origins to fire late, as short telomeres replicate subtelomeric origins early [57, 58]. Telomeres and subtelomeric regions are frequently silenced in transcription, and origin firing is delayed to late in S phase by telomeric suppression of origin activation by the histone deacetylase (HDAC) Sir2 [52, 59], or by the Ku complex [58, 60] (Fig. 2.1b). Loss of function of Sir2 suppresses the *cdc6-4* mutation, and rescues DNA synthesis and plasmid stability of other pre-RC mutants, implying that Sir2 regulates initiation of DNA replication [61]. The Sir2-dependent inhibition of origin activity is differential among origins suggesting that it relies on origin sequences or structure, and mechanistically it could be explained by the presence of an inhibitory sequence (I^s) on ARSs that requires Sir2 [62] (Fig. 2.1b). Further supporting the idea that histone acetylation regulates the selection of origins for firing timing, the loss of the HDAC Rpd3 causes advanced firing timing and Cdc45 recruitment of the subset of late

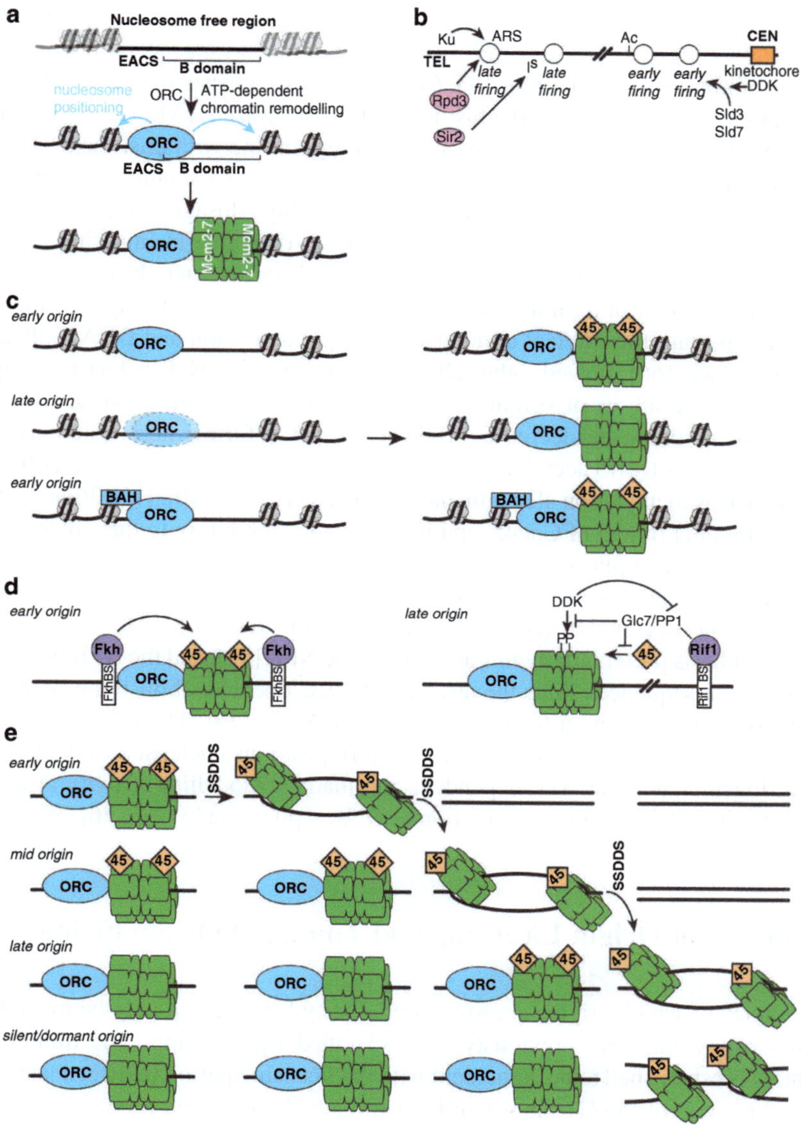

Fig. 2.1 Factors determining origin specification and choice in the control of replication timing. (**a**) *Cis*-acting elements at origins (EACS and B domain) influence origin efficiency and also the surrounding chromatin structure by maintaining a nucleosome-free region window that facilitates ORC recruitment; ORC–DNA binding and an ATP-dependent chromatin remodelling activity position flanking nucleosomes for pre-RC formation in the G1 phase. (**b**) Local and global chromatin determinants influence the choice and firing timing of origins, including centromeres and telomeres that influence the replication timing (early or late) of surrounding origins, and HDACs like Rpd3 and Sir2. Ac, histone acetylation. (**c**) The strength and/or the cell cycle timing of the ORC–chromatin interaction influence pre-RC formation and the timing of origin firing. Earlier or more stable ORC binding (*solid line*), or the presence of the BAH domain, associates with earlier pre-RC formation and with early origins; more labile or later ORC binding (*dashed line*) associates with late origins. The diamond represents Cdc45. (**d**) Apart from the general initiation machinery, specialised factors like Fkh1/2 and Rif1 influence the timing of origin activation by modulating the maturation of pre-RCs into active replisomes during origin firing. FkhBS, Fkh1/2-binding site; Rif1 BS, Rif1-binding site. (**e**) Rate-limiting availability of firing factors and sequential usage by origins govern the distributed timing of origin firing during S phase in the budding yeast. SSDDS, Sld2, Sld3, Dbf4, Dpb11, Sld7

origins of non-telomeric regions [63, 64] (Fig. 2.1b). Consistently, targeted histone acetylation by recruitment of the histone acetyltransferase (HAT) Gcn5 to the late-firing origin ARS1412 advances the firing timing. In line with this, the deletion of Gcn5 compromises minichromosome maintenance, alters the chromatin structure, decreases the level of Mcms at origins and finally a high dosage of Gcn5 suppresses the thermosensibility of *ORC* and *MCM* mutants defective in initiation of DNA replication [64, 65]. Acetylation of H3 and H4 has been shown to be present around a replication origin in a minichromosome, and acetylation of multiple lysine residues is important for efficient chromosomal origin activation and DNA replication during S phase [66]. Indeed, other HDAC, such as the Sum1-Rfm1-Hst1 complex that binds to a subset of origins, is required for normal initiation activity, and removal of a binding site decreases origin activity [67]; furthermore, deletion of this HDAC increases H4K5 acetylation and decreases origin activity [68]. Other histone marks such as methylation also influence origin activity [69, 70]. H3K36me by Set2 aids the binding timing of Cdc45, and H3K36me3 is inhibitory to this process [69].

Regarding the implication of chromatin structure and remodelling in origin activation and replication initiation, almost all origin sequences in *S. cerevisiae* maintain a nucleosome-free region (NFR) which starts from the ACS [71, 72]. Leading-strand synthesis preferentially initiates within the NFR [73], and the NFR is directed by the origin sequences since the absence of ORC does not abolish the NFR [37]. Nucleosome positioning affects the function of ARS1 [74]. The NFR presumably provides access for ORC binding, which in turn positions nucleosomes flanking the origin together with an ATP-dependent chromatin remodelling activity at almost every origin and facilitates the initiation of DNA replication [37, 75, 76] (Fig. 2.1a).

Dynamics of Origin Licensing and Timing of Origin Firing

The choice of which origins are to be fired during S phase can only be made from those origins that have previously been specified by licensing. The timing programme is established to be coincident with the licensing period (between late mitosis and the end of the G1 phase), at least for subtelomeric late-replication regions where a subtelomeric late origin excised in G2/M, and not in the G1 phase, switches the activation time to early firing [77]. Hence, the schedule of origin licensing can contribute to the timing of origin activation.

The chromatin structure may influence the activity of origins by regulating the accessibility of initiation factors to origins, both globally at entire chromatin regions because origin activity correlates with nuclear positioning [78] and origins organise into foci of multiple origins that fire at similar times [79], and locally at specific origins under the influence of the chromatin environment. The NFR at origins can accommodate Mcm2-7 hexamers, and disruption of nucleosome positioning by ORC interferes with pre-RC formation [76]. Significantly, the nucleosome positioning which is established during the G1 phase differs between early and late origins, and is modulated during origin activation in the cell cycle [80]. Supporting the view that the association of ORC with chromatin during the cell cycle is a determinant of origin efficiency, a detailed analysis of DNA proteins by MNase footprint mapping at origins [81] has shown a cell cycle regulation of ORC binding to origins and consistent nucleosome remodelling. This approach discriminates between 30 % of origins showing a detectable ORC-dependent footprint in G2, and another 15 % of origins having a footprint detected only in G1 (consistent with transient ORC binding in G2 or with no binding until G1) (Fig. 2.1c). Significantly, ORC binding in G2 is a determinant for efficient or early origin activation, although the effect is not global [81]. In *S. pombe* the timing of ORC binding and pre-RC formation during mitosis and G1 influences origin efficiency and firing timing during S phase [82]. Origins with earlier Mcm2-7 loading could have further time for additional Mcm2-7 recruitment, thus increasing the probability of attracting firing timing factors and therefore of firing earlier during S phase [21, 26] (Fig. 2.1c).

However, the determinants of ORC binding to specific origins and during the cell cycle are not known. In metazoans, the conserved chromatin-binding module bromo-adjacent homologous (BAH) domain of Orc1 [83] recognises and binds to H4K20me2 methylated histones, but this function is not conserved in Orc1BAH in yeasts [84]. Instead, in *S. cerevisiae* the BAH domain of Orc1 is important for origin selection within chromatin [85] (Fig. 2.1c). *orc1bahΔ* cells show reduced ORC and Mcm2-7 association with chromatin. Consistent with the BAH domain not being a general regulator of origin activation, the effect is differential among origins so that some origins are *orc1bahΔ* sensitive and others *orc1bahΔ* resistant. Furthermore, there are differential responses among sensitive origins, and the loss of the BAH domain does not completely remove ORC/Mcm2-7 binding to *orc1bahΔ*-sensitive origins, but replication initiation, efficiency of origin firing and plasmid maintenance are compromised in sensitive and not in resistant origins [85]. Importantly, the BAH domain is not the determinant of origin efficiency, as efficient and inefficient origins have been found among *orc1bahΔ*-sensitive and -resistant origins [85]. Significantly, ORC binds more stably to origin-containing chromatin than to naked DNA suggesting that ORC at origins is stabilised through the interaction with nucleosomes, and is independent of the BAH domain [86]. Origins relying more on local chromatin determinants, defined as chromatin dependent, are enriched in early-firing origins [47].

Hence, it seems that origin selection can be viewed as the intrinsic origin sequence capability modified by several local chromatin determinants that differentially merge at each origin and modulate the characteristic origin probabilities of efficiency or firing timing during S phase.

Factors Regulating the Timing of Origin Activation

Although firing at all origins occurs by the maturation of licensed origins from pre-RCs to pre-ICs and replisomes, there are conserved distinctive activation times and efficiencies between different origins [13–15], and more origins are licensed than are actually selected to initiate replication [50], evidencing a timely choice of origins for firing. Identification of the factors that regulate timing of origin firing is of relevance considering that the replication profile results predominantly from the kinetics of origin firing [73]. The determinant factors of firing timing had remained elusive until recently, and key discoveries have now shed light on the process.

The transcription factors Fkh1 and Fkh2 are determinants of the origin firing timing programme because they promote early firing to a subset of early origins through ORC binding, clustering of early origins and association with Cdc45 during pre-RC maturation in the G1 phase [87] (Fig. 2.1d). Fkh1/2-binding sites on origins are limited to early origins, although the presence of these sites at origins is insufficient to confer early firing (which is also dependent on the close proximity to the ACS). Furthermore, not all early origins contain Fkh1/2-binding sites, and the introduction of Fkh1/2-binding sites at late origins is insufficient to confer early replication [87, 88]. The position and number of Fkh1/2 sites relative to the ACS seem to be important for origin activity and only a subset of origins contain two sites in a position which flanks the NFR in a precise localisation relative to ACSs [80, 88] (Fig. 2.1d). This regulation by Fkh1/2 in tethering early origins together is consistent with these clusters being poles of attraction for firing factors, including Cdc45, which concentrate spatially and temporally leading to early replication [87] (Fig. 2.1d), and also with evidence that early origins frequently interact [89].

The telomere-binding protein Rif1 is also a conserved regulator of the replication timing programme in normal cell cycles from yeast to human cells [90–93]. In its absence there is a premature activation of origins at telomeres and earlier replication [58]. Rif1 regulates the firing timing of late/dormant origins in internal and subtelomeric chromosome regions in *S. cerevisiae* [92] (Fig. 2.1d). Both in fission and budding yeast Rif1 binds to telomeres and along chromosomes, and although binding is close to some Rif1-regulated origins there is no specific enrichment at origins [91, 92, 94]. The details of the mechanism by which Rif1 controls the firing timing of origins have been elucidated recently. In *S. pombe* the binding of Cdc45, but not of Mcm4, to origins is affected in *rif1*Δ cells, where it was shown that Cdc45 was bound to late origins in contrast to wild-type cells, suggesting that Rif1 influences the steps after pre-RC assembly [91]. In *S. cerevisiae* Rif1 contains two Glc7/protein phosphatase 1 (PP1) interaction motifs at the N-terminus, which enable Rif1 to target PP1 activity to pre-RCs to counteract the DDK phosphorylation of Mcm4 that is critical for the Rif1-repressive effect on the firing timing of late origins [95–97] (Fig. 2.1d). Importantly, Rif1 is also regulated in its binding to Glc7 by interaction with DDK and by DDK-dependent phosphorylation [95–97], and by Tel1 phosphorylation at short telomeres [98] (Fig. 2.1d).

A more global determinant of origin firing timing seems to be the limiting step of pre-RC maturation towards active replisomes. Indeed, while the earliest origins recruit Cdc45 (although loosely during the G1 phase), late origins remain unbound until late S phase [99] (Fig. 2.1e). Work in *S. pombe* has indicated that the recruitment of rate-limiting initiation factors to origins controls origin efficiency by ordered ORC and Mcm2-7 binding, and of firing timing by limiting DDK [19, 82]. Furthermore, Fkh1/2 and Rif1 influence the schedule of Cdc45 recruitment and of

DDK phosphorylation to origins to regulate the firing timing of origins [87, 95–97] (Fig. 2.1d). Work in budding yeast has shown that the essential initiation factors for pre-RC maturation Sld2, Sld3, Sld7, Cdc45, Dpb11 and Dbf4 are rate limiting for origin association and influence the timing of origin firing as their combined overexpression advances the firing timing of late origins to earlier in S phase [100, 101]. Hence, sequential origin firing timing is ordered by the binding of rate-limiting factors to early origins, and further release and subsequent recycling by mid and late origins until complete replication (Fig. 2.1e). In this context, the prevention of late origin activation under replication stress or DNA damage by the S-phase checkpoint also operates. New firing events at licensed origins are prevented upon activation of the S-phase checkpoint [102, 103] mediated by phosphorylation of Dbf4 and Sld3, and inhibition of Cdc45 recruitment to late origins [99, 104–106].

Significantly, those determinants of firing timing actually operate in parallel. Fkh1/2-binding sites are excluded from the subset of origins regulated by Rpd3L [87], while the combined lack of Rpd3L-dependent late firing of dormant origins and overexpression of the rate-limiting firing factors is required for early firing of dormant origins [100].

Significance of Spatio-Temporal Programmes of Origin Activation

Replication timing patterns are more conserved across eukaryotic evolution than strict origin positioning (mainly of dormant origins), even in closely related species [107]. The function of performing regulated temporal programmes of origin choice and replication therefore seems important but is, as yet, incompletely understood, and a number of suggestions have been posed and modelled mathematically [21, 22, 24, 26]. The identification of factors that determine this control has allowed the consequences of its mutation for genome integrity to be addressed experimentally.

The replication timing programme can influence the mutagenic landscape of chromosomes (see [108]). Chromosomes contain an irregular distribution of distinct elements whose replication pattern provides evidence of preferred replication dynamics. The function or homeostasis of these elements may require specific replication control, which could explain the existence of replication timing programming. This is the case with centromeres that replicate early in budding and fission yeasts [109], presumably to ensure optimal chromosome segregation and prevention of aneuploidy [110]. It is also the case with fragile sites (where chromosomes break more frequently), which are present from yeasts to human cells, that frequently have specific chromatin structure or composition, display difficult replication, break under defective or slow replication dynamics, and correlate with a paucity of dormant origins along large chromosome regions and retarded replication in human cells [111–113]. Also of relevance is that mutagenesis is non-random across the genome. Replicative polymerases have distinct error rates and contribute differently to mutation rates by inducing compositional biases along DNA associated with the asymmetry of DNA replication, and accordingly active origins establish a strand bias for mutagenesis [114, 115]. Late-replicating regions have been shown to have a higher incidence of mutagenesis than early regions [116]. Inserting a sequence at distinct replication-time positions along a chromosome reveals a strong correlation with timing and rates of mutagenesis, and consistently the deletion of an early origin leads to a mutagenic increase presumably by retarding replication of nearby regions [117].

Excessive firing could be restricted to proceeding sequentially during S phase if replication proteins or other factors are rate limiting so that the progression during S phase must accommodate the rates of synthesis or recycling of those factors. This is the case of the rate-limiting firing factors (Sld2, Sld3, Dbf4, Dpb11, Cdc45 and Sld7) that impede inappropriate origin activation during S phase and control S-phase length in budding yeast [100, 101]. Otherwise, simultaneous firing can be deleterious. Indeed, dNTP pools are rate limiting and balanced for precise genome duplication by the ribonucleotide reductase [118]. The simultaneous firing of early and late origins in S phase by the overexpression of the rate-limiting firing factors in budding yeast imbalances replication by the elevated numbers of replication forks: dNTPs are deprived, replication stress arises, and the checkpoint kinase Rad53 is activated depending on dNTP levels [100].

Deregulated origin usage could alter the optimal distribution of initiation events along chromosomes needed to ensure replication completion according to the proposed 'random completion' or 'replication gap' problem [27, 29, 119]. For timely completion of replication, the 'origin redundancy' model [27] proposes two solutions: first, that a large excess of licensed origins are selected to fire during S phase in a regular distribution, and second that unreplicated regions retain initiation potential at licensed unfired origins whose activation would facilitate replication completion. A compatible proposed solution is that the efficiency of origin firing increases as S phase progresses at unreplicated regions [27, 29]. Indeed, the replication programme displays exceeding numbers of origins used below saturation during S phase, and non-random origin distribution. The features of regulated activation timing of origins (providing origin diversity) and some allowed stochastic origin selection (providing flexibility) together lead to strong origin redundancy in replication. This replication programme could thus provide the optimal organisation for completion of replication [120]. This is particularly important for two reasons: firstly, replisome progression is normally highly irregular due to eventual fork stalling or collapse [121], regulated pausing at programmed fork barriers [122–124] or delayed replication progression across chromatin regions that display difficult replication like fragile sites [111], and secondly, considering that every chromosomal sequence has a maximum of two opportunities of replication by incoming replisomes from each flank, new origin firing within the region can easily rescue irreversible fork arrest [125]. Consistently, reducing origin numbers compromises chromosome maintenance and integrity, and is further aggravated upon reducing origin diversity by the simultaneous deletion of dormant origins [126, 127]. Similarly, a paucity of origins delays replication completion leading to the expression of fragile sites in human cells [112]. Mutants of licensing factors also reduce the efficiency of origin activation and cause the loss of minichromosomes and elevated rates of chromosomal rearrangements [45, 49, 128]; consistent upregulation of licensing-inhibitory CDK kinases in the G1 phase reduces origin licensing and the efficiency of origin firing, compromising the dynamics of S phase and genome stability possibly by incomplete genome duplication before the initiation of anaphase [129, 130]. Strongly linking genome instability to defective origin usage, the elevated rate of gross chromosomal rearrangements (GCR) caused by the deregulation of CDK activity in the G1 phase at a chromosome region is suppressed by increasing the concentration or distribution of origins in that region presumably by increasing the density of initiation events [130]. And consistently, the rates of GCR reflect the paucity of initiation events from active origins in that region [131]. Hence, compromising the number or choice of origins available during replication could reduce the flexibility of initiation and the robustness of S phase towards replication completion and genome maintenance.

Acknowledgments I apologize to those colleagues whose work could not be discussed or cited in the text owing to space constraints. Work in the laboratory was supported by the Spanish Ministry of Economy and Competitiveness (grant no. BFU2010-18225).

References

1. Leonard AC, Méchali M. DNA replication origins. Cold Spring Harb Perspect Biol. 2013;5 (10):a010116.
2. Gilbert DM. In search of the holy replicator. Nat Rev Mol Cell Biol. 2004;5(10):848–55.
3. Siddiqui K, On KF, Diffley JFX. Regulating DNA replication in eukarya. Cold Spring Harb Perspect Biol 2013; 5(9).
4. Tanaka S, Araki H. Helicase activation and establishment of replication forks at chromosomal origins of replication. Cold Spring Harb Perspect Biol. 2013;5(12):a010371.
5. Diffley JF, Cocker JH, Dowell SJ, Rowley A. Two steps in the assembly of complexes at yeast replication origins in vivo. Cell. 1994;78(2):303–16.
6. Zou L, Stillman B. Formation of a preinitiation complex by S-phase cyclin CDK-dependent loading of Cdc45p onto chromatin. Science. 1998;280(5363):593–6.
7. Muramatsu S, Hirai K, Tak Y-S, Kamimura Y, Araki H. CDK-dependent complex formation between replication proteins Dpb11, Sld2, Pol, and GINS in budding yeast. Genes Dev. 2010;24(6):602–12.
8. Tanaka S, Umemori T, Hirai K, Muramatsu S, Kamimura Y, Araki H. CDK-dependent phosphorylation of Sld2 and Sld3 initiates DNA replication in budding yeast. Nature. 2007; 445(7125):328–32.
9. Zegerman P, Diffley JFX. Phosphorylation of Sld2 and Sld3 by cyclin-dependent kinases promotes DNA replication in budding yeast. Nature. 2007;445(7125):281–5.
10. Gambus A, Jones RC, Sanchez-Diaz A, Kanemaki M, van Deursen F, Edmondson RD, et al. GINS maintains association of Cdc45 with MCM in replisome progression complexes at eukaryotic DNA replication forks. Nat Cell Biol. 2006;8(4):358–66.
11. Moyer SE, Lewis PW, Botchan MR. Isolation of the Cdc45/Mcm2-7/GINS (CMG) complex, a candidate for the eukaryotic DNA replication fork helicase. Proc Natl Acad Sci U S A. 2006;103(27):10236–41.
12. Diffley JFX. Regulation of early events in chromosome replication. Curr Biol. 2004;14(18): R778–86.
13. Friedman KL, Brewer BJ, Fangman WL. Replication profile of Saccharomyces cerevisiae chromosome VI. Genes Cells. 1997;2(11):667–78.

14. Yamashita M, Hori Y, Shinomiya T, Obuse C, Tsurimoto T, Yoshikawa H, et al. The efficiency and timing of initiation of replication of multiple replicons of Saccharomyces cerevisiae chromosome VI. Genes Cells. 1997;2(11):655–65.
15. Poloumienko A, Dershowitz A, De J, Newlon CS. Completion of replication map of Saccharomyces cerevisiae chromosome III. Mol Biol Cell. 2001;12(11):3317–27.
16. Raghuraman MK, Winzeler EA, Collingwood D, Hunt S, Wodicka L, Conway A, et al. Replication dynamics of the yeast genome. Science. 2001;294(5540):115–21.
17. Wyrick JJ, Aparicio JG, Chen T, Barnett JD, Jennings EG, Young RA, et al. Genome-wide distribution of ORC and MCM proteins in S. cerevisiae: high-resolution mapping of replication origins. Science. 2001;294(5550):2357–60.
18. Yabuki N, Terashima H, Kitada K. Mapping of early firing origins on a replication profile of budding yeast. Genes Cells. 2002;7(8):781–9.
19. Patel PK, Arcangioli B, Baker SP, Bensimon A, Rhind N. DNA replication origins fire stochastically in fission yeast. Mol Biol Cell. 2006;17(1):308–16.
20. Czajkowsky DM, Liu J, Hamlin JL, Shao Z. DNA combing reveals intrinsic temporal disorder in the replication of yeast chromosome VI. J Mol Biol. 2008;375(1):12–9.
21. Yang SC-H, Rhind N, Bechhoefer J. Modeling genome-wide replication kinetics reveals a mechanism for regulation of replication timing. Mol Syst Biol. 2010;6:404.
22. Bechhoefer J, Rhind N. Replication timing and its emergence from stochastic processes. Trends Genet. 2012;28(8):374–81.
23. Méchali M, Yoshida K, Coulombe P, Pasero P. Genetic and epigenetic determinants of DNA replication origins, position and activation. Curr Opin Genet Dev. 2013;23(2):124–31.
24. Raghuraman MK, Brewer BJ. Molecular analysis of the replication program in unicellular model organisms. Chromosome Res. 2010;18(1):19–34.
25. Renard-Guillet C, Kanoh Y, Shirahige K, Masai H. Temporal and spatial regulation of eukaryotic DNA replication: from regulated initiation to genome-scale timing program. Semin Cell Dev Biol. 2014;30:110–20.
26. Rhind N, Gilbert DM. DNA replication timing. Cold Spring Harb Perspect Biol. 2013; 5(8):a010132.
27. Hyrien O, Marheineke K, Goldar A. Paradoxes of eukaryotic DNA replication: MCM proteins and the random completion problem. Bioessays. 2003;25(2):116–25.
28. Rhind N, Yang SC-H, Bechhoefer J. Reconciling stochastic origin firing with defined replication timing. Chromosome Res. 2010;18(1):35–43.
29. Rhind N. DNA replication timing: random thoughts about origin firing. Nat Cell Biol. 2006;8(12):1313–6.
30. Brewer BJ, Fangman WL. The localization of replication origins on ARS plasmids in S. cerevisiae. Cell. 1987;51(3):463–71.
31. Celniker SE, Campbell JL. Yeast DNA replication in vitro: initiation and elongation events mimic in vivo processes. Cell. 1982;31(1):201–13.
32. Huberman JA, Spotila LD, Nawotka KA, el-Assouli SM, Davis LR. The in vivo replication origin of the yeast 2 microns plasmid. Cell. 1987;51(3):473–81.
33. Stinchcomb DT, Thomas M, Kelly J, Selker E, Davis RW. Eukaryotic DNA segments capable of autonomous replication in yeast. Proc Natl Acad Sci U S A. 1980;77(8):4559–63.
34. Marahrens Y, Stillman B. A yeast chromosomal origin of DNA replication defined by multiple functional elements. Science. 1992;255(5046):817–23.
35. Theis JF, Newlon CS. The ARS309 chromosomal replicator of Saccharomyces cerevisiae depends on an exceptional ARS consensus sequence. Proc Natl Acad Sci U S A. 1997; 94(20):10786–91.
36. Xu W, Aparicio JG, Aparicio OM, Tavaré S. Genome-wide mapping of ORC and Mcm2p binding sites on tiling arrays and identification of essential ARS consensus sequences in S. cerevisiae. BMC Genomics. 2006;7:276.
37. Eaton ML, Galani K, Kang S, Bell SP, MacAlpine DM. Conserved nucleosome positioning defines replication origins. Genes Dev. 2010;24(8):748–53.

38. Siow CC, Nieduszynska SR, Müller CA, Nieduszynski CA. OriDB, the DNA replication origin database updated and extended. Nucleic Acids Res. 2012;40(Database issue):D682–6.
39. Diffley JF, Cocker JH. Protein-DNA interactions at a yeast replication origin. Nature. 1992;357(6374):169–72.
40. Rao H, Marahrens Y, Stillman B. Functional conservation of multiple elements in yeast chromosomal replicators. Mol Cell Biol. 1994;14(11):7643–51.
41. Wilmes GM, Bell SP. The B2 element of the Saccharomyces cerevisiae ARS1 origin of replication requires specific sequences to facilitate pre-RC formation. Proc Natl Acad Sci U S A. 2002;99(1):101–6.
42. Remus D, Beuron F, Tolun G, Griffith J, Morris E, Diffley J. Concerted loading of Mcm2-7 double hexamers around DNA during DNA replication origin licensing. Cell. 2009;139:719.
43. On KF, Beuron F, Frith D, Snijders AP, Morris EP, Diffley JFX. Prereplicative complexes assembled in vitro support origin-dependent and independent DNA replication. EMBO J. 2014;33(6):605–20.
44. Crevel G, Cotterill S. Forced binding of the origin of replication complex to chromosomal sites in Drosophila S2 cells creates an origin of replication. J Cell Sci. 2012;125(Pt 4):965–72.
45. Huang RY, Kowalski D. Multiple DNA elements in ARS305 determine replication origin activity in a yeast chromosome. Nucleic Acids Res. 1996;24(5):816–23.
46. Theis JF, Newlon CS. Domain B of ARS307 contains two functional elements and contributes to chromosomal replication origin function. Mol Cell Biol. 1994;14(11):7652–9.
47. Hoggard T, Shor E, Müller CA, Nieduszynski CA, Fox CA. A link between ORC-origin binding mechanisms and origin activation time revealed in budding yeast. PLoS Genet. 2013;9(9):e1003798.
48. Donato JJ, Chung SCC, Tye BK. Genome-wide hierarchy of replication origin usage in Saccharomyces cerevisiae. PLoS Genet. 2006;2(9):e141.
49. Nieduszynski CA, Blow JJ, Donaldson AD. The requirement of yeast replication origins for pre-replication complex proteins is modulated by transcription. Nucleic Acids Res. 2005;33(8):2410–20.
50. Tanny RE, MacAlpine DM, Blitzblau HG, Bell SP. Genome-wide analysis of re-replication reveals inhibitory controls that target multiple stages of replication initiation. Mol Biol Cell. 2006;17(5):2415–23.
51. Brewer BJ, Fangman WL. Initiation at closely spaced replication origins in a yeast chromosome. Science. 1993;262(5140):1728–31.
52. Ferguson BM, Fangman WL. A position effect on the time of replication origin activation in yeast. Cell. 1992;68(2):333–9.
53. Pohl TJ, Kolor K, Fangman WL, Brewer BJ, Raghuraman MK. A DNA sequence element that advances replication origin activation time in Saccharomyces cerevisiae. G3. 2013;3(11):1955–63.
54. Koren A, Tsai H-J, Tirosh I, Burrack LS, Barkai N, Berman J. Epigenetically-inherited centromere and neocentromere DNA replicates earliest in S-phase. PLoS Genet. 2010;6(8):e1001068.
55. Pohl TJ, Brewer BJ, Raghuraman MK. Functional centromeres determine the activation time of pericentric origins of DNA replication in Saccharomyces cerevisiae. PLoS Genet. 2012;8(5):e1002677.
56. Hayashi MT, Takahashi TS, Nakagawa T, Nakayama J-I, Masukata H. The heterochromatin protein Swi6/HP1 activates replication origins at the pericentromeric region and silent mating-type locus. Nat Cell Biol. 2009;11(3):357–62.
57. Bianchi A, Shore D. Early replication of short telomeres in budding yeast. Cell. 2007; 128(6):1051–62.
58. Lian H-Y, Robertson ED, Hiraga S-I, Alvino GM, Collingwood D, McCune HJ, et al. The effect of Ku on telomere replication time is mediated by telomere length but is independent of histone tail acetylation. Mol Biol Cell. 2011;22(10):1753–65.
59. Stevenson JB, Gottschling DE. Telomeric chromatin modulates replication timing near chromosome ends. Genes Dev. 1999;13(2):146–51.
60. Cosgrove AJ, Nieduszynski CA, Donaldson AD. Ku complex controls the replication time of DNA in telomere regions. Genes Dev. 2002;16(19):2485–90.

61. Pappas DL, Frisch R, Weinreich M. The NAD(+)-dependent Sir2p histone deacetylase is a negative regulator of chromosomal DNA replication. Genes Dev. 2004;18(7):769–81.
62. Crampton A, Chang F, Pappas DL, Frisch RL, Weinreich M. An ARS element inhibits DNA replication through a SIR2-dependent mechanism. Mol Cell. 2008;30(2):156–66.
63. Knott SRV, Viggiani CJ, Tavaré S, Aparicio OM. Genome-wide replication profiles indicate an expansive role for Rpd3L in regulating replication initiation timing or efficiency, and reveal genomic loci of Rpd3 function in Saccharomyces cerevisiae. Genes Dev. 2009;23(9):1077–90.
64. Vogelauer M, Rubbi L, Lucas I, Brewer BJ, Grunstein M. Histone acetylation regulates the time of replication origin firing. Mol Cell. 2002;10(5):1223–33.
65. Espinosa MC, Rehman MA, Chisamore-Robert P, Jeffery D, Yankulov K. GCN5 is a positive regulator of origins of DNA replication in Saccharomyces cerevisiae. PLoS One. 2010;5(1):e8964.
66. Unnikrishnan A, Gafken PR, Tsukiyama T. Dynamic changes in histone acetylation regulate origins of DNA replication. Nat Struct Mol Biol. 2010;17(4):430–7.
67. Irlbacher H, Franke J, Manke T, Vingron M, Ehrenhofer-Murray AE. Control of replication initiation and heterochromatin formation in Saccharomyces cerevisiae by a regulator of meiotic gene expression. Genes Dev. 2005;19(15):1811–22.
68. Weber JM, Irlbacher H, Ehrenhofer-Murray AE. Control of replication initiation by the Sum1/Rfm1/Hst1 histone deacetylase. BMC Mol Biol. 2008;9:100.
69. Pryde F, Jain D, Kerr A, Curley R, Mariotti FR, Vogelauer M. H3 k36 methylation helps determine the timing of cdc45 association with replication origins. PLoS One. 2009;4(6):e5882.
70. Rizzardi LF, Dorn ES, Strahl BD, Cook JG. DNA replication origin function is promoted by H3K4 di-methylation in Saccharomyces cerevisiae. Genetics. 2012;192(2):371–84.
71. Lee W, Tillo D, Bray N, Morse RH, Davis RW, Hughes TR, et al. A high-resolution atlas of nucleosome occupancy in yeast. Nat Genet. 2007;39(10):1235–44.
72. Mavrich TN, Ioshikhes IP, Venters BJ, Jiang C, Tomsho LP, Qi J, et al. A barrier nucleosome model for statistical positioning of nucleosomes throughout the yeast genome. Genome Res. 2008;18(7):1073–83.
73. Mcguffee SR, Smith DJ, Whitehouse I. Quantitative, genome-wide analysis of eukaryotic replication initiation and termination. Mol Cell. 2013;50:123–35.
74. Simpson RT. Nucleosome positioning can affect the function of a cis-acting DNA element in vivo. Nature. 1990;343(6256):387–9.
75. Berbenetz NM, Nislow C, Brown GW. Diversity of eukaryotic DNA replication origins revealed by genome-wide analysis of chromatin structure. PLoS Genet. 2010;1:6(9).
76. Lipford JR, Bell SP. Nucleosomes positioned by ORC facilitate the initiation of DNA replication. Mol Cell. 2001;7(1):21–30.
77. Raghuraman MK, Brewer BJ, Fangman WL. Cell cycle-dependent establishment of a late replication program. Science. 1997;276(5313):806–9.
78. Heun P, Laroche T, Raghuraman MK, Gasser SM. The positioning and dynamics of origins of replication in the budding yeast nucleus. J Cell Biol. 2001;152(2):385–400.
79. Pope BD, Aparicio OM, Gilbert DM. SnapShot: replication timing. Cell. 2013;152(6):1390–1.
80. Soriano I, Morafraile EC, Vázquez E, Antequera F, Segurado M. Different nucleosomal architectures at early and late replicating origins in Saccharomyces cerevisiae. BMC Genomics. 2014;15:791.
81. Belsky JA, MacAlpine HK, Lubelsky Y, Hartemink AJ, MacAlpine DM. Genome-wide chromatin footprinting reveals changes in replication origin architecture induced by pre-RC assembly. Genes Dev. 2015;29(2):212–24.
82. Wu P-YJ, Nurse P. Establishing the program of origin firing during S phase in fission Yeast. Cell. 2009;136(5):852–64.
83. Callebaut I, Courvalin JC, Mornon JP. The BAH (bromo-adjacent homology) domain: a link between DNA methylation, replication and transcriptional regulation. FEBS Lett. 1999;446(1):189–93.
84. Kuo AJ, Song J, Cheung P, Ishibe-Murakami S, Yamazoe S, Chen JK, et al. The BAH domain of ORC1 links H4K20me2 to DNA replication licensing and Meier-Gorlin syndrome. Nature. 2012;484(7392):115–9.

85. Müller P, Park S, Shor E, Huebert DJ, Warren CL, Ansari AZ, et al. The conserved bromo-adjacent homology domain of yeast Orc1 functions in the selection of DNA replication origins within chromatin. Genes Dev. 2010;24(13):1418–33.
86. Hizume K, Yagura M, Araki H. Concerted interaction between origin recognition complex (ORC), nucleosomes and replication origin DNA ensures stable ORC-origin binding. Genes Cells. 2013;18(9):764–79.
87. Knott SRV, Peace JM, Ostrow AZ, Gan Y, Rex AE, Viggiani CJ, et al. Forkhead transcription factors establish origin timing and long-range clustering in S. cerevisiae. Cell. 2012;148(1-2):99–111.
88. Lõoke M, Kristjuhan K, Värv S, Kristjuhan A. Chromatin-dependent and -independent regulation of DNA replication origin activation in budding yeast. EMBO Rep. 2013;14(2): 191–8.
89. Duan Z, Andronescu M, Schutz K, Mcilwain S, Kim YJ, Lee C, et al. A three-dimensional model of the yeast genome. Nature. 2010;465(7296):363–7.
90. Cornacchia D, Dileep V, Quivy J-P, Foti R, Tili F, Santarella-Mellwig R, et al. Mouse Rif1 is a key regulator of the replication-timing programme in mammalian cells. EMBO J. 2012;31(18):3678–90.
91. Hayano M, Kanoh Y, Matsumoto S, Renard-Guillet C, Shirahige K, Masai H. Rif1 is a global regulator of timing of replication origin firing in fission yeast. Genes Dev. 2012;26(2):137–50.
92. Peace JM, Ter-Zakarian A, Aparicio OM. Rif1 regulates initiation timing of late replication origins throughout the S. cerevisiae genome. PLoS One. 2014;9(5):e98501.
93. Yamazaki S, Ishii A, Kanoh Y, Oda M, Nishito Y, Masai H. Rif1 regulates the replication timing domains on the human genome. EMBO J. 2012;31(18):3667–77.
94. Smith CD, Smith DL, DeRisi JL, Blackburn EH. Telomeric protein distributions and remodeling through the cell cycle in Saccharomyces cerevisiae. Mol Biol Cell. 2003;14(2):556–70.
95. Davé A, Cooley C, Garg M, Bianchi A. Protein phosphatase 1 recruitment by Rif1 regulates DNA replication origin firing by counteracting DDK activity. Cell Rep. 2014;7(1):61.
96. Hiraga S-I, Alvino GM, Chang F, Lian H-Y, Sridhar A, Kubota T, et al. Rif1 controls DNA replication by directing Protein Phosphatase 1 to reverse Cdc7-mediated phosphorylation of the MCM complex. Genes Dev. 2014;28(4):372–83.
97. Mattarocci S, Shyian M, Lemmens L, Damay P, Altintas DM, Shi T, et al. Rif1 controls DNA replication timing in yeast through the PP1 phosphatase Glc7. Cell Rep. 2014;7(1):69.
98. Sridhar A, Kedziora S, Donaldson AD. At short telomeres Tel1 directs early replication and phosphorylates Rif1. PLoS Genet. 2014;10(10):e1004691.
99. Aparicio OM, Stout AM, Bell SP. Differential assembly of Cdc45p and DNA polymerases at early and late origins of DNA replication. Proc Natl Acad Sci U S A. 1999;96(16):9130–5.
100. Mantiero D, Mackenzie A, Donaldson A, Zegerman P. Limiting replication initiation factors execute the temporal programme of origin firing in budding yeast. EMBO J. 2011;30(23): 4805–14.
101. Tanaka T, Umemori T, Endo S, Muramatsu S, Kanemaki M, Kamimura Y, et al. Sld7, an Sld3-associated protein required for efficient chromosomal DNA replication in budding yeast. EMBO J. 2011;30(10):2019–30.
102. Santocanale C, Diffley JF. A Mec1- and Rad53-dependent checkpoint controls late-firing origins of DNA replication. Nature. 1998;395(6702):615–8.
103. Shirahige K, Hori Y, Shiraishi K, Yamashita M, Takahashi K, Obuse C, et al. Regulation of DNA-replication origins during cell-cycle progression. Nature. 1998;395(6702):618–21.
104. Duncker BP, Shimada K, Tsai-Pflugfelder M, Pasero P, Gasser SM. An N-terminal domain of Dbf4p mediates interaction with both origin recognition complex (ORC) and Rad53p and can deregulate late origin firing. Proc Natl Acad Sci U S A. 2002;99(25):16087–92.
105. Lopez-Mosqueda J, Maas NL, Jonsson ZO, Defazio-Eli LG, Wohlschlegel J, Toczyski DP. Damage-induced phosphorylation of Sld3 is important to block late origin firing. Nature. 2010;467(7314):479–83.
106. Zegerman P, Diffley JFX. Checkpoint-dependent inhibition of DNA replication initiation by Sld3 and Dbf4 phosphorylation. Nature. 2010;467(7314):474–8.
107. Müller CA, Nieduszynski CA. Conservation of replication timing reveals global and local regulation of replication origin activity. Genome Res. 2012;22(10):1953–62.

108. Donley N, Thayer MJ. DNA replication timing, genome stability and cancer: late and/or delayed DNA replication timing is associated with increased genomic instability. Semin Cancer Biol. 2013;23(2):80–9.
109. McCarroll RM, Fangman WL. Time of replication of yeast centromeres and telomeres. Cell. 1988;54(4):505–13.
110. Feng W, Bachant J, Collingwood D, Raghuraman MK, Brewer BJ. Centromere replication timing determines different forms of genomic instability in Saccharomyces cerevisiae checkpoint mutants during replication stress. Genetics. 2009;183(4):1249–60.
111. Durkin SG, Glover TW. Chromosome fragile sites. Annu Rev Genet. 2007;41:169–92.
112. Letessier A, Millot GA, Koundrioukoff S, Lachagès A-M, Vogt N, Hansen RS, et al. Cell-type-specific replication initiation programs set fragility of the FRA3B fragile site. Nature. 2011;470(7332):120–3.
113. Ozeri-Galai E, Lebofsky R, Rahat A, Bester AC, Bensimon A, Kerem B. Failure of origin activation in response to fork stalling leads to chromosomal instability at fragile sites. Mol Cell. 2011;43(1):122–31.
114. Kunkel TA. Evolving views of DNA replication (in)fidelity. Cold Spring Harb Symp Quant Biol. 2009;74:91–101.
115. Marsolier-Kergoat M-C, Goldar A. DNA replication induces compositional biases in yeast. Mol Biol Evol. 2012;29(3):893–904.
116. Agier N, Fischer G. The mutational profile of the yeast genome is shaped by replication. Mol Biol Evol. 2012;29(3):905–13.
117. Lang GI, Murray AW. Mutation rates across budding yeast chromosome VI are correlated with replication timing. Genome Biol Evol. 2011;3:799–811.
118. Kumar D, Viberg J, Nilsson AK, Chabes A. Highly mutagenic and severely imbalanced dNTP pools can escape detection by the S-phase checkpoint. Nucleic Acids Res. 2010;38(12):3975–83.
119. Blow JJ, Gillespie PJ, Francis D, Jackson DA. Replication origins in Xenopus egg extract Are 5-15 kilobases apart and are activated in clusters that fire at different times. J Cell Biol. 2001;152(1):15–25.
120. Newman TJ, Mamun MA, Nieduszynski CA, Blow JJ. Replisome stall events have shaped the distribution of replication origins in the genomes of yeasts. Nucleic Acids Res. 2013;41(21): 9705–18.
121. Branzei D, Foiani M. Maintaining genome stability at the replication fork. Nat Rev Mol Cell Biol. 2010;11(3):208–19.
122. Calzada A, Hodgson B, Kanemaki M, Bueno A, Labib K. Molecular anatomy and regulation of a stable replisome at a paused eukaryotic DNA replication fork. Genes Dev. 2005;19(16):1905–19.
123. Ivessa AS, Lenzmeier BA, Bessler JB, Goudsouzian LK, Schnakenberg SL, Zakian VA. The Saccharomyces cerevisiae helicase Rrm3p facilitates replication past nonhistone protein-DNA complexes. Mol Cell. 2003;12(6):1525–36.
124. Tourrière H, Versini G, Cordón-Preciado V, Alabert C, Pasero P. Mrc1 and Tof1 promote replication fork progression and recovery independently of Rad53. Mol Cell. 2005;19(5):699–706.
125. Blow JJ, Ge XQ, Jackson DA. How dormant origins promote complete genome replication. Trends Biochem Sci. 2011;36(8):405–14.
126. Dershowitz A, Newlon CS. The effect on chromosome stability of deleting replication origins. Mol Cell Biol. 1993;13(1):391–8.
127. Dershowitz A, Snyder M, Sbia M, Skurnick JH, Ong LY, Newlon CS. Linear derivatives of Saccharomyces cerevisiae chromosome III can be maintained in the absence of autonomously replicating sequence elements. Mol Cell Biol. 2007;27(13):4652–63.
128. Maine GT, Sinha P, Tye BK. Mutants of S. cerevisiae defective in the maintenance of minichromosomes. Genetics. 1984;106(3):365–85.
129. Lengronne A, Schwob E. The yeast CDK inhibitor Sic1 prevents genomic instability by promoting replication origin licensing in late G(1). Mol Cell. 2002;9(5):1067–78.
130. Tanaka S, Diffley JFX. Deregulated G1-cyclin expression induces genomic instability by preventing efficient pre-RC formation. Genes Dev. 2002;16(20):2639–49.
131. Ayuda-Durán P, Devesa F, Gomes F, Sequeira-Mendes J, Avila-Zarza C, Gómez M, et al. The CDK regulators Cdh1 and Sic1 promote efficient usage of DNA replication origins to prevent chromosomal instability at a chromosome arm. Nucleic Acids Res. 2014;42(11):7057–68.

Chapter 3
Epigenetic vs. Sequence-Dependent Control of Eukaryotic Replication Timing

Kyle N. Klein and David M. Gilbert

Abstract Eukaryotic DNA replication follows a reproducible temporal pattern throughout S phase known as the replication timing (RT) program. RT is correlated with gene expression, chromatin structure, and 3D chromatin folding states; it helps to maintain genome integrity, correlates with mutation frequencies, and is altered in many diseases. However, the mechanisms regulating RT remain poorly defined. Studies over the last three decades have attempted to identify specific DNA sequences that regulate this program from yeasts to humans. Recent studies have implicated defined protein-binding motifs in yeasts. In mammals, there is indisputable evidence that epigenetic mechanisms regulate homologue-specific differences in RT, while artificial constructs have been shown to influence RT in a sequence-dependent manner and genomics approaches find compelling correlations of sequence variation to RT. However, the mechanisms linking these features to RT remain elusive.

Keywords DNA replication • Replication timing • DNA replication origin • Replication domains • Epigenetics • Primary DNA sequence • Transcription • Metazoans • *Cis*-acting • *Trans*-acting • Replication variation

Introduction

Eukaryotic chromosomes are replicated in a specific temporal pattern throughout S phase known as the "replication timing" (RT) program. Four hundred to eight hundred kilobase regions of the genome termed "replication domains" are replicated coordinately due to the synchronous firing of replication origins within each domain. In metazoans, gene-rich chromatin is replicated early and located at the nuclear interior, while gene-poor chromatin is replicated late and enriched at the nuclear and nucleolar periphery [1]. RT programs are regulated during development, often associated with changes in gene expression, conferring cell type-specific RT programs [2].

K.N. Klein • D.M. Gilbert (✉)
Department of Biological Science, Florida State University, Tallahassee, FL 32306-4295, USA
e-mail: gilbert@bio.fsu.edu

D.L. Kaplan (ed.), *The Initiation of DNA Replication in Eukaryotes*,
DOI 10.1007/978-3-319-24696-3_3

The biological significance of this phenomenon remains unclear. RT is evolutionarily conserved between closely related species and there is evidence for roles in maintaining genome integrity and genome evolution, and regulating programs of gene expression, which are the subject of several recent reviews [3–7].

Here we discuss evidence for the role of DNA sequence vs. chromatin and epigenetic mechanisms in regulating the RT program. First we develop the concepts that specific DNA sequences are not necessary for initiation of replication in eukaryotes to the degree that they are in prokaryotes. We outline the sequence-independent nature of the molecular mechanisms by which replication is licensed, initiated, and regulated to ensure the complete and once-per-cell-cycle duplication of the genome. We delve further into the independence of regulatory mechanisms determining origin usage vs. replication timing, taking many examples from yeasts while providing evidence for similarly independent mechanisms in metazoans. This provides the background for a discussion of the evidence for sequence-dependent versus epigenetic mechanisms regulating replication timing in both yeasts and metazoans. We remark on the importance of replication timing control in development and its possible links to transcription. We also investigate the more abstract concept of stochasticity and its role in determining origin firing patterns. We conclude with ideas on where the field of replication timing needs to progress in order to establish higher resolution replication timing profiles and more accurate characterization of replication dynamics in populations versus single cells.

Initiation of Eukaryotic Chromosomal DNA Replication Does Not Require Specific Consensus DNA Sequences

Eukaryotic DNA replication initiates at origins of replication interspersed along the length of each chromosome, but what determines the location of these origins is still poorly understood. Origins of replication were originally defined in budding yeast as autonomously replicating sequence (ARS) elements that could confer replicative capacity to minichromosomes or plasmid DNA [8]. These observations led to the notion that eukaryotic origins, like bacterial, plasmid, and many viral origins, are specified by specific DNA sequences that serve as binding sites for sequence-specific initiator proteins, conforming to the replicon model for bacterial plasmid replication originally proposed by Jacob and Cuzin [9]. However, it was later discovered that yeast is one of the only eukaryotic organisms to have a consensus sequence at which replication initiates [10]. Moreover, even in budding yeast, not all origins harbor a clear consensus [11, 12], and when the consensus origin sites are deleted, noncanonical sites can initiate replication [13]. Fission yeast preferentially initiate DNA replication within AT-rich stretches of DNA but those sites lack a consensus DNA sequence [14, 15]. Metazoans appear to lack any specific sequence requirements. In fact, during the rapid early cleavage stages of *Drosophila* and *Xenopus* development, when transcription is silent, replication initiates at random with respect to DNA sequence [16–19]. However, using extracts from these early embryos, artificial systems have been developed that can achieve site-specific initiation by manipulating features such as DNA methylation [20] or transcription factor targeting [21], leading many to hypothesize that the local chromatin environment contributes more strongly than DNA sequence to defining the locations of origins in higher organisms [22]. In mammalian cells, any DNA sequence above a critical size can replicate autonomously as a circular minichromosome [23], and cases of ectopic insertion into chromosomes have been identified where replication is found to initiate within the bacterial vector sequences [24–27].

Since functional assays failed to identify sequence elements that confer origin activity in metazoans, many turned to genomics to identify any features, sequence-dependent or chromatin-associated, which are correlated with origin activity. Methods such as isolation of small nascent strands (SNS), deemed "nascent" by virtue of being labeled during a brief BrdU pulse or by having a short RNA primer on their 5′ ends [28–31], or trapping DNA bubble structures ("bubble trap") characteristic of nascent replicons [32], have identified chromatin features such as DNaseI hypersensitive sites (DHSs), and even features of the primary DNA sequence such as those that can form a four stranded DNA structure known as the G4 quadruplex, that correlate with the positions of origins [33, 34]. In fact, in the case of the G4 quadruplex structures, transplantation studies in chicken DT40 cells demonstrated that G4 structure and orientation were necessary but not sufficient for origin activity at an ectopic site [34]. In the case of DHSs, a mathematical model whose input was based solely on the assumption that DHSs are the determinants of replication initiation sites was able to accurately model the temporal program for DNA replication during S phase. However, only a fraction of DHSs or G4 quadruplexes align with origins and vice versa, so the extent to which these and similar correlations (e.g., H3K4me1, H3K27me3, and others) [35, 36] are causatively linked to the activity of particular sets of origins remains to be determined.

Several limitations must be taken into account with the interpretation of genomic approaches to origin mapping, particularly in metazoan genomes. First, the concordance of bubble trap and SNS datasets is only about 50 % [35], suggesting that the two methods capture different populations of origins (Fig. 3.1) [37]. Even the concordance of SNS datasets across laboratories is rather weak [38]. Contamination of SNS preparations with lambda exonuclease (lexo)-resistant unreplicated DNA (e.g., GC rich and G4 quadruplex-containing DNA) [39] is unlikely to account for this lack of concordance, since some datasets were collected by enriching for BrdU-substituted SNS or by releasing captured small molecules with RNaseI without lexo treatment [31, 40]. Rather, the fact that concordance is improved by sequencing SNS to saturation suggests that each dataset is capturing only a subset of potential initiation sites [37, 41, 42]. A second problem is that, like all methods that begin by pooling cell populations, cell-to-cell heterogeneity is lost (Fig. 3.1). This is particularly important in the case of replication origins, as it is clear from studies that map sites of DNA synthesis on individual isolated DNA fibers that sites of initiation vary tremendously from molecule to molecule (even in yeasts) (Fig. 3.1), with different

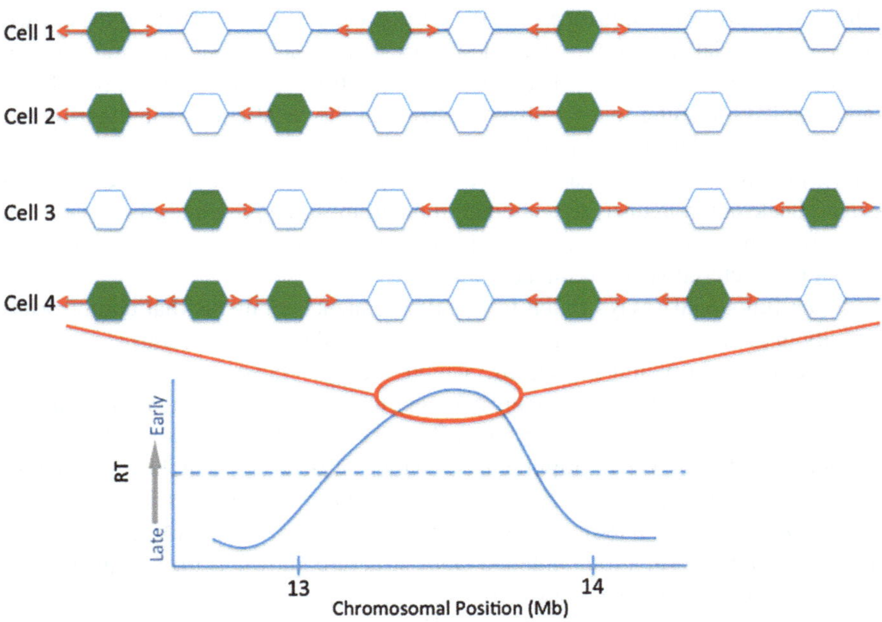

Fig. 3.1 Replication origin usage is heterogeneous in higher eukaryotes. Origins of replication (*hexagons*) are activated heterogeneously in a population of cells. *White hexagons* represent inactive origins while *green hexagons* represent activated origins. Metazoan RT profile (*bottom*) averages origin usage within population of cells to create a smooth curve around replication origin clusters. *X*-axis indicates chromosome position. *Y*-axis is RT on $\log_2$ scale; regions with positive values (above *dashed zero line*) are early replicating and regions with negative values are late replicating. Different techniques used to map origins may only be capturing a subset of these origins

sites having widely different probabilities of firing [43–48]. This means that population-based methods are likely only detecting the most frequently utilized origins, at their respective resolution (i.e., clusters of inefficient origins might be detected as a single efficient origin by bubble trap but go undetected by SNS) [37]. Moreover, the only study that compared ensemble methods to DNA fiber methods directly demonstrated that many more origins are detected in population-based methods than are activated in a single S phase [49].

Taken together, it is clear that there is a great deal of flexibility in the sites that can be selected to initiate DNA replication, particularly in metazoans (Fig. 3.1). To date, high-throughput methods capture only the most frequently utilized sites of initiation, and even those sites appear to lack any single-consensus DNA sequence. A complete picture of the relative degree and efficiencies with which specific sites are utilized as origins of replication will require improvements in the throughput of single-molecule methods [37].

Once-per-Cell-Cycle Regulation of Initiation Does Not Require Specific Consensus DNA Sequences

Despite differences in sequence conservation, all eukaryotes assemble the basic molecular machinery necessary for DNA replication in a similar fashion, and utilize similar mechanisms to ensure that the genome is duplicated once and exactly once per cell cycle. This includes preventing any molecule from firing more than once, as well as mechanisms to ensure that all DNA is replicated before cell division. Here, we would like to make clear that none of these well-accepted mechanisms invoke a requirement for specific DNA sequences.

First, to prevent multiple rounds of replication within a single cell cycle, eukaryotes have devised a system of two non-overlapping windows in which origins are first licensed and then activated. In late M phase and early G1 origins are bound by the origin recognition complex (ORC), which along with cell division cycle 6 (Cdc6)- and cdc10-dependent transcript 1 (Cdt1) ultimately load the double-hexamer helicase minichromosomal maintenance (Mcm) complex to form a pre-replicative complex (pre-RC) [50]. Sites occupied by pre-RCs are "licensed"; they have the potential to be activated during S phase [51]. As G1 phase progresses, the S phase-specific kinases cyclin-dependent kinase (S-CDK) and Dbf4-dependent kinase (DDK) become activated. First, strong DDK phosphorylation of the N-terminus of Mcm4 alleviates the repressive activity of this subunit [52] allowing for recruitment of Sld3 and Cdc45 and the activation of the helicase complex [53, 54]. S-CDK then acts to recruit a second suit of essential replication factors Sld2, Dpb11, Mcm10, GINS, and DNA polymerase ε [53–57]. Further steps, possibly dependent upon RecQ4 [58], unwind the DNA and recruit DNA polymerase α to initiate DNA replication. Thus, pre-RCs can only form under conditions of low CDK and DDK activity, which do not permit initiation, while initiation can occur only under conditions of high CDK and DDK, which prevent pre-RC formation [53, 54]. These two mutually exclusive periods of the cell cycle strictly prevent any molecules from reinitiating DNA replication. Although multiple redundant mechanisms exist that vary slightly in different species, this basic underlying mechanism is highly conserved [59]. Importantly, this well-studied and universally accepted mechanism does not invoke any requirement for specific DNA sequences. Regardless of where pre-RCs assemble, initiation at any given site will occur exactly once per cell cycle.

Equally important as preventing more than one initiation per cell cycle is ensuring that all DNA completes replication before cell division. Given that there is always a finite chance that any given pre-RC might not be activated in a timely manner, eukaryotes assemble considerably more pre-RCs than necessary. As mentioned, each pre-RC has a probability of firing, and in yeasts only the highest probability origins are utilized in most normal cell cycles, while additional "dormant" origins can be activated if the cell experiences conditions of replication stress that stall replication forks [60–64]. Normally, these dormant pre-RCs are destroyed when forks emanating from the activated replicons pass through them, but they can be recruited if fork arrest substantially delays the replication of downstream DNA containing these dormant origins. In metazoans, it has been more difficult to assess which origins should be classified as dormant, likely due to the broader spectrum of origin efficiencies (Fig. 3.1). Nonetheless, it is clear that a vast excess of origins are assembled and that these excess origins become necessary for genome integrity under conditions of replication stress [60–64]. As each cell progresses through S phase, the probability of initiation at any given pre-RC increases. Mathematical models that invoke the existence of recycled limiting factors whose free concentration is dependent upon the amount of actively replicating DNA find the best fit to experimental data, suggesting that as DNA replication approaches completion the entire reservoir of limiting factors is made available to initiate at low-probability (i.e., dormant) origins [65].

Taken together, the complex distribution of replication origin firing in mammalian cells results from a combination of stochastic activation of origins at an approximate distribution of one per 125 kb (40–60 min of DNA replication time) and the inactivation of other origins by passing replication forks. Pre-RCs assembled at high-probability sites are more likely to initiate and are thus more easily detected by mapping methods, but any one of the many sites of pre-RC assembly may or may not fire in any particular cell cycle. Overall, the regulatory logic for the completion problem is to assemble many more pre-RCs than needed to ensure that no large genomic segment is devoid of initiation sites [66, 67]. Importantly, as with mechanisms to prevent reinitiation, the mechanism to ensure complete duplication does not invoke a requirement for any specific sequences.

Replication Timing Control Is Uncoupled from Origin Specification

In eukaryotic genomes, not all origins are activated at the same time. Some origins are fired early during S phase and others are fired later, resulting in a defined replication timing (RT) program [7]. Although this program is frequently misrepresented as being mediated by origins that are intrinsically programmed to fire early or late during S phase (i.e., "early origins" or "late origins"), the determinants of RT are clearly separate from the origins themselves. Identification of specific sites of initiation in budding and fission yeasts allowed for some of the first studies of the elements controlling RT. For example, ectopic positioning of an early-firing origin to the late-replicating telomeric region resulted in delayed firing of the origin [68], and late origins cloned on ARS plasmids generally replicated early [68], providing some of the first evidence that RT determinants can be uncoupled from the sequences that confer origin activity. Subsequently, telomeric late origins in fission yeast were shown to replicate early on circular ARS plasmids unless they also contained separate telomeric sequences flanking the origins [69]. Telomeres are clustered at the nuclear periphery in yeast, but simply artificially tethering an early origin to the nuclear periphery in budding yeast was not sufficient to confer late initiation onto this origin [70]. Yeast genomes also contain late-firing origins distal from the telomere (internal late origins) and their delayed firing is also influenced by *cis* elements in the regions flanking the ARS in both budding [71] and fission [72] yeasts. In budding yeast, the late element could delay origin firing when inserted up to 6 kb away [71]. In fission yeast, a 10-bp G-rich late consensus sequence (LCS) was identified that, when inserted up to 800 bp away, could force normally early origins to fire later in S phase [72].

DNA sequences have also been identified that are necessary for early RT [73–75]. Centromeres can confer early replication of origins located up to 20 kb away [73, 76]. In fission yeast, this has been shown to be due to the ability of the heterochromatin protein-1 (HP1) homolog Swi6 to recruit the DDK subunit Dfp1 (homolog of Dbf4) to advance replication of pericentromeric origins [77]. In budding yeast, a computational screen for protein-binding sites near Rpd3L-delayed late replication origins revealed significant depletion of DNA-binding sites for the forkhead transcription factors (Fkh1 and Fkh2), leading to the finding that Fkh1 and Fkh2 proteins are required for early replication of 30 % of the early-firing origins and appear to mediate this effect by clustering early origins and enhancing their association with initiation protein Cdc45, a step that is controlled by DDK (Fig. 3.2) [74]. Additional sequences have been identified that can confer early replication on ARS plasmids through as-yet unidentified mechanisms [75]. Altogether, these data make

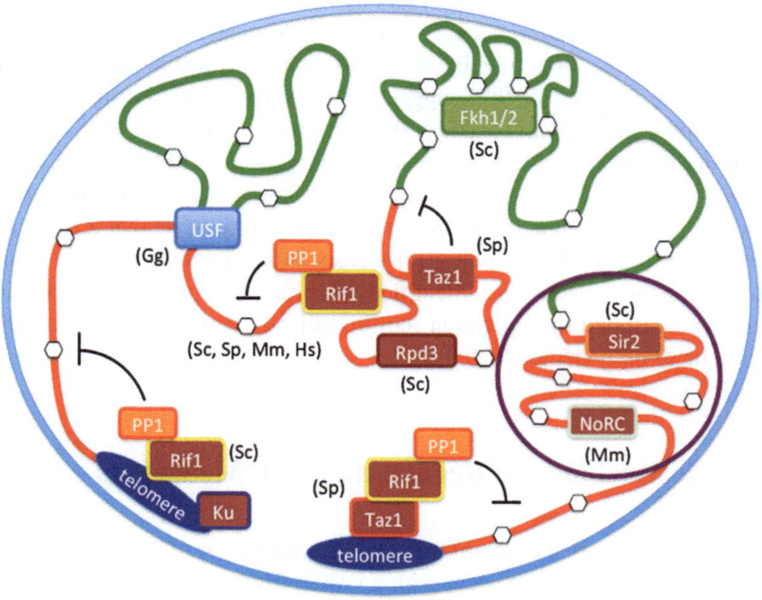

Fig. 3.2 RT control by *trans*-acting factors. Factors that promote the firing of origins (*white hexagons*) are colored in *green* and bind within early (*green*) domains. Factors that inhibit origin firing are colored in *red* and bind within late (*red*) domain. The *violet circle* denotes the nucleolus. Species specificity of particular *trans*-acting factors is indicated by abbreviation. *Sc Saccharomyces cerevisiae*, *Sp Schizosaccharomyces pombe*, *Gg Gallus gallus*, *Mm Mus musculus*, *Hs Homo sapiens* (adapted from Fig. 2, Mechali et al. [170])

it clear that *cis*-acting DNA sequence elements that are separate from replication origins themselves influence the RT of nearby origins in yeasts.

Comparative genome-wide studies of origin locations and RT also support the conclusion that RT is independent of replication origin specification. Comparisons of different species of budding [78] or fission [43] yeasts demonstrated poor conservation of the precise positions of replication origins, but strong conservation of RT across these species. In mammals as well, RT is highly conserved between species [79, 80] but when the same origin mapping method was applied to a conserved locus (the beta-globin locus) in both mouse and human cells, the pattern of initiation was found to be highly divergent, with human cells initiating in a defined (2 kb) region [81] and mouse cells initiating at many sites distributed throughout a 40-kb initiation zone [82].

Finally, while the determinants of mammalian origin specification remain poorly defined, several studies provide indirect evidence supporting the uncoupling of origin sites and RT. First, by initiating DNA replication in mammalian nuclei isolated from cells synchronized at various times during G1 phase it was shown that the RT of domains is established distinctly prior to the selection of replication origin sites [83]. Second, studies in cycling *Xenopus* egg extracts [84] and cultured mammalian cells [85, 86] demonstrated that domains labeled early in one S phase were labeled early in the second S phase, whereas sites of initiation labeled on DNA fibers did not coincide. Third, as proposed many years ago [87], the S-phase replication checkpoint responds to replication stress by activating dormant origins within replication domains, while at the same time inhibiting any initiation within later firing domains that have yet to initiate [88–90], suggesting that origin firing is subordinate to the time of domain activation. Together, these results suggest that RT is determined early during G1 phase within the context of large-scale (400–800 kb) replication domains, while selection of the specific origin sites is a downstream event that is considerably more flexible. In fact, recent findings demonstrate that RT in mammals is regulated at the level of topologically associating domains (TADs), which are structural units of chromosomes identified by chromatin conformation capture [91] that likely correspond to replication foci visualized cytogenetically (although this has yet to be directly demonstrated).

Epigenetic Mechanisms Regulating RT in Metazoans

Ever since the discovery that one of the two X chromosomes of mammals is randomly chosen to be inactivated coincident with a switch from early to late replication [92, 93], it has been presumed that epigenetic mechanism regulate RT. In fact, genomic imprinting in mammals is also associated with silencing and delayed replication of the imprinted allele [94, 95]. In addition, some non-imprinted loci are mono-allelically expressed, with the active allele again replicated earlier [96–98], including different copies of rDNA genes, reinforcing the notion that RT is epigenetically regulated. Exactly how replication is delayed by these allele-specific mechanisms is not clear, but several examples reveal a common theme of ncRNA expression. For example, approximately 50 % of rDNA copies are expressed in somatic cells, and those expressed copies are early replicating while the silent copies are late [96]. In mouse, this differential regulation appears to be established early in development, possibly at a time similar to X inactivation [99]. Interestingly, overexpression of the ATP-dependent chromatin remodeling complex NoRC is sufficient to shift a large percentage of the early-replicating rDNA copies to late replication, coincident with their transcriptional silencing (Fig. 3.2). The mechanism by which NoRC silences rDNA is by the binding of its TIP5 subunit to a small ncRNA that is complementary to the rDNA promoter and transcribed during mid-S phase [100], implicating but by no means directly demonstrating a link between ncRNA and late RT. During X-inactivation, the switch to late replication requires the synthesis of a long ncRNA termed Xist [101, 102]. By contrast early replication of the Dlk1-Dio3 imprinted locus requires bidirectional transcription of ncRNAs from the imprinting control region [103]. Other mechanisms to distinguish imprinted alleles also exist and may influence RT, such as the binding of CTCF near the H19 imprinting control region [104, 105]. Overall, these results point to specific epigenetic mechanisms that control homologue-specific RT of large chromosomal domains.

A Closer Look at X Inactivation

Using the copy number method for RT determination genome wide [37, 106, 107] and deep sequencing, it is possible to identify a sufficient density of SNPs to distinguish the RT of individual homologous chromosomes of phased diploid genomes. One such study investigated the dramatic allelic difference in RT between the active and inactive X chromosomes (Xa and Xi) in female mammals. This difference had been well established by cytogenetic studies [92]. Moreover, cytogenetic and live-cell imaging studies demonstrated that all detectable Xi DNA replication occurs within a 1–2-h period of time in mid-late S phase [108, 109], suggesting a near-synchronous firing of origins throughout the chromosome, while the Xa was replicated in an autosomal-like domain pattern. Genome-wide analysis of SNPs between the Xa and Xi was able to provide molecular confirmation of this distinction [106].

Unfortunately, while the resolution of Xa/Xi replication timing differential was certainly improved by the SNP analysis, several conclusions of this study were overstated. First, the authors over-interpreted from this study that the inactive X chromosome is replicated "randomly" in a fashion similar to the replication patterns of frog and fly embryos [106, 110]. Of course such a synchronous replication pattern at a narrowly defined time during S phase is far from random. Randomly replicating sequences in such ensemble molecular studies would appear to replicate throughout S phase due to their replication at different times in different cells, such as is observed when cells in M and G1 phase are coerced to initiate replication before RT is established at the timing decision point (TDP) [111] or when G2-phase cells that have lost their replication program are coerced to re-replicate [112].

Random replication would also be detected cytogenetically as a large degree of cell-to-cell heterogeneity in Xi RT, which is not what is observed. Second, the authors also conclude from this study that origin specification is random on the Xi and that origins are closely spaced [106, 110], but their methodology did not have the resolution to map replication origin specification or spacing. Moreover, this conclusion contradicts prior origin-mapping studies indicating that both the Xa and Xi use the same specific origin sites to initiate replication of their DNA regardless of their differing RT patterns [113, 114].

Over-interpretations aside, the Koren et al. study confirmed that inactivation and heterochromatinization of the Xi is associated with synchronous late replication across most of the Xi. This cannot be due to structural arrangements that cause the Xi to form one very large TAD, as TAD structure seems to be conserved on the Xa and Xi [115]. Rather, it is likely due to a consolidation of many TADs into one larger subnuclear compartment, as occurs on autosomes coincident with X-inactivation [2, 116]. Since X-inactivation is random with respect to parent of origin, it is unquestionably an epigenetic mechanism.

Evidence for Sequence-Dependent Mechanisms Regulating RT in Metazoans

Despite the compelling evidence for epigenetic mechanisms controlling RT, there is also intriguing evidence for a role of the primary DNA sequence in regulating RT, both in artificial and more native contexts. Experiments with transgenic mice indicated that the locus control region controlling expression of the mouse beta-globin gene was able to dictate local replication timing in a transcription-independent manner [117]. Targeting histone acetylases and deacetylases near a mapped initiation site at the human beta globin locus was shown to accelerate or delay, respectively, the timing of replication of the local region [118]. When a replication origin near the avian beta-globin gene was flanked by insulator sequences and inserted into a late-replicating region in avian cells, a shift to early replication occurred that was dependent upon the presence of both flanking insulators, nearby transcription, and binding sites for the USF transcription factor (Fig. 3.2) [119]. Together, these artificial systems demonstrate that specific DNA sequences can influence RT in the right context.

Some endogenous *cis*-acting regions have also been shown to influence RT on a chromosome wide scale in mouse and human cells. Although Xist is implicated in homologue-specific epigenetic silencing and late replication of the inactive X chromosome (Xi) [92, 101, 102], deletion of Xist in somatic cells leads to a chromosome-wide further delay in the replication of the inactive chromosome [101]. Similarly, long ncRNAs have been identified on human chromosomes 6 and 15 (with evidence that they exist on all human and mouse chromosomes) whose deletion results in severely delayed RT of the entire chromosome while apparently retaining the relative replication times of domains along the length of the chromosome [120–123].

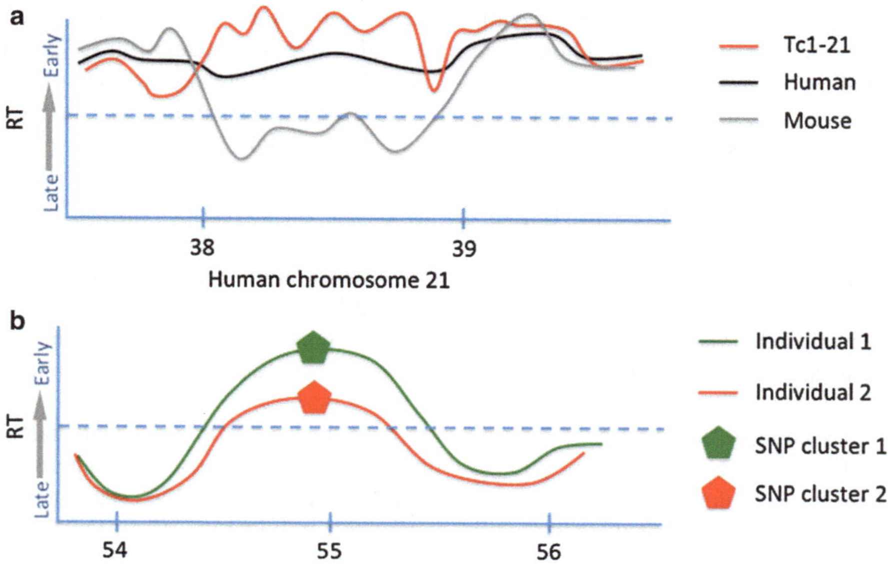

Fig. 3.3 Evidence that primary DNA sequence regulates mammalian RT. (**a**) The Tc1-21 human chromosome retains portions of its original human RT pattern in multiple mouse cell types. RT profile of human chromosome 21 stably maintained in mouse cells shows regions that retain human chromosome 21 RT patterns. RT profiles of Tc1-21 (*red*), human (*black*), and syntenic mouse (*grey*) region from fibroblasts. RT profile axes as in Fig. 3.1 (adapted with permission from Fig. 2C, Pope et al. [124]). (**b**) Clusters of SNPs correlate with changes in RT patterns in human cells. Variations in RT between individuals (red versus green RT profiles) on the Mb scale correlate with SNP clusters and small indels within or in the direct vicinity of varied regions. Shown is a variation in replication peak height, but other variations such as peak presence and slope of RT profile have been observed. RT profile axes as in Fig. 3.1

A different approach to the question of whether primary DNA sequence can dictate RT was taken by Pope et al. by studying mice that carry a copy of human chromosome 21 [124]. These researchers found that the ectopic human Chr21 maintained a human-specific RT profile in two different mouse tissues, including human-specific developmental changes in RT (Fig. 3.3a). At several regions where RT did not match either human RT or the RT of regions of conserved synteny in mouse, it was found that intrachromosomal rearrangements introduced during the construction of the mouse strain had juxtaposed early- and late-replicating DNA, allowing for a preliminary glance of the sequences sufficient to retain RT in an ectopic context. Segments that retained a replication domain boundary (defined as transitions in replication timing previously identified with genome-wide methods) retained their native RT at the ectopic site, while segments that were separated from a domain boundary adopted the RT of the neighboring sequences or insertion site. Together, these results provided compelling evidence that interspecies differences in RT result from the underlying DNA sequence, and that DNA sequences establish the boundaries between differentially replicating regions [124].

Next-generation sequencing techniques have allowed for the study of how DNA sequence variation between homologues and individuals correlates with RT. Such studies, focused on the RT differences at polymorphic autosomal sites either between autosomes or between individuals, have revealed a small number of subtle RT differences linked to sequence variation. In the first study, approximately 9.5 % of the autosomal genome showed allelic variation in RT in primary human erythroblasts, with each segment differing by less than 30 % of the length of S phase [125]. Differences in SNPs or small indels correlated poorly with RT differences between alleles on a genome-wide scale, but large structural variations (10 kb to 1 Mb) were associated with allelic asynchrony indicating that large-scale sequence divergence may cause RT variations between alleles. Asynchronous RT regions between alleles were also enriched in G-quadruplex motifs, CpG islands, and transcription start sites [125].

By contrast, another study has shown that clusters of SNPs and small indels within areas of RT variation correlate with RT variations between individuals or between homologues, indicating a role for smaller variations in DNA sequence in RT control; these areas were termed RT quantitative trait loci (rtQTLs) (Fig. 3.3b) [126]. The stretches of DNA encompassing the clusters of SNPs for each of the rtQTLs (2–160 kb; median 20 kb) were generally smaller than those reported by the study discussed above [125]; yet the changes in RT between individuals at rtQTLs (median 0.66 Mb) matched the size of replication domains (400–800 kb; see below), suggesting that sequence variation at the 2–160 kb scale can influence the RT of entire replication domains. The authors found that rtQTLs were within a median distance of 52 kb from the very broad computationally defined replication timing peaks. From this, they suggested that rtQTLs affect RT by affecting replication origins [126, 127], but of course the resolution was not sufficient to draw any conclusions about replication origins. Moreover, given what is known about RT regulation (summarized above), particularly the heterogeneity of origin usage in cell populations, if these rtQTLs are causally linked to RT, they are likely influencing the RT of domains by increasing the probability of firing within the domain, rather than, as the authors and commentator's suggest, activating or silencing specific origins. In fact, a recent study of allelic replication origin usage concluded that there is little variation in origin site selection between the two homologues, even in locations where the rtQTLs were found [128]. Rather, in cases of asynchronous replication, these authors found differences in the efficiencies of usage of the same origins. Finally, since >95 % of RT variants were not associated with detectable sequence variation, it remains to be determined whether the rtQTLs are causally linked to the RT variation. However, in one published case a single SNP was found to be associated with origin activity 53 kb upstream of fragile X repeats [129] accompanied by delayed RT of regions both up- and downstream of the repeats [130], suggesting that differential origin usage and timing can emerge from a single SNP, with profound phenotypic consequences.

Altogether, these studies provide compelling correlative evidence that specific DNA sequences may have important influences on RT, but the causal linkages and mechanisms remain elusive.

Trans-Acting Factors Affecting Replication Timing

Studies of *trans*-acting factors provide insight into both sequence-dependent and epigenetic mechanisms regulating replication timing. For example, the function of the various late-conferring sequence (LCS) elements in yeasts is beginning to become clear, primarily through the identification of *trans*-acting factors regulating RT. Telomere and subtelomere binding factors Ku, Taz1, and Rif1 have all been shown to maintain late replication of chromosome ends in yeasts. The disruption of any of these proteins advances RT of telomeric and subtelomeric regions [131–133]. Taz1 and Rif1 are known to interact at yeast telomeres to control telomere length and stability [134]. Taz1 homodimers bind two tandem telomeric repeats at both telomeres and internal late origins in fission yeast. These sequences are essential to delay origin firing until late S phase (Fig. 3.2) [131]. Fission yeast Rif1 binds at telomeres in a Taz1-dependent fashion, but also binds centromeres and chromosome arms in a Taz1-independent manner (Fig. 3.2). Rif1 binding in fission yeast is less specific than Taz1 binding but 40 % of Rif1-binding sites contain the LCS motif [133]. Budding yeast Rif1 is also required for late replication of internal origins, but it remains unclear from genome-wide ChIP analysis whether budding yeast Rif1 binds to loci other than telomeres (A. Donaldson, personal communication). In both budding and fission yeasts, Rif1 targets a PP1 phosphatase to areas of late replication, counteracting DDK-mediated phosphorylation of MCM required to fire origins (Fig. 3.2). Disruption of Rif1's PP1-binding domain mimics the advanced RT of telomeric origins observed in a complete Rif1 knockout [135–138]. Interplay between Rif1, which controls telomere length in yeasts, and the telomerase recruiting Ku complex has also been hypothesized to regulate RT of telomeric origins as telomeric origins are fired early in both Rif1 and Ku mutants [139]. Rif1 is thought to "measure" the number of telomeric repeats by direct binding and then delaying RT of nearby origins once a threshold of Rif1 binding has been achieved (Fig 3.2). Without Ku-mediated telomerase recruitment telomeres are shortened and cannot reach the Rif1-binding threshold that delays firing of nearby origins. Telomere length and late replication of telomeric origins can be rescued in Ku-negative cells by further deletion of Pif1, a negative regulator of telomere length, confirming the role of telomere length in RT regulation at proximal origins [139]. Sir complex proteins that establish a repressive chromatin state have also been shown to delay RT of telomeric regions (Fig. 3.2) [140]. Telomere-binding proteins have been thought to recruit these chromatin modifiers to telomeres to establish a condensed and thus late-replicating chromatin state. Deletion of the histone deacetylase Rpd3 or its binding partner Sin3 advances the RT of over 100 internal late origins in budding yeast without affecting telomeric origins (Fig. 3.2) [141, 142]. Others have hypothesized that the tethering of telomeres to the nuclear periphery, as by the Ku complex, may sequester late-replicating regions away from limiting essential replication factors (Fig. 3.2) [132]. Sir4 also contributes to tethering of silent telomeres to the nuclear envelope and may prevent early firing of origins through both chromatin structure and peripheral sequestration [143]. However, as mentioned above, artificially tethering an internally localized origin to the periphery is not sufficient to delay its initiation of replication [70]. Hence, a combination of epigenetic state, nuclear localization, and biochemical modification status is likely contributing to the maintenance telomeric regions as late replicating in yeasts.

Trans-acting factors influencing RT control in metazoans have proven very difficult to identify. In fact, ablation of most chromatin regulators and histone post-translational modifications have very little influence on RT, even those that correlate very strongly to RT [108, 118, 144, 145]. As mentioned above, the chromatin remodeler NoRC is necessary for late rDNA replication [100], while the embryonic stem cell-specific esBAF chromatin remodeler is necessary to maintain the RT of a small number of replication domains [146]. A recent breakthrough identified Rif1 as the first *trans*-acting factor that is responsible for the RT of approximately 30 % of the mouse and human genome [147, 148]. Disruption of human Rif1 led to increased levels of MCM phosphorylation by DDK [147], suggesting that human Rif1 may antagonize DDK in a mechanism similar to that in yeast. Indeed mammalian and yeast Rif1 are structurally very similar [149] and human Rif1 interacts with PP1 [150]. Disruption of Rif1 led to an increase in the sizes of chromatin loops, suggesting that Rif1 may also organize chromatin spatially in the nucleus. Rif1 appears to localize to mammalian heterochromatin [147, 148], so a current working model (Fig. 3.2) is that Rif1 may bind late domains and help tether them to the nuclear envelope and other late-replicating compartments of the nucleus, creating zones of high PP1 activity that antagonize DDK. Rif1 knockout cells exhibit both early to late and late to early replication timing alterations. Depletion of Rif1 may both disrupt chromatin organization and redistribute PP1 throughout the nucleus, lowering PP1 near late-replicating regions and increasing PP1 near early-replicating domains. This model is supported by recent work in yeast showing that de-repression of normally late origins in rDNA repeats (~30 % of yeast origins) by deletion of the histone deacetylase (HDAC) Sir2 causes earlier firing of these origins in conjunction with delayed firing of normally early origins, suggesting that de-repressed rDNA origins compete with normally early origins for limiting replication factors [151].

Replication and Transcription During Development

As alluded to in the introduction, in mammalian cells approximately half of the genome changes RT during development in segments of 400–800 kb accompanied by changes in transcription of genes within the changing regions [2, 79]. In fact, the genome can be divided into constitutively early, constitutively late, and developmentally regulated domains. Constitutively early and late domains have distinct sequence compositions (AT content, repetitive sequence family composition, and gene densities), subnuclear interaction compartments, and chromatin accessibility to nuclease attack, while developmentally regulated domains have intermediate or unusual sequence composition [2, 152–154] and are particularly resistant to nuclease attack regardless of their RT [155, 156] and are less confined to particular subnuclear compartments [157]. Hence, while it is currently unclear how these factors influence the regulation of RT, it is clear that the individual replication domains differ dramatically in many properties that are associated with their RT behavior during development.

Developmental control of RT and its correlation with transcriptional changes offer the opportunity to study mechanisms regulating changes in RT, which may be influenced by specific regulatory DNA sequences. As discussed above, in the case of imprinting it is clear that one of the two homologues is chosen to be later replicating by epigenetic mechanisms during development. However, there are cases where specific DNA sequences can induce RT changes either through histone modifications or possibly transcriptional induction [118]. In addition, induction of transcription from a Gal4/UAS element in *Drosophila* is accompanied by earlier replication of a broad area of chromatin neighboring the newly active gene and changes in histone modifications and chromatin structure to a more "open" state [158]. Recently, targeting of a strong transactivator to induce the transcription of a gene within a developmentally regulated domain was able to partially advance the RT of that domain [159]. Contrary to the conclusions of these authors, however, transcription is clearly not sufficient to regulate RT as the observed shift in RT upon artificial gene induction did not reproduce the normal developmental RT change. Moreover, many genes are induced without RT changes, and there are even a small number of genes that are induced upon a switch from early to late RT [153, 160]. There are also domains that change RT without any detectable changes in transcription [153], although it is conceivable that changes in transcription have occurred but have escaped detection. It may very well be the case that each domain is regulated by a complex combination of mechanisms, similar to regulation of gene expression itself.

Stochastic vs. Deterministic Mechanisms Regulating RT

Underpinning all of these observations is the recent debate between two general models of replication control that have been used to explain the RT program. The first model posits that origin usage is deterministic, mediated by origins that are programed to fire at specific times during S phase. The second model posits that the probability an origin will fire at any given time during S phase, and the resulting timing program is an average of the stochastic probabilities within a population; origins with a higher probability of firing will tend to fire earlier. The well-defined and often highly efficient origins in budding yeast naively appear to suggest a more deterministic model [161–163]. However, DNA fiber experiments have shown that origin usage is stochastic in both budding and fission yeasts, with different origins being used in different S phases [44, 164]. Competition between origins for limiting replication factors has been shown to influence the RT and firing efficiency of yeast origins [48, 165–167], and the differential ability to compete for such factors likely underlies the probability of an origin firing early vs. late. In mammals, although general accessibility does not seem to accurately predict replication timing [155], computer models that can accurately predict the RT of entire genomes from DHS data suggest that local regions of hypersensitive chromatin may influence the probability of firing [36]. Another possible contributing factor is that the activation of one origin may increase the probability of neighboring origins firing [168–170].

Future Directions

The mechanisms that control the ordered program of DNA replication remain enigmatic. One emerging theme is that the probability of an origin firing at any given time is a function of the ability of its surrounding environment to recruit or antagonize DDK activity, but the individual mechanisms at different chromosome loci are likely to be complex and variable. One of the single greatest concerns that needs to be addressed if we are to generate a complete picture of replication control is to understand the heterogeneity of replication between single cells and homologues within cells. With the exception of DNA fiber analyses, all current replication analysis methods rely on large numbers of cells to determine an average of replication dynamics over a population of cells, precluding such insight. Even DNA fiber analyses, while providing molecule-to-molecule heterogeneity, still pool DNA fibers from large populations and do not retain cell of origin information. New efforts are needed to generate RT profiles and origin maps in single cells and on single molecules to assess the positions of all potential initiation sites and their probabilities of firing in different metabolic and developmental contexts. Such analyses will also allow deeper insight into the influence of deterministic vs. stochastic mechanisms, with intracellular (inter-homologue) heterogeneity reflecting intrinsic influences and cell-to-cell heterogeneity reflecting extrinsic influences.

A second area of needed research is to identify the *cis* and *trans* factors that regulate large-scale replication domain structure. Chromosomes are organized nonrandomly into stable TADs that serve as the units of DNA replication timing regulation. The boundaries of these units are generally stable during differentiation, and are even detectable as preserved in single-cell chromatin conformation capture methods [171]. Given the one-to-one structure–function relationship of TADs and their 3D interaction profiles with replication domains and their RT, respectively, understanding the elements that organize these self-associating units and their 3D arrangement in the nucleus will be certain to reveal important insights into the elements regulating RT.

Acknowledgements We would like to acknowledge J.C. Rivera Mulia for critical reading of the manuscript. Work in the Gilbert lab is supported by Grants GM083337 and GM085354.

References

1. Küpper K, Kölbl A, Biener D, Dittrich S, von Hase J, Thormeyer T, et al. Radial chromatin positioning is shaped by local gene density, not by gene expression. Chromosoma. 2007;116: 285–306.
2. Hiratani I, Ryba T, Itoh M, Yokochi T, Schwaiger M, Chang CW, et al. Global reorganization of replication domains during embryonic stem cell differentiation. PLoS Biol. 2008;6: 2220–36.
3. Renard-Guillet C, Kanoh Y, Shirahige K, Masai H. Temporal and spatial regulation of eukaryotic DNA replication: from regulated initiation to genome-scale timing program. Semin Cell Dev Biol. 2014;30:110–20.
4. Sima J, Gilbert DM. Complex correlations: replication timing and mutational landscapes during cancer and genome evolution. Curr Opin Genet Dev. 2014;25:93–100.
5. Yoshida K, Poveda A, Pasero P. Time to be versatile: regulation of the replication timing program in budding yeast. J Mol Biol. 2013;425:4696–705.
6. Pope BD, Gilbert DM. The replication domain model: regulating replicon firing in the context of large-scale chromosome architecture. J Mol Biol. 2013;425:4690–5.
7. Rhind N, Gilbert DM. DNA replication timing. Cold Spring Harb Perspect Biol. 2013;5.
8. Stinchcomb DT, Struhl K, Davis RW. Isolation and characterisation of a yeast chromosomal replicator. Nature. 1979;282:39–43.
9. Jacob F, Brenner S, Cuzin F. On the regulation of DNA replication in bacteria. Cold Spring Harb Symp [Internet]. 1963. http://symposium.cshlp.org/content/28/329.full.pdf\npapers2://publication/uuid/5D4D31D1-C062-4C84-AE47-1DA6A02A3001.
10. Celniker SE, Sweder K, Srienc F, Bailey JE, Campbell JL. Deletion mutations affecting autonomously replicating sequence ARS1 of Saccharomyces cerevisiae. Mol Cell Biol. 1984;4:2455–66.
11. Theis JF, Yang C, Schaefer CB, Newlon CS. DNA sequence and functional analysis of homologous ARS elements of Saccharomyces cerevisiae and S. carlsbergensis. Genetics. 1999;152: 943–52.
12. Theis JF, Newlon CS. Two compound replication origins in Saccharomyces cerevisiae contain redundant origin recognition complex binding sites. Mol Cell Biol. 2001;21:2790–801.
13. Bogenschutz NL, Rodriguez J, Tsukiyama T. Initiation of DNA replication from non-canonical sites on an origin-depleted chromosome. PLoS One [Internet]. 2014;9(12):e114545. http://www.pubmedcentral.nih.gov/articlerender.fcgi?artid=4259332&tool=pmcentrez&rendertype=abstract. Accessed 14 Feb 2015.
14. Hayashi M, Katou Y, Itoh T, Tazumi A, Yamada Y, Takahashi T, et al. Genome-wide localization of pre-RC sites and identification of replication origins in fission yeast. EMBO J. 2007;26: 1327–39.
15. Heichinger C, Penkett CJ, Bähler J, Nurse P, Bahler J. Genome-wide characterization of fission yeast DNA replication origins. EMBO J. 2006;25:5171–9.
16. Hyrien O, Méchali M. Chromosomal replication initiates and terminates at random sequences but at regular intervals in the ribosomal DNA of Xenopus early embryos. EMBO J. 1993; 12:4511–20.
17. Sasaki T, Sawado T, Yamaguchi M, Shinomiya T. Specification of regions of DNA replication initiation during embryogenesis in the 65-kilobase DNApolα-dE2F locus of Drosophila melanogaster. Mol Cell Biol. 1999;19:547–55.
18. Walter J, Newport JW. Regulation of replicon size in Xenopus egg extracts. Science. 1997; 275:993–5.
19. Harland RM, Laskey RA. Regulated replication of DNA microinjected into eggs of Xenopus laevis. Cell. 1980;21:761–71.
20. Harvey KJ, Newport J. CpG methylation of DNA restricts prereplication complex assembly in Xenopus egg extracts. Mol Cell Biol. 2003;23:6769–79.

21. Cox LS, Hupp T, Midgley CA, Lane DP. A direct effect of activated human p53 on nuclear DNA replication. EMBO J. 1995;14:2099–105.
22. Ding Q, MacAlpine DM. Defining the replication program through the chromatin landscape. Crit Rev Biochem Mol Biol. 2011;46:165–79.
23. Heinzel SS, Krysan PJ, Tran CT, Calos MP. Autonomous DNA replication in human cells is affected by the size and the source of the DNA. Mol Cell Biol [Internet]. 1991;11(4):2263–72. http://www.pubmedcentral.nih.gov/articlerender.fcgi?artid=359926&tool=pmcentrez&rendertype=abstract. Accessed 25 Feb 2015
24. Ishimi Y, Matsumoto K, Ohba R. DNA replication from initiation zones of mammalian cells in a model system. Mol Cell Biol. 1994;14:6489–96.
25. Berberich S, Trivedi A, Daniel DC, Johnson EM, Leffak M. In vitro replication of plasmids containing human c-myc DNA. J Mol Biol. 1995;245:92–109.
26. McWhinney C, Leffak M. Autonomous replication of a DNA fragment containing the chromosomal replication origin of the human c-myc gene. Nucleic Acids Res. 1990;18:1233–42.
27. Krysan PJ, Calos MP. Replication initiates at multiple locations on an autonomously replicating plasmid in human cells. Mol Cell Biol. 1991;11:1464–72.
28. Giacca M, Zentilin L, Norio P, Diviacco S, Dimitrova D, Contreas G, et al. Fine mapping of a replication origin of human DNA. Proc Natl Acad Sci U S A. 1994;91:7119–23.
29. Gerbi S. Replication initiation point mapping. Methods [Internet]. 1997;13:271–80. http://linkinghub.elsevier.com/retrieve/pii/S1046-2023(97)90526-0.
30. Bielinsky AK, Gerbi SA. Discrete start sites for DNA synthesis in the yeast ARS1 origin. Science [Internet]. 1998;279:95–8. http://www.sciencemag.org/cgi/doi/10.1126/science.279.5347.95, http://www.ncbi.nlm.nih.gov/pubmed/9417033.
31. Kunnev D, Freeland A, Qin M, Leach RW, Wang J, Shenoy RM, et al. Effect of minichromosome maintenance protein 2 deficiency on the locations of DNA replication origins. Genome Res [Internet]. 2015;25(4):558–69. http://www.ncbi.nlm.nih.gov/pubmed/25762552. Accessed 26 May
32. Mesner LD, Valsakumar V, Karnani N, Dutta A, Hamlin JL, Bekiranov S. Bubble-chip analysis of human origin distributions demonstrates on a genomic scale significant clustering into zones and significant association with transcription. Genome Res. 2011;21:377–89.
33. Besnard E, Babled A, Lapasset L, Milhavet O, Parrinello H, Dantec C, et al. Unraveling cell type-specific and reprogrammable human replication origin signatures associated with G-quadruplex consensus motifs. Nat Struct Mol Biol. 2012;19(8):837–44.
34. Valton AL, Hassan-Zadeh V, Lema I, Boggetto N, Alberti P, Saintomé C, et al. G4 motifs affect origin positioning and efficiency in two vertebrate replicators. EMBO J. 2014;33:732–46.
35. Picard F, Cadoret JC, Audit B, Arneodo A, Alberti A, Battail C, et al. The spatiotemporal program of DNA replication is associated with specific combinations of chromatin marks in human cells. PLoS Genet. 2014;10.
36. Gindin Y, Valenzuela MS, Aladjem MI, Meltzer PS, Bilke S. A chromatin structure-based model accurately predicts DNA replication timing in human cells. Mol Syst Biol. 2014;10.
37. Gilbert DM. Evaluating genome-scale approaches to eukaryotic DNA replication. Nat Rev Genet. 2010;11:673–84.
38. Gilbert DM. Replication origins run (ultra) deep. Nat Struct Mol Biol [Internet]. Nature Publishing Group; 2012;19(8):740–2. http://www.nature.com/nsmb/journal/v19/n8/full/nsmb.2352.html#ref2. Accessed 5 Mar 2015.
39. Foulk MS, Urban JM, Casella C, Gerbi SA. Characterizing and controlling intrinsic biases of Lambda exonuclease in nascent strand sequencing reveals phasing between nucleosomes and G-quadruplex motifs around a subset of human replication origins. Genome Res [Internet]. 2015. http://www.ncbi.nlm.nih.gov/pubmed/25695952. Accessed 20 Feb 2015.
40. Aladjem MI, Wahl GM. Mapping replication fork direction by leading strand analysis. Methods. 1997;13:281–92.
41. Cadoret J-C, Meisch F, Hassan-Zadeh V, Luyten I, Guillet C, Duret L, et al. Genome-wide studies highlight indirect links between human replication origins and gene regulation. Proc Natl Acad Sci U S A. 2008;105:15837–42.

42. Karnani N, Taylor CM, Malhotra A, Dutta A. Genomic study of replication initiation in human chromosomes reveals the influence of transcription regulation and chromatin structure on origin selection. Mol Biol Cell. 2010;21:393–404.
43. Xu J, Yanagisawa Y, Tsankov AM, Hart C, Aoki K, Kommajosyula N, et al. Genome-wide identification and characterization of replication origins by deep sequencing. Genome Biol. 2012;13:27.
44. Patel P, Arcangioli B, Baker S. DNA replication origins fire stochastically in fission yeast. Mol Biol Cell [Internet]. 2006;17:308–16. http://www.ncbi.nlm.nih.gov/pubmed/16251353, http://www.molbiolcell.org/cgi/content/abstract/17/1/308.
45. Dijkwel PA, Wang S, Hamlin JL. Initiation sites are distributed at frequent intervals in the Chinese hamster dihydrofolate reductase origin of replication but are used with very different efficiencies. Mol Cell Biol. 2002;22:3053–65.
46. Guan Z, Hughes CM, Kosiyatrakul S, Norio P, Sen R, Fiering S, et al. Decreased replication origin activity in temporal transition regions. J Cell Biol. 2009;187:623–35.
47. Kaykov A, Nurse P. The spatial and temporal organization of origin firing during the S-phase of fission yeast. Genome Res [Internet]. 2015. http://www.ncbi.nlm.nih.gov/pubmed/25650245. Accessed 6 Feb 2015.
48. Wu PYJ, Nurse P. Establishing the program of origin firing during S phase in fission yeast. Cell. 2009;136:852–64.
49. Cayrou C, Coulombe P, Vigneron A, Stanojcic S, Ganier O, Peiffer I, et al. Genome-scale analysis of metazoan replication origins reveals their organization in specific but flexible sites defined by conserved features. Genome Res. 2011;21:1438–49.
50. Remus D, Beuron F, Tolun G, Griffith JD, Morris EP, Diffley JFX. Concerted loading of Mcm2-7 double hexamers around DNA during DNA replication origin licensing. Cell. 2009;139:719–30.
51. Bell SP, Dutta A. DNA replication in eukaryotic cells. Annu Rev Biochem. 2002;71:333–74.
52. Sheu Y-J, Stillman B. The Dbf4-Cdc7 kinase promotes S phase by alleviating an inhibitory activity in Mcm4. Nature. 2010;463:113–7.
53. Heller RC, Kang S, Lam WM, Chen S, Chan CS, Bell SP. Eukaryotic origin-dependent DNA replication in vitro reveals sequential action of DDK and S-CDK kinases. Cell. 2011;146:80–91.
54. Yeeles JTP, Deegan TD, Janska A, Early A, Diffley JFX. Regulated eukaryotic DNA replication origin firing with purified proteins. Nature [Internet]. Nature Publishing Group, a division of Macmillan Publishers Limited. All Rights Reserved. 2015. doi:10.1038/nature14285. Accessed 6 Mar 2015.
55. Moyer SE, Lewis PW, Botchan MR. Isolation of the Cdc45/Mcm2-7/GINS (CMG) complex, a candidate for the eukaryotic DNA replication fork helicase. Proc Natl Acad Sci U S A. 2006;103: 10236–41.
56. Ilves I, Petojevic T, Pesavento JJ, Botchan MR. Activation of the MCM2-7 helicase by association with Cdc45 and GINS proteins. Mol Cell. 2010;37:247–58.
57. Labib K. How do Cdc7 and cyclin-dependent kinases trigger the initiation of chromosome replication in eukaryotic cells? Genes Dev. 2010;24(12):1208–19.
58. Sanuki Y, Kubota Y, Kanemaki MT, Takahashi TS, Mimura S, Takisawa H. RecQ4 promotes the conversion of the pre-initiation complex at a site-specific origin for DNA unwinding in Xenopus egg extracts. Cell Cycle [Internet]. 2015. http://www.ncbi.nlm.nih.gov/pubmed/25602506. Accessed 25 Feb 2015.
59. Remus D, Diffley JF. Eukaryotic DNA replication control: lock and load, then fire. Curr Opin Cell Biol. 2009;21(6):771–7.
60. Ge XQ, Jackson DA, Blow JJ. Dormant origins licensed by excess Mcm2-7 are required for human cells to survive replicative stress. Genes Dev. 2007;21:3331–41.
61. Blow JJ, Ge XQ, Jackson DA. How dormant origins promote complete genome replication. Trends Biochem Sci. 2011;36:405–14.
62. Woodward AM, Göhler T, Luciani MG, Oehlmann M, Ge X, Gartner A, et al. Excess Mcm2-7 license dormant origins of replication that can be used under conditions of replicative stress. J Cell Biol. 2006;173:673–83.
63. McIntosh D, Blow JJ. Dormant origins, the licensing checkpoint, and the response to replicative stresses. Cold Spring Harb Perspect Biol. 2012;4.

64. Ibarra A, Schwob E, Méndez J. Excess MCM proteins protect human cells from replicative stress by licensing backup origins of replication. Proc Natl Acad Sci U S A. 2008;105:8956–61.
65. Goldar A, Labit H, Marheineke K, Hyrien O. A dynamic stochastic model for DNA replication initiation in early embryos. PLoS One. 2008;3.
66. Méchali M. Eukaryotic DNA, replication origins: many choices for appropriate answers. Nat Rev Mol Cell Biol. 2010;11:728–38.
67. Cvetic C, Walter JC. Eukaryotic origins of DNA replication: could you please be more specific? Sem Cell Dev Biol. 2005;16:343–53.
68. Ferguson BM, Fangman WL. A position effect on the time of replication origin activation in yeast. Cell. 1992;68:333–9.
69. Chaudari A, Huberman J. Identification of two telomere-proximal fission yeast DNA replication origins constrained by nearby cis-acting sequences to replicate in late S phase. F1000Research [Internet]. 2012;1:58. http://www.pubmedcentral.nih.gov/articlerender.fcgi?artid=3790605&tool=pmcentrez&rendertype=abstract.
70. Ebrahimi H, Robertson ED, Taddei A, Gasser SM, Donaldson AD, Hiraga S. Early initiation of a replication origin tethered at the nuclear periphery. J Cell Sci. 2010;123:1015–9.
71. Friedman KL, Diller JD, Ferguson BM, Nyland SVM, Brewer BJ, Fangman WL. Multiple determinants controlling activation of yeast replication origins late in S phase. Genes Dev. 1996;10:1595–607.
72. Yompakdee C, Huberman JA. Enforcement of late replication origin firing by clusters of short G-rich DNA sequences. J Biol Chem. 2004;279:42337–44.
73. Pohl TJ, Brewer BJ, Raghuraman MK. Functional centromeres determine the activation time of pericentric origins of DNA replication in Saccharomyces cerevisiae. PLoS Genet. 2012;8.
74. Knott SRV, Peace JM, Ostrow AZ, Gan Y, Rex AE, Viggiani CJ, et al. Forkhead transcription factors establish origin timing and long-range clustering in S. cerevisiae. Cell. 2012;148:99–111.
75. Pohl TJ, Kolor K, Fangman WL, Brewer BJ, Raghuraman MK. A DNA sequence element that advances replication origin activation time in Saccharomyces cerevisiae. G3 (Bethesda) [Internet]. 2013;3:1955–63. http://www.pubmedcentral.nih.gov/articlerender.fcgi?artid=3815058&tool=pmcentrez&rendertype=abstract.
76. Natsume T, Müller CA, Katou Y, Retkute R, Gierliński M, Araki H, et al. Kinetochores coordinate pericentromeric cohesion and early DNA replication by Cdc7-Dbf4 kinase recruitment. Mol Cell. 2013;50:661–74.
77. Hayashi MT, Takahashi TS, Nakagawa T, Nakayama J, Masukata H. The heterochromatin protein Swi6/HP1 activates replication origins at the pericentromeric region and silent mating-type locus. Nat Cell Biol [Internet]. 2009;11(3):357–62. http://www.ncbi.nlm.nih.gov/pubmed/19182789. Accessed 25 Feb 2015.
78. Di Rienzi SC, Lindstrom KC, Mann T, Noble WS, Raghuraman MK, Brewer BJ. Maintaining replication origins in the face of genomic change. Genome Res. 2012;22:1940–52.
79. Ryba T, Hiratani I, Lu J, Itoh M, Kulik M, Zhang J, et al. Evolutionarily conserved replication timing profiles predict long-range chromatin interactions and distinguish closely related cell types. Genome Res. 2010;20:761–70.
80. Yue F, Cheng Y, Breschi A, Vierstra J, Wu W, Ryba T, et al. A comparative encyclopedia of DNA elements in the mouse genome. Nature [Internet]. 2014;515(7527):355–64. http://www.ncbi.nlm.nih.gov/pubmed/25409824. Accessed 19 Nov 2014.
81. Kitsberg D, Selig S, Keshet I, Cedar H. Replication structure of the human beta-globin gene domain. Nature. 1993;366:588–90.
82. Aladjem MI, Rodewald LW, Lin CM, Bowman S, Cimbora DM, Brody LL, et al. Replication initiation patterns in the beta-globin loci of totipotent and differentiated murine cells: evidence for multiple initiation regions. Mol Cell Biol [Internet]. 2002;22(2):442–52. http://www.pubmedcentral.nih.gov/articlerender.fcgi?artid=139749&tool=pmcentrez&rendertype=abstract. Accessed 1 Mar 2015.
83. Gilbert DM, Takebayashi SI, Ryba T, Lu J, Pope BD, Wilson KA, et al. Space and time in the nucleus developmental control of replication timing and chromosome architecture. Cold Spring Harb Symp Quant Biol. 2010;75:143–53.

84. Labit H, Perewoska I, Germe T, Hyrien O, Marheineke K. DNA replication timing is deterministic at the level of chromosomal domains but stochastic at the level of replicons in Xenopus egg extracts. Nucleic Acids Res. 2008;36:5623–34.
85. Jackson DA, Pombo A. Replicon clusters are stable units of chromosome structure: evidence that nuclear organization contributes to the efficient activation and propagation of S phase in human cells. J Cell Biol. 1998;140:1285–95.
86. Sadoni N, Cardoso MC, Stelzer EHK, Leonhardt H, Zink D. Stable chromosomal units determine the spatial and temporal organization of DNA replication. J Cell Sci. 2004;117: 5353–65.
87. Gilbert DM. Replication origin plasticity, Taylor-made: inhibition vs recruitment of origins under conditions of replication stress. Chromosoma. 2007;116:341–7.
88. Gillespie PJ, Blow JJ. Clusters, factories and domains: the complex structure of S-phase comes into focus. Cell Cycle. 2010;9:3218–26.
89. Thomson AM, Gillespie PJ, Blow JJ. Replication factory activation can be decoupled from the replication timing program by modulating Cdk levels. J Cell Biol. 2010;188:209–21.
90. Ge XQ, Blow JJ. Chk1 inhibits replication factory activation but allows dormant origin firing in existing factories. J Cell Biol. 2010;191:1285–97.
91. Pope BD, Ryba T, Dileep V, Yue F, Wu W, Denas O, et al. Topologically associating domains are stable units of replication-timing regulation. Nature [Internet]. 2014;515(7527):402–5. http://www.ncbi.nlm.nih.gov/pubmed/25409831. Accessed 19 Nov 2014.
92. TAYLOR JH. Asynchronous duplication of chromosomes in cultured cells of Chinese hamster. J Biophys Biochem Cytol. 1960;7:455–64.
93. Lyon MF. Lyonisation of the X chromosome. Lancet [Internet]. 1963;2(7317):1120–1. http://www.ncbi.nlm.nih.gov/pubmed/14063435. Accessed 1 Nov 2015.
94. Kitsberg D, Selig S, Brandeis M, Simon I, Keshet I, Driscoll DJ, et al. Allele-specific replication timing of imprinted gene regions. Nature. 1993;364:459–63.
95. Simon I, Tenzen T, Reubinoff BE, Hillman D, McCarrey JR, Cedar H. Asynchronous replication of imprinted genes is established in the gametes and maintained during development. Nature. 1999;401:929–32.
96. Li J, Santoro R, Koberna K, Grummt I. The chromatin remodeling complex NoRC controls replication timing of rRNA genes. EMBO J. 2005;24:120–7.
97. Mostoslavsky R, Singh N, Tenzen T, Goldmit M, Gabay C, Elizur S, et al. Asynchronous replication and allelic exclusion in the immune system. Nature. 2001;414:221–5.
98. Farago M, Rosenbluh C, Tevlin M, Fraenkel S, Schlesinger S, Masika H, et al. Clonal allelic predetermination of immunoglobulin-κ rearrangement. Nature. 2012;490:561–5.
99. Schlesinger S, Selig S, Bergman Y, Cedar H. Allelic inactivation of rDNA loci. Genes Dev. 2009;23:2437–47.
100. Bierhoff H, Schmitz K, Maass F, Ye J, Grummt I. Noncoding transcripts in sense and antisense orientation regulate the epigenetic state of ribosomal RNA genes. Cold Spring Harb Symp Quant Biol. 2010;75:357–64.
101. Diaz-Perez S V, Ferguson DO, Wang C, Csankovszki G, Wang C, Tsai S-C, et al. A deletion at the mouse Xist gene exposes *trans*-effects that alter the heterochromatin of the inactive X chromosome and the replication time and DNA stability of both X chromosomes. Genetics [Internet]. 2006;174(3):1115–33. http://www.pubmedcentral.nih.gov/articlerender.fcgi?artid=1667074&tool=pmcentrez&rendertype=abstract. Accessed 1 Mar 2015.
102. Diaz-Perez S, Ouyang Y, Perez V, Cisneros R, Regelson M, Marahrens Y. The element(s) at the nontranscribed Xist locus of the active X chromosome controls chromosomal replication timing in the mouse. Genetics. 2005;171:663–72.
103. Kota SK, Llères D, Bouschet T, Hirasawa R, Marchand A, Begon-Pescia C, et al. ICR noncoding RNA expression controls imprinting and DNA replication at the Dlk1-Dio3 domain. Dev Cell [Internet]. 2014;31(1):19–33. http://www.ncbi.nlm.nih.gov/pubmed/25263792. Accessed 1 Mar 2015.
104. Rand E, Ben-Porath I, Keshet I, Cedar H. CTCF elements direct allele-specific undermethylation at the imprinted H19 locus. Curr Biol. 2004;14:1007–12.

105. Bergström R, Whitehead J, Kurukuti S, Ohlsson R. CTCF regulates asynchronous replication of the imprinted H19/Igf2 domain. Cell Cycle. 2007;6:450–4.
106. Koren A, McCarroll SA. Random replication of the inactive X chromosome. Genome Res. 2014;24:64–9.
107. Dimitriadou E, Van der Aa N, Cheng J, Voet T, Vermeesch JR. Single cell segmental aneuploidy detection is compromised by S phase. Mol Cytogenet [Internet]. 2014;7:46. http://www.pubmedcentral.nih.gov/articlerender.fcgi?artid=4114140&tool=pmcentrez&rendertype=abstract.
108. Wu R, Singh PB, Gilbert DM. Uncoupling global and fine-tuning replication timing determinants for mouse pericentric heterochromatin. J Cell Biol. 2006;174:185–94.
109. Casas-Delucchi CS, Brero A, Rahn H-P, Solovei I, Wutz A, Cremer T, et al. Histone acetylation controls the inactive X chromosome replication dynamics. Nat Commun. 2011;2:222.
110. Koren A. DNA replication timing: coordinating genome stability with genome regulation on the X chromosome and beyond. Bioessays [Internet]. 2014;36(10):997–1004. http://www.ncbi.nlm.nih.gov/pubmed/25138663. Accessed 12 Jan 2015.
111. Dimitrova DS, Gilbert DM. The spatial position and replication timing of chromosomal domains are both established in early G1 phase. Mol Cell. 1999;4:983–93.
112. Lu J, Li F, Murphy CS, Davidson MW, Gilbert DM. G2 phase chromatin lacks determinants of replication timing. J Cell Biol. 2010;189:967–80.
113. Gómez M, Brockdorff N. Heterochromatin on the inactive X chromosome delays replication timing without affecting origin usage. Proc Natl Acad Sci U S A. 2004;101:6923–8.
114. Cohen SM, Brylawski BP, Cordeiro-Stone M, Kaufman DG. Same origins of DNA replication function on the active and inactive human X chromosomes. J Cell Biochem. 2003;88:923–31.
115. Nora EP, Lajoie BR, Schulz EG, Giorgetti L, Okamoto I, Servant N, et al. Spatial partitioning of the regulatory landscape of the X-inactivation centre. Nature. 2012;485:381–5.
116. Hiratani I, Gilbert DM. Autosomal lyonization of replication domains during early mammalian development. Adv Exp Med Biol. 2010;695:41–58.
117. Simon I, Tenzen T, Mostoslavsky R, Fibach E, Lande L, Milot E, et al. Developmental regulation of DNA replication timing at the human beta globin locus. EMBO J. 2001;20:6150–7.
118. Goren A, Tabib A, Hecht M, Cedar H. DNA replication timing of the human beta-globin domain is controlled by histone modification at the origin. Genes Dev. 2008;22:1319–24.
119. Hassan-Zadeh V, Chilaka S, Cadoret JC, Ma MKW, Boggetto N, West AG, et al. USF binding sequences from the HS4 insulator element impose early replication timing on a vertebrate replicator. PLoS Biol. 2012;10.
120. Donley N, Smith L, Thayer MJ. ASAR15, a cis-acting locus that controls chromosome-wide replication timing and stability of human chromosome 15. PLoS Genet [Internet]. 2015;11(1):e1004923. http://journals.plos.org/plosgenetics/article?id = 10.1371/journal.pgen.1004923. Accessed 18 Jan 2015.
121. Donley N, Stoffregen EP, Smith L, Montagna C, Thayer MJ. Asynchronous replication, mono-allelic expression, and long range cis-effects of ASAR6. PLoS Genet. 2013;9.
122. Stoffregen EP, Donley N, Stauffer D, Smith L, Thayer MJ. An autosomal locus that controls chromosomewide replication timing and mono-allelic expression. Hum Mol Genet. 2011;20:2366–78.
123. Thayer MJ. Mammalian chromosomes contain *cis*-acting elements that control replication timing, mitotic condensation, and stability of entire chromosomes. Bioessays. 2012;34:760–70.
124. Pope BD, Chandra T, Buckley Q, Hoare M, Ryba T, Wiseman FK, et al. Replication-timing boundaries facilitate cell-type and species-specific regulation of a rearranged human chromosome in mouse. Hum Mol Genet. 2012;21:4162–70.
125. Mukhopadhyay R, Lajugie J, Fourel N, Selzer A, Schizas M, Bartholdy B, et al. Allele-specific genome-wide profiling in human primary erythroblasts reveal replication program organization. PLoS Genet. 2014;10.
126. Koren A, Handsaker RE, Kamitaki N, Karlić R, Ghosh S, Polak P, et al. Genetic variation in human DNA replication timing. Cell [Internet]. 2014;159(5):1015–26. http://www.ncbi.nlm.nih.gov/pubmed/25416942. Accessed 17 Nov 2014.

127. Hause RJ, Shendure J. Genetic variation meets replication origins. Cell [Internet]. Elsevier; 2014;159(5):973–4. http://www.ncbi.nlm.nih.gov/pubmed/25416936. Accessed 20 Nov 2014.
128. Bartholdy B, Mukhopadhyay R, Lajugie J, Aladjem MI, Bouhassira EE. Allele-specific analysis of DNA replication origins in mammalian cells. Nat Commun [Internet]. 2015;6:7051. http://www.ncbi.nlm.nih.gov/pubmed/25987481. Accessed 20 May 2015.
129. Gerhardt J, Zaninovic N, Zhan Q, Madireddy A, Nolin SL, Ersalesi N, et al. *Cis*-acting DNA sequence at a replication origin promotes repeat expansion to fragile X full mutation. J Cell Biol [Internet]. 2014;206(5):599–607. http://jcb.rupress.org/content/206/5/599.abstract. Accessed 12 Mar 2015.
130. Hansen RS, Canfield TK, Lamb MM, Gartler SM, Laird CD. Association of fragile X syndrome with delayed replication of the FMR1 gene. Cell. 1993;73:1403–9.
131. Tazumi A, Fukuura M, Nakato R, Kishimoto A, Takenaka T, Ogawa S, et al. Telomere-binding protein Taz1 controls global replication timing through its localization near late replication origins in fission yeast. Genes Dev. 2012;26:2050–62.
132. Cosgrove AJ, Nieduszynski CA, Donaldson AD. Ku complex controls the replication time of DNA in telomere regions. Genes Dev. 2002;16:2485–90.
133. Hayano M, Kanoh Y, Matsumoto S, Renard-Guillet C, Shirahige K, Masai H. Rif1 is a global regulator of timing of replication origin firing in fission yeast. Genes Dev. 2012;26:137–50.
134. Kanoh J, Ishikawa F. spRap1 and spRif1, recruited to telomeres by Taz1, are essential for telomere function in fission yeast. Curr Biol. 2001;11:1624–30.
135. Mattarocci S, Shyian M, Lemmens L, Damay P, Altintas DM, Shi T, et al. Rif1 Controls DNA replication timing in yeast through the PP1 Phosphatase Glc7. Cell Rep. 2014;7:62–9.
136. Davé A, Cooley C, Garg M, Bianchi A. Protein phosphatase 1 recruitment by Rif1 regulates DNA replication origin firing by counteracting DDK activity. Cell Rep. 2014;7:53–61.
137. Hiraga SI, Alvino GM, Chang F, Lian HY, Sridhar A, Kubota T, et al. Rif1 controls DNA replication by directing protein phosphatase 1 to reverse Cdc7-mediated phosphorylation of the MCM complex. Genes Dev. 2014;28:372–83.
138. Peace JM, Ter-Zakarian A, Aparicio OM. Rif1 regulates initiation timing of late replication origins throughout the S. cerevisiae genome. PLoS One. 2014;9(5):98501.
139. Lian H-Y, Robertson ED, Hiraga S, Alvino GM, Collingwood D, McCune HJ, et al. The effect of Ku on telomere replication time is mediated by telomere length but is independent of histone tail acetylation. Mol Biol Cell. 2011;22:1753–65.
140. Stevenson JB, Gottschling DE. Telomeric chromatin modulates replication timing near chromosome ends. Genes Dev. 1999;13:146–51.
141. Aparicio JG, Viggiani CJ, Gibson DG, Aparicio OM. The Rpd3-Sin3 histone deacetylase regulates replication timing and enables intra-S origin control in Saccharomyces cerevisiae. Mol Cell Biol. 2004;24:4769–80.
142. Knott SRV, Viggiani CJ, Tavaré S, Aparicio OM. Genome-wide replication profiles indicate an expansive role for Rpd3L in regulating replication initiation timing or efficiency, and reveal genomic loci of Rpd3 function in Saccharomyces cerevisiae. Genes Dev. 2009;23:1077–90.
143. Taddei A, Gasser SM. Structure and function in the budding yeast nucleus. Genetics. 2012;192:107–29.
144. Yokochi T, Poduch K, Ryba T, Lu J, Hiratani I, Tachibana M, et al. G9a selectively represses a class of late-replicating genes at the nuclear periphery. Proc Natl Acad Sci U S A. 2009;106:19363–8.
145. Jørgensen HF, Azuara V, Amoils S, Spivakov M, Terry A, Nesterova T, et al. The impact of chromatin modifiers on the timing of locus replication in mouse embryonic stem cells. Genome Biol. 2007;8:R169.
146. Takebayashi S-I, Lei I, Ryba T, Sasaki T, Dileep V, Battaglia D, et al. Murine esBAF chromatin remodeling complex subunits BAF250a and Brg1 are necessary to maintain and reprogram pluripotency-specific replication timing of select replication domains. Epigenetics Chromatin [Internet]. 2013;6:42. http://www.pubmedcentral.nih.gov/articlerender.fcgi?artid=3895691&tool=pmcentrez&rendertype=abstract.

147. Yamazaki S, Ishii A, Kanoh Y, Oda M, Nishito Y, Masai H. Rif1 regulates the replication timing domains on the human genome. EMBO J. 2012;31:3667–77.
148. Cornacchia D, Dileep V, Quivy J-P, Foti R, Tili F, Santarella-Mellwig R, et al. Mouse Rif1 is a key regulator of the replication-timing programme in mammalian cells. EMBO J. 2012;31:3678–90.
149. Sreesankar E, Senthilkumar R, Bharathi V, Mishra RK, Mishra K. Functional diversification of yeast telomere associated protein, Rif1, in higher eukaryotes. BMC Genomics. 2012;13:255.
150. Moorhead GBG, Trinkle-Mulcahy L, Nimick M, De Wever V, Campbell DG, Gourlay R, et al. Displacement affinity chromatography of protein phosphatase one (PP1) complexes. BMC Biochem. 2008;9:28.
151. Yoshida K, Bacal J, Desmarais D, Padioleau I, Tsaponina O, Chabes A, et al. The histone deacetylases Sir2 and Rpd3 Act on ribosomal DNA to control the replication program in budding yeast. Mol Cell. 2014;54:691–7.
152. Hiratani I, Leskovar A, Gilbert DM. Differentiation-induced replication-timing changes are restricted to AT-rich/long interspersed nuclear element (LINE)-rich isochores. Proc Natl Acad Sci U S A [Internet]. 2004;101(48):16861–6. http://www.pnas.org/content/101/48/16861. Accessed 27 May 2015.
153. Hiratani I, Ryba T, Itoh M, Rathjen J, Kulik M, Papp B, et al. Genome-wide dynamics of replication timing revealed by in vitro models of mouse embryogenesis. Genome Res. 2010;20: 155–69.
154. Hansen RS, Thomas S, Sandstrom R, Canfield TK, Thurman RE, Weaver M, et al. Sequencing newly replicated DNA reveals widespread plasticity in human replication timing. Proc Natl Acad Sci U S A. 2010;107:139–44.
155. Takebayashi S-i, Dileep V, Ryba T, Dennis JH, Gilbert DM. Chromatin-interaction compartment switch at developmentally regulated chromosomal domains reveals an unusual principle of chromatin folding. Proc Natl Acad Sci. 2012;109:12574–9.
156. Takebayashi S, Ryba T, Gilbert DM. Developmental control of replication timing defines a new breed of chromosomal domains with a novel mechanism of chromatin unfolding. Nucleus [Internet]. 2012;3:500–7. http://www.pubmedcentral.nih.gov/articlerender.fcgi?artid=3515532&tool=pmcentrez&rendertype=abstract.
157. Dileep V, Ay F, Sima J, Vera DL, Noble WS, Gilbert DM. Topologically-associating domains and their long-range contacts are established during early G1 coincident with the establishment of the replication timing program. Genome Res [Internet]. 2015. http://www.ncbi.nlm.nih.gov/pubmed/25995270. Accessed 25 May 2015.
158. Koryakov DE, Pokholkova GV, Maksimov DA, Belyakin SN, Belyaeva ES, Zhimulev IF. Induced transcription results in local changes in chromatin structure, replication timing, and DNA polytenization in a site of intercalary heterochromatin. Chromosoma. 2012;121:573–83.
159. Therizols P, Illingworth RS, Courilleau C, Boyle S, Wood AJ, Bickmore WA. Chromatin decondensation is sufficient to alter nuclear organization in embryonic stem cells. Science [Internet]. 2014;346(6214):1238–42. http://www.ncbi.nlm.nih.gov/pubmed/25477464. Accessed 5 Dec 2014.
160. Rivera-Mulia J, Buckley Q, Sasaki T, Zimmerman J, Didier R, Nazor K et al. Dynamic changes in replication timing and gene expression during lineage specification of human pluripotent stem cells. Genome Research. 2015;25(8):1091–1103.
161. Newlon CS, Lipchitz LR, Collins I, Deshpande A, Devenish RJ, Green RP, et al. Analysis of a circular derivative of Saccharomyces cerevisiae chromosome III: a physical map and identification and location of ARS elements. Genetics. 1991;129:343–57.
162. Brewer BJ, Fangman WL. Mapping replication origins in yeast chromosomes. Bioessays. 1991;13:317–22.
163. Czajkowsky DM, Liu J, Hamlin JL, Shao Z. DNA combing reveals intrinsic temporal disorder in the replication of yeast chromosome VI. J Mol Biol. 2008;375:12–9.
164. Patel PK, Kommajosyula N, Rosebrock A, Bensimon A, Leatherwood J, Bechhoefer J, et al. The Hsk1(Cdc7) replication kinase regulates origin efficiency. Mol Biol Cell. 2008;19:5550–8.
165. Mantiero D, Mackenzie A, Donaldson A, Zegerman P. Limiting replication initiation factors execute the temporal programme of origin firing in budding yeast. EMBO J. 2011;30:4805–14.

166. Tanaka S, Nakato R, Katou Y, Shirahige K, Araki H. Origin association of Sld3, Sld7, and Cdc45 proteins is a key step for determination of origin-firing timing. Curr Biol. 2011;21:2055–63.
167. Shaw A, Olivares-Chauvet P, Maya-Mendoza A, Jackson DA. S-phase progression in mammalian cells: modelling the influence of nuclear organization. Chromosome Res. 2010;18:163–78.
168. Audit B, Zaghloul L, Vaillant C, Chevereau G, d'Aubenton-Carafa Y, Thermes C, et al. Open chromatin encoded in DNA sequence is the signature of "master" replication origins in human cells. Nucleic Acids Res. 2009;37:6064–75.
169. Guilbaud G, Rappailles A, Baker A, Chen CL, Arneodo A, Goldar A, et al. Evidence for sequential and increasing activation of replication origins along replication timing gradients in the human genome. PLoS Comput Biol. 2011;7.
170. Nagano T, Lubling Y, Stevens TJ, Schoenfelder S, Yaffe E, Dean W, et al. Single-cell Hi-C reveals cell-to-cell variability in chromosome structure. Nature [Internet]. 2013;502:59–64. http://www.ncbi.nlm.nih.gov/pubmed/24067610\nfile:///Users/Daniel/Documents/Mendeley Desktop/Nagano et al/Nagano et al._2013_Single-cell Hi-C reveals cell-to-cell variability in chromosome structure._Nature.pdf.
171. Méchali M, Yoshida K, Coulombe P, Pasero P. Genetic and epigenetic determinants of DNA replication origins, position and activation. Curr Opin Genet Dev. 2013;23:124–31.

Chapter 4
Up and Down the Slope: Replication Timing and Fork Directionality Gradients in Eukaryotic Genomes

Olivier Hyrien

Abstract Modern techniques allow the genome-wide determination of the replication time (RT) of any sequence in eukaryotic cell populations. Because origin firing is stochastic, the mean replication time (MRT) of a locus in a cell population depends on the firing time probability distribution of both neighboring and distant origins as well as on replication fork progression rates. Interpreting MRT profiles in terms of origin firing is therefore delicate. Theory predicts a simple relationship between the derivative (slope) of MRT profiles, the speed of replication forks, and the proportions of rightward- and leftward-moving forks replicating that locus (replication fork directionality; RFD). RFD profiles have been obtained by several independent methods: derivative of MRT profiles; nucleotide compositional skew analysis; sequencing of purified Okazaki fragments; and analysis of biased ribonucleotide incorporation in the two strands of the DNA. Using mathematical models, both MRT and RFD profiles allow quantitative inferences about the location and timing of replication initiation and termination events genome-wide. We summarize results and models of the replication program obtained by these approaches and their potential links with replication foci, chromatin states, and globular chromosomal domains.

Keywords Replication origins • Replication termini • Replication fork • Chromatin structure • Mathematical modelling

Introduction: From Single Replication Origins to Replication Domains and Genome-Wide Replication Dynamics

Eukaryotic organisms replicate their genome from multiple initiation sites, termed replication origins, that are activated (fire) at different times in S phase [1–3]. Replication forks then progress at a relatively constant rate until they merge with

O. Hyrien (✉)
Institut de Biologie de l'Ecole Normale Superieure (IBENS), CNRS UMR8197, Inserm U1024, 46 rue d'Ulm, Paris 75005, France
e-mail: hyrien@biologie.ens.fr

D.L. Kaplan (ed.), *The Initiation of DNA Replication in Eukaryotes*,
DOI 10.1007/978-3-319-24696-3_4

converging forks. The genome-wide RT program is mainly determined by the kinetics of origin firing [4, 5]. Recent years have seen an explosion of genome-wide replication profiles, which potentially allow to connect the dynamics of DNA replication at different scales to the properties of individual replication origins.

Eukaryotic replication origins are best understood in budding yeast [6–8]. First identified by genetic means [9], yeast origins correspond to highly specific sites and are used in a variable but often large fraction of the cell cycles [6, 10]. Origins bind the origin recognition complex (ORC) [11]. During the G1 phase of the cell cycle, ORC loads the MCM complex, a core component of the replicative helicase, in an inactive double hexameric (DH) form around double-stranded DNA. MCM DHs can be potentially activated during S phase by protein kinases and firing factors that split the DH to assemble a pair of divergent replication forks, or can be removed from chromatin by passive replication [7]. The current spatiotemporal resolution of yeast genome-wide replication profiles is sufficient to identify the full complement of origins and characterize their firing-time distribution [4]. As detailed below, mathematical analysis of these profiles suggests that the reproducible RT pattern observed in cell populations can be explained by the stochastic parameters governing the independent firing of individual origins.

The nature and location of mammalian replication origins remain much less clear [1]. Initiation events are clearly more dispersive than in yeast, are developmentally plastic, and can occur at different times in S phase, but no definitive information is available about the extent of initiation zones, their specification by genetic or epigenetic elements, their firing efficiency, or their firing time distribution. Unlike in yeast, genome-wide RT profiles in mammals are not sufficiently resolutive to identify individual origins [5, 12, 13]. They highlight early- or late-replicating megabase domains that presumably contain multiple origins firing at roughly similar times. Single-molecule DNA replication mapping techniques (detailed below) have long suggested spatial and temporal correlations between neighboring origins in mammals [3]. More recently, mammalian RT gradients have been proposed to reflect either the sequential activation of multiple origins [14] or the unidirectional progression of single forks [15]. Together with chromatin interaction data, replication kinetics suggest that mammalian genomes are segmented into megabase units that form either flat or U-shaped domains of RT [2, 13–15]. Elucidating the determinants of mammalian origin location and firing efficiency, and understanding whether origin synchrony or sequentiality reflect fork propagation effects, origin cross-talk mechanisms, or co-regulation of independent origins by higher order chromatin structures, are important issues to address in the future.

Origins and Replication Forks as Seen by Single-Molecule Techniques

The DNA fiber autoradiographic studies of Huberman and Riggs [16] established that replication of eukaryotic chromosomes initiated at multiple origins. In mammals, single DNA fibers often showed tandem arrays of replicons spaced at 20–400 kb intervals, which initiated at similar times and completed replication within ~1 h. Since S phase typically lasts 8–10 h in mammalian cells, these results implied a sequential activation of origin clusters through S phase, with only ~10 % of all replicons active at any time in S phase. How origin clusters were connected together was unclear. As measurements of interorigin distances were limited to replicon arrays, replicons larger than the typical fiber length were automatically excluded. By recording grain tracks which could be attributed to the progression of a single fork or pair of forks in the absence of signal from adjacent replicons, Yurov and Liapunova [17] proposed the existence of ~1–2 Mb long replicons that replicated through a large window of S phase. To date, the proportions of mammalian replicons of various sizes and the total number of origins activated per cell cycle in mammalian genomes remain uncertain [3].

Modern DNA fiber techniques (DNA combing and single-molecule analysis of replicated DNA (SMARD) combine fluorographic detection of labelled replication tracts with identification of specific 0.1–1.5 Mb genomic segments by fluorescent in situ hybridization (FISH). DNA combing and SMARD were used to reveal the distribution and efficiency of multiple initiation and termination zones in a few specific mammalian loci spanning up to 1.5 Mb [18–21]. Clusters of initiation zones as well as originless regions up to 700 kb long were identified. One class of originless regions connect early- and late-replicating domains and replicate in a strictly unidirectional manner [19]. Other originless regions can replicate in either direction from flanking origins to form late RT troughs, which can cause chromosome fragility in conditions of replicative stress [21]. Originless regions identified in one cell type may show initiation events in other cell types, confirming the epigenetic nature of origin specification.

DNA autoradiography revealed that the fork progression rate in mammals was not strictly uniform but ranged from 0.6 to 3.6 kbp/min even for individual cells at a single time during S phase [22]. DNA combing and SMARD essentially confirmed these data and did not reveal marked differences in fork rate distributions between the bulk genome and specific loci [14, 23]. Local or long-range differences in DNA sequence, DNA-bound proteins and chromatin structure, transient changes in replisome components, dNTP concentrations, and stochasticity inherent to the biochemical cycle of its molecular motors may all contribute to fork speed variation within and between cells. It was also reported that fork speed changes during S phase but these results were obtained using artificially synchronized cells [24, 25]. A more recent DNA combing study using retroactive (FACS) synchronization, which does not perturb the cell cycle, found instead a constancy of fork speed distribution through S phase [14]. This study also suggested that the synchrony of neighboring origins decays with their distance, consistent with the propagation of an initiation wave, and increases as S phase progresses. The latter finding is consistent with an increasing rate of initiation during S phase, which was detected in many eukaryotes [26–29] and may represent a universal feature of eukaryotic DNA replication dynamics [26]. Models that can quantitatively account for the observed increase assume that the firing of potential origins is governed by their encounter with a limiting initiation factor that can be recycled between early and late replicons, and whose concentration increases during S phase [26, 30–32].

From Stochastic Origin Firing to Replication Timing and Fork Directionality Gradients

In a single chromosomal copy, the time at which a replication fork reaches a locus is the sum of the time taken by the origin emitting this fork to fire, and the time taken by the fork to reach the locus. In the corresponding single-chromosome RT profile (Fig. 4.1, red curve; note that the time axis is oriented from top to bottom), peaks and troughs mark origins and termini, respectively, and the slope (or timing gradient) of the segments between origins and termini, dt/dx, represents the inverse of the replication fork velocity ($1/v$). The slope is linear if v is constant along the segment.

The first eukaryotic genome-wide RT profile (obtained from budding yeast cell populations) showed peaks and troughs connected by lines of various slopes [33]. The peak and troughs allowed identification of origins and termini and the peak heights were taken to precisely reflect origin activation times. The broad range of slopes suggested a corresponding range of fork velocities, compatible with data gathered with DNA fiber and other techniques. As detailed below, some conclusions of this pioneering experiment needed to be revisited to take into account cell-to-cell variability in replication origin usage.

Experimental, mean replication time (MRT) profiles are population averages. Of critical importance, bulk (e.g., 2D gel electrophoresis [34, 35]) and single-molecule (e.g., DNA combing [10]) techniques to analyze replicating DNA revealed that in a cell population, origins fire in only a fraction of the chromosomal copies (termed origin efficiency) and over a broad window of time rather than at a precise time. Owing to this stochasticity, a locus can be replicated with some probability by forks originated from *any* of the origins and moving in *both* directions [36–38]. Consequently (Fig. 4.1), (1) the peaks and troughs associated with active origins and termini in single-copy profiles may weaken or disappear in a population-average MRT profile; (2) an origin's MRT may not necessarily reflect its activation time but may also depend on the behavior of other origins and fork speeds; and (3) the slopes in MRT profiles depend on cell-to-cell heterogeneity in replication direction as much as on fork velocity [14, 36, 38, 39].

The difference of the proportions of rightward-moving (R) and leftward-moving (L) forks replicating a locus in a population of chromosomes, ($R-L$), is defined as replication fork directionality (RFD). Assuming a constant v, it can be shown that the derivative of the MRT profile with respect to chromosomal coordinate (x) is equal to $\mathrm{dMRT}/\mathrm{d}x=(R-L)/v$ [14, 39]. In words, loci replicated equally often in both directions ($R-L=0$) have a flat MRT profile whereas loci replicated in a more predominant direction ($|R-L|>0$) show a correspondingly steeper MRT gradient. Variations in the MRT gradient reflect variations in RFD, due to initiation or termination inside the considered chromosomal segment. RFD increases across an initiation site, and decreases across a termination site, in proportion to the origin or terminus efficiency [40]. More specifically, the spatial derivative of RFD (i.e., the second derivative of MRT) equals twice the difference between the number of initiation (N_i) and termination (N_t) per unit length: $\mathrm{dRFD}/\mathrm{d}x=2\ (N_i-N_t)$ [41]. In the general

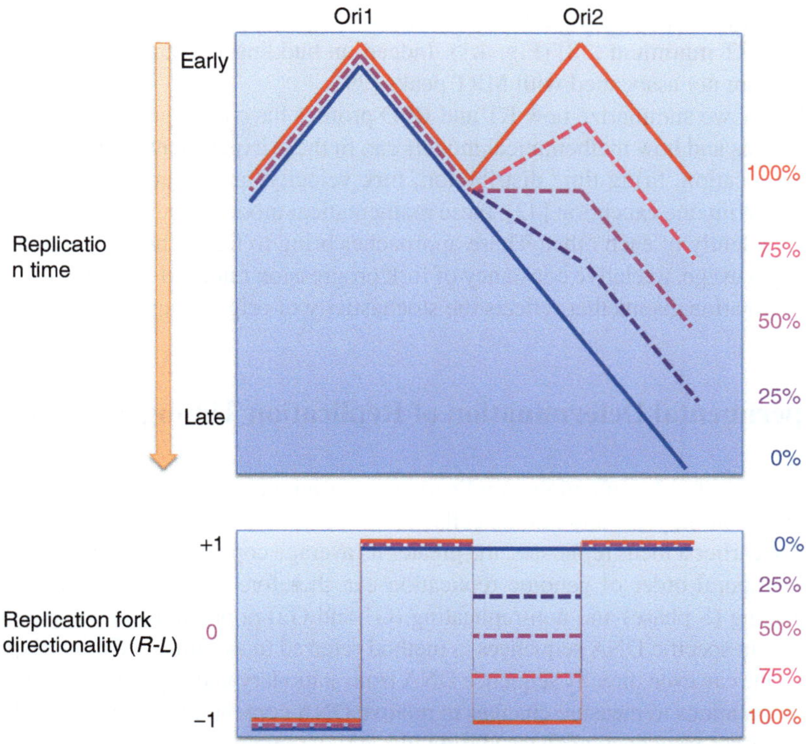

Fig. 4.1 Effect of origin efficiency of replication timing (*top*) and fork directionality (*bottom*) profiles. The figure depicts a chromosome segment containing two synchronous replication origins. Ori1 fires in 100 % of the chromosomal copies whereas Ori2 fires in a variable fraction of the chromosomal copies (indicated by *percentages* on *right side*). The *red curves* show the replication timing (RT) and fork directionality (RFD) profiles expected for a single chromosome (or a homogeneous chromosomal population) in which both Ori1 and Ori2 fire. The *blue curves* correspond to a single chromosome (or a homogeneous chromosomal population) in which Ori1 fires but Ori2 does not. In both these homogeneous cases, active origins are associated with RT peaks and the slope of the RT profile can only take two values, $+1/v$ and $-1/v$, corresponding to RFD values of +1 and −1. The *intermediate color curves* correspond to heterogeneous chromosomal populations containing various percentages of both types of chromosomes. In this case, the MRT is the weighted average of the *red* and *blue* RT. Since means and derivatives commute, the slope of the MRT profile is the weighted average of the *red* and *blue* slopes, and the RFD in the population is the weighted average of the *red* and *blue* RFDs. Thus, Ori2 is associated with a peak only if it fires in >50 % of the chromosomal copies and the height of the peak does not reflect its firing time. The amplitude of the RFD shift at Ori2 is proportional to its firing efficiency

case, an ascending RFD (or convex MRT) segment reflects a predominance of initiation whereas a descending RFD (concave MRT) segment reflects a predominance of termination. Note that a change in RFD can occur without a sign shift (i.e., a change in MRT convexity can occur without forming a local extremum), which explains why origins and termini are not necessarily associated with RT peaks and troughs. For example, a weak origin that is frequently passively replicated by a nearby strong origin may not sufficiently affect the RFD to change its sign, and so may not appear as an MRT minimum [38] (Fig. 4.1). Indeed, in budding yeast, about a third of all origins are not associated with MRT peaks [29].

Below we summarize how RT and RFD profiles have been generated in various organisms and how mathematical models can fit these experimental data and extract origin location, firing time distribution, fork velocity, and location of termination events. With one exception [42], these mathematical models assume that origins fire independently of each other. These approaches bring to light a broad heterogeneity of origin usage, a relative constancy of fork progression rate, and a wide distribution of termination events that reflects the stochasticity of origin usage.

Experimental Determination of Replication Timing Profiles

The temporal order of genome replication has been measured by different methodologies, both in yeast and human cells.

The earlier a locus replicates, the greater its average copy number in S-phase cells. The temporal order of genome replication can therefore be determined by sorting replicating (S phase) and non-replicating (G1 and G2) populations to compare their content in specific DNA sequences, a method referred to as TimEx (Timing Express) [43]. One can even directly sequence DNA from stationary and exponentially growing cell populations to measure changes in relative DNA copy number, a method referred to as marker frequency analysis (MFA) [44, 45]. Copy number has a negative linear relationship with MRT [38]; therefore plots of copy number along a genome are comparable with MRT profiles. Although these data give no direct information about the dispersion of replication time around the mean, a signature of replication stochasticity is apparent in the morphology of the profile, as explained below.

The distribution of RT in a population can be observed by monitoring the incorporation of detectable nucleotides [8], the occupancy of replication fork proteins [46], the presence of single-stranded DNA [47], or the copy number of DNA sequences [48] in timed samples during a synchronous S phase. Alternatively, asynchronous cells can be "post-sorted" into multiple S-phase fractions by FACS or elutriation prior to tracking replicative label incorporation by microarray hybridization (Repli-Chip) [15] or sequencing (Repli-Seq) [49, 50]. Post-sorting avoids potential replication perturbations introduced by the synchronization procedure. On the other hand, since the multiple S-phase fractions are ordered by total DNA content, rather than time during S phase, calculating the absolute replication time from post-sorted cell populations requires to calibrate the fraction of replicated DNA as a function of time in S phase. This function is not linear; it can be mathematically extracted from cell cycle and FACS data [14], or determined by independent experimental means [32]. In both time-course and post-sort methods, the observed distributions of RT provide direct, genome-wide information about cell-to-cell variability in RT. Obviously, the temporal resolution of these data increases with the number of time points or post-sort compartments.

Determination of RFD Profiles

The directionality of replication forks has been estimated by direct and indirect methods.

The first indirect method, knowing fork speed v, consists in extracting RFD by spatial derivation of MRT profiles: RFD = $v\, d\mathrm{MRT}/dx$ [14, 39].

The second indirect method is nucleotide compositional skew analysis. It was first reported in bacteria that the two DNA strands have an asymmetric nucleotide composition, with enrichment of the leading strand in G over C and T over A [51]. The strict correlation of the GC and TA skews $S_{GC} = (G-C)/(G+C)$ and $S_{TA} = (T-A)/(T+A)$ with replication fork direction suggested that the leading and lagging strands experience different rates of nucleotide substitution leading to accumulation of nucleotide compositional skew over evolutionary times [51]. Upward skew jumps ((+) *S*-jumps) similar to bacterial origins were subsequently detected at 1546 sites in the human genome [52, 53]. Between upward jumps, the skew decreased in a linear manner suggesting a progressive inversion of RFD across megabase-sized "N-domains," which together cover one-third of the genome [54]. Comparison of skew profiles with Repli-Seq data from several somatic cell lines strongly supported the notion that the GC and TA skews are a direct reflection of RFD in germline cells [39, 55]. Concerns were raised that N-shaped patterns may be caused by mutational strand asymmetry associated with transcription rather than replication [56]. However, a detailed study of the mutational profile of S-jumps established the existence of replication-associated mutational asymmetries and showed that S-jump profiles could only be explained by the additive effect of transcription- and replication-associated mutational asymmetries [55]. Techniques have been developed to deconvolute the transcription-associated and replication-associated skews [57].

Replication-associated GC and TA skews were more recently detected in the yeasts *S. cerevisiae*, *K. lactis*, and *L. kluyveri*, but only when all interorigin intervals were analyzed together [58–60]. Interestingly, the leading strand was enriched in C and A, whereas it is enriched in G and T in mammalian and most eubacterial genomes. In contrast, no convincing replication-associated skew could be detected in the fission yeast *S. pombe* [58]. In *L. kluyveri*, a lack of replication-associated skew was specifically observed in a chromosomal arm that has a much higher GC content and replicates earlier and faster than the rest of the genome [60]. It was proposed that the lack of skew may be caused by a random fork direction. However, the RT profile of this chromosomal arm showed peaks and troughs rather than the flat curve expected if RFD was null. It remains possible that mutational patterns for this chromosomal arm differ from the rest of the genome.

The third indirect method to map RFD is based on the fact that ribonucleotides are covalently incorporated into genomic DNA at different rates by Pol ε, the primary leading strand replicase, and Pols α and δ, which are primarily responsible for lagging strand synthesis [61]. Ribonucleotides are normally removed by ribonucleotide excision repair (RER) but are well tolerated in yeast RER mutants [61, 62]. Moreover, polymerase mutants that incorporate ribonucleotides at higher rates than their wild-type counterparts have been obtained [61, 63]. Four recently developed methods (dubbed EmRiboSeq [64], Pu-Seq [65], HydEn-Seq [66], and Ribose-Seq [67]) allowed to determine the genome-wide distribution of embedded ribonucleotides in RER and polymerase mutants in *S. cerevisiae* and *S. pombe*. These patterns support the previous assignments of leading and lagging strand polymerases inferred from the use of mutator polymerase alleles [63, 68–70], and thereby allow indirect but generally precise determination of RFD across the genome. They also identify regions of the genome in which ribonucleotide incorporation patterns deviate from the expectations of a simple division of labor among the three replicases. Deviations specifically observed at origins were proposed to result from occasional leading strand initiation by Pol δ followed by exchange with Pol ε [65]. Note that a recent study [71] challenged the "established" distribution of DNA polymerases at the replication fork and proposed other explanations for the observed distribution of rNMP incorporation in polymerase mutants.

The last and so far only direct method to analyze RFD genome-wide is to purify and sequence Okazaki fragments, using strand identity to discern fragments replicated as the Watson strand (*L* forks) or Crick strand (*R* forks). This was first achieved in *S. cerevisiae* using conditional lethal mutants lacking DNA ligase and DNA damage checkpoint activity, which allow massive accumulation of unligated Okazaki fragments and continuation of S-phase progression despite the presence of unligated nicks in replicated DNA [40, 72]. More recently, genome-wide RFD profiles from both yeast and human cells were obtained using a novel Okazaki-fragment purification technique that does not require overproducing mutant cells (unpublished). RFD profiles obtained by Okazaki-fragment sequencing are highly consistent with RFD and MRT profiles obtained by other techniques and allow a direct and quantitative, genome-wide analysis of initiation and termination events in yeast and mammals.

Mathematical Analysis of Budding Yeast Replication Profiles

The recognition that origin initiation, fork progression, and fork merge at termini is formally analogous to nucleation, growth, and coalescence in crystallization kinetics allowed to adapt to DNA replication [28, 73–75], a formalism developed long ago to describe crystallization kinetics [76–80]. In this formalism, an initiation function $I(x,t)$ describes the rate of initiation at position x at time t. If fork velocity v is assumed to be constant, then $I(x,t)$ entirely determines the replication fraction $f(x,t)$, which has been estimated in the most precise time-course experiments in budding yeast [48] at 1 kb resolution in space and 5-min resolution in time. The goal is to estimate the $I(x,t)$ that best explains such experimental $f(x,t)$ data. This is usually performed by curve-fitting strategies [29] requiring some a priori knowledge of the initiation function, although more general strategies such as Bayesian inference [42] or direct analytical inversion [81] seem achievable.

Yang et al. [29] performed the first detailed mathematical analysis of budding yeast genome replication time-course data. This study revealed that the RT distribution of origins could be fitted by sigmoid curves whose width and mean were correlated. In other words, RT was more precise for early than for late origins. An origin's *replication* time does not simply reflect its *firing* time, unless this origin fires in all copies of the chromosomal population. Therefore, an analytical model that incorporated passive replication and a sigmoid shape for the cumulative firing time distribution of origins was used to fit to the data and extract these distributions. Their widths (10–40 min) were comparable to S-phase length (60 min), and indeed correlated with median firing time, as expected for stochastic activation. The distributions allowed to compute *potential* origin efficiencies, defined as the efficiency of initiation that would be observed over the length of S phase in the absence of passive replication, and to compare them with their *observed* efficiencies, whose computation relied on the firing time distribution of *all* origins of the same chromosome and on fork speed. A constant fork speed (1.9 kb/min) was used since this gave as good a fit as allowing variable velocities. Potential efficiencies were quite high (>0.9 for one-half of the origins, >0.5 for most of the rest). In contrast, observed efficiencies were more evenly spread from 0 to 1, owing to frequent passive replication from neighboring origins. Finally, potential efficiencies decreased monotonically with median firing time and temporally alike origins did not appear to form clusters along the chromosomes. All these results suggested that stochastic activation of origins, firing independently of each other and with variable efficiency, suffice to explain the replication timing "program" of budding yeast with no need to invoke regulation by external triggers.

Parallel work in yeasts [82–85] and metazoans [86] has shown that the firing of origins is regulated by competition for initiation factors less abundant than potential origins, which explains why origins cannot all fire at the same time. Stochastic interaction of these factors with origins would explain simply why origins fire stochastically. However, interorigin differences in accessibility or response to the limiting factors must be invoked to explain why some origins fire with higher probability than others. The fact that origins cannot all fire at the same time implies that origins are not entirely independent of each other, contrary to the assumption of several mathematical models. The extent of origin correlation induced by these limiting amounts remains to be addressed.

In the "multiple initiator model" (MIM) [29], a variable number of initiator proteins is loaded at each origin, so that origins with more initiators fire with an earlier and sharper time distribution. This model can quantitatively recapitulate the variety of origin firing time distributions extracted from the replication time-course data in yeast. One obvious candidate for the initiator was the MCM complex, because origin efficiency correlated with MCM occupancy in some ChIP-chip experiments [87] and because in vitro, a single ORC can load multiple MCM double hexamers that can passively translocate along DNA prior to activation in S phase [88–90]. However, recent in vivo data integrating nucleotide-resolution "footprints" of origin architecture with MCM ChIP-seq suggested that, unlike in vitro, only one MCM double hexamer is loaded per origin [91]. Further work is therefore required to validate or refute the MIM model.

The stochasticity of origin activation in budding yeast was confirmed and further explored in several other studies, which also analyzed termination in more detail.

Hawkins et al. [48] analyzed high-resolution replication time-course data using a mathematical model in which the probability that an origin achieves licensing (termed origin competence) was explicitly formulated [38]. The median and width of origin firing time distributions were again correlated, and single-cell fluorescence microscopy measurements of a locus replication time confirmed the estimated variability in origin activity. The extracted origin competences, however, showed no correlation with firing time, suggesting that the probabilities of origin licensing and firing are governed by distinct mechanisms. Hawkins et al. [48] also used the equation $dMRT/dx = (R - L)/v$ to derive a genome-wide RFD profile from the experimental MRT. An experimental RFD profile was independently obtained by McGuffee et al. [40] by sequencing Okazaki fragments purified from an asynchronous culture. The inferred RFD profile by Hawkins et al. [48] and this experimental RFD were quite similar. The MRT profile computed from the RFD was also reasonably consistent with experiments [40].

Inferred and experimental RFD profiles were used to quantitate initiation and termination events genome-wide [40, 48]. The observed origin efficiencies decreased with MRT, but the correlation was looser than between potential efficiencies and MRT, consistent with the confounding effects of passive replication. Termination sites were widely distributed, covering >75 % of the genome, and inefficient (<4 % per kb even at the most efficient sites). A cruder, previous study identified 71 termination (TER) sites covering only ~3 % of the genome [92]. The new RFD data revealed that these TER sites, although more efficient than average, together represented <5 % of all termination events. Inactivation of specific origins flanking a previously described TER site suppressed termination in this region, without delaying fork progression [48]. In a strain overexpressing limiting factors for origin activation [84], origins fired at similar times genome-wide and with less temporal precision than normal, and the termination zones broadened and their midpoints moved toward the midpoints of interorigin intervals [40]. These results show that the location of termination events is determined by the firing time of origins rather than by specific *cis*-acting elements.

The link between the temporal distribution of origin firing and the spatial distribution of termination events (Fig. 4.2) suggests that the former can be extracted from the latter. Retkute et al. [37, 38] have mathematically studied the case when two adjacent origins fire within a sufficiently sharp time window, Δt, that a fork from one origin can reach the other origin only if it is not competent. In this case, the second derivative of the MRT curve at its maximum (i.e., the slope of the decreasing RFD profile around 0) is inversely proportional to Δt, with little dependence on the exact shape of the origin firing time distribution. Application to experimental MRT data for suitable interorigin intervals in the *S. cerevisiae* genome yielded estimates of 10-15 min for Δt, in good agreement with time-course analyses. Thus, stochastic origin parameters can be extracted without recourse to time-course experiments, using a single MRT or RFD profile from non-synchronized cells.

Quite recently, RFD profiles have been determined by monitoring ribonucleotide incorporation in *S. cerevisiae* [66] and *S. pombe* [65]. Their analysis led to similar conclusions regarding the stochasticity of replication initiation and termination and the utility of RFD data to infer reliable MRT profiles.

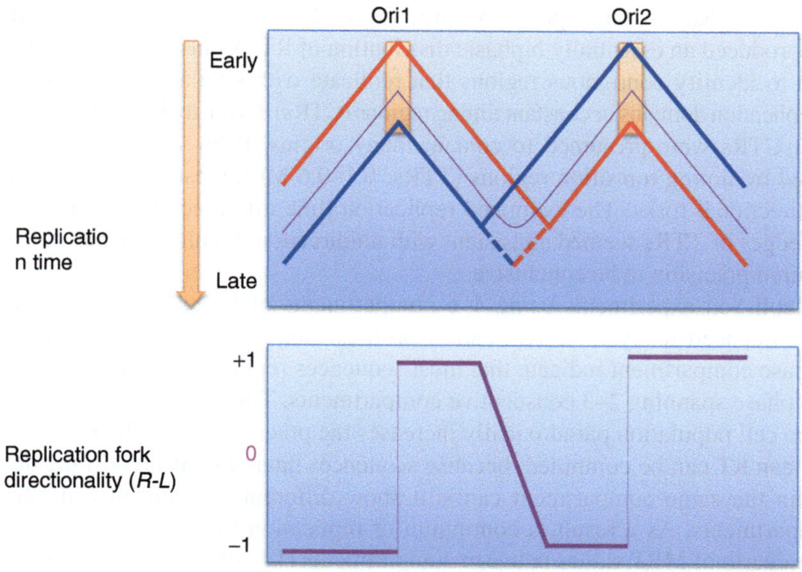

Fig. 4.2 Effect of origin firing time distribution on the dispersion of termination events. The figure depicts a chromosome segment containing two efficient replication origins that can fire independently over a temporal window symbolized by the *beige boxes* in the *top panel*. Forks emanating from these two origins meet at variable positions due to the variable firing time difference between the two origins. The extremes of these ranges are indicated by the *red* and *blue curves*, and their mean by the *purple curve*. The RFD profile in the population shows the range of corresponding termination positions as a central descending RFD segment. The extent of the termination zone is proportional to the width of the firing time window and the slope of this descending segment (or curvature of the MRT profile minimum) is inversely proportional to this time window

Mammalian Replication Profiles

In contrast to yeast RT profiles, the resolution of mammalian RT profiles is generally not sufficient to identify single-replication origins. The spatial resolution of copy number profiles may in theory attain the single-nucleotide resolution of sequencing. In practice, however, the low signal-to-noise ratio of most experiments imposed the use of smoothing algorithms that limited resolution to about 50 kb [43]. Nevertheless, increasing sequencing depth can reduce the need for smoothing so that identification of individual origins in human TimEx profiles may become feasible [93]. In post-sort methods (Repli-Chip [15, 94] or Repli-Seq [49, 50]), it is the number of consecutive S-phase samples that is limiting temporal resolution. Only 2–6 S-phase compartments are used to analyze the typical 8–10-h S phase of mouse or human cells, precluding sharp distinction of origins and termini that replicate within ~1 h of each other. Consistently with this limitation, the profiles are not improved by pushing spatial resolution of microarrays below 5 kb or by using sequencing.

The first Repli-Chip experiments [15, 94] used only two S-phase compartments and produced an essentially biphasic distribution of RT. A clustering algorithm was used to identify contiguous regions that replicate with similar timing, referred to as replication domains or constant timing regions (CTRs). Given their size (0.2–2.0 Mb), most CTRs were presumed to contain many origins. Early and late CTRs were joined by timing transition regions (TTRs; 0.1–0.6 Mb), presumed to replicate by unidirectional forks. The estimated replication time difference between early and late edges of TTRs seemed consistent with unidirectional replication but lacked the required precision to be conclusive.

Repli-Seq experiments using 4–6 compartments of S phase [49, 50] led to a more nuanced analysis of replication profiles. Sequence tag densities for each S-phase compartment indicate that most sequences replicate over a broad window of S phase spanning 2–3 consecutive compartments. This broad distribution of RT in the cell population paradoxically increases the precision with which the median or mean RT can be computed, because sequences having their highest tag density within the same compartment can still show different tag densities in adjacent compartments. As a result, a continuum of replication times was observed, with little dearth of MRT values between compartments [14, 49]. The smoothed profiles (50–100 kb resolution) traced a landscape of sharp peaks and shallow valleys, only occasionally interspersed with extended domains of constant RT. An objective analysis of RT gradients at multiple scales revealed a broad and continuous (not biphasic) distribution of slopes [14]. Thus, there was no sharp demarcation between CTRs and TTRs. Given that the range of fork speeds along the genome was much too narrow to account for the range of MRT gradients, the equation $dMRT/dx = RFD/v$ implied a broad distribution of RFD along the genome, at least when averaged at 50–100 kb spatial resolution. In most cell lines, only the 1–5 % steepest segments had a slope compatible with unidirectional replication ($|RFD| = 1$). Most of the genome had a smaller |RFD| and was therefore replicated by variable proportions of forks moving in both directions.

The valleys were U-shaped rather than V-shaped, suggesting that |RFD| decreased in later replicating regions. Automated detection identified in multiple cell lines 800–1500 megabase-scale U-shaped domains of replication timing covering 40–60 % of the genome [39]. Their averaged profile was parabolic, and the derived RFD profile was therefore an N-shape, strikingly similar to the skew profile of N-domains (the derivative of a parabola is a straight line). Significant overlaps were observed between U-domains of different cell lines and between U-domains and N-domains. These results suggested that U-domains and N-domains represent the same fundamental unit of replication but are developmentally plastic and that N-domains are the U-domains of the germline.

U-shaped MRT profiles [48] and N-shaped RFD profiles [40, 65] have also been observed in interorigin intervals of budding and fission yeast. In these cases, the sharpness of the MRT peaks bordering the U reflects the precise location of origins, whereas the curvature at the bottom of the U (i.e., the slope of the descending branch of the N) reflects the firing time variability of flanking origins. The greater this variability, the larger the intervening segment of mixed replication directionality (Fig. 4.2 and top panel of Fig. 4.3).

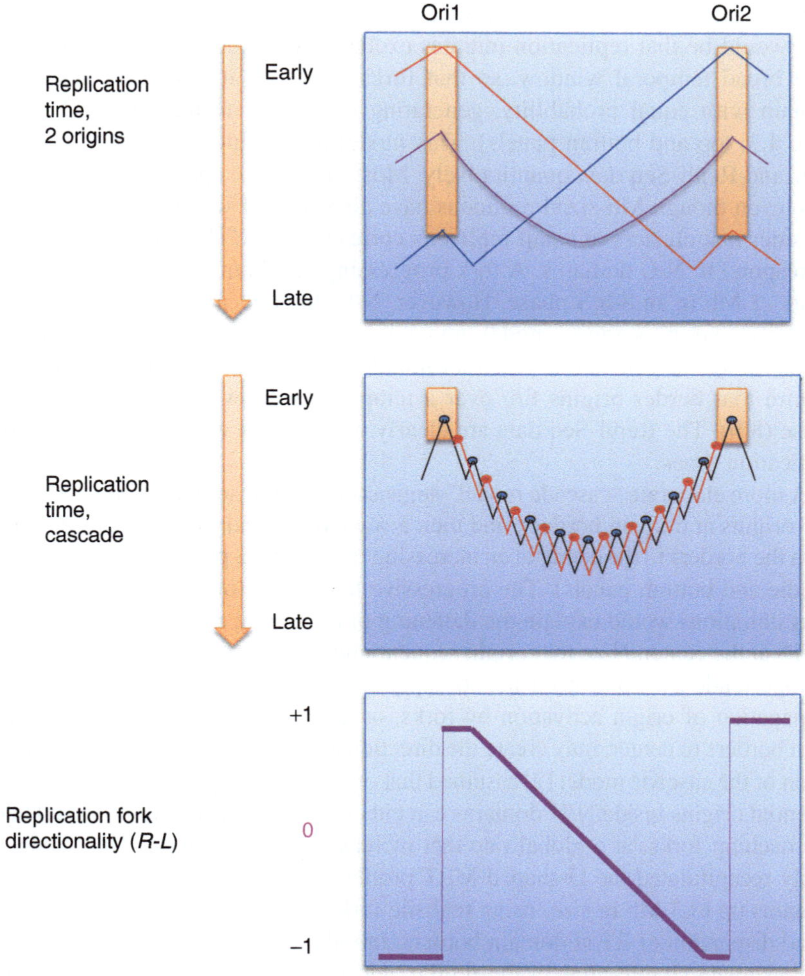

Fig. 4.3 Two-origin and cascade models for replication of domains with U-shaped MRT and N-shaped RFD profiles. The *top panel* shows that an extended termination zone can form between two origins if their firing time window (*beige boxes*) is as long as the time needed by a single fork to traverse the entire interorigin interval (refer to Fig. 4.2). The *middle panel* shows temporally ordered cascades of origin activation from domain borders to centers. *Dark* and *red lines* represent replication timing profiles of two different cells activating the same border origins but different internal origins in the intradomain cascade. Due to sequential origin firing the two forks emanating from each origin travel unequal distances before merging with converging forks from upstream and downstream origins. As origins become more synchronous at the domain center, forks travelling in opposite orientations replicate more equal lengths of DNA and |RFD| decreases from borders to center. Both the two-origin and the cascade models predict an N-shaped pattern of RFD across the domain (*bottom panel*). In the cascade model, however, origins at domain borders need not fire over a broad time window

Likewise, a simple two-origin model for mammalian N/U-domain replication [54] would be that replication initiates exclusively at domain borders, but over a very broad temporal window, so that forks converge at any position along the domain with equal probability, generating a linearly decreasing RFD segment (Fig. 4.3, top and bottom panels). This model is difficult to reconcile with DNA fiber and Repli-Seq data quantitatively. Most origins are spaced at ~100 kb intervals, even though Mb-sized replicons have also been observed [16, 17]. One may consider that clusters of small replicons correspond to CTRs while long replicons correspond to N/U domains. A fork progressing at 2 kb/min may indeed replicate up to ~1 Mb in an 8-h S phase. However, N/U-domains are up to 3 Mb long [39, 54], and the two-origin model appears insufficient for large domain sizes. Furthermore, to generate a 1 Mb termination zone, the two-origin model would require that border origins fire over a temporal window as long as the entire S phase (8 h). The Repli-Seq data are clearly inconsistent with such a dispersion of replication times.

A more elaborate "cascade model" suggests that replication first initiates at "master" origins at domain borders, and then a wave of secondary initiations propagates from the borders to the center at an increasing rate during S phase [2, 14, 39] (Fig. 4.3, middle and bottom panels). The progressive transition from sequential to synchronous initiations would explain the flattening of the timing curve and the decrease of |RFD| at the center. Note that origin sequentiality and increasing synchrony are only average tendencies and need not to apply strictly at the single-molecule level. A propagation of origin activation by forks, or a gradient of chromatin accessibility from borders to center, may create the directional initiation wave. A numerical simulation of the cascade model [2] assumed that once master origins have fired, abundant potential origins inside N/U-domains can either fire on their own or be stimulated by approaching forks, at a global rate that increases during S phase. This model precisely recapitulated the U-shaped MRT profile and N-shaped RFD profile of N/U-domains up to 3 Mb in size, using realistic fork velocities and no requirement for a broad dispersion of RT at domain borders. Intradomain origin firing was too dispersive to create singularities in the profile, and the changing balance between fork-stimulated and autonomous origin firing during S phase generated the smooth inversion of RFD between the two borders.

More recent analyses of the human genome have highlighted a novel skew structure, termed the "split-N" domain, that has a shape reminiscent of an N, but split in half, leaving in the center a region of null skew [95]. This central region appears when the distance between domain borders is >3 Mb, and its length, which can reach several Mb, increases with domain size. Split N-domains together cover 13 % of the genome. The null skew of the central regions suggests a null RFD. Indeed, the MRT of the central region is homogeneously late, consistent with late and spatially random replication. The split-N pattern cannot be explained by a two-origin model. Strikingly, however, the cascade model predicts a transition from the standard N to the split-N profile of RFD when the domain size increases above a critical threshold (unpublished). This is because autonomous activation of intradomain origins becomes highly efficient late in S phase in long central regions before they can be reached by stimulating forks progressing from domain sides. These results support the notion that mammalian genomes replicate by a superposition of early and efficient initiation at specific sites or zones followed by more random initiation later in S phase [1, 2].

Human genome-wide RFD profiles have been recently obtained directly by sequencing of highly purified Okazaki fragments (unpublished). These profiles are highly consistent with published MRT profiles but offer much higher resolution of initiation and termination events. They reveal in multiple cell lines 6000–10,000 ascending RFD segments representing initiation zones (mean size ~30 kb), alternating with descending RFD segments spanning up to 3 Mb. In some cases ascending and descending RFD segments are joined by high RFD segments of unidirectional replication (representing <10 % of the genome). Thus, RFD ranges broadly and N-shaped patterns of RFD are observed over the entire genome, with clusters of small Ns discernible in regions previously annotated as CTRs. These observations provide further support for the widespread existence of linear RFD gradients, for spatial and temporal stochasticity in replication initiation and termination, and for the cascade model of human genome replication.

Mammalian Replication Timing Program and Chromatin Architecture

Nuclear organization has appeared as a strong determinant of DNA replication kinetics. A correlation has long been observed between heterochromatin and late replication [96]. Early-replicating regions tend to be enriched in active genes and open chromatin marks, while late-replicating regions show opposite features [5]. Moreover, early- and late-replicating regions tend to be located in the interior and periphery of the nucleus, respectively [5]. A discrete point during G1 phase was discovered at which the replication time of specific sequences is established for each cell cycle, coincident with their intranuclear repositioning following mitosis [97]. This result pointed to a close link between the spatial organization of the genome in interphase and the temporal regulation of its replicons.

Cytological analysis of the pulse-labeled intranuclear sites of DNA synthesis revealed the existence of replication "foci" that, on average, encompass ~1 Mb of DNA and replicate in 45–60 min [3]. As S phase progresses, their morphology, number, and intranuclear position change [98], with new foci appearing in close vicinity to earlier ones [99]. Early foci tend to occupy the interior volume of the nucleus whereas later foci colocalize with nucleoli and the nuclear periphery, and still later foci with a few blocks of constitutive heterochromatin. It was proposed that foci are chromatin structural units equivalent to replicon clusters and/or CTRs [3, 13] and that their ordered activation reflects the sequential synthesis of genetically continuous chromatin domains, suggesting some sort of propagation effect [99, 100].

N/U-domain borders are enriched in DNaseI hypersensitive sites (HS) and prone to transcription, whereas N/U-domain central regions appear transcriptionally silent

[39, 54, 101]. A recent integrative analysis of 13 epigenetic marks led to definition of four distinct chromatin states (C1–C4) in the human genome [102]. C1 is a transcriptionally active chromatin state, C2 a repressive state associated with polycomb complexes, C3 a silent state lacking characteristic marks, and C4 a gene-poor, HP-1-bound heterochromatic state. It was found that U-domain replication proceeds from C1 at borders to C2, C3, and C4 at centers, whereas early and late CTRs consist entirely of C1 + C2 and C3 + C4 states, respectively.

High-throughput chromatin conformation capture (Hi-C) technology allows to measure the frequency of pairwise contacts between any pair of loci genome-wide [103]. When averaged over the genome, the frequency of interactions decays rapidly with distance, but some interactions are more or less frequent than expected. This led to definition of two sets of loci (named A and B) within which contacts are enriched and between which contacts are depleted [103]. Comparison with RT profiles revealed a striking correlation of A and B compartments with early- and late-replicating DNA, respectively, consistent with their different intranuclear localization [94]. Moreover, when the chromatin folding of U-domains was analyzed, it was found that the sequences within a single U-domain preferentially interact between themselves rather than with outside sequences, even if they are located closer [39]. Thus, sequences belonging to A and B compartments within a single U-domain interact more strongly with each other than with their cognate partners outside the domain. This self-interaction pattern suggested a tight connection between U-domains and "topological domains" (TADs) revealed by higher resolution Hi-C data [104]. TADs are delimited by sharp boundaries containing housekeeping genes and insulator sites, and N/U-domains also share these features. Furthermore, TADs are abolished in mitosis and must reform in interphase of each cell cycle [105], an additional similarity with RT regulation.

The developmental plasticity of RT profiles [5, 13] is in apparent contrast with the developmental stability of TADs [104]. Developmental changes in RT primarily occur in 400–800 kb units, suggesting that CTRs observed in any cell type comprise multiple, potentially switchable units. Examination of RT across TADs [106] revealed that some TADs are either entirely early or late replicating, some span all or part of a single TTR, and others contain convergent TTRs forming the previously described U-domains [39]. When TTRs were mapped in many cell types, their early borders coincided almost one to one with TAD boundaries, whereas their late borders had no detectable relationship to TAD structure [106]. Higher resolution MRT or RFD profiles are needed to elucidate whether "master" origins are active at TAD boundaries whatever their replication timing and chromatin composition, or whether this is a unique feature of U-domains.

A chromatin-based replication model was recently developed to look for genomic markers able to predict MRT profiles [107]. In this model, rate-limiting activators probabilistically select genomic locations and initiate replication and then remain engaged with forks until they merge and terminate. The probability of initiation thus depends on both the density of the tested genomic landmark and the number of unengaged factors at time t, and is therefore negatively regulated by fork density, which depends on the collective behavior of all origins. Remarkably, this simple model predicts MRT profiles with an accuracy rivaling experimental repeats when the initiation probability landscape is defined by the density of DNase-HS sites. The recycling of the limiting factor at fork collision was a required model feature to predict RT. This model assumes no other interdependency of origin firing and does not require any explicit reference to the spatial organization of the genome. The limiting factor hypothesis de facto introduces an interdependency of origins. It is also possible that the distribution of DNase-HS sites contains in a non-explicit form some information about chromatin interactions and origin correlations. Emerging higher resolution datasets and advanced mathematical models will undoubtedly promote further progress.

References

1. Hyrien O. Peaks cloaked in the mist: the landscape of mammalian replication origins. J Cell Biol. 2015;208(2):147–60.
2. Hyrien O, Rappailles A, Guilbaud G, Baker A, Chen CL, Goldar A, et al. From simple bacterial and archaeal replicons to replication n/u-domains. J Mol Biol. 2013;425(23):4673–89.
3. Berezney R, Dubey DD, Huberman JA. Heterogeneity of eukaryotic replicons, replicon clusters, and replication foci. Chromosoma. 2000;108(8):471–84.
4. Bechhoefer J, Rhind N. Replication timing and its emergence from stochastic processes. Trends Genet. 2012;28:374–81.
5. Rhind N, Gilbert DM. DNA replication timing. Cold Spring Harbor Perspect Med. 2013;3(7): 1–26.
6. Fangman WL, Brewer BJ. Activation of replication origins within yeast chromosomes. Annu Rev Cell Biol. 1991;7:375–402.
7. Tognetti S, Riera A, Speck C. Switch on the engine: how the eukaryotic replicative helicase MCM2-7 becomes activated. Chromosoma. 2014.
8. Raghuraman MK, Brewer BJ. Molecular analysis of the replication program in unicellular model organisms. Chromosome Res. 2010;18(1):19–34.
9. Newlon CS, Theis JF. The structure and function of yeast ARS elements. Curr Opin Genet Dev. 1993;3(5):752–8.
10. Czajkowsky DM, Liu J, Hamlin JL, Shao Z. DNA combing reveals intrinsic temporal disorder in the replication of yeast chromosome VI. J Mol Biol. 2008;375(1):12–9.
11. Bell SP, Stillman B. ATP-dependent recognition of eukaryotic origins of DNA replication by a multiprotein complex [see comments]. Nature. 1992;357(6374):128–34.
12. Gilbert DM. Evaluating genome-scale approaches to eukaryotic DNA replication. Nat Rev Genet. 2010;11(10):673–84.
13. Pope BD, Gilbert DM. The replication domain model: regulating replicon firing in the context of large-scale chromosome architecture. J Mol Biol. 2013;425(23):4690–5.
14. Guilbaud G, Rappailles A, Baker A, Chen CL, Arneodo A, Goldar A, et al. Evidence for sequential and increasing activation of replication origins along replication timing gradients in the human genome. PLoS Comput Biol. 2011;7(12):e1002322.
15. Hiratani I, Ryba T, Itoh M, Yokochi T, Schwaiger M, Chang CW, et al. Global reorganization of replication domains during embryonic stem cell differentiation. PLoS Biol. 2008;6(10), e245.
16. Huberman JA, Riggs AD. On the mechanism of DNA replication in mammalian chromosomes. J Mol Biol. 1968;32(2):327–41.
17. Yurov YB, Liapunova NA. The units of DNA replication in the mammalian chromosomes: evidence for a large size of replication units. Chromosoma. 1977;60(3):253–67.

18. Lebofsky R, Heilig R, Sonnleitner M, Weissenbach J, Bensimon A. DNA replication origin interference increases the spacing between initiation events in human cells. Mol Biol Cell. 2006;17(12):5337–45.
19. Norio P, Kosiyatrakul S, Yang Q, Guan Z, Brown NM, Thomas S, et al. Progressive activation of DNA replication initiation in large domains of the immunoglobulin heavy chain locus during B cell development. Mol Cell. 2005;20(4):575–87.
20. Demczuk A, Gauthier MG, Veras I, Kosiyatrakul S, Schildkraut CL, Busslinger M, et al. Regulation of DNA replication within the immunoglobulin heavy-chain locus during B cell commitment. PLoS Biol. 2012;10(7), e1001360.
21. Letessier A, Millot GA, Koundrioukoff S, Lachages AM, Vogt N, Hansen RS, et al. Cell-type-specific replication initiation programs set fragility of the FRA3B fragile site. Nature. 2011;470(7332):120–3.
22. Yurov YB. Rate of DNA replication fork movement within a single mammalian cell. J Mol Biol. 1980;136(3):339–42.
23. Anglana M, Apiou F, Bensimon A, Debatisse M. Dynamics of DNA replication in mammalian somatic cells: nucleotide pool modulates origin choice and interorigin spacing. Cell. 2003;114(3):385–94.
24. Takebayashi S, Sugimura K, Saito T, Sato C, Fukushima Y, Taguchi H, et al. Regulation of replication at the R/G chromosomal band boundary and pericentromeric heterochromatin of mammalian cells. Exp Cell Res. 2005;304(1):162–74.
25. Housman D, Huberman JA. Changes in the rate of DNA replication fork movement during S phase in mammalian cells. J Mol Biol. 1975;94(2):173–81.
26. Goldar A, Marsolier-Kergoat MC, Hyrien O. Universal temporal profile of replication origin activation in eukaryotes. PLoS One. 2009;4(6), e5899.
27. Lucas I, Chevrier-Miller M, Sogo JM, Hyrien O. Mechanisms ensuring rapid and complete DNA replication despite random initiation in Xenopus early embryos. J Mol Biol. 2000;296(3):769–86.
28. Herrick J, Jun S, Bechhoefer J, Bensimon A. Kinetic model of DNA replication in eukaryotic organisms. J Mol Biol. 2002;320(4):741–50.
29. Yang SC, Rhind N, Bechhoefer J. Modeling genome-wide replication kinetics reveals a mechanism for regulation of replication timing. Mol Syst Biol. 2010;6:404.
30. Goldar A, Labit H, Marheineke K, Hyrien O. A dynamic stochastic model for DNA replication initiation in early embryos. PLoS One. 2008;3(8), e2919.
31. Gauthier MG, Bechhoefer J. Control of DNA replication by anomalous reaction-diffusion kinetics. Phys Rev Lett. 2009;102(15):158104.
32. Ma E, Hyrien O, Goldar A. Do replication forks control late origin firing in Saccharomyces cerevisiae? Nucleic Acids Res. 2012;40(5):2010–9.
33. Raghuraman MK, Winzeler EA, Collingwood D, Hunt S, Wodicka L, Conway A, et al. Replication dynamics of the yeast genome. Science. 2001;294(5540):115–21.
34. Friedman KL, Brewer BJ, Fangman WL. Replication profile of Saccharomyces cerevisiae chromosome VI. Genes Cells. 1997;2(11):667–78.
35. Yamashita M, Hori Y, Shinomiya T, Obuse C, Tsurimoto T, Yoshikawa H, et al. The efficiency and timing of initiation of replication of multiple replicons of Saccharomyces cerevisiae chromosome VI. Genes Cells. 1997;2(11):655–65.
36. de Moura AP, Retkute R, Hawkins M, Nieduszynski CA. Mathematical modelling of whole chromosome replication. Nucleic Acids Res. 2010;38:5623–63.
37. Retkute R, Nieduszynski CA, de Moura A. Dynamics of DNA replication in yeast. Phys Rev Lett. 2011;107(6):068103.
38. Retkute R, Nieduszynski CA, de Moura A. Mathematical modeling of genome replication. Phys Rev E Stat Nonlin Soft Matter Phys. 2012;86(3 Pt 1):031916.
39. Baker A, Audit B, Chen C-L, Moindrot B, Leleu A, Guilbaud G, et al. Replication fork polarity gradients revealed by megabase-sized U-shaped replication timing domains in human cell lines. PLoS Comput Biol. 2012;8(4), e1002443.
40. McGuffee SR, Smith DJ, Whitehouse I. Quantitative, genome-wide analysis of eukaryotic replication initiation and termination. Mol Cell. 2013;50(1):123–35.

41. Audit B, Baker A, Chen CL, Rappailles A, Guilbaud G, Julienne H, et al. Multiscale analysis of genome-wide replication timing profiles using a wavelet-based signal-processing algorithm. Nat Protoc. 2013;8(1):98–110.
42. Baker A, Bechhoefer J. Inferring the spatiotemporal DNA replication program from noisy data. Phys Rev E Stat Nonlin Soft Matter Phys. 2014;89(3):032703.
43. Desprat R, Thierry-Mieg D, Lailler N, Lajugie J, Schildkraut C, Thierry-Mieg J, et al. Predictable dynamic program of timing of DNA replication in human cells. Genome Res. 2009;19(12):2288–99.
44. Muller CA, Hawkins M, Retkute R, Malla S, Wilson R, Blythe MJ, et al. The dynamics of genome replication using deep sequencing. Nucleic Acids Res. 2014;42(1), e3.
45. Koren A, Handsaker RE, Kamitaki N, Karli R, Ghosh S, Polak P, et al. Genetic variation in human DNA replication timing. Cell. 2014;159(5):1015–26.
46. Sekedat MD, Fenyo D, Rogers RS, Tackett AJ, Aitchison JD, Chait BT. GINS motion reveals replication fork progression is remarkably uniform throughout the yeast genome. Mol Syst Biol. 2010;6:353.
47. Feng W, Collingwood D, Boeck ME, Fox LA, Alvino GM, Fangman WL, et al. Genomic mapping of single-stranded DNA in hydroxyurea-challenged yeasts identifies origins of replication. Nat Cell Biol. 2006;8(2):148–55.
48. Hawkins M, Retkute R, Muller CA, Saner N, Tanaka TU, de Moura AP, et al. High-resolution replication profiles define the stochastic nature of genome replication initiation and termination. Cell Rep. 2013;5(4):1132–41.
49. Chen CL, Rappailles A, Duquenne L, Huvet M, Guilbaud G, Farinelli L, et al. Impact of replication timing on non-CpG and CpG substitution rates in mammalian genomes. Genome Res. 2010;20:447–57.
50. Hansen RS, Thomas S, Sandstrom R, Canfield TK, Thurman RE, Weaver M, et al. Sequencing newly replicated DNA reveals widespread plasticity in human replication timing. Proc Natl Acad Sci U S A. 2010;107(1):139–44.
51. Lobry JR. Asymmetric substitution patterns in the two DNA strands of bacteria. Mol Biol Evol. 1996;13(5):660–5.
52. Touchon M, Nicolay S, Audit B, Brodie Of Brodie EB, d'Aubenton-Carafa Y, Arneodo A, et al. Replication-associated strand asymmetries in mammalian genomes: toward detection of replication origins. Proc Natl Acad Sci U S A. 2005;28:28.
53. Brodie Of Brodie EB, Nicolay S, Touchon EB, Audit B, d'Aubenton-Carafa Y, Thermes C, et al. From DNA sequence analysis to modeling replication in the human genome. Phys Rev Lett. 2005;94(24):248103.
54. Huvet M, Nicolay S, Touchon M, Audit B, d'Aubenton-Carafa Y, Arneodo A, et al. Human gene organization driven by the coordination of replication and transcription. Genome Res. 2007;17(9):1278–85.
55. Chen CL, Duquenne L, Audit B, Guilbaud G, Rappailles A, Baker A, et al. Replication-associated mutational asymmetry in the human genome. Mol Biol Evol. 2011;28(8):2327–37.
56. Necsulea A, Guillet C, Cadoret JC, Prioleau MN, Duret L. The relationship between DNA replication and human genome organization. Mol Biol Evol. 2009;26(4):729–41.
57. Baker A, Nicolay S, Zaghloul L, d'Aubenton-Carafa Y, Thermes A, d'Aubenton-Carafa Y, Audit B, et al. Wavelet-based method to disentangle transcription- and replication-associated strand asymmetries in mammalian genomes. Appl Comput Harmon Anal. 2010;28:150–70.
58. Marsolier-Kergoat MC, Goldar A. DNA replication induces compositional biases in yeast. Mol Biol Evol. 2012;29(3):893–904.
59. Agier N, Fischer G. The mutational profile of the yeast genome is shaped by replication. Mol Biol Evol. 2012;29(3):905–13.
60. Agier N, Romano OM, Touzain F, Cosentino Lagomarsino M, Fischer G. The spatiotemporal program of replication in the genome of Lachancea kluyveri. Genome Biol Evol. 2013;5(2):370–88.
61. Nick McElhinny SA, Kumar D, Clark AB, Watt DL, Watts BE, Lundstrom EB, et al. Genome instability due to ribonucleotide incorporation into DNA. Nat Chem Biol. 2010;6(10):774–81.

62. Sparks JL, Chon H, Cerritelli SM, Kunkel TA, Johansson E, Crouch RJ, et al. RNase H2-initiated ribonucleotide excision repair. Mol Cell. 2012;47(6):980–6.
63. Miyabe I, Kunkel TA, Carr AM. The major roles of DNA polymerases epsilon and delta at the eukaryotic replication fork are evolutionarily conserved. PLoS Genet. 2011;7(12), e1002407.
64. Reijns MA, Kemp H, Ding J, de Proce SM, Jackson AP, Taylor MS. Lagging-strand replication shapes the mutational landscape of the genome. Nature. 2015;518(7540):502–6.
65. Daigaku Y, Keszthelyi A, Muller CA, Miyabe I, Brooks T, Retkute R, et al. A global profile of replicative polymerase usage. Nat Struct Mol Biol. 2015;22(3):192–8.
66. Clausen AR, Lujan SA. Tracking replication enzymology in vivo by genome-wide mapping of ribonucleotide incorporation. Nat Struct Mol Biol. 2015;22(3):185–91.
67. Koh KD, Balachander S, Hesselberth JR, Storici F. Ribose-seq: global mapping of ribonucleotides embedded in genomic DNA. Nat Methods. 2015;12(3):251–7.
68. Pursell ZF, Isoz I, Lundstrom EB, Johansson E, Kunkel TA. Yeast DNA polymerase epsilon participates in leading-strand DNA replication. Science. 2007;317(5834):127–30.
69. Nick McElhinny SA, Gordenin DA, Stith CM, Burgers PM, Kunkel TA. Division of labor at the eukaryotic replication fork. Mol Cell. 2008;30(2):137–44.
70. Lujan SA, Clausen AR, Clark AB, MacAlpine HK, MacAlpine DM, Malc EP, et al. Heterogeneous polymerase fidelity and mismatch repair bias genome variation and composition. Genome Res. 2014;24(11):1751–64.
71. Johnson RE, Classen R, Prakash L, Prakash S. A major role of DNA polymerase delta in replication of both the leading and lagging DNA strands. Mol. Cell 2015; 59(2):163-175.
72. Smith DJ, Whitehouse I. Intrinsic coupling of lagging-strand synthesis to chromatin assembly. Nature. 2012;483(7390):434–8.
73. Hyrien O, Goldar A. Mathematical modelling of eukaryotic DNA replication. Chromosome Res. 2010;18(1):147–61.
74. Gauthier MG, Norio P, Bechhoefer J. Modeling inhomogeneous DNA replication kinetics. PLoS One. 2012;7(3), e32053.
75. Yang SC, Bechhoefer J. How Xenopus laevis embryos replicate reliably: investigating the random-completion problem. Phys Rev E Stat Nonlin Soft Matter Phys. 2008;78(4 Pt 1):041917.
76. Kolmogorov AN. On the statistical theory of crystallization in metals. Izv Akad Nauk SSSR Ser Fiz. 1937;1:355–9.
77. Johnson WA, Mehl PA. Reaction kinetics in processes of nucleation and growth. Trans AIMME. 1939;135:416–42.
78. Avrami M. Kinetics of phase change. I. General theory. J Chem Phys. 1939;7:1103–12.
79. Avrami M. Kinetics of phase change. II. Transformation-time relations for random distributions of nuclei. J Chem Phys. 1940;8:212–24.
80. Avrami M. Kinetics of phase change. III. Granulation, phase change, and microstructure. J Chem Phys. 1941;9:177–84.
81. Baker A, Audit B, Yang SC-H, Bechhoefer J, Arneodo A. Inferring where and when replication initiates from genome-wide replication timing data. Phys Rev Lett. 2012;108:268101.
82. Patel PK, Kommajosyula N, Rosebrock A, Bensimon A, Leatherwood J, Bechhoefer J, et al. The Hsk1/Cdc7 replication kinase regulates origin efficiency. Mol Biol Cell. 2008;12:5550–8.
83. Wu PY, Nurse P. Establishing the program of origin firing during S phase in fission Yeast. Cell. 2009;136(5):852–64.
84. Mantiero D, Mackenzie A, Donaldson A, Zegerman P. Limiting replication initiation factors execute the temporal programme of origin firing in budding yeast. EMBO J. 2011;30(23): 4805–14.
85. Tanaka S, Nakato R, Katou Y, Shirahige K, Araki H. Origin association of Sld3, Sld7, and Cdc45 proteins is a key step for determination of origin-firing timing. Curr Biol. 2011;21(24):2055–63.
86. Wong PG, Winter SL, Zaika E, Cao TV, Oguz U, Koomen JM, et al. Cdc45 limits replicon usage from a low density of preRCs in mammalian cells. PLoS One. 2011;6(3), e17533.
87. Wyrick JJ, Aparicio JG, Chen T, Barnett JD, Jennings EG, Young RA, et al. Genome-wide distribution of ORC and MCM proteins in S. cerevisiae: high-resolution mapping of replication origins. Science. 2001;294(5550):2357–60.

88. Remus D, Beuron F, Tolun G, Griffith JD, Morris EP, Diffley JF. Concerted loading of Mcm2-7 double hexamers around DNA during DNA replication origin licensing. Cell. 2009;139(4):719–30.
89. Bowers JL, Randell JC, Chen S, Bell SP. ATP hydrolysis by ORC catalyzes reiterative Mcm2-7 assembly at a defined origin of replication. Mol Cell. 2004;16(6):967–78.
90. Evrin C, Clarke P, Zech J, Lurz R, Sun J, Uhle S, et al. A double-hexameric MCM2-7 complex is loaded onto origin DNA during licensing of eukaryotic DNA replication. Proc Natl Acad Sci U S A. 2009;106(48):20240–5.
91. Belsky JA, MacAlpine HK, Lubelsky Y, Hartemink AJ, MacAlpine DM. Genome-wide chromatin footprinting reveals changes in replication origin architecture induced by pre-RC assembly. Genes Dev. 2015;29(2):212–24.
92. Fachinetti D, Bermejo R, Cocito A, Minardi S, Katou Y, Kanoh Y, et al. Replication termination at eukaryotic chromosomes is mediated by Top2 and occurs at genomic loci containing pausing elements. Mol Cell. 2010;39(4):595–605.
93. Mukhopadhyay R, Lajugie J, Fourel N, Selzer A, Schizas M, Bartholdy B, et al. Allele-specific genome-wide profiling in human primary erythroblasts reveal replication program organization. PLoS Genet. 2014;10(5), e1004319.
94. Ryba T, Hiratani I, Lu J, Itoh M, Kulik M, Zhang J, et al. Evolutionarily conserved replication timing profiles predict long-range chromatin interactions and distinguish closely related cell types. Genome Res. 2010;20(6):761–70.
95. Zaghloul L, Drillon G, Boulos RE, Argoul F, Thermes C, Arneodo A, et al. Large replication skew domains delimit GC-poor gene deserts in human. Comput Biol Chem. 2014;53 Pt A:153–65.
96. Lima-de-Faria A, Jaworska H. Late DNA synthesis in heterochromatin. Nature. 1968;217(5124):138–42.
97. Dimitrova DS, Gilbert DM. The spatial position and replication timing of chromosomal domains are both established in early G1 phase. Mol Cell. 1999;4(6):983–93.
98. O'Keefe RT, Henderson SC, Spector DL. Dynamic organization of DNA replication in mammalian cell nuclei: spatially and temporally defined replication of chromosome-specific alpha-satellite DNA sequences. J Cell Biol. 1992;116(5):1095–110.
99. Sporbert A, Gahl A, Ankerhold R, Leonhardt H, Cardoso MC. DNA polymerase clamp shows little turnover at established replication sites but sequential de novo assembly at adjacent origin clusters. Mol Cell. 2002;10(6):1355–65.
100. Maya-Mendoza A, Olivares-Chauvet P, Shaw A, Jackson DA. S phase progression in human cells is dictated by the genetic continuity of DNA foci. PLoS Genet. 2010;6(4), e1000900.
101. Audit B, Zaghloul L, Vaillant C, Chevereau G, d'Aubenton-Carafa Y, Thermes C, et al. Open chromatin encoded in DNA sequence is the signature of 'master' replication origins in human cells. Nucleic Acids Res. 2009;37(18):6064–75.
102. Julienne H, Zoufir A, Audit B, Arneodo A. Human genome replication proceeds through four chromatin states. PLoS Comput Biol. 2013;9(10), e1003233.
103. Lieberman-Aiden E, van Berkum NL, Williams L, Imakaev M, Ragoczy T, Telling A, et al. Comprehensive mapping of long-range interactions reveals folding principles of the human genome. Science. 2009;326(5950):289–93.
104. Dixon JR, Selvaraj S, Yue F, Kim A, Li Y, Shen Y, et al. Topological domains in mammalian genomes identified by analysis of chromatin interactions. Nature. 2012;485(7398):376–80.
105. Naumova N, Imakaev M, Fudenberg G, Zhan Y, Lajoie BR, Mirny LA, et al. Organization of the mitotic chromosome. Science. 2013;342(6161):948–53.
106. Pope BD, Ryba T, Dileep V, Yue F, Wu W, Denas O, et al. Topologically associating domains are stable units of replication-timing regulation. Nature. 2014;515(7527):402–5.
107. Gindin Y, Valenzuela MS, Aladjem MI, Meltzer PS, Bilke S. A chromatin structure-based model accurately predicts DNA replication timing in human cells. Mol Syst Biol. 2014;10:722.

Chapter 5
Chromatin Determinants of Origin Selection and Activation

Mónica P. Gutiérrez and David M. MacAlpine

Abstract DNA replication is an essential cell cycle-regulated process necessary for the accurate duplication of the genome. DNA replication begins at *cis*-acting replicator loci (replication origins) that are distributed throughout each of the eukaryotic chromosomes. The first factor to bind to the replicator is the origin recognition complex (ORC). ORC directs the recruitment of the Mcm2-7 helicase complex to form the pre-replication complex (pre-RC), licensing the origin for activation. Origin selection and activation are dependent on both DNA sequence and epigenetic features. The *cis*-acting sequence elements that function as replicators are well defined in *Saccharomyces cerevisiae*; in contrast, metazoan replicators are not defined by primary sequence, but rather by secondary structural features like G-quadruplexes. In both yeast and higher eukaryotes, however, *cis*-acting sequences or G-quadruplexes are not sufficient for origin function, implying the necessity for epigenetic mechanisms in regulating the selection and activation of DNA replication origins. In higher eukaryotes, the chromatin landscape surrounding origins of replication is important for the plasticity of the DNA replication program, allowing it to adapt and respond to developmental and environmental signals. Here we describe the role of chromatin structure and histone modifications in specifying and regulating eukaryotic DNA replication origins.

Keywords DNA replication • Chromatin • ORC • Pre-RC • Origin • G-quadruplex • Nucleosome • Epigenetics

---

M.P. Gutiérrez
University Program in Genetics and Genomics, Duke University Medical Center, Durham, NC, USA
e-mail: monica.gutierrez@duke.edu

D.M. MacAlpine (✉)
Department of Pharmacology and Cancer Biology, Duke University Medical Center, Durham, NC, USA
e-mail: david.macalpine@duke.edu

D.L. Kaplan (ed.), *The Initiation of DNA Replication in Eukaryotes*,
DOI 10.1007/978-3-319-24696-3_5

Introduction

The entire genome must be duplicated in an accurate and complete manner every cell cycle. This process must be tightly regulated to ensure the complete duplication of the genome within the confines of S-phase [1]. To accomplish this, higher eukaryotes employ tens of thousands of DNA replication start sites distributed throughout the genome. Failure to properly regulate the DNA replication program may lead to under- or over-replication which may cause significant genomic instability [2]. In 1963, Jacob, Brenner, and Cuzin proposed a simple and elegant model for the regulation of prokaryotic DNA replication [3]. This model is based on the premise that a *cis*-acting sequence of DNA defined as a *replicator*, is acted upon, in *trans*, by an *initiator* factor to direct the duplication of the replicon. In prokaryotic systems like *Escherichia coli*, the replicator sequence *oriC* is recognized by DnaA, the initiator, which results in localized unwinding of the DNA and the recruitment and loading of the DnaB helicase by the helicase loader DnaC [4]. The fundamental principles of the replicon model, first described in prokaryotes, are conserved across prokaryotic and eukaryotic systems. The increased complexity of eukaryotic genomes requires that the replicon model and the selection and activation of DNA replication origins be inherently dynamic. This is necessary to establish and maintain cell-, tissue-, and developmental-specific DNA replication programs. For example, many more origins are required during early development, a stage where S-phase lasts only a few minutes, while in differentiated cells fewer origins are activated and S-phase can progress for longer than 6 h [5, 6]. The plasticity of the DNA replication program is likely driven by changes in the concentration of *trans*-activating initiator factors [7, 8] and epigenetic features that contribute to the recognition of *cis*-acting replicator sequences [9–11].

Trans-Acting Initiators of Eukaryotic DNA Replication

The *trans*-activating initiators that direct DNA replication are functionally conserved between prokaryotes and eukaryotes. Analogous to DnaA in prokaryotes, the origin recognition complex (ORC) associates with replicator sequences throughout the majority of the cell cycle. ORC is an essential and conserved protein complex composed of six subunits (Orc1-6) [12], and serves as a scaffold to mark potential DNA replication origins for helicase loading in G1 of the cell cycle (Fig. 5.1). With the exception of the AT-hook domain of Orc4 in *Schizosaccharomyces pombe*, ORC does not interact with the DNA via a specific DNA-binding domain. Instead, the DNA is threaded through a central channel in the complex which makes contact with the Orc1-5 subunits [13]. In G1, ORC recruits Cdc6 which, together with Cdt1, coordinates the loading of the Mcm2-7 replicative helicase to form the pre-replicative complex (pre-RC) [14]. Formation of the pre-RC at the origin serves to "license" the origin for initiation of DNA replication in the following S-phase [15].

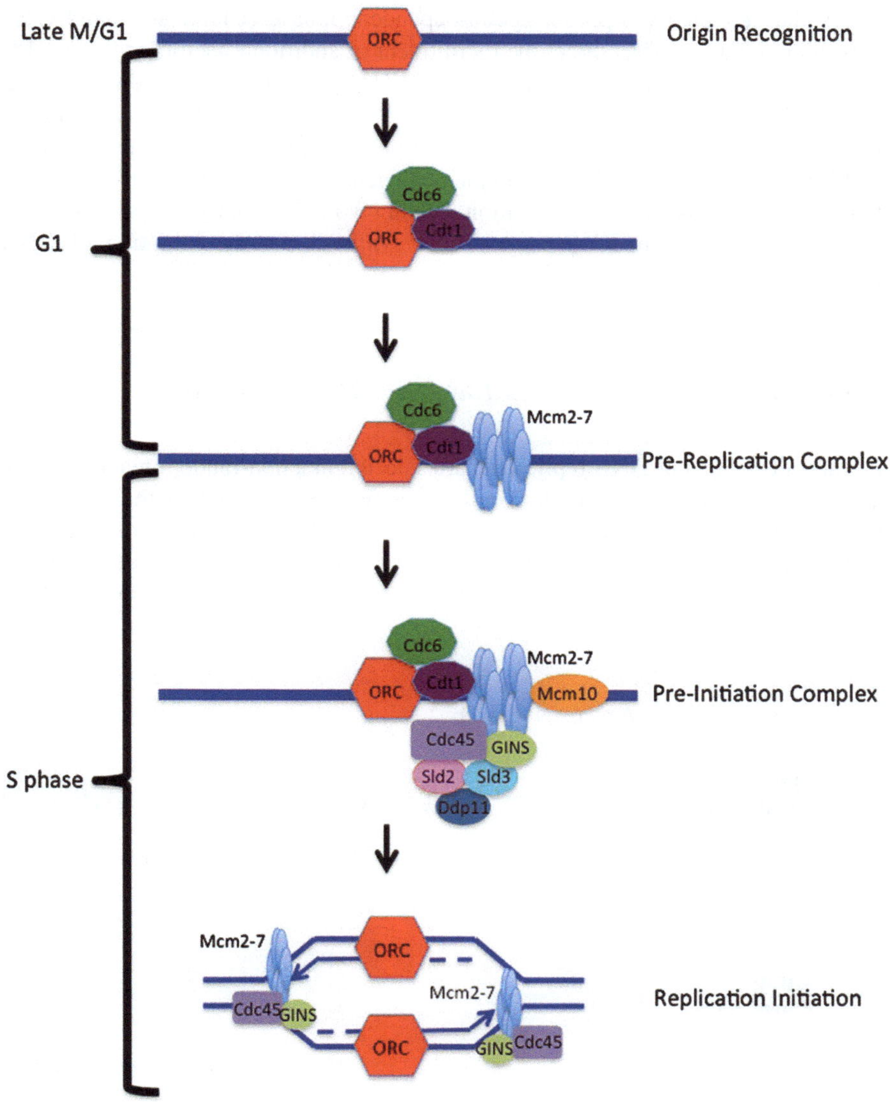

Fig. 5.1 Pre-RC assembly. Selection of replication origins starts with binding of the initiator of eukaryotic replication, the origin recognition complex (ORC), during late M phase and early G1 phase of the cell cycle. ORC binding to replicator sites leads to recruitment of Cdc6 and Cdt1, which are necessary for loading of the minichromosome maintenance complex (Mcm2-7) during G1 phase; together, these factors form the pre-replication complex (pre-RC). Recruitment of Mcm2-7 licenses the pre-RC, leading to replication origin activation in the subsequent S-phase. After origin licensing, recruitment of pre-initiation factors is promoted by cyclin- and Dbf4-dependent kinases (CDK and DDK, respectively) during the transition to S-phase, leading to formation of a pre-initiation complex (pre-IC) and subsequent replication initiation

As the cell enters S-phase, cyclin-dependent kinase (CDK) and Dbf4-Cdc7 dependent kinase (DDK) direct the recruitment of additional factors to the pre-RC to form the pre-initiation complex (pre-IC) [16]. In a DDK-dependent manner, Sld3, Sld7, and Cdc45 associate with the pre-RC [8, 17], and CDK activity directs the recruitment of Sld2, Dbp11, GINS, and polymerase ε [18–20]. Cdc45, GINS, and Mcm2-7 form the CMG holocomplex which has robust helicase activity in vitro [21, 22] and travels with the replication fork in vivo [23]. Finally, the primase Pol α is recruited to the pre-IC to start primer synthesis for the initiation of DNA replication.

Cis-Acting Replicators of Eukaryotic DNA Replication

Eukaryotic origins of DNA replication are selected in a non-random manner by the association of ORC at specific loci in the genome. ORC interacts with specific *cis*-acting replicator sequences distributed throughout the *Saccharomyces cerevisiae* genome [24]; in contrast, a conserved primary sequence resembling the sequence composition of *S. cerevisiae* origins has yet to emerge in higher eukaryotes (Fig. 5.2).

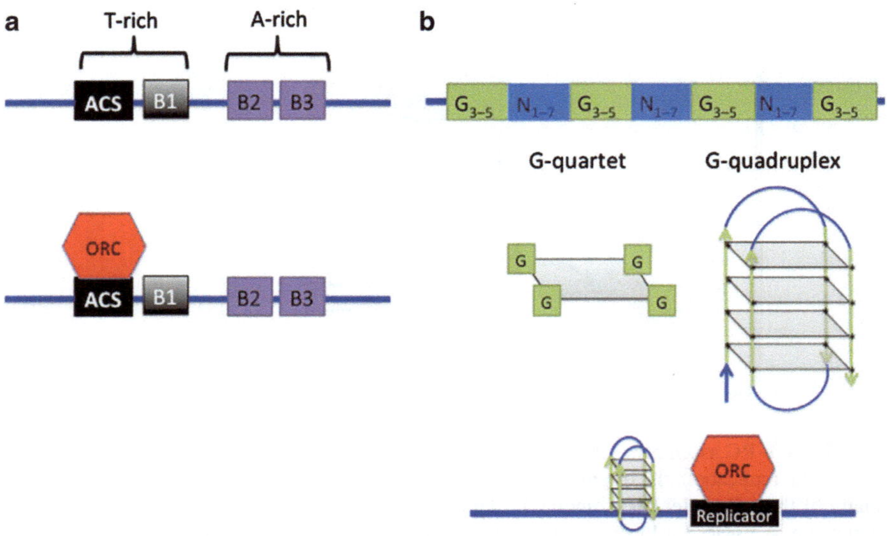

Fig. 5.2 Sequence determinants of DNA replication in higher eukaryotes. (**a**) In *S. cerevisiae*, replication origins are defined by the ARS consensus sequence (ACS), which is necessary, but not sufficient, for replication origin site selection. B elements are components within the ACS that are thought to function as sites of DNA unwinding and strand separation. (**b**) In the genomes of higher eukaryotes, guanine-rich sequences with a specific motif (G_{3-5}-N_{1-7}-G_{3-5}-N_{1-7}-G_{3-5}-N_{1-7}-G_{3-5}) can form four-stranded DNA structures, called G-quadruplexes. A significant fraction of replication origins in mammalian cell lines have been mapped in the vicinity of these structures

ORC purified from metazoans exhibits little sequence specificity [25]; however, recent evidence suggests that secondary sequence structures such as G-quadruplexes may function as part of replicators [26].

Cis-Acting Replicators in S. cerevisiae

Studies in the budding yeast led to the identification of the first eukaryotic *cis*-acting sequences with replicator function (Fig. 5.2a). Autonomously replicating sequences (ARSs) were identified in *S. cerevisiae* based on their ability to confer stable inheritance to an episome [27]. Genetic dissection of the ARS1 replication origin revealed an 11 base pair T-rich ARS consensus sequence (ACS) that is required for origin function [28] by promoting ORC binding [12]. Additional sequence features in the ARS, referred to as B elements, also contribute to origin function and are thought to facilitate helicase loading [29] and DNA unwinding [30]. A systematic genome-wide screen for DNA fragments with replicator activity across the *S. cerevisiae* genome identified 366 unique loci with replicator potential [31]. Although the ACS is necessary for origin function, there are many more ACS motif matches in the *S. cerevisiae* genome ($\sim$10,000) than functional origins [32]. Together, these results suggest that, in addition to the ACS, other *cis*-acting chromosomal features are required to specify origins of replication in the genome.

The sequences that function as replicator elements in the budding yeast are conserved among other *sensu stricto Saccharomyces* species [33]. Despite this evolutionary conservation for not only the ACS sequence but also its location and distribution throughout the genome, the DNA replication program is remarkably tolerant of losing specific origins. Deletion of the majority of active origins on an extra copy of chromosome III was remarkably stable across yeast generations, only resulting in a minimal S-phase delay [34]. Cryptic origins near the telomeres accounted for the cell's ability to faithfully replicate and segregate a mutant chromosome lacking nearly all origins. A more recent study investigated the consequence of deleting seven highly characterized origins from the left arm of chromosome VI [35]. As in the earlier study, loss of origins on chromosome VI had a minimal impact on growth even in the presence of replicative stress [35]. Surprisingly, despite loss of ORC association at the deleted origins, cryptic initiation events in the vicinity of the original origins still occurred. It is unlikely that these were ORC-independent initiation events, but rather pre-RC assembly and subsequent initiation may have been facilitated by transient ORC interactions with the DNA [36]. This phenomenon underscores the plasticity of the replication program, allowing the activation of non-canonical origins in order to complete DNA replication and maintain genomic stability.

Cis-Acting Replicators in Higher Eukaryotes

The identification of *cis*-acting replicator sequences in higher eukaryotes has been hampered, in part, by the size and increased complexity of the genome and the lack of reliable plasmid-based assays for origin function. Early attempts to identify *cis*-acting replicators using plasmid-based assays found that plasmid maintenance was most strongly correlated with the size but not the specific sequence composition of the replicator element [37]. In addition, purified ORC from higher eukaryotes also exhibited little sequence specificity in vitro [25, 38]. Despite the apparent lack of a conserved *cis*-acting element, origin selection in higher eukaryotes is not an entirely stochastic process as there are many examples of specific loci functioning as replication origins [39–41].

The rapid proliferation of genome-wide approaches to map ORC binding sites and replication intermediates in higher eukaryotes has identified genomic features frequently associated with origins of DNA replication. ChIP-seq analysis of ORC binding in *Drosophila* [42], human [43], and mouse [44] studies found that ORC was frequently associated at promoter regions. Similarly, replication intermediates (nascent strands) arising from origin activity were also enriched at promoter elements [45].

Nascent strand analysis revealed an enrichment of guanine-rich sequences at origins in a variety of model systems [46–48]. These stretches of guanine nucleotides have the potential to form a four-stranded molecule, called a G-quadruplex, when single strands of DNA are exposed during replication and transcription (Fig. 5.2b). Although G-quadruplexes can form in vitro, their propensity to form in vivo is less clear [49]. Recent reports have mapped their location and determined that these structures are long-lived [50], suggesting that their presence and stability could have a biological role; however, it is unclear exactly how these sequences function in DNA replication. In vitro, purified ORC has a high affinity for G-rich single-stranded synthetic oligomers [51]; paradoxically, in vivo ORC is found downstream of G-rich structures, approximately 160 bp in *Drosophila* cells and 280 bp in mouse cells [47]. The location and function of these sequences suggest that they may not function strictly as replicators, but perhaps in other aspects of replication initiation, or by stalling or blocking active replication forks. Finally, sequences with the potential to form a G-quadruplex are very abundant in the genome and only a small fraction of all potential G-quadruplexes are associated with enriched replication intermediates.

A defining feature of multicellular eukaryotes is the considerable cellular plasticity required for normal development and tissue-specific function in the organism. Just as there are developmental and tissue-specific transcriptional programs, DNA replication must also be dynamic and respond to developmental, tissue-specific, and environmental cues. For example, during early embryogenesis in the fruit fly, when S-phase is only a matter of minutes, there needs to be many more origins of replication than in a differentiated tissue with a significantly longer S-phase [5]. Unlike in yeast cells where each potential origin is established every S-phase, multicellular eukaryotes exhibit cell type-specific patterns of origin selection and activation. For example, the human β-globin locus is duplicated by a single bidirectional origin during early S-phase in erythrocytes, while in other cell types it is a late replicating origin [52]. Together, these and numerous other experiments demonstrate that sequence alone is insufficient to define the location or activation properties of replicators in higher eukaryotes.

Epigenetic Determinants of Pre-RC Assembly and Replication Initiation

Chromatin is the macromolecular complex of DNA, RNA, and proteins within the nucleus. The fundamental organizing unit of chromatin is the nucleosome, composed of two copies each of H2A, H2B, H3, and H4 that form a histone octamer to which 147 bp of DNA wrap around approximately 1.7 times [53]. Arrays of nucleosomes are able to form higher order structures of compact chromatin which facilitate the organization and packaging of the genome in the nucleus. The degree of chromatin compaction dictates the accessibility of regulatory DNA sequences to *trans*-acting factors required for transcription and DNA replication.

Nucleosome Positioning and Chromatin Remodelers

Nucleosome positioning is critical for origin function. Pioneering studies identified well-positioned nucleosomes flanking the ARS1 origin in *S. cerevisiae* [54]. A nucleosome-free region at the origin is critical for function as the forced encroachment of a nucleosome into the *cis*-acting ACS replicator element impaired origin function on a plasmid [55]. Nucleosome occupancy at the ACS likely prevented ORC association rendering the origin non-functional. Given that in *S. cerevisiae* the ACS is necessary but not sufficient for origin function and that metazoan ORC exhibits little sequence specificity, an appealing hypothesis is that local chromatin structure and nucleosome positioning serve as important determinants for ORC binding and origin selection in eukaryotic genomes (Fig. 5.3).

Advances in genomic technologies like microarrays and next-generation sequencing have made it possible to systematically and comprehensively catalog nucleosome positions in a number of eukaryotic organisms [56–58]. Nucleosome positioning throughout the genome is not random, and characteristic patterns of nucleosome occupancy have emerged for many genomic features. For example, there is a nucleosome depleted region at transcription start sites with well-positioned nucleosomes at the +1 position [59, 60]. Similar patterns are also evident at DNA replication origins in *S. cerevisiae*, with well-positioned nucleosomes flanking the vast majority of active DNA replication origins [61, 62]. Consistent with the

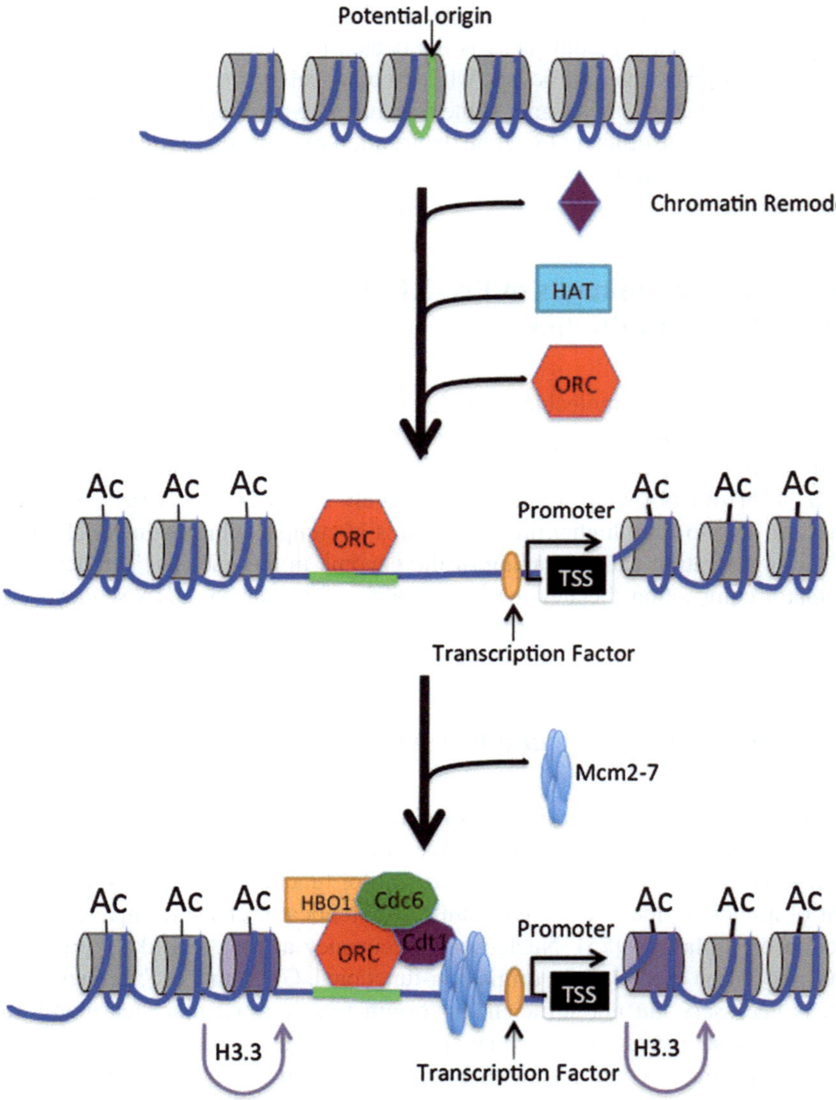

Fig. 5.3 Chromatin determinants of DNA replication. Specific origins of replication are selected throughout the eukaryotic genome. They are defined by areas of open chromatin marked by histone acetylation, and most, but not all, are located near transcription start sites. The activity of histone acetylases (HBO1) and chromatin remodelers (ISWI and SWI/SNF) facilitates ORC binding and pre-RC assembly at replication origins. Nucleosomes flanking replication origins undergo constant turn-over and contribute to a dynamic chromatin landscape. HBO1 also acts with Cdt1 and aids in the recruitment of the Mcm2-7 replicative helicase complex forming the pre-RC, likely by means of its role in histone H4 acetylation

hypothesis that chromatin structure and nucleosome positioning may contribute to ORC localization, nucleosome-free regions at ORC binding sites have been observed in a number of model organisms [42, 63, 64].

Precise nucleosome positioning at the origin is also important for pre-RC assembly and origin activation. Displacement of the nucleosomes away from the ACS at ARS1 negatively impairs origin function by preventing helicase loading and pre-RC formation [65]. These results suggest that the proximity of the flanking nucleosome to ORC may facilitate helicase loading either through an Mcm2-7-histone interaction [66, 67] or perhaps by stabilizing ORC via the $ORC1_{BAH}$ domain [68]. Nucleosome positioning at the origin is dynamic with a Cdc6-dependent expansion of the nucleosome-free region occurring between G2 and G1 [66]. However, it is unclear if the expansion of the nucleosome-free region is due to Cdc6 loading or subsequent recruitment of the Mcm2-7 helicase. Importantly, cell cycle dependent nucleosome remodeling at the origin is associated with efficient and early activating origins of DNA replication.

ATP-dependent chromatin remodelers are required to slide and evict nucleosomes from DNA. In vitro ORC and the activity of an ATP-dependent chromatin remodeler (ISWI) are necessary and sufficient to establish precisely positioned nucleosomes flanking an ACS on template DNA [62]. There is also an abundance of evidence linking chromatin remodeling to origin usage in higher eukaryotes. ORC binding sites in *Drosophila* are enriched for dynamic nucleosomes that are frequently turned over [69] and exchanged for histone H3.3 in a replication-independent manner [42]. H3.3-containing nucleosomes are deposited on DNA outside of S-phase and mark genomic regions undergoing active chromatin turn-over and remodeling [70]. Not surprisingly, specific ATP-dependent chromatin remodelers are enriched at replication origins. For example, *Drosophila* ISWI and NURF chromatin remodeling complexes are among the top predictive features for specifying ORC binding [71]. Similarly, nearly a third of human origins of replication were located in close proximity to regions of SWI/SNF activity [72]. The exact mechanisms by which chromatin remodelers are targeted to specific origins of replication are poorly understood. However, at a subset of origins, chromatin remodelers may be recruited in a cell cycle-dependent manner via interactions with specific pre-RC components. For example, the ATPase SNF2H of the ISWI chromatin remodeling complex interacts with Cdt1 to promote Mcm2-7 loading in human cell lines [73].

Histone Post-translational Modifications

The non-structured N-terminal tail of each of the histones contains a high concentration of lysine residues that act as a substrate for a variety of different covalent post-translational modifications (PTMs), including methylation, acetylation, sumoylation, ribosylation, and ubiquitination [74]. Together, these patterns of histone PTMs form the basis of a complex "histone code" that regulates the compaction of chromatin and recruitment of DNA-binding proteins [75]. Importantly, these

PTMs are reversible. They provide dynamic chromatin states capable of responding to external, developmental, and cell type-specific signals, which in turn modulate and regulate DNA-templated processes including transcription and DNA replication.

Numerous histone PTMs have been linked to the DNA replication program. Early observations noted that gene-rich euchromatin environments were typically copied in S-phase prior to gene-poor heterochromatin environments [76–78]. More recently, studies from ENCODE (Encyclopedia of DNA Elements) and modENCODE (model organism ENCODE) have extended these observations genome-wide and correlated the average time of DNA replication with chromatin states that correspond to specific histone PTMs [79, 80]. Early replicating regions of the genome are enriched for chromatin marks associated with active transcription, including histone acetylation (H3K9ac, H3K14ac, H3K18ac) and methylation (H3K4me1/2/3). Conversely, late replicating regions of the genome are coupled with repressive chromatin modifications frequently associated with constitutive and facultative heterochromatin (H3K9me2/3, H3K27me3). Elegant experiments at the human β-globin locus demonstrated that tethering a histone acetyltransferase (HAT) near the origin was sufficient to promote early replication and, similarly, tethering a histone deacetylase (HDAC) delayed the time of replication activation [11]. Broadly speaking, these results suggest that the DNA replication program responds to similar epigenetic cues that regulate transcription; however, only a few histone modifications and their respective modifying enzymes have been mechanistically linked to regulation of origin selection or activation.

Methylation of Histone H4 on Lysine 20

A major question in the DNA replication field is how metazoan ORC localizes to specific loci in the absence of any apparent sequence specificity. Human ORC1 contains a bromo-adjacent homology (BAH) domain which specifically recognizes and binds to H4K20me2 with micromolar affinity [81]. The $ORC1_{BAH}$ domain is conserved in eukaryotes and exhibits affinity for H4K20me2 across a wide range of metazoan species. Interestingly, in *S. cerevisiae*, where ORC exhibits sequence-specific interactions with the ACS, the yeast $ORC1_{BAH}$ domain does not interact with H4K20me2. This suggests that H4K20me2 may function as a specificity factor for ORC in higher eukaryotes. Consistent with this model, H4K20me2 is enriched at select origins in the human genome, and mutations in the BAH domain that disrupt the recognition of H4K20me2 cause a decrease in ORC occupancy at replication origins and a cell cycle delay [81]. However, it is difficult to reconcile this chromatin-mediated model for ORC specificity with H4K20me2 being the most abundant histone modification. More than 85 % of histone H4 is dimethylated on lysine 20 genome-wide [82], and thus nearly all nucleosomes contain at least one histone H4 with H4K20me2. It seems more likely that H4K20me2 may function to stabilize ORC binding on chromatin via interaction with the $ORC1_{BAH}$ domain.

In humans, mutations in the $ORC1_{BAH}$ domain as well as other pre-RC components have been linked to Meier–Gorlin Syndrome (MGS), a primordial form of dwarfism [83, 84]. This group of diseases is characterized by intrauterine reduced growth, postnatal short stature, and microcephaly, with normal intellectual development [84]. Functional studies demonstrated that MGS mutations in pre-RC components lead to defects in origin licensing, causing a prolonged G1 phase with a delayed transition into S-phase [84]. Several of the MGS phenotypes can be recapitulated in zebrafish models by mutating ORC1 [81, 84], or, alternatively, by depleting the conserved methyltransferases (Suv4-20h1/2) required for H4K20me2/3 [81]. Together, these data demonstrate that the $ORC1_{BAH}$ domain and its interaction with H4K20me2 are critical components of origin selection necessary for proper organismal development.

In addition to the methyltransferases Suv4-20h1 and Suv4-20h2, which catalyze di- and trimethylation of H4K20, PR-Set7, also known as Set8, is the sole methyltransferase responsible for monomethylation of H4K20 [85, 86]. PR-Set7 is cell cycle-regulated, and its targeted destruction by the proteasome is tightly coupled to S-phase by the E3 ubiquitin ligase, Crl4, in a PCNA-dependent manner [87–89]. PR-Set7 and H4K20me1 levels peak during late S-phase and remain elevated through G2/M. PR-Set7 and, presumably, H4K20me1 are critical for maintaining genome stability as loss of PR-Set7 activity results in chromosome decondensation, S-phase delay, centrosome amplification, and activation of the DNA damage checkpoint [90–92]. Stabilization or overexpression of PR-Set7 also leads to genomic instability resulting from re-replication of DNA. Tethering of PR-Set7 to specific loci results in a local increase in H4K20me1 and the promotion of pre-RC assembly [93]. Subsequent experiments argue that efficient pre-RC assembly is dependent not only on PR-Set7 but also on Suv4-20h1/2 [94]. This suggests that it is not the de novo deposition of H4K20me1, but rather the subsequent conversion to H4K20me2/3 that is important for ORC binding and pre-RC assembly. However, somewhat paradoxically, only the loss of PR-Set7, and not Suv4-20h1/2, led to DNA damage and cell cycle arrest [92, 94, 95]. Future experiments will be needed to further establish the role of PR-Set7 in pre-RC formation and genome stability. It is likely that the methylation state of H4K20 is critical not only for ORC binding and pre-RC assembly, but also in maintaining genome stability.

Acetylation of Histone H4 by HBO1

The HAT binding to ORC1 (HBO1) is responsible for the bulk of histone H4 acetylation in mammals [96]. HBO1 was initially identified in a two-hybrid screen for factors that interact with human ORC1 [97]. Subsequently, HBO1 was shown to interact with multiple pre-RC components including Mcm2 [98] and Cdt1 [99]. HBO1 is targeted to origins of replication in G1 by a direct interaction with the licensing factor Cdt1 [100]. Artificially tethering HBO1 or its *Drosophila* homolog, Chameau, to origins promotes pre-RC assembly and origin function [10, 101].

Presumably, HBO1 acetylates histone H4 on nucleosomes proximal to the origin to facilitate Mcm2-7 loading; however, it remains possible that the target of the HBO1 HAT activity is not origin-proximal nucleosomes, but rather that HBO1 facilitates pre-RC assembly via the direct acetylation of pre-RC components [99].

Additional Histone PTMs Involved in Origin Function

A number of post-translational histone modifications have been implicated in regulating the selection and activation of DNA replication origins from a variety of eukaryotic organisms. H3K36 methylation mediated by Set2 in *S. cerevisiae* promotes the recruitment of Cdc45 and early origin activation [102]. H3K4me2 is also enriched at yeast origins, and mutations in either Set1 or Bre3, both regulators of H3K4 methylation, impair plasmid maintenance [103]. Ubiquitination of H2B (H2Bub1) is enriched at *S. cerevisiae* origins; however, it does not regulate initiation, but instead appears to promote fork elongation [104]. In human cells, H3K79me2 enrichment at replication origins may negatively regulate origin licensing, as depletion of the methyltransferase Dot1L results in re-replication and genome instability [105]. Similarly, in *Arabidopsis*, the loss of H3K27 methylation in the heterochromatin also leads to re-replication [106]. Together, these data demonstrate the importance of the chromatin landscape in origin selection and activation, ensuring that DNA replication occurs once and only once per cell cycle. Hyperacetylation of H4K16 promotes transcription and origin activation in *Drosophila*. In *Drosophila* males, the single X chromosome is hyperacetylated on H4K16 by the dosage compensation complex (DCC), which upregulates transcription of the X chromosome twofold to balance gene expression with the autosomes [107]. In addition to having elevated gene expression, the X chromosome is also replicated earlier than the autosomes or the two female X chromosomes [108], suggesting a link between dosage compensation and origin function. This linkage was confirmed by genome-wide experiments correlating male-specific H4K16Ac on the X chromosome with replication during early S-phase [109]. Inactivation of the HAT, MOF, an integral component of the DCC, prevented the male-specific early replication of the X chromosome [110]. The H4K16Ac-mediated early replication of the male X chromosome was due to an increase in origin activation, not origin selection (ORC binding). Together, these results suggest that transcription and replication initiation are regulated by the same epigenetic cues.

PTM of histones modulates almost all DNA-templated processes including DNA replication. However, given the broad distribution and potential secondary effects due to transcriptional regulation, care must be taken in assigning a direct causal effect of a particular chromatin state on the DNA replication program. For example, the time of origin activation in *S. cerevisiae* was thought to be regulated, in part, by various HDACs including Rpd3 [9, 111, 112] and Sir2 [113]. A correlation was found between activation of late replication origins and Rpd3; similarly, Sir2 was found to both repress select origins and promote the activation of early origins [114]. Paradoxically, however, despite the substantial impact they exhibited on replication origin activation, there was little evidence for origin-specific changes in histone acetylation. Instead, it was found that Rpd3 and Sir2 modulated origin function not by deacetylating origin-specific chromatin, but rather by regulating rDNA function [114]. In *S. cerevisiae*, the rDNA exists as hundreds of copies of tandem repeats with each repeat containing an origin of replication. Rpd3 and Sir2 promoted or repressed origin activation at the rDNA locus, which resulted in the recruitment or sequestration of key replication initiation factors. Thus, the activation of non-rDNA origins was not dependent on local chromatin changes, but rather the availability of rate limiting replication initiation factors [7].

Acknowledgements We thank members of the MacAlpine laboratory for critical comments and suggestions. This work is supported by the National Institute Of General Medical Sciences of the National Institutes of Health under Award Number R01-GM104079 to D.M.M and F31-GM115158 to M.P.G.

References

1. Leonard AC, Méchali M. DNA replication origins. Cold Spring Harb Perspect Biol. 2013; 5(10):a010116.
2. Green BM, Finn KJ, Li JJ. Loss of DNA replication control is a potent inducer of gene amplification. Science. 2010;329(5994):943–6.
3. Jacob F, Brenner S, Cuzin F. On the regulation of DNA replication in bacteria. Cold Spring Harb Symp Quant Biol. 1963;28:329–48.
4. O'Donnell M, Langston L, Stillman B. Principles and concepts of DNA replication in bacteria, archaea, and eukarya. Cold Spring Harb Perspect Biol. 2013;5(7):a010108.
5. Blumenthal AB, Kriegstein HJ, Hogness DS. The units of DNA replication in *Drosophila melanogaster* chromosomes. Cold Spring Harb Symp Quant Biol. 1974;38:205–23.
6. Hyrien O, Maric C, Méchali M. Transition in specification of embryonic metazoan DNA replication origins. Science. 1995;270(5238):994–7.
7. Mantiero D, Mackenzie A, Donaldson A, Zegerman P. Limiting replication initiation factors execute the temporal programme of origin firing in budding yeast. EMBO J. 2011;30(23): 4805–14.
8. Tanaka S, Nakato R, Katou Y, Shirahige K, Araki H. Origin association of Sld3, Sld7, and Cdc45 proteins is a key step for determination of origin-firing timing. Curr Biol. 2011;21(24):2055–63.
9. Vogelauer M, Rubbi L, Lucas I, Brewer BJ, Grunstein M. Histone acetylation regulates the time of replication origin firing. Mol Cell. 2002;10(5):1223–33.
10. Aggarwal BD, Calvi BR. Chromatin regulates origin activity in *Drosophila* follicle cells. Nature. 2004;430(6997):372–6.
11. Goren A, Tabib A, Hecht M, Cedar H. DNA replication timing of the human beta-globin domain is controlled by histone modification at the origin. Genes Dev. 2008;22(10):1319–24.
12. Bell SP, Stillman B. ATP-dependent recognition of eukaryotic origins of DNA replication by a multiprotein complex. Nature. 1992;357(6374):128–34.
13. Bleichert F, Botchan MR, Berger JM. Crystal structure of the eukaryotic origin recognition complex. Nature. 2015;519(7543):321–6.
14. Bell SP, Kaguni JM. Helicase loading at chromosomal origins of replication. Cold Spring Harb Perspect Biol. 2013;5(6):a010124.
15. Blow JJ, Laskey RA. A role for the nuclear envelope in controlling DNA replication within the cell cycle. Nature. 1988;332(6164):546–8.

16. Tanaka S, Araki H. Helicase activation and establishment of replication forks at chromosomal origins of replication. Cold Spring Harb Perspect Biol. 2013;5(12):a010371.
17. Yabuuchi H, Yamada Y, Uchida T, Sunathvanichkul T, Nakagawa T, Masukata H. Ordered assembly of Sld3, GINS and Cdc45 is distinctly regulated by DDK and CDK for activation of replication origins. EMBO J. 2006;25(19):4663–74.
18. Heller RC, Kang S, Lam WM, Chen S, Chan CS, Bell SP. Eukaryotic origin-dependent DNA replication in vitro reveals sequential action of DDK and S-CDK kinases. Cell. 2011;146(1): 80–91.
19. Muramatsu S, Hirai K, Tak YS, Kamimura Y, Araki H. CDK-dependent complex formation between replication proteins Dpb11, Sld2, Pol, and GINS in budding yeast. Genes Dev. 2010;24(6):602–12.
20. Zegerman P, Diffley JFX. Phosphorylation of Sld2 and Sld3 by cyclin-dependent kinases promotes DNA replication in budding yeast. Nature. 2007;445(7125):281–5.
21. Ilves I, Petojevic T, Pesavento JJ, Botchan MR. Activation of the MCM2-7 helicase by association with Cdc45 and GINS proteins. Mol Cell. 2010;37(2):247–258.
22. Moyer SE, Lewis PW, Botchan MR. Isolation of the Cdc45/Mcm2-7/GINS (CMG) complex, a candidate for the eukaryotic DNA replication fork helicase. Proc Natl Acad Sci U S A. 2006;103(27):10236–41.
23. Aparicio OM, Weinstein DM, Bell SP. Components and dynamics of DNA replication complexes in *S. cerevisiae*: redistribution of MCM proteins and Cdc45p during S phase. Cell. 1997;91(1):59–69.
24. Celniker SE, Campbell JL. Yeast DNA replication in vitro: initiation and elongation events mimic in vivo processes. Cell. 1982;31(1):201–13.
25. Vashee S, Cvetic C, Lu W, Simancek P, Kelly TJ, Walter JC. Sequence-independent DNA binding and replication initiation by the human origin recognition complex. Genes Dev. 2003;17(15):1894–908.
26. Valton AL, Hassan-Zadeh V, Lema I, Boggetto N, Alberti P, Saintomé C, et al. G4 motifs affect origin positioning and efficiency in two vertebrate replicators. EMBO J. 2014;33(7): 732–46.
27. Stinchcomb DT, Struhl K, Davis RW. Isolation and characterisation of a yeast chromosomal replicator. Nature. 1979;282(5734):39–43.
28. Marahrens Y, Stillman B. A yeast chromosomal origin of DNA replication defined by multiple functional elements. Science. 1992;255(5046):817–23.
29. Wilmes GM, Bell SP. The B2 element of the *Saccharomyces cerevisiae* ARS1 origin of replication requires specific sequences to facilitate pre-RC formation. Proc Natl Acad Sci U S A. 2002;99(1):101–6.
30. Huang RY, Kowalski D. A DNA unwinding element and an ARS consensus comprise a replication origin within a yeast chromosome. EMBO J. 1993;12(12):4521–31.
31. Liachko I, Youngblood RA, Keich U, Dunham MJ. High-resolution mapping, characterization, and optimization of autonomously replicating sequences in yeast. Genome Res. 2013;23(4):698–704.
32. Breier AM, Chatterji S, Cozzarelli NR. Prediction of *Saccharomyces cerevisiae* replication origins. Genome Biol. 2004;5(4):R22.
33. Nieduszynski CA, Knox Y, Donaldson AD. Genome-wide identification of replication origins in yeast by comparative genomics. Genes Dev. 2006;20(14):1874–79.
34. Dershowitz A, Newlon CS. The effect on chromosome stability of deleting replication origins. Mol Cell Biol. 1993;13(1):391–98.
35. Bogenschutz NL, Rodriguez J, Tsukiyama T. Initiation of DNA replication from non-canonical sites on an origin-depleted chromosome. PLoS One. 2014;9(12):e114545.
36. Gros J, Devbhandari S, Remus D. Origin plasticity during budding yeast DNA replication in vitro. EMBO J. 2014;33(6):621–36.
37. Heinzel SS, Krysan PJ, Tran CT, Calos MP. Autonomous DNA replication in human cells is affected by the size and the source of the DNA. Mol Cell Biol. 1991;11(4):2263–72.

38. Remus D, Beall EL, Botchan MR. DNA topology, not DNA sequence, is a critical determinant for *Drosophila* ORC–DNA binding. EMBO J. 2004;23(4):897–907.
39. Ariga H, Imamura Y, Iguchi-Ariga SM. DNA replication origin and transcriptional enhancer in c-myc gene share the c-myc protein binding sequences. EMBO J. 1989;8(13):4273–9.
40. Delidakis C, Kafatos FC. Amplification enhancers and replication origins in the autosomal chorion gene cluster of *Drosophila*. EMBO J. 1989;8(3):891–901.
41. Kitsberg D, Selig S, Keshet I, Cedar H. Replication structure of the human beta-globin gene domain. Nature. 1993;366(6455):588–90.
42. MacAlpine HK, Gordan R, Powell SK, Hartemink AJ, MacAlpine DM. *Drosophila* ORC localizes to open chromatin and marks sites of cohesin complex loading. Genome Res. 2010;20(2):201–11.
43. Dellino GI, Cittaro D, Piccioni R, Luzi L, Banfi S, Segalla S, et al. Genome-wide mapping of human DNA-replication origins: levels of transcription at ORC1 sites regulate origin selection and replication timing. Genome Res. 2013;23(1):1–11.
44. Cayrou C, Coulombe P, Vigneron A, Stanojcic S, Ganier O, Peiffer I, et al. Genome-scale analysis of metazoan replication origins reveals their organization in specific but flexible sites defined by conserved features. Genome Res. 2011;21(9):1438–49.
45. Cadoret JC, Meisch F, Hassan-Zadeh V, Luyten I, Guillet C, Duret L, et al. Genome-wide studies highlight indirect links between human replication origins and gene regulation. Proc Natl Acad Sci U S A. 2008;105(41):15837–42.
46. Martin MM, Ryan M, Kim R, Zakas AL, Fu H, Lin CM, et al. Genome-wide depletion of replication initiation events in highly transcribed regions. Genome Res. 2011;21(11):1822–32.
47. Cayrou C, Coulombe P, Puy A, Rialle S, Kaplan N, Segal E, et al. New insights into replication origin characteristics in metazoans. Cell Cycle. 2012;11(4):658–67.
48. Besnard E, Babled A, Lapasset L, Milhavet O, Parrinello H, Dantec C, et al. Unraveling cell type-specific and reprogrammable human replication origin signatures associated with G-quadruplex consensus motifs. Nat Struct Mol Biol. 2012;19(8):837–44.
49. Bochman ML, Paeschke K, Zakian VA. DNA secondary structures: stability and function of G-quadruplex structures. Nat Rev Genet. 2012;13(11):770–80.
50. Lam EY, Beraldi D, Tannahill D, Balasubramanian S. G-quadruplex structures are stable and detectable in human genomic DNA. Nat Commun. 2013;4:1796.
51. Hoshina S, Yura K, Teranishi H, Kiyasu N, Tominaga A, Kadoma H, et al. Human origin recognition complex binds preferentially to G-quadruplex-preferable RNA and single-stranded DNA. J Biol Chem. 2013;288(42):30161–71.
52. Epner E, Forrester WC, Groudine M. Asynchronous DNA replication within the human beta-globin gene locus. Proc Natl Acad Sci U S A. 1988;85(21):8081–5.
53. Luger K, Mäder AW, Richmond RK, Sargent DF, Richmond TJ. Crystal structure of the nucleosome core particle at 2.8 A resolution. Nature. 1997;389(6648):251–60.
54. Thoma F, Bergman LW, Simpson RT. Nuclease digestion of circular TRP1ARS1 chromatin reveals positioned nucleosomes separated by nuclease-sensitive regions. J Mol Biol. 1984;177(4):715–33.
55. Simpson RT. Nucleosome positioning can affect the function of a cis-acting DNA element in vivo. Nature. 1990;343(6256):387–9.
56. Mavrich TN, Jiang C, Ioshikhes IP, Li X, Venters BJ, Zanton SJ, et al. Nucleosome organization in the *Drosophila* genome. Nature. 2008;453(7193):358–62.
57. Jiang C, Pugh BF. A compiled and systematic reference map of nucleosome positions across the *Saccharomyces cerevisiae* genome. Genome Biol. 2009;10(10):R109.
58. Gaffney DJ, McVicker G, Pai AA, Fondufe-Mittendorf YN, Lewellen N, Michelini K, et al. Controls of nucleosome positioning in the human genome. PLoS Genet. 2012;8(11):e1003036.
59. Mavrich TN, Ioshikhes IP, Venters BJ, Jiang C, Tomsho LP, Qi J, et al. A barrier nucleosome model for statistical positioning of nucleosomes throughout the yeast genome. Genome Res. 2008;18(7):1073–83.
60. Shivaswamy S, Bhinge A, Zhao Y, Jones S, Hirst M, Iyer VR. Dynamic remodeling of individual nucleosomes across a eukaryotic genome in response to transcriptional perturbation. PLoS Biol. 2008;6(3):e65.

61. Berbenetz NM, Nislow C, Brown GW. Diversity of eukaryotic DNA replication origins revealed by genome-wide analysis of chromatin structure. PLoS Genet. 2010;6(9):e1001092.
62. Eaton ML, Galani K, Kang S, Bell SP, MacAlpine DM. Conserved nucleosome positioning defines replication origins. Genes Dev. 2010;24(8):748–53.
63. Lubelsky Y, Sasaki T, Kuipers MA, Lucas I, Le Beau MM, Carignon S, et al. Pre-replication complex proteins assemble at regions of low nucleosome occupancy within the Chinese hamster dihydrofolate reductase initiation zone. Nucleic Acids Res. 2011;39(8):3141–55.
64. Xu J, Yanagisawa Y, Tsankov AM, Hart C, Aoki K, Kommajosyula N, et al. Genome-wide identification and characterization of replication origins by deep sequencing. Genome Biol. 2012;13(4):R27.
65. Lipford JR, Bell SP. Nucleosomes positioned by ORC facilitate the initiation of DNA replication. Mol Cell. 2001;7(1):21–30.
66. Belsky JA, MacAlpine HK, Lubelsky Y, Hartemink AJ, MacAlpine DM. Genome-wide chromatin footprinting reveals changes in replication origin architecture induced by pre-RC assembly. Genes Dev. 2015;29(2):212–24.
67. Groth A, Corpet A, Cook AJL, Roche D, Bartek J, Lukas J, et al. Regulation of replication fork progression through histone supply and demand. Science. 2007;318(5858):1928–31.
68. Müller P, Park S, Shor E, Huebert DJ, Warren CL, Ansari AZ, et al. The conserved bromo-adjacent homology domain of yeast Orc1 functions in the selection of DNA replication origins within chromatin. Genes Dev. 2010;24(13):1418–33.
69. Deal RB, Henikoff JG, Henikoff S. Genome-wide kinetics of nucleosome turnover determined by metabolic labeling of histones. Science. 2010;328(5982):1161–4.
70. Ahmad K, Henikoff S. The histone variant H3.3 marks active chromatin by replication-independent nucleosome assembly. Mol Cell. 2002;9(6):1191–200.
71. Eaton ML, Prinz JA, MacAlpine HK, Tretyakov G, Kharchenko PV, MacAlpine DM. Chromatin signatures of the *Drosophila* replication program. Genome Res. 2011;21(2):164–74.
72. Euskirchen GM, Auerbach RK, Davidov E, Gianoulis TA, Zhong G, Rozowsky J, et al. Diverse roles and interactions of the SWI/SNF chromatin remodeling complex revealed using global approaches. PLoS Genet. 2011;7(3):e1002008.
73. Sugimoto N, Yugawa T, Iizuka M, Kiyono T, Fujita M. Chromatin remodeler sucrose nonfermenting 2 homolog (SNF2H) is recruited onto DNA replication origins through interaction with Cdc10 protein-dependent transcript 1 (Cdt1) and promotes pre-replication complex formation. J Biol Chem. 2011;286(45):39200–10.
74. Kouzarides T. Chromatin modifications and their function. Cell. 2007;128(4):693–705.
75. Jenuwein T, Allis CD. Translating the histone code. Science. 2001;293(5532):1074–80.
76. Hsu TC, Schmid W, Stubblefield E. DNA replication sequences in higher animals. In: Locke M, editor. The role of chromosomes in development. New York: Academic; 1964. p. 83–112.
77. Lyon MF. Chromosomal and subchromosomal inactivation. Annu Rev Genet. 1968;2(1):31–52.
78. Stambrook PJ, Flickinger RA. Changes i chromosomal DNA replication tterns in developing frog embryos. J Exp Zool. 1970;174(1):101–13.
79. ENCODE Project Consortium, Birney E, Stamatoyannopoulos JA, Dutta A, Guigó R, Gingeras TR, et al. Identification and analysis of functional elements in 1 % of the human genome by the ENCODE pilot project. Nature. 2007;447(7146):799–816.
80. The modENCODE Consortium, Roy S, Ernst J, Kharchenko PV, Kheradpour P, Negre N, et al. Identification of functional elements and regulatory circuits by *Drosophila* modENCODE. Science. 2010;330(6012):1787–97.
81. Kuo AJ, Song J, Cheung P, Ishibe-Murakami S, Yamazoe S, Chen JK, et al. The BAH domain of ORC1 links H4K20me2 to DNA replication licensing and Meier-Gorlin syndrome. Nature. 2012;484(7392):115–9.
82. Schotta G, Sengupta R, Kubicek S, Malin S, Kauer M, Callén E, et al. A chromatin-wide transition to H4K20 monomethylation impairs genome integrity and programmed DNA rearrangements in the mouse. Genes Dev. 2008;22(15):2048–61.

83. Bicknell LS, Bongers EMHF, Leitch A, Brown S, Schoots J, Harley ME, et al. Mutations in the pre-replication complex cause Meier-Gorlin syndrome. Nat Genet. 2011;43(4):356–59.
84. Bicknell LS, Walker S, Klingseisen A, Stiff T, Leitch A, Kerzendorfer C, et al. Mutations in ORC1, encoding the largest subunit of the origin recognition complex, cause microcephalic primordial dwarfism resembling Meier-Gorlin syndrome. Nat Genet. 2011;43(4):350–55.
85. Fang J, Feng Q, Ketel CS, Wang H, Cao R, Xia L, et al. Purification and functional characterization of SET8, a nucleosomal histone H4-lysine 20-specific methyltransferase. Curr Biol. 2002;12(13):1086–99.
86. Nishioka K, Rice JC, Sarma K, Erdjument-Bromage H, Werner J, Wang Y, et al. PR-Set7 is a nucleosome-specific methyltransferase that modifies lysine 20 of histone H4 and is associated with silent chromatin. Mol Cell. 2002;9(6):1201–13.
87. Abbas T, Shibata E, Park J, Jha S, Karnani N, Dutta A. CRL4(Cdt2) regulates cell proliferation and histone gene expression by targeting PR-Set7/Set8 for degradation. Mol Cell. 2010;40(1):9–21.
88. Centore RC, Havens CG, Manning AL, Li JM, Flynn RL, Tse A, et al. CRL4(Cdt2)-mediated destruction of the histone methyltransferase Set8 prevents premature chromatin compaction in S phase. Mol Cell. 2010;40(1):22–33.
89. Oda H, Hübner MR, Beck DB, Vermeulen M, Hurwitz J, Spector DL, et al. Regulation of the histone H4 monomethylase PR-Set7 by CRL4(Cdt2)-mediated PCNA-dependent degradation during DNA damage. Mol Cell. 2010;40(3):364–76.
90. Jørgensen S, Elvers I, Trelle MB, Menzel T, Eskildsen M, Jensen ON, et al. The histone methyltransferase SET8 is required for S-phase progression. J Cell Biol. 2007;179(7): 1337–45.
91. Houston SI, McManus KJ, Adams MM, Sims JK, Carpenter PB, Hendzel MJ, et al. Catalytic function of the PR-Set7 histone H4 lysine 20 monomethyltransferase is essential for mitotic entry and genomic stability. J Biol Chem. 2008;283(28):19478–88.
92. Oda H, Okamoto I, Murphy N, Chu J, Price SM, Shen MM, et al. Monomethylation of histone H4-Lysine 20 is involved in chromosome structure and stability and is essential for mouse development. Mol Cell Biol. 2009;29(8):2278–95.
93. Tardat M, Brustel J, Kirsh O, Lefevbre C, Callanan M, Sardet C, et al. The histone H4 Lys 20 methyltransferase PR-Set7 regulates replication origins in mammalian cells. Nat Cell Biol. 2010;12(11):1086–93.
94. Beck DB, Burton A, Oda H, Ziegler-Birling C, Torres-Padilla ME, Reinberg D. The role of PR-Set7 in replication licensing depends on Suv4-20h. Genes Dev. 2012;26(23):2580–89.
95. Sakaguchi A, Karachentsev D, Seth-Pasricha M, Druzhinina M, Steward R. Functional characterization of the *Drosophila* Hmt4-20/Suv4-20 histone methyltransferase. Genetics. 2008;179(1):317–22.
96. Doyon Y, Cayrou C, Ullah M, Landry AJ, Côté V, Selleck W, et al. ING tumor suppressor proteins are critical regulators of chromatin acetylation required for genome expression and perpetuation. Mol Cell. 2006;21(1):51–64.
97. Iizuka M, Stillman B. Histone acetyltransferase HBO1 interacts with the ORC1 subunit of the human initiator protein. J Biol Chem. 1999;274(33):23027–34.
98. Burke TW, Cook JG, Asano M, Nevins JR. Replication factors MCM2 and ORC1 interact with the histone acetyltransferase HBO1. J Biol Chem. 2001;276(18):15397–408.
99. Iizuka M, Matsui T, Takisawa H, Smith MM. Regulation of replication licensing by Acetyltransferase Hbo1. Mol Cell Biol. 2006;26(3):1098–108.
100. Miotto B, Struhl K. HBO1 histone acetylase is a coactivator of the replication licensing factor Cdt1. Genes Dev. 2008;22(19):2633–38.
101. Chen X, Liu G, Leffak M. Activation of a human chromosomal replication origin by protein tethering. Nucleic Acids Res. 2013;41(13):6460–74.
102. Pryde F, Jain D, Kerr A, Curley R, Mariotti FR, Vogelauer M. H3 k36 methylation helps determine the timing of cdc45 association with replication origins. PLoS One. 2009;4(6):e5882.
103. Rizzardi LF, Dorn ES, Strahl BD, Cook JG. DNA replication origin function is promoted by H3K4 di-methylation in *Saccharomyces cerevisiae*. Genetics. 2012;192(2):371–84.

104. Trujillo KM, Osley MA. A role for H2B ubiquitylation in DNA replication. Mol Cell. 2012;48(5):734–46.
105. Fu H, Maunakea AK, Martin MM, Huang L, Zhang Y, Ryan M, et al. Methylation of histone H3 on lysine 79 associates with a group of replication origins and helps limit DNA replication once per cell cycle. PLoS Genet. 2013;9(6):e1003542.
106. Jacob Y, Stroud H, Leblanc C, Feng S, Zhuo L, Caro E, et al. Regulation of heterochromatic DNA replication by histone H3 lysine 27 methyltransferases. Nature. 2010; 466(7309):987–91.
107. Straub T, Becker PB. Dosage compensation: the beginning and end of generalization. Nat Rev Genet. 2007;8(1):47–57.
108. Berendes HD. Differential replication of male and female X-chromosomes in *Drosophila*. Chromosoma. 1966;20(1):32–43. doi:10.1007/BF00331896
109. Schwaiger M, Stadler MB, Bell O, Kohler H, Oakeley EJ, Schübeler D. Chromatin state marks cell-type- and gender-specific replication of the *Drosophila* genome. Genes Dev. 2009;23(5):589–601.
110. Lubelsky Y, Prinz JA, DeNapoli L, Li Y, Belsky JA, MacAlpine DM. DNA replication and transcription programs respond to the same chromatin cues. Genome Res. 2014;24(7): 1102–14.
111. Aparicio JG, Viggiani CJ, Gibson DG, Aparicio OM. The Rpd3-Sin3 histone deacetylase regulates replication timing and enables intra-S origin control in *Saccharomyces cerevisiae*. Mol Cell Biol. 2004;24(11):4769–80.
112. Knott SRV, Viggiani CJ, Tavaré S, Aparicio OM. Genome-wide replication profiles indicate an expansive role for Rpd3L in regulating replication initiation timing or efficiency, and reveal genomic loci of Rpd3 function in *Saccharomyces cerevisiae*. Genes Dev. 2009;23(9):1077–90.
113. Crampton A, Chang F, Pappas J Donald L, Frisch RL, Weinreich M. An ARS element inhibits DNA replication through a SIR2-dependent mechanism. Mol Cell. 2008;30(2):156–66.
114. Yoshida K, Bacal J, Desmarais D, Padioleau I, Tsaponina O, Chabes A, et al. The histone deacetylases sir2 and rpd3 act on ribosomal DNA to control the replication program in budding yeast. Mol Cell. 2014;54(4):691–97.

Chapter 6
Global and Local Regulation of Replication Origin Activity

Conrad A. Nieduszynski

Abstract Eukaryotic genomes are replicated from multiple initiation sites called DNA replication origins. Different origins fire at different times during S phase, giving rise to a characteristic temporal order to genome replication. However, the physiological role for temporal regulation of the order of genome replication remains largely unknown. Powerful genomic approaches have allowed genome replication dynamics to be characterised in various mutants and a range of species. Work in several organisms has revealed that limiting levels of *trans*-acting replication initiation factors are likely to play a role in determining origin firing time. This raises the question of how the initiation factors distinguish between origins. Recent work has started to identify *cis*-acting elements at origins that might be responsible for characteristic firing times. The identification of mechanisms that regulate the temporal order of genome replication is starting to allow investigation of potential physiological roles for temporally regulated replication.

Keywords Replication origins • Origin licensing • Origin firing • DNA replication timing • Replication fork

Introduction

DNA replication is controlled primarily by the regulated activation of replication origins. Eukaryotic genomes are replicated from multiple origins to help ensure completion of DNA replication [1, 2]. Origins activate at characteristic times during S phase to produce bidirectional replication forks that progress to replicate the flanking DNA. Consequently, genomes are replicated in a highly conserved and characteristic temporal order—some sequences replicate early in S phase, and others late [3]. This is of critical importance to genome stability, as illustrated by the disruption of replication timing in cancer cells that contributes to chromosomal breakage, translocations and aneuploidy [4].

C.A. Nieduszynski (✉)
Sir William Dunn School of Pathology, University of Oxford, South Parks Road, Oxford OX1 3RE, UK
e-mail: conrad.nieduszynski@path.ox.ac.uk

D.L. Kaplan (ed.), *The Initiation of DNA Replication in Eukaryotes*,
DOI 10.1007/978-3-319-24696-3_6

From late mitosis until the end of G1 phase, replication origins are 'licensed' for subsequent use by the loading of Mcm2-7 double hexamers [5]. During S phase, the activity of two S phase kinases, cyclin-dependent kinase (CDK) and Dbf4-dependent kinase (DDK), activates Mcm2-7 hexamers to form the core of the replicative helicase that unwinds template DNA [6]. CDK has a dual role: it inhibits the licensing reaction and activates replication origins. This prevents re-replication of the DNA by ensuring that each origin can only activate once per cell cycle [7].

Replication origin sites are determined by the binding specificity of ORC and nucleosome positioning around potential ORC-binding sites [8, 9]. In budding yeast, ORC recognises a specific sequence motif that is non-transcribed and nucleosome depleted. Metazoan ORC is also recruited to non-transcribed and nucleosome-depleted regions, but with little sequence specificity. DNA-bound ORC, together with Cdc6 and Cdt1, loads the Mcm2-7 complex to form the pre-replication complex [10]. The origin is now licensed to fire upon entry into S phase.

As cells enter into S phase, inactive Mcm2-7 helicases must be converted to mature, processive replication forks (Fig. 6.1). Helicase activation is DDK and CDK dependent and requires the Mcm2-7 core to be joined by Cdc45 and GINS to give the CMG complex [11]. DDK directly phosphorylates several Mcm2-7 subunits to relieve an inhibitory effect of the N-terminal tail of Mcm4 and to recruit additional

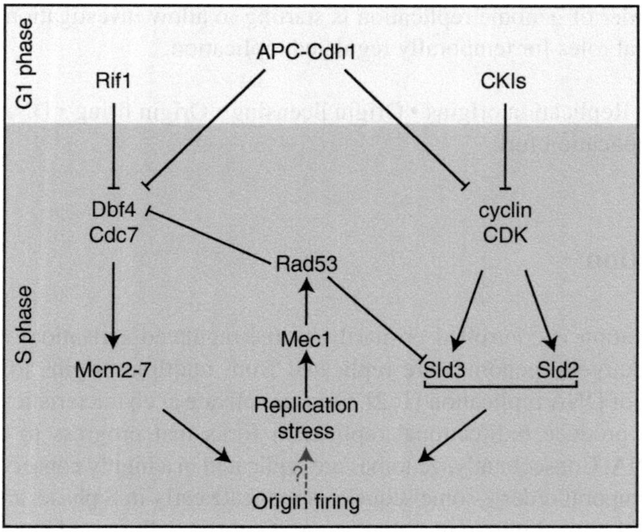

Fig. 6.1 Pathways regulating replication origin firing. During G1 phase multiple pathways inhibit origin firing via inhibition of Dbf4-dependent kinase (DDK) and cyclin-dependent kinase (CDK) activities. Upon entry to S phase increasing cyclin and Dbf4 concentrations coupled with double-negative feedback loops allow activation of both DDK and CDK whose concerted activities permit origin firing. If resulting replication forks stall due to encountering DNA damage or limited dNTP supply this will activate the Mec1-Rad53 checkpoint pathway. Targets of Rad53 include the origin firing factors Dbf4 and Sld3, thereby inhibiting both of the pathways required for origin activation. It remains to be determined how many of these pathways play a role in executing the temporal ordering of genome replication

subunits Cdc45, Sld3 and Sld7 [12]. CDK phosphorylation of Sld2 and Sld3 is necessary for an Sld2-Dpb11-Sld3 interaction that recruits GINS to give the CMG complex [13]. Additional factors are recruited that crucially include Polα–primase, Polε (leading strand polymerase) and Polδ (lagging strand polymerase) to form the replisome and allow DNA replication initiation. Sld2, Sld3, Sld7 and Dpb11 are not part of the replisome and, like CDK and DDK, behave as origin 'firing' factors.

The G1–S phase transition results from the switch-like activation of CDK activity as the inhibitory roles of both the APC^{Cdh1} and CDK inhibitors are removed (Fig. 6.1). However, not all replication origins are activated immediately upon entry into S phase. Instead origins have characteristic activation times, with some origins activating early whereas others activate later in S phase. In addition, each origin is only active in a proportion of the cell population (termed the efficiency). Some origins are active in the majority of cells (>80 % in budding and fission yeasts [14–16]) whereas others are rarely used and are termed dormant or backup origins [17]. Dormant origins allow completion of DNA replication under conditions of replicative stress. If replication fork progression is inhibited, proximal dormant origins activate to help ensure complete chromosome replication [18]. Conversely, a sparsity of replication initiation sites has been observed at certain common fragile sites [19]. These observations underline the physiological importance of understanding the regulation of origin activity. This chapter reviews some of the mechanisms responsible for these differences in origin activity. I focus on global *trans*-acting and local *cis*-acting regulators, but first review the methodologies that have revealed a genome-wide view of DNA replication.

Measuring DNA Replication Genome-Wide

The progress of DNA replication during S phase can be measured directly in one of two ways. First, nascent strands can be marked and then detected, for example by the use of heavy versus light isotopes or by the incorporation of nucleotide analogues such as BrdU. Second, the change in DNA copy number, from one to two as a sequence replicates, can be measured. Both approaches require enrichment of S phase cells that can be achieved either using cell cycle synchronisation or by fluorescence-activated cell sorting (FACS). A pair of landmark papers used the powerful cell cycle synchronisation available in *Saccharomyces cerevisiae* to determine DNA replication dynamics genome-wide [20, 21]. Raghuraman et al. used a switch from dense to light isotopes to enrich replicated DNA followed by detection on a microarray [21]. Yabuki et al. measured the increase in DNA copy number by microarray [20]. Both studies used multiple S phase time points to calculate the median time that each genomic location replicates, called the Trep.

Subsequent studies have extended these approaches via the use of nucleotide analogues [22], enrichment of S phase cell by FACS [23] and most recently replacement of microarrays by deep sequencing [24, 25]. In addition, the use of replication inhibitors (such as hydroxyurea) can be used to slow replication [20, 26] and/or dissect the role of replication checkpoint pathways [27]. These experimental techniques have now been used to determine the dynamics of genome replication in a range of organisms, cell types and mutants [3].

Measurements of genome replication dynamics are presented in replication timing profiles, where the x-axis represents chromosomal coordinate and the y-axis the median replication time (Trep) or fraction replicated (Fig. 6.2). By convention, Trep is presented with time running down the y-axis. Although this is initially counter-intuitive it has the advantage that profiles of Trep resemble those of fraction replicated. Therefore, in timing profiles peaks represent the earliest replicating loci and valleys later replicating loci. In eukaryotes with small genomes, such as yeasts, the defined nature of replication origins coupled with high-resolution data permit the association of profile peaks with individual origins [24]. By contrast, in organisms with larger genomes the reduction in resolution currently precludes the association of individual origins with profile peaks. Instead, replication profiles are divisible into constant timing regions (CTRs) separated by timing transition regions (TTRs) [3]. Each CTR is replicated from clusters of multiple origins that activate at a similar time.

Stochastic replication origin usage complicates the interpretation of replication timing profiles. Some naïve interpretations of timing profiles have erroneously assumed that the population average (ensemble) data represents the dynamics of replication in individual cells—i.e. the cell population is completely homogeneous. Common misinterpretations include associating the Trep at origins with the origin firing time and differences in timing profile gradients with variability in replication fork speed [28]. It is well established that origins are only used in a subset of cells [29, 30] and therefore it follows that different cells within a population will activate different cohorts of origins. As such, ensemble timing profiles cannot represent the situation within individual cells [31]. Mathematical models have demonstrated that variable origin usage is consistent with the ensemble replication data [16, 28, 32, 33]. In addition, these models have shown that ensemble timing profiles are consistent with a constant average fork velocity throughout the genome. That is not to say that there is no variability in fork velocity between cells, but on average the replication profiles do not provide evidence for variability by position in the genome (to the level of resolution provided by current replication timing studies).

What then can explain the differences in gradient across replication timing profiles? Mathematical models indicate that this variability can be explained by variability in the proportion of replication forks moving in each direction. Adjacent to a highly active origin (efficiency approaching 100 %), virtually all forks will be moving away from the origin and the gradient of the replication profile will be close to the velocity of an individual fork. However, more commonly there will be a significant number of forks moving in both directions and here the replication profile gradient will represent the average velocity of leftward and rightward moving forks. This was a highly valuable observation, since it allowed the gradient of replication timing profiles to be transformed into fork direction data across the genome (Fig. 6.2c) [16, 34]. In turn, changes in fork direction allow the estimation of origin efficiencies and the distribution of replication termination events [16]. The reciprocal calculation allows the estimation of relative replication time from genome-wide

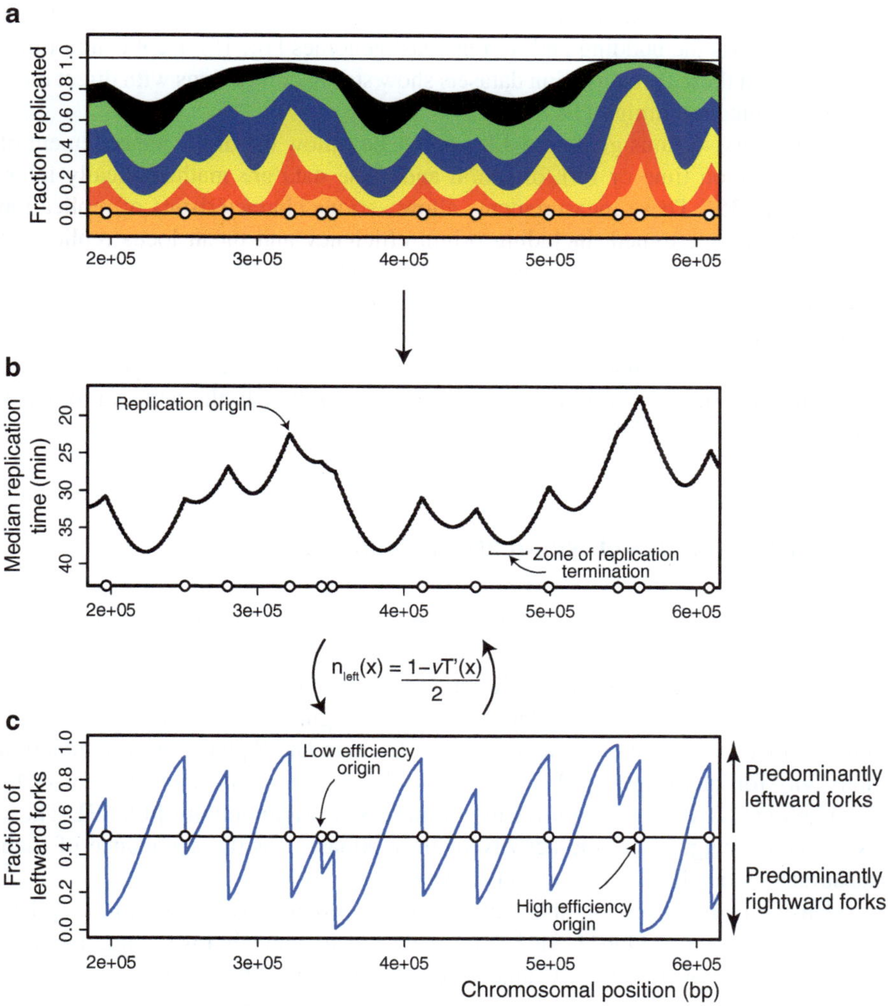

Fig. 6.2 Quantitative measurement of genome replication dynamics. (**a**) The population average dynamics of genome replication can be determined by measuring the fraction of cells in which each locus is replicated. The panel shows a window of *S. cerevisiae* chromosome 14 with representative data for 5-min time points through S phase [16]. White circles represent the locations of replication origins. (**b**) Such a time series allows calculation of the median replication time (Trep) for each locus. By convention Trep profiles are plotted with time running down the *y*-axis. As such, peaks in replication timing profiles represent the location of chromosomally active origins while the valleys indicate the location of zones of replication termination. (**c**) For any particular genomic locus (x) the fraction of leftward moving forks (n_{left}) can be calculated from the fork velocity (v) and the gradient of the replication profile (T') [35]. This relationship allows calculation of replication fork direction, from Trep and vice versa [14, 16]. In plots of replication fork direction, sharp changes in fork direction are seen at the sites of active origins—the magnitude of the transition indicates the proportion of cells in which the origin is active (efficiency)

measures of fork direction. Recent experimental approaches have determined fork direction across the budding and fission yeast genomes [14, 15]. Replication times inferred from these fork direction datasets show strong correlations with direct measures of replication timing [14, 16].

In summary, various genome-wide approaches allow the direct measurement of DNA replication time or fork direction and these data are mathematically interchangeable [35]. These measurements allow various characteristics of replication origins to be determined, including origin efficiency and mean locus replication time. However, sophisticated mathematical models and/or single-cell/molecule approaches are required to determine the underlying properties of the origins. These properties include the proportion of cells in which an origin is licensed (origin *competence* [36]) and the time at which the origin activates. It is these origin properties that determine observed origin characteristics and ultimately the temporal order in which the genome is replicated.

Global Regulators of Replication Origin Activity

Replication origins are activated by the concerted activity of cyclin-dependent kinase (CDK) and Dbf4-dependent kinase (DDK). Both kinases are required throughout S phase. Transient DDK activity at the start of S phase is sufficient to activate early origins, but not later activating origins [37, 38]. Likewise, yeast mutants that have a decreased abundance of CDK in late S phase show a reduction in activity for later origins [39–41]. In *S. pombe* increasing the abundance of DDK or targeting DDK to an origin results in increased origin activity [42, 43]. These experiments suggest that the activity of these kinases (or the abundance of their substrates) is rate limiting for origin activation.

Measurements of protein abundance identified potentially rate-limiting replication initiation factors. Dpb11, Sld3, Sld7, Sld2, Cdc45 and Dbf4 were found to have concentrations lower than ORC and therefore could limit the rate of origin activation [44, 45]. Simultaneous overexpression of various subsets of these proteins (for example, Sld2 + Sld3 + Dpb11 + Dbf4 or Sld3 + Sld7 + Cdc45) was able to advance origin activation times with (1) normally early origins activating slightly earlier and (2) normally late origins activating much earlier in S phase. Therefore, a small distinction between early- and late-activating origins remained, but the time between early and late origin activation was greatly reduced—there was a compaction in the temporal order of origin activation. A presumed consequence of overexpressing these initiation factors is the simultaneous activation of many more origins than normal, resulting in more replication forks and a greater demand for dNTPs. Consistent with this, simultaneous overexpression of Sld2, Sld3, Dpb11, Dbf4, Cdc45 and Sld7 gave transient activation of the checkpoint kinase Rad53 that could be suppressed by increasing dNTP supply by *SML1* deletion [45].

Epigenetic mechanisms may influence the accessibility of rate-limiting initiation factors to replication origins. For example, in budding yeast deletion of the histone deacetylase *RPD3* resulted in earlier activation of many normally late-activating origins [46–48], suggesting that histone tail acetylation may stimulate origin activity. Targeting a histone acetyltransferase to a single origin advanced the firing timing, whereas targeting a histone deacetylase resulted in delayed activation [49–51]. However, in budding yeast the genomic pattern of histone acetylation does not display a simple correlation with origin activation times [52]. At least in budding yeast, it seems that the histone deacetylases Rpd3 and Sir2 do not act directly at individual origins, but rather regulate the ability of the ~200 rDNA origins to compete with single-copy origins (~400) for the rate-limiting initiation factors [46]. Consistent with this interpretation, a natural polymorphism within the rDNA ORC-binding site that reduces origin activity frees up limiting initiation factors and thereby stimulates the activity of single-copy origins [53]. It is unknown whether a similar mechanism operates in human cells; however mammalian genomes contain an abundance of repetitive elements within heterochromatin. Therefore it is possible that origins in these regions of repetitive sequence could be repressed to prevent them from competing for initiation factors with single-copy origins.

Together the above data support the model that replication initiation factors are rate limiting for origin activation. Under normal cellular conditions those origins that have greatest affinity for the initiation factors will activate first with activation of other origins requiring recycling of the initiation factors. With the notable exception of Cdc45, the other rate-limiting initiation factors are not thought to be associated with elongating replication forks and therefore after origin activation can be recycled ready to activate other origins. It remains to be determined whether the kinetics of initiation factor recycling are sufficient to explain the temporal difference between early and later origin activation times or whether additional mechanisms are at play. Notably, most mutants that slow DNA replication (and dNTP depletion by hydroxyurea) also proportionately scale origin activation times, such that the relative order of genome replication is maintained [23, 26], consistent with active regulatory mechanisms in addition to the *hardwired* rate-limiting factors. This is of crucial importance given the replication stress and potential for genome instability that results from the activation of too many or too few replication origins. Below I consider two further mechanisms that have the potential to enforce a delay in late origin activation and thus modulate origin activation time.

First, protein phosphatases oppose the activities of the kinases responsible for origin activation. Recent studies in yeasts have identified a role for protein phosphatase 1 (PP1), targeted by Rif1, in inhibiting origin activation via the dephosphorylation of replication factors [54–57]. Thus PP1-Rif1 can contribute to the inhibition of DNA replication in G1 phase by reversing precocious kinase activity (Fig. 6.1). In S phase, PP1-Rif1 may contribute to the temporal order of origin activation by slowing down the recycling of replication initiation factors, thereby introducing a delay between early and late origin activation. This could be at a global level for all origins or via a targeting mechanism that could specifically inhibit subsets of origins and thereby contribute to the differentiation of early- and late-activating origins [54, 58].

Rif1 is proposed to be one of a range of targeting subunits that provides PP1 substrate specificity. In the absence of Rif1 there are elevated levels of Mcm4 and Sld3 phosphorylation in G1 phase and this phosphorylation is DDK dependent. Deletion of Rif1 is able to rescue an *S. pombe hsk1* deletion (*hsk1* is the fission yeast orthologue of *CDC7*) and rescue growth at a semi-permissive temperature in *S. cerevisiae CDC7* mutants [55–58]. Together these experiments suggest that Rif1 directs PP1 to dephosphorylate DDK substrates, including Mcm4 and potentially Sld3, and thereby inhibit origin activation. However, the mechanisms by which Rif1 targets PP1 to DDK substrates or specific subsets of replication origins remain to be determined.

A second mechanism that can regulate origin activity is inhibition by checkpoint kinases (Fig. 6.1). In the presence of replicative stress, such as DNA damage, bulk genome replication is slowed due to inhibition of late origin activation. This response to stress is dependent upon the checkpoint kinases Mec1 (the yeast orthologue of human ATM-related kinase, ATR) and Rad53 (the yeast analogue of CHK1). The checkpoint kinase Rad53 inhibits both the CDK- and DDK-dependent steps of origin activation. DDK is directly inhibited, via Dbf4 phosphorylation, while the CDK-dependent step is inhibited via phosphorylation of its substrate, Sld3. Thus, in response to replicative stress the checkpoint provides a 'double lock' preventing origin activation while maintaining high CDK levels to prevent cell cycle reversal. While it is clear that DNA damage can induce the checkpoint to inhibit late origin activation, it is less clear whether the checkpoint controls origin activity in an unperturbed S phase. In the absence of exogenous replicative stress, a *rad53* mutant showed significantly earlier activation of a normally late origin [59]. However this study only looked at two loci (*ARS607* and *ARS609*) and therefore it is not possible to determine whether the relative dynamics of DNA replication remained intact. Recent genome-wide analyses of DNA replication dynamics suggest that in the absence of DNA damage the Mec1-Rad53 checkpoint is not involved in maintaining relative replication times, for example by inhibiting late origin firing [60].

In summary, cell cycle transitions must be irreversible and this is achieved by the switch-like activation of CDKs. However, within a cell cycle stage more nuanced regulation is required, for example, the gradual activation of replication origins rather than a switch-like simultaneous activation of all origins. The utilisation of two kinases may provide for both a switch-like change (in CDK activity) and a more graduated increase (in DDK activity). The requirement for two origin-activating kinases has been proposed to help prevent re-licensing and hence prevent re-replication. However, a further role may be to provide both the global control that limits origin activation to S phase and the fine-level control that prevents excess origin activation.

Local Regulators of Replication Origin Activity

The global *trans*-acting regulators of DNA replication described above need to distinguish between early and later activating origins to give rise to the observed characteristic temporal patterns of genome replication. Mathematical models indicate that the temporal pattern of genome replication can be accounted for by stochastic origin firing, where origins have different firing probabilities. Whether an origin fires is determined by its firing probability coupled with the proximity and firing probability of neighbouring origins that give rise to forks that may passively replicate and thereby inactivate the origin. Therefore, in these models the temporal order of genome replication is determined by differences in origin firing probabilities, the molecular basis for which is unclear.

Replication timing correlates with, and has been proposed to regulate or be regulated by, transcription levels, chromatin state, cellular differentiation and chromosome structure/positioning. Furthermore, the activity of individual replication origins has been proposed to be determined by the number of MCM double hexamers loaded during the replication licensing step and/or by various *cis*-acting elements. These models are not necessarily mutually exclusive; for example *cis*-acting elements could influence the MCM load. A role for *cis*-acting elements in influencing origin firing probabilities is supported by a range of studies, including transplanting of origins [61, 62], analysis of origin activity in hybrid species [63] and via mutation of candidate elements [64, 65].

Early studies in yeast support a model by which replication origins activate early in S phase by default, with multiple *cis*-acting elements located over a region of several kilobases imposing a late replication time [62]. These experiments were based upon origins carried on plasmids and supported by transplanting chromosomal origins. The transplanted origins acquired the replication characteristics of the new location. However, in these experiments the proximity of the origin to a centromere was altered and centromeres have since been shown to influence origin activity *in cis* (see below) [66, 67]. This permits an alternative explanation: some origins activate late in S phase by default, but local elements (such as centromere proximity or the binding of specific factors) can impose an early replication time. Indeed, recent studies have found that some early-activating origins when transplanted to normally late-replicating regions of the genome retain their early-activating characteristics [61]. Therefore origin activity may be regulated *in cis*, with some factors stimulating while others delay origin firing. The *cis*-acting elements may be close to the origin or in the case of the centromere may influence the activity of origins up to 20 kb away.

One mechanism by which *cis*-acting elements could determine origin firing probabilities is via modulating the affinity of the origin and/or pre-RC for the rate-limiting initiation factors. Difference in origin initiation factor affinity could be a consequence of accessibility (for example, as a consequence of DNA packaging [46]), specific recruitment of initiation factors [46] (or inhibitors) or the nature of the replication licensing reaction [68, 69]. It may be that each of these plays a role in determining origin activity and hence genome replication dynamics.

Perhaps the most straightforward proposed mechanism is that the efficiency of the origin licensing step dictates the subsequent origin firing probability. This could involve the nature of the ORC-DNA interaction at a particular origin determining the efficiency or manner of MCM loading [42, 68, 69]. For example, origins at which many MCM double hexamers are loaded would have higher affinity for the rate-limiting initiation factors and therefore activate earlier than those origins with fewer MCMs [32]. In addition, there is a further potential contribution of the licensing step to origin replication time. A low-competence origin may be licensed and therefore able to activate in only a subset of cell within the population. On average within the population, such an origin would replicate later than a highly competent origin despite there being no difference between the origin firing times [28]. This would be anticipated to be particularly apparent in experiments where the time available for origin licensing is short. Consistent with this, different methods of cell cycle synchronisation in fission yeast (which has a short G1 phase) result in significant differences in the temporal order of genome replication [70]. Therefore, in both the MCM load and origin competence hypotheses, aspects of replication timing are a direct consequence of the efficiency of the licensing step. Although these models are attractive, to date genome-wide MCM ChIP data has not identified a clear correlation between MCM load and origin activation time [15], with the caveat that ChIP datasets may not be sufficiently quantitative.

A second mechanism, by which the affinity of origins for limiting initiation factors could be varied, is via *cis* recruitment of these factors or their inhibitors (Fig. 6.3). Such mechanisms may be responsible for the characteristic replication times of centromeric and telomeric regions. In *S. cerevisiae*, *Candida albicans*, *S. pombe*, *Trypanosoma brucei* and *Drosophila* cells the centromere (or the core centromeric

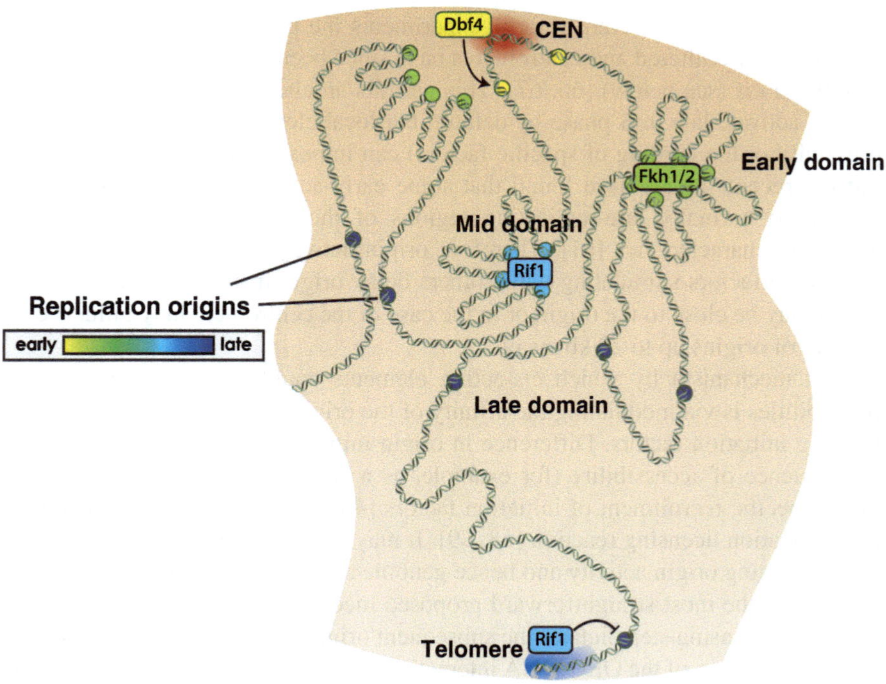

Fig. 6.3 Temporal and spatial organisation of genome replication. There are strong correlations between replication time and higher order chromosomal structure. Thus positive (e.g. Dbf4 and Fkh1/2) and negative (e.g. Rif1) regulators of replication origin firing may in part function via organisation of the genome into temporally and spatially coordinated domains

region) replicates early in S phase [71–75]. By contrast, the telomeric regions in yeast replicate late in S phase [71]. In *S. cerevisiae* and *S. pombe* the centromere contributes to the early activation of neighbouring replication origins [67, 72] via recruitment of one of the rate-limiting initiation factors, Dbf4 [66, 76, 77]. In budding yeast, this recruitment is to the kinetochore and is required and sufficient for the early activation of origins within ~20 kb [66]. By contrast, the late replication of telomeres depends upon Rif1 binding, which in turn may locally inhibit the activity of Dbf4-dependent kinase [56, 78, 79]. Cells with shorter telomeres bind less Rif1, and thus are likely to have a lower level of local DDK inhibition, permitting telomereproximal origins to fire earlier in S phase. Therefore, both centromere and telomere replication times are regulated by direct recruitment of replication origin activators (Dbf4 at centromeres) or inhibitors (Rif1 at telomeres). It remains to be determined whether the replication time of other genomic loci is regulated in an analogous manner. However, in the fission yeast genome Rif1-binding sites are found closer to late than to early-firing origins consistent with *cis*-acting sequences recruiting Rif1 to delay origin activation [58].

There is accumulating evidence that the spatial organization of chromosomes within the nucleus is related to the dynamics of genome replication. In mammalian cells, replication timing correlates better with chromosome interaction maps than any of the other chromosomal feature analysed to date [80]. In budding yeast, global chromosomal interaction maps revealed contacts between early-firing replication origins [81] and single-cell analyses revealed that replication forks from neighbouring origins could stochastically associate in replication factories [82]. In fission yeast, single-cell and single-molecule data are consistent with a model where clusters of neighbouring origins fire at similar times within a replication factory—visualised as replication foci within single cells [83]. Mechanistically, the budding yeast forkhead transcription factors, Fkh1 and Fkh2, have been implicated in organizing early-replicating regions. Inactivation of Fkh1/2 reduced interactions between two early-firing origins and resulted in a global change to the temporal dynamics of genome replication. Consensus binding sites for Fkh1/2 have been found close to many early-firing origins and mutation of these binding sites delayed origin firing, demonstrating a role for these factors in *cis* regulation of origin activity [61, 65]. However, the molecular mechanism by which Fkh1/2 stimulates origin activity remains to be fully elucidated. Therefore these experiments suggest an important role for chromosomal interactions in the regulation of genome replication dynamics, but with much still to be learnt about the molecular mechanisms involved.

In summary, current experimental data point towards the involvement of multiple mechanisms in the regulation of the temporal dynamics of genome replication. Even in the small and predominantly euchromatic genome of budding yeast there are stimulators (e.g. Fkh1/2 and Dbf4) and repressors (e.g. Rif1) of origin firing. In organisms with larger genomes the differences in accessibility between euchromatin and heterochromatin likely add an additional level of control over origin activation time. Finally, it remains an open question as to whether the efficiency of origin licensing during G1 phase contributes to subsequent origin activity in S phase.

Physiological Role for Replication Timing Control

The number and velocity of replication forks determine the rate of genome replication with limits imposed by the supply of dNTPs and histones. Consequently, a null hypothesis for a physiological role for the temporal ordering of genome replication is to match supply of precursors with demand at replication forks. Rate-limiting replication initiation factors *hardwire* a limit on the number of active forks to ensure that supply can meet demand. In experiments where additional replication origins are permitted to fire, demand outstrips supply with resultant stalling of forks and activation of the intra-S phase checkpoint. Limiting the number of active replication forks does not necessitate a reproducible temporal order for genome replication. However, if replication origins differ in affinity for the rate-limiting initiation factors, this could be sufficient to impose a characteristic temporal order to genome replication with no further physiological role. Therefore, by this hypothesis the temporal order of genome replication may be an indirect measure of a higher order chromosomal state such as 3D conformation or packaging without any direct physiological role.

Is there any evidence for a physiological role for the temporal regulation of genome replication? One hint at a physiological role comes from comparisons of genome replication dynamics in related organisms. In the closely related *Saccharomyces* species, the location and efficiency of active replication origins are conserved, while dormant origins are poorly conserved [63]. Consequently the temporal order of genome replication is conserved between these species. Comparisons of replication timing between mouse and human have revealed strong conservation of replication timing in similar cell types [84]. In more distantly related yeast species certain features of genome replication dynamics are conserved, including early centromere and late telomere replication [85]. These data are consistent with a physiological requirement for regulation of the replication times of particular genomic loci.

Budding yeast offers a relatively unique experimental system, since individual replication origins on the chromosome can be inactivated by point mutations within the ORC-binding consensus sequence. Such origin inactivation allows the replication time of a particular locus to be altered from early to late S phase and hence to test for physiological consequences [16, 66]. The conservation of early centromere replication points towards a physiological role in ensuring faithful chromosome segregation. Inactivation of centromere proximal origins specifically increased the rate of loss for the chromosome with the delayed centromere replication [66]. At least three mechanisms contribute to faithful chromosome segregation: early centromere replication, the S phase checkpoint and the spindle assembly checkpoint [66]. Therefore, there is a clear physiological role for early centromere replication.

In many organisms telomeres are among the last genomic regions to replicate. In budding yeast, the length of the telomere has been discovered to influence their replication time: long telomeres replicate late while short telomeres replicate earlier in S phase [86, 87]. Short telomeres need to be prioritised for elongation by telomerase; therefore this hints at a physiological role for regulated telomere replication time. Telomerase is observed at telomeres after passage of the canonical replication machinery [88, 89]. Consequently, early replication of short telomeres allows earlier recruitment of telomerase, allowing greater opportunity for telomere lengthening during S phase. Therefore, telomere replication time may be part of the signalling mechanism that contributes to telomere length homeostasis [86].

It remains to be determined how many other genomic loci have an analogous physiological role for a particular replication time. The correlations between replication time, epigenetic states and levels of gene transcription have led to suggestions that there could be co-dependencies. For example, the time at which a locus replicates could be important for the maintenance of the epigenetic state that in turn would influence gene expression levels [90, 91]. Such a mechanism could self-maintain: chromatin state influences replication time, while the replication time helps to re-establish the chromatin state.

Regulation of centromere and telomere replication time is crucial for genome stability by ensuring faithful chromosome segregation and chromosome end protection. This link between replication time and genome stability may be more general, due to the emerging links between replication time and the rate of mutation [4, 92]. In yeast, flies and mammalian cells [93, 94] there is a positive correlation between mutation rate (or single-nucleotide polymorphism (SNP) density) and replication time [95–97] that may be a consequence of temporal variation in repair pathways [92]. Therefore, genes in early-replicating regions may evolve more slowly or be mutated less often than those in later replicating regions. Late-replicating regions of the human genome are depleted of cancer-related genes whose mutation could result in tumor formation [93]. In contrast, late-replicating regions of many genomes contain genes whose mutation is more likely to be tolerated or even potentially advantageous. For example, the late-replicating telomeric regions of yeast genomes lack essential genes but are enriched for genes involved in niche adaptation [98]. Therefore, elevated mutation rates in such late-replicating regions may help support rapid adaptation to novel niches.

In summary, the physiological requirements for replication timing control remain poorly characterised. However, recent studies have started to elucidate requirements for regulated DNA replication time in ensuring genome stability. These discoveries allow us to rule out the null hypothesis that replication timing control is solely a read-out of higher order chromosomal states. Further work is required to determine how general these examples are and to help our understanding of whether the deregulation of genome replication timing control observed in early stages of cancer development contributes to the resulting pathology.

Conclusions

Recently, significant advances have been made in isolating the global factors involved in regulating the dynamics of genome replication. The identification of rate-limiting *trans*-acting replication initiation factors supports a mechanism by which early and late origins are distinguished by their accessibility to these initiation factors. However, the *cis*-acting elements that contribute to the distinction between origins remain poorly understood with the exception of well-characterised positive and negative *cis*-acting mechanisms at centromeres and telomeres, respectively. A further major challenge exists in determining the physiological requirements for temporal regulation of genome replication. The elucidation of further *cis*-acting regulatory mechanisms will offer the opportunity to determine the physiological consequences of their loss.

References

1. Newman TJ, Mamun MA, Nieduszynski CA, Blow JJ. Replisome stall events have shaped the distribution of replication origins in the genomes of yeasts. Nucleic Acids Res. 2013; 41(21):9705–18.
2. Siow CC, Nieduszynska SR, Müller CA, Nieduszynski CA. OriDB, the DNA replication origin database updated and extended. Nucleic Acids Res. 2012;40(Database issue):D682–6.
3. Rhind N, Gilbert DM. DNA replication timing. Cold Spring Harb Perspect Biol. 2013;5(8): a010132.
4. Donley N, Thayer MJ. DNA replication timing, genome stability and cancer: late and/or delayed DNA replication timing is associated with increased genomic instability. Semin Cancer Biol. 2013;23(2):80–9.
5. DePamphilis ML, Blow JJ, Ghosh S, Saha T, Noguchi K, Vassilev A. Regulating the licensing of DNA replication origins in metazoa. Curr Opin Cell Biol. 2006;18(3):231–9.
6. Labib K. How do Cdc7 and cyclin-dependent kinases trigger the initiation of chromosome replication in eukaryotic cells? Genes Dev. 2010;24(12):1208–19.
7. Diffley JF. Quality control in the initiation of eukaryotic DNA replication. Philos Trans R Soc Lond B Biol Sci. 2011;366(1584):3545–53.
8. Leonard AC, Mechali M. DNA replication origins. Cold Spring Harb Perspect Biol. 2013;5(10):a010116.
9. Ding Q, MacAlpine DM. Defining the replication program through the chromatin landscape. Crit Rev Biochem Mol Biol. 2011;46(2):165–79.
10. Riera A, Tognetti S, Speck C. Helicase loading: how to build a MCM2-7 double-hexamer. Semin Cell Dev Biol. 2014;30:104–9.
11. Ilves I, Petojevic T, Pesavento JJ, Botchan MR. Activation of the MCM2-7 helicase by association with Cdc45 and GINS proteins. Mol Cell. 2010;37(2):247–58.
12. Sheu YJ, Stillman B. The Dbf4-Cdc7 kinase promotes S phase by alleviating an inhibitory activity in Mcm4. Nature. 2010;463(7277):113–7.
13. Zegerman P, Diffley JF. Phosphorylation of Sld2 and Sld3 by cyclin-dependent kinases promotes DNA replication in budding yeast. Nature. 2007;445(7125):281–5.
14. Daigaku Y, Keszthelyi A, Muller CA, Miyabe I, Brooks T, Retkute R, et al. A global profile of replicative polymerase usage. Nat Struct Mol Biol. 2015;22(3):192–8.
15. McGuffee SR, Smith DJ, Whitehouse I. Quantitative, genome-wide analysis of eukaryotic replication initiation and termination. Mol Cell. 2013;50(1):123–35.
16. Hawkins M, Retkute R, Müller CA, Saner N, Tanaka TU, de Moura AP, et al. High-resolution replication profiles define the stochastic nature of genome replication initiation and termination. Cell Rep. 2013;5(4):1132–41.
17. Alver RC, Chadha GS, Blow JJ. The contribution of dormant origins to genome stability: from cell biology to human genetics. DNA Repair (Amst). 2014;19:182–9.
18. McIntosh D, Blow JJ. Dormant origins, the licensing checkpoint, and the response to replicative stresses. Cold Spring Harb Perspect Biol. 2012;4(10):a012955.

19. Letessier A, Millot GA, Koundrioukoff S, Lachages AM, Vogt N, Hansen RS, et al. Cell-type-specific replication initiation programs set fragility of the FRA3B fragile site. Nature. 2011;470(7332):120–3.
20. Yabuki N, Terashima H, Kitada K. Mapping of early firing origins on a replication profile of budding yeast. Genes Cells. 2002;7(8):781–9.
21. Raghuraman MK, Winzeler EA, Collingwood D, Hunt S, Wodicka L, Conway A, et al. Replication dynamics of the yeast genome. Science. 2001;294(5540):115–21.
22. Viggiani CJ, Knott SR, Aparicio OM. Genome-wide analysis of DNA synthesis by BrdU immunoprecipitation on tiling microarrays (BrdU-IP-chip) in Saccharomyces cerevisiae. Cold Spring Harb Protoc. 2010;2010(2):pdbprot5385.
23. Koren A, Soifer I, Barkai N. MRC1-dependent scaling of the budding yeast DNA replication timing program. Genome Res. 2010a;20(6):781–90.
24. Müller CA, Hawkins M, Retkute R, Malla S, Wilson R, Blythe MJ, et al. The dynamics of genome replication using deep sequencing. Nucleic Acids Res. 2014;42(1), e3.
25. Gilbert DM. Evaluating genome-scale approaches to eukaryotic DNA replication. Nat Rev Genet. 2010;11(10):673–84.
26. Alvino GM, Collingwood D, Murphy JM, Delrow J, Brewer BJ, Raghuraman MK. Replication in hydroxyurea: it's a matter of time. Mol Cell Biol. 2007;27(18):6396–406.
27. Crabbe L, Thomas A, Pantesco V, De Vos J, Pasero P, Lengronne A. Analysis of replication profiles reveals key role of RFC-Ctf18 in yeast replication stress response. Nat Struct Mol Biol. 2010;17(11):1391–7.
28. de Moura AP, Retkute R, Hawkins M, Nieduszynski CA. Mathematical modelling of whole chromosome replication. Nucleic Acids Res. 2010;38(17):5623–33.
29. Yamashita M, Hori Y, Shinomiya T, Obuse C, Tsurimoto T, Yoshikawa H, et al. The efficiency and timing of initiation of replication of multiple replicons of Saccharomyces cerevisiae chromosome VI. Genes Cells. 1997;2:655–66.
30. Friedman KL, Brewer BJ, Fangman WL. Replication profile of Saccharomyces cerevisiae chromosome VI. Genes Cells. 1997;2(11):667–78.
31. Rhind N. DNA replication timing: random thoughts about origin firing. Nat Cell Biol. 2006;8(12):1313–6.
32. Yang SC, Rhind N, Bechhoefer J. Modeling genome-wide replication kinetics reveals a mechanism for regulation of replication timing. Mol Syst Biol. 2010;6:404.
33. Sekedat MD, Fenyo D, Rogers RS, Tackett AJ, Aitchison JD, Chait BT. GINS motion reveals replication fork progression is remarkably uniform throughout the yeast genome. Mol Syst Biol. 2010;6:353.
34. Retkute R, Nieduszynski CA, de Moura A. Dynamics of DNA replication in yeast. Phys Rev Lett. 2011;107(6):068103.
35. Retkute R, Nieduszynski CA, de Moura A. Mathematical modeling of genome replication. Phys Rev E. 2012;86(3 Pt 1):031916.
36. Raghuraman MK, Brewer BJ. Molecular analysis of the replication program in unicellular model organisms. Chromosome Res. 2010;18(1):19–34.
37. Donaldson AD, Fangman WL, Brewer BJ. Cdc7 is required throughout the yeast S phase to activate replication origins. Genes Dev. 1998a;12(4):491–501.
38. Bousset K, Diffley JF. The Cdc7 protein kinase is required for origin firing during S phase [published erratum appears in Genes Dev 1998 Apr 1;12(7):1072]. Genes Dev. 1998;12(4): 480–90.
39. Jackson LP, Reed SI, Haase SB. Distinct mechanisms control the stability of the related S-phase cyclins Clb5 and Clb6. Mol Cell Biol. 2006;26(6):2456–66.
40. Gibson DG, Aparicio JG, Hu F, Aparicio OM. Diminished S-phase cyclin-dependent kinase function elicits vital Rad53-dependent checkpoint responses in Saccharomyces cerevisiae. Mol Cell Biol. 2004;24(23):10208–22.
41. Donaldson AD, Raghuraman MK, Friedman KL, Cross FR, Brewer BJ, Fangman WL. CLB5-dependent activation of late replication origins in S. cerevisiae. Mol Cell. 1998b;2(2): 173–82.

42. Wu PY, Nurse P. Establishing the program of origin firing during S phase in fission Yeast. Cell. 2009;136(5):852–64.
43. Patel PK, Kommajosyula N, Rosebrock A, Bensimon A, Leatherwood J, Bechhoefer J, et al. The Hsk1(Cdc7) replication kinase regulates origin efficiency. Mol Biol Cell. 2008;19(12):5550–8.
44. Tanaka S, Nakato R, Katou Y, Shirahige K, Araki H. Origin association of Sld3, Sld7, and Cdc45 proteins is a key step for determination of origin-firing timing. Curr Biol. 2011;21(24):2055–63.
45. Mantiero D, Mackenzie A, Donaldson A, Zegerman P. Limiting replication initiation factors execute the temporal programme of origin firing in budding yeast. EMBO J. 2011;30(23):4805–14.
46. Yoshida K, Bacal J, Desmarais D, Padioleau I, Tsaponina O, Chabes A, et al. The histone deacetylases sir2 and rpd3 act on ribosomal DNA to control the replication program in budding yeast. Mol Cell. 2014;54(4):691–7.
47. Knott SR, Viggiani CJ, Tavare S, Aparicio OM. Genome-wide replication profiles indicate an expansive role for Rpd3L in regulating replication initiation timing or efficiency, and reveal genomic loci of Rpd3 function in Saccharomyces cerevisiae. Genes Dev. 2009;23(9):1077–90.
48. Aparicio JG, Viggiani CJ, Gibson DG, Aparicio OM. The Rpd3-Sin3 histone deacetylase regulates replication timing and enables intra-S origin control in Saccharomyces cerevisiae. Mol Cell Biol. 2004;24(11):4769–80.
49. Goren A, Tabib A, Hecht M, Cedar H. DNA replication timing of the human beta-globin domain is controlled by histone modification at the origin. Genes Dev. 2008;22(10):1319–24.
50. Aggarwal BD, Calvi BR. Chromatin regulates origin activity in Drosophila follicle cells. Nature. 2004;430(6997):372–6.
51. Vogelauer M, Rubbi L, Lucas I, Brewer BJ, Grunstein M. Histone acetylation regulates the time of replication origin firing. Mol Cell. 2002;10(5):1223–33.
52. Nieduszynski CA, Knox Y, Donaldson AD. Genome-wide identification of replication origins in yeast by comparative genomics. Genes Dev. 2006;20(14):1874–9.
53. Kwan EX, Foss EJ, Tsuchiyama S, Alvino GM, Kruglyak L, Kaeberlein M, et al. A natural polymorphism in rDNA replication origins links origin activation with calorie restriction and lifespan. PLoS Genet. 2013;9(3):e1003329.
54. Peace JM, Ter-Zakarian A, Aparicio OM. Rif1 regulates initiation timing of late replication origins throughout the S. cerevisiae genome. PLoS One. 2014;9(5):e98501.
55. Mattarocci S, Shyian M, Lemmens L, Damay P, Altintas DM, Shi T, et al. Rif1 controls DNA replication timing in yeast through the PP1 phosphatase Glc7. Cell Rep. 2014;7(1):62–9.
56. Hiraga S, Alvino GM, Chang F, Lian HY, Sridhar A, Kubota T, et al. Rif1 controls DNA replication by directing Protein Phosphatase 1 to reverse Cdc7-mediated phosphorylation of the MCM complex. Genes Dev. 2014;28(4):372–83.
57. Dave A, Cooley C, Garg M, Bianchi A. Protein phosphatase 1 recruitment by Rif1 regulates DNA replication origin firing by counteracting DDK activity. Cell Rep. 2014;7(1):53–61.
58. Hayano M, Kanoh Y, Matsumoto S, Renard-Guillet C, Shirahige K, Masai H. Rif1 is a global regulator of timing of replication origin firing in fission yeast. Genes Dev. 2012;26(2):137–50.
59. Shirahige K, Hori Y, Shiraishi K, Yamashita M, Takahashi K, Obuse C, et al. Regulation of DNAreplication origins during cell-cycle progression. Nature. 1998;395(6702):618–21.
60. Gispan A, Carmi M, Barkai N. Checkpoint-independent scaling of the Saccharomyces cerevisiae DNA replication program. BMC Biol. 2014;12:79.
61. Looke M, Kristjuhan K, Varv S, Kristjuhan A. Chromatin-dependent and -independent regulation of DNA replication origin activation in budding yeast. EMBO Rep. 2013;14(2):191–8.
62. Friedman KL, Diller JD, Ferguson BM, Nyland SV, Brewer BJ, Fangman WL. Multiple determinants controlling activation of yeast replication origins late in S phase. Genes Dev. 1996;10(13):1595–607.

63. Müller CA, Nieduszynski CA. Conservation of replication timing reveals global and local regulation of replication origin activity. Genome Res. 2012;22(10):1953–62.
64. Pohl TJ, Kolor K, Fangman WL, Brewer BJ, Raghuraman MK. A DNA sequence element that advances replication origin activation time in Saccharomyces cerevisiae. G3 (Bethesda). 2013;3(11):1955–63.
65. Knott SR, Peace JM, Ostrow AZ, Gan Y, Rex AE, Viggiani CJ, et al. Forkhead transcription factors establish origin timing and long-range clustering in S. cerevisiae. Cell. 2012;148(1–2): 99–111.
66. Natsume T, Müller CA, Katou Y, Retkute R, Gierlinski M, Araki H, et al. Kinetochores coordinate pericentromeric cohesion and early DNA replication by Cdc7-Dbf4 kinase recruitment. Mol Cell. 2013;50(5):661–74.
67. Pohl TJ, Brewer BJ, Raghuraman MK. Functional centromeres determine the activation time of pericentric origins of DNA replication in Saccharomyces cerevisiae. PLoS Genet. 2012;8(5):e1002677.
68. Hoggard T, Shor E, Müller CA, Nieduszynski CA, Fox CA. A link between ORC-origin binding mechanisms and origin activation time revealed in budding yeast. PLoS Genet. 2013;9:e1003798.
69. Belsky JA, MacAlpine HK, Lubelsky Y, Hartemink AJ, MacAlpine DM. Genome-wide chromatin footprinting reveals changes in replication origin architecture induced by pre-RC assembly. Genes Dev. 2015;29(2):212–24.
70. Xu J, Yanagisawa Y, Tsankov AM, Hart C, Aoki K, Kommajosyula N, et al. Genome-wide identification and characterization of replication origins by deep sequencing. Genome Biol. 2012;13(4):R27.
71. McCarroll RM, Fangman WL. Time of replication of yeast centromeres and telomeres. Cell. 1988;54(4):505–13.
72. Koren A, Tsai HJ, Tirosh I, Burrack LS, Barkai N, Berman J. Epigenetically-inherited centromere and neocentromere DNA replicates earliest in s-phase. PLoS Genet. 2010b;6(8):e1001068.
73. Kim SM, Dubey DD, Huberman JA. Early-replicating heterochromatin. Genes Dev. 2003;17(3):330–5.
74. Tiengwe C, Marcello L, Farr H, Dickens N, Kelly S, Swiderski M, et al. Genome-wide analysis reveals extensive functional interaction between DNA replication initiation and transcription in the genome of Trypanosoma brucei. Cell Rep. 2012;2(1):185–97.
75. Ahmad K, Henikoff S. Centromeres are specialized replication domains in heterochromatin. J Cell Biol. 2001;153(1):101–10.
76. Li PC, Chretien L, Cote J, Kelly TJ, Forsburg SL. S. pombe replication protein Cdc18 (Cdc6) interacts with Swi6 (HP1) heterochromatin protein: region specific effects and replication timing in the centromere. Cell Cycle. 2011;10(2):323–36.
77. Hayashi MT, Takahashi TS, Nakagawa T, Nakayama J, Masukata H. The heterochromatin protein Swi6/HP1 activates replication origins at the pericentromeric region and silent mating-type locus. Nat Cell Biol. 2009;11(3):357–62.
78. Sridhar A, Kedziora S, Donaldson AD. At short telomeres Tel1 directs early replication and phosphorylates Rif1. PLoS Genet. 2014;10(10):e1004691.
79. Cooley C, Dave A, Garg M, Bianchi A. Tel1ATM dictates the replication timing of short yeast telomeres. EMBO Rep. 2014;15(10):1093–101.
80. Ryba T, Hiratani I, Lu J, Itoh M, Kulik M, Zhang J, et al. Evolutionarily conserved replication timing profiles predict long-range chromatin interactions and distinguish closely related cell types. Genome Res. 2010;20(6):761–70.
81. Duan Z, Andronescu M, Schutz K, McIlwain S, Kim YJ, Lee C, et al. A three-dimensional model of the yeast genome. Nature. 2010;465(7296):363–7.
82. Saner N, Karschau J, Natsume T, Gierlinski M, Retkute R, Hawkins M, et al. Stochastic association of neighboring replicons creates replication factories in budding yeast. J Cell Biol. 2013;202(7):1001–12.

83. Kaykov A, Nurse P. The spatial and temporal organization of origin firing during the S-phase of fission yeast. Genome Res. 2015;25(3):391–401.
84. Yaffe E, Farkash-Amar S, Polten A, Yakhini Z, Tanay A, Simon I. Comparative analysis of DNA replication timing reveals conserved large-scale chromosomal architecture. PLoS Genet. 2010;6(7):e1001011.
85. Di Rienzi SC, Lindstrom KC, Mann T, Noble WS, Raghuraman MK, Brewer BJ. Maintaining replication origins in the face of genomic change. Genome Res. 2012;22(10):1940–52.
86. Bianchi A, Shore D. Early Replication of Short Telomeres in Budding Yeast. Cell. 2007;128(6):1051–62.
87. Cosgrove AJ, Nieduszynski CA, Donaldson AD. Ku complex controls the replication time of DNA in telomere regions. Genes Dev. 2002;16(19):2485–90.
88. Dionne I, Wellinger RJ. Processing of telomeric DNA ends requires the passage of a replication fork. Nucleic Acids Res. 1998;26(23):5365–71.
89. Marcand S, Brevet V, Mann C, Gilson E. Cell cycle restriction of telomere elongation. Curr Biol. 2000;10(8):487–90.
90. Zhang J, Xu F, Hashimshony T, Keshet I, Cedar H. Establishment of transcriptional competence in early and late S phase. Nature. 2002;420(6912):198–202.
91. Lande-Diner L, Zhang J, Cedar H. Shifts in replication timing actively affect histone acetylation during nucleosome reassembly. Mol Cell. 2009;34(6):767–74.
92. Sima J, Gilbert DM. Complex correlations: replication timing and mutational landscapes during cancer and genome evolution. Curr Opin Genet Dev. 2014;25:93–100.
93. Woo YH, Li WH. DNA replication timing and selection shape the landscape of nucleotide variation in cancer genomes. Nat Commun. 2012;3:1004.
94. Stamatoyannopoulos JA, Adzhubei I, Thurman RE, Kryukov GV, Mirkin SM, Sunyaev SR. Human mutation rate associated with DNA replication timing. Nat Genet. 2009;41(4):393–5.
95. Agier N, Fischer G. The mutational profile of the yeast genome is shaped by replication. Mol Biol Evol. 2012;29(3):905–13.
96. Lang GI, Murray AW. Mutation rates across budding yeast chromosome VI are correlated with replication timing. Genome Biol Evol. 2011;3:799–811.
97. Weber CC, Pink CJ, Hurst LD. Late-replicating domains have higher divergence and diversity in Drosophila melanogaster. Mol Biol Evol. 2012;29(2):873–82.
98. Brown CA, Murray AW, Verstrepen KJ. Rapid expansion and functional divergence of subtelomeric gene families in yeasts. Curr Biol. 2010;20(10):895–903.

Chapter 7
Sequence Determinants of Yeast Replication Origins

M.K. Raghuraman and Ivan Liachko

Abstract The first eukaryotic replication origin was isolated from *S. cerevisiae* in the late 1970s using a plasmid maintenance assay. Combined with Sanger sequencing, this assay gave valuable insights into origin structure in *S. cerevisiae* and a few other yeast species. Fast-forward to this millennium, and the same simple assay in conjunction with modern "next-generation" sequencing and other high-throughput techniques testing origin structure and activity has led to an explosion of powerful approaches for dissecting origin sequence and function. Although such studies are still in their infancy, they have already uncovered a surprising diversification of origin sequences over a relatively short evolutionary time span. In this chapter we focus on how these approaches are being applied to understand origin structure and evolution in diverse species of budding yeasts. These approaches hold out the hope that through a comprehensive analysis of origin function across the budding yeast lineage, we can begin to understand the evolutionary forces that shape the replication landscape.

Keywords Replication • Replication origin • ARS • Initiation • ACS • B-element • ARS-seq • Yeast

Introduction

Some 30-plus years after the identification of the first eukaryotic replication origin, computational identification of origins still escapes us in most eukaryotes, including the well-studied *Saccharomyces cerevisiae*, where the first eukaryotic origin was identified. However, the advent of "next-generation" sequencing enables massively parallel screening methods that help us investigate the sequence determinants of origin function at an unprecedented scale and resolution. These technologies hold out the promise that, to the extent that underlying nucleotide sequence specifies

M.K. Raghuraman (✉) • I. Liachko
Department of Genome Sciences, University of Washington,
Seattle, WA 98195, USA
e-mail: raghu@uw.edu; liachko.ivan@gmail.com

D.L. Kaplan (ed.), *The Initiation of DNA Replication in Eukaryotes*,
DOI 10.1007/978-3-319-24696-3_7

origin function, it should be possible to uncover such sequences. We and others have embarked on such studies not only in *S. cerevisiae* but also in other yeasts, one goal being to gain insights into how replication origins evolve. Replication origins are unlike other *cis*-acting elements such as transcription factor-binding sites and centromeres in that loss of function of any one origin is unlikely to have deleterious consequences for the cell [1, 2]. Origins as a class therefore are essential for genome maintenance, but individually are dispensable and redundant—raising the question of whether or to what extent they would show evolutionary sequence conservation. Furthermore, regardless of whether origin sequences are conserved, the fact that individual origins may be lost with no apparent fitness cost raises the possibility that they would show a more rapid turnover than would genes in their neighborhood. Indeed, comparison of *S. cerevisiae* with *Lachancea waltii*—a pre-whole-genome duplication (WGD) yeast—has revealed just such a plasticity of origin locations [3].

In this chapter we summarize the state of our knowledge of origin sequences and highlight some themes that have emerged in recent years. Because there are several excellent recent reviews of origins in mammalian systems [4, 5], our focus will be on replication origins in yeasts. At the core of these yeast studies is the availability of two classes of techniques: a plasmid-based genetic test for origin function, and molecular tests that query the activity of those sequences in initiating DNA synthesis. The genetic test identifies sequences that have the potential to act as chromosomal origins of replication; the molecular tests reveal variation in their performance as origins in the chromosomal context—such as the extent to which they are used in any given cell cycle (their efficiency of firing) and in the time within S phase when they are most likely to become active.

The ARS Assay: A Genetic Test for Origin Function

Based on studies in *Escherichia coli*, Jacob et al. [6] proposed the replicon hypothesis: each replication unit or replicon contains a *cis*-acting element, the replicator, that is the target of a *trans*-activator, the initiator, to begin replication. Depending on the species, replicons in eukaryotes follow this model more or less closely. The replicator is what we would currently refer to as an origin of replication. However, it wasn't until the late 1970s that the first eukaryotic replication origin was discovered in *S. cerevisiae* [7, 8]. At the time, several possibilities—that were not mutually exclusive—for the nature of eukaryotic replication origins were being considered [9]. One possibility was that as in *E. coli*, origins were sequence specific, with perhaps many different origin sequences within any given species. An alternative possibility was that origins weren't sequence specific, but were restricted to particular chromosomal regions, either as a result of transcription or through different chromatin states. A breakthrough came from the discovery by Stinchcomb et al. that a plasmid carrying the yeast *TRP1* gene as an insert was able to transform *trp1* mutant yeast cells to Trp+ with an efficiency at least a 1000-fold higher than, for example, *HIS3* or *LEU2* plasmids were able to transform *his3* or *leu2* mutants, respectively [7, 8, 10]. Furthermore, the His+ and Leu+ transformants all had integrated the plasmid into the genome. In contrast, the *TRP1* plasmid was maintained episomally and therefore clearly was able to replicate as an autonomous genetic element. Stinchcomb et al. surmised that the *TRP1* genomic fragment must fortuitously also carry an origin of replication; by deletion analysis, they mapped the *autonomous replication sequence* (ARS) to a sequence of ~850 bp and named it *ARS1* [7]. As expected for a replication origin, this sequence acted in *cis*: in co-transformation experiments, the sequence boosted the transformation efficiency only of those plasmids that carried it.

The ARS assay—i.e., the ability of a sequence to confer autonomous maintenance on a recombinant plasmid—was a quick and easy genetic test for putative origin sequences in yeast, and it wasn't long before additional ARS elements were discovered [11]. Direct proof that ARS elements were DNA synthesis initiation sites came with the advent of two-dimensional (2-D) agarose gel electrophoresis techniques to examine replication intermediates [12, 13].

ARS Structure

From sequence comparison and mutational analysis of a handful of ARS elements, a few patterns emerged [11, 14–17].

- ARS elements are small, A/T-rich sequence of 100–150 bp.
- Although not identical in sequence, they do contain a loose match (at least 9/11 [18, 19]) to an 11-base pair sequence, the ARS consensus sequence (ACS) or "A" element: $\genfrac{}{}{0pt}{}{\mathrm{A}}{\mathrm{T}}\mathrm{TTTA}\genfrac{}{}{0pt}{}{\mathrm{T}}{\mathrm{C}}\genfrac{}{}{0pt}{}{\mathrm{A}}{\mathrm{G}}\mathrm{TTT}\genfrac{}{}{0pt}{}{\mathrm{A}}{\mathrm{T}}$
- The ACS is necessary but not sufficient for ARS activity—sequences called "B" elements 3′ to the to T-rich strand of the ACS are also needed, different ARSs having different numbers of B elements with little sequence conservation.
- A third, accessory element, the "C" element, has also been described [20], although its nature is even less understood or explored than that of B elements.

The contribution of the accessory elements to ARS function is highlighted by the fact that, as we now know, there are >15,000 matches of 10+ bp matches to the ACS in the *S. cerevisiae* genome but only a few hundred ARSs. Considerable effort has gone into defining ARS sequences, the ultimate goal being to predict origin location computationally. Initially, continued comparative analysis of additional ARS sequences yielded an expanded, 17-bp ACS [21]. Systematic mutational analysis of *ARS1* [22] and a few other ARS sequences confirmed the identity of the ACS and revealed substructure within the ACS, wherein certain positions within the ACS were more tolerant of mutation than were others. Furthermore, the same analyses also revealed the diversity of sequence and structure of the B elements ([23]; Fig. 7.1). Most ARS sequences examined by mutational analysis appear to have a B1 element, but they diverge in structure beyond that (see Fig. 7.1). Such analyses

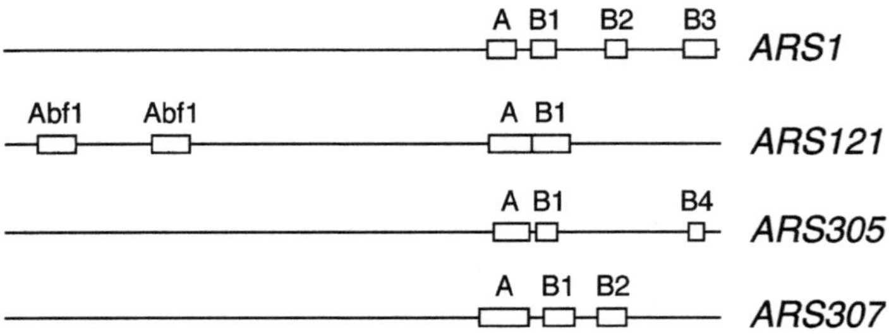

Fig. 7.1 ARS structure in *S. cerevisiae*: A (ACS) and B elements as deduced by sequence comparison and mutational analysis are shown for *ARS1* [22], *ARS121* [75], *ARS305* [76], and *ARS307* [32, 77]. All four ARSs are depicted with the T-rich strand of their A element as the Watson strand. Binding sites for transcription factor Abf1p are shown for *ARS121*; note that the B3 element of *ARS1* also is a binding site for Abf1p, but the sites in *ARS121* are not annotated as B elements because they occur upstream of the T-rich strand of the ACS, unlike other known B elements

went into high gear with the advent of high-throughput "deep" sequencing technologies (see below). Nevertheless, as discussed below, the nature of the B elements remains poorly understood, and we still are unable to predict them based on sequence alone.

Any given ARS may have between one and three nonidentical B elements. The ACS and B1 element together form a bipartite binding site for the origin recognition complex (ORC, [24]), the highly conserved heterohexamer that then recruits Mcm2p-Mcm7p, the double-hexameric *mini*chromosome *m*aintenance component of the replicative helicase, thereby establishing an initiation complex that is "licensed" to fire [25, 26]. Other B elements occur as well, also 3′ of the T-rich strand of the ACS. In *ARS1*, the B2 element corresponds to a region of high helical instability. It was thought that this region of instability—called the duplex unwinding element or DUE, found at most ARSs—would provide the site of initial unwinding of DNA as part of the initiation process [27–30]. The B2 element of *ARS1* can substitute for the B2 elements of *ARS305* and *ARS307*, supporting the view that the B2 element contributed to a general property of ARS elements [31, 32]. However, detailed mutational analysis of the *ARS1* B2 element showed that its function does not appear to be that of duplex unwinding, as no correlation was found between helical instability and the ability of the B2 variant to promote ARS activity [33]. Rather, the data supported a protein recruitment role for the B2 element: first, consistent with prior reports [34, 35], loading of the MCM complex was reduced in B2− but not in B2+ variants of the B2 sequence, and second, overexpression of Cdc6p, needed for MCM complex loading, suppressed the defect in ARS function of the B2− variants [33]. Nevertheless, as a DUE does appear to be associated with most if not all ARSs, these two possible roles of the B2 element (structural and chromatin) are not necessarily mutually exclusive, and the relative contribution of these two roles may depend on the context and identity of the ARS.

Additional B elements were also described—e.g., B3 in *ARS1* and B4 in *ARS305* (Fig. 7.1). In some instances, these elements are transcription factor-binding sites—

e.g., the transcription factor Abf1, which took its name (*A*RS-*b*inding *f*actor *1*) from its identification as a protein that binds to the B3 element in *ARS1*, binds to many but not all ARS elements ([36] and references therein), and Rap1 binds to and contributes to the activity of ARS elements associated with transcriptional silencers [37, 38]. The transcription factors involved are thought not to have specific interactions with ORC and associated machinery; rather, they are thought to recruit chromatin modifiers or remodelers, which in turn alter the nucleosome environment to promote origin activity [39]. Consistent with this idea is the observation that the Rap1p-binding site or the binding site for the Gal4p transcriptional activator can functionally substitute for the Abf1p site in *ARS1* [22].

ARS and Origin Mapping in the Genomic Era

Identification and fine-structure mapping of ARS elements and origins took a giant step forward with the arrival of genomics and massively parallel DNA sequencing technologies, marked in 1996 by the release of the *S. cerevisiae* genome sequence [40] and subsequently by the release of genome sequences of additional *Saccharomyces* and related budding yeast species as well as that of the fission yeast species *Schizosaccharomyces pombe*. Shortly after the turn of the millennium, genome-wide replication studies in *S. cerevisiae* profiled various aspects of origin function in the genome (Fig. 7.2; summarized in [41]). Replication profiles capturing genome-wide locations of active origins and termination zones, relative times of replication within S phase, and relative fork migration rates were obtained using a modernized version of the classic dense isotope Meselson/Stahl experiment [42]. "ChIP-chip"—chromatin immunoprecipitation of ORC and Mcm complexes

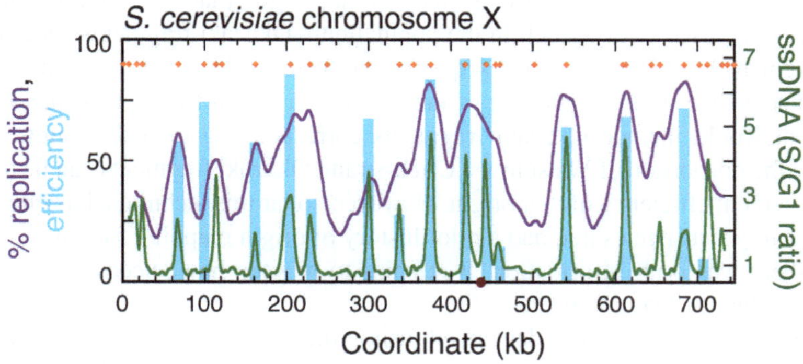

Fig. 7.2 Replication profiling: The replication pattern of *S. cerevisiae* chromosome X revealed by four methods: percent replication in mid-S phase as judged by density transfer analysis (*purple*), origin activity revealed by ssDNA accumulation during S phase in a *rad53* mutant strain (*green*), ORC/MCM binding detected by ChIP-chip analysis (*orange diamonds*), and origin efficiency inferred from Okazaki fragment abundance of Watson vs. Crick strands (*cyan bars*). The brown dot on *x*-axis marks the location of the centromere. Based on data from [44, 47, 78, 79]

cross-linked to the underlying genomic DNA, followed by microarray hybridization (now replaced by deep sequencing, ChIP-seq) generated a genomic map of pre-replicative complex locations [43, 44]. Genomic sites of bromodeoxyuridine incorporation by cells entering S phase in the presence of the ribonucleotide reductase inhibitor hydroxyurea (HU) showed the locations of origins that predominantly become active in the first half of S phase [45]. Likewise, mapping of single-stranded DNA associated with replication forks in the presence of HU in *rad53* checkpoint-deficient cells mapped, at high resolution, the sites of all active origins in the genome [46]. The conclusions from these and other such studies are available online at http://cerevisiae.oridb.org/.

Each of those studies had its strengths and limitations. For example, dense isotope-based replication profiling revealed a wealth of information about patterns of chromosome replication—including origin locations—but the resolution of those origin assignments was not high enough to map origin sequences precisely. ChIP-chip analysis of ORC- and Mcm-binding sites gave high-resolution information about the underlying DNA sequence (and hence, origin sequence), but those studies did not include information on which of those sites were actually used as origins. And none of the studies solved the thorny problem of untangling the replication time of an origin—the time when that sequence is replicated on average in a population of cells—and origin efficiency, the percent of cells in which the origin is actually used. Nevertheless, the aggregate of these studies gave enough detailed information about origin usage across the genome to serve as a benchmark against which predictions of origin location and function could be tested. A relatively recent approach of mapping Okazaki fragments across the genome has been used to not only identify origin locations in *S. cerevisiae* but also measure their firing efficiency [47]. This approach detects origins by mapping discontinuities between Watson-strand and Crick-strand Okazaki fragments and uses the ratio of Watson-strand to Crick-strand Okazaki fragments on either side of the origin to estimate firing efficiency (Fig. 7.2). The premise is that for an origin that has an efficiency of 100 %—i.e., used in every cell in the population—Okazaki fragments should consist exclusively of Watson-strand sequence on the left side of the origin and Crick-strand sequence on the right side of the origin. An origin that is less than 100 % efficient would include Okazaki fragments corresponding to both strands on its flanks; the proportion of Watson- vs. Crick-strand Okazaki fragments can be related to the percent efficiency of that origin. However, so far this technique has only been applied to yeast species that had a prior history of origin mapping for which, therefore, there was prior information. It remains to be seen how successful it will be when applied to "naïve" species.

Computational methods developed using the genome-wide maps described above yielded some gains in the success rate of predicting origin locations (e.g., [48]). Initially, cheaper Sanger sequencing enabled scaled-up testing and analysis of ARS elements, both from *S. cerevisiae* and from other budding yeasts. For example, through an iterative process of successive rounds of ARS identification using a plasmid library, computational sequence analysis of the ARSs so cloned, prediction of other ARSs in the genome, and testing of those predictions, ARSs were mapped and analyzed on a large scale in *Kluyveromyces lactis* and *Lachancea kluyveri* [49, 50]. Using a combination of comparative genomic analysis and molecular biology/genetics, Nieduszynski et al. [51] mapped *S. cerevisiae* origin locations by looking for conserved islands of DNA sequence in intergenic regions from four members of the sensu stricto group of yeasts—*S. cerevisiae*, *S. paradoxus*, *S. mikatae*, and *S. bayanus*. Their assumption was that the extent of conservation in origin sequences of closely related species would be sufficient to produce islands of sequence identity in the otherwise divergent intergenic regions. As proof of their findings they performed ARS assays on a few hundred predicted *S. cerevisiae* origins. Their work expanded and refined the locations and sequence motif of the *S. cerevisiae* ACS; however, their work did not address the nature of the non-ACS essential elements, nor did it address the question of whether these conserved regions actually act as ARSs (or origins) in the other sensu stricto species. Such questions are much more amenable to analysis using the "next-gen" sequencing approaches described below.

A Systematic, High-Resolution Approach for Studying Origin Sequence and Function

High-throughput, "next-generation" sequencing has brought an additional level of power to investigations of replication origins. As outlined below, we now have a near-routine pipeline for rapid identification of potential origins, analysis of sequence determinants of origin function, and examining usage of those potential origins in the chromosomal context (Fig. 7.3).

1. First is "ARS-seq" (Fig. 7.3), a high-throughput ARS assay that identifies all sequences in a genome that have the potential to act as origins on plasmids [52]. Genomic DNA from the species of interest is fragmented, either by mechanical shearing or by digestion with a combination of restriction enzymes with short (4 bp) recognition sites. The DNA fragments are size-selected and cloned into a non-ARS yeast vector that has a selectable marker suitable for the yeast species being studied (e.g., a nutritional marker *URA3*, or resistance to G418) as well as sequences for propagation in *E. coli*. Cells of the species of interest are then transformed with the library of plasmids and grown under conditions selecting for the presence of the plasmid, thereby selecting for cells with plasmids that can be replicated—i.e., containing ARSs in their inserts. All resulting colonies are pooled, DNA is isolated, and the plasmid sequences recovered using paired-end next-generation sequencing. This process yields a comprehensive set of ARS fragments that should comprise most if not all of the ARS elements present in the source genome. Mapping the sequences of those fragments back to the genome gives the precise genomic locations of most or all potential origins in the genome and sets the stage for tests and further analysis of origin activity at those sites using the replication profiling methods outlined above. Additionally, since most

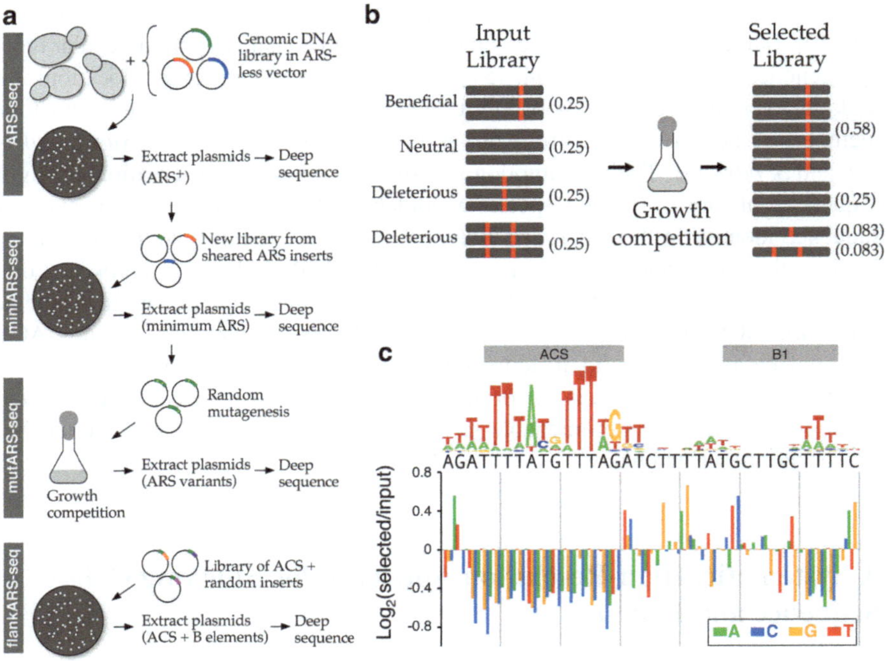

Fig. 7.3 High-throughput identification and dissection of ARS structure and function. (**a**) Overview of a pipeline for analysis of ARS structure and function. ARS-seq: A genomic DNA library made from a species of interest and constructed in a vector lacking an ARS element for that species is introduced back into the species under conditions that select for the presence of the plasmid. A cell that is able to form a colony under these conditions must harbor a plasmid that has acquired a functional ARS in its genomic DNA insert. The plasmids are extracted from the pool of transformants and their inserts are sequenced by next-gen sequencing. The deduced insert sequences are mapped back to the genome to identify the genomic sites of the putative origins. See [52] for details. MiniARS-seq: ARS elements identified and cloned using ARS-seq are sheared into smaller pieces, recloned, and retested in the ARS assay to yield the minimal ARS sequence for each member ARS in the collection. MutARS-seq: The functional contribution of each possible base within the minimal ARS sequence for any given ARS can be assayed in parallel by competitive co-culture of a pool of all possible single-base substitutions within the core ARS (plus some proportion of multiply mutated variants). FlankARS-seq: Starting with a vector that has one ARS element (e.g., the ACS) but lacks the others (e.g., B elements), a library of small inserts is created to ask: what are the sequences of short inserts that can, together with the element already present on the vector, reconstitute a functional ARS. (**b**) Data collection from MutARS-seq: A library of ARS sequence variants is introduced into yeast and the collection is cultured as a pool under conditions selecting for maintenance of the plasmid. Mutations that compromise ARS function are progressively depleted from the pool, whereas mutations that improve ARS function are enriched. Sequencing of the pools at intervals during the growth identifies the sequence of each allele as well as the relative abundance of each allele. Numbers in parentheses, relative abundance of each variant in the total pool before and after co-culture. Note that the input library need not have all variants in equal proportions; what matters is the comparison of proportions in the input relative to the selected library. (**c**) Interpretation of MutARS-seq for the A and B1 elements of *ARS1* of *S. cerevisiae*. The wild-type sequence is shown in *black*; the effect of each single-base change, color-coded by the variant base, is shown as a relative change in the proportion of that variant. Most substitutions within the core ACS and B1 sequence clearly are detrimental, whereas certain base changes (e.g., in the region between the ACS and the B1 element) improve ARS function. Based on [52]

ARSs are isolated more than once, on overlapping fragments, this technique can define minimal ARS regions and separate ARSs that are too close to each other to be delineated using other assays.

2. Second is miniARS-seq (Fig. 7.3). Once a library of ARS fragments has been obtained, the minimal sequences that constitute the ARSs in the library can be delineated by amplifying the ARS inserts from the ARS library clones en masse, shearing and re-cloning the ARS sub-fragments, recovering those that still retain ARS function by passing them again through the host species, and finally sequencing the set of minimal ARSs (Fig. 7.3). This process allows us to delineate the minimal functional ARS regions with single-nucleotide resolution for numerous ARSs in a single experiment. Subsequent sequence comparison of these regions (e.g., using MEME; [53]) reveals any conserved sequence motifs within the collection of minimal ARS sequences. Yeast transformed with miniARS plasmids can be competed in large populations to rank ARS sequences based on their individual effectiveness in plasmid maintenance (IL and M. Dunham, unpublished).
3. Third is mutARS-seq (Fig. 7.3). Minimal ARS elements can be subjected to a modified version of deep mutational scanning [54] to evaluate simultaneously the functional contribution of each base in the sequence ([52, 55]; mutARS-seq, Fig. 7.3). Deep mutational scanning is the massively parallel embodiment of scanning mutagenesis, where sequences are systematically mutagenized and tested for function. The ARS insert is randomly mutagenized in vitro, the mutagenized DNA is cloned into an ARS-less vector, and the host species is transformed with this pool of variant ARS plasmids. Large populations of transformed cells are co-cultured as a pool under conditions that select for the presence of the plasmid; deep sequencing of the ARS insert at different times during propagation of the culture gives a quantitative measure of relative abundance of each variant ARS sequence in the pool. Because propagation of any given plasmid in the pool will depend on its ability to be replicated, the change in relative abundance of particular ARS variants is a readout of ARS performance—mutations that alter base pairs required for ARS function will be depleted from the pool over time, whereas mutations that improve ARS function will be enriched. This approach allows the simultaneous measurement of the effects of all mutations on ARS function. When applied to *ARS1* as proof of principle, not only did this technique yield information on the structure of the ACS and B1 element at an unprecedented level of resolution, but it also provided some surprises: for example, some base substitution improved *ARS1* function above that of the wild-type starting sequence and deletion of particular base pairs between the ACS and B1 elements improved *ARS1* activity. Furthermore, variant *ARS1* inserts that had two or more mutations revealed epistatic interactions that previously would have been difficult to identify. Based on these results, it was possible to construct a variant of *ARS1*—*ARS1max*—that doesn't exist in nature, consisting of a 100 bp sequence in which each position had the nucleotide that individually performed best in the mutARSseq assay. Although this combination of "best" nucleotides need not necessarily have produced an ARS with improved function, this synthetic variant did in fact vastly outperform the wild-type *ARS1* fragment [52].

4. Fourth is flankARS-seq (Fig. 7.3). Additional variations on the ARS assay can be applied to further dissect the structure of ACS-flanking elements. For example, we are currently in the process of completing a "B-element screen" in *S. cerevisiae* (Fig. 7.3). A plasmid that has an ACS, but no B elements (and therefore is incapable of autonomous replication), is used as the vector for a library of short genomic DNA fragments. An ARS assay performed on this library yields plasmids that now have acquired a functional ARS—and therefore must have a functional B element in the insert. We hope to use this approach to better understand the properties of B elements.
5. Fifth are cross-species ARS tests. To understand species-specific aspects of ARS function, an ARS library created from one species can be tested en masse in a second species. These experiments provide some unique insights into the diversity of sequences that contribute to origin function. For example, about two-thirds of the ARS elements from *S. cerevisiae* support plasmid maintenance in *L. waltii* and vice versa [3]. And whereas testing for ARS function in *Pichia pastoris* (*Komagataella phaffii*) gave 311 ARS loci in the *P. pastoris* genome, testing of *S. cerevisiae* genomic fragments in *P. pastoris* surprisingly yielded >1800 unique *S. cerevisiae* fragments that could function as ARSs in *P. pastoris* (IL and M. Dunham, unpublished).
6. Lastly, in parallel with the above approaches, replication profiling can be done to monitor chromosomal replication dynamics. A combined outcome of ARS mapping using ARS-seq and replication mapping using dense isotope transfer as well as single-stranded DNA mapping for *L. waltii* is shown in Fig. 7.4.

What has emerged from such studies is that although there certainly are obvious similarities between species, there also is a surprising diversity of origin sequences across yeast lineages (Fig. 7.5). In the budding yeast species studied thus far, ARS

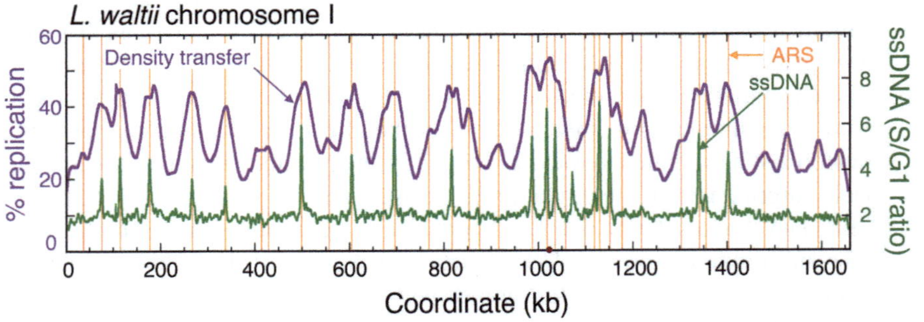

Fig. 7.4 Replication map for *L. waltii* chromosome I. Density transfer replication profiling (*purple*), ssDNA mapping (*green*), and ARS-seq (*orange*) mapping of ARS elements are shown. Peaks in the density transfer profile indicate the locations of origins; taller peaks are earlier replicating. Peaks in the ssDNA map indicate early-firing origins. Notice that peaks in the two replication assays are centered on ARSs identified by ARS-seq. Individual peaks predict origin locations that can be tested by 2-D gel electrophoresis. Based on [3]

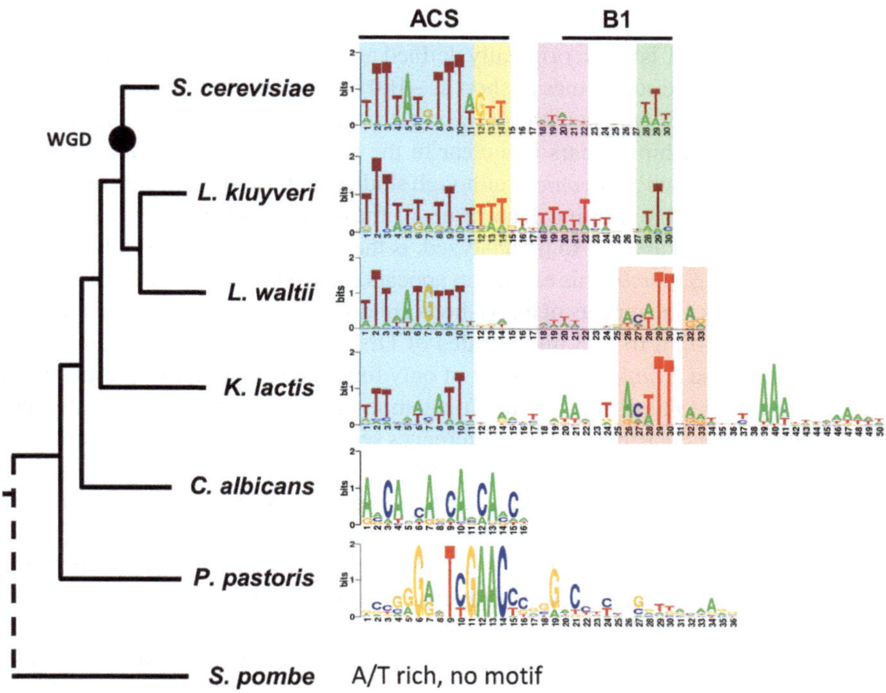

Fig. 7.5 ARS sequence structure in six budding yeast species. Sequence conservation is depicted as information content in the form of a sequence logo [80, 81]. *Cyan* and *yellow shading*: 11-bp and extended 15-bp A element, respectively; *pink*, *green*, and *pale red shading* matches to the B1 element. Sequence logos based on [3, 49, 50, 55, 71] and generated using http://weblogo.berkeley.edu. Tree phylogeny based on Dujon [82, 83]; branch lengths are arbitrary. WGD, whole -genome duplication

elements almost overwhelmingly are found in intergenic regions. Although studies in the sensu stricto species *S. bayanus* are ongoing, comparative sequence and replication analysis suggests that the ACS in this species is very similar to that in *S. cerevisiae*. Rather surprisingly, the pre-WGD species *L. waltii* also has an ACS that is very similar to that of *S. cerevisiae* but with some clear differences [3]. As mentioned above, ARSs show partial cross-functionality between the two species, so the differences in the ACS or flanking elements likely contribute to the overlapping species specificities. *L. kluyveri* has an ACS that consists mostly of Ts on one strand and As on the other [50]. From the observation that 98 % of *S. cerevisiae* ARSs can support plasmid maintenance in *L. kluyveri* (whereas only about 50 % of *L. kluyveri* ARSs function in *S. cerevisiae* hosts) we can infer that the ORC and associated replication licensing apparatus in *L. kluyveri* have much more relaxed specificity than that of *S. cerevisiae*. In contrast, the ACS in *K. lactis* is a long and partially palindromic structure [49, 56]. Unlike in *S. cerevisiae* and *L. waltii*, this ACS sequence is necessary and sufficient for ARS function, and genomic ARS locations can therefore be predicted computationally with high accuracy [49].

Going further afield, origins in the fission yeast *S. pombe* consist of A/T-rich sequences called AT islands, originally defined as being >500 bp in length but more recently shown to be sometimes as short as ~100 bp [57, 58]. It is possible to predict origin locations in *S. pombe* with >90 % success based just on this simple criterion [58]. This relationship appears less clear in the related *Schizosaccharomyces* species *S. octosporus* and *S. japonicus* although studies in those species are still in their infancy [58, 59]. One striking feature of all these *Schizosaccharomyces* species, unique amongst eukaryotes so far analyzed, is the presence of an N-terminal extension on Orc4. In *S. pombe* this extension appears to be necessary for origin recognition and contains nine repeats of an AT-hook subdomain [60, 61], each repeat having the capacity to bind to 6–8 nucleotides of DNA [62]. The AT-hook domain in *S. octosporus* and *S. japonicus* consist of only four and five AT-hooks, respectively [58], as opposed to nine in *S. pombe*. It is tempting to speculate that this difference in AT-hook count contributes to the differences seen between the species as to which sequences are associated with origin function.

Perhaps the biggest surprise thus far to come from ARS studies was found in *P. pastoris* (*K. phaffii*). There, ARS-seq and mutARS-seq analysis revealed two classes of origin sequences. One-third of the origins, called GC-ARSs, contain a relatively G/C-rich sequence motif and replicate early in S phase. Mutational analysis (mutARS-seq) confirmed that the motif is necessary for ARS activity. The other class of origins, called AT-ARSs, lacks the G/C-rich motif but instead has an A/T-rich region and fires significantly later in S phase (or is significantly less efficient than the G/C-rich class) [55]. Particularly intriguing is the observation that the G/C-rich motif matches the binding site for the human version of Hsf1, a transcriptional activator, but not that of the *S. cerevisiae* version of Hsf1. The fact that *P. pastoris* has four homologs of Hsf1 while *S. cerevisiae* only has one copy of the gene raises the very interesting possibility that the G/C-rich motif, so unlike the *S. cerevisiae* Hsf1 site, is recognized by at least one of the *P. pastoris* Hsf1 proteins, and that this interaction promotes recruitment of ORC to the GC-ARSs. In fact, the GC-rich ARS motifs in *P. pastoris* are located in promoters and show a nucleosome depletion pattern similar to transcription start sites, suggesting a link between transcription start and origin firing as seen in metazoans [55]. Taken further, this line of reasoning would suggest that the use of G/C-rich motifs for replication is perhaps an ancestral trait that was lost in the lineage leading to *Saccharomyces*, *Kluyveromyces*, and *Lachancea* [55]. Characterization of *P. pastoris* Hsf1 proteins and their DNA recognition properties would help clarify some of these questions.

One hypothesis for origin evolution in the budding yeast lineage is that the massive genomic upheaval that occurred in the aftermath of the whole-genome duplication (WGD) may have been a time of rapid change not only in genome structure but also in origin structure. Although our information across the budding yeast lineage is still very sparse, what has been observed so far is not in keeping with this hypothesis: species that span the whole-genome duplication divide may vary greatly in origin location while sharing very similar ACSs (e.g., *S. cerevisiae* and *L. waltii*) while among pre-WGD species there may be substantial diversification of origin sequences. It will obviously be illuminating to examine several more clades on either side of the WGD event to see if any patterns emerge as to origin sequence change in the context of other genomic changes amongst these species.

Chromatin as a Determinant of Origin Selection

ARS activity—i.e., the ability of a sequence to support replication of a plasmid—is a measure of the genetic potential of that sequence to act as an origin of replication. However, examples abound of ARSs that show high activity on a plasmid but are rarely used in their native chromosomal context—e.g., *ARS301* [17, 63] and *ARS604* [64, 65]. An obvious explanation for this context-dependent difference in origin activity is that the local chromatin structure must be influencing the ability of the origin to function, perhaps by limiting its accessibility to initiation factors. Early studies on the chromatin structure of *ARS1* had in fact pointed to this possibility. Mapping of chromatin at *ARS1* detected a nucleosome-free region over the core ARS sequence and a set of phased nucleosomes flanking the ARS [66]. Insertion and deletion mutations that shifted a nucleosome to be positioned over one or more of the A or B elements of the ARS reduced ARS function, and ARS function was restored when the α2 operator was used to shift the nucleosome back to its original position [67]. A subsequent study concluded that the nucleosome-free region at *ARS1* was not needed for ORC association per se, but was needed for pre-RC formation (i.e., recruitment of the Mcm complex) [34]. Thus, it is not surprising that the activity of an ARS might be strongly influenced by its chromosomal context.

The generality of the nucleosome-free nature of the core *ARS1* sequence became apparent from the mapping of a few hundred ARSs in *S. cerevisiae*, which revealed a nucleosome-free region (NFR) positioned asymmetrically over and extending 3′ to the ACS [68–70]. Genomic ACS matches not associated with ARSs as a group also have an NFR, but the ACS in those cases is positioned symmetrically within the NFR (Fig. 7.6). Furthermore, whereas the NFR in ARSs is flanked by tightly positioned nucleosomes, the non-ARS matches to the ACS do not show this arrangement of nucleosomes. The association of ACS with NFR suggests that the ACS sequence itself is able to establish a nucleosome-free region. In support of this idea, nucleosomes reconstituted in vitro on DNA fragments containing a match to the ACS also were excluded from the ACS; as with non-ARS ACS sequences in vivo, nucleosomes did not show phased positioning flanking the NFR [69, 70].

This property of ARS-associated NFRs is a useful "reality check" in mapping origins in species that have not been studied before. For example, NFRs were found at GC-ARS as well as AT-ARS sequences in *P. pastoris*, but the nucleosome profiles looked markedly different between these two classes of origins: the NFR was much wider at GC-ARSs when compared to those at AT-ARSs or to ARSs of *S. cerevisiae*. However, whereas the NFR at GC-ARSs was flanked by well-positioned nucleosomes, such an arrangement was much less clear at AT-ARSs, again underscoring

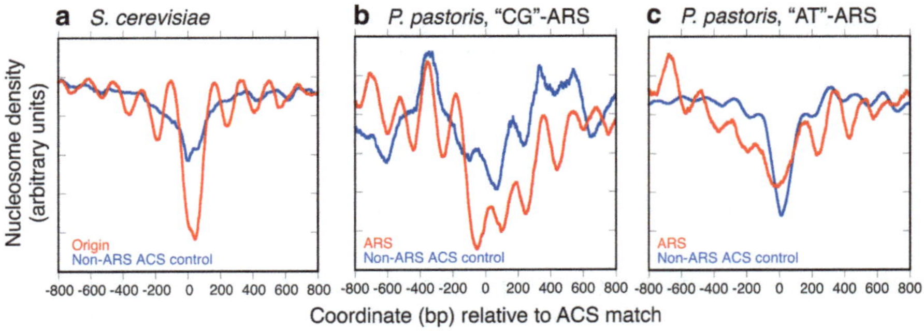

Fig. 7.6 Nucleosome profiling of ARSs. (**a**) Nucleosome density profiles centered around ACS elements (position 0) that map to confirmed origins or ARSs (*red*) and to intergenic regions not associated with origins or ARSs (*blue*) are shown for *S. cerevisiae*. Nucleosome densities derived from data of Lee et al. [84]. (**b, c**) Similar plots for CG-rich and AT-rich ARS elements in *P. pastoris*. Based on [55]

the difference in the nature and perhaps mechanism of function of these classes of origins [55]. The combination of the signature nucleosome depletion region and ORC binding has been used to predict origin locations in *Candida albicans* [71].

The contribution of chromatin and nucleosome context to origin function in *S. cerevisiae* was further revealed in a recent study comparing the relative strength of ORC-origin interaction in vivo with that seen in vitro [72]. Based on the premise that functional intracellular ORC concentration becomes limiting for origin function in a mutant *orc2-1* strain [73], the authors compared origin occupancy by ORC in wild-type and *orc2-1* mutant cells to deduce a relative ranking of ORC-origin interaction strength genome-wide. In parallel, they performed in vitro gel mobility shift assays coupled to microarray hybridization to determine the innate binding strength of those origin sequences to ORC. From this analysis, origins were classified as "DNA dependent" or "chromatin dependent." The former category comprised origins whose relative ORC binding affinity in vivo matched that seen in vitro, whereas the latter category comprised origins for which in vivo binding deviated from that expected based on in vitro binding. This classification revealed a hitherto unsuspected correlation: the chromatin-dependent group was enriched for origins that fire early in S phase, while the DNA-dependent group was enriched for late-replicating origins. These assays lend themselves to further analysis (e.g., examining the sequence contribution at nucleotide level in DNA-dependent vs. chromatin-dependent origins) using high-throughput approaches of the sort outlined in Fig. 7.3. The crystal structure of *Drosophila melanogaster* ORC has been published recently [74]; one hopes that in the near future the field will make inroads into understanding the interaction of ORC with origin sequences in the context of chromatin.

Concluding Thoughts

Some 35+ years after the identification of the first eukaryotic replication origin, we are entering a new golden age of replication studies. It is now possible to generate comprehensive maps of potential origins in a species and profile the major features of chromosomal replication within a month or two. Such studies, while still in their infancy, have already uncovered an unsuspected diversity of replication origin sequences and types within a relatively narrow slice of the tree of life. This exploration of diverse yeasts has already shaken the long-held paradigm that, contrary to metazoans, yeasts utilize A/T-rich DNA sequences for origin function. And although origins in metazoans appear not to share the defined sequence structure of yeast origins, what we are learning from yeasts will surely inform us as to the evolutionary mechanism behind the distribution and control of origins amongst all eukaryotes.

Acknowledgments We thank Bonny Brewer for critical reading of this manuscript. I.L. was supported by a grant from the NSF (1243710). M.K.R. was supported by grants from the NIH (GM018926) and the NSF (1243710).

References

1. Bogenschutz NL, Rodriguez J, Tsukiyama T. Initiation of DNA replication from non-canonical sites on an origin-depleted chromosome. PLoS One. 2014;9(12):e114545.
2. Dershowitz A, Snyder M, Sbia M, Skurnick JH, Ong LY, Newlon CS. Linear derivatives of Saccharomyces cerevisiae chromosome III can be maintained in the absence of autonomously replicating sequence elements. Mol Cell Biol. 2007;27(13):4652–63.
3. Di Rienzi SC, Lindstrom KC, Mann T, Noble WS, Raghuraman MK, Brewer BJ. Maintaining replication origins in the face of genomic change. Genome Res. 2012;22:1940–52.
4. Hyrien O. Peaks cloaked in the mist: the landscape of mammalian replication origins. J Cell Biol. 2015;208(2):147–60.
5. Smith OK, Aladjem MI. Chromatin structure and replication origins: determinants of chromosome replication and nuclear organization. J Mol Biol. 2014;426(20):3330–41.
6. Jacob F, Brenner S, Cuzin F. On the regulation of DNA replication in bacteria. Cold Spring Harb Symp Quant Biol. 1963;28:329–48.
7. Stinchcomb DT, Struhl K, Davis RW. Isolation and characterisation of a yeast chromosomal replicator. Nature. 1979;282(5734):39–43.
8. Struhl K, Stinchcomb DT, Scherer S, Davis RW. High-frequency transformation of yeast: autonomous replication of hybrid DNA molecules. Proc Natl Acad Sci U S A. 1979; 76(3):1035–9.
9. Hand R. Eucaryotic DNA: organization of the genome for replication. Cell. 1978;15:317–25.
10. Hinnen A, Hicks JB, Fink GR. Transformation of yeast. Proc Natl Acad Sci U S A. 1978; 75(4):1929–33.
11. Broach JR, Li YY, Feldman J, Jayaram M, Abraham J, Nasmyth KA, et al. Localization and sequence analysis of yeast origins of DNA replication. Cold Spring Harb Symp Quant Biol. 1983;2:1165–73.
12. Brewer BJ, Fangman WL. The localization of replication origins on ARS plasmids in S. cerevisiae. Cell. 1987;51(3):463–71.

13. Huberman JA, Spotila LD, Nawotka KA, el-Assouli SM, Davis LR. The in vivo replication origin of the yeast 2 microns plasmid. Cell. 1987;51(3):473–81.
14. Bouton AH, Smith MM. Fine-structure analysis of the DNA sequence requirements for autonomous replication of Saccharomyces cerevisiae plasmids. Mol Cell Biol. 1986;6(7):2354–63.
15. Kearsey S. Analysis of sequences conferring autonomous replication in baker's yeast. EMBO J. 1983;2(9):1571–5.
16. Kearsey S. Structural requirements for the function of a yeast chromosomal replicator. Cell. 1984;37(1):299–307.
17. Newlon CS, Theis JF. The structure and function of yeast ARS elements. Curr Opin Genet Dev. 1993;3(5):752–8.
18. Shirahige K, Iwasaki T, Rashid MB, Ogasawara N, Yoshikawa H. Location and characterization of autonomously replicating sequences from chromosome VI of Saccharomyces cerevisiae. Mol Cell Biol. 1993;13:5043–56.
19. Walker SS, Francesconi SC, Eisenberg S. A DNA replication enhancer in Saccharomyces cerevisiae. Proc Natl Acad Sci U S A. 1990;87(12):4665–9.
20. Celniker SE, Sweder K, Srienc F, Bailey JE, Campbell JL. Deletion mutations affecting autonomously replicating sequence ARS1 of Saccharomyces cerevisiae. Mol Cell Biol. 1984;4(11):2455–66.
21. Theis JF, Newlon CS. The ARS309 chromosomal replicator of Saccharomyces cerevisiae depends on an exceptional ARS consensus sequence. Proc Natl Acad Sci U S A. 1997;94(20):10786–91.
22. Marahrens Y, Stillman B. A yeast chromosomal origin of DNA replication defined by multiple functional elements. Science. 1992;255(5046):817–23.
23. Lucas IA, Raghuraman MK. The dynamics of chromosome replication in yeast. Curr Top Dev Biol. 2003;55:1–73.
24. Li H, Stillman B. The origin recognition complex: a biochemical and structural view. Subcell Biochem. 2012;62:37–58.
25. Coverley D, Laskey RA. Regulation of eukaryotic DNA replication. Annu Rev Biochem. 1994;63:745–76.
26. Riera A, Tognetti S, Speck C. Helicase loading: how to build a MCM2-7 double-hexamer. Semin Cell Dev Biol. 2014;30:104–9.
27. Huang RY, Kowalski D. A DNA unwinding element and an ARS consensus comprise a replication origin within a yeast chromosome. EMBO J. 1993;12(12):4521–31.
28. Natale DA, Schubert AE, Kowalski D. DNA helical stability accounts for mutational defects in a yeast replication origin. Proc Natl Acad Sci U S A. 1992;89(7):2654–8.
29. Umek RM, Kowalski D. Thermal energy suppresses mutational defects in DNA unwinding at a yeast replication origin. Proc Natl Acad Sci U S A. 1990;87(7):2486–90.
30. Umek RM, Kowalski D. The ease of DNA unwinding as a determinant of initiation at yeast replication origins. Cell. 1988;52(4):559–67.
31. Lin S, Kowalski D. Functional equivalency and diversity of cis-acting elements among yeast replication origins. Mol Cell Biol. 1997;17(9):5473–84.
32. Rao H, Marahrens Y, Stillman B. Functional conservation of multiple elements in yeast chromosomal replicators. Mol Cell Biol. 1994;14(11):7643–51.
33. Wilmes GM, Bell SP. The B2 element of the Saccharomyces cerevisiaeARS1 origin of replication requires specific sequences to facilitate pre-RC formation. Proc Natl Acad Sci U S A. 2002;99(1):101–6.
34. Lipford JR, Bell SP. Nucleosomes positioned by ORC facilitate the initiation of DNA replication. Mol Cell. 2001;7(1):21–30.
35. Zou L, Stillman B. Assembly of a complex containing Cdc45p, replication protein A, and Mcm2p at replication origins controlled by S-phase cyclin-dependent kinases and Cdc7p-Dbf4p kinase. Mol Cell Biol. 2000;20(9):3086–96.
36. Rhode PR, Sweder KS, Oegema KF, Campbell JL. The gene encoding ARS-binding factor I is essential for the viability of yeast. Genes Dev. 1989;3(12A):1926–39.

37. Diffley JF, Stillman B. Similarity between the transcriptional silencer binding proteins ABF1 and RAP1. Science. 1989;246(4933):1034–8.
38. Shore D, Nasmyth K. Purification and cloning of a DNA binding protein from yeast that binds to both silencer and activator elements. Cell. 1987;51(5):721–32.
39. Knott SR, Peace JM, Ostrow AZ, Gan Y, Rex AE, Viggiani CJ, et al. Forkhead transcription factors establish origin timing and long-range clustering in S. cerevisiae. Cell. 2012;148(1–2):99–111.
40. Goffeau A, Barrell BG, Bussey H, Davis RW, Dujon B, Feldmann H, et al. Life with 6000 genes. Science. 1996;274(5287):546. 63–7.
41. Raghuraman MK, Brewer BJ. Molecular analysis of the replication program in unicellular model organisms. Chromosome Res. 2010;18(1):19–34.
42. Raghuraman MK, Winzeler EA, Collingwood D, Hunt S, Wodicka L, Conway A, et al. Replication dynamics of the yeast genome. Science. 2001;294(5540):115–21.
43. Wyrick JJ, Aparicio JG, Chen T, Barnett JD, Jennings EG, Young RA, et al. Genome-wide distribution of ORC and MCM proteins in S. cerevisiae: high- resolution mapping of replication origins. Science. 2001;294(5550):2357–60.
44. Xu W, Aparicio JG, Aparicio OM, Tavare S. Genome-wide mapping of ORC and Mcm2p binding sites on tiling arrays and identification of essential ARS consensus sequences in S. cerevisiae. BMC Genomics. 2006;7:276.
45. Yabuki N, Terashima H, Kitada K. Mapping of early firing origins on a replication profile of budding yeast. Genes Cells. 2002;7(8):781–9.
46. Feng W, Collingwood D, Boeck ME, Fox LA, Alvino GM, Fangman WL, et al. Genomic mapping of single-stranded DNA in hydroxyurea-challenged yeasts identifies origins of replication. Nat Cell Biol. 2006;8(2):148–55.
47. McGuffee SR, Smith DJ, Whitehouse I. Quantitative, genome-wide analysis of eukaryotic replication initiation and termination. Mol Cell. 2013;50(1):123–35.
48. Breier AM, Chatterji S, Cozzarelli NR. Prediction of Saccharomyces cerevisiae replication origins. Genome Biol. 2004;5(4):R22.
49. Liachko I, Bhaskar A, Lee C, Chung SC, Tye BK, Keich U. A comprehensive genome-wide map of autonomously replicating sequences in a naive genome. PLoS Genet. 2010; 6(5):e1000946.
50. Liachko I, Tanaka E, Cox K, Chung SC, Yang L, Seher A, et al. Novel features of ARS selection in budding yeast Lachancea kluyveri. BMC Genomics. 2011;12:633.
51. Nieduszynski CA, Knox Y, Donaldson AD. Genome-wide identification of replication origins in yeast by comparative genomics. Genes Dev. 2006;20(14):1874–9.
52. Liachko I, Youngblood RA, Keich U, Dunham MJ. High-resolution mapping, characterization, and optimization of autonomously replicating sequences in yeast. Genome Res. 2013; 23(4):698–704.
53. Bailey TL, Boden M, Buske FA, Frith M, Grant CE, Clementi L, et al. MEME SUITE: tools for motif discovery and searching. Nucleic Acids Res. 2009;37(Web Server issue):W202–8.
54. Fowler DM, Fields S. Deep mutational scanning: a new style of protein science. Nat Methods. 2014;11(8):801–7.
55. Liachko I, Youngblood RA, Tsui K, Bubb KL, Queitsch C, Raghuraman MK, et al. GC-rich DNA elements enable replication origin activity in the methylotrophic yeast Pichia pastoris. PLoS Genet. 2014;10(3):e1004169.
56. Fabiani L, Frontali L, Newlon CS. Identification of an essential core element and stimulatory sequences in a Kluyveromyces lactis ARS element, KARS101. Mol Microbiol. 1996;19(4):756–66.
57. Antequera F. Genomic specification and epigenetic regulation of eukaryotic DNA replication origins. EMBO J. 2004;23(22):4365–70.
58. Cotobal C, Segurado M, Antequera F. Structural diversity and dynamics of genomic replication origins in Schizosaccharomyces pombe. EMBO J. 2010;29(5):934–42.

59. Xu J, Yanagisawa Y, Tsankov AM, Hart C, Aoki K, Kommajosyula N, et al. Genome-wide identification and characterization of replication origins by deep sequencing. Genome Biol. 2012;13(4):R27.
60. Chuang RY, Kelly TJ. The fission yeast homologue of Orc4p binds to replication origin DNA via multiple AT-hooks. Proc Natl Acad Sci U S A. 1999;96(6):2656–61.
61. Lee JK, Moon KY, Jiang Y, Hurwitz J, Chuang RY, Kelly TJ. The Schizosaccharomyces pombe origin recognition complex interacts with multiple AT-rich regions of the replication origin DNA by means of the AT-hook domains of the spOrc4 protein. The fission yeast homologue of Orc4p binds to replication origin DNA via multiple AT-hooks. Proc Natl Acad Sci U S A. 2001;98(24):13589–94.
62. Reeves R, Nissen MS. The A.T-DNA-binding domain of mammalian high mobility group I chromosomal proteins. A novel peptide motif for recognizing DNA structure. J Biol Chem. 1990;265(15):8573–82.
63. Santocanale C, Sharma K, Diffley JF. Activation of dormant origins of DNA replication in budding yeast. Genes Dev. 1999;13(18):2360–4.
64. Friedman KL, Brewer BJ, Fangman WL. Replication profile of Saccharomyces cerevisiae chromosome VI. Genes Cells. 1997;2(11):667–78.
65. Yamashita M, Hori Y, Shinomiya T, Obuse C, Tsurimoto T, Yoshikawa H, et al. The efficiency and timing of initiation of replication of multiple replicons of Saccharomyces cerevisiae chromosome VI. Genes Cells. 1997;2(11):655–65.
66. Thoma F, Bergman LW, Simpson RT. Nuclease digestion of circular TRP1ARS1 chromatin reveals positioned nucleosomes separated by nuclease-sensitive regions. J Mol Biol. 1984;177(4):715–33.
67. Simpson RT. Nucleosome positioning can affect the function of a cis-acting DNA element in vivo. Nature. 1990;343(6256):387–9.
68. Belsky JA, MacAlpine HK, Lubelsky Y, Hartemink AJ, MacAlpine DM. Genome-wide chromatin footprinting reveals changes in replication origin architecture induced by pre-RC assembly. Genes Dev. 2015;29(2):212–24.
69. Eaton ML, Galani K, Kang S, Bell SP, MacAlpine DM. Conserved nucleosome positioning defines replication origins. Genes Dev. 2010;24(8):748–53.
70. Berbenetz NM, Nislow C, Brown GW. Diversity of eukaryotic DNA replication origins revealed by genome-wide analysis of chromatin structure. PLoS Genet. 2010;6(9):e1001092.
71. Tsai HJ, Baller JA, Liachko I, Koren A, Burrack LS, Hickman MA, et al. Origin replication complex binding, nucleosome depletion patterns, and a primary sequence motif can predict origins of replication in a genome with epigenetic centromeres. MBio. 2014;5(5):e01703–14.
72. Hoggard T, Shor E, Muller CA, Nieduszynski CA, Fox CA. A link between ORC-origin binding mechanisms and origin activation time revealed in budding yeast. PLoS Genet. 2013;9(9):e1003798.
73. Shimada K, Pasero P, Gasser SM. ORC and the intra-S-phase checkpoint: a threshold regulates Rad53p activation in S phase. Genes Dev. 2002;16(24):3236–52.
74. Bleichert F, Botchan MR, Berger JM. Crystal structure of the eukaryotic origin recognition complex. Nature. 2015;519(7543):321–6.
75. Walker SS, Malik AK, Eisenberg S. Analysis of the interactions of functional domains of a nuclear origin of replication from Saccharomyces cerevisiae. Nucleic Acids Res. 1991;19(22):6255–62.
76. Huang RY, Kowalski D. Multiple DNA elements in ARS305 determine replication origin activity in a yeast chromosome. Nucleic Acids Res. 1996;24(5):816–23.
77. Theis JF, Newlon CS. Domain B of ARS307 contains two functional elements and contributes to chromosomal replication origin function. Mol Cell Biol. 1994;14(11):7652–9.
78. Alvino GM, Collingwood D, Murphy JM, Delrow J, Brewer BJ, Raghuraman MK. Replication in hydroxyurea: it's a matter of time. Mol Cell Biol. 2007;27(18):6396–406.
79. Feng W, Bachant J, Collingwood D, Raghuraman MK, Brewer BJ. Centromere replication timing determines different forms of genomic instability in Saccharomyces cerevisiae checkpoint mutants during replication stress. Genetics. 2009;183(4):1249–60.

80. Schneider TD. Information content of individual genetic sequences. J Theor Biol. 1997;189(4):427–41.
81. Schneider TD, Stephens RM. Sequence logos: a new way to display consensus sequences. Nucleic Acids Res. 1990;18(20):6097–100.
82. Dujon B. Yeasts illustrate the molecular mechanisms of eukaryotic genome evolution. Trends Genet. 2006;22(7):375–87.
83. Dujon B. Yeast evolutionary genomics. Nat Rev Genet. 2010;11(7):512–24.
84. Lee W, Tillo D, Bray N, Morse RH, Davis RW, Hughes TR, et al. A high-resolution atlas of nucleosome occupancy in yeast. Nat Genet. 2007;39(10):1235–44.

Chapter 8
Rif1, a Conserved Chromatin Factor Regulating DNA Replication, DNA Repair, and Transcription

Naoko Yoshizawa-Sugata, Satoshi Yamazaki, and Hisao Masai

Abstract Rif1, originally discovered as a telomere binding factor in yeast, is evolutionally conserved and regulates various aspects of chromosome reactions including repair, DNA replication, and transcription in addition to telomere regulation. In mammals, Rif1 suppresses homologous recombination-dependent repair and stimulates non-homologous end-joining repair of double-stranded DNA breaks. Rif1 plays a crucial role in regulating timing of genome replication during S phase. It also affects the transcription profiles. Rif1, composed of the N-terminal HEAT repeat domain and the C-terminal DNA binding/oligomerization domain, tightly binds to chromatin and may facilitate the formation of chromatin domains that may be repressive for initiation of replication as well as for transcription. Rif1 also binds to many factors including protein phosphatase 1, which plays a role in suppression of origin firing. Rif1 is expressed at a high level in mouse embryonic stem (ES) cells and is involved in regulation of differentiation processes.

Keywords Rif1 • DNA replication • DNA repair • Transcription • Chromatin • Telomere

Introduction

DNA replication is spatially and temporally regulated by high-order chromosomal structures to coordinate various chromosome transactions during cell cycle. Cells appear to have tens of thousands of potential DNA replication origins on chromatin to maintain and inherit genome information [1]. At late-M/early-G1 phase and probably throughout G1 phase, origin licensing process permits the formation of pre-RC complex on chromatin, establishing potential replication origins. Once cells enter S phase, replication origins are fired coordinately along the chromosomes so

N. Yoshizawa-Sugata • S. Yamazaki • H. Masai (✉)
Department of Genome Medicine, Tokyo Metropolitan Institute of Medical Science, Tokyo, Japan
e-mail: masai-hs@igakuken.or.jp

D.L. Kaplan (ed.), *The Initiation of DNA Replication in Eukaryotes*,
DOI 10.1007/978-3-319-24696-3_8

that the entire genome is duplicated once and only once. This process is partly under the cells' inherent chromatin program, but choice of fired origins could be stochastic. Generally, replication origins are present in excess and not all the origins are utilized in a given cell or in a given cell cycle. Chromatin conditions including epigenomic features, transcriptional activity, and higher order chromatin configuration may affect the choice of replication origins to be fired.

Recently, a conserved factor, Rif1, was reported to profoundly affect the genome-wide replication timing program both in fission yeast and mammalian cells [2]. Rif1 is a chromatin-bound factor that may define the replication timing domains on the chromosomes. Rif1 was also shown to affect the choice of repair pathways for double-stranded DNA breaks. Rif1, originally discovered as a telomere binding factor, regulates telomere length in yeasts, but roles of Rif1 at telomeres in higher eukaryotes are less obvious. Instead, it may function as a more general chromatin factor that regulates various chromosome reactions including replication, repair, and transcription through organizing local chromatin compartments and recruiting various factors. Here we summarize the potential roles and mechanisms of Rif1 in regulation of various chromosome transactions.

Roles of Rif1 in Normal and Dysfunctional Telomeres

Telomeres, the terminal segments of chromosome in eukaryotes, are composed of unique repetitive DNA sequences. The maintenance of telomere structures is essential for genome stability and the loss of its homeostasis is implicated in aging in mammalian cells. A group of telomere-binding proteins function to guard the chromosome ends [3], and the unprotected chromosome terminus is subject to aberrant recombination reactions that would lead to rearrangement and fusions of chromosomes (Table 8.1).

In budding yeast, Rap1 is a core telomeric double-stranded DNA (dsDNA)-binding protein and its inactivation leads to telomere uncapping and resection, which is mainly executed by Exo1 nuclease. Rif1, together with Rif2, is recruited to telomeres through the C-terminus of Rap1 and also has roles in telomere stability. The deletion of *rif1* causes extensive telomere elongation [4, 5] both in Rap1-dependent and -independent manners [6]. In addition to the Rap1-Rif1-Rif2 complex on dsDNA region, the telomeric single-stranded DNA (ssDNA) overhang is protected by the Cdc13-Stn1-Ten1 (CST) complex. Inactivation of the CST complex causes uncapped telomeres, ultimately leading to the accumulation of extensive tracts of ssDNA and lethality [7, 8]. Rif1 and Rif2 are similar in their functions and protein structures, although there are several differences. The lack of Rif1, but not that of Rif2, caused dramatic reduction in viability of *cdc13-1* or *cdc13-5*ts mutant cells [9, 10]. Rif1 also competes with histone deacetyltransferase Sir3 and Sir4 for the binding to Rap1, and loss of Rif1 causes enhanced silencing in telomeres [5]. Not only the telomere-binding proteins but also telomeric repeat-containing RNA (TERRA), a large noncoding RNA found in animals and fungi, are integral components of

Table 8.1 Comparison of features and Rif1-mediated regulation of telomeres in yeasts and vertebrates [54–58]

Organism		*Saccharomyces cerevisiae*	*Schizosaccharomyces pombe*	Vertebrates
Telomeric repeat	Sequence	$T(G)_{2-3}(TG)_{1-6}$ [54]	GGTTAC [55]	TTAGGG [57]
	Length (nt)	300–350 [58]	200–300 [56]	15,000 [58]
Telomere-binding proteins	Telomeric dsDNA	Rap1-Rif1-Rif2	Taz1-Rap1-Rif1	TRF1;TRF2-Rap1
	ssDNA overhang	Cdc13-Stn1-Ten1	Pot1	Pot1
	Ku proteins	Ku70-Ku80	pku70-pku80	Ku70/Xrcc6-Ku86/Xrcc5
	Linkers of Shelterin	–	Tpz1	TIN2-TPP1
Phenotype of *rif1* depletion in telomere homeostasis		Elongated telomere length [4, 5]	Elongated telomere length [12]	Heterogeneity of telomeres Elongated telomere length, telomere loss, telomere end fusion (Mouse embryonic stem cells) [17] Normal (Mouse embryonic fibroblast cells) [18]

telomeric heterochromatin. Recently, Rif1/2 was reported to affect the transcription of TERRA [11]. At telomeres that contain only X-elements, the Rap1 carboxy-terminal domain recruits the Sir2/3/4 and Rif1/2 complexes to repress transcription, whereas, at telomeres containing Y′ elements, Rap1 represses TERRA mainly through recruitment of Rif1 and Rif2.

In fission yeast, Rap1 also binds to telomeric DNA, although this binding is through Taz1 [12], which serves as the platform of Rif1 binding to telomere. Unlike in budding yeast, the Taz1-Rap1-Rif1 complex on dsDNA and Pot1 (functional analog of CST complex) on ssDNA are connected together by the linker adaptor protein Tpz1 [13], and potentially form a large, stable "Shelterin" complex at the chromosome ends. Deletion of Rif1 leads to moderate elongation of telomere length in a manner dependent on Taz1 [12]. Genome-wide mapping of Rif1-binding sites by ChIP-chip and ChIP-seq revealed extensive accumulation of Rif1 at the telomeric regions, as expected. It also showed that Rif1 binds to chromosome arms in a manner independent of Taz1, while telomere binding depends on Taz1 [14].

In mammalian cells, TRF1 and TRF2 are the platform for assembly of the telomeric dsDNA-binding proteins, and Pot1 binds to ssDNA overhang. These proteins are linked by TIN2/TPP1 to form the Shelterin complex. Although Rap1 is recruited to telomeres through TRF2, mammalian Rif1 does not localize to normal telomeres. When the uncapped and dysfunctional telomeres were induced by overexpression of TRF2 or by that of a mutant telomerase in human cells, Rif1 protein localizes to damaged telomeres in a 53BP1- and ATM-dependent manner [15, 16]. This is likely to be a consequence of DNA damage response rather than reflection of its function in telomere maintenance. However, recently, potential roles of Rif1 in telomere homeostasis were reported. In mouse ES cells, depletion of Rif1 induced the elongation and abnormalities of the telomere structures including telomere heterogeneity, telomere loss, and end fusions [17], although these defects were not detected in Rif1-depleted MEF cells [18]. One of the reasons of telomere defects caused by Rif1 depletion might be due to the up-regulation of Zscan4, a positive regulator of telomere length in mouse ES cells [19], since the codepletion of Zscan4 partially rescues the telomere elongation caused by Rif1 depletion [17]. In Drosophila, Rif1, localized in nuclei, does not bind to telomere, and does not appear to have a role in telomere regulation [20].

Roles of Rif1 in Regulation of DNA Replication Timing in Yeasts

In yeasts, Rif1 has a function in telomere maintenance, as described above. We and other groups have reported that Rif1 is a critical regulator of DNA replication timing in yeasts [14, 21–24]. In the absence of *hsk1* (homologue of Cdc7 kinase), the origin firing efficiency is decreased during S phase, leading to impairment of DNA replication and growth defect. However, genetic analyses and systematic screening for mutants that can suppress the growth defect of *hsk1Δ* cells led to identification of *mrc1*, *cds1*, and *rif1* as mutants capable of bypassing *hsk1* requirement for cell growth [14, 22, 24]. The deletion mutant of *rif1* can restore the growth of *hsk1* deletion, and this may be due to firing of many dormant origins in the absence of *rif1*. Origin firing analyses by BrdU-IP showed that not only origins at telomere (which are normally not fired in early S phase) but also late-firing/dormant origins on the arm regions are activated at early S phase by Rif1 depletion. Interestingly, some of the early-S firing origins are suppressed by loss of Rif1. This suggests a possibility that Rif1 regulates origin firing program both negatively and positively. Alternatively, excess firing at normally late/dormant origins may deplete limiting initiation factors, leading to suppression of some of the early-firing origins.

Taz1 that regulates telomere maintenance in conjunction with Rif1 also regulates the origin firing program in fission yeasts [25]. The origins, suppressed by Taz1, are located at telomere and a part of arm regions. Affected origins are usually fired at late-S phase. Interestingly, Taz1-mediated origin regulation appears to depend on Rif1 functions, although Taz1-binding and Rif1-binding sites do not precisely overlap [21], [59].

Rif1 binds not only at telomere regions but also at many locations on the arm of the chromosomes during G1 phase. Rif1 gradually dissociates from chromatin during S phase. The Rif1-binding sites on the arms tend to be present near late-firing/dormant origins in the intergenic regions, but not precisely overlap with the pre-RC locations [11]. Depletion of Rif1 facilitates the loading of Cdc45 at affected origins, but does not affect the pre-RC formation at G1 phase. More recent studies revealed a part of mechanism underlying Rif1-mediated origin suppression [22–24]. Protein phosphatase 1 (PP1) interaction motifs are conserved in Rif1 proteins from yeast to mammalian cells. PP1 can directly interact with the yeast Rif1 N-terminal segment, and the recruited PP1 dephosphorylates MCM complex to counteract the Cdc7-mediated phosphorylation. During S phase, Cdc7/Dbf4 interacts with the Rif1 C-terminal regions and phosphorylates residues surrounding the PP1-binding motifs, facilitating the dissociation of PP1 from Rif1 for regulated origin firing program.

Pfa4 (palmitoyltransferase) contributes to Rif1 localization at nuclear membrane via enhancing palmitoylation of Rif1 in budding yeast [26]. Deletion of *pfa4* leads to dissociation of Rif1 from nuclear periphery in nuclei. However, *pfa4*Δ did not affect replication timing profiles [21]. This result suggests that anchoring of Rif1 at nuclear periphery may not be essential for control of replication timing, consistent with the report that forced tethering of an origin at nuclear periphery did not change its replication timing [27].

Roles of Rif1 in Regulation of DNA Replication Timing Domains in Mammalian Cells

As stated above, mammalian Rif1 dose not directly regulate telomere maintenance. Recent studies showed two major functions for mammalian Rif1. First, it is a major regulator of DNA replication timing, as reported in yeast [14, 28, 29] (Fig. 8.1). Genome-wide DNA replication timing analyses revealed that depletion of Rif1 results in major changes of DNA replication timing profiles in cancer cells and mouse embryonic fibroblast (MEF) cells. The changes occur throughout the genome, and timing regulation is generally lost and the entire domains are replicated toward the mid-S. Replication timing domains, defined as the segments containing origins coregulated so that they fire simultaneously, switched from early to late and vice verse, by loss of Rif1 and were fragmented into smaller domains [28]. In the absence of Rif1, Cdc7-mediated phosphorylation events (phosphorylation of Mcms) and chromatin loading of replication factors including Cdc45 and PCNA temporally increased at early S phase, reflecting the temporal burst of origin firing in Rif1-depleted cells due to deregulation of normally suppressed origins [29].

Rif1 is tightly bound to nuclease-resistant nuclear structure all through the interphase. Bulk chromatin loop sizes increased in the absence of Rif1, suggesting a possibility that Rif1 may facilitate chromatin loop formation at nuclear matrix-like structures near the nuclear periphery [29].

Rif1-binding sites on mammalian genomes are still unknown. However, immunofluorescence analyses of Rif1 protein showed that some population of Rif1 localizes at nuclear and nucleoli periphery, which coincides with the distribution of mid S-phase replication foci. In the absence of Rif1, cells with mid-S replication foci patterns (strong signals near the nuclear and nucleoli periphery) disappeared and

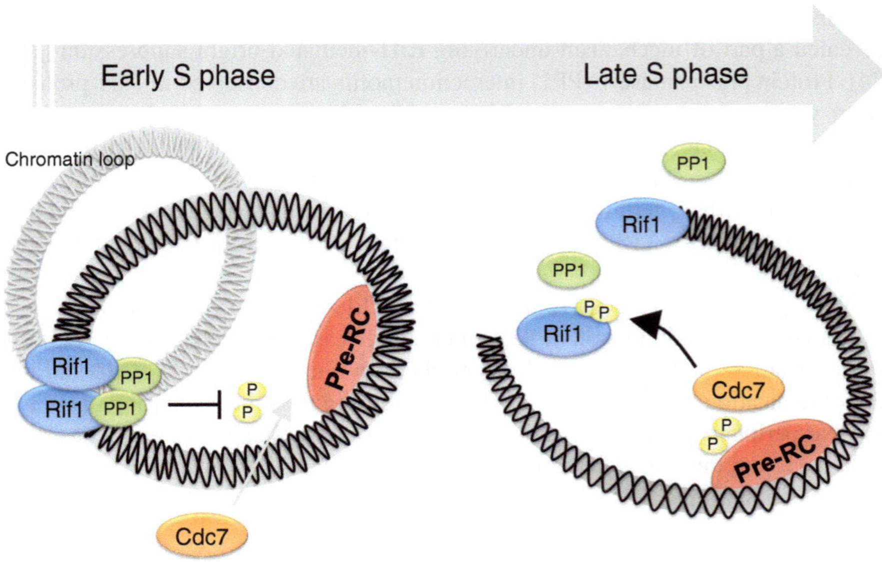

Fig. 8.1 A model of regulation of late replication origins. Oligomeric Rif1 proteins bind to DNA and generate chromatin loop/compartment. The replication origins present in the compartments or close to the Rif1-binding site are inhibited from firing at early S phase by protein phosphatase 1 (PP1) which would counteract the phosphorylation by Cdc7 kinase, essential for initiation. Chromatin loop/compartment formation may facilitate the coregulation of the origins in the same loop/compartment. At late S phase, chromatin loop structures may be disassembled and PP1 dissociates from Rif1 to permit the activation of the suppressed late-firing origins

smaller and dispersed foci, reminiscent of early-S foci, were maintained until mid-S phase. This result suggests that Rif1 may be required for generation of mid-S replication foci structures [29].

During mitosis, Rif1 is strongly phosphorylated and dissociates from chromatin. The dissociated Rif1 localizes to midzone microtubules in early anaphase and then rebinds to chromatin at nuclease-resistant structures [16, 29]. It has been proposed that, during late M/early-G1 phase, chromatin repositioning occurs at timing decision point (TDP) which is coincident with determination of the origin firing program [30]. It is an intriguing possibility that the chromatin binding of Rif1 at late M/early G1 constitutes a part of TDP.

Roles of Rif1 in DNA Double-Strand Break Repair

When DNA damage occurs, cells activate DNA damage response (DDR) to allow activation of checkpoint pathways and to repair broken DNA ends. The second function of mammalian Rif1 is to promote repair of DNA souble-strand break (DSB) by protecting the 5′ end from resection at DSB sites. This would inhibit homologous recombination (HR)-dependent repair and promote non-homologous end-joining (NHEJ)-mediated repair. DSBs arise upon exposure to X-ray or replication fork stall/collapse during S phase.

Recently, Rif1 in mammalian cells was shown to accumulate at DNA lesions to promote 53BP1-dependent NHEJ during G1 phase [31–35]. 53BP1 is a key regulator that dictates the DSB repair pathway choice. During G1 phase, Ku70/80-mediated canonical NHEJ pathway preferentially operates, while 5′-end resection of DSB is generally suppressed. On the other hand, cells in S phase permit the resection of DSBs for HR-dependent repair pathway by BRCA1/CtIP. Rif1 interacts with 53BP1 upon DNA damage or at dysfunctional telomeres. A 53BP1 mutant, in which the S/TQ residues, ATM-dependent phosphorylation sites, in the N-terminal region were changed to AQ, did not permit accumulation of Rif1 at DSB sites, whereas Rif1 can be recruited to DSB sites in the 53BP1 C-terminal BRCT domain mutant. These results indicate that Rif1 is recruited to DNA lesion in a manner dependent on the phosphorylation of ATM target sites in the N-terminal region of 53BP1. In contrast to inhibitory roles of Rif1 against DNA end resection in mammalian cells, it was recently reported that budding yeast Rif1 promotes end resection to facilitate DSB repair by homologous recombination. This activity becomes important especially when resection activities are suboptimal [36].

Rif1, in conjunction with BLM helicase, is important also for the recovery of replication forks [37]. Indeed, Rif1 interacts with BLM through its C-terminal segment and is recruited to stalled replication forks in the presence of HU or aphidicolin. Recently, it was reported that BLM counteracts DSB-derived, CtIP/Mre11-dependent deletions generated by alternative end joining, in a manner epistatic with 53BP1 and Rif1 [38]. The results implicate the BLM-53BP1-Rif1 complex in the third pathway of DNA damage repair.

In Rif1-depleted or knockout cells under ionizing radiation (IR) treatment, 5′-end is resected at DSB sites and BRCA1 accumulates even during G1 phase. Furthermore, RPA phosphorylation is elevated. These were not further enhanced in Rif1/53BP1 double-knockout cells. By contrast, Rif1 damage foci increased during S/G2 phase by depletion of BRCA1 or CtIP. These opposing results indicate that Rif1/53BP1 may counteract BRCA1 in DNA damage repair in G1 phase, while BRCA1 may prevent recruitment of Rif1 to DSB sites to promote HR during S/G2 phase [33, 34]. The formation of Rif1 damage foci requires the N-terminal HEAT repeats, to which phosphorylated 53BP1 binds, whereas deletion of the C-terminal 200 amino acids only partially reduces the foci. In contrast, suppression of BRCA damage foci during G1 requires both N-terminal and C-terminal regions of Rif1. BARD1, a binding partner of BRCA1, also counteracts Rif1 in S phase, possibly through the interaction with HP1γ in the heterochromatin compartment [39] (Fig. 8.2).

In immunogenic reactions in vivo, V(D)J recombination and class switch recombination (CSR) promote genetic rearrangement to increase immune diversity. These events involve repair of DSBs generated by increased activity of recombination-activating gene (RAG) and activation-induced deaminase (AID), respectively. In these events, the canonical NHEJ pathway is activated to join the broken DNA ends. The factors, involved in the NHEJ pathway, are essential for V(D)J recombination

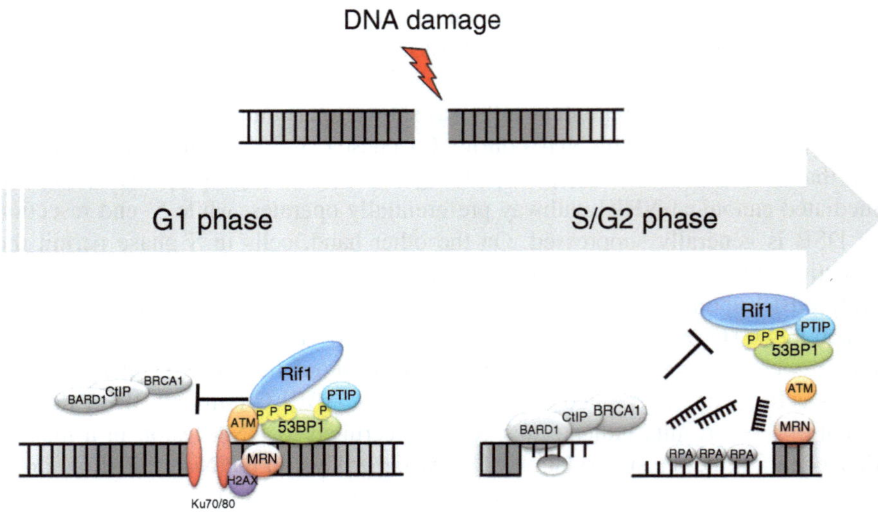

Fig. 8.2 Rif1 accumulates at DNA lesions during the 53BP1-dependent NHEJ repair in G1 phase. In response to DSB, 53BP1 is phosphorylated during G1 phase in its N-terminal segment by ATM kinase. Rif1 is recruited to DSB via phosphorylated 53BP1 and inhibits 5′-end resection to promote the NHEJ repair pathway. In S/G2 phase, in contrast, Rif1 is inhibited from accessing at DSB sites by the BRCA1/BARD1/CtIP complex to facilitate the HR repair pathway

and for CSR. In CSR, the production of IgG/IgA from IgM is reduced by depletion of Rif1, although proliferation and cell cycle progression are not affected [31–33]. This phenotype is similar to that observed in 53BP1-deficient B cells, supporting the notion that Rif1 functions in the same epistasis pathway as 53BP1 in immunoglobulin heavy-chain gene rearrangements through the NHEJ pathway. Similarly, 53BP1 participates in V(D)J recombination that is associated with long-range recombination events in T cells but not in that with short-range recombination [40]. However, long-range V(D)J recombination events that require 53BP1 do occur almost normally in Rif1 knockout T cells (our unpublished data). Thus, Rif1 may be involved differentially in NHEJ for V(D)J recombination and for CSR.

Roles of Rif1 in Mammalian Pluripotent Stem Cells

Rif1 is highly expressed in mouse embryos at the preimplantation developmental stage, in primordial germ cells, and in ES cells [41–43]. Rif1-deficient mice are embryonic lethal at the postimplantation stage in the C57BL6 genetic background [18]. Given that 53BP1- or ATM-deficient mice are viable, Rif1 may have a role in embryogenesis in addition to that in DNA damage response pathway downstream of the ATM-53BP1 pathway. However, truncated Rif1 mutant mice generated by the gene trap system, producing only the N-terminal 778 aa of Rif1 fused to beta-Geo, are viable in the same background, although the frequency of offspring and fertility rate are significantly reduced [18]. Consistent with this finding, ENU mutagenesis mice, truncated at the 1669th aa with a stop codon, are also viable [44]. These observations indicate that the Rif1 N-terminal regions may be sufficient for embryonic development. Intriguingly, Rif1-deficient male mice in the CD-1 genetic background are viable and do not show developmental abnormality, although female mutant mice are lethal [31]. Significant variability in Rif1 dependency during embryogenesis among mouse strains may be caused by the differences of expression levels of Rif1, a repertoire of its splicing variants, and/or posttranslational modification of Rif1 during the early development (our unpublished data).

In mouse ES cells, the pluripotency is maintained by the core circuit of the transcription factors including Oct4, Sox2, Nanog, and Rex1, which activate themselves by a positive feedback. Genome-wide chromatin immunoprecipitation analysis by ChIP-seq of these factors to screen the target genes revealed that Rif1 is one of the top targets of multiple key pluripotent factors [45]. In agreement with this observation, the expression of Rif1 is rapidly downregulated along with other pluripotent markers, when mouse ES cells are differentiated by depletion of LIF and addition of retinoic acid [42]. In spite of obvious correlation between Rif1 expression and stemness, roles of Rif1 in maintenance of pluripotency of ES cells are still unclear. In one report, when Rif1 was depleted in E14 ES cells by siRNA, the undifferentiated state could not be maintained, as indicated by alkaline phosphatase staining [42]. In another report, depletion of Rif1 led to almost no obvious morphology phenotype in the same ES cell line [46], as well as in other cell lines [17, our unpublished data]. The reason for these differences in phenotypes of Rif1-depleted mouse ES cells is unclear. In contrast, Rif1 depletion seems to affect the differentiation process of ES cells both in vivo and in vitro [17, our unpublished data], indicating that Rif1 might be required for proper induction of differentiation from the pluripotent state, rather than self-renewal.

The results of ChIP-seq analysis indicate that ES cell-specific transcription factors, bound at the promoter of *the Rif1* gene, may play a major role in regulation of Rif1 transcription [45]. Recently, transcription regulation of Rif1 by Oct4 and Smad3 in ES cells was reported [47]. Depletion of Smad3 led to reduced recruitment of PRC2 (Polycomb repressive complex 2) at the Rif1 promoter and increased Rif1 expression, while Oct4 is still bound at the Rif1 promoter. PRC2 recruits repressive histone modification complex on the Rif1 promoter [48, 49]. On the contrary, depletion of Oct4 leads to suppression of the Rif1 gene expression in ES cells, but at the same time resulted in reduced recruitment of Smad3. Thus, Oct4 and Smad3 may act together to maintain the expression of Rif1 at a proper level in ES cells. The roles of other ES cell-specific transcription factors bound to the Rif1 promoter need to be clarified in the future.

It was recently reported that Rif1 represses the expression of *zscan4*, *tsctv1*, and other two-cell embryo-specific genes in ES cells [17] (Fig. 8.3). In their report, at least two potential mechanisms were discussed. First, the promoter activity of Zscan4 may be directly suppressed by Rif1. ChIP analyses indicate binding of Rif1 upstream of the *Zscan4* gene. Secondly, Rif1 may somehow increase histone modification for silent chromatin in the regions containing target genes. Indeed, although the overall change of the H3K9me3 level in Rif1-depleted cells seems to be subtle, changes are more evident in the telomeric and centromeric regions. In addition, H3K9 methylase,

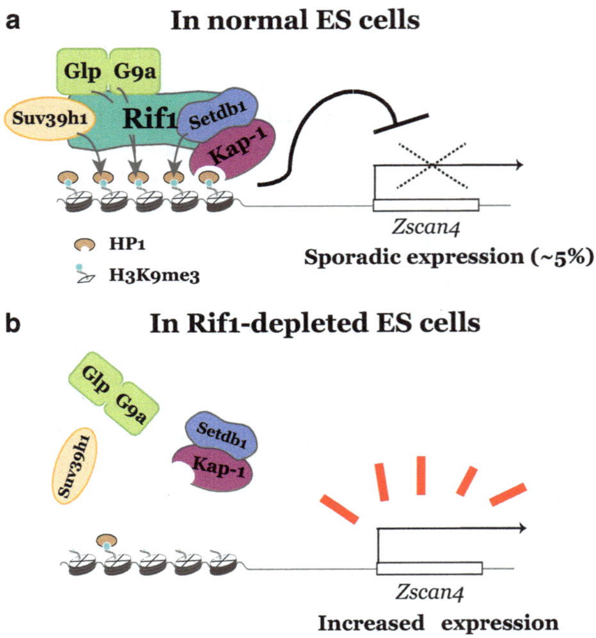

Fig. 8.3 Regulation of transcription of the zscan4 gene by Rif1. (**a**) In normal ES cells, *Zscan4* gene is sporadically expressed in approximate 5 % population with constant fluctuation between ON/OFF states. Rif1 binds to the promoter region of Zscan4 and forms complexes composed of HP1α, Kap-1 (transcription repressor), and histone H3K9 methyltransferase including Glp-G9a heterodimer complex, Setbd1, and Suv39h1. Rif1 may maintain the inactive states by ensuring persistent H3K9 methylation. (**b**) When Rif1 is depleted from ES cells, the H3K9 methyltransferase complex becomes unstable and dissociates from the Zscan4 promoter region, resulting in its increased expression

Glp, was identified as a Rif1-binding protein and its binding to the *Zscan4* gene is substantially reduced by Rif1 knockdown. Interestingly, most of the genes repressed by Rif1 in normal ES cells are also transcriptionally deregulated by depletion of H3K4 demethylase Lsd1 [50]. Although interaction of Rif1 with methylated H3K4 or its catalytic factors has not been reported, histone modifications other than H3K9me might be involved in Rif1-mediated regulation of two-cell specific transcripts.

Possible involvement of Rif1 in cell reprogramming process is intriguing. In an early study, it was reported that the expression of Rif1 in embryogenesis is high during the preimplantation stages, decreased afterwards in late blastocysts and epiblasts, and emerged again in primordial cells at 11.5 dpc [41]. Notably, Rif1 is not abundantly expressed in ICM, and is induced again during ES cell derivation [41]. These results are in keeping with the idea that Rif1 might be required for reprogramming of the differentiated cells into the highly pluripotent states. To examine this hypothesis, the effect of Rif1 depletion on reprogramming process is to be precisely evaluated both in vivo and in vitro.

Structure-Functions of Rif1 Protein

Rif1 protein consists of HEAT/Armadillo-like repeat structures and DNA-binding domain present in the N-terminal and C-terminal regions, respectively. It is predicted that there are 21 HEAT repeats in the N-terminal domain of human and mouse Rif1 [37], and the 133aa residues (from 172 to 304) are particularly well conserved among Rif1 homologs of diverse eukaryotic species and constitute the core conserved region of the Rif1 HEAT repeat [20]. In vertebrate Rif1, there is a long stretch of amino acids between the N-terminal and C-terminal domains that are predicted to constitute intrinsically disordered polypeptide (IDP) structure [51]. The functional significance of the IDP is unknown, although a number of phosphorylation sites have been mapped in this domain. The C-terminal domain of Rif1 can interact with dsDNA with higher affinity than with ssDNA [37]. More recently, it was shown that the DNA-binding domain of mouse Rif1 has higher affinity to structured DNAs including cruciform structures [51]. Direct interaction of Rif1 with DNA through the C-terminal domain may play a pivotal role in chromosome regulation including DNA replication timing and DNA damage repair.

It was also reported that the C-terminal domain of human Rif1 carries sequences capable of forming oligomers [37]. Recently, a crystal structure of the budding yeast Rif1 C-terminal segment (from 1857 to 1916aa), well conserved among eukaryotes, revealed a tetramerization function [52]. The budding yeast Rif1 is anchored to membrane through lipid modification, and this feature could be conserved also in Rif1 from higher eukaryotes [26].

Another conserved feature of Rif1 is the presence of protein phosphatase (PP1)-interaction domain. All the Rif1 homologues have both SILK and RVxF/W motifs [20], the PP1 docking motif [53], with a short stretch of amino acids in between. Functional significance of the PP1-binding sequences has been proven in budding and fission yeasts [22–24].

Rif1 as a General Regulator of Chromosome Dynamics

Rif1 affects DNA replication, DNA repair, and transcription. Involvement of Rif1 in various chromosome reactions, the basic machinery for which is not shared, is intriguing. Rif1 may have many faces that would interact with factors regulating each process. Indeed, Rif1 seems to interact with 53BP1, phosphatase, and BLM through different domains; in DNA damage response, human Rif1 binds to phosphorylated 53BP1 in ATM-dependent manner possibly through N-terminal HEAT repeat domain, which is required for its recruitment to DNA damage sites [33, 35]. Fungi Rif1 has RVxF/W-SILK motifs near the N-terminus, whereas Rif1 from multicellular organisms carries similar motifs in the C-terminal segment [20]. Recruitment of PP1 to Rif1 is a part of the mechanism with which firing of origins is inhibited by Rif1 in yeasts [22–24]. A conserved domain near the C-terminus serves also as the BLM-binding site, enhancing the interaction of Rif1 with BLM-Rmi1-Top3α complex [37].

A portion of Rif1 is associated with chromatin at nuclease-insoluble compartments. Localization of nuclease-insoluble Rif1 at nuclear and nucleoli periphery could be important for establishment of mid-S replication timing domains. This interaction in conjunction with its binding to DNA and oligomerization may contribute to chromatin loop formation, leading to generation of chromatin compartments that may be related to replication timing domains. Furthermore, Rif1 may regulate the locations of chromatin within nuclei through its ability to be tethered to nuclear membrane. Thus, Rif1-mediated regulation of chromatin architecture and nuclear localization could simultaneously and coordinately affect replication, repair, and transcription and potentially other chromosome transactions.

Concluding Remarks and Future Perspectives

Rif1, originally discovered as a telomere binding factor, is conserved from yeasts to human. Although roles of Rif1 in telomere regulation are important in lower eukaryotes, those appear to be less significant in higher eukaryotes. Rif1 is a critical regulator of DNA replication timing, the mechanisms that determine the temporal order of DNA replication. In fission yeast, Rif1 negatively regulates initiation of DNA replication by binding to the vicinity of late-firing/dormant origins. One of the mechanisms of the suppression is recruitment of a phosphatase by Rif1, which may counteract the phosphorylation events required for firing of origins. In mammalian cells, depletion of Rif1 results in genome-wide alteration of replication timing domain structures, similar to the phenotypes in fission yeast. Rif1 also plays important roles in DSB repair, suppressing HR-dependent repair and promoting 53BP1-dependent NHEJ pathway in mammalian cells. Rif1 is recruited to phosphorylated 53BP1, and inhibits resection of DSB ends required for HR-dependent repair. Rif1 also affects transcription profiles. In addition to affecting the expression of isolated genes, Rif1 knockout appears to deregulate clusters of amplified genes encompassing nearly a Mb.

Rif1 is a chromatin factor that appears to be bound at nuclease-resistant nuclear structures. Rif1 affects chromatin loop formation in vivo. In vitro, the C-terminal domain binds to DNA and oligomerizes. Therefore, it may be an interesting possibility that Rif1 may hold together multiple chromatin fibers through its oligomerization activity, generating a chromatin compartment that may be inhibitory (or stimulatory) for various chromosome reactions. This may explain rather long-range effect of Rif1 knockdown in both replication and transcription. Rif1 is abundantly expressed in embryonic stem cells, and its expression level goes down upon differentiation. The level of Rif1 is expected to affect the chromatin networks which would have profound impact on replication timing, DNA repair, or transcription.

The effect of Rif1 on chromatin compartments needs to be experimentally investigated in future both in yeasts and mammals. Future works would also include the clarification of nature of Rif1-DNA interaction, including the precise targets of Rif1 and impact of Rif1 binding on DNA/chromatin structures. In this regard, it should be noted that Rif1 was recently reported to specifically recognize and bind to G-quadruplex structures present in the intergeninc regions on the fission yeast genome [59]. How interaction of Rif1 with G-quadruplex structures generates higher-order chromatin structure would be an interesting issue that needs to be explored in the future. Rif1 could also act as a landing pad for many regulatory proteins including phosphatase and histone modification enzymes. Search of other interacting factors would be needed to clarify the actions of Rif1 in various settings. Rif1 has a unique structure composed of N-terminal HEAT repeats and C-terminal DNA-binding/oligomerization domains. In addition, Rif1 from higher eukaryotes carries IDP segment of more than 1000 aa. Functional contribution of each domain needs to be biochemically analyzed. The roles of the extreme long IDP in functions of Rif1 are of particular interest. Finally, biological roles of Rif1 as an epigenomic regulator during development and differentiation as well as possible involvement in human diseases would be fascinating issues that could be addressed through various mutant or gene-modified model animals.

Acknowledgements We would like to thank members of our laboratory for collaboration and useful discussion.

References

1. Masai H, Matsumoto S, You Z, Yoshizawa-Sugata N, Oda M. Eukaryotic chromosome DNA replication: where, when, and how? Annu Rev Biochem. 2010;79:89–130.
2. Yamazaki S, Hayano M, Masai H. Replication timing regulation of eukaryotic replicons: Rif1 as a global regulator of replication timing. Trends Genet. 2013;29(8):449–60.
3. Smogorzewska A, de Lange T. Regulation of telomerase by telomeric proteins. Annu Rev Biochem. 2004;73:177–208.
4. Hardy CF, Balderes D, Shore D. Dissection of a carboxy-terminal region of the yeast regulatory protein RAP1 with effects on both transcriptional activation and silencing. Mol Cell Biol. 1992;12(3):1209–17.
5. Wotton D, Shore D. A novel Rap1p-interacting factor, Rif2p, cooperates with Rif1p to regulate telomere length in Saccharomyces cerevisiae. Genes Dev. 1997;11(6):748–60.
6. Levy DL, Blackburn EH. Counting of Rif1p and Rif2p on Saccharomyces cerevisiae telomeres regulates telomere length. Mol Cell Biol. 2004;24(24):10857–67.
7. Lydall D, Weinert T. Yeast checkpoint genes in DNA damage processing: implications for repair and arrest. Science. 1995;270(5241):1488–91.
8. Garvik B, Carson M, Hartwell L. Single-stranded DNA arising at telomeres in cdc13 mutants may constitute a specific signal for the RAD9 checkpoint. Mol Cell Biol. 1995;15(11):6128–38.
9. Xue Y, Rushton MD, Maringele L. A novel checkpoint and RPA inhibitory pathway regulated by Rif1. PLoS Genet. 2011;7(12):e1002417.
10. Anbalagan S, Bonetti D, Lucchini G, Longhese MP. Rif1 supports the function of the CST complex in yeast telomere capping. PLoS Genet. 2011;7(3):e1002024.
11. Iglesias N, Redon S, Pfeiffer V, Dees M, Lingner J, Luke B. Subtelomeric repetitive elements determine TERRA regulation by Rap1/Rif and Rap1/Sir complexes in yeast. EMBO Rep. 2011;12(6):587–93.
12. Kanoh J, Ishikawa F. spRap1 and spRif1, recruited to telomeres by Taz1, are essential for telomere function in fission yeast. Curr Biol. 2001;11(20):1624–30.

13. Miyoshi T, Kanoh J, Saito M, Ishikawa F. Fission yeast Pot1-Tpp1 protects telomeres and regulates telomere length. Science. 2008;320(5881):1341–4.
14. Hayano M, Kanoh Y, Matsumoto S, Renard-Guillet C, Shirahige K, Masai H. Rif1 is a global regulator of timing of replication origin firing in fission yeast. Genes Dev. 2012;26(2):137–50.
15. Silverman J, Takai H, Buonomo SB, Eisenhaber F, de Lange T. Human Rif1, ortholog of a yeast telomeric protein, is regulated by ATM and 53BP1 and functions in the S-phase checkpoint. Genes Dev. 2004;18(17):2108–19.
16. Xu L, Blackburn EH. Human Rif1 protein binds aberrant telomeres and aligns along anaphase midzone microtubules. J Cell Biol. 2004;167(5):819–30.
17. Dan J, Liu Y, Liu N, Chiourea M, Okuka M, Wu T, et al. Rif1 maintains telomere length homeostasis of ESCs by mediating heterochromatin silencing. Dev Cell. 2014;29(1):7–19.
18. Buonomo SB, Wu Y, Ferguson D, de Lange T. Mammalian Rif1 contributes to replication stress survival and homology-directed repair. J Cell Biol. 2009;187(3):385–98.
19. Zalzman M, Falco G, Sharova LV, Nishiyama A, Thomas M, Lee SL, et al. Zscan4 regulates telomere elongation and genomic stability in ES cells. Nature. 2010;464(7290):858–63.
20. Sreesankar E, Senthilkumar R, Bharathi V, Mishra RK, Mishra K. Functional diversification of yeast telomere associated protein, Rif1, in higher eukaryotes. BMC Genomics. 2012;13:255.
21. Peace JM, Ter-Zakarian A, Aparicio OM. Rif1 regulates initiation timing of late replication origins throughout the S. cerevisiae genome. PLoS One. 2014;9(5):e98501.
22. Davé A, Cooley C, Garg M, Bianchi A. Protein phosphatase 1 recruitment by Rif1 regulates DNA replication origin firing by counteracting DDK activity. Cell Rep. 2014;7(1):53–61.
23. Hiraga S, Alvino GM, Chang F, Lian HY, Sridhar A, Kubota T, et al. Rif1 controls DNA replication by directing protein phosphatase 1 to reverse Cdc7-mediated phosphorylation of the MCM complex. Genes Dev. 2014;28(4):372–83.
24. Mattarocci S, Shyian M, Lemmens L, Damay P, Altintas DM, Shi T, et al. Rif1 controls DNA replication timing in yeast through the PP1 phosphatase Glc7. Cell Rep. 2014;7(1):62–9.
25. Tazumi A, Fukuura M, Nakato R, Kishimoto A, Takenaka T, Ogawa S, et al. Telomere-binding protein Taz1 controls global replication timing through its localization near late replication origins in fission yeast. Genes Dev. 2012;26(18):2050–62.
26. Park S, Patterson EE, Cobb J, Audhya A, Gartenberg MR, Fox CA. Palmitoylation controls the dynamics of budding-yeast heterochromatin via the telomere-binding protein Rif1. Proc Natl Acad Sci U S A. 2011;108(35):14572–7.
27. Ebrahimi H, Robertson ED, Taddei A, Gasser SM, Donaldson AD, Hiraga S. Early initiation of a replication origin tethered at the nuclear periphery. J Cell Sci. 2010;123(Pt 7):1015–9.
28. Cornacchia D, Dileep V, Quivy JP, Foti R, Tili F, Santarella-Mellwig R, et al. Mouse Rif1 is a key regulator of the replication-timing programme in mammalian cells. EMBO J. 2012;31(18):3678–90.
29. Yamazaki S, Ishii A, Kanoh Y, Oda M, Nishito Y, Masai H. Rif1 regulates the replication timing domains on the human genome. EMBO J. 2012;31(18):3667–77.
30. Gilbert DM. Cell fate transitions and the replication timing decision point. J Cell Biol. 2010;191(5):899–903.
31. Chapman JR, Barral P, Vannier JB, Borel V, Steger M, Tomas-Loba A, et al. RIF1 is essential for 53BP1-dependent nonhomologous end joining and suppression of DNA double-strand break resection. Mol Cell. 2013;49(5):858–71.
32. Di Virgilio M, Callen E, Yamane A, Zhang W, Jankovic M, Gitlin AD, et al. Rif1 prevents resection of DNA breaks and promotes immunoglobulin class switching. Science. 2013;339(6120):711–5.
33. Escribano-Diaz C, Orthwein A, Fradet-Turcotte A, Xing M, Young JT, Tkac J, et al. A cell cycle-dependent regulatory circuit composed of 53BP1-RIF1 and BRCA1-CtIP controls DNA repair pathway choice. Mol Cell. 2013;49(5):872–83.
34. Feng L, Fong KW, Wang J, Wang W, Chen J. RIF1 counteracts BRCA1-mediated end resection during DNA repair. J Biol Chem. 2013;288(16):11135–43.
35. Zimmermann M, Lottersberger F, Buonomo SB, Sfeir A, de Lange T. 53BP1 regulates DSB repair using Rif1 to control 5′ end resection. Science. 2013;339(6120):700–4.

36. Martina M, Bonetti D, Villa M, Lucchini G, Longhese MP. Saccharomyces cerevisiae Rif1 cooperates with MRX-Sae2 in promoting DNA-end resection. EMBO Rep. 2014;15(6):695–704. PMCID: PMC4197880. eng.
37. Xu D, Muniandy P, Leo E, Yin J, Thangavel S, Shen X, et al. Rif1 provides a new DNA-binding interface for the Bloom syndrome complex to maintain normal replication. EMBO J. 2010;29(18):3140–55.
38. Grabarz A, Guirouilh-Barbat J, Barascu A, Pennarun G, Genet D, Rass E, et al. A role for BLM in double-strand break repair pathway choice: prevention of CtIP/Mre11-mediated alternative nonhomologous end-joining. Cell Rep. 2013;5(1):21–8.
39. Wu W, Nishikawa H, Fukuda T, Vittal V, Asano M, Miyoshi Y, et al. Interaction of BARD1 and HP1 is required for BRCA1 retention at sites of DNA damage. Cancer Res. 2015;75(7): 1311–21.
40. Difilippantonio S, Gapud E, Wong N, Huang CY, Mahowald G, Chen HT, et al. 53BP1 facilitates long-range DNA end-joining during V(D)J recombination. Nature. 2008;456(7221):529–33.
41. Adams IR, McLaren A. Identification and characterisation of mRif1: a mouse telomere-associated protein highly expressed in germ cells and embryo-derived pluripotent stem cells. Dev Dyn. 2004;229(4):733–44.
42. Loh YH, Wu Q, Chew JL, Vega VB, Zhang W, Chen X, et al. The Oct4 and Nanog transcription network regulates pluripotency in mouse embryonic stem cells. Nat Genet. 2006;38(4):431–40.
43. Wang J, Rao S, Chu J, Shen X, Levasseur DN, Theunissen TW, et al. A protein interaction network for pluripotency of embryonic stem cells. Nature. 2006;444(7117):364–8.
44. Daxinger L, Harten SK, Oey H, Epp T, Isbel L, Huang E, et al. An ENU mutagenesis screen identifies novel and known genes involved in epigenetic processes in the mouse. Genome Biol. 2013;14(9):R96.
45. Kim J, Chu J, Shen X, Wang J, Orkin SH. An extended transcriptional network for pluripotency of embryonic stem cells. Cell. 2008;132(6):1049–61.
46. Fazzio TG, Huff JT, Panning B. An RNAi screen of chromatin proteins identifies Tip60-p400 as a regulator of embryonic stem cell identity. Cell. 2008;134(1):162–74.
47. Li P, Ma X, Adams IR, Yuan P. A tight control of Rif1 by Oct4 and Smad3 is critical for mouse embryonic stem cell stability. Cell Death Dis. 2015;6:e1588.
48. Margueron R, Reinberg D. The Polycomb complex PRC2 and its mark in life. Nature. 2011;469(7330):343–9. PMCID: PMC3760771. eng.
49. Simon JA, Kingston RE. Occupying chromatin: polycomb mechanisms for getting to genomic targets, stopping transcriptional traffic, and staying put. Mol Cell. 2013;49(5):808–24. PMCID: PMC3628831. eng.
50. Macfarlan TS, Gifford WD, Agarwal S, Driscoll S, Lettieri K, Wang J, et al. Endogenous retroviruses and neighboring genes are coordinately repressed by LSD1/KDM1A. Genes Dev. 2011;25(6):594–607.
51. Sukackaite R, Jensen MR, Mas PJ, Blackledge M, Buonomo SB, Hart DJ. Structural and biophysical characterization of murine rif1 C terminus reveals high specificity for DNA cruciform structures. J Biol Chem. 2014;289(20):13903–11.
52. Shi T, Bunker RD, Mattarocci S, Ribeyre C, Faty M, Gut H, et al. Rif1 and Rif2 shape telomere function and architecture through multivalent Rap1 interactions. Cell. 2013;153(6):1340–53.
53. Moorhead GB, Trinkle-Mulcahy L, Nimick M, De Wever V, Campbell DG, Gourlay R, et al. Displacement affinity chromatography of protein phosphatase one (PP1) complexes. BMC Biochem. 2008;9:28.
54. McEachern MJ, Blackburn EH. A conserved sequence motif within the exceptionally diverse telomeric sequences of budding yeasts. Proc Natl Acad Sci U S A. 1994;91(8):3453–7. PMCID: PMC43595. eng.
55. Sugawara N. DNA sequences at the telomeres of fission yeast *S. pombe* [PhD thesis]. Harvard University; 1989.
56. Joseph I, Lustig AJ. Telomeres in meiotic recombination: the yeast side story. Cell Mol Life Sci. 2007;64(2):125–30.

57. Meyne J, Ratliff RL, Moyzis RK. Conservation of the human telomere sequence (TTAGGG)n among vertebrates. Proc Natl Acad Sci U S A. 1989;86(18):7049–53. PMCID: PMC297991. eng.
58. Dewar JM, Lydall D. Similarities and differences between "uncapped" telomeres and DNA double-strand breaks. Chromosoma. 2012;121(2):117–30.
59. Kanoh Y, Matsumoto S, Fukatsu R, Kakusho N, Kono N, Renard-Guillet C et al. Rif1 binds to G quadruplexes and suppresses replication over long distances. Nat Struct Mol Biol. *in press*.

Chapter 9
The Origin Recognition Complex in the Initiation of DNA Replication

Timothy Hoggard and Catherine A. Fox

Abstract The origin recognition complex (ORC) is the six-subunit heteromeric protein complex that binds directly to the positions on eukaryotic chromosomes where the protein-DNA complex required for the activation of DNA replication origins assembles. This protein-DNA complex, named the MCM complex, contains the replicative helicases and performs the actual unwinding of the parental DNA duplex (a.k.a. origin activation or firing) at the origin. While there is no evidence that ORC is required for origin unwinding directly, it is nevertheless essential for origin activation because of its key role in the MCM complex loading reaction. ORC loads the MCM complex onto chromosomal DNA during G1-phase, thus "licensing" origins for activation in the subsequent S-phase. To perform this role, ORC directly contacts double-stranded DNA, positioning the substrate onto which the MCM complex must load as well as providing a protein interaction surface to guide MCM complex loading. ORC binds and hydrolyzes ATP, and uses these steps to alter its conformation. These conformational changes in turn alter ORC's ability to interact with both DNA and partner proteins to achieve a complete MCM complex loading cycle. Other chapters in this volume focus on the steps of the MCM complex loading reaction (a.k.a. pre-RC formation; origin licensing). While a discussion of the ORC's role in replication initiation naturally requires consideration of this reaction, this chapter emphasizes the current understanding of the structural and functional features of ORC that must impinge on MCM complex loading, and perhaps other as yet undiscovered roles for ORC in DNA replication.

Keywords ORC • Origins • pre-RC • Cell cycle • MCM

T. Hoggard • C.A. Fox (✉)
Department of Biomolecular Chemistry, School of Medicine and Public Health, University of Wisconsin, Madison, WI 53706, USA
e-mail: cfox@wisc.edu

D.L. Kaplan (ed.), *The Initiation of DNA Replication in Eukaryotes*,
DOI 10.1007/978-3-319-24696-3_9

A Brief History

The ORC was identified in *Saccharomyces cerevisiae* (herein referred to as *Sc*ORC) as a stable, six-subunit heteromer that binds to an essential element within budding yeast origin DNA [1]. This element includes a conserved sequence called the ACS, for *A*RS *c*onsensus *s*equence, because it can be found in all chromosomal fragments that can provide for autonomous replication of bacterial plasmids in budding yeast [2, 3]. The ACS is required for yeast origin function in both plasmid and native chromosomal contexts [4–8]. Thus by analogy to the prokaryotic cellular initiator dnaA binding to the *E. coli* chromosomal origin, oriC, and several viral initiators binding to their cognate sites, there was strong reason to hypothesize that the ACS was a prime target for the binding of the eukaryotic initiator. The importance of having a defined, functional sequence linked to origins cannot be overemphasized—in no other eukaryote were origins defined at a level to be useful for this type of direct biochemistry and this fact remains true. The identification of *Sc*ORC as a protein complex that protected the ACS, by a DNaseI footprinting assay, and required it for DNA binding therefore provided compelling evidence that it was, or was strongly linked to, the eukaryotic initiator that functioned directly at the DNA replication origins of the cellular chromosomes of eukaryotes. *Sc*ORC was also intriguing because it bound ATP, indicating another similarity to previously characterized bacterial and viral initiators (reviewed in [9]). Following *Sc*ORC's biochemical discovery, temperature-sensitive mutant alleles of two *Sc*ORC genes, *ScORC2* and *ScORC5*, were isolated in forward genetic screens, allowing for their rapid cloning [10–12]. *ScORC6* was identified and cloned using a one-hybrid protein-DNA interaction assay, and the remaining yeast ORC genes were cloned using degenerate primers designed from protein sequence [13, 14]. Orthologs of each of the six *Sc*ORC genes have since been identified in many other organisms, and sequence and structural information about ORC subunits from archeal and eukaryotic model organisms, as described in the following sections, extend the analogy between ORC and other initiators of DNA replication to the structural level (discussed in several reviews [15–17]).

Primary and Secondary Structural Features of ORC Subunits

The ORC is a complex comprised of six distinct subunits. To understand how these individual subunits interact to form a functional ORC, this section will first consider what is known about some of the key primary and secondary features of the subunits. Five of the six *Sc*ORC subunits, Orc1-5 contain a region with substantial sequence similarities to the defining domain of the AAA+ family of ATP-binding and hydrolysis proteins and these subunits will be discussed together [16, 18]. Orc6, which has a distinct structure, will be discussed separately.

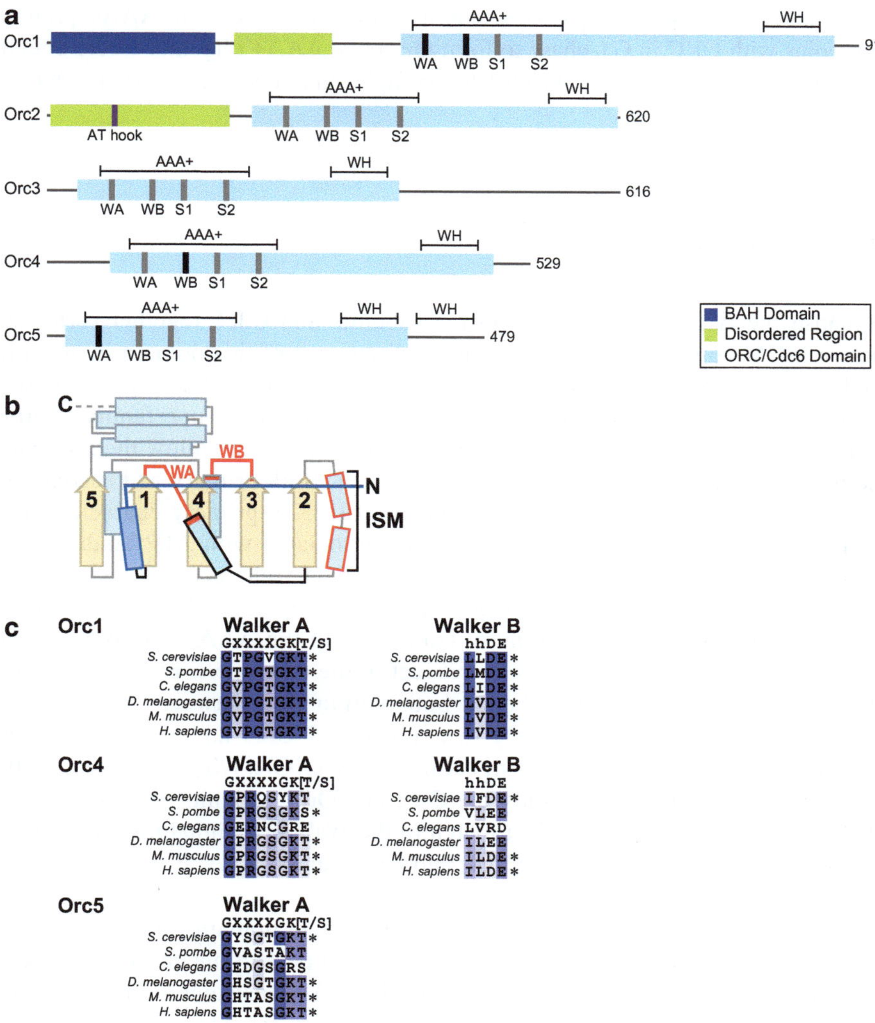

Fig. 9.1 (**a**) Schematics of primary structure of *S. cerevisiae* Orc1-5 subunits adapted from [16]. Perfects matches to Walker A (WA) and B (WB) motifs are colored in *black*. Deviations from consensus motif signatures of the AAA+ domain are colored in *grey*. (**b**) Schematic of the AAA+ secondary fold for the Initiator clade adapted from [19]. WA and WB motif positions indicated in *red*, and the Initiator Specific Motif (ISM) (two similarly sized alpha helices following beta strand 2) is outlined in *red*. (**c**) Alignments of the WA and WB motifs of Orc1, Orc4, and Orc5

The primary structure of the five AAA+ *Sc*ORC subunits is diagrammed in Fig. 9.1a. Each subunit is named based on its relative molecular weight, so Orc1 is the largest subunit and Orc6 is the smallest subunit of ORC. This size differential is generally conserved, though in metazoans Orc3 is similar in size or slightly larger than Orc2, and metazoan Orc6 is considerably smaller than *Sc*Orc6. Orc1-5, and particularly Orc1, share sequence similarity with Cdc6, a seventh AAA+ protein that joins with ORC in G1-phase to form the active AAA+ oligomeric core bound to origin DNA. It is this ORC-Cdc6-DNA complex that functions to load the MCM complex [18].

The AAA+ Domain

Because the AAA+ domains of Orc1-5 subunits and Cdc6 together form the active MCM complex loading machine, a discussion of this domain in general and its variation within individual Orc subunits is warranted. AAA+ (*A*TPases *A*ssociated with diverse cellular *A*ctivities) proteins define a superfamily of proteins that share features of ATP binding and hydrolysis that allow them to participate in a vast array of cellular processes [18]. Both primary and secondary structural characteristics define this domain [19]. Key primary characteristics include the Walker A and Walker B motifs, associated with ATP binding and hydrolysis, respectively (Fig. 9.1). The key secondary characteristic is the core sheet of parallel beta strands arranged in a 51432 topology (Fig. 9.1b). While Orc1-5 share secondary and primary features with the AAA+ family, there is variation among these subunits with respect to the catalytic core motifs within the Walker A and Walker B boxes both within and between species [18] (Fig. 9.1c). In budding yeast, only two subunits, *Sc*Orc1 and *Sc*Orc5, contain perfect matches to the Walker A signature (GXXXXGK[T/S]), consistent with the experimental finding that only the Orc1 and Orc5 subunits of *Sc*ORC bind ATP [20]. In metazoans, a perfect match to the Walker A signature can also be found in Orc4. Metazoan Orc1, Orc4, and Orc5 each bind ATP (Figs. 9.1c and 9.2) [21, 22].

While *Sc*Orc1-5 each contain a recognizable AAA+ domain, and both *Sc*Orc1 and *Sc*Orc5 can bind ATP, only *Sc*Orc1 contains a perfect match to the Walker B signature, hydrophobic-hydrophobic-DE (hhDE). Consistent with this observation, Orc1 is the subunit within *Sc*ORC that hydrolyzes ATP and in fact is the major ATPase of ORC in all eukaryotes examined. Interestingly, Orc2 contains a highly conserved sequence at the predicted position of the Walker A motif, though it does not match the canonical Walker A motif, suggesting that it performs an important but as yet undefined function of ORC [18]. Orc3 is the most diverged of the Orc1-5 subunits [16, 18]. In summary, experimental work, including a recently solved crystal structure of *Drosophila melanogaster* ORC (*Dm*ORC), and sequence conservation provide strong evidence that the AAA+ domains of the Orc1-5 subunits drive the assembly of a functional complex [23]. The fundamental role of ATP binding and hydrolysis by ORC that is essential to MCM complex loading onto chromosomal DNA is performed by the Orc1 subunit.

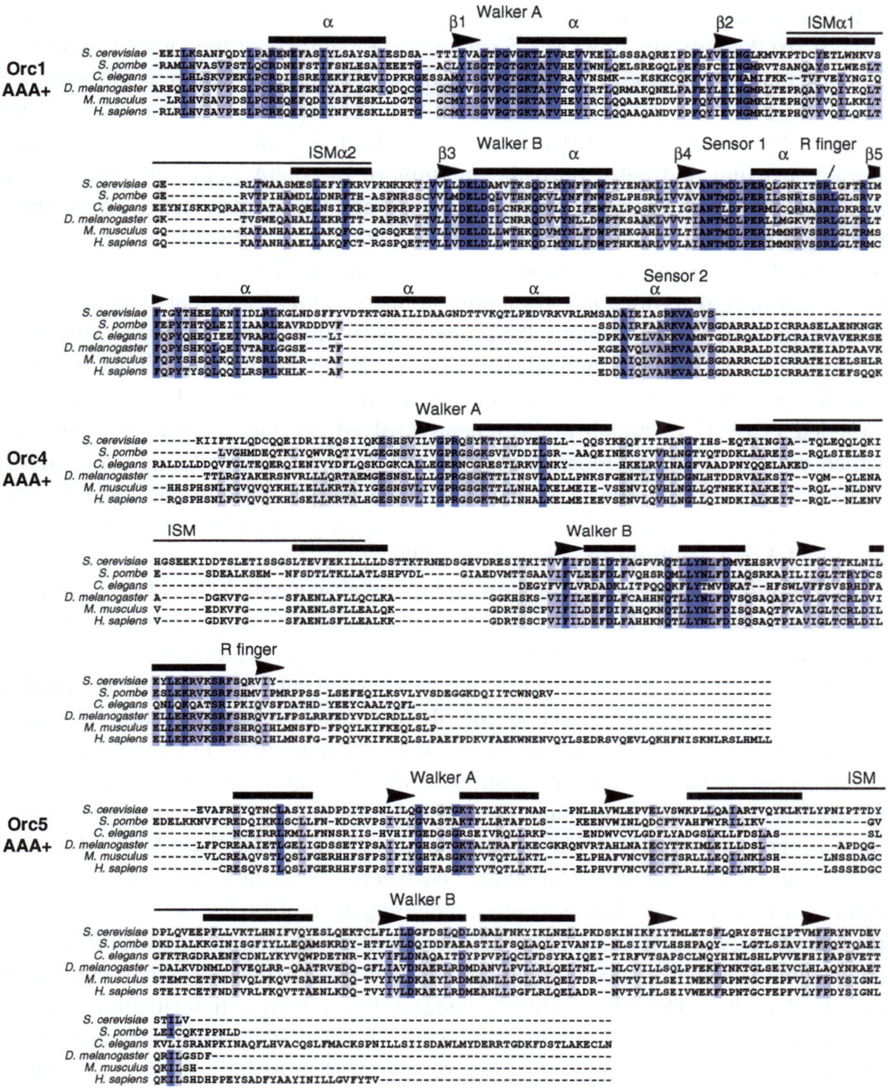

Fig. 9.2 Alignments of the AAA+ domains of Orc1, Orc4, and Orc5: Positions of relative amino acids are indicated. For Orc1 the beta strands are numbered to aid in comparison to Fig. 9.1b

Initiator Specific Motif (ISM) and Winged-Helix (WH) Domain

Fundamental to ORC's catalytic role in loading the MCM complex is its ability to directly contact double-stranded DNA. By binding DNA, ORC serves to position it for MCM complex loading. Two lines of evidence reveal that these ORC-DNA contacts, which do not necessarily play a role in the selection of *specific* chromosomal sites by ORC (see below), are likely performed by two motifs present in the Orc1-5 subunits. One motif is the ISM that is part of the AAA+ fold in these subunits (Fig. 9.1b). Specifically, the Orc1-5 subunits are members of the DnaA/Cdc6/ORC Initiator clade within the AAA+ superfamily whose defining feature is the presence of an extra helical insertion, resulting in two similarly sized alpha helices after strand 2 [9, 10, 19] (Fig. 9.1b). In dnaA, the bacterial initiator, the ISM forms a wedge structure that helps push the subunits of the dnaA homo-oligomer into a spiral that binds single-stranded DNA (ssDNA) and thus contributes to origin DNA melting [9, 24, 25]. In an archeal Orc1 (*S. solfataricus*) the ISM forms a different, more parallel arrangement that contacts double-stranded DNA [26]. It has been proposed that the eukaryotic Orc1-5 subunits contain similar ISMs that adopt a conformation in which each subunit ISM interfaces with the duplex DNA (dsDNA) substrate [9, 26, 27]. This conformation may be adapted for the role of ORC/Cdc6 in stabilizing both the DNA substrate and the MCM helicase for loading onto double-stranded DNA [28]. Notably, the recently solved crystal structure of *Dm*ORC, which will be discussed more in sections that follow, and modeling using the *S. solfataricus* Orc1-1 structure, reveals a central channel through the *Dm*ORC that forms from the ISMs of Orc2-5 and the β-hairpin wings of the WH domains of Orc1 and Orc3-5 that can fully accommodate dsDNA [23].

It is worth noting here that in vitro and in the absence of ATP, *Sc*ORC can bind single-stranded DNA (ssDNA) and does so in a distinct conformation, indicating that there are forms of *Sc*ORC that favor ssDNA binding [29]. However, no biological role for this interaction has been demonstrated and current models do not predict that such an activity is necessary for the MCM complex loading reaction [30–32].

The second motif likely involved in catalytically relevant DNA contacts (i.e., regulation of ATPase activity; regulation of MCM complex loading) is the winged-helix (WH) DNA binding domain. The Orc1-5 subunits contain regions predicted to form winged-helix (WH) DNA binding domains C-terminal to their AAA+ domains (Fig. 9.1a). The WH domains, predicted based on a secondary structural signature, can be found in multiple Orc orthologs, including archeal Orc1/Cdc6 orthologs. For archeal Orc1s, the WH domains participate in extensive interactions with DNA and contribute to sequence-specific binding, although an unexpectedly small number of protein side-chain-DNA-base contacts suggest that context or deformability plays a larger role in origin specificity than sequence [26, 27, 33]. Regardless, these observations led to the obvious hypothesis that the WH domains of eukaryotic ORC subunits contact DNA in a similar manner to archeal Orc1 that forms a lobster claw-like structure, with the ISM and WH domains forming the two halves of the claw that bind DNA. However, the WH domains of eukaryotic ORCs have not been studied in great detail, perhaps because they do not appear to contribute substantially to ORC-origin specificity across species. Interestingly, while the isolated WH domains of human Orc1 (*Hs*Orc1) bind DNA, the WH domain of *Sc*Orc1 does not but instead forms oligomers in solution [34]. It is posited that the WH domain of *Sc*Orc1 may have evolved instead for protein-protein interactions important to the mechanism of transcriptional silencing. After the genome duplication in budding yeast, the *SIR3* gene, an *ORC1* paralog, evolved for dedicated transcriptional silencing, and indeed the Sir3 WH domain is important for both Sir3 oligomerization and silencing in yeast [34, 35]. Nevertheless, these observations indicate that a WH domain of an ORC subunit can function in mediating protein-protein interactions. The recent crystal structure of *Dm*ORC reveals that the WH domains of the Orc1-5 subunits are integral to the stability of the complex because they mediate extensive contacts with the AAA+ domains on adjacent subunits [23]. In fact these contacts rely on the canonical DNA recognition helices of the WH domains interacting extensively with the AAA+ domains and thus made unavailable for DNA binding.

Other DNA Binding Domains in ORC

The ORC-DNA contacts mediated by the ISM and WH domains discussed above are likely conserved and critical for the catalytic role of ORC in loading the MCM complex onto DNA. However, ORC is also important for selecting specific positions on chromosomes to serve as MCM complex loading sites. It is likely that for most eukaryotic ORCs, with the possible exception of budding yeast, that these ORC-DNA contacts are quite distinct from the contacts discussed above that are intimately connected to ORC's mechanism in MCM complex loading. In addition, it appears that these types of contacts are not well conserved. For example, *S. pombe* Orc4 (*Sp*Orc4) contains a distinct N-terminal extension with nine AT-hook motifs important for ORC-origin binding and replication function in vivo, but the binding mediated by the AT-hook appears to be separable from tighter ORC-DNA interactions needed for ORC's catalytic function [36, 37]. *Sc*ORC is the only eukaryotic ORC that shows sequence-specific DNA binding to origins in vitro and this binding, but not nonspecific DNA binding, can regulate the Orc1 ATPase activity [20]. Thus for *Sc*ORC, there appears to exist a connection or overlap between ORC-DNA interactions critical for ORC's selection of chromosomal origins and ORC-DNA interactions with functional roles in MCM complex loading. However even in budding yeast, recent in vitro studies indicate that the specific sequence of yeast origin DNA, that is the conserved ORC binding site, is not intrinsically essential to ORC's ability to establish functional replication origins [38, 39]. Regardless the mechanism by which particular ORC-DNA interactions function "catalytically" in the MCM complex loading reaction and their relationship to ORC-origin interactions required primarily for localizing ORC to appropriate chromosomal regions for this reaction to occur remain open questions. These questions have been raised to a new level of interest by the crystal structure of *Dm*ORC that suggests a stable form of the ATP-bound complex exists in an auto-inhibited form that prevents ORC from engaging with the double-stranded DNA in a catalytically useful way [23]. In particular, perhaps ORC-DNA, or ORC-chromatin contacts important for origin selection by ORC help relieve this auto-inhibited form of ORC such that it can encircle dsDNA.

For metazoans, the most prominent models for origin selection by ORC invoke specific interactions between certain types of chromatin structures and ORC, as described below and later in this review. Thus the prominent model is that origin selection mechanisms in metazoan cells do not involve substantial and direct sequence-specific contacts between ORC and origin DNA. However, a recent study of *Hs*ORC-DNA binding challenges an overly rigid take on this model [40]. In particular, *Hs*ORC, via discrete region of the Orc1 protein N-terminal to the AAA+ domain, shows in vitro binding preference for G-quadruplex (G4)-preferable G-rich ssDNA or RNA. Notably this observation is not the first example of ORC-origin interactions involving nucleic acids other than dsDNA [41–44]. However, what is particularly significant about these biochemical observations is that G4-preferable elements are common features in human DNA replication origins [45–47]. In summary there appear to be multiple acceptable mechanisms for localizing ORC to chromosomal regions in a way that will allow for catalytically relevant ORC-DNA contacts to form subsequently that direct the MCM complex loading reaction. It is possible that these differing selection mechanisms influence ORC dynamics with the DNA in a manner that is relevant to the efficiency of MCM complex loading, as discussed later.

Orc1 Bromo Adjacent Homology (BAH) Domain

Many observations suggest that metazoan ORC relies on specific contacts with chromatin to select and bind to the chromosomal positions that will serve as MCM complex loading sites. The largest subunit of Orc1 contains a BAH domain, a protein module found in many nuclear proteins that likely functions, at least in part, through direct interactions with nucleosomes [48–51]. This domain is defined primarily at the level of secondary structure. While it can be found in most Orc1 orthologs, from yeast to humans, there is limited conservation between yeast and metazoan Orc1BAH domains at the sequence level (Fig. 9.3). Nevertheless accumulating evidence supports a role for the Orc1BAH domain in ORC-chromatin interactions in both metazoans and yeast.

In human cells the Orc1BAH domain contributes to re-binding of Orc1 to chromosomes in G1-phase and the replication of a plasmid that depends on the Epstein-Barr virus oriP [52]. The metazoan Orc1BAH domain binds a histone H3 N-terminal peptide dimethylated on lysine 20 (H3K20me2) [51] (Fig. 9.3). Mutations that cause defects in this interaction reduce ORC binding to an origin in vivo and cause growth defects in a zebrafish model that mimic Meier-Gorlin Syndrome (MGS), a form of human dwarfism associated with defects in several proteins that function in pre-RC assembly [53]. These data provide a direct link between ORC and specific chromatin modifications relevant to ORC-origin binding. More recent studies demonstrate that the MGS-associated substitution in the Orc1BAH domain, R105Q, reduces an Orc1BAH-nucleosome interaction in vitro [54]. All together, these data provide evidence that a specific metazoan Orc1-BAH-nucleosome interaction is important for ORC-chromatin binding and ORC's replication function in metazoans.

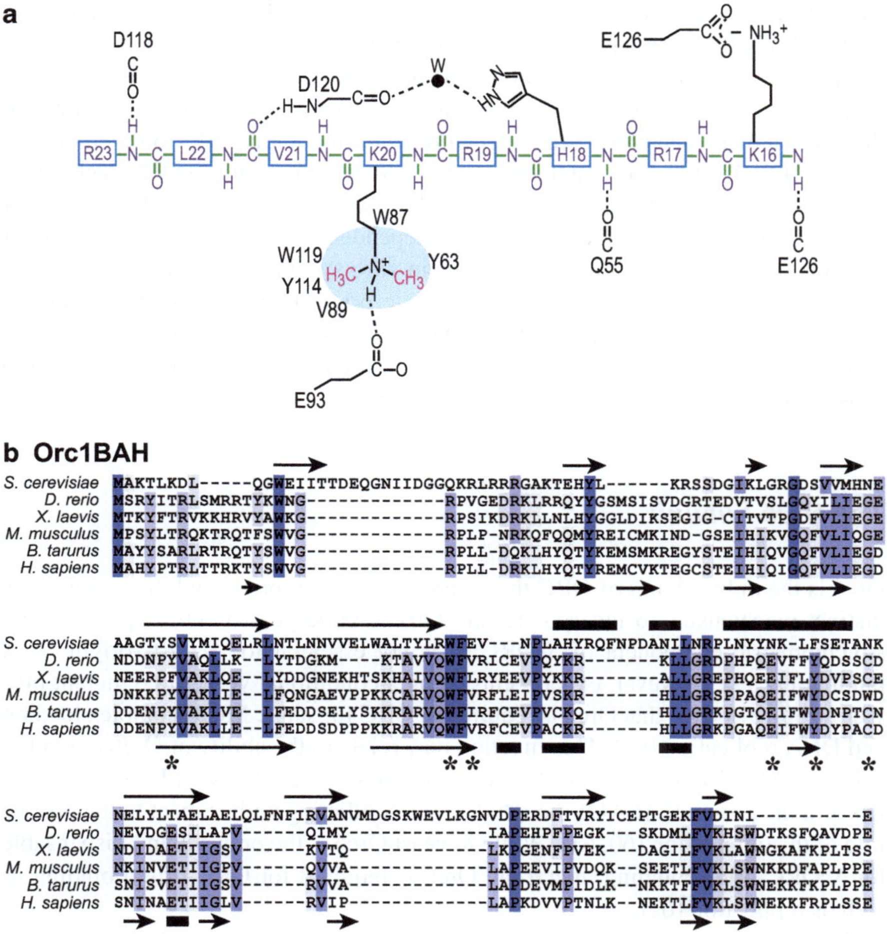

Fig. 9.3 (**a**) Schematic of the H3K20me2 peptide interactions with the Orc1BAH domain adapted from [51]. (**b**) Alignments of the Orc1BAH domains from the indicated species, *asterisks* refer to key residues involved in forming the hydrophobic cage that binds K20me2 on histone H3 peptide

The paradigm for *Sc*ORC-origin binding is that interactions between ORC and specific sequences within yeast origin DNA are key determinants. While this model is generally correct and contributes predictive power with respect to identifying yeast origins, accumulating evidence reveals that ORC-chromatin interactions likely play a positive role in yeast ORC-origin binding just as they do in metazoans [55, 56]. While the yeast Orc1BAH domain does not bind H3K20me2, it does bind chromatin [50]. Given recent structural studies of the similar Sir3BAH domain bound to nucleosomes, it is likely that the yeast Orc1BAH domain binds nucleosomes directly [48]. In addition, while a yeast mutant lacking the Orc1BAH domain (i.e., *ORC-bah*Δ) shows no growth defects, ORC association with chromatin is reduced, and multiple individual origins show substantially reduced ORC binding and defects in activation in vivo [14, 56]. Thus the *Sc*Orc1BAH domain is important for ORC-origin binding, but ORC-origin binding capability must exist in excess of what is required for yeast cell viability. Consistent with this interpretation, viable mutations in genes encoding proteins that participate in pre-RC assembly are synthetically lethal when combined with an *orc1bahΔ* mutation [56]. However, no specific *Sc*Orc1BAH-histone interaction has been reported, and *Sc*ORC lacking an Orc1BAH domain can still bind origins in chromatin in vitro, though this result may indicate that the appropriate *Sc*Orc1BAH-dependent nucleosome configuration or modification is currently missing from the in vitro system and/or that other regions of *Sc*ORC may contact chromatin to facilitate origin binding [56–58]. While the specifics of the molecular interactions may differ, these observations provide evidence that the Orc1BAH domain is a conserved chromatin-binding module from yeast to humans important for ORC-origin binding.

The Orc6 Subunit

Orc6 is the only subunit of ORC that is not an AAA+ protein, and thus it must use distinct mechanisms to incorporate into ORC. While yeast Orc6 co-purifies as a stoichiometric component of *Sc*ORC, it is not essential for *Sc*ORC-origin DNA binding [1, 59]. However, experiments using a conditional yeast Orc6 mutant provide strong evidence that Orc6 is required in ORC for stable MCM complex loading and DNA replication [60]. This function may relate to the ability of Orc6 to bind to Cdt1, an MCM chaperone needed for stable loading of the MCM complex onto DNA in the presence of ATP, though data are conflicting with respect to the precise role of this interaction [60, 61]. While in yeast Orc6 is the only subunit dispensable for *Sc*ORC-origin binding, metazoan Orc6 *is* required for ORC-origin binding as well as replication [62].

In comparison to the AAA+-domain-containing ORC subunits, Orc6 is less conserved between yeast and metazoans [16]. However, a recent study presents revised Orc6 sequence alignments to reveal greater similarity in Orc6 domain structure between yeast and metazoans than previously thought [63]. In addition, experiments show that yeast and metazoan Orc6 actually use a similar mechanism for binding to ORC [41].

Orc6 is conserved among metazoan species and contains three distinct domains. Two of these domains are similar to regions in Transcription Initiation Factor II B (TFIIB, domains A and B, Fig. 9.4a) [62]. The third region of similarity occurs in an otherwise novel C-terminal domain (Orc6 CTD). Importantly, similarity to all three of these regions can also be located within *Sc*Orc6, once a large insertion between the TFIIB domains is correctly accounted for [41]. The region of highest conservation within the Orc6 CTD contains a missense substitution in human Orc6 (Y232S) associated with MGS. The corresponding substitution in *Dm*Orc6 (Y225S) reduces Orc6 binding to Orc1-5 in vitro and MCM loading in vivo [63]. This metazoan Orc6-(Orc1-5) interaction is mediated by an interface between a highly

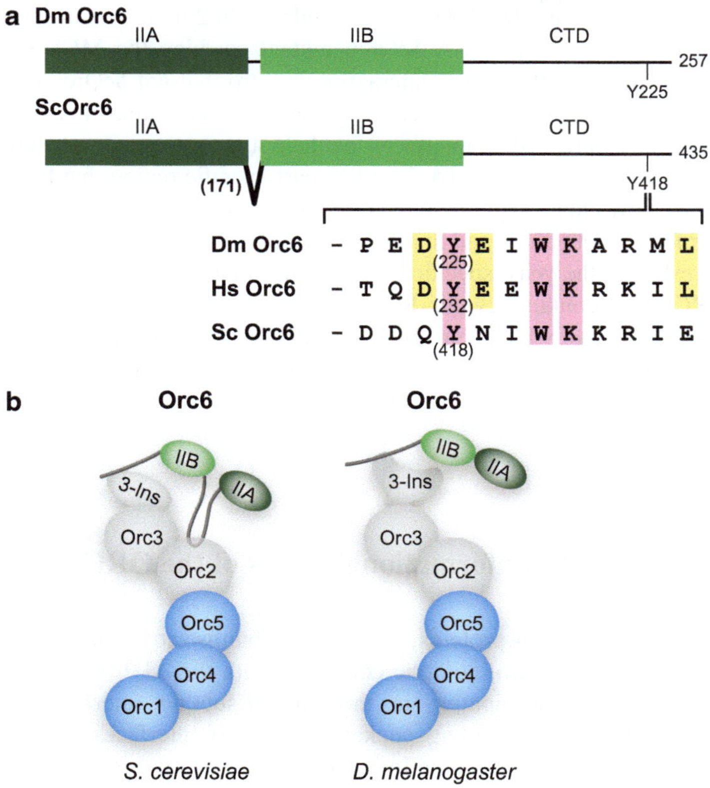

Fig. 9.4 (**a**) Schematics of the primary structure for *Dm*Orc6 and *Sc*Orc6 adapted from [63]. Region of the Orc6 CTDs that contain the "MGS" implicated Y residue and contact Orc3 are highlighted for *Homo sapiens*, *Drosophila melanogaster*, and *Saccharomyces cerevisiae*. (**b**) Cartoon models depicting the basic subunit organization of *Sc*ORC and *Dm*ORC as adapted from [63]. These models are based on the analyses of EM structures as described in the text [63]; a recent crystal structure of *Dm*ORC revised the order of the subunits from Orc1-Orc4-Orc5-Orc2-Orc3 in the AAA+ core to Orc1-Orc4-Orc5-Orc3-Orc2, and shows that ATPγS bound *Dm*ORC forms a ring with an enclosed central channel that could accommodate dsDNA [23]

conserved C-terminal region of Orc6 and a distinct region of Orc3 (Fig. 9.4b). These results align well with data that show that human Orc6 and Orc3 interact [64, 65]. While these data are consistent with the human Orc6 MGS-mutant's (Y232S)) effects on DNA replication causing MGS-associated phenotypes, it must be noted that Orc6, including the Orc6CTD (and other subunits of ORC), has additional non-replicative functions in metazoans that may also contribute to MGS phenotypes [66–74].

*Sc*Orc6's interaction with the Orc1-5 AAA+ core complex appears to involve both yeast-specific features and metazoan-like features. Independent studies provide evidence that *Sc*Orc6 interacts with *Sc*Orc2 or *Sc*Orc3 to incorporate into ORC. In particular the C-terminal region of *Sc*Orc6 binds ORC through an interaction with *Sc*Orc3, but *Sc*Orc6 also binds *Sc*Orc2 under conditions where it cannot bind *Sc*Orc3 [60, 75]. A *Sc*Orc6 mutant carrying the MGS substitution (Y418S) abolishes an Orc6-Orc3 interaction in vitro, but not *Sc*Orc6's interaction with the Orc1-5 AAA+ core [63]. Together these studies suggest that *Sc*Orc6 contains at least two independent interfaces that help it incorporate into *Sc*ORC, a conserved Orc6 CTD that binds *Sc*Orc3 and a yeast-specific region that binds *Sc*Orc2 (Fig. 9.4b). Nevertheless, these studies identify strong similarities between Orc6 and its structure, both alone and in the context of ORC, between yeast and metazoans. Interestingly, the yeast-specific insertion plays an important functional role in yeast ORC, as it contains cyclin-dependent kinase (CDK) sites and an S-cyclin binding motif, both of which help regulate ORC's activities in yeast [76, 77]. While the core structure and mechanical functions of ORC are probably highly conserved from yeast to metazoans, strategies to regulate its activity appear to vary considerably (e.g., [64]).

Architecture and Conformational States of the ORC

As is true for many AAA+ proteins, the Orc1-5 subunits acquire their functionality by coming together in an oligomeric complex, ,with the ATP-binding pockets of individual subunits participating in the subunit-subunit interfaces [19]. The arrangement of the ATP-binding pockets at subunit interfaces plays a role in ATPase activity of AAA+ oligomeric complexes [78]. In particular, X-ray crystal structures of AAA+ oligomers suggest an arginine of one subunit, referred to as an "arginine finger," interacts with the ATP-binding pocket of the neighboring subunit forming interactions with the bound ATP, promoting ATP hydrolysis [78]. *Sc*Orc1 and *Sc*Orc4 follow this paradigm, with the R267 residue of *Sc*Orc4 being essential for ATP hydrolysis by *Sc*Orc1 in ORC [79]. Interestingly, while both EM structures of *Sc*ORC and *Dm*ORC and the recent crystal structure of *Dm*ORC reveal that Orc1 and Orc4 are adjacent subunits, consistent with these experimental findings (Fig. 9.4b), the recent crystal structure of *Dm*ORC indicates that the arginine finger of *Dm*Orc4 is 40 angstroms away from nucleotide binding cleft of *Dm*Orc1. This and additional observations strongly suggest that *Dm*Orc1 must undergo a large conformational shift so that *Dm*Orc4 can trigger its ATPase activity [23]. Such changes could be achieved in part by ORC binding to DNA and/or other proteins that participate in the MCM complex loading reaction, including Cdc6 and MCM itself. In addition the *Dm*ORC crystal structure suggests that ATP binding and hydrolysis by *Dm*Orc4, regulated by an arginine finger from the neighboring subunit of *Dm*Orc5, may execute important *Dm*ORC functions in some organisms, consistent with metazoan Orc4s containing both consensus Walker A and B motifs. Because *Sc*Orc4 lacks a Walker A motif, whatever this role of metazoan Orc4 may be, it is not essential in budding yeast and therefore likely more critical for regulation than core mechanism.

Substantial efforts to define the structures of *Sc*ORC and *Dm*ORC by electron microscopy (EM), and more recently an X-ray crystal structure of *Dm*ORC, have generated images of ORC that provide clues about how it may function in the MCM complex loading reaction [18, 23, 28, 63, 75, 80, 81]. Single-particle reconstructions of negative-stained EM complexes in which the heterologous Maltose Binding Protein is fused to individual Orc subunits allowed the subunit organization within *Sc*ORC and *Dm*ORC to be determined with impressive accuracy, though the recent crystal structure of *Dm*ORC reveals that the subunit arrangement of Orc2-Orc3 is actually Orc3-Orc2 [23, 63, 81] (Fig. 9.4b). Regardless, all together these studies indicate that both the subunit organization and basic shape of *Sc*ORC and *Dm*ORC are similar, as predicted based on the fundamental role ORC plays at DNA replication origins (Fig. 9.4b) [63].

For *Dm*ORC, high levels of ATPγS were required to provide the level of resolution necessary to locate the subunits in the EM images and make a useful comparison to *Sc*ORC, suggesting that the ATP-bound *Dm*ORC forms a more stable and homogenous conformation. Based on EM analysis the ORCs from both organisms form a two-lobed structure in which roughly one half contains the subunits with ATP-binding and/or hydrolysis capabilities in the following order: Orc1-Orc4-Orc5, and the other half contains the remaining subunits in the following order: (Orc2-Orc3)-Orc6 [63] (Fig. 9.4b). The crystal structure of *Dm*ORC shows that ATP-bound ORC reveals a ring structure with a clear central channel for duplex DNA and, as mentioned above, revises the placement of Orc2 and Orc3 within ORC1-5 to Orc1-Orc4-Orc5-Orc3-Orc2 [23]. Interestingly, this structure provides no entry point for duplex DNA, suggesting metazoan ORC must undergo conformational changes even in the ATP-bound state to allow for DNA binding. There are two distinct tiers to *Dm*ORC, with one being formed by both canonical and noncanonical interactions between adjacent AAA+ domains and the other being formed by WH domain interactions between the AAA+ domains of *neighboring* subunits. These interactions lead to the WH domains forming a "collar" that is offset from the AAA+ ring. It is the flexible linker regions between the AAA+ and WH domains on most subunits that allow for these interactions to occur in this way. While there is as yet no reported crystal structure for *Sc*ORC, and potentially important differences may exist between *Dm*ORC and *Sc*ORC, the overall similarity in shape and organization derived from completely independent EM studies is substantial and comforting. In addition, all of these data are consistent with previous examinations of ORC organization as well as subunit-subunit interactions in these and other systems, including human ORC [59, 64, 65, 82, 83].

EM structural studies to address possible conformational changes in ORC have been performed with *Sc*ORC and include examining the effect of relevant *Sc*ORC partners, including Cdc6 and origin DNA and, most recently, a partially assembled putative intermediate in the MCM loading reaction, the OCCM (for ORC-Cdc6-Cdt1-MCM) on ORC structure [18, 28, 75, 81]. In each of these situations ORC's structure is determined in the presence of ATPγS to stabilize intermediates. The ORC-dependent MCM loading machine consists of ORC and Cdc6 together on origin DNA, such that the loading complex contains six related AAA+ protein subunits and Orc6. In yeast, ORCs remain bound to most origins throughout the cell cycle, and Cdc6 joins ORC only in G1-phase when the MCM complex is loaded onto chromosomal DNA [84, 85]. In vitro, Cdc6 preferentially binds ORC when ORC is bound to origin DNA, and upon doing so creates an extended ORC-Cdc6 DNaseI footprint that includes sequences adjacent to the ORC binding site but not required for ORC binding per se [18]. In fact this extended footprint contains an origin element required for MCM complex loading and looks similar to the extended pre-RC (G1-phase) footprint initially defined at yeast origins by in vivo footprinting [85–87]. Together these observations suggest that distinguishing features of the in vivo G1-footprint result from the ORC-Cdc6 complex participating in MCM complex loading, rather than the loaded MCM complex itself. The formation of this ORC-Cdc6-origin DNA complex, including the diagnostic extended DNaseI footprint, requires ATP and a *functional* ORC binding site, that is defined as an ORC site that can support origin function in vivo and not merely ORC binding in vitro [18]. ORC-1A is an ORC containing a defective Orc1 subunit (a K485A substitution in Orc1). ORC-1A is lethal in vivo and does not support formation of the diagnostic ORC-Cdc6-origin DNA complex in vitro, providing evidence that this distinctive ternary complex is functionally relevant to the essential MCM complex loading reaction [18, 20].

A single-particle cryo-EM-derived structure of the *Sc*ORC-origin DNA and *Sc*ORC-Cdc6-origin DNA complexes in the presence of ATPγS indicates that conformational changes in ORC accompany these assemblies [75]. First, ORC binding to origin DNA stabilizes the ORC structure considerably and induces a rotation of the Orc1-Orc4-Orc5 region relative to the rest of ORC. Second, upon Cdc6 binding, additional conformational changes in ORC are observed, including a shift in the Orc1 N-terminal domain, which contains the nucleosome-binding BAH module. This shift may accommodate new Cdc6-Orc1 contacts. Orc6 is also rearranged in the complex upon binding Cdc6 such that it juts out of the central portion of the ring's surface and comes into contact with Orc1, which is supported by additional Orc1-Orc6 interaction data. This change is proposed to position Orc6 to help recruit the MCM complex via direct contacts with MCM-bound Cdt1. Recent studies with purified proteins suggest that this interaction is not intrinsically essential for recruitment of MCM to ORC-Cdc6-DNA, though both Orc6 and Cdt1 are required for a complete reaction that culminates in stable loading of the MCM complex [60, 61]. These major conformational changes to Orc1 and Orc6 within *Sc*ORC observed from the EM studies are consistent with earlier limited protease digestion patterns of ORC compared to ORC-Cdc6 [88]. The bottom line is that *Sc*ORC conformations are altered considerably by nucleotide binding, origin DNA, and the key protein partner Cdc6.

More recently, *Sc*ORC structure by cryo-EM has been examined in a trapped and putative intermediate in the MCM complex loading reaction on origin DNA termed the OCCM for ORC-Cdc6-Cdt1-MCM [28]. The formation of this intermediate is accompanied by additional changes in the conformation of ORC-Cdc6 in which the relatively flat complex changes to a more dome-like shape, with the concave face directed toward the Cdt1-MCM complex. While the initial interpretation of this structure was that the AAA+ domains reach toward the MCM subunits within the Cdt1-MCM complex, the recent model based on the crystal structure of *Dm*ORC suggests that the WH collar on ORC contacts the AAA+ ring on the MCM [23]. Based on the EM structural analysis of the OCCM, the movement of *Sc*Orc6 from the middle of the *Sc*ORC-Cdc6 complex seems to be essential to allow for the binding of the Cdt1-MCM complex and to help promote additional transitions in *Sc*Orc3 and *Sc*Cdc6. Interestingly, one outcome of these transitions is that Orc4 and Orc1 come into closer contact, consistent with the notion that these conformational changes promote Orc4's ability to stimulate ATP hydrolysis by Orc1, a reaction essential for efficient MCM complex loading in vitro as well as cell viability [79]. Together with the crystal structure of *Dm*ORC that shows *Dm*Orc4 too far away from *Dm*Orc1 to stimulate Orc1 ATPase, these data suggest that ORC's catalytically important ATPase activity is highly regulated by steps in the MCM complex loading reaction. Another conformational change is that the ORC-Cdc6 complex forms a right-handed spiral structure within the OCCM that has a 34 angstrom helical rise from bottom (Orc3) to top (Cdc6), matching the helical pitch of B-form dsDNA. Thus, an attractive hypothesis is that these matching structures allow ORC-Cdc6 to hold the DNA substrate for loading of the MCM complex [28].

As an AAA+ initiator protein, the observed conformational changes in ORC in response to ATP binding and interactions with DNA and other proteins would be predicted to drive the mechanical work that culminates in the loading of a replication-competent MCM complex onto DNA. The challenge, of course, is to link these conformational states to discrete biochemical steps in the MCM loading reaction, and currently a strong opportunity for meeting this challenge exists in the yeast system where the MCM complex loading reaction, and most recently the entire origin reaction, from *Sc*ORC binding to origin activation, can be reconstituted with recombinant proteins [61, 89–92]. While it makes sense that the structural changes described to date cause the changes in protein-protein and protein-DNA interaction affinities necessary to drive the G1-phase MCM complex loading cycle, precisely how ORC's conformational states fit into this scheme remains an area of intense research and debate [93, 94]. The recent reports about both *Sc*ORC in the OCCM and the high resolution crystal structure of *Dm*ORC described above raise the possibility that the MCM complex loading step(s) promote ORC-Cdc6 architectural changes that modulate both the ORC and Cdc6 ATPases. Interestingly, recent studies show that the ATP hydrolysis by subunits of MCM complex substrate itself is required for MCM complex loading [89, 90]. Thus, while there may be an intrinsic rate of ATP hydrolysis by ATP-bound ORC-Cdc6 on DNA that could conceivably function in proofreading (i.e., dissolution of any incomplete MCM complexes that may form) as proposed, perhaps another and not mutually exclusive possibility is that a fully loaded MCM complex (i.e., double hexamer of MCMs) signals its "maturation" to its ORC-Cdc6 loading complex and promotes ORC-Cdc6 ATPase activity and release from DNA [61]. In this model, a loaded MCM complex is made competent for S-phase activation (i.e., now released from ORC-Cdc6 and free to move away for the "loading site" on double-stranded DNA) and, at the same time, ORC and Cdc6 are "reset" such that they can now participate in another independent MCM complex loading reaction at the same or another origin [79]. The end result would be a cycle that could continue to load MCM complexes onto chromosomes through G1-phase as long as Cdt1-MCM complexes were available. For a cycle to work efficiently and to load multiple MCM complexes, a "mature" MCM complex would have to move out of the way to allow another loading event. This movement on double-stranded DNA is not hard to envision given the demonstrated passive sliding of the MCM complex on naked DNA in vitro and the dynamic behavior of nucleosomes observed near origins [30, 32, 95]. The bottom line based on several lines of accumulating evidence, in terms of ORC, is that ORC must *bind* and *release* its origin DNA substrate to establish one complete MCM complex loading reaction. Mechanisms that facilitate re-binding of released ORC would promote multiple MCM complex loading reactions in a given G1-phase. The recent models based on the crystal structure of *Dm*ORC raise the possibility that origin binding dynamics could be regulated by the rate that ATP-bound *Dm*ORC is able to alter its conformation to access double-stranded DNA (i.e., the "on" reaction) [23]. Specific interactions between ORC subunits and chromatin structures neighboring origins are suggested as one possible mechanism for achieving such a conformational alteration.

Roles for ATP Binding and Hydrolysis by ORC

The specific roles for ATP binding and hydrolysis events catalyzed by ORC have been studied extensively in yeast using genetic tools, purified proteins and extracts. In one basic approach, a defined mutant form of a *Sc*ORC subunit is generated. The mutant subunit is then expressed in vivo as the sole source of that subunit and/or as a component of an overproduced ORC (i.e., where all subunits are overproduced to make an excess of the desired mutant ORC complex). In addition, the mutant subunit is reconstituted with the remaining *Sc*ORC subunits in an expression system (Baculovirus infected Sf9 cells or a yeast strain optimized for protein expression and purification) to facilitate purification of a defined mutant form of *Sc*ORC. The biochemical functions of the mutant *Sc*ORC are then compared to that of a wild-type *Sc*ORC using a variety of in vitro assays. Similar in vitro expression approaches can be used to study metazoan ORCs in vitro as well, as mentioned below.

Because only Orc1 and Orc5 were shown to bind ATP within *Sc*ORC, an early study addressed their ATP-binding functions in the context of ORC by generating mutant versions of each of these subunits in which their Walker A motifs were altered to abolish their ATP-binding function [20]. Two distinct mutant forms of *Sc*ORC can be assembled that either cannot bind ATP at the Orc1 subunit (*Sc*ORC-1A) or cannot bind ATP at the Orc5 subunit (*Sc*ORC-5A). *Sc*ORC-1A but not *Sc*ORC-5A cannot bind a yeast origin in vitro, indicating that ATP binding by ORC via the Orc1 subunit but not the Orc5 subunit is essential for specific *Sc*ORC-origin binding. In addition an *orc1-A* mutant allele fails to complement an *orc1Δ* mutation in vivo, while an *orc5-A* mutant allele complements an *orc5Δ* mutation.

Interestingly, the biochemical role of ATP binding by *Sc*Orc5 remains unknown, even though ATP binding by this subunit is conserved. A clue comes from a genetic screen to identify mutant alleles that enhance gross chromosomal rearrangements (GCRs) in yeast that isolated *orc5-70,* an allele that encodes a G42R substitution in *Sc*Orc5's Walker A motif [96]. While *orc5-70* causes temperature-sensitive growth defects, it causes no detectable defects in origin activation per se, in contrast to the temperature-sensitive *orc5-1* allele [11, 84, 97]. However, an increase in origin number per chromosome exacerbates the elevated GCR caused by *orc5-70*, suggesting that whatever defect in *Sc*ORC this allele causes is related in some way to *Sc*ORC behavior at origins.

To examine the role of ATP hydrolysis by *Sc*ORC, ,which occurs through the *Sc*Orc1 subunit, a similar approach was taken. One type of allele was identified in *ORC1* through a dominant negative overexpression screen that assessed the effect of mutations in the region encoding the Orc1 Walker B motif predicted to be important for ATP hydrolysis [98]. Two alleles were isolated that affect the same amino acid, *orc1-d1*(D569Y) and *orc1-d2*(D569F). The goal of this screen was to isolate a "clean" defective version of *Sc*ORC in which Orc1 ATPase was nonfunctional but ATP binding was unaffected. While this goal was not achieved—an *Sc*ORC-d1 mutant complex has a 10-fold reduction in the Km for ATP and a 16-fold defect in kcat for ATPase activity—the analyses of these mutants provide important insights into ORC function. Interestingly the *orc1-d* alleles support viability and complement an *orc1Δ* mutation, indicating that a substantial reduction in intrinsic ATPase activity of *Sc*ORC is tolerated in vivo. Consistent with this observation, under saturating levels of ATP, a mutant *Sc*ORC-d1 binds yeast origin DNA indistinguishably from wild-type *Sc*ORC. Thus the fundamental role for ORC in origin binding appears unperturbed in vivo, indicating that MCM complexes must be loaded onto origin DNA by *Sc*ORC-d mutants. However, overproduction of an orc1-d subunit *together with overproduction of each of the other Orc subunits* produces a dominant negative growth phenotype. Thus an excess of ORC-d1 complexes is dominant negative over wild-type ORC. (Overexpression of *orcd-1* alone is *not* dominant negative because an ORC-d1 complex can provide enough ATPase activity to support viability—only when excess ORC-d1 is made, such that it accumulates in the nucleus is a dominant negative effect observed.) However, this dominant negative effect of ORC-d1 can be bypassed by overexpressing CDC6. The interpretation is that ORC-d1, essentially acting as an excess of ORC in the ATP-bound state, is titrating limiting levels of Cdc6 away from any ORC (wild type or ORC-d1) bound to origins. Cdc6 has the highest affinity for the ORC-DNA-ATP complex [18]. Thus this observation is consistent with a model whereby ORC's ATPase is required to release ORC-Cdc6 from DNA and possibly each other, thus recycling the proteins for further rounds of MCM complex loading [79, 99].

While the Cdc6 overexpression rescue of the dominant negative effect of *Sc*ORC-d1 makes sense, an unanswered question is why *Sc*ORC-d1 or *Sc*ORCd2, when expressed at normal levels and as the sole source of ORC, provides for normal levels of cell growth. At least two different models may explain this observation. The first is that, in the context of Cdc6, origin DNA, and MCM-Cdt1, the

ORC-d1 (or -d2)-Cdc6 complex has no substantial defect in whatever level of ATPase activity is required by this complex to participate in a normal MCM complex loading reaction. Related to this point, it must be acknowledged that the requirement and roles of ATP hydrolysis by the ORC-Cdc6 complex for MCM complex loading remain incompletely understood. Indeed, recent reports show that, in a purified system, the Cdc6-ATPase is not required for loading a replication-competent MCM complex in vitro, even though the Cdc6-ATPase is required for viability *and* MCM complex loading in a crude extract [89, 90, 100]. Nevertheless, this first model posits that, in the biologically relevant context, ORC-d1 or ORC-d2 has little to no defect in ORC ATPase activity despite showing weak intrinsic ATPase activities as isolated complexes in vitro. A second possibility is that, while MCM complex loading at origins can still occur in cells in which ORC-d1 or ORC-d2 is the only source of ORC, the MCM loading cycle may be substantially less efficient. However, perhaps in yeast cells, as in mammalian cells, levels of loaded MCM complexes (i.e., numbers of *potential* origins) are in excess of what is needed to complete an unchallenged S-phase [101–103].

A second mutant of version ORC defective in ATPase activity is called ORC-4R to signify that the gene encoding the Orc4 subunit contains a missense mutation in the codon for arginine 267 (i.e., *orc4-R267A*, *orc4-R267K*, *orc4-R267E*), the arginine "finger" discussed above needed for Orc1's ATPase activity [79]. In contrast to the *orc1-d* alleles, *orc4-R* alleles do not support yeast viability. In addition, in vitro ORC-4R complexes show no ORC ATPase activity but can bind ATP similarly to wild-type ORC, indicating Orc1's ability to bind ATP remains intact. As predicted by this result, ORC-4R binds origin DNA similarly to WT ORC both in vivo and in vitro. However, ORC-4R reduces MCM complex loading efficiency in an in vitro MCM complex loading extract in which wild-type ORC is depleted and replaced with recombinant ORC-4R. Interestingly, in vitro, MCM complex loading, as measured by salt-resistant origin DNA-associated MCM, is not abolished by ORC4-R, only reduced. These and other data lead to a model that posits that ATP hydrolysis by ORC is required for multiple rounds of MCM complex loading reactions onto DNA in vitro, but is not essential for loading per se. These data also support a model in which the ORC ATPase activity functions to recycle the ORC-Cdc6 loading machine and, perhaps, also release the loaded MCM complex for "activation competence" in S-phase, as discussed above. With respect to these ideas, it would be useful to know the cell cycle arrest point of orc-4R mutants, and whether ORC and/or MCM are associated with origins at this arrest stage. ORC4-R may remain "stuck" at origins, or it may leave, but load inadequate levels of MCM complexes to support replication. Suppressors of orc-4R lethality might also be informative, though not necessarily trivial (or even possible) to obtain.

In terms of ATP binding and ATPase activity, metazoan ORC, as represented by studies of *Dm*ORC, shows strong similarities to *Sc*ORC even though *Dm*ORC DNA binding shows considerably lower preference for any particular DNA sequence [21, 104, 105]. In particular, both tight DNA binding and ATPase activity of ORC depend on *Dm*Orc1 [21]. Furthermore, chromatin binding by ORC as well as DNA replication in a cell-free extract depends on an intact Orc1, i.e., an Orc1 containing wild type and functional Walker A and Walker B motifs. Analogous defects in Orc4 or Orc5 have only minimal effects. In substantial contrast to *Sc*ORC, however, *Dm*Orc6 is essential for DNA and chromatin binding by ORC. In general, however, there are strong functional similarities between *Dm*ORC and *Sc*ORC in terms of ATP binding and hydrolysis.

Studies of human ORC (*Hs*ORC),), however, suggest that, while mechanistic aspects of Orc1-ATP binding and hydrolysis are conserved once ORC is bound to an origin and presumably participating in the MCM complex loading reaction, differences in terms of regulation of *Hs*ORC are apparent [22, 64, 65]. As with *Sc*ORC and *Dm*ORC, Orc1 is required to support ATP hydrolysis of *Hs*ORC. However, in contrast to *Dm*ORC, recombinant *Hs*ORC requires the Walker A motifs of Orc1, Orc4, and Orc5 to associate with chromatin and support replication in a cell-free extract derived from *Xenopus laevis* [22]. A partial explanation for these somewhat unexpected observations may be provided by the subsequent studies of recombinant *Hs*ORC that show its assembly into a stable complex requires ATP binding by the Orc4 and Orc5 subunits and follows an ordered pathway [64, 65]. In particular, *Hs*Orc4 binds to a stable subcomplex of *Hs*Orc2-3-5 only in the presence of ATP, and intact Walker A motifs in Orc4 and Orc5 are necessary. Moreover, Orc1 association with an Orc2-5 subcomplex also requires ATP, though Orc1 does not interact with Orc4 on its own, regardless of ATP. Thus a model is proposed whereby ATP binding by the Orc4 and Orc5 subunits controls an ordered assembly of *Hs*ORC, with Orc1 joining the complex last in G1-phase to allow ORC to participate in the MCM complex loading reaction. This assembly process may proceed in reverse during S-phase, after an origin is activated, and be important for the S-phase mediated degradation of Orc1 in human cells and the inactivation of ORC activity until the following cell division is complete and the cells re-enter G1-phase [106, 107].

Regulation of ORC Activity

The eukaryotic cell division cycle demands a precise coordination between chromosome duplication in S-phase and chromosome segregation in M-phase. In particular, every chromosome has to be duplicated completely and only once during a normal S-phase or chromosome breakage and aneuploidy can result after M-phase. In terms of DNA replication origins, this rule means that any given region of a chromosome, including a region containing a potential origin, is replicated only once during a given cell cycle, regardless of whether the origin-containing region is replicated from the functional activation of that origin or a replication fork emanating from a neighboring origin. In either situation, the origin must be inactivated and made incapable of firing until cell division is complete and the new daughter cells enter their own S-phases. To achieve this goal, multiple mechanisms exist that inhibit the MCM complex loading reaction during S-phase and promote it only during G1-phase. Thus, the proteins required for the MCM complex loading reaction, including ORC, Cdc6, Cdt1, and MCM itself, are prime targets for these inhibitory mechanisms. While the specific mechanisms vary between organisms, and the pathways emphasized for inhibition vary as well, the ultimate result is the same—the MCM complex loading reaction is strictly confined to G1-phase. A complete discussion of this topic is beyond this chapter's scope, but ORC activity, because of its central role in the MCM complex loading reaction, is one relevant target and mechanisms for its regulation, some of which are mentioned above, are briefly summarized here.

As mentioned above, *Sc*ORC binds origins throughout most of the cell cycle, though recently a class of origins was discovered that lacks robust ORC association in G2-phase [108–110]. Nevertheless, *Sc*ORC does not appear to be regulated in general or substantially at the level of ORC-origin binding during the cell cycle, at least compared to metazoans. Instead, as cells enter S-phase, *Sc*ORC is phosphorylated by S-phase cyclin-dependent kinases (CDKs) on the Orc2 and Orc6 subunits [76]. In addition, one S-phase cyclin, Clb5, binds directly to an RXL motif in Orc6 [111]. While phosphorylation does not appear to reduce ORC-origin binding, it inhibits *Sc*ORC's function in the MCM complex loading reaction [61, 77]. Thus while S-phase CDKs activate targets needed for origin activation during S-phase, they simultaneously phosphorylate and inhibit targets that function in the MCM complex loading reaction.

Evidence from studies of *Dm*ORC suggests both similarities and differences in terms of ORC regulation by CDKs. In particular *Dm*ORC is also a target CDK-dependent phosphorylation, though the target subunits are Orc1 and Orc2. In vitro studies show that *Dm*ORC's ATPase activity but not ATP-binding affinity is reduced by this phosphorylation. Moreover, ATP-dependent ORC-DNA binding is also inhibited, suggesting a mechanism different from that used by *Sc*ORC. Finally, as mentioned above, *Hs*ORC activity in MCM complex loading is regulated by the degradation of Orc1 during S-phase [106]. Thus the assembly of a complete *Hs*ORC functional for MCM complex loading is confined to G1-phase by this mechanism and possibly ATP-dependent ordered assembly of the remainder of the complex as well [64, 65].

Selecting Chromosomal Origins

While ORC binding is not sufficient to establish an origin, it is obviously an essential step. The mechanisms that recruit ORC to specific chromosomal regions and their relationship to the catalytic role of ORC in MCM complex loading remain incompletely understood. For example, while it is reasonable to think that key aspects of the ISM and WH domain contacts with origin DNA have fundamental catalytic roles and therefore are similar at all origins, the mechanisms responsible for localizing ORC to a chromosomal region where it has the opportunity to make such contacts in the first place appear to vary considerably between organisms and even between origins within a single cell. This variability creates substantial challenges to understanding precisely how ORC selects origins across species, and how these selection mechanisms might affect origin activity. However, this variability may also create opportunities for achieving cell-type-specific modulation of ORC function as well as increasing our basic understanding of the relationships between chromatin structure and origin activity.

Because *Sc*ORC shows binding specificity for a conserved sequence element found within yeast origins, origin selection is relatively well understood in *S. cerevisiae*. Indeed, an ORC binding site can be identified in most yeast origins [3]. However, even in *S. cerevisiae*, an ORC binding site is not sufficient to identify a yeast origin. For example, a tight ORC binding site in vitro does not necessarily predict an efficient origin in vivo [57, 112]. In fact, a large number of budding yeast origins actually have surprisingly weak ORC binding sites such that the intrinsic strength of the ORC-DNA interaction at these origins is insufficient to explain the strength of ORC-origin binding in vivo [57]. In addition, a number of highly efficient ORC binding sites can be found in the yeast genome that do not function as origins or even bind ORC in vivo [57]. Thus, while ORC binding site specificity certainly aids in mechanistic studies of yeast origins, there is as yet no clear understanding of ORC-DNA contacts critical for ORC's catalytic roles versus contacts used "simply" to localize or stabilize ORC to a particular chromosomal region. In budding yeast this issue is challenging because the same DNA element that localizes ORC to origins appears to have catalytic functions as well.

Moreover, accumulating evidence in budding yeast suggests that, as in metazoans, ORC accessory factors, in the form of specific chromatin structures and/or other DNA binding proteins, may have direct and positive roles in promoting selective *Sc*ORC-origin binding. In fact, the best predictor of a functional *Sc*ORC binding site in vivo is not the closest sequence match to a consensus ORC site but rather an ORC binding site contained within a defined nucleosome configuration in which the predicted ORC site is relatively free of nucleosomes [55]. As mentioned above, the *Sc*Orc1BAH domain, a putative nucleosome-binding module, is needed for wild-type levels of *Sc*ORC-origin binding at most yeast origins, and origins that are particularly dependent on the Orc1BAH domain for *Sc*ORC binding are enriched for a distinctive local nucleosome configuration that places two nucleosomes closer to the Orc1 subunit side of ORC [56]. Recently, another distinct Orc1BAH-independent group of yeast origins have been identified that likely rely on the presence of additional proteins, including chromatin, to either properly remodel otherwise weak ORC sites or to provide for direct ORC-protein interactions [57]. While the specific ORC-accessory interactions that promote ORC-origin binding at these yeast origins have not been defined, a recent study demonstrates a physical interaction between an evolutionarily conserved transcription factor, a member of the forkhead transcription factor family conserved from yeast to metazoans, named Fkh1 and *Sc*ORC [113]. In addition, forkhead protein binding sites are enriched near many *S. cerevisiae* DNA replication origins and contribute to their normal early activation during S-phase [113, 114]. Therefore, while sequence specificity of *Sc*ORC-double-stranded origin DNA interactions still makes it exceptional among eukaryotes, accumulating evidence makes it clear that, as for ORCs in other organisms, accessory factors, including nucleosomes, play a role in promoting functional ORC-origin binding.

In metazoans, including human cells and excluding the rapidly dividing cells of early embryos, the guiding paradigm for origin selection by ORC is that ORC-origin binding specificity is achieved by ORC accessory factors, including specific modification states of chromatin that directly bind regions of ORC as discussed above (e.g., K3K20me2 binding by Orc1BAH). Consistent with this model, ORC interacting proteins that have a key role in recruiting *Hs*ORC to origins have been identified. HMGA1a, a high-mobility group chromatin-associated protein, interacts directly with *Hs*ORC and, when tethered to a heterologous DNA binding domain, can recruit ORC to a heterologous DNA region and establish a functional origin [115]. Thus the HMGA1a-ORC contacts govern the specificity of ORC-origin binding, but establishing the functional MCM complex loading site has little, if any, requirement for specific ORC-DNA contacts. The interaction between HMGA1a and ORC is robust within AT-rich heterochromatin, consistent with a proposal that such ORC-protein interactions may be particularly critical in difficult-to-replicate heterochromatin, an abundant component of metazoan genomes [116]. ORCA/LRWD1 (ORC associated/leucine rich repeats and WD containing 1) is a more recently described ORC accessory factor with similar roles to HMGA1a in recruiting ORC to DNA [117, 118]. Again, the most abundant co-localization between ORCA and ORC occurs within heterochromatic regions. ORCA also interacts with other regulators of MCM complex loading such as Cdt1. Depletion experiments in embryonic stem cells and human primary cells show that ORCA is required for normal levels of chromatin-associated MCM. A recent review discusses many other ORC accessory factors that have been described in metazoans, several of which appear to function after the ORC-origin binding step and facilitate the MCM complex loading reaction [119]. What is clear from these studies is that metazoan ORC relies on a number of accessory factors to perform its role in establishing origins in complex genomes. The relatively stringent requirement for a specific sequence for ORC-origin binding in the small compact genome of budding yeast may have evolved to insure origins would not be disrupted by or interfere with gene transcription, though even in budding yeast robust gene transcription makes some origins more sensitive to defects in MCM complex loading components compared to others [120]. Regardless, in larger genomes with substantial gene-free regions, there is an increased opportunity for ORC-origin interactions that may have relieved some of the evolutionary pressure for sequence-specific binding by ORC. Despite differences between ORC-origin binding in budding yeast and metazoans, it is worth noting that ORC binding regions in chromatin occur within chromatin structures that appear, at least in terms of local nucleosome organization, similar to the structure described for yeast origins [55, 121–123].

An important question is whether these varied ORC-origin selection mechanisms have any impact on ORC's catalytic role in loading MCM complexes onto chromosomal DNA and/or other steps relevant to origin activation. Interestingly, independent studies in *S. pombe* and *S. cerevisiae* indicate a link between ORC-origin binding dynamics and origin activation time during S-phase, even though no functional role for ORC in origin activation during S-phase is known or proposed [57, 110, 124]. The multiple origins that replicate eukaryotic chromosomes are not activated simultaneously during S-phase, but rather individual origins each activate with a distinct probability and at a characteristic point during S-phase, with some origins activating soon after S-phase begins and others activating later after much of the genome has been duplicated. In *S. pombe*, origins that bind to ORC early after M-phase, presumably because they have a higher affinity for ORC, also show earlier MCM recruitment *and* are activated earlier in the subsequent S-phase compared to origins that bind ORC later [124]. One interpretation is that an origin with high affinity for ORC increases the probability of a successful MCM complex loading event occurring, thus allowing for multiple MCM complexes to be loaded at such an origin in a given G1-phase. Because the MCM complex is the target for limiting S-phase activators, origins that have received more MCM complexes during G1-phase can more effectively compete for these limiting activators and hence activate soon after S-phase begins [125, 126]. In *S. cerevisiae*, a recent study provides a complimentary correlative observation [110, 127, 128]. In particular, while budding yeast origins are thought to be bound by *Sc*ORC throughout the cell cycle, a recent study used a more sensitive measure for ORC binding in vivo than used in prior genome-wide studies to identify a subset of yeast origins that fail to associate with ORC immediately following the completion of replication in G2-phase. The origins that show this delay in ORC binding are enriched for origins that activate relatively late during S-phase. Of course, other models that do not invoke reiterative loading of MCM complexes are reasonable. For example, whatever chromatin structure promotes early ORC binding to an origin may also promote access of that origin to limiting S-phase factors, and the correlation between ORC binding and activation time may merely reflect a "shared response" to the same chromatin cues. However, some experimental evidence argues against this interpretation of the correlative data [124].

Another series of studies focused on *Sc*ORC-origin binding interactions may seem to contradict these findings at first glance, but on closer consideration all of these data may be reconciled [57, 112]. In these studies, yeast origins that bind ORC tightly because they possess an ORC DNA binding site with a high-intrinsic affinity for *Sc*ORC, as measured in vitro, are enriched for later activating origins, a result that is the opposite of that described for *Sp*ORC above. However, it is important to recall that *Sp*ORC selects origins via an AT-hook on the Orc4 subunit acting essentially as an ORC-origin tether. Therefore, the *Sp*ORC catalytic DNA contacts, presumably mediated by the ISM and WH domains conserved in all ORCs, are distinct from the contacts *Sp*ORC uses to select an origin site (more accurately thought of as selecting an MCM complex loading site). Based on the role of accessory ORC factors in ORC-origin binding in metazoans as discussed above, it seems likely that a similar division of labor occurs for metazoan ORC, except that the ORC-origin tethering may be achieved through ORC-protein or ORC-chromatin contacts as opposed to an AT-hook-DNA interaction. In contrast to these situations, *Sc*ORC, at least at these high-affinity DNA-dependent origins, appears to rely on the same DNA element for both origin selection and for ORC's catalytic role in MCM complex loading. Thus for *Sc*ORC an intrinsically tight ORC binding site may inhibit the kind of ORC-DNA dynamics required for efficient MCM complex loading (i.e., efficient binding *and* release). In support of this interpretation, budding yeast origins that also bind *Sc*ORC tightly in vivo but instead show weak intrinsic ORC-origin DNA interactions are enriched for origins that are activated early and efficiently in S-phase. These origins are referred to as "chromatin-dependent" because they use factor(s) extrinsic to the ORC binding site, such as ORC-chromatin contacts, to "tether" ORC to the origin. Thus at these origins *Sc*ORC is positioned to have a dynamic relationship with the DNA that controls its catalytic function, similarly to how the *Sp*Orc4 AT-hook positions *Sp*ORC. The bottom line is that ORC may require a dynamic interaction with its catalytically relevant DNA contacts (i.e., it must bind and release and rebind efficiently) to either maximize the rate of an MCM complex loading cycle in G1-phase, as suggested by biochemical data or, alternatively, "release" a loaded MCM complex for activation during S-phase [99]. An ORC-origin tethering mechanism distinct from catalytically important ORC-DNA interactions may facilitate such dynamics.

While the mechanisms that control origin activation timing are mechanistically interesting and relevant to genome stability and cell differentiation in their own right, several studies now provide evidence that the *number* of MCM complexes loaded onto a chromosome directly reflects the number of *potential* origins on that chromosome, and these potential origins are essential for chromosomal stability under conditions of replicative stress, including the levels of stress encountered during the typical lifespan of a multicellular organism [101–103, 129–133]. Thus specific ORC-origin selection mechanisms may facilitate distinct ORC-DNA dynamics that in turn establish the rate for an MCM complex loading cycle within a particular chromosomal region. An efficient cycle would help establish the "excess" number of potential origins, in the form of loaded and competent MCM complexes, which appears so critical to the stable inheritance of eukaryotic genomes.

References

1. Bell SP, Stillman B. ATP-dependent recognition of eukaryotic origins of DNA replication by a multiprotein complex. Nature. 1992;357(6374):128–34.
2. Stinchcomb DT, Struhl K, Davis RW. Isolation and characterisation of a yeast chromosomal replicator. Nature. 1979;282(5734):39–43.
3. Siow CC, Nieduszynska SR, Muller CA, Nieduszynski CA. OriDB, the DNA replication origin database updated and extended. Nucleic Acids Res. 2012;40(Database issue):D682–6. Pubmed Central PMCID: 3245157.
4. Deshpande AM, Newlon CS. The ARS consensus sequence is required for chromosomal origin function in Saccharomyces cerevisiae. Mol Cell Biol. 1992;12(10):4305–13.
5. Newlon CS, Theis JF. The structure and function of yeast ARS elements. Curr Opin Genet Dev. 1993;3(5):752–8.
6. Marahrens Y, Stillman B. A yeast chromosomal origin of DNA replication defined by multiple functional elements. Science (New York, NY). 1992;255(5046):817–23.
7. Marahrens Y, Stillman B. Replicator dominance in a eukaryotic chromosome. EMBO J. 1994;13(14):3395–400.
8. Brewer BJ, Fangman WL. The localization of replication origins on ARS plasmids in S. cerevisiae. Cell. 1987;51(3):463–71.

9. Duderstadt KE, Berger JM. A structural framework for replication origin opening by AAA+ initiation factors. Curr Opin Struct Biol. 2013;23(1):144–53. Pubmed Central PMCID: 3572333.
10. Foss M, McNally FJ, Laurenson P, Rine J. Origin recognition complex (ORC) in transcriptional silencing and DNA replication in S. cerevisiae. Science (New York, NY). 1993;262(5141):1838–44.
11. Loo S, Fox CA, Rine J, Kobayashi R, Stillman B, Bell S. The origin recognition complex in silencing, cell cycle progression, and DNA replication. Mol Biol Cell. 1995;6(6):741–56.
12. Micklem G, Rowley A, Harwood J, Nasmyth K, Diffley JF. Yeast origin recognition complex is involved in DNA replication and transcriptional silencing. Nature. 1993;366(6450):87–9.
13. Li JJ, Herskowitz I. Isolation of ORC6, a component of the yeast origin recognition complex by a one-hybrid system. Science (New York, NY). 1993;262(5141):1870–4.
14. Bell SP, Mitchell J, Leber J, Kobayashi R, Stillman B. The multidomain structure of Orc1p reveals similarity to regulators of DNA replication and transcriptional silencing. Cell. 1995;83(4):563–8.
15. Bell SP. The origin recognition complex: from simple origins to complex functions. Genes Dev. 2002;16(6):659–72.
16. Duncker BP, Chesnokov IN, McConkey BJ. The origin recognition complex protein family. Genome Biol. 2009;10(3):214. Pubmed Central PMCID: 2690993.
17. Li H, Stillman B. The origin recognition complex: a biochemical and structural view. Subcell Biochem. 2012;62:37–58. Pubmed Central PMCID: 3779782.
18. Speck C, Chen Z, Li H, Stillman B. ATPase-dependent cooperative binding of ORC and Cdc6 to origin DNA. Nat Struct Mol Biol. 2005;12(11):965–71.
19. Iyer LM, Leipe DD, Koonin EV, Aravind L. Evolutionary history and higher order classification of AAA+ ATPases. J Struct Biol. 2004;146(1-2):11–31.
20. Klemm RD, Austin RJ, Bell SP. Coordinate binding of ATP and origin DNA regulates the ATPase activity of the origin recognition complex. Cell. 1997;88(4):493–502.
21. Chesnokov I, Remus D, Botchan M. Functional analysis of mutant and wild-type Drosophila origin recognition complex. Proc Natl Acad Sci U S A. 2001;98(21):11997–2002. Pubmed Central PMCID: 59756.
22. Giordano-Coltart J, Ying CY, Gautier J, Hurwitz J. Studies of the properties of human origin recognition complex and its Walker A motif mutants. Proc Natl Acad Sci U S A. 2005;102(1):69–74. Pubmed Central PMCID: 544074.
23. Bleichert F, Botchan MR, Berger JM. Crystal structure of the eukaryotic origin recognition complex. Nature. 2015;519(7543):321–6. Pubmed Central PMCID: 4368505.
24. Duderstadt KE, Chuang K, Berger JM. DNA stretching by bacterial initiators promotes replication origin opening. Nature. 2011;478(7368):209–13. Pubmed Central PMCID: 3192921.
25. Erzberger JP, Mott ML, Berger JM. Structural basis for ATP-dependent DnaA assembly and replication-origin remodeling. Nat Struct Mol Biol. 2006;13(8):676–83.
26. Dueber EL, Corn JE, Bell SD, Berger JM. Replication origin recognition and deformation by a heterodimeric archaeal Orc1 complex. Science (New York NY). 2007;317(5842):1210–3.
27. Gaudier M, Schuwirth BS, Westcott SL, Wigley DB. Structural basis of DNA replication origin recognition by an ORC protein. Science (New York NY). 2007;317(5842):1213–6.
28. Sun J, Evrin C, Samel SA, Fernandez-Cid A, Riera A, Kawakami H, et al. Cryo-EM structure of a helicase loading intermediate containing ORC-Cdc6-Cdt1-MCM2-7 bound to DNA. Nat Struct Mol Biol. 2013;20(8):944–51. Pubmed Central PMCID: 3735830.
29. Lee DG, Makhov AM, Klemm RD, Griffith JD, Bell SP. Regulation of origin recognition complex conformation and ATPase activity: differential effects of single-stranded and double-stranded DNA binding. EMBO J. 2000;19(17):4774–82.
30. Remus D, Beuron F, Tolun G, Griffith JD, Morris EP, Diffley JF. Concerted loading of Mcm2-7 double hexamers around DNA during DNA replication origin licensing. Cell. 2009;139(4):719–30.
31. Remus D, Diffley JF. Eukaryotic DNA replication control: lock and load, then fire. Curr Opin Cell Biol. 2009;21(6):771–7.

32. Evrin C, Clarke P, Zech J, Lurz R, Sun J, Uhle S, et al. A double-hexameric MCM2-7 complex is loaded onto origin DNA during licensing of eukaryotic DNA replication. Proc Natl Acad Sci U S A. 2009;106(48):20240–5.
33. Wigley DB. ORC proteins: marking the start. Curr Opin Struct Biol. 2009;19(1):72–8.
34. Oppikofer M, Kueng S, Keusch JJ, Hassler M, Ladurner AG, Gut H, et al. Dimerization of Sir3 via its C-terminal winged helix domain is essential for yeast heterochromatin formation. EMBO J. 2013;32(3):437–49. Pubmed Central PMCID: 3567499.
35. Hickman MA, Rusche LN. Transcriptional silencing functions of the yeast protein Orc1/Sir3 subfunctionalized after gene duplication. Proc Natl Acad Sci U S A. 2010;107(45):19384–9. Pubmed Central PMCID: 2984160.
36. Houchens CR, Lu W, Chuang RY, Frattini MG, Fuller A, Simancek P, et al. Multiple mechanisms contribute to Schizosaccharomyces pombe origin recognition complex-DNA interactions. J Biol Chem. 2008;283(44):30216–24. Pubmed Central PMCID: 2573090.
37. Chuang RY, Chretien L, Dai J, Kelly TJ. Purification and characterization of the Schizosaccharomyces pombe origin recognition complex: interaction with origin DNA and Cdc18 protein. J Biol Chem. 2002;277(19):16920–7.
38. On KF, Beuron F, Frith D, Snijders AP, Morris EP, Diffley JF. Prereplicative complexes assembled in vitro support origin-dependent and independent DNA replication. EMBO J. 2014;33(6):605–20. Pubmed Central PMCID: 3989654.
39. Gros J, Devbhandari S, Remus D. Origin plasticity during budding yeast DNA replication in vitro. EMBO J. 2014;33(6):621–36. Pubmed Central PMCID: 3989655.
40. Hoshina S, Yura K, Teranishi H, Kiyasu N, Tominaga A, Kadoma H, et al. Human origin recognition complex binds preferentially to G-quadruplex-preferable RNA and single-stranded DNA. J Biol Chem. 2013;288(42):30161–71. Pubmed Central PMCID: 3798484.
41. Norseen J, Thomae A, Sridharan V, Aiyar A, Schepers A, Lieberman PM. RNA-dependent recruitment of the origin recognition complex. EMBO J. 2008;27(22):3024–35. Pubmed Central PMCID: 2585170.
42. Mohammad MM, Donti TR, Sebastian Yakisich J, Smith AG, Kapler GM. Tetrahymena ORC contains a ribosomal RNA fragment that participates in rDNA origin recognition. EMBO J. 2007;26(24):5048–60. Pubmed Central PMCID: 2140106.
43. Norseen J, Johnson FB, Lieberman PM. Role for G-quadruplex RNA binding by Epstein-Barr virus nuclear antigen 1 in DNA replication and metaphase chromosome attachment. J Virol. 2009;83(20):10336–46. Pubmed Central PMCID: 2753104.
44. Deng Z, Norseen J, Wiedmer A, Riethman H, Lieberman PM. TERRA RNA binding to TRF2 facilitates heterochromatin formation and ORC recruitment at telomeres. Mol Cell. 2009;35(4):403–13. Pubmed Central PMCID: 2749977.
45. Besnard E, Babled A, Lapasset L, Milhavet O, Parrinello H, Dantec C, et al. Unraveling cell type-specific and reprogrammable human replication origin signatures associated with G-quadruplex consensus motifs. Nat Struct Mol Biol. 2012;19(8):837–44.
46. Cayrou C, Coulombe P, Puy A, Rialle S, Kaplan N, Segal E, et al. New insights into replication origin characteristics in metazoans. Cell Cycle. 2012;11(4):658–67. Pubmed Central PMCID: 3318102.
47. Cayrou C, Coulombe P, Vigneron A, Stanojcic S, Ganier O, Peiffer I, et al. Genome-scale analysis of metazoan replication origins reveals their organization in specific but flexible sites defined by conserved features. Genome Res. 2011;21(9):1438–49. Pubmed Central PMCID: 3166829.
48. Armache KJ, Garlick JD, Canzio D, Narlikar GJ, Kingston RE. Structural basis of silencing: Sir3 BAH domain in complex with a nucleosome at 3.0 A resolution. Science (New York, NY). 2011;334(6058):977–82.
49. Callebaut I, Courvalin JC, Mornon JP. The BAH (bromo-adjacent homology) domain: a link between DNA methylation, replication and transcriptional regulation. FEBS Lett. 1999;446(1):189–93.
50. Onishi M, Liou GG, Buchberger JR, Walz T, Moazed D. Role of the conserved Sir3-BAH domain in nucleosome binding and silent chromatin assembly. Mol Cell. 2007;28(6):1015–28.

51. Kuo AJ, Song J, Cheung P, Ishibe-Murakami S, Yamazoe S, Chen JK, et al. The BAH domain of ORC1 links H4K20me2 to DNA replication licensing and Meier-Gorlin syndrome. Nature. 2012;484(7392):115–9.
52. Noguchi K, Vassilev A, Ghosh S, Yates JL, DePamphilis ML. The BAH domain facilitates the ability of human Orc1 protein to activate replication origins in vivo. EMBO J. 2006;25(22):5372–82.
53. Bicknell LS, Walker S, Klingseisen A, Stiff T, Leitch A, Kerzendorfer C, et al. Mutations in ORC1, encoding the largest subunit of the origin recognition complex, cause microcephalic primordial dwarfism resembling Meier-Gorlin syndrome. Nat Genet. 2011;43(4):350–5.
54. Zhang W, Sankaran S, Gozani O, Song J. A Meier-Gorlin Syndrome mutation impairs the ORC1-nucleosome association. ACS Chem Biol. 2015;10:1176–80.
55. Eaton ML, Galani K, Kang S, Bell SP, MacAlpine DM. Conserved nucleosome positioning defines replication origins. Genes Dev. 2010;24(8):748–53. Pubmed Central PMCID: 2854390.
56. Muller P, Park S, Shor E, Huebert DJ, Warren CL, Ansari AZ, et al. The conserved bromo-adjacent homology domain of yeast Orc1 functions in the selection of DNA replication origins within chromatin. Genes Dev. 2010;24(13):1418–33.
57. Hoggard T, Shor E, Muller CA, Nieduszynski CA, Fox CA. A link between ORC-origin binding mechanisms and origin activation time revealed in budding yeast. PLoS Genet. 2013;9(9), e1003798. Pubmed Central PMCID: 3772097.
58. Hizume K, Yagura M, Araki H. Concerted interaction between origin recognition complex (ORC), nucleosomes and replication origin DNA ensures stable ORC-origin binding. Genes Cells. 2013;18(9):764–79.
59. Lee DG, Bell SP. Architecture of the yeast origin recognition complex bound to origins of DNA replication. Mol Cell Biol. 1997;17(12):7159–68.
60. Chen S, de Vries MA, Bell SP. Orc6 is required for dynamic recruitment of Cdt1 during repeated Mcm2-7 loading. Genes Dev. 2007;21(22):2897–907.
61. Frigola J, Remus D, Mehanna A, Diffley JF. ATPase-dependent quality control of DNA replication origin licensing. Nature. 2013;495(7441):339–43.
62. Liu S, Balasov M, Wang H, Wu L, Chesnokov IN, Liu Y. Structural analysis of human Orc6 protein reveals a homology with transcription factor TFIIB. Proc Natl Acad Sci U S A. 2011;108(18):7373–8. Pubmed Central PMCID: 3088625.
63. Bleichert F, Balasov M, Chesnokov I, Nogales E, Botchan MR, Berger JM. A Meier-Gorlin syndrome mutation in a conserved C-terminal helix of Orc6 impedes origin recognition complex formation. Elife. 2013;2, e00882. Pubmed Central PMCID: 3791464.
64. Siddiqui K, Stillman B. ATP-dependent assembly of the human origin recognition complex. J Biol Chem. 2007;282(44):32370–83.
65. Ranjan A, Gossen M. A structural role for ATP in the formation and stability of the human origin recognition complex. Proc Natl Acad Sci U S A. 2006;103(13):4864–9. Pubmed Central PMCID: 1458761.
66. Hossain M, Stillman B. Meier-Gorlin syndrome mutations disrupt an Orc1 CDK inhibitory domain and cause centrosome reduplication. Genes Dev. 2012;26(16):1797–810. Pubmed Central PMCID: 3426759.
67. Stiff T, Alagoz M, Alcantara D, Outwin E, Brunner HG, Bongers EM, et al. Deficiency in origin licensing proteins impairs cilia formation: implications for the aetiology of Meier-Gorlin syndrome. PLoS Genet. 2013;9(3), e1003360. Pubmed Central PMCID: 3597520.
68. Prasanth SG, Prasanth KV, Siddiqui K, Spector DL, Stillman B. Human Orc2 localizes to centrosomes, centromeres and heterochromatin during chromosome inheritance. EMBO J. 2004;23(13):2651–63. Pubmed Central PMCID: 449767.
69. Prasanth SG, Prasanth KV, Stillman B. Orc6 involved in DNA replication, chromosome segregation, and cytokinesis. Science (New York NY). 2002;297(5583):1026–31.
70. Hemerly AS, Prasanth SG, Siddiqui K, Stillman B. Orc1 controls centriole and centrosome copy number in human cells. Science (New York, NY). 2009;323(5915):789–93.
71. Chesnokov IN, Chesnokova ON, Botchan M. A cytokinetic function of Drosophila ORC6 protein resides in a domain distinct from its replication activity. Proc Natl Acad Sci U S A. 2003;100(16):9150–5. Pubmed Central PMCID: 170887.

72. Balasov M, Huijbregts RP, Chesnokov I. Functional analysis of an Orc6 mutant in Drosophila. Proc Natl Acad Sci U S A. 2009;106(26):10672–7. Pubmed Central PMCID: 2705598.
73. Akhmetova K, Balasov M, Huijbregts RP, Chesnokov I. Functional insight into the role of Orc6 in septin complex filament formation in Drosophila. Mol Biol Cell. 2015;26(1):15–28. Pubmed Central PMCID: 4279225.
74. Bernal JA, Venkitaraman AR. A vertebrate N-end rule degron reveals that Orc6 is required in mitosis for daughter cell abscission. J Cell Biol. 2011;192(6):969–78. Pubmed Central PMCID: 3063140.
75. Sun J, Kawakami H, Zech J, Speck C, Stillman B, Li H. Cdc6-induced conformational changes in ORC bound to origin DNA revealed by cryo-electron microscopy. Structure. 2012;20(3):534–44. Pubmed Central PMCID: 3299985.
76. Nguyen VQ, Co C, Li JJ. Cyclin-dependent kinases prevent DNA re-replication through multiple mechanisms. Nature. 2001;411(6841):1068–73.
77. Chen S, Bell SP. CDK prevents Mcm2-7 helicase loading by inhibiting Cdt1 interaction with Orc6. Genes Dev. 2011;25(4):363–72. Pubmed Central PMCID: 3042159.
78. Ogura T, Whiteheart SW, Wilkinson AJ. Conserved arginine residues implicated in ATP hydrolysis, nucleotide-sensing, and inter-subunit interactions in AAA and AAA+ ATPases. J Struct Biol. 2004;146(1-2):106–12.
79. Bowers JL, Randell JC, Chen S, Bell SP. ATP hydrolysis by ORC catalyzes reiterative Mcm2-7 assembly at a defined origin of replication. Mol Cell. 2004;16(6):967–78.
80. Clarey MG, Erzberger JP, Grob P, Leschziner AE, Berger JM, Nogales E, et al. Nucleotide-dependent conformational changes in the DnaA-like core of the origin recognition complex. Nat Struct Mol Biol. 2006;13(8):684–90.
81. Chen Z, Speck C, Wendel P, Tang C, Stillman B, Li H. The architecture of the DNA replication origin recognition complex in Saccharomyces cerevisiae. Proc Natl Acad Sci U S A. 2008;105(30):10326–31.
82. Chastain 2nd PD, Bowers JL, Lee DG, Bell SP, Griffith JD. Mapping subunit location on the Saccharomyces cerevisiae origin recognition complex free and bound to DNA using a novel nanoscale biopointer. J Biol Chem. 2004;279(35):36354–62.
83. Matsuda K, Makise M, Sueyasu Y, Takehara M, Asano T, Mizushima T. Yeast two-hybrid analysis of the origin recognition complex of Saccharomyces cerevisiae: interaction between subunits and identification of binding proteins. FEMS Yeast Res. 2007;7(8):1263–9.
84. Liang C, Weinreich M, Stillman B. ORC and Cdc6p interact and determine the frequency of initiation of DNA replication in the genome. Cell. 1995;81(5):667–76.
85. Cocker JH, Piatti S, Santocanale C, Nasmyth K, Diffley JF. An essential role for the Cdc6 protein in forming the pre-replicative complexes of budding yeast. Nature. 1996;379(6561):180–2.
86. Santocanale C, Diffley JF. ORC- and Cdc6-dependent complexes at active and inactive chromosomal replication origins in Saccharomyces cerevisiae. EMBO J. 1996;15(23):6671–9.
87. Wilmes GM, Bell SP. The B2 element of the Saccharomyces cerevisiae ARS1 origin of replication requires specific sequences to facilitate pre-RC formation. Proc Natl Acad Sci U S A. 2002;99(1):101–6.
88. Mizushima T, Takahashi N, Stillman B. Cdc6p modulates the structure and DNA binding activity of the origin recognition complex in vitro. Genes Dev. 2000;14(13):1631–41.
89. Coster G, Frigola J, Beuron F, Morris EP, Diffley JF. Origin licensing requires ATP binding and hydrolysis by the MCM replicative helicase. Mol Cell. 2014;55(5):666–77. Pubmed Central PMCID: 4157578.
90. Kang S, Warner MD, Bell SP. Multiple functions for Mcm2-7 ATPase motifs during replication initiation. Mol Cell. 2014;55(5):655–65. Pubmed Central PMCID: 4156824.
91. Heller RC, Kang S, Lam WM, Chen S, Chan CS, Bell SP. Eukaryotic origin-dependent DNA replication in vitro reveals sequential action of DDK and S-CDK kinases. Cell. 2011;146(1):80–91.
92. Yeeles JT, Deegan TD, Janska A, Early A, Diffley JF. Regulated eukaryotic DNA replication origin firing with purified proteins. Nature. 2015;519(7544):431–5.

93. Samson RY, Bell SD. MCM loading--an open-and-shut case? Mol Cell. 2013;50(4):457–8.
94. Yardimci H, Walter JC. Prereplication-complex formation: a molecular double take? Nat Struct Mol Biol. 2014;21(1):20–5.
95. Dion MF, Kaplan T, Kim M, Buratowski S, Friedman N, Rando OJ. Dynamics of replication-independent histone turnover in budding yeast. Science (New York NY). 2007;315(5817):1405–8.
96. Huang D, Koshland D. Chromosome integrity in Saccharomyces cerevisiae: the interplay of DNA replication initiation factors, elongation factors, and origins. Genes Dev. 2003;17(14):1741–54. Pubmed Central PMCID: 196182.
97. Fox CA, Loo S, Dillin A, Rine J. The origin recognition complex has essential functions in transcriptional silencing and chromosomal replication. Genes Dev. 1995;9(8):911–24.
98. Klemm RD, Bell SP. ATP bound to the origin recognition complex is important for preRC formation. Proc Natl Acad Sci U S A. 2001;98(15):8361–7.
99. Tsakraklides V, Bell SP. Dynamics of pre-replicative complex assembly. J Biol Chem. 2010;285(13):9437–43. Pubmed Central PMCID: 2843193.
100. Randell JC, Bowers JL, Rodriguez HK, Bell SP. Sequential ATP hydrolysis by Cdc6 and ORC directs loading of the Mcm2-7 helicase. Mol Cell. 2006;21(1):29–39.
101. Ge XQ, Jackson DA, Blow JJ. Dormant origins licensed by excess Mcm2-7 are required for human cells to survive replicative stress. Genes Dev. 2007;21(24):3331–41.
102. Ibarra A, Schwob E, Mendez J. Excess MCM proteins protect human cells from replicative stress by licensing backup origins of replication. Proc Natl Acad Sci U S A. 2008;105(26):8956–61.
103. Kawabata T, Luebben SW, Yamaguchi S, Ilves I, Matise I, Buske T, et al. Stalled fork rescue via dormant replication origins in unchallenged S phase promotes proper chromosome segregation and tumor suppression. Mol Cell. 2011;41(5):543–53.
104. Remus D, Beall EL, Botchan MR. DNA topology, not DNA sequence, is a critical determinant for Drosophila ORC-DNA binding. EMBO J. 2004;23(4):897–907.
105. Austin RJ, Orr-Weaver TL, Bell SP. Drosophila ORC specifically binds to ACE3, an origin of DNA replication control element. Genes Dev. 1999;13(20):2639–49.
106. Mendez J, Zou-Yang XH, Kim SY, Hidaka M, Tansey WP, Stillman B. Human origin recognition complex large subunit is degraded by ubiquitin-mediated proteolysis after initiation of DNA replication. Mol Cell. 2002;9(3):481–91.
107. Ohta S, Tatsumi Y, Fujita M, Tsurimoto T, Obuse C. The ORC1 cycle in human cells: II. Dynamic changes in the human ORC complex during the cell cycle. J Biol Chem. 2003;278(42):41535–40.
108. Diffley JF, Cocker JH, Dowell SJ, Rowley A. Two steps in the assembly of complexes at yeast replication origins in vivo. Cell. 1994;78(2):303–16.
109. Aparicio OM, Weinstein DM, Bell SP. Components and dynamics of DNA replication complexes in S. cerevisiae: redistribution of MCM proteins and Cdc45p during S phase. Cell. 1997;91(1):59–69.
110. Belsky JA, MacAlpine HK, Lubelsky Y, Hartemink AJ, MacAlpine DM. Genome-wide chromatin footprinting reveals changes in replication origin architecture induced by pre-RC assembly. Genes Dev. 2015;29(2):212–24. Pubmed Central PMCID: 4298139.
111. Wilmes GM, Archambault V, Austin RJ, Jacobson MD, Bell SP, Cross FR. Interaction of the S-phase cyclin Clb5 with an "RXL" docking sequence in the initiator protein Orc6 provides an origin-localized replication control switch. Genes Dev. 2004;18(9):981–91.
112. Palacios DeBeer MA, Muller U, Fox CA. Differential DNA affinity specifies roles for the origin recognition complex in budding yeast heterochromatin. Genes Dev. 2003;17(15):1817–22.
113. Knott SR, Peace JM, Ostrow AZ, Gan Y, Rex AE, Viggiani CJ, et al. Forkhead transcription factors establish origin timing and long-range clustering in S. cerevisiae. Cell. 2012;148(1-2):99–111. Pubmed Central PMCID: 3266545.
114. Looke M, Kristjuhan K, Varv S, Kristjuhan A. Chromatin-dependent and -independent regulation of DNA replication origin activation in budding yeast. EMBO Rep. 2013;14(2):191–8. Pubmed Central PMCID: 3596130.

115. Thomae AW, Pich D, Brocher J, Spindler MP, Berens C, Hock R, et al. Interaction between HMGA1a and the origin recognition complex creates site-specific replication origins. Proc Natl Acad Sci U S A. 2008;105(5):1692–7. Pubmed Central PMCID: 2234206.
116. Leatherwood J, Vas A. Connecting ORC and heterochromatin: why? Cell Cycle. 2003;2(6):573–5.
117. Shen Z, Sathyan KM, Geng Y, Zheng R, Chakraborty A, Freeman B, et al. A WD-repeat protein stabilizes ORC binding to chromatin. Mol Cell. 2010;40(1):99–111.
118. Shen Z, Chakraborty A, Jain A, Giri S, Ha T, Prasanth KV, et al. Dynamic association of ORCA with prereplicative complex components regulates DNA replication initiation. Mol Cell Biol. 2012;32(15):3107–20. Pubmed Central PMCID: 3434513.
119. Shen Z, Prasanth SG. Emerging players in the initiation of eukaryotic DNA replication. Cell Div. 2012;7(1):22. Pubmed Central PMCID: 3520825.
120. Nieduszynski CA, Blow JJ, Donaldson AD. The requirement of yeast replication origins for pre-replication complex proteins is modulated by transcription. Nucleic Acids Res. 2005;33(8):2410–20. Pubmed Central PMCID: 1087785.
121. Eaton ML, Prinz JA, MacAlpine HK, Tretyakov G, Kharchenko PV, MacAlpine DM. Chromatin signatures of the Drosophila replication program. Genome Res. 2010;21(2): 164–74.
122. Lubelsky Y, Sasaki T, Kuipers MA, Lucas I, Le Beau MM, Carignon S, et al. Pre-replication complex proteins assemble at regions of low nucleosome occupancy within the Chinese hamster dihydrofolate reductase initiation zone. Nucleic Acids Res. 2011;39(8):3141–55. Pubmed Central PMCID: 3082903.
123. Papior P, Arteaga-Salas JM, Gunther T, Grundhoff A, Schepers A. Open chromatin structures regulate the efficiencies of pre-RC formation and replication initiation in Epstein-Barr virus. J Cell Biol. 2012;198(4):509–28. Pubmed Central PMCID: 3514025.
124. Wu PY, Nurse P. Establishing the program of origin firing during S phase in fission Yeast. Cell. 2009;136(5):852–64.
125. Mantiero D, Mackenzie A, Donaldson A, Zegerman P. Limiting replication initiation factors execute the temporal programme of origin firing in budding yeast. EMBO J. 2011;30(23):4805–14. Pubmed Central PMCID: 3243606.
126. Tanaka S, Nakato R, Katou Y, Shirahige K, Araki H. Origin association of Sld3, Sld7, and Cdc45 proteins is a key step for determination of origin-firing timing. Curr Biol. 2011;21(24):2055–63.
127. Rhind N, Yang SC, Bechhoefer J. Reconciling stochastic origin firing with defined replication timing. Chromosome Res. 2010;18(1):35–43. Pubmed Central PMCID: 2862975.
128. Yang SC, Rhind N, Bechhoefer J. Modeling genome-wide replication kinetics reveals a mechanism for regulation of replication timing. Mol Syst Biol. 2010;6:404. Pubmed Central PMCID: 2950085.
129. Ryba T, Hiratani I, Sasaki T, Battaglia D, Kulik M, Zhang J, et al. Replication timing: a fingerprint for cell identity and pluripotency. PLoS Comput Biol. 2011;7(10), e1002225. Pubmed Central PMCID: 3197641.
130. Debatisse M, Le Tallec B, Letessier A, Dutrillaux B, Brison O. Common fragile sites: mechanisms of instability revisited. Trends Genet. 2012;28(1):22–32.
131. Watanabe Y, Maekawa M. Spatiotemporal regulation of DNA replication in the human genome and its association with genomic instability and disease. Curr Med Chem. 2010;17(3):222–33.
132. Smith L, Plug A, Thayer M. Delayed replication timing leads to delayed mitotic chromosome condensation and chromosomal instability of chromosome translocations. Proc Natl Acad Sci U S A. 2001;98(23):13300–5. Pubmed Central PMCID: 60865.
133. Woodward AM, Gohler T, Luciani MG, Oehlmann M, Ge X, Gartner A, et al. Excess Mcm2-7 license dormant origins of replication that can be used under conditions of replicative stress. J Cell Biol. 2006;173(5):673–83.

Chapter 10
Licensing of Replication Origins

Alberto Riera and Christian Speck

Abstract All living organisms need to duplicate their genetic material prior to cell division in order to maintain genomic-stability. Cells have evolved sophisticated DNA replication mechanisms to ensure that this process is as faithful as possible. Eukaryotic initiation of DNA replication is a two-step process, where the replicative DNA helicase becomes loaded onto DNA to license DNA replication during late M-phase of the cell cycle prior to helicase-activation in S-phase. Importantly, helicase loading is entirely blocked in S-phase, which is a crucial regulatory mechanism that hinders re-replication of DNA and is crucial for genomic stability. Moreover, multiple copies of the replicative helicase become loaded at each origin to serve as backup-helicases in case a fork becomes terminally arrested. For these reasons it is imperative that helicase loading is as efficient as possible. MCM2–7 represent the core of the replicative helicase, which becomes loaded in an ATP-hydrolysis-dependent process as a double-hexamer onto double-stranded DNA. Current data suggest a model where ORC, Cdc6, and Cdt1 load in a stepwise process the MCM2–7 double-hexamer onto DNA. In this review we discuss the emerging mechanism of ATP-hydrolysis-driven helicase loading, the regulation of this process, and the structure and function of the MCM2–7 double-hexamer.

Keywords MCM2–7 • Pre-RC • Origin • AAA+ • DNA licensing • Cancer • Cell cycle • ORC • Cdc6 • Cdt1 • Helicase

Overview

All organisms living on earth need to precisely duplicate their genetic material prior to cell division in order to maintain genomic stability. Thus, cells have evolved sophisticated mechanisms to control DNA replication, which ensure that this process is as faithful as possible. The more complex a cell is, the more complex these mechanisms are. Indeed, the highest level of complexity has been observed in eukaryotes, especially in multicellular organisms.

A. Riera • C. Speck (✉)
DNA Replication Group, Institute of Clinical Science, Faculty of Medicine, Imperial College London, London, W12 0NN, UK
e-mail: chris.speck@imperial.ac.uk

D.L. Kaplan (ed.), *The Initiation of DNA Replication in Eukaryotes*,
DOI 10.1007/978-3-319-24696-3_10

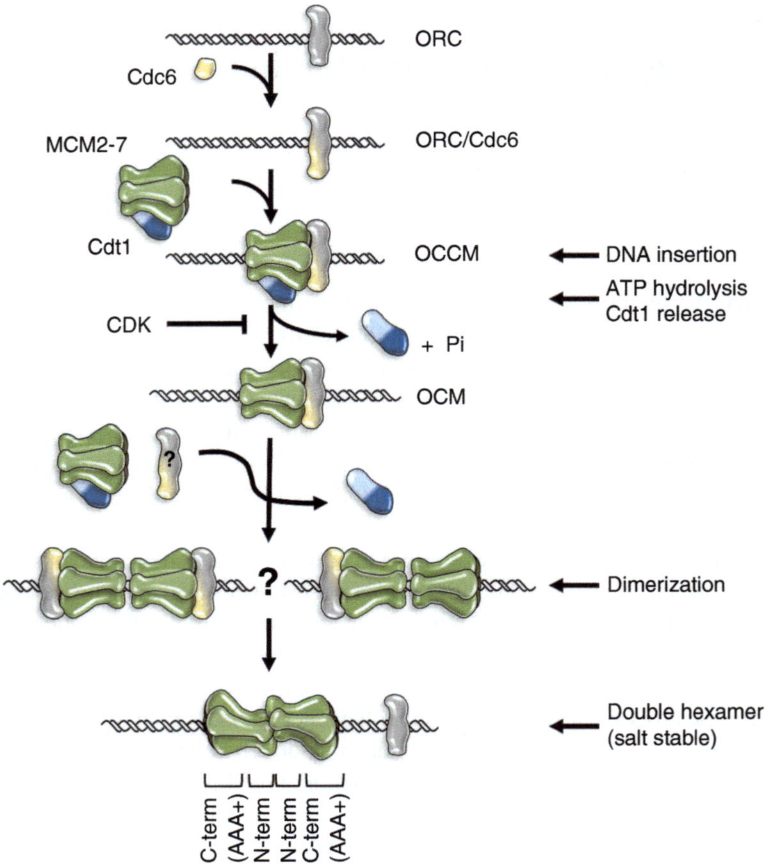

Fig. 10.1 Licensing of replication origins in *S. cerevisiae*. ORC binds specifically to replication origins. In late M phase of the cell cycle, Cdc6 joins ORC on DNA. The ORC/Cdc6 complex constitutes a landing pad for a Cdt1/MCM2–7 heptamer. In a fast reaction, an ORC/Cdc6/Cdt1/MCM2–7 (OCCM) complex is formed, with double stranded DNA already inserted into the central pore of the helicase. Upon ATP hydrolysis, Cdt1 is released, resulting in an ORC/Cdc6/MCM2–7 (OCM) complex. The OCM is competent to recruit a second MCM2–7/Cdt1, but the mechanism is unclear. The final product of DNA licensing is a salt stable MCM2–7 double hexamer, which requires activation in S-phase to function as part of the replicative helicase

Replication is initiated at specific regions called origins of DNA replication [1]. In eukaryotes these origins are recognized by the six-subunit Origin Recognition Complex (ORC) (Fig. 10.1). In budding yeast this complex binds to DNA in a strictly ATP-dependent process [2]. During the entire cell cycle ORC is chromatin bound [3], but late in M phase Cdc6 binds to ORC [4]. This ORC/Cdc6 complex is the landing pad for the Mini Chromosome Maintenance (MCM) complex. MCM2–7 forms the core of the eukaryotic replicative DNA helicase and the six proteins are arranged in a ring shaped complex [5–7]. However, MCM2–7 cannot interact directly with ORC/Cdc6, but requires an adaptor protein, Cdt1 [8–10]. Cooperative interactions between all these proteins allow the assembly of a transient ORC/Cdc6/

Cdt1/MCM2–7 (OCCM) complex, which promotes insertion of double-stranded DNA into the MCM2–7 ring [10–13]. Yet, upon ATP hydrolysis, Cdt1 is quickly ejected, resulting in an ORC/Cdc6/MCM2–7 (OCM) complex [10, 11]. This OCM contains a single copy of ORC, Cdc6, and MCM2–7 and is competent to recruit a second Cdt1/MCM2–7 heptamer, which leads to formation of a stable MCM2–7 double hexamer that encircles double stranded DNA [10, 14–16].

During the G1 phase of the cell cycle, multiple copies of the replicative helicase are loaded onto DNA at hundreds or thousands of replication origins, but remain inactive until S-phase. Conversely, in S-phase, helicase loading is prevented by multiple redundant mechanisms that block re-replication of DNA [17–19]. Ultimately, numerous protein factors and kinases are required in S-phase to promote the formation of the Cdc45/MCM2–7/GINS (CMG) complex, which represents the active form of the helicase [20–24].

Licensing Is a Tale of AAA+ Proteins

ORC, Cdc6, and MCM2–7 belong to the same family of ancient proteins: ATPases associated with various cellular activities (AAA+) [25]. In order to understand their function during the licensing of DNA replication it is necessary to look at their common characteristics.

Conserved Features of the AAA+ Domain

The AAA+ protein family is composed of a diverse group of enzymes, which can be found throughout all the kingdoms of life and appeared early during evolution [26]. Their hallmark is a highly conserved 200–250 amino acid ATP-binding domain (AAA+ domain) that contains conserved motifs essential for ATP binding and hydrolysis, while a short C-terminal helix-bundle mediates interprotomer contacts and distinguishes the AAA+ protein family from other ATPases [25, 27]. In particular, short insertions into the ATPase core define seven specific AAA+ clades. ORC and Cdc6 belong to the initiator clade and contain a helical insertion supporting filament formation (Fig. 10.2a). On the other hand, MCM2–7 belongs to the PS-II insert clade, which forms hexameric ring structures and is characterized by three insertions; two beta-hairpins important for DNA unwinding and an alpha-helix that is unique to this clade, affecting the positioning of the C-terminal helix bundle [26, 28].

ATP-binding and hydrolysis depends on several conserved motifs (Fig. 10.2b): The Walker A motif—GXXXXGK[T/S]—directly interacts with the phosphate groups of ATP. The conserved lysine in this motif is essential for interaction with ATP and its mutation abrogates nucleotide binding, inactivating the protein, while an adjacent sensor 2 motif functions in a similar manner. The Walker B motif—hhhhDE—(h stands for hydrophobic) contacts the nucleotide as well and facilitates ATP-hydrolysis by coordinating a magnesium ion and several water molecules.

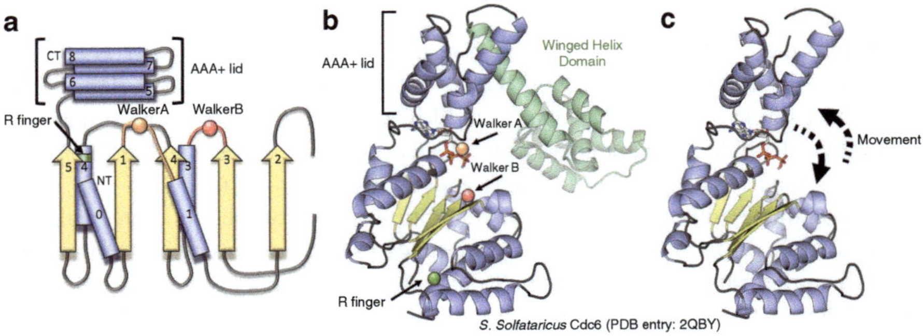

Fig. 10.2 AAA+ ATPases: general organization and essential motifs. (**a**) Arrangement of a classical AAA+ ATPase core. Walker A—*orange*, Walker B—*red*, arginine finger—*green*, and C-terminal AAA+ lid—*blue* are displayed. (**b**) *Sulfolobus solfataricus* Cdc6 crystal structure [PDB 2QBY], highly related to eukaryotic Cdc6 and Orc1. Essential motifs are highlighted using the same color code as in (**a**), as well as the Winged Helix Domain, involved in DNA binding and protein interactions—*pale green*. (**c**) Example of motions caused by ATP binding/hydrolysis. This movement is transmitted to the rest of the AAA+ protein and/or associated AAA+ proteins

Mutations in the Walker B motif, or the nearby sensor 1 motif, typically impair ATP hydrolysis, but do not abrogate ATP binding. In proximity to the Walker B motif is an arginine finger, which is essential for nucleotide hydrolysis. Within larger AAA+ protein assemblies, the arginine finger of one subunit constitutes part of the nucleotide hydrolysis motif of the adjacent protomer. This arrangement allows inter-subunit crosstalk and coordination of ATP hydrolysis, an essential feature for processive DNA unwinding in hexameric helicases.

It is well known that almost all AAA+ proteins undergo dramatic conformational changes upon ATP binding and/or hydrolysis. The conformational changes are then transmitted to the insertions attached to the AAA+ core or to adjacent domains (Fig. 10.2c). These movements make the enzymes especially suitable for a large array of diverse cellular processes involving DNA replication, protein degradation, or movement of microtubule motors [25]. The movements of AAA+ proteins function in different ways—either as switches or motors. A classic example of an AAA+ switch is the bacterial DNA replication initiator DnaA. Here, ATP-hydrolysis is solely required to inactivate DnaA, and thus the time point of DNA replication initiation is controlled by the cellular ratio of ATP-DnaA to ADP-DnaA [29]. On the other hand, the bacterial DnaB helicase is an AAA+ motor, where ATP-hydrolysis propels the rapid DNA unwinding at the replication fork [30].

Origin Recognition Complex

The main function of the origin recognition complex is to promote helicase loading during pre-RC formation [11, 31]. ORC is composed of six subunits (Orc1-6) and was first identified in budding yeast, where it binds to the budding yeast replication origin, termed Autonomous Replication Sequence (ARS) [2]. Consequently it was found that ORC is conserved within all eukaryotes [32] where it universally functions for DNA licensing [33, 34]. Additionally it was shown that ORC has important roles in epigenetic gene-silencing, cytokinesis, chromosome segregation, and dendrite formation in post-mitotic neurons [35–41]. ORC's ability to interact with chromatin is central to most, if not all, of its functions. In *Saccharomyces cerevisiae*, ORC binds to a specific DNA sequence [2], while in higher eukaryotes no consensus sequences were identified. Instead, in vertebrates and mammals a multitude of chromatin recruitment mechanisms have been described, including epigenetic marks, recruitment factors, or DNA structures [1].

Five out of the six ORC subunits (Orc1 to Orc5) are structurally related, belonging to the AAA+ protein family, while Orc6 has an unrelated structure [42, 43]. The overall structure of *S. cerevisiae* and *Drosophila* has been studied by electron microscopy [42–47], and a recent crystal structure of the *Drosophila* ORC has provided more in depth insights [48]. Although Orc1 to Orc5 adopt an ATP-binding fold, only Orc1 actively hydrolyzes ATP. Orc5, and in higher eukaryotes also Orc4, are able to bind the nucleotide, but ATP hydrolysis is not important to their function, instead it may play a structural role [49–51]. Although budding yeast Orc4 cannot bind ATP, it contains an arginine finger that is important for Orc1 ATP-hydrolysis on DNA [52]. On the other hand, Orc2 and Orc3 contain recognizable AAA+ signatures in their sequence, but several conserved structural motifs are absent, hence they cannot bind or hydrolyze ATP [32, 42, 43]. However, by maintaining an overall AAA+-like fold the Orc1–5 proteins together with Cdc6 can be arranged in an AAA+ typical ring shaped complex with DNA passing through the center of the ring [43, 45, 48]. We suggest that an Orc1 ATP-hydrolysis power-stroke is transmitted to the entire ORC complex, which in turn facilitates pre-RC formation. C-terminal to the AAA+ fold Orc1–5 contain conserved winged-helix-domains (WHDs), which allow the complex to form tight interactions with DNA. Interestingly, the AAA+ core of Orc1 and its WHD are functionally linked, as binding of ORC to origin DNA downregulates its ATPase activity [50]. On the other hand, Orc6, the only non AAA+ protein, is composed of two small domains that are connected by a loop with structural similarity to transcription factor TFIIB [53]. Orc6 associates tightly with *Drosophila* or *S. cerevisiae* Orc1–5, but only weakly with human Orc1–5 and not at all with *Xenopus* Orc1–5, consistent with the idea that Orc6 has acquired additional functions outside of DNA replication [2, 38, 49, 54–56].

Cdc6

Cell division cycle 6 (Cdc6) was first identified by Lee Hartwell, based on temperature sensitive mutants that arrest the budding yeast cell cycle [57]. Consequently it was found that Cdc6 and ORC are structurally related [58, 59], and that during late M-phase of the cell cycle, Cdc6 becomes recruited to chromatin in an

ORC-dependent manner, involving specific contacts with Orc1 and Orc3 [12, 60]. The ORC–Cdc6–DNA interaction is ATP dependent and results in a remodeling of the protein–DNA complex, causing a rearrangement of Orc1 and an extended protein–DNA interface [43, 61]. Importantly, the ORC-Cdc6 complex adopts a ring-shaped structure, which can embrace double stranded DNA, that is ideally suited to recruit the ring shaped MCM2–7/Cdt1 complex [11, 12, 54]. Cdc6, from a structural point of view, is a conserved AAA+ protein with a central ATPase core, which allows it to integrate into Orc1–5 to form a hexameric AAA+ assembly [28, 43]. The Cdc6 C-terminus contains a winged-helix-domain that interacts with DNA, similar as in Orc1–5 [42, 43, 62], and at the N-terminus Cdc6 carries a small extension, which becomes phosphorylated by CDK and regulates its stability [63]. Crucially, Cdc6, like Orc1, can hydrolyze ATP, causing structural changes that can be transmitted by the other AAA+ proteins to the rest of the complex, which are important for pre-RC assembly; see section "The Role of ATP Hydrolysis During OCM Formation" [11, 45, 61].

MCM2–7: The Core of the Replicative Helicase

The hexameric MCM2–7 complex forms the core of the replicative helicase [64]. These proteins were first identified by Bik Tye's laboratory while screening for genes that affect plasmid maintenance in budding yeast [65]. The six subunits of the eukaryotic MCM2–7 (Fig. 10.3) share a common core and are highly conserved in all eukaryotes. In archaea the hexamer is usually constituted by a single Mcm protein, which is structurally similar to the core of eukaryotic subunits [66] (Fig. 10.3a). All six proteins are essential for DNA licensing [67], replication initiation [68, 69] and DNA synthesis [70]. Indeed the proteins travel with the replication fork [71]. Surprisingly, MCM2–7 by itself is only a weak helicase [7]. However, when the complex is joined by Cdc45 and GINS the MCM2–7 helicase activity becomes strongly activated [20]. Crucial for MCM2–7 complex function is its ability to assemble into a ring, with a subunit arrangement of Mcm5-3-7-4-6-2 [72]. ATP-binding sites are located at the interface between adjacent subunits, meaning that residues from flanking protomers coordinate ATP binding and hydrolysis [72] (Fig. 10.3b). The inter-subunit nature of the active sites in conjunction with the ring-shaped assembly of the helicase is ideally suited for a highly coordinated ATP-hydrolysis and DNA unwinding activity. Consistent with this model, the incorporation of a single ATP binding mutant into the MCM2–7 complex results in a strong reduction of ATP-hydrolysis for the entire complex, causing lethality in *S. cerevisiae* [73, 74]. This shows that defects in ATP-binding influence the overall conformation of the MCM2–7 complex, potentially leading to MCM2–7 ring opening and a complete block of the ATP-hydrolysis chain. Yet introduction of ATP-hydrolysis mutants have a less severe effect [74, 75], since they likely cause a smaller structural change. Overall, it was shown in budding yeast that the purified MCM2–7 hexamer exhibits in vitro a low but robust idle ATPase activity, which

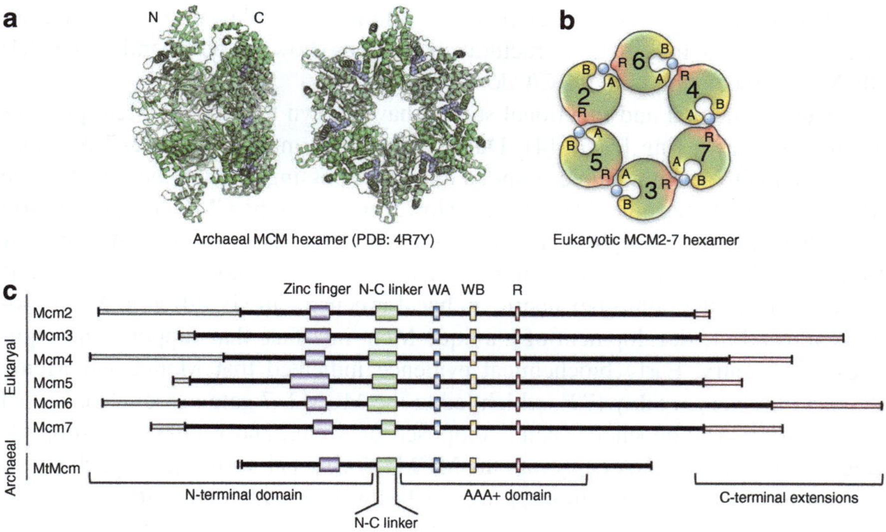

Fig. 10.3 MCM2–7 replicative helicase. (**a**) Crystal structure of an active archaeal MCM hexamer displaying the central channel that accommodates double stranded DNA [PDB: 4R7Y]. (**b**) Arrangement of AAA+ subunits in the MCM2–7 helicase. One monomer provides Walker A (**a**) and Walker B (**b**) motifs and the next one the arginine finger (R). This distribution allows inter-subunit coordination of ATP hydrolysis and inter-subunit cross talk. The ATP is represented by a blue sphere. (**c**) Diagram showing the N- and C-terminal extensions of eukaryotic MCM subunits. All Mcm proteins share a core common to archaeal and eukaryotic protomers, but the extensions allow a multitude of additional functions that are not present in lower organisms (loading quality control, histone interaction, etc.)

gets suppressed when MCM2–7 assembles into the double-hexamer [19]. In *Drosophila* it was found that once MCM2–7 becomes integrated into the CMG, its ATPase activity is increased by about 100-fold [76], which explains in part why this complex is highly active in DNA unwinding.

The Mcm proteins, from a structural point of view, are highly related, with the highest homology in a central section that contains several functional elements (Fig. 10.3c). The N-terminal domain contains a conserved zinc finger and an oligonucleotide/oligosaccharide binding fold (OB-fold), which is connected to a C-terminal domain, which comprises the AAA+ domain and a winged helix domain [64]. In contrast to archaeal Mcm, eukaryotic Mcm2–7 have characteristic N- and C-terminal extensions of various lengths, which have important regulatory functions. For instance, the long N-termini of Mcm2, Mcm4, and Mcm6 contain phosphorylation sites for the S-phase kinase Dbf4-Cdc7 (DDK) [77]. Phosphorylation of Mcm4 by DDK promotes S-phase progression by alleviating an inhibitory activity in the amino terminal region of Mcm4 [78]. Then again, the Mcm C-termini of Mcm3 and Mcm6 are important for recruitment of Cdt1/MCM2–7 to ORC/Cdc6. The Mcm3 C-terminus targets ORC/Cdc6 [12, 79], while Mcm6 contains an inhibitory domain that blocks MCM2–7 association with ORC/Cdc6. Importantly, the binding of Cdt1 to MCM2–7 induces a structural change in the helicase alleviating the inhibitory activity in Mcm6. Indeed, this structural change was shown to be essential for Cdt1/MCM2–7 association with ORC/Cdc6 [10].

Several structural and functional studies have shown that the six Mcm proteins are arranged as a ring [5, 7, 14]. During pre-RC formation, MCM2–7 becomes loaded onto DNA, with double stranded (ds) DNA passing through the center of the MCM2–7 double-hexamer [14, 16, 25]. However, within the CMG, the active form of the helicase, it is thought that single-stranded DNA passes through the complex [80]. Thus, MCM2–7 ring opening represents a unique topological challenge. The evolution from the homo-hexameric archaeal precursor to six different Mcm proteins allowed the development of a unique Mcm interface that supports and regulates DNA entry. Early biochemical evidence indicated that Mcm2 and Mcm5 interact only very weakly [72], which led to the MCM2–7 gate model. This model postulates an ATP-dependent gate, composed of Mcm2 and Mcm5, that supports single stranded (ss) DNA entry into the MCM2–7 ring [7]. Consequently, chemical biology tools and synthetic biology approaches were used to demonstrate the existence of a unique DNA entry gate during pre-RC formation. Interestingly, the Mcm2/5 gate was shown to operate in the absence of ATP-hydrolysis at the stage of OCCM formation, with dsDNA becoming inserted into the MCM2–7 ring [5]. It is still unclear if this gate also functions for CMG formation or during DNA synthesis.

Cdt1

Cdt1 was originally identified in *S. pombe*, where its transcription was shown to be regulated by Cdc10, a factor which controls the G1-S transition of the cell cycle [81]. Then, work by the Nurse and Mechali laboratories demonstrated that *S. pombe* and *Xenopus* Cdt1 are essential for DNA licensing [8, 9], which is now universally accepted for all eukaryotes. In *Xenopus* and humans Cdt1 associates via ORC/Cdc6 with chromatin [8, 54, 82], while in budding yeast Cdt1 becomes recruited to replication origins in the form of the Cdt1/MCM2–7 complex [83, 84].

The structure of Cdt1 has been analyzed [10, 84–86]. The protein can be divided into three sections: an N-terminal domain whose structure is not known, and two Winged Helix Domains (WHD) localized in the central and C-terminal domains (Fig. 10.4) [87, 88]. In metazoans, the N-terminal domain becomes post-transcriptionally modified in a cell cycle-dependent manner, which regulates protein stability. Ubiquitination of this domain promotes proteolysis during S and G2 phases, while acetylation hinders ubiquitination [89]. In budding yeast an interaction between the N-terminal domain and MCM2–7 was observed, but this is not sufficient to recruit MCM2–7 to origins in vitro [10, 85]. Moreover, it was found that the Cdt1 N-terminus is important for the assembly of a functional MCM2–7 double-hexamer [85]. On the other hand, both the central and the C-terminal WHD of *Mus musculus* Cdt1 have been crystallized [87, 88]. Winged

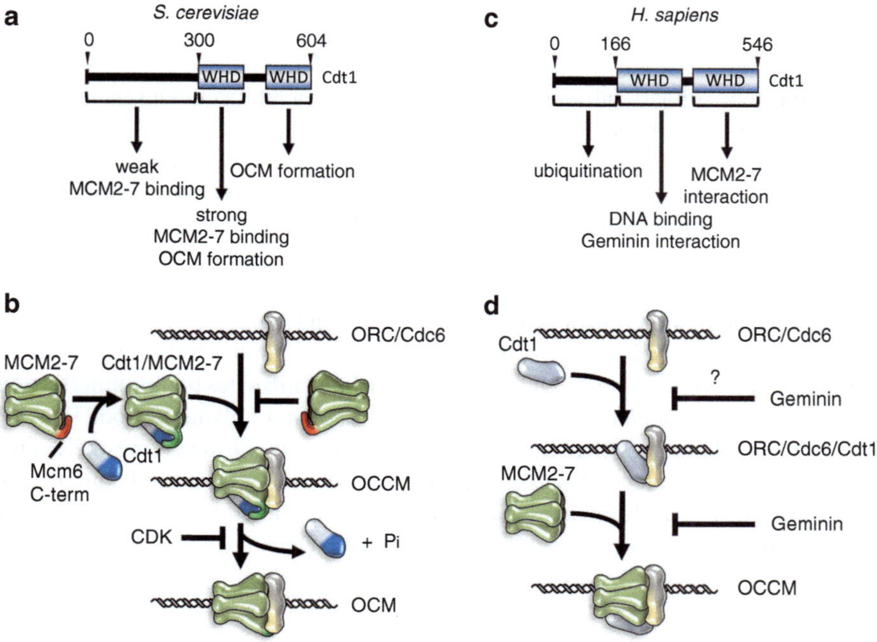

Fig. 10.4 Cdt1's and geminin's roles in pre-RC formation. (**a**) The *Saccharomyces cerevisiae* Cdt1 protein has an N-terminal domain involved in MCM2–7 binding and two Wing Helix Domains (WHD) in the central and C-terminal regions. (**b**) OCM formation in *Saccharomyces cerevisiae*. MCM2–7 contains an inhibitory domain in Mcm6 that blocks interactions with ORC/Cdc6. Cdt1 alleviates this inhibitory activity and allows the binding of MCM2–7/Cdt1 to ORC/Cdc6. Upon OCCM formation the Cdt1–Mcm6 interaction is essential to induce ATP-hydrolysis, which results in Cdt1 ejection and OCM formation. (**c**). *Homo sapiens* Cdt1 has an N-terminal regulatory domain, and two Wing Helix Domains (WHD) in the central and C-terminal regions. (**d**) Human Cdt1 associates with ORC/Cdc6 on chromatin to promote MCM2–7 loading. Geminin interferes with pre-RC formation in a concentration-dependent manner

helix domains are often associated with DNA recognition and protein–protein interactions. Indeed, in mice, the first WHD has been implicated in DNA binding [90], while in budding yeast both the first and second WHD have been shown to participate in the recruitment of MCM2–7 to replication origins, with the first WHD being particularly important for stable OCCM formation and induction of ORC/Cdc6 ATP-hydrolysis during pre-RC formation [10, 85] (Fig. 10.4). Consistently, budding yeast Cdt1 mutants that affect either WHD completely fail to promote MCM2–7 double-hexamer assembly and cause lethality [10, 85]. Additionally, N- and C-terminal extensions to the central WHD contain binding sites for geminin (see also section "Geminin"), the primary inhibitor of DNA licensing in higher eukaryotes [91, 92]. Interestingly, overexpression of hCdt1 is oncogenic, probably by promoting re-licensing of replicated DNA in S-phase, which results in re-replication and genomic instability [93], highlighting the functional importance of Cdt1.

Geminin

Geminin is the primary inhibitor of DNA replication in multicellular organisms and was first identified by the Kirschner laboratory [94]. This small 33 kDa protein was initially thought to work by binding and sequestering Cdt1, preventing MCM complex loading during S-phase [91, 92]. However, we know today that geminin has a far more complex mechanism of inhibition. Indeed, experiments in *Xenopus* egg extracts showed that geminin significantly stabilizes the binding of ORC, Cdc6, and Cdt1 on plasmid DNA [54, 95].

Looking at the structure of geminin allows us to understand its function. Geminin consists of an N-terminal region that is important for the interaction with Cdt1, a central coiled-coil domain that interacts with Cdt1 and mediates homo-dimerization, and a C-terminal domain, the function of which is still not fully understood [88, 96]. Current models predict that Geminin-Cdt1 oligomerisation involving the central coiled-coiled region regulates DNA licensing: lower order oligomers being permissive for licensing, while greater concentrations of geminin promote the formation of higher order Cdt1-geminin complexes, which block DNA licensing [89, 94, 95].

Licensing of Replication Origins

During DNA licensing ORC, Cdc6, Cdt1, and MCM2–7 assemble into pre-replication complexes, which lead to the loading of head-to-head MCM2–7 double-hexamers onto origin DNA, with double stranded DNA running through the central channel of the complex [97, 98]. At each origin repeated rounds of MCM2–7 loading occur, yielding additional copies of MCM2–7 double-hexamers that function as dormant origins. When a fork becomes terminally arrested due to severe DNA damage, dormant origins become activated to reestablish new replication forks. Therefore, efficient and repetitive MCM2–7 loading represents a mechanism that is particularly important for genomic stability [99].

ORC/Cdc6/Cdt1/MCM2–7 (OCCM) Complex Establishment and ORC/Cdc6/MCM2–7 (OCM) Formation

Xenopus, human, and budding yeast pre-RC formation has been reconstituted with extracts and purified proteins, and these systems have generated significant insights into the process [14, 16, 52, 54, 100, 101]. Employing the budding yeast system, ORC, Cdc6, Cdt1, and MCM2–7 were shown to assemble rapidly, within 30 s, into an ORC/Cdc6/Cdt1/MCM2–7 (OCCM) complex in an ATP hydrolysis independent way [10–12]. The experiments employed ATPγS, an ATP analogue that can be hydrolyzed only slowly, which was crucial for capturing the very short-lived intermediate. Biochemical analysis demonstrated that the OCCM contains only a single MCM2–7 hexamer, while a crosslinking analysis supported the idea that the complex contains a single copy of ORC, Cdc6, and Cdt1. Then, a cryo-EM analysis visualized for the first time the 3D structure of the entire 14 poly-peptide complex, revealing its overall architecture [13]. Biochemical and electron-microscopy data together proved that MCM2–7 double-hexamer formation is a step-wise process: initially only a single MCM2–7 hexamer associates with ORC, Cdc6, and Cdt1. However, upon ATP-hydrolysis Cdt1 becomes quickly released, resulting in an ORC/Cdc6/MCM2–7 (OCM) complex that contains a single copy of ORC, Cdc6, and MCM2–7 [10, 11]. In addition, very recent single-molecule experiments indicate that ATP-hydrolysis not only promotes Cdt1 release, but also Cdc6 release, which is followed by rebinding of a second Cdc6 molecule to form the OCM complex [101]. Importantly, the OCM, but not the OCCM, can recruit a second MCM2–7 hexamer, demonstrating that the OCM is a true intermediate of MCM2–7 double-hexamer formation [10, 102]. Crucially, while the OCM is formed within 30 s, MCM2–7 double-hexamer formation takes between 8 and 15 min, suggesting that the second MCM2–7 hexamer becomes recruited in a different way (for more details see section "MCM2–7 Dimerization").

Establishment of the OCCM is strictly Cdt1 dependent (Fig. 10.4). In the absence of Cdt1, MCM2–7 is not able to associate with ORC/Cdc6 due to the presence of an inhibitory activity in the C-terminal part of Mcm6. Specific Cdt1–Mcm6 interactions overcome this inhibition. Structural work with human proteins in combination with yeast genetics [84] identified five amino acids in *S. cerevisiae* Mcm6-5A (E945, D947, L951, E953, and Y954), plus three amino acids in *S. cerevisiae* Cdt1-3A (R486, L487, and R490) that are important for Cdt1–Mcm6 association. In budding yeast, the Mcm6–Cdt1 interaction is essential for the nuclear import of MCM2–7 and, as a consequence, mutations affecting this interaction block chromatin binding of MCM2–7 [83, 103]. In vitro work in *S. cerevisiae* showed that an MCM2–7 complex, containing Mcm6-5A, is able to interact with Cdt1, while Cdt1-3A was able to bind to MCM2–7, which is consistent with the finding that MCM2–7 has several Cdt1 binding sites. However, both Mcm6-5A and Cdt1-3A containing Cdt1/MCM2–7 complexes fail to bind to ORC/Cdc6, as these mutants are not competent to promote productive Cdt1–MCM2–7 interactions [10]. Yet *S. cerevisiae* MCM2–7 that lacks the C-terminal domain of Mcm6 (MCM2–7-Δ6C) is able to interact with ORC/Cdc6, even in the absence of Cdt1, demonstrating that the Mcm6 C-terminus contains an auto-inhibitory domain. Importantly, although MCM2–7-Δ6C it is able to interact with Cdt1, it does not support MCM2–7 double-hexamer formation, as MCM2–7-Δ6C fails to induce pre-RC ATP-hydrolysis or to eject Cdt1. In summary, the budding yeast Mcm6–Cdt1 interaction governs nuclear import, pre-RC assembly and ATP-hydrolysis [10, 83, 103]. Yet it is still unclear if this interaction functions in a similar way during DNA licensing in higher eukaryotes.

Orc6 and OCM Establishment

Orc6 is the smallest Orc subunit, interacts with Orc3 and has a peripheral localization within the *D. melanogaster* ORC complex [47]. In vitro analysis in budding yeast has shown that Orc6 interacts with Cdt1, while in vivo depletion of the protein inhibits pre-RC formation [104]. More detailed analysis using a fully reconstituted pre-RC system revealed that Orc1–5 is competent for OCCM formation, but that the complex fails during OCM formation. Indeed, Cdc6 ATP hydrolysis leads to disassembly of the Orc1–5 containing OCCM, which in turn blocks MCM2–7 double-hexamer formation. In summary, these data demonstrate that Orc6 plays a role during the establishment of the OCM [10, 79].

The Role of ATP Hydrolysis During OCM Formation

Orc1 and Cdc6 ATP-Hydrolysis

ATP-hydrolysis is crucial for pre-RC formation. Indeed, in the absence of ATP-hydrolysis, MCM2–7 double-hexamer formation cannot occur. Yet investigating the role of ATP-hydrolysis during pre-RC formation proved highly complex. Early analysis showed that overexpression of Orc1 and Cdc6 Walker B mutants in yeast caused dominant lethality and MCM2–7 loading defects [105, 106], while microinjection of *H. sapiens* Cdc6 Walker B mutant protein into cells inhibited DNA replication [107]. Then again, a mutant in the arginine finger of budding yeast Orc4 (ORC4R), which affects Orc1 ATP hydrolysis in the context of origin DNA, caused lethality in budding yeast, while its overexpression resulted in MCM2–7 loading defects [52]. Moreover, depletion experiments in *Xenopus* egg extracts coupled with add back of Cdc6 Walker B mutant protein revealed that the mutant causes a small decrease in MCM2–7 loading, but blocked DNA synthesis substantially. In summary, the in vivo data indicate an important role for Orc1 and Cdc6 ATP hydrolysis for the viability of cells, especially in the context of overexpression, but it is difficult to conclude a specific mechanism. In vitro analysis of budding yeast Cdc6 ATP-hydrolysis mutants by several groups indicated a Cdt1 release defect [10, 11, 108, 109]. However, the effect of these Cdc6 mutants on MCM2–7 double-hexamer formation was less clear, with two studies observing no effect [74, 109], while two other studies reported a reduction in MCM2–7 double-hexamer formation [10, 13]. ATPase measurements in the context of pre-RC assays demonstrated that Cdc6 Walker B or Sensor 1 mutants reduced ATP-hydrolysis significantly. An ORC4R (mutation in the arginine finger of ORC4) mutant had no effect on pre-RC ATP-hydrolysis or MCM2–7 double-hexamer formation [10, 13], therefore the essential role of the Orc4 arginine finger in yeast is currently unclear. On the other hand, an Orc1 Walker B mutant reduced pre-RC ATP-hydrolysis, affected Cdt1 release and MCM2–7 double-hexamer formation, similar as observed with a Cdc6 ATPase mutant. The combination of the Cdc6 and Orc1 ATPase mutants blocked Cdt1 release and MCM2–7 double-hexamer formation to a similar level as observed with ATPγS [10]. Therefore, it was suggested that Orc1 and Cdc6 synergize during pre-RC formation to promote Cdt1 release.

MCM2–7 ATP Hydrolysis

In *Xenopus* it has been shown that the ATPase activity of MCM2–7 is dispensable for pre-RC assembly, although required for DNA unwinding [110]. In *S. cerevisiae* it was reported that an MCM2–7 complex carrying a mutation in the arginine finger of Mcm3 greatly reduces its idle ATPase activity, but it is still competent for MCM2–7 double-hexamer formation and has no effect on ORC/Cdc6 ATP hydrolysis or Cdt1 release [10]. Then two studies analyzed all possible *S. cerevisiae* MCM2–7 helicase arginine-finger (R to A) mutants. The most severe phenotypes were observed with Mcm5 and Mcm6 mutants, which blocked MCM2–7 double-hexamer formation, while the other mutants affected the process only to a smaller extent [74, 109]. In particular the arginine finger mutant in Mcm5 affected Cdt1 release, probably because it interferes with Orc1 or Cdc6 ATP hydrolysis, but this has not yet been analyzed. MCM2–7 ATP-hydrolysis per se may not be essential for pre-RC formation, as a number of mutants supported double-hexamer formation, although at somewhat reduced levels. Moreover, the arginine finger is positioned at the interface of neighboring MCM2–7 subunits; therefore, it appears possible that arginine finger mutations could also affect the MCM2–7 structure. One possibility is that during DNA insertion, when the MCM2–7 ring is opened, these mutations stress the complex, resulting in structural defects that interfere with OCM formation. Alternatively, MCM2–7 ATP hydrolysis could be actively induced during pre-RC formation to propel Cdt1 release. However, as the Mcm3 arginine finger mutant blocks MCM2–7 ATP hydrolysis strongly while still promoting MCM2–7 double-hexamer formation [10], this model appears unlikely.

The MCM2–7 DNA Entry Gate

Work in *S. cerevisiae* indicated that in-between Mcm2 and Mcm5 a discontinuity exists, and it was suggested that this gate could open in a nucleotide-dependent manner [7, 72]. However, helicase loading is a highly regulated process that depends on ORC, Cdc6, and Cdt1 in vitro and in vivo. Therefore it can be assumed that the MCM2–7 ring does not open by itself inside the cell. Indeed, budding yeast MCM2–7 was observed by electron microscopy in a closed circular conformation [5, 7]. To test which Mcm subunits in the context of helicase loading represent the DNA entry gate, conditional linkages between neighboring Mcm subunits were designed employing FKRB and FRB insertions, which control in a rapamycin-dependent manner DNA entry into the MCM2–7 ring [5].

In vitro analysis showed that one specific fusion, MCM2–7-M2/M5 (MCM2–7 with the conditional gate created between the subunits 2 and 5) showed efficient MCM2–7 loading in the absence of rapamycin (Fig. 10.5). In contrast, in the presence of the drug this construct displayed efficient OCCM assembly, but significantly reduced OCM formation and a complete block in salt resistant double

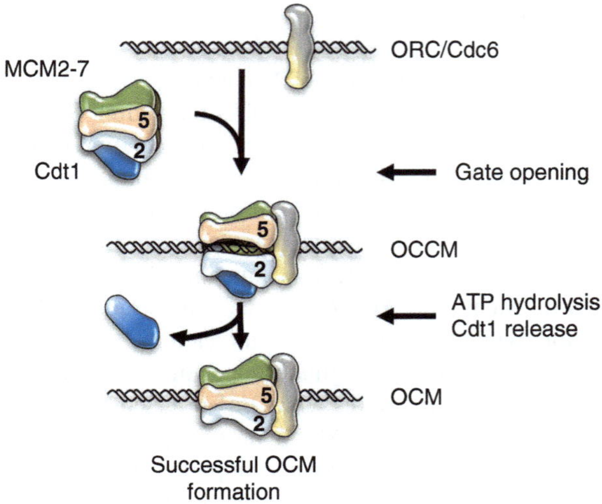

Fig. 10.5 The MCM2–7 DNA entry gate. The gate is located in the interface between the Mcm2 and Mcm5 subunits. MCM2–7 is a closed hexameric ring, but upon interaction of Cdt1/MCM2–7 with ORC/Cdc6 a gate between Mcm2 and Mcm5 opens and DNA is inserted into the MCM2–7 ring. This loading process does not require ATP hydrolysis, but ATPase activity is required for Cdt1 ejection

hexamer formation. Yet this effect was not observed with conditional links between any other neighboring Mcm subunit combinations. Consistent with the in vitro work, a yeast strain expressing MCM2–7-M2/M5 displayed a rapamycin-dependent block in MCM2–7 recruitment to chromatin. Using this mutant it was demonstrated that DNA insertion occurs at the stage of OCCM formation, as rapamycin-dependent closing of the gate during ORC/Cdc6/Cdt1/MCM2–7 complex formation was sufficient to promote the DNA loading of salt stable single MCM2–7 hexamers. Indeed, this observation is consistent with the cryo-EM structure of the OCCM, which indicated the presence of DNA inside the MCM2–7 ring [12].

Do One or Two OCMs Participate in Double Hexamer Formation?

For budding yeast, it is clear that ORC binds to replication origins via a unique binding site, the ARS consensus sequence (ACS). Consequently, the OCM complex is formed at the ACS. Nonetheless, it is currently being investigated whether the OCM directly recruits a second MCM2–7 hexamer, or whether a second OCM is formed at the replication origin in an ACS independent process. Answers to these questions will reveal the molecular basis of eukaryotic replication origins. Indeed, several models have been suggested already [98, 111].

Although we do not know the exact process yet, here we describe the current state of knowledge. Recently, it was found that the second MCM2–7 hexamer is recruited in a Cdt1-dependent manner [10]. In addition, it was observed that both hexamers require an intact Mcm3 C-terminus, which represents a binding site for ORC/Cdc6 [12, 79]. Therefore one can conclude that each hexamer is recruited in a Cdt1-dependent manner. However, it still awaits hard proof whether the recruitment of the second MCM2–7 hexamer also requires ORC/Cdc6. On the other hand, electron microscopy has identified complexes containing a single ORC in complex with an MCM2–7 double-hexamer, while MCM2–7 double-hexamers with two ORC attached were not detected [19]. Moreover, very recent single molecule analysis of pre-RC formation has failed to detect a second ORC complex, suggesting that one ORC complex loads two MCM2–7 hexamers [101, 112]. However, this would require different loading mechanism for the first and second MCM2–7 hexamer. Either complexes that contain two ORCs are too short lived, or they do not exist at all. In order to conclude on the mechanism of MCM2–7 double-hexamer formation additional investigations will be needed.

MCM2–7 Dimerization

Although there is an ongoing discussion about the presence of one or more OCMs, it is clear that the final product of the loading reaction is the salt resistant head-to-head MCM2–7 double hexamer with double stranded DNA passing through the central pore. This means that at some point the two hexameric MCM2–7 rings must be joined via their N-terminal domains.

The double-hexamer interface of archaeal MCM2–7 has been crystallized and spans a large surface that is localized between the zinc binding domains of the Mcm subunits [113]. Recently a systematic mutation analysis of this double-hexamer interface was performed with the aim to arrest pre-RC formation just prior to double-hexamer formation. It was found that budding yeast tolerates single short insertion mutations in this region, while the introduction of three or more of these insertions causes dominant lethality and cell cycle arrest. In particular, mutations in the three neighboring subunits (Mcm2/5/3) have a strong effect, which could indicate that these subunits have an important function during MCM2–7 dimerization [102]. In vitro analysis demonstrated that several triple insertion mutants support OCCM formation and ATP-hydrolysis-dependent OCM formation, but arrest DNA licensing prior to MCM2–7 double-hexamer formation. Interestingly, biochemical experiments proved that these complexes already contain two MCM2–7 hexamers. However, in contrast to wild type MCM2–7 double-hexamers, these MCM2–7 dimers are salt sensitive and inefficient substrates for the S-phase kinase Cdc7-Dbf4. In summary, these experiments discovered important features of MCM2–7 double-hexamer formation. Initially, two MCM2–7 hexamers join and this is followed by a dramatic reorganization of the complex, which makes the Mcm4, Mcm6, and Mcm2 N-termini accessible for DDK and promotes the establishment of the salt-stable double-hexamer interface [102].

The MCM2–7 Double Hexamer

Eukaryotes have hundreds to thousands of replication origins, while bacteria only have a single one. Therefore, initiation of DNA replication in eukaryotes needs to be regulated extremely tightly, otherwise re-replication of DNA might occur. For this reason eukaryotes evolved a two-stage initiation process, which separates helicase loading from helicase activation. In consequence, eukaryotes, in contrast to bacteria, have an entirely different helicase loading mechanism, which yields a helicase that is completely inactive until it is activated in S-phase. Hence, the detailed structural and functional characterization of the MCM2–7 double-hexamer was particularly relevant to understand the molecular basis of eukaryotic DNA replication initiation [14, 16, 19] (Fig. 10.6). An initial EM analysis showed that within the double hexamer the two MCM2–7 rings interact via their N-terminal domains and form a salt-resistant stable structure with a central channel that accommodates double stranded DNA [14, 16]. A more refined EM analysis then identified important structural features and the subunit organization of the complex. One striking observation was that in contrast to the single hexamer the Mcm subunits in the double hexamer are tilted by 30° (Fig. 10.6a). This structural change has significant consequences on the function of the complex, as it leads to an uncoupling of the arginine finger that interconnect the neighboring subunits for ATP hydrolysis (Fig. 10.6b). Indeed, it was observed that the ATPase rate of the double hexamer, in contrast of the single-hexamer was markedly reduced. As a consequence of their interaction the two single hexamers are also tilted and staggered within the double-hexamer. This arrangement, nonetheless, does not disturb the continuity of the central channel that accommodates DNA. Clearly, uncoupling of the Mcm ATPases represents a powerful mechanism to restrict Mcm2–7 helicase activity prior to activation in S-phase [19].

Furthermore, the identification of the subunit organization revealed how the two MCM2–7 rings are aligned with respect to the Mcm2/5 DNA entry gate in the double hexamer (Fig. 10.6a). It was found that the 2/5 gates are offset within the double-hexamer, with the opening of the ring being seemingly impossible, as inter-hexamer connections keep each gate in a closed conformation. Moreover, this subunit organization could constitute another safety mechanism to restrict MCM2–7 ring opening and premature helicase unloading or activation [19].

The architecture of the double hexamer does not only ensure complex stability, but also revealed the structural basis of kinase-dependent helicase activation in S-phase. Cdc7-Dbf4 (DDK) is essential for MCM2–7 helicase activation in all eukaryotes. In budding yeast it was shown that DDK interacts directly with the N-terminal domains of Mcm2 and Mcm4 [78, 114, 115], while genetic experiments have shown that a specific mutation in mcm5 (bob-1) promotes structural changes in MCM2–7, which allow DDK independent helicase activation [116]. Then again, DDK targets not only Mcm2 and Mcm4, but also Mcm6 [77]. The identification of the MCM2–7 double hexamer organization revealed that the two opposing Mcm hexamers generate an integrated surface for DDK-dependent helicase activation, as

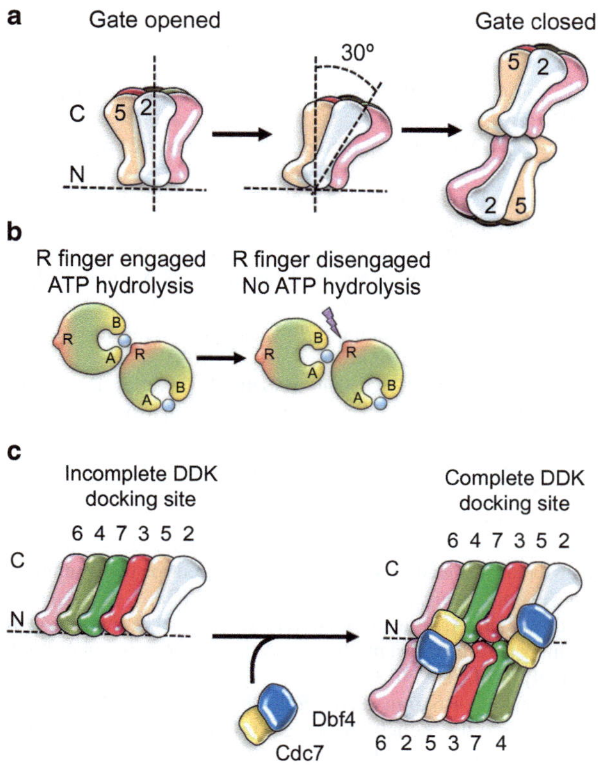

Fig. 10.6 The MCM2–7 double hexamer. (**a**) When the double hexamer is formed, MCM2–7 subunits are tilted by 30°. Within the MCM2–7 double-hexamer the Mcm2/Mcm5 gates are offset and strong inter-hexamer interactions block ring opening. (**b**) The Mcm subunit organization within the single MCM2–7 hexamer support ATP-hydrolysis, but tilting of the subunits during MCM2–7 double-hexamer formation disengages the arginine fingers, which blocks MCM2–7 ATP-hydrolysis. (**c**) The subunit organization within the double hexamer creates a Cdc7/Dbf (DDK) docking site consisting of Mcm2 and Mcm4. Therefore the MCM2–7 double-hexamer can be efficiently phosphorylated by DDK, which is essential for MCM2–7 helicase activation

they bring the Mcm2/4/5/6 subunits in proximity, which can be much better phosphorylated by DDK than within the single hexamer. However, DDK phosphorylation of the two hexamers does not result in large conformational changes in the MCM2–7 double-hexamer [19, 108]. In this regard, phosphorylation is likely to affect the surface of the proteins, generating binding sites to promote recruitment of additional factors required for helicase activation.

Acknowledgements We are grateful to Nick Kennedy, Almut Caspary, and members of the Speck laboratory for helpful discussions and critical reading of the manuscript. Work in the Speck laboratory is supported by the Biotechnology and Biological Sciences Research Council (BBSRC), UK. The authors declare that they have no competing financial interest.

References

1. Leonard AC, Mechali M. DNA replication origins. Cold Spring Harb Perspect Biol. 2013;5(10):a010116.
2. Bell SP, Stillman B. ATP-dependent recognition of eukaryotic origins of DNA replication by a multiprotein complex. Nature. 1992;357(6374):128–34.
3. Liang C, Stillman B. Persistent initiation of DNA replication and chromatin-bound MCM proteins during the cell cycle in cdc6 mutants. Genes Dev. 1997;11(24):3375–86.
4. Weinreich M, Liang C, Stillman B. The Cdc6p nucleotide-binding motif is required for loading mcm proteins onto chromatin. Proc Natl Acad Sci U S A. 1999;96(2):441–6.
5. Samel SA, Fernandez-Cid A, Sun J, Riera A, Tognetti S, Herrera MC, et al. A unique DNA entry gate serves for regulated loading of the eukaryotic replicative helicase MCM2-7 onto DNA. Genes Dev. 2014;28(15):1653–66.
6. Hesketh EL, Parker-Manuel RP, Chaban Y, Satti R, Coverley D, Orlova EV, et al. DNA induces conformational changes in a recombinant human minichromosome maintenance complex. J Biol Chem. 2015;290(12):7973–9.
7. Bochman ML, Schwacha A. The Mcm2-7 complex has in vitro helicase activity. Mol Cell. 2008;31(2):287–93.
8. Nishitani H, Lygerou Z, Nishimoto T, Nurse P. The Cdt1 protein is required to license DNA for replication in fission yeast. Nature. 2000;404(6778):625–8.
9. Maiorano D, Moreau J, Mechali M. XCDT1 is required for the assembly of pre-replicative complexes in Xenopus laevis. Nature. 2000;404(6778):622–5.
10. Fernandez-Cid A, Riera A, Tognetti S, Herrera MC, Samel S, Evrin C, et al. An ORC/Cdc6/MCM2-7 complex is formed in a multistep reaction to serve as a platform for MCM double-hexamer assembly. Mol Cell. 2013;50(4):577–88.
11. Randell JC, Bowers JL, Rodriguez HK, Bell SP. Sequential ATP hydrolysis by Cdc6 and ORC directs loading of the Mcm2-7 helicase. Mol Cell. 2006;21(1):29–39.
12. Sun J, Evrin C, Samel SA, Fernandez-Cid A, Riera A, Kawakami H, et al. Cryo-EM structure of a helicase loading intermediate containing ORC-Cdc6-Cdt1-MCM2-7 bound to DNA. Nat Struct Mol Biol. 2013;20(8):944–51.
13. Evrin C, Fernandez-Cid A, Zech J, Herrera MC, Riera A, Clarke P, et al. In the absence of ATPase activity, pre-RC formation is blocked prior to MCM2-7 hexamer dimerization. Nucleic Acids Res. 2013;41(5):3162–72.
14. Remus D, Beuron F, Tolun G, Griffith JD, Morris EP, Diffley JF. Concerted loading of Mcm2-7 double hexamers around DNA during DNA replication origin licensing. Cell. 2009;139(4):719–30.
15. Gambus A, Khoudoli GA, Jones RC, Blow JJ. MCM2-7 form double hexamers at licensed origins in Xenopus egg extract. J Biol Chem. 2011;286(13):11855–64.
16. Evrin C, Clarke P, Zech J, Lurz R, Sun J, Uhle S, et al. A double-hexameric MCM2-7 complex is loaded onto origin DNA during licensing of eukaryotic DNA replication. Proc Natl Acad Sci U S A. 2009;106(48):20240–5.
17. Diffley JF. Once and only once upon a time: specifying and regulating origins of DNA replication in eukaryotic cells. Genes Dev. 1996;10(22):2819–30.
18. Arias EE, Walter JC. Strength in numbers: preventing rereplication via multiple mechanisms in eukaryotic cells. Genes Dev. 2007;21(5):497–518.
19. Sun J, Fernandez-Cid A, Riera A, Tognetti S, Yuan Z, Stillman B, et al. Structural and mechanistic insights into Mcm2-7 double-hexamer assembly and function. Genes Dev. 2014;28(20):2291–303.
20. Moyer SE, Lewis PW, Botchan MR. Isolation of the Cdc45/Mcm2-7/GINS (CMG) complex, a candidate for the eukaryotic DNA replication fork helicase. Proc Natl Acad Sci U S A. 2006;103(27):10236–41.
21. Gambus A, Jones RC, Sanchez-Diaz A, Kanemaki M, van Deursen F, Edmondson RD, et al. GINS maintains association of Cdc45 with MCM in replisome progression complexes at eukaryotic DNA replication forks. Nat Cell Biol. 2006;8(4):358–66.

22. Tanaka S, Araki H. Helicase activation and establishment of replication forks at chromosomal origins of replication. Cold Spring Harb Perspect Biol. 2013;5(12):a010371.
23. Heller RC, Kang S, Lam WM, Chen S, Chan CS, Bell SP. Eukaryotic origin-dependent DNA replication in vitro reveals sequential action of DDK and S-CDK kinases. Cell. 2011; 146(1):80–91.
24. Yeeles JT, Deegan TD, Janska A, Early A, Diffley JF. Regulated eukaryotic DNA replication origin firing with purified proteins. Nature. 2015;519(7544):431–5.
25. Neuwald AF, Aravind L, Spouge JL, Koonin EV. AAA+: a class of chaperone-like ATPases associated with the assembly, operation, and disassembly of protein complexes. Genome Res. 1999;9(1):27–43.
26. Erzberger JP, Berger JM. Evolutionary relationships and structural mechanisms of AAA+ proteins. Annu Rev Biophys Biomol Struct. 2006;35:93–114.
27. Lenzen CU, Steinmann D, Whiteheart SW, Weis WI. Crystal structure of the hexamerization domain of N-ethylmaleimide-sensitive fusion protein. Cell. 1998;94(4):525–36.
28. Iyer LM, Leipe DD, Koonin EV, Aravind L. Evolutionary history and higher order classification of AAA+ ATPases. J Struct Biol. 2004;146(1-2):11–31.
29. Skarstad K, Katayama T. Regulating DNA replication in bacteria. Cold Spring Harb Perspect Biol. 2013;5(4):a012922.
30. Caruthers JM, McKay DB. Helicase structure and mechanism. Curr Opin Struct Biol. 2002; 12(1):123–33.
31. Rowley A, Cocker JH, Harwood J, Diffley JF. Initiation complex assembly at budding yeast replication origins begins with the recognition of a bipartite sequence by limiting amounts of the initiator, ORC. EMBO J. 1995;14(11):2631–41.
32. Duncker BP, Chesnokov IN, McConkey BJ. The origin recognition complex protein family. Genome Biol. 2009;10(3):214.
33. Romanowski P, Madine MA, Rowles A, Blow JJ, Laskey RA. The Xenopus origin recognition complex is essential for DNA replication and MCM binding to chromatin. Curr Biol. 1996;6(11):1416–25.
34. Mendez J, Stillman B. Chromatin association of human origin recognition complex, cdc6, and minichromosome maintenance proteins during the cell cycle: assembly of prereplication complexes in late mitosis. Mol Cell Biol. 2000;20(22):8602–12.
35. Bell SP, Kobayashi R, Stillman B. Yeast origin recognition complex functions in transcription silencing and DNA replication. Science. 1993;262(5141):1844–9.
36. Prasanth SG, Shen Z, Prasanth KV, Stillman B. Human origin recognition complex is essential for HP1 binding to chromatin and heterochromatin organization. Proc Natl Acad Sci U S A. 2010;107(34):15093–8.
37. Pak DT, Pflumm M, Chesnokov I, Huang DW, Kellum R, Marr J, et al. Association of the origin recognition complex with heterochromatin and HP1 in higher eukaryotes. Cell. 1997; 91(3):311–23.
38. Prasanth SG, Prasanth KV, Stillman B. Orc6 involved in DNA replication, chromosome segregation, and cytokinesis. Science. 2002;297(5583):1026–31.
39. Sasaki T, Gilbert DM. The many faces of the origin recognition complex. Curr Opin Cell Biol. 2007;19(3):337–43.
40. Chesnokov IN. Multiple functions of the origin recognition complex. Int Rev Cytol. 2007;256: 69–109.
41. Huang Z, Zang K, Reichardt LF. The origin recognition core complex regulates dendrite and spine development in postmitotic neurons. J Cell Biol. 2005;170(4):527–35.
42. Clarey MG, Erzberger JP, Grob P, Leschziner AE, Berger JM, Nogales E, et al. Nucleotide-dependent conformational changes in the DnaA-like core of the origin recognition complex. Nat Struct Mol Biol. 2006;13(8):684–90.
43. Speck C, Chen Z, Li H, Stillman B. ATPase-dependent cooperative binding of ORC and Cdc6 to origin DNA. Nat Struct Mol Biol. 2005;12(11):965–71.
44. Chen Z, Speck C, Wendel P, Tang C, Stillman B, Li H. The architecture of the DNA replication origin recognition complex in Saccharomyces cerevisiae. Proc Natl Acad Sci U S A. 2008;105(30):10326–31.

45. Sun J, Kawakami H, Zech J, Speck C, Stillman B, Li H. Cdc6-induced conformational changes in ORC bound to origin DNA revealed by cryo-electron microscopy. Structure. 2012;20(3): 534–44.
46. Clarey MG, Botchan M, Nogales E. Single particle EM studies of the Drosophila melanogaster origin recognition complex and evidence for DNA wrapping. J Struct Biol. 2008;164(3): 241–9.
47. Bleichert F, Balasov M, Chesnokov I, Nogales E, Botchan MR, Berger JM. A Meier-Gorlin syndrome mutation in a conserved C-terminal helix of Orc6 impedes origin recognition complex formation. eLife. 2013;2, e00882.
48. Bleichert F, Botchan MR, Berger JM. Crystal structure of the eukaryotic origin recognition complex. Nature. 2015;519(7543):321–6.
49. Chesnokov I, Remus D, Botchan M. Functional analysis of mutant and wild-type Drosophila origin recognition complex. Proc Natl Acad Sci U S A. 2001;98(21):11997–2002.
50. Klemm RD, Austin RJ, Bell SP. Coordinate binding of ATP and origin DNA regulates the ATPase activity of the origin recognition complex. Cell. 1997;88(4):493–502.
51. Giordano-Coltart J, Ying CY, Gautier J, Hurwitz J. Studies of the properties of human origin recognition complex and its Walker A motif mutants. Proc Natl Acad Sci U S A. 2005;102(1):69–74.
52. Bowers JL, Randell JC, Chen S, Bell SP. ATP hydrolysis by ORC catalyzes reiterative Mcm2-7 assembly at a defined origin of replication. Mol Cell. 2004;16(6):967–78.
53. Liu S, Balasov M, Wang H, Wu L, Chesnokov IN, Liu Y. Structural analysis of human Orc6 protein reveals a homology with transcription factor TFIIB. Proc Natl Acad Sci U S A. 2011;108(18):7373–8.
54. Gillespie PJ, Blow JJ. Nucleoplasmin-mediated chromatin remodelling is required for Xenopus sperm nuclei to become licensed for DNA replication. Nucleic Acids Res. 2000;28(2):472–80.
55. Ghosh S, Vassilev AP, Zhang J, Zhao Y, DePamphilis ML. Assembly of the human origin recognition complex occurs through independent nuclear localization of its components. J Biol Chem. 2011;286(27):23831–41.
56. Thomae AW, Baltin J, Pich D, Deutsch MJ, Ravasz M, Zeller K, et al. Different roles of the human Orc6 protein in the replication initiation process. Cell Mol Life Sci. 2011;68(22):3741–56.
57. Hartwell LH. Three additional genes required for deoxyribonucleic acid synthesis in Saccharomyces cerevisiae. J Bacteriol. 1973;115(3):966–74.
58. Bell SP, Mitchell J, Leber J, Kobayashi R, Stillman B. The multidomain structure of Orc1p reveals similarity to regulators of DNA replication and transcriptional silencing. Cell. 1995;83(4):563–8.
59. Muzi-Falconi M, Kelly TJ. Orp1, a member of the Cdc18/Cdc6 family of S-phase regulators, is homologous to a component of the origin recognition complex. Proc Natl Acad Sci U S A. 1995;92(26):12475–9.
60. Wang B, Feng L, Hu Y, Huang SH, Reynolds CP, Wu L, et al. The essential role of Saccharomyces cerevisiae CDC6 nucleotide-binding site in cell growth, DNA synthesis, and Orc1 association. J Biol Chem. 1999;274(12):8291–8.
61. Speck C, Stillman B. Cdc6 ATPase activity regulates ORC x Cdc6 stability and the selection of specific DNA sequences as origins of DNA replication. J Biol Chem. 2007;282(16):11705–14.
62. Liu J, Smith CL, DeRyckere D, DeAngelis K, Martin GS, Berger JM. Structure and function of Cdc6/Cdc18: implications for origin recognition and checkpoint control. Mol Cell. 2000;6(3):637–48.
63. Perkins G, Drury LS, Diffley JF. Separate SCF(CDC4) recognition elements target Cdc6 for proteolysis in S phase and mitosis. EMBO J. 2001;20(17):4836–45.
64. Bochman ML, Schwacha A. The Mcm complex: unwinding the mechanism of a replicative helicase. Microbiol Mol Biol Rev. 2009;73(4):652–83.
65. Maine GT, Sinha P, Tye BK. Mutants of S. cerevisiae defective in the maintenance of minichromosomes. Genetics. 1984;106(3):365–85.

66. Costa A, Onesti S. Structural biology of MCM helicases. Crit Rev Biochem Mol Biol. 2009;44(5):326–42.
67. Chong JP, Mahbubani HM, Khoo CY, Blow JJ. Purification of an MCM-containing complex as a component of the DNA replication licensing system. Nature. 1995;375(6530):418–21.
68. Pacek M, Walter JC. A requirement for MCM7 and Cdc45 in chromosome unwinding during eukaryotic DNA replication. EMBO J. 2004;23(18):3667–76.
69. Shechter D, Ying CY, Gautier J. DNA unwinding is an Mcm complex-dependent and ATP hydrolysis-dependent process. J Biol Chem. 2004;279(44):45586–93.
70. Labib K, Tercero JA, Diffley JF. Uninterrupted MCM2-7 function required for DNA replication fork progression. Science. 2000;288(5471):1643–7.
71. Aparicio OM, Weinstein DM, Bell SP. Components and dynamics of DNA replication complexes in S. cerevisiae: redistribution of MCM proteins and Cdc45p during S phase. Cell. 1997; 91(1):59–69.
72. Davey MJ, Indiani C, O'Donnell M. Reconstitution of the Mcm2-7p heterohexamer, subunit arrangement, and ATP site architecture. J Biol Chem. 2003;278(7):4491–9.
73. Schwacha A, Bell SP. Interactions between two catalytically distinct MCM subgroups are essential for coordinated ATP hydrolysis and DNA replication. Mol Cell. 2001;8(5): 1093–104.
74. Coster G, Frigola J, Beuron F, Morris EP, Diffley JF. Origin licensing requires ATP binding and hydrolysis by the MCM replicative helicase. Mol Cell. 2014;55(5):666–77.
75. Bochman ML, Bell SP, Schwacha A. Subunit organization of Mcm2-7 and the unequal role of active sites in ATP hydrolysis and viability. Mol Cell Biol. 2008;28(19):5865–73.
76. Ilves I, Petojevic T, Pesavento JJ, Botchan MR. Activation of the MCM2-7 helicase by association with Cdc45 and GINS proteins. Mol Cell. 2010;37(2):247–58.
77. Randell JC, Fan A, Chan C, Francis LI, Heller RC, Galani K, et al. Mec1 is one of multiple kinases that prime the Mcm2-7 helicase for phosphorylation by Cdc7. Mol Cell. 2010;40(3): 353–63.
78. Sheu YJ, Stillman B. The Dbf4-Cdc7 kinase promotes S phase by alleviating an inhibitory activity in Mcm4. Nature. 2010;463(7277):113–7.
79. Frigola J, Remus D, Mehanna A, Diffley JF. ATPase-dependent quality control of DNA replication origin licensing. Nature. 2013;495(7441):339–43.
80. Costa A, Renault L, Swuec P, Petojevic T, Pesavento JJ, Ilves I, et al. DNA binding polarity, dimerization, and ATPase ring remodeling in the CMG helicase of the eukaryotic replisome. eLife. 2014;3, e03273.
81. Hofmann JF, Beach D. cdt1 is an essential target of the Cdc10/Sct1 transcription factor: requirement for DNA replication and inhibition of mitosis. EMBO J. 1994;13(2):425–34.
82. Cook JG, Chasse DA, Nevins JR. The regulated association of Cdt1 with minichromosome maintenance proteins and Cdc6 in mammalian cells. J Biol Chem. 2004;279(10):9625–33.
83. Tanaka S, Diffley JF. Interdependent nuclear accumulation of budding yeast Cdt1 and Mcm2-7 during G1 phase. Nat Cell Biol. 2002;4(3):198–207.
84. Liu C, Wu R, Zhou B, Wang J, Wei Z, Tye BK, et al. Structural insights into the Cdt1-mediated MCM2-7 chromatin loading. Nucleic Acids Res. 2012;40(7):3208–17.
85. Takara TJ, Bell SP. Multiple Cdt1 molecules act at each origin to load replication-competent Mcm2-7 helicases. EMBO J. 2011;30(24):4885–96.
86. Wei Z, Liu C, Wu X, Xu N, Zhou B, Liang C, et al. Characterization and structure determination of the Cdt1 binding domain of human minichromosome maintenance (Mcm) 6. J Biol Chem. 2010;285(17):12469–73.
87. Khayrutdinov BI, Bae WJ, Yun YM, Lee JH, Tsuyama T, Kim JJ, et al. Structure of the Cdt1 C-terminal domain: conservation of the winged helix fold in replication licensing factors. Protein Sci. 2009;18(11):2252–64.
88. Lee C, Hong B, Choi JM, Kim Y, Watanabe S, Ishimi Y, et al. Structural basis for inhibition of the replication licensing factor Cdt1 by geminin. Nature. 2004;430(7002):913–7.
89. Caillat C, Perrakis A. Cdt1 and geminin in DNA replication initiation. Subcell Biochem. 2012;62:71–87.

90. Yanagi K, Mizuno T, You Z, Hanaoka F. Mouse geminin inhibits not only Cdt1-MCM6 interactions but also a novel intrinsic Cdt1 DNA binding activity. J Biol Chem. 2002;277(43): 40871–80.
91. Wohlschlegel JA, Dwyer BT, Dhar SK, Cvetic C, Walter JC, Dutta A. Inhibition of eukaryotic DNA replication by geminin binding to Cdt1. Science. 2000;290(5500):2309–12.
92. Tada S, Li A, Maiorano D, Mechali M, Blow JJ. Repression of origin assembly in metaphase depends on inhibition of RLF-B/Cdt1 by geminin. Nat Cell Biol. 2001;3(2):107–13.
93. Liontos M, Koutsami M, Sideridou M, Evangelou K, Kletsas D, Levy B, et al. Deregulated overexpression of hCdt1 and hCdc6 promotes malignant behavior. Cancer Res. 2007;67(22):10899–909.
94. McGarry TJ, Kirschner MW. Geminin, an inhibitor of DNA replication, is degraded during mitosis. Cell. 1998;93(6):1043–53.
95. Lutzmann M, Maiorano D, Mechali M. A Cdt1-geminin complex licenses chromatin for DNA replication and prevents rereplication during S phase in Xenopus. EMBO J. 2006; 25(24):5764–74.
96. Saxena S, Yuan P, Dhar SK, Senga T, Takeda D, Robinson H, et al. A dimerized coiled-coil domain and an adjoining part of geminin interact with two sites on Cdt1 for replication inhibition. Mol Cell. 2004;15(2):245–58.
97. Siddiqui K, On KF, Diffley JF. Regulating DNA replication in eukarya. Cold Spring Harb Perspect Biol. 2013;5(9):012930.
98. Riera A, Tognetti S, Speck C. Helicase loading: how to build a MCM2-7 double-hexamer. Semin Cell Dev Biol. 2014;30:104–9.
99. Alver RC, Chadha GS, Blow JJ. The contribution of dormant origins to genome stability: from cell biology to human genetics. DNA Repair. 2014;19:182–9.
100. Wu M, Lu W, Santos RE, Frattini MG, Kelly TJ. Geminin inhibits a late step in the formation of human pre-replicative complexes. J Biol Chem. 2014;289(44):30810–21.
101. Ticau S, Friedman LJ, Ivica NA, Gelles J, Bell SP. Single-molecule studies of origin licensing reveal mechanisms ensuring bidirectional helicase loading. Cell. 2015;161(3):513–25.
102. Evrin C, Fernandez-Cid A, Riera A, Zech J, Clarke P, Herrera MC, et al. The ORC/Cdc6/ MCM2-7 complex facilitates MCM2-7 dimerization during prereplicative complex formation. Nucleic Acids Res. 2014;42(4):2257–69.
103. Wu R, Wang J, Liang C. Cdt1p, through its interaction with Mcm6p, is required for the formation, nuclear accumulation and chromatin loading of the MCM complex. J Cell Sci. 2012; 125(Pt 1):209–19.
104. Chen S, de Vries MA, Bell SP. Orc6 is required for dynamic recruitment of Cdt1 during repeated Mcm2-7 loading. Genes Dev. 2007;21(22):2897–907.
105. Klemm RD, Bell SP. ATP bound to the origin recognition complex is important for preRC formation. Proc Natl Acad Sci U S A. 2001;98(15):8361–7.
106. Perkins G, Diffley JF. Nucleotide-dependent prereplicative complex assembly by Cdc6p, a homolog of eukaryotic and prokaryotic clamp-loaders. Mol Cell. 1998;2(1):23–32.
107. Herbig U, Marlar CA, Fanning E. The Cdc6 nucleotide-binding site regulates its activity in DNA replication in human cells. Mol Biol Cell. 1999;10(8):2631–45.
108. On KF, Beuron F, Frith D, Snijders AP, Morris EP, Diffley JF. Prereplicative complexes assembled in vitro support origin-dependent and independent DNA replication. EMBO J. 2014; 33(6):605–20.
109. Kang S, Warner MD, Bell SP. Multiple functions for Mcm2-7 ATPase motifs during replication initiation. Mol Cell. 2014;55(5):655–65.
110. Ying CY, Gautier J. The ATPase activity of MCM2-7 is dispensable for pre-RC assembly but is required for DNA unwinding. EMBO J. 2005;24(24):4334–44.
111. Yardimci H, Walter JC. Prereplication-complex formation: a molecular double take? Nat Struct Mol Biol. 2014;21(1):20–5.
112. Duzdevich D, Warner MD, Ticau S, Ivica NA, Bell SP, Greene EC. The dynamics of eukaryotic replication initiation: origin specificity, licensing, and firing at the single-molecule level. Mol Cell. 2015;58(3):483–94.

113. Fletcher RJ, Shen J, Gomez-Llorente Y, Martin CS, Carazo JM, Chen XS. Double hexamer disruption and biochemical activities of Methanobacterium thermoautotrophicum MCM. J Biol Chem. 2005;280(51):42405–10.
114. Ramer MD, Suman ES, Richter H, Stanger K, Spranger M, Bieberstein N, et al. Dbf4 and Cdc7 proteins promote DNA replication through interactions with distinct Mcm2-7 protein subunits. J Biol Chem. 2013;288(21):14926–35.
115. Lei M, Kawasaki Y, Young MR, Kihara M, Sugino A, Tye BK. Mcm2 is a target of regulation by Cdc7-Dbf4 during the initiation of DNA synthesis. Genes Dev. 1997;11(24):3365–74.
116. Hardy CF, Dryga O, Seematter S, Pahl PM, Sclafani RA. mcm5/cdc46-bob1 bypasses the requirement for the S phase activator Cdc7p. Proc Natl Acad Sci U S A. 1997;94(7):3151–5.

Chapter 11
Coordination of DNA Replication and Histone Synthesis During S Phase

Johanna Paik, Serena Giovinazzi, and Akash Gunjan

Abstract The coordination of DNA replication with histone synthesis is of utmost importance as any imbalance between the two processes results in genomic instability and may even cause lethality. Hence, to maintain genome stability, histone synthesis is regulated at multiple levels—transcriptionally, posttranscriptionally and by modulating protein stability. This tight regulation facilitates the creation of a very transient histone pool for replication-coupled chromatin assembly and ensures that histone synthesis is downregulated when DNA replication is completed or stalled due to replication inhibition. As illustrated in this chapter, the bulk of histone synthesis during S phase is activated by the same cell cycle signals that initiate DNA replication and downregulated by the same DNA damage response pathways that arrest the DNA replication machinery upon DNA damage. Conversely, the availability of histone proteins and their chaperones that help package the newly replicated DNA into chromatin in turn regulate replication fork progression. Further, in senescent cells, the histone chaperone *Hi*stone *R*egulatory Homolog *A* (HIRA), a co-repressor of histone gene transcription, plays an important role in the formation of transcriptionally silent heterochromatin that incorporates replication-dependent histone genes as well as many genes needed for DNA replication to concomitantly shut down both histone and DNA synthesis. This chapter discusses the current state of knowledge on the coregulation of histone and DNA synthesis during S phase.

Keywords Histone • Histone gene regulation • Chromatin assembly • Histone chaperone • Replication-dependent histones • MCM • HIRA • Cyclin • Cdk • NPAT • SLBP • DDR • SAHF • Hir

---

J. Paik (✉) • S. Giovinazzi • A. Gunjan
Department of Biomedical Sciences, College of Medicine, Florida State University, 1115 West Call Street, Tallahassee, FL 32306, USA
e-mail: johanna.paik@med.fsu.edu; serena.giovinazzi@med.fsu.edu; akash.gunjan@med.fsu.edu

D.L. Kaplan (ed.), *The Initiation of DNA Replication in Eukaryotes*,
DOI 10.1007/978-3-319-24696-3_11

Introduction

The genomic DNA in eukaryotic cells exists in the form of nucleoprotein filaments collectively known as chromatin. The basic repeating unit of chromatin comprises the nucleosome in which 147 bp of DNA is wrapped around an octameric protein core consisting of two molecules each of the core histone proteins H2A, H2B, H3, and H4 [1]. The binding of linker histone H1 to the nucleosomes helps fold the chromatin further by forming more condensed higher order structures [2]. Histone chaperones help package the DNA into nucleosomes rapidly behind the replication fork as soon as a sufficient amount (~200 bp) of newly synthesized DNA emerges from the replication fork [3]. DNA replication and histone synthesis are strictly coordinated and it has been known for decades that when DNA replication is inhibited upon exposure to genotoxic agents, histone synthesis is rapidly downregulated in an evolutionary conserved response [4–7]. Conversely, a deficiency of histones or histone chaperones also has a strong inhibitory effect on both DNA replication initiation and elongation in higher eukaryotes [8–12], demonstrating that histone and DNA synthesis are intimately interlinked processes. This chapter discusses our current understanding of how histone and DNA synthesis are coordinately regulated during S phase of the cell cycle in budding yeast and mammalian cells.

Coordination of Histone Synthesis and DNA Replication in Budding Yeast

The coupling of histone gene expression and DNA replication in the budding yeast *Saccharomyces cerevisiae* is believed to be mainly achieved through transcriptional control as histone transcripts have a relatively short half-life [13]. There are 11 histone genes in budding yeast: single genes encoding H2A.Z (*HTZ1*) variant, centromeric H3 variant (*CSE4*), the linker histone H1 (*HHO1*) and two genes for each of the core histone proteins H2A, H2B, H3 and H4. The core histone genes occur in pairs, with the genes for H3 (*H*istone *H* *T*hree 1 and 2, *HHT1* and *HHT2*) being paired with H4 (*H*istone *H* *F*our 1 and 2, *HHF1* and *HHF2*) and that of H2A (*H*istone *T*wo *A* 1 and 2, *HTA1* and *HTA2*) with H2B (*H*istone *T*wo *B* 1 and 2, *HTB1* and *HTB2*), which presumably helps maintain the right stoichiometry of individual core histone proteins relative to each other [14].

Each core histone gene pair is transcribed divergently from a common promoter at the G1/S boundary and repressed outside of S phase. This cell cycle regulation is conferred by activating and repressing elements present in the replication-dependent core histone (H2A, H2B, H3, and H4) promoters (Fig. 11.1a). Common to all the four histone promoters are *U*pstream *A*ctivating *S*equence (UAS) elements that are bound by the *S*u*p*pressor of *T*y 10 (Spt10), the only histone promoter specific transcriptional co-activator known so far in the budding yeast. It has been suggested that Spt10 dimerizes and binds to pairs of histone UAS elements in late G1 [15].

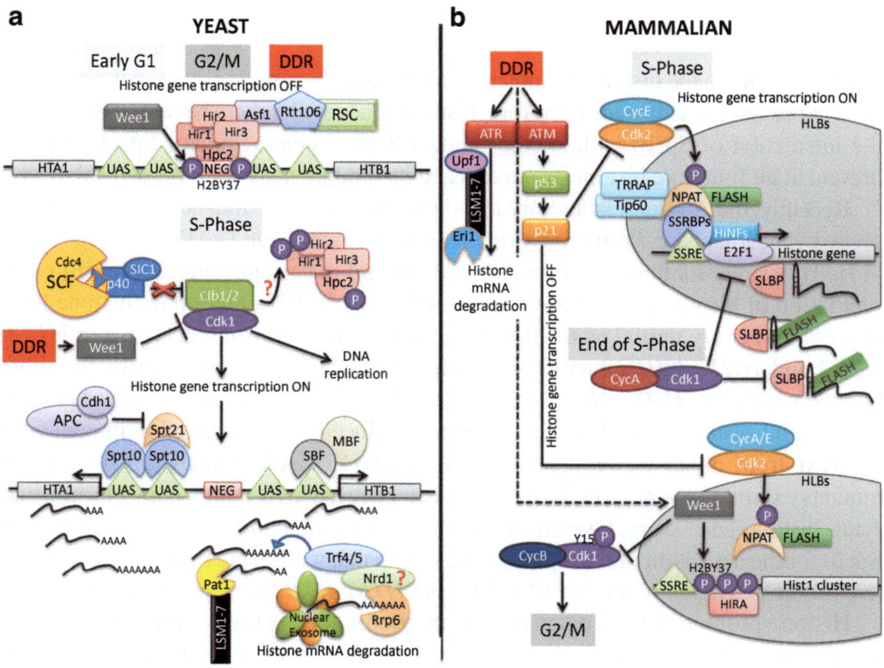

Fig. 11.1 Regulation of histone mRNA levels in budding yeast and mammals. Although the actual regulatory factors involved are different, in principle similar mechanisms govern histone gene transcription as well as histone mRNA levels in yeast and mammals. (**a**) Yeast histone genes are repressed outside of S phase and upon DNA damage in S phase by the Hir complex (Hir1–3, Hpc2) with help from the chaperones Asf1 and Rtt106. These factors mediate their repressive effects via the negative regulatory element (NEG). In S phase, SCF^{Cdc4} promotes the activation and accumulation of the mitotic cyclins Clb1/2. We speculate that Clb1/2 may prompt Cdk1 mediated Hir complex phosphorylation, thereby relieving the repression of histone genes. Wee1 kinase phosphorylates and inhibits Cdk1 as part of the G2/M cell cycle checkpoint and the DNA Damage Response (DDR). Additionally, Wee1 phosphorylation of H2B tyrosine 37 has been shown to facilitate Hir complex binding to repress histone gene transcription. The Upstream Activating Sequences (UAS) in the histone gene promoters are bound by coactivators Spt10 and SBF (a heterodimer of Swi4 and Swi6), in a mutually exclusive manner to activate histone gene transcription in early S phase. Polyadenylation of histone mRNAs by Trf4/5 may contribute to their posttranscriptional regulation by modulating their half-lives. Histone mRNAs are degraded by the nuclear exosome in a process possibly mediated by Nrd1. The Lsm1-7-Pat1 complex may also degrade histone mRNAs based on the length of their polyA tails. (**b**) Mammalian histone genes require NPAT for transcription which is recruited to histone promoters by SSRE binding proteins (SSRBPs). Numerous transcription factors (HiNFs) confer tissue specificity or transcriptional control for different replication-dependent core histone subtypes. Mammalian replication-dependent core histone mRNAs do not possess poly-A tails but a stem loop in their 3′ UTR which is recognized by Stem Loop Binding Protein (SLBP) for posttranscriptional and translational processing. Mature histone mRNAs are stabilized and targeted to the cytoplasm by SLBP. At the end of S phase, the cell cycle machinery represses histone synthesis by inhibiting SLBP and E2F1, whilst activating the yeast Hir1/2 protein homolog HIRA. As in yeast, Wee1 phosphorylation of H2B tyrosine 37 also facilitates HIRA binding and termination of histone synthesis at the end of S phase. In addition, upon DNA damage, Wee1 inhibits Cdk1 activity through specific residues phosphorylation; this conserved mechanism prevents mitotic entry in presence of DNA lesions. In a complex signaling network involving checkpoint kinases, the DDR triggers both replication arrest and downregulation of histone synthesis through inhibition of histone gene transcription via NPAT inactivation and histone mRNA degradation via the LSM1–7 complex. Question marks indicate speculative pathways that currently lack direct experimental evidence. *P* phosphate. *HLBs* histone locus bodies

Spt10 is bound to the histone promoters in alpha factor arrested G1 cells before histone gene expression is activated. Therefore, histone genes are likely to be turned on through the interaction of Spt10 with another known histone gene regulator Spt21 which is present at all four core histone promoters and its binding peaks in S phase [16].

Recently, the Andrews laboratory has shown that Spt21 abundance is restricted to S phase by the *A*naphase-*P*romoting *C*omplex *Cdc*20 *h*omologue 1 (APC^{Cdh1}) (Fig. 11.1a) [17]. This suggests that Spt21 levels oscillate and may contribute to the cell cycle regulation of histone genes. However, the effect of *SPT21* deletion on histone gene expression is only partial as the deletion mutants show a strong decrease in *HTA2-HTB2* and *HHT2-HHF2* histone transcript levels but the expression of the other histone gene pairs *HHT1-HHF1* and *HTA1-HTB1* are only modestly reduced [17]. In contrast, all histone mRNA levels are greatly reduced in *spt10* deletion mutant cells released from alpha factor mediated G1 arrest [18]. Additionally, *spt10* deletion mutants exhibit a severe growth defect that can be rescued by a high copy plasmid carrying histone gene pairs. Overall, these observations demonstrate the importance of Spt10 for the activation of multiple histone gene pairs, while Spt21 may be more selective for the activation of the *HHT2-HHF2* and *HTA2-HTB2* gene pairs (Fig. 11.1a).

Histone promoters have putative binding motifs for the transcription factor Swi4 and genome-wide *Ch*romatin *I*mmuno*p*recipitation (ChIP) analysis shows that they are bound by SBF and possibly MBF transcription factors [19–22]. MBF and SBF are heterodimers containing Swi6 (a regulatory factor) and a sequence specific DNA binding factor Mbp1 (in MBF) and Swi4 (in SBF) respectively [23]. SBF and Spt10 bind to the same UAS elements (UAS2 and UAS3) at the *HTA1-HTB1* promoter and their binding is mutually exclusive in vitro (Fig. 11.1a). SBF binding peaks in early S phase and concomitantly a SBF mediated small peak in histone gene expression is detectable. By the end of S phase, the binding of SBF and Spt10 at the histone promoters is completely abolished [15]. Overall, histone gene expression is only modestly reduced in *swi4* or *mbp1* deletion strains with some histone genes showing a small but significant reduction [16]. Spt10 and SBF may together control the expression from the *HTA1-HTB1* promoter in G1/S as *HTA1-HTB1* expression is completely abolished when all four UAS elements are removed from the promoter [15].

Inappropriate expression of histone genes outside of S phase or when DNA replication is arrested is associated with toxicity. To prevent this, histone genes are also subjected to negative regulation [8]. A group of four functionally related *Hi*stone *R*egulators (Hir1, Hir2, and Hir3) and the *H*istone *P*romoter *C*ontrol 2 (Hpc2) proteins are shown to act in a complex as transcriptional co-repressors for three of the four histone gene pairs (Fig. 11.1a), *HTA1-HTB1*, *HHT1-HHF1*, and *HHT2-HHF2*, both outside of S phase and in response to replication arrest [14, 24]. A fifth protein, Asf1 (Anti-Silencing Function 1) also co-purifies with Hir proteins and *asf1* mutants are partially defective in the repression of transcription from *HTA1-HTB1*, *HHT1-HHF1*, and *HHT2-HHF2* gene pairs upon replication arrest [25]. Asf1/Hir-mediated repression of transcription relies on a specific DNA sequence, the negative (NEG) regulatory element which is present on the promoters of the three Hir-regulated gene pairs (Fig. 11.1a), but is absent from that of the *HTA2-HTB2* promoter [4]. The *HTA2-HTB2* gene pair also undergoes cell cycle regulation and its transcripts disappear outside of S phase as well as in response to replication arrest, but in a Hir-independent manner [26]. Asf1 and Hir are also chromatin assembly factors that bind to histones and help assemble and disassemble chromatin [27]. Asf1 stimulates the replication-independent nucleosome assembly activity of the Hir complex in vitro [28]. More recently, the H3/H4 chromatin assembly factor Rtt106 was identified as an additional regulator of histone gene expression [29]. Rtt106 functions downstream of Asf1 and the Hir complex and localizes to the histone gene promoters in a Hir- and Asf1-dependent manner (Fig. 11.1a) [29]. Further, the chromatin remodeling complex Rsc is also recruited to the histone gene promoters in a Hir and Rtt106-dependent manner (Fig. 11.1a) [30]. Rsc recruitment is mediated by Rtt106 and its recruitment to the histone gene promoters is compromised in *rtt106* mutant cells, although the Hir complex and Asf1 remain bound [30].

Histone promoters are devoid of nucleosomes in either *hir* or *asf1* or *rtt106* mutant cells, suggesting that these histone chaperones form a repressive chromatin structure at the histone gene promoters to shut down transcription [29]. However, the defect in histone gene repression in an *rtt106* or *asf1* mutant is relatively mild compared to the nearly complete derepression of histone genes observed in *hir/hpc* mutants [29]. This raises the intriguing question whether a different pathway/factor exists through which the Hir complex exerts its repressive role at the histone gene promoters. It is also possible that part of the repressive activity of the Hir complex is mediated via Asf1 and Rtt106 such as under conditions of replication stress in S phase, while the remainder is carried out directly by the Hir complex itself in G2, M, and G1 in an Asf1/Rtt106 independent manner [31].

Previous work from the Osley and Johnston laboratories has demonstrated the requirement of a functional *C*ell *D*ivision *C*ycle 4 (Cdc4) protein for the transcription of the *HTA1* gene in late G1 [32, 33] (Fig. 11.2). Cdc4 is critical for DNA replication initiation at G1/S and for meiotic nuclear division at G2/M [34, 35]. Cdc4, Cdc34, Cdc53, and Skp1 proteins form the *S*kp, *C*ullin, *F*-box containing SCF-Cdc4 complex which is required for the ubiquitin-dependent proteolysis of a number of genes that regulate cell cycle progression such as Sic1, Cdc6, Clb6, Cln3 [36–41]. Based on this, Amin et al tested if Cdc4 coordinates DNA replication with histone synthesis in G1/S. Using *cdc4 ts* (temperature sensitive) mutants they demonstrated that the activation of the *HTA1-HTB1* gene pair is dependent on Cdc4 [31]. Since the major role of SCF-Cdc4 complex at the G1/S transition is the activation of all six Clb/Cdk1 complexes, this raises the question whether any of the six Clb/Cdk1 complexes are involved in activating histone gene expression [31]. Unexpectedly, S phase cyclins Clb5 and Clb6 which are involved in DNA initiation did not have any major impact on histone gene activation, although their activation was slightly delayed. Instead, the activation of histone genes *HTA1-HTB1* and *HTA2* showed a strong dependence on the mitotic Clb1 and Clb2 cyclins [31]. Further, Clb1/2 cyclin dependency can be bypassed by the disruption of the Hir complex which negatively regulates histone gene expression [31]. This suggests that the Cdks may negatively regulate the Hir-complex and as a result histone genes are activated in G1/S (Fig. 11.1a). *Hi*stone *R*egulatory homolog *A* (HIRA), the mammalian homolog of Hir1 and Hir2 in yeast is a direct Cdk target and as such it is entirely possible that the Hir-complex is also

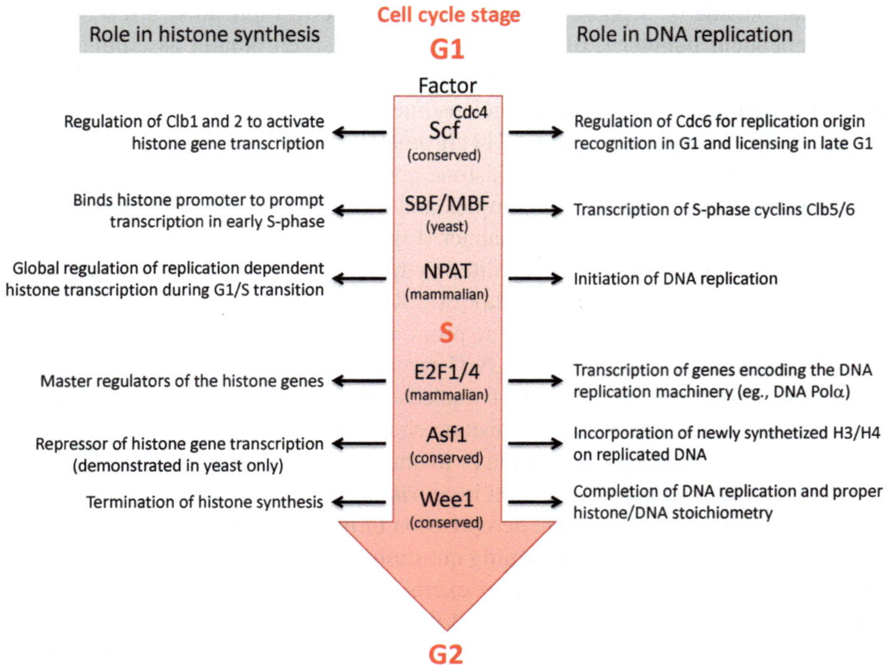

Fig. 11.2 Cross talk between the regulation of DNA and histone synthesis. Regulation of histone levels and DNA replication are tightly coupled. Several components of the cell cycle machinery cooperate with common factors involved in regulating both DNA and histone synthesis to ensure that enough histones are synthetized to satisfy chromatin assembly requirements following DNA replication, since both free histones and naked DNA are toxic for the cell. Some of these regulatory factors for DNA/histone synthesis are evolutionarily conserved while others are unique to either yeast or mammalian cells

regulated by Cdk phosphorylation [42]. Indeed, Hpc2 was identified as a direct target for Cdk1 in a global screen to identify new Clb2/Cdk1 substrates [43]. Further, Clb2 associates with the Hir complex *in vivo* [31] and the Clb1/Clb2-associated Cdk1 kinase is active at the G1/S transition when histone gene transcription is activated, although its kinase activity is lower than that observed in G2 [44–46].

Histone genes in yeast are also subjected to posttranscriptional regulation and this mechanism contributes to the cell cycle regulation of histone genes. Unlike their mammalian counterparts that end in a stem loop, yeast histone mRNAs are polyadenylated (Fig. 11.1a). The region that confers cell cycle regulation even when histone genes are expressed from a constitutive heterologous promoter has been mapped and lies in the 3′ *U*ntranslated *R*egion (UTR). The 3′ UTR of the *HTB1* gene contains a *D*istal *D*ownstream *E*lement (DDE) that lies about 100 nt downstream of the 3′-end cleavage site [47]. In addition, the DDE contains a transcription termination site and mutations in the DDE prevent the degradation of a reporter mRNA [48]. Work from the Bond and Campbell laboratories has demonstrated the involvement of the nuclear exosome (a multi-subunit complex of 3′ to 5′ exonucleases) and the 3′-end processing machinery consisting of the *Tr*f4/*A*ir2/*M*tr4p *P*olyadenylation (TRAMP) complex in the cell cycle regulation of histone mRNAs [48, 49]. The nuclear exosome processes many RNAs in the nucleus that need to be first polyadenylated by the TRAMP complex comprising the PolyA polymerases Trf4 and Trf5, a putative ATP-dependent RNA helicase Mtr4p/Dob1p and the zinc knuckle-binding proteins Air1 and Air2 (TRAMP) [50–52]. Deletion of the nuclear exosome component Rrp6 as well as the TRAMP components Trf4 and Trf5 result in high levels of histone mRNAs [49]. It is thought that Trf4 and Trf5 components of the TRAMP complex add short polyA tails and tag the mRNA to be degraded by the nuclear exosome [50, 53]. In the case of histone mRNA regulation, the Air1 and Air2 RNA-binding proteins of the TRAMP complex are not needed and therefore another RNA-binding protein such as Nrd1 may support the recruitment of the nuclear exosome via interaction with Rrp6. Indeed, Nrd1 has been shown to interact with the TRAMP complex and stimulate the 3′-end processing of RNA Pol II transcripts by the exosome [54]. The transcriptional termination site which lies in the DDE of the *HTB1* gene seems to be critical for the interaction with TRAMP and the nuclear exosome (Table 11.1).

The Role of Transcriptional Regulation in Coordinating Histone Synthesis and DNA Replication in S Phase in Mammalian Cells

Transcriptional Regulation of Core Histones in Mammalian Cells

The tremendous increase in histone protein levels during S phase to meet the demand for new chromatin assembly is a direct result of the accumulation of histone mRNAs in S phase, which in turn is controlled at the level of transcription, pre-mRNA processing and mRNA stability in mammalian cells [55–57]. We will first consider the contribution of histone gene transcription in ensuring an adequate supply of histones during DNA replication. There are multiple copies of replication-dependent or canonical histone H2A, H2B, H3, and H4 genes, the majority of which are clustered in a small number of chromosomal loci in mammals and are expressed only during S phase in a replication-dependent manner [57, 58]. The remainder of the histones genes is mostly single copy replication-*in*dependent histone genes scattered across the genome and these express primary sequence variants of the canonical histone genes throughout the cell cycle and appear to possess unique functions [57]. Replication-dependent histone gene promoters generally contain a core promoter for initiation by RNA polymerase II, a *S*ubtype *S*pecific *C*onsensus *E*lement (SSCE) and a distal activation domain. In the case of histone H4, whose regulation is the best studied among mammalian histone genes, cell cycle dependency comes from the SSCE element. The regulation of each of the mammalian histone genes involves different *cis*-acting elements and multiple transcription factors. Over the years, studies have uncovered many of these transcription factors and named them

Table 11.1 List of genes

Gene		Description	Species
Air2	Arginine methyltransferase-Interacting RING finger 2	RNA-binding subunit of the TRAMP complex	Yeast
APC	Anaphase-promoting complex	Ubiquitin-protein ligase degrades substrates at the metaphase/anaphase transition	Yeast; mammal
Asf1	Anti-silencing function 1	Corepressor; stimulates the replication-independent nucleosome assembly	Yeast
ATM	Ataxia Telangiectasia mutated	DNA damage response (DDR) kinase	Mammal; yeast (Tel1)
ATR	ATM and Rad3 related	DDR kinase	Mammal; yeast (Mec1)
Cdc	Cell division cycle	Group of proteins that timely regulate cell cycle	Yeast; mammal
Cdks	Cyclin-dependent kinases	Ser/Thr kinases regulating the cell cycle; active only in a complex with cyclins	Yeast; mammal
CDP	CCAAT displacement protein	HiNF-D	Mammal
CHK2	Checkpoint kinase 2	Protein kinase; DDR activated to arrest the cell-cycle and DNA replication and to initiate DNA repair	Mammal/ yeast (Rad53)
Clb	B type cyclin	Cell cycle progression/DNA replication	Yeast; mammal
E2F1/4	E2F transcription factor 1/4	Transcription factor with crucial role in cell cycle control	Mammal
FLASH	Fadd Like IL1B converting enzyme associated Huge protein	NPAT mediated regulator of histones transcription	Mammal
HiNF-P	Histone nuclear factor-P	Transcription factor specific for H4	Mammal
HiNFs	Histone nuclear factors	Transcription factors for histone genes	Mammal
Hir1/2/3	HIstone regulation 1/2/3	Subunits of the Hir corepressor complex; nucleosome assembly	Yeast; mammal (HIRA)
HMG-I	High mobility group-I	HiNF-A	Mammal
HPC2	Histone periodic control 2	Subunit of the Hir complex	Yeast
IRF2	Interferon regulatory factor 2	HiNF-M	Mammal
LSM1–7	Sm-like proteins 1–7	Heptameric complex with role in mRNA degradation	Yeast; mammal
MBF	Mbp1-Swi6 heterodimer binding factor	Transcription factor	Yeast
MCM	Mini-chromosome maintenance	The MCM2–7 complex is a putative replicative helicase	Mammal; yeast (Cdc47)

(continued)

Table 11.1 (continued)

Gene		Description	Species
Mtr4	mRNA transport 4	ATP-dependent 3′–5′ RNA helicase of the TRAMP complex	Yeast
NPAT	Nuclear protein Ataxia Telangiectasia	Essential factor for S phase-dependent histone gene transcription	Mammal
Nrd1	Nuclear pre-mRNA downregulation 1	RNA-binding subunit of the exosome complex	Yeast
OCA-S	Oct-1 co-activator in S phase	Cell cycle regulator of H2B transcription	Mammal
Oct-1	Octamer binding factor 1	Transcription factor	Mammal
PCNA	Proliferating cell nuclear antigen	Protein essential for DNA replication and repair; sliding clamp	Mammal; yeast (Pol30)
Rad53	RADiation sensitive 53	Protein kinase; DDR activated to arrest the cell-cycle and DNA replication and to initiate DNA repair	Yeast; mammal (Chk2)
Rrp6	Ribosomal RNA processing 6	Nuclear exosome exonuclease	Yeast
Rtt106	Regulator of Ty1 transposition	Histone chaperone	Yeast
SBF	SBC binding factor	Swi4-Swi6 Heterodimer transcription factor	Yeast
SLBP	Stem loop binding protein	Protein required for histone pre-mRNA processing and mRNA export to cytoplasm	Mammal
SMAD	Small and mothers against decapentaplegic	Transcription factor	Mammal
SP1	Specificity protein 1	HiNF-C	Mammal
Spt10	Suppressor of Ty 10	Sequence-specific activator of histone genes	Yeast
Spt21	Suppressor of Ty 21	Transcriptional silencer	Yeast
Swi4	SWItching deficient 4	DNA binding of SCB	Yeast
Swi6	SWItching deficient 6	Regulatory factor	Yeast
Tip-60	60 kDa Tat-interactive protein	Histone acetyltransferase	Mammal
TRAMP	Trf4/Air2/Mtr4p polyadenylation complex	Nuclear RNA surveillance complex	Yeast
Trf4/5	Trf poly(A) polymerase 4/5	Pol beta-like nucleotidyltransferases	Yeast
TRRAP	TRansformation/tRanscription domain associated protein	Transcriptional histone acetyltransferase cofactor/Adaptor	Mammal
Wee1	Wee (Scottish word for small)-1G2-checkpoint kinase	Nuclear tyrosine kinase regulator of cell cycle progression	Yeast; mammal
YY1	Yin Yang 1	HiNF-I	Mammal

HiNFs (*H*istone *N*uclear *F*actors), although these mostly correspond to known general mammalian transcription factors. For example, histone H4 genes contain two major protein–DNA interaction sites in vivo [59]. One of these is an enhancer element for activating histone H4 transcription via interactions with a variety of general transcription factors such as SP1 (*S*pecificity *P*rotein 1)/HiNF-C, Yin Yang 1 (YY1)/HiNF-I, *H*igh *M*obility *G*roup-I (HMG-I)/HiNF-A) [60–62]. Cell cycle control in G1/S is conferred by the second regulatory site via interactions with IRF2 (*I*nterferon *R*egulatory *F*actor 2)/HiNF-M, CDP (*C*CAAT *D*isplacement *P*rotein)/HiNF-D and a H4 specific transcription factor HiNF-P [63, 64], all of which contribute to the cell cycle regulation of histone H4 expression [65].

Transcriptional control plays an important role in regulating the expression of the replication-dependent histone genes at the G1/S boundary. As cells enter S phase, histone mRNA transcription increases three- to fivefold compared to basal G1-levels [66, 67]. This cell cycle-dependent S phase expression of histone genes is conferred by the action of Cyclin E/Cdk2, the key regulator of S phase entry which also initiates DNA replication (Fig. 11.1b). The activity of Cyclin E/Cdk2 is required for phosphorylation of Cdc6 which protects it from degradation by the *A*naphase-*P*romoting *C*omplex (APC), which in turn allows the licensing of origins in late G1 [68, 69]. Another essential substrate for Cyclin E/Cdk2 is the *N*uclear *P*rotein *A*taxia–*T*elangiectasia locus (NPAT) which is localized in a cell cycle-dependent manner in subnuclear structures known as the *H*istone *L*ocus *B*odies (HLBs) that contain the clustered histone genes [70–73]. The histone locus bodies also contain many factors required for histone mRNA synthesis and processing [74]. NPAT is an essential protein for normal mammalian development and enhances histone gene transcription as overexpression of NPAT activates promoters of multiple histone genes via the SSREs within the promoters [72]. The suppression of NPAT expression through RNA interference or conditional knockout impedes expression of all replication-dependent core histone genes. The promoter DNA sequences of different core histone subtypes are quite divergent, and direct DNA binding by NPAT has not been detected. Therefore, it has been proposed that coordination of the transcription of multiple core histone subtypes by NPAT occurs through the interaction of NPAT with factors that regulate transcription of the individual core histone subtypes. Indeed, after phosphorylation by Cyclin E/Cdk2, NPAT co-activates the transcription of histone genes by interacting with the H4 promoter specific transcription factor HiNF-P, a conserved Zn finger protein that binds to a histone H4 promoter regulatory element [75]. Although a large number of histone gene transcription factors have been characterized, HiNF-P is unique because it is the only known histone H4 promoter-specific factor that interacts directly with NPAT and they both reside in HLBs. Although the mechanism of HiNF-P action is not fully understood, it is clear that the HiNF-P/NPAT complex mediates a unique cell cycle regulatory mechanism that controls the G1/S phase transition.

NPAT also interacts with transcription factor Oct1 (*Oct*amer Binding Factor 1) and its coactivator complex OCA-S (*O*ct-1 *C*o-*A*ctivator in *S* phase) that bind to the octamer binding element in the H2B promoter SSRE to activate H2B transcription [76]. Remarkably, components of OCA-S include nuclear p38/*glyceraldehyde-3-*

phosphate dehydrogenase (GAPDH) and lactate dehydrogenase. Further, the activity of OCA-S is regulated by NAD (*N*icotinamide *A*denine *D*inucleotide) and NADH, suggesting a link between the histone gene transcription and the cellular metabolic state/redox status [76]. The molecular mechanism underlying the regulation of H2B transcription by Oct1/OCA-S in accordance with the cellular metabolic/redox states is not yet clear. Nevertheless, the pattern of Oct-1 phosphorylation changes throughout the cell cycle, and may be CDK mediated, thereby potentially regulating Oct1 function at histone promoters [77].

Studies aimed at further elucidating the molecular mechanisms by which NPAT regulates histone gene activation at G1/S revealed an interaction of NPAT with components of the Tip60 *H*istone *A*cetyl*t*ransferase (HAT) complex [78]. The association of the Tip60-TRRAP (*Tr*ansformation/T*r*anscription Domain-*A*ssociated *P*rotein) complex increases histone H4 acetylation at the H4 promoter during G1/S transition in a NPAT-dependent manner (Fig. 11.1b). Another factor, FLASH (FADD-like IL-1β-converting enzyme associated huge protein), has been implicated in the NPAT-mediated regulation of histone gene transcription [79]. Initially, FLASH was shown to associate with NPAT as well as histone gene promoters and was required for histone transcription and S phase progression [80]. Subsequently, FLASH was shown to be required for 3′-end processing of histone pre-mRNAs [81–83].

A number of transcription factors such as YY1 (Yin Yang 1) and SMAD (*S*mall and *M*others *A*gainst *D*ecapentaplegic) that have not been shown to interact with NPAT so far also contribute to the transcription of histone genes, although the detailed molecular mechanisms involved are largely unknown. YY1 is a ubiquitous transcription factor involved in activating histone genes upon binding specific sequences both in the coding sequences [84] and promoter regions [62] of replication-dependent histone genes. A recent study suggests that YY1 may be primarily contributing to tissue specific activation of histone genes upon differentiation [85]. The same study also reported that in embryonic stem cells, histone genes were the major targets for repression by SMAD1 and SMAD2, which are effectors of extracellular signaling by BMP (*B*one *M*orphogenetic *P*rotein) and TGF-β (*T*ransforming *G*rowth *F*actor-β). Most surprisingly, this study also finds E2F1 and E2F4 transcription factors to be highly enriched on nearly all replication-dependent histone genes, which is contrary to the general belief that most histone genes are not regulated by the E2F family of proteins [86]. Additional studies are required to determine the molecular mechanisms involved in the regulation of histone genes by these transcription regulators.

Wee1 is a conserved nuclear tyrosine kinase and known to negatively regulate the activity of Cdk1 (Cdc28 in yeast) in the Cyclin B/Cdk1 complex by phosphorylating tyrosine 15 of Cdk1 and inhibiting it throughout S phase, thereby preventing cells from entering mitosis until DNA replication is completed [87, 88]. Recently, Wee1 has been shown to be playing an important role in shutting off histone gene transcription at the end of S phase [89]. Wee1 phosphorylates histone H2B at tyrosine 37 in the nucleosomes found upstream of the major histone gene cluster *HIST1* and suppresses histone transcription in late S phase [89]. Inhibition of the WEE1 kinase results in the loss of H2B tyrosine 37 phosphorylation (pH2B Y37) and an increase in histone transcript levels. The nucleosomes upstream of the *HIST1* cluster containing pH2B Y37 may hinder the recruitment of NPAT and thereby prevent the activation of histone gene transcription (Fig. 11.1a) [89]. Additionally pH2B Y37 has been shown to recruit HIRA to the histone gene cluster, where it may help to enforce a repressive chromatin structure [89]. This mode of transcriptional repression seems to be conserved as pH2B Y37 phosphorylation is enriched on the yeast histone promoters containing the negative regulatory (NEG) element as well (Fig. 11.1a) [89]. Thus, Wee1 couples the completion of DNA replication with the termination of histone synthesis at the end of S phase, thereby ensuring a proper histone–DNA stoichiometry before the cell proceeds to mitosis (Fig. 11.1b).

Taken together, all the available evidence suggests that NPAT functions as a key global regulator of coordinated transcriptional activation of multiple core histone subtypes during the G1/S phase transition and links the regulation of histone gene expression to the cell cycle machinery. In addition to histone gene expression, NPAT has been shown to play a critical role in S phase entry [70, 90, 91]. NPAT deletion cells do not enter S phase despite high CDK activity, demonstrating that it has a CDK-independent role in cell cycle progression. Therefore, while Cyclin E/Cdk2 is at the top of the cascade which leads to both the initiation of DNA replication and histone synthesis in mammalian cells, NPAT may serve as the crucial link that couples histone gene activation and DNA replication. On the other hand, the CDK regulator Wee1 coordinates the shutdown of histone gene transcription with the completion of replication at the end of S phase.

Linker Histone H1 and DNA Replication

There are at least three major groups of H1-encoding genes in vertebrates—cleavage stage H1, replication-dependent H1, and differentiation-specific H1. The promoter of each group consists of a specific combination of regulatory elements that are evolutionarily well conserved.

Among invertebrates, the best-studied species is the sea urchin, where these different types of H1-encoding genes have been described and their promoter functionally dissected [92–94]. The replication-dependent histone promoter contains a CAAT box upstream from the TATA box, and is followed by a GC-rich region and the characteristic H1 box (also named AC box). In addition, there is a near perfect inverted repeat of the H1/AC box named UCE (*U*pstream *C*onserved *E*lement) that was first found in the differentiation-associated $H1^0$ promoter in *Xenopus*. This element was also found later in vertebrate H1 promoters and named the TG box [95]. The role of these regulatory elements in the expression of replication-dependent histone H1 has been extensively studied [94]. The first clue regarding the link between the cell cycle and the expression of H1s came from the work of van Wijnen and colleagues showing the involvement of the HiNF-D complex in the control of the expression of several S phase histones, including H1 [96]. It is likely that H1 and core histone genes become activated by the same S phase signaling initiated by the

CyclinE/Cdk2 complex [32]. However, H1 is transcribed throughout S phase and core histones are only transcribed in a short pulse during early S phase. Indeed, work from the Tijan laboratory using single-cell imaging in *Drosophila* suggests that alternative *P*re*i*nitiation *C*omplex (PIC) subunits are used for H1 transcription initiation [97].

Phosphorylation is an important cell cycle dependent modification for histone H1. A low level of H1 phosphorylation can be detected in G1 which increases steadily through S phase until it reaches its maximum in mitotic cells. S phase histone H1 exist in both unphosphorylated and hypo-phosphorylated forms and there is evidence that H1 phosphorylation is important for the process of DNA replication [98–100]. Replicating DNA and phosphorylated H1 colocalize in vivo [101], suggesting that this H1 phosphorylation may promote DNA decondensation to facilitate DNA replication. The effects of histone H1 on DNA replication has been studied directly using *Xenopus* egg extracts. Lu et al showed that DNA replication is inhibited to equal extent in extracts supplemented with either of the somatic variants H1c or $H1^0$ [102]. Additionally, somatic H1s inhibit replication initiation by limiting the assembly of pre-replication complexes on sperm chromatin [103]. Together, these studies suggest that on one hand the interaction of preexisting histone H1 may be weakened via phosphorylation to facilitate replication fork progression, while on the other hand the deposition of histone H1 on newly synthesized DNA may contribute to block replication licensing and prevent re-replication.

The Role of Posttranscriptional Regulation in Coordinating Histone Synthesis and DNA Replication during S Phase in Mammalian Cells

Since the half-life of the mammalian histone transcript is generally much greater compared to that in yeast, it is important to consider the contribution of posttranscriptional regulation in coordinating histone and DNA synthesis. Mammalian histone mRNAs are unique as they do not end in a polyA tail but instead have a conserved stem loop at their 3′ end that is bound by the *S*tem *L*oop *B*inding *P*rotein (SLBP) [104]. SLPB is involved in the histone pre-mRNA processing and its degradation at the end of S phase (Fig. 11.1b). SLBP is subjected to cell cycle regulation and is stable only in S phase. Accumulation of SLBP before the beginning of S phase is crucial to allow the accumulation of histone mRNAs necessary for histone protein synthesis. At the end of S phase, SLBP needs to be degraded to shut off histone synthesis (Fig. 11.1b). The degradation of SLBP is initiated by Cyclin A/Cdk1 mediated phosphorylation of SLBP on Threonine 61 (T61) which is followed by *C*asein *K*inase 2 (CK2) mediated phosphorylation of T60 [105]. Both sites need to be phosphorylated for subsequent SLBP degradation via the ubiquitin–proteasome pathway, although the ubiquitin ligase responsible for its degradation is not known.

Recent work from the Marzluff laboratory suggests that the major S phase transcription factor, E2F1 is regulated in parallel with SLBP, possibly by Cyclin

A/Cdk1 (Fig. 11.1b) [106]. In HeLa cells, E2F1 is cell cycle regulated and the level of E2F1 protein increases as cells enter S phase and decreases at the end of S phase just like SLBP. Further, deletion of the amino terminus of E2F1 that removes the Cyclin A binding site results in the stabilization of E2F1 protein at the end of S phase. A mutant E2F1 carrying a deletion of the C-terminal residues 300–379 was also not degraded at the end of S phase, suggesting that two regions are needed for degradation of E2F1 protein. E2F1 is a critical transcription factor that regulates the transcription of many genes involved in DNA replication including the MCM (*M*ini *C*hromosome *M*aintenance) and ORC (*O*rigin *R*ecognition *C*omplex) proteins, components of the replication apparatus such as RFC (*R*eplication *F*actor *C*), PCNA (*P*roliferating *C*ell *N*uclear *A*ntigen), and RPA (*R*eplication *P*rotein *A*) to name a few [107]. Further, genes encoding enzymes involved in deoxyribonucleotide metabolism also require the E2F1 transcription factor. At the beginning of S phase, members of the retinoblastoma tumor suppressor pRb family (pRb, p130, and p107) are phosphorylated by Cyclin E/Cdk2 and this results in their dissociation from the transcription factor E2F1 and consequently in the upregulation of the genes involved in DNA replication [108]. At the end of S phase, E2F1 is very likely to be inactivated along with SLBP by Cyclin A/Cdk1, which also phosphorylates and inactivates proteins required for DNA replication such as ORC1, FEN1, and the CDP (*C*CAAT *D*isplacement *P*rotein) transcription factor that is needed for the expression of the DNA polymerase alpha gene [109, 110]. Thus, Cyclin A/Cdk1 may signal the end of S phase by both inactivating pre-replication complexes to prevent re-replication and simultaneously inhibiting histone synthesis by degrading SLBP.

Coupling of DNA Synthesis and Histone Synthesis upon Replication Inhibition or DNA Damage in S Phase

The previous sections have focused on how the activation of DNA and histone synthesis by Cyclin/Cdk2 activity ensures the coordination between them. This coupling is maintained throughout S phase and as such the amount of free histones is very small. This is not surprising given the known effects of histone imbalance on the fidelity of diverse nuclear processes and genomic stability [8, 9, 111, 112]. The close coordination of histone and DNA synthesis is exemplified in the rapid and concerted downregulation of histone mRNAs in response to DNA replication inhibiting drugs such as hydroxyurea (HU) [14]. This evolutionarily conserved response reflects the existence of S phase controls which ensure that the rate of histone production exactly matches the rate of their incorporation into chromatin. In mammalian cells, the *D*NA *d*amage *r*esponse (DDR) kinase ATM (*A*taxia *T*elangiectasia *M*utated) is directly involved in the activation of mechanisms that ultimately lead to the repression of histone expression in *i*onizing *r*adiation (IR) treated cells [113]. The DDR activates a checkpoint when a cell senses DNA damage and the checkpoint response leads to a sequential activation of kinases that slow down the cell cycle to allow time for repair [114]. Following activation of ATM, its downstream targets p53 and p21 are consequently upregulated. p21 blocks the activation of the key regulator Cyclin E/Cdk2 complex which leads the dephosphorylation of NPAT. Unphosphorylated NPAT can no longer localize to the histone clusters in the HLB to activate histone gene transcription [113]. In fact, both NPAT and FLASH are degraded upon DNA damage mediated by UV-C irradiation [115]. This leads to the disruption of HLBs and the subsequent drop in histone transcription may contribute to the cell cycle arrest. Histone mRNAs are also rapidly degraded when DNA replication is inhibited with drugs such as hydroxyurea during S phase. The degradation is induced by the addition of untemplated uridines at the 3′ end of histone transcripts by terminal uridyl transferases [5, 116]. Such 3′ oligouridylated transcripts are preferentially recognized and bound by the hetero-heptameric LSM1–7 complex. The LSM1–7 complex then triggers the degradation of histone mRNAs through the interaction with Eri1, a 3′–5′ exoribonuclease and the RNA helicase Upf1. They all contribute to the subsequent histone mRNA degradation through the 3′ stem loop [117]. Although the mechanism of histone mRNA degradation is now known in significant detail, the important question of how the stalling of DNA replication in the nucleus is signaled to the cytoplasmic histone mRNA degradation machinery is still not clear. An attractive hypothesis is that changes in Upf1 protein phosphorylation in response to DNA replication stress modulate histone mRNA stability. Upf1 is a direct target of the *p*hosphatidyl*i*nositol-*3* (PI-3) kinases ATR (*AT*M and *R*ad3 related) and hSMG-1 [118].

In S. cerevisiae, the Hir complex plays a major role in the DNA damage induced transcriptional repression of histone genes. In addition, other histone chaperones such as Asf1 and Rtt106 and the chromatin remodeler Rsc are also implicated in the formation of a repressive chromatin structure at the histone promoters to block transcription [29]. Our laboratory has obtained evidence suggesting that the DDR kinases trigger the downregulation of histone mRNAs by targeting the Hir-complex (Paik et al., unpublished data). Interestingly, the yeast DDR kinase Rad53 (homolog of the human tumor suppressor CHK2) targets excess histone proteins for degradation via the ubiquitin–proteasome pathway both at the end of a normal S phase, as well as upon DNA damage or replication arrest in S phase, thereby contributing to genome stability [8, 111].

Recent evidence suggests that histone mRNAs in *S. cerevisiae* are also posttranscriptionally regulated and Herrero and Moreno have reported the importance of Lsm1 (as part of the yeast Lsm1–7-Pat1 complex) in histone mRNA degradation [119]. Lsm1 mutants accumulate high levels of histone mRNAs and are unable to degrade them following HU treatment. How the Lsm1–7 complex recognizes the histone mRNAs for degradation is not clear as yeast histone transcripts are not found to be uridylated. The Lsm1–7-Pat1 complex has been found to bind preferentially to U-tracts carrying mRNAs in humans and to oligoadenylated over polyadenylated mRNAs in yeast [116]. It has been reported that the average length of the poly A tail of yeast histone H2B mRNA is quite short compared to other transcripts. Further, the H2B poly A tail length varies during the cell cycle and shortens as cells progress from G1 to S phase [120]. Hence, it is attractive to speculate that the Lsm1–7-Pat1 complex may recognize histone mRNAs based on the length of their poly A tails and selectively degrade them. Future research should unravel the relative contributions and the molecular details of how the Lsm1–7-Pat1 complex and the TRAMP complex regulate histone mRNAs in yeast.

Is MCM2 a Sensor of Histone Supply and Demand at the Replication Fork?

To maintain a proper histone–DNA stoichiometry, the amount of histones assembled onto chromatin is exactly matched with the rate of ongoing DNA replication. The assembly of chromatin is facilitated by the interaction of certain components of the DNA replication machinery with chromatin assembly factors or histone chaperones which mediate the deposition of histones onto the newly replicated DNA. Accordingly, the pace of DNA replication is tightly coupled to chromatin assembly. This is illustrated by the fact that depletion of chromatin assembly factors such as ASF1 and CAF-1 in mammalian cells [121], as well as diminished histone supply due to knockdown of SLBP result in inhibition of replication [80, 122]. A study from the Groth laboratory suggests that replication fork progression may be coupled to nucleosome assembly through a feedback mechanism from CAF-1 to the MCM replicative helicase and/or the unloading of PCNA from newly synthesized DNA upon nucleosome assembly [123].

There is evidence that MCMs might play a role in sensing histone supply and demand at the replication fork. The amino terminus of MCM2 has been shown to interact with several chromatin proteins [124–126]. Histone H3 binds to the amino-terminal domain of MCM2 with high affinity and can likewise interact with the replicative helicase large T antigen encoded by the SV40 DNA tumor virus. In human cells, the histone chaperone ASF1 forms a complex with MCM2–7 [127], and a fraction of ASF1 colocalizes with MCM2 on chromatin. This interaction is bridged by an H3–H4 dimer, indicating that MCM2–7 loaded onto chromatin can bind non-nucleosomal histone H3–H4 dimers [126]. Histones in complex with ASF1 carry modifications that are typical of new histones, but chromatin-specific marks that would be present on parental histones can also be detected in association with MCM2 [126]. These interactions suggest that MCM2 may be ideally placed to sense the rate of replication fork progression, and in response to replication inhibition, it can potentially detect the accumulation of histones and transduce this signal to the histone synthesis machinery which can then be downregulated accordingly.

Do Senescent Cells Undergo Coordinated Loss of Replicative Potential and Histone Synthesis?

Replicative senescence is defined as an irreversible cell cycle arrest typically in G1 phase of the cell cycle. It is triggered by a variety of endogenous and exogenous cellular insults including eroded telomeres, DNA damage, oxidative stress, and/or oncogene activation [128]. Senescent cells show striking changes in gene expression—two cell cycle inhibitors p21 and p16 are usually upregulated, whereas genes that encode proteins to facilitate cell cycle progression (replication-dependent histones, c-FOS, Cyclin A, Cyclin B, and PCNA) are incorporated into transcriptionally silent heterochromatin [129, 130]. The molecular basis of the G1 arrest is thought to be the result of accumulation of the Cdk inhibitors p21 and p16 that block the inactivating phosphorylation of pRb. In the absence of inactivating Cdk phosphorylation, the Rb family of proteins (Rb, p107, and p130) binds E2F transcription factors and blocks S-phase entry [131, 132]. Many of the senescence associated Rb-specific targets are DNA replication factors as well as Cyclin E1 that induces Cdk2 activity and further phosphorylates and inactivates Rb family members, among others [133]. Interestingly, Cyclin E1 is also believed to stimulate the formation of the pre-replication complexes through the recruitment of MCMs to DNA replication origins that are Cdk-independent [134]. Thus, in senescent cells Rb may inhibit replication by repressing the genes associated with DNA replication—including Cyclin E1 which participates in pre-RC formation.

An additional consequence of Cyclin E1 inhibition in senescent cells is that the synthesis of replication-dependent histones will be inhibited as the transcriptional coactivator NPAT is not phosphorylated and fails to localize to the HLBs where histone gene transcription takes place. The Karlseder laboratory has shown that chronic DNA damage signals caused by telomere shortening decrease new histone synthesis and reduce expression of SLBP and histone chaperones ASF1 and CAF1 during replicative aging [135]. This downregulation of histone synthesis is independent of p53 and pRb, suggesting that the histone synthesis is posttranscriptionally regulated.

Senescence is often characterized by domains of facultative heterochromatin, called *S*enescence-*A*ssociated *H*eterochromatin *F*oci (SAHF), which repress expression of proliferation-promoting genes such as E2F target genes. The formation of these SAHF is dependent on the p16-pRb pathway and requires several days to develop [136]. During this time, a complex of histone chaperones, comprising HIRA/UBN1/CABIN and ASF1a and HP1 transiently interacts and drives the formation of the histone variant macroH2A-containing SAHF and senescence-associated cell cycle exit. At the end of this process, each SAHF contains portions of a single condensed chromosome in which the linker histone H1 is replaced by HP1 and enriched for macroH2A [137, 138]. MacroH2A containing chromatin is resistant to ATP-dependent chromatin remodeling proteins and binding of transcription factors [139].

HIRA/UBN1/CABIN is a functional homolog of the yeast Hir complex and its function in human cells is mostly associated with chromatin assembly of replication-independent histones. However, several studies suggest that this complex may also play an important role in histone regulation. Overexpression of HIRA is able to repress histone gene transcription in human cells and block S-phase progression [42]. Cyclin E/Cdk2 and Cyclin A-Cdk2 can phosphorylate HIRA and this phosphorylation in inhibited by the cyclin inhibitor p21 [121, 140]. Thus, HIRA could act as a repressor for histone gene expression and at the same time as a repressor of genes needed for DNA replication in cells undergoing senescence.

References

1. Richmond TJ, Davey CA. The structure of DNA in the nucleosome core. Nature. 2003;423(6936):145–50.
2. Thoma F, Koller T, Klug A. Involvement of histone H1 in the organization of the nucleosome and of the salt-dependent superstructures of chromatin. J Cell Biol. 1979;83(2 Pt 1):403–27. Pubmed Central PMCID: 2111545.
3. Sogo JM, Stahl H, Koller T, Knippers R. Structure of replicating simian virus 40 minichromosomes. The replication fork, core histone segregation and terminal structures. J Mol Biol. 1986;189(1):189–204.
4. Lycan DE, Osley MA, Hereford LM. Role of transcriptional and posttranscriptional regulation in expression of histone genes in Saccharomyces cerevisiae. Mol Cell Biol. 1987;7(2):614–21. Pubmed Central PMCID: 365116.
5. Kaygun H, Marzluff WF. Regulated degradation of replication-dependent histone mRNAs requires both ATR and Upf1. Nat Struct Mol Biol. 2005;12(9):794–800.
6. Dominski Z, Yang XC, Kaygun H, Dadlez M, Marzluff WF. A 3′ exonuclease that specifically interacts with the 3′ end of histone mRNA. Mol Cell. 2003;12(2):295–305.
7. Sittman DB, Graves RA, Marzluff WF. Histone mRNA concentrations are regulated at the level of transcription and mRNA degradation. Proc Natl Acad Sci U S A. 1983;80(7):1849–53. Pubmed Central PMCID: 393707.
8. Gunjan A, Verreault A. A Rad53 kinase-dependent surveillance mechanism that regulates histone protein levels in S. cerevisiae. Cell. 2003;115(5):537–49.
9. Meeks-Wagner D, Hartwell LH. Normal stoichiometry of histone dimer sets is necessary for high fidelity of mitotic chromosome transmission. Cell. 1986;44(1):43–52.
10. Prado F, Aguilera A. Impairment of replication fork progression mediates RNA polII transcription-associated recombination. EMBO J. 2005;24(6):1267–76. Pubmed Central PMCID: 556405.
11. Prado F, Cortes-Ledesma F, Aguilera A. The absence of the yeast chromatin assembly factor Asf1 increases genomic instability and sister chromatid exchange. EMBO Rep. 2004; 5(5):497–502. Pubmed Central PMCID: 1299049.
12. Myung K, Pennaneach V, Kats ES, Kolodner RD. Saccharomyces cerevisiae chromatin-assembly factors that act during DNA replication function in the maintenance of genome stability. Proc Natl Acad Sci U S A. 2003;100(11):6640–5. Pubmed Central PMCID: 164500.
13. Herrick D, Parker R, Jacobson A. Identification and comparison of stable and unstable mRNAs in Saccharomyces cerevisiae. Mol Cell Biol. 1990;10(5):2269–84. Pubmed Central PMCID: 360574.
14. Osley MA. The regulation of histone synthesis in the cell cycle. Annu Rev Biochem. 1991;60:827–61.
15. Eriksson PR, Ganguli D, Nagarajavel V, Clark DJ. Regulation of histone gene expression in budding yeast. Genetics. 2012;191(1):7–20. Pubmed Central PMCID: 3338271.
16. Hess D, Liu B, Roan NR, Sternglanz R, Winston F. Spt10-dependent transcriptional activation in Saccharomyces cerevisiae requires both the Spt10 acetyltransferase domain and Spt21. Mol Cell Biol. 2004;24(1):135–43. Pubmed Central PMCID: 303362.
17. Kurat CF, Lambert JP, Petschnigg J, Friesen H, Pawson T, Rosebrock A, et al. Cell cycle-regulated oscillator coordinates core histone gene transcription through histone acetylation. Proc Natl Acad Sci U S A. 2014;111(39):14124–9. Pubmed Central PMCID: 4191790.
18. Xu F, Zhang K, Grunstein M. Acetylation in histone H3 globular domain regulates gene expression in yeast. Cell. 2005;121(3):375–85.
19. Iyer VR, Horak CE, Scafe CS, Botstein D, Snyder M, Brown PO. Genomic binding sites of the yeast cell-cycle transcription factors SBF and MBF. Nature. 2001;409(6819):533–8.
20. Simon I, Barnett J, Hannett N, Harbison CT, Rinaldi NJ, Volkert TL, et al. Serial regulation of transcriptional regulators in the yeast cell cycle. Cell. 2001;106(6):697–708.

21. Ng HH, Robert F, Young RA, Struhl K. Genome-wide location and regulated recruitment of the RSC nucleosome-remodeling complex. Genes Dev. 2002;16(7):806–19. Pubmed Central PMCID: 186327.
22. Osley MA, Gould J, Kim S, Kane MY, Hereford L. Identification of sequences in a yeast histone promoter involved in periodic transcription. Cell. 1986;45(4):537–44.
23. Koch C, Moll T, Neuberg M, Ahorn H, Nasmyth K. A role for the transcription factors Mbp1 and Swi4 in progression from G1 to S phase. Science. 1993;261(5128):1551–7.
24. Spector MS, Raff A, DeSilva H, Lee K, Osley MA. Hir1p and Hir2p function as transcriptional corepressors to regulate histone gene transcription in the Saccharomyces cerevisiae cell cycle. Mol Cell Biol. 1997;17(2):545–52. Pubmed Central PMCID: 231779.
25. Sutton A, Bucaria J, Osley MA, Sternglanz R. Yeast ASF1 protein is required for cell cycle regulation of histone gene transcription. Genetics. 2001;158(2):587–96. Pubmed Central PMCID: 1461693.
26. Xu H, Kim UJ, Schuster T, Grunstein M. Identification of a new set of cell cycle-regulatory genes that regulate S-phase transcription of histone genes in Saccharomyces cerevisiae. Mol Cell Biol. 1992;12(11):5249–59. Pubmed Central PMCID: 360458.
27. De Koning L, Corpet A, Haber JE, Almouzni G. Histone chaperones: an escort network regulating histone traffic. Nat Struct Mol Biol. 2007;14(11):997–1007.
28. Green EM, Antczak AJ, Bailey AO, Franco AA, Wu KJ, Yates 3rd JR, et al. Replication-independent histone deposition by the HIR complex and Asf1. Curr Biol. 2005;15(22):2044–9. Pubmed Central PMCID: 2819815.
29. Fillingham J, Kainth P, Lambert JP, van Bakel H, Tsui K, Pena-Castillo L, et al. Two-color cell array screen reveals interdependent roles for histone chaperones and a chromatin boundary regulator in histone gene repression. Mol Cell. 2009;35(3):340–51.
30. Ferreira ME, Flaherty K, Prochasson P. The Saccharomyces cerevisiae histone chaperone Rtt106 mediates the cell cycle recruitment of SWI/SNF and RSC to the HIR-dependent histone genes. PLoS One. 2011;6(6), e21113. Pubmed Central PMCID: 3115976.
31. Amin AD, Dimova DK, Ferreira ME, Vishnoi N, Hancock LC, Osley MA, et al. The mitotic Clb cyclins are required to alleviate HIR-mediated repression of the yeast histone genes at the G1/S transition. Biochim Biophys Acta. 2012;1819(1):16–27. Pubmed Central PMCID: 3249481.
32. White JH, Green SR, Barker DG, Dumas LB, Johnston LH. The CDC8 transcript is cell cycle regulated in yeast and is expressed coordinately with CDC9 and CDC21 at a point preceding histone transcription. Exp Cell Res. 1987;171(1):223–31.
33. Sherwood PW, Tsang SV, Osley MA. Characterization of HIR1 and HIR2, two genes required for regulation of histone gene transcription in Saccharomyces cerevisiae. Mol Cell Biol. 1993;13(1):28–38. Pubmed Central PMCID: 358881.
34. Goh PY, Surana U. Cdc4, a protein required for the onset of S phase, serves an essential function during G(2)/M transition in Saccharomyces cerevisiae. Mol Cell Biol. 1999;19(8):5512–22. Pubmed Central PMCID: 84393.
35. Drury LS, Perkins G, Diffley JF. The Cdc4/34/53 pathway targets Cdc6p for proteolysis in budding yeast. EMBO J. 1997;16(19):5966–76. Pubmed Central PMCID: 1170227.
36. Feldman RM, Correll CC, Kaplan KB, Deshaies RJ. A complex of Cdc4p, Skp1p, and Cdc53p/cullin catalyzes ubiquitination of the phosphorylated CDK inhibitor Sic1p. Cell. 1997;91(2):221–30.
37. Mathias N, Johnson SL, Winey M, Adams AE, Goetsch L, Pringle JR, et al. Cdc53p acts in concert with Cdc4p and Cdc34p to control the G1-to-S-phase transition and identifies a conserved family of proteins. Mol Cell Biol. 1996;16(12):6634–43. Pubmed Central PMCID: 231665.
38. Schneider BL, Yang QH, Futcher AB. Linkage of replication to start by the Cdk inhibitor Sic1. Science. 1996;272(5261):560–2.
39. Schwob E, Bohm T, Mendenhall MD, Nasmyth K. The B-type cyclin kinase inhibitor p40SIC1 controls the G1 to S transition in S. cerevisiae. Cell. 1994;79(2):233–44.

40. Skowyra D, Craig KL, Tyers M, Elledge SJ, Harper JW. F-box proteins are receptors that recruit phosphorylated substrates to the SCF ubiquitin-ligase complex. Cell. 1997;91(2):209–19.
41. DeSalle LM, Pagano M. Regulation of the G1 to S transition by the ubiquitin pathway. FEBS Lett. 2001;490(3):179–89.
42. Hall C, Nelson DM, Ye X, Baker K, DeCaprio JA, Seeholzer S, et al. HIRA, the human homologue of yeast Hir1p and Hir2p, is a novel cyclin-cdk2 substrate whose expression blocks S-phase progression. Mol Cell Biol. 2001;21(5):1854–65. Pubmed Central PMCID: 86753.
43. Ubersax JA, Woodbury EL, Quang PN, Paraz M, Blethrow JD, Shah K, et al. Targets of the cyclin-dependent kinase Cdk1. Nature. 2003;425(6960):859–64.
44. Amon A, Surana U, Muroff I, Nasmyth K. Regulation of p34CDC28 tyrosine phosphorylation is not required for entry into mitosis in S. cerevisiae. Nature. 1992;355(6358):368–71.
45. Kellogg DR, Murray AW. NAP1 acts with Clb1 to perform mitotic functions and to suppress polar bud growth in budding yeast. J Cell Biol. 1995;130(3):675–85. Pubmed Central PMCID: 2120541.
46. Hereford L, Bromley S, Osley MA. Periodic transcription of yeast histone genes. Cell. 1982;30(1):305–10.
47. Campbell SG, Li Del Olmo M, Beglan P, Bond U. A sequence element downstream of the yeast HTB1 gene contributes to mRNA 3′ processing and cell cycle regulation. Mol Cell Biol. 2002;22(24):8415–25. Pubmed Central PMCID: 139887.
48. Canavan R, Bond U. Deletion of the nuclear exosome component RRP6 leads to continued accumulation of the histone mRNA HTB1 in S-phase of the cell cycle in Saccharomyces cerevisiae. Nucleic Acids Res. 2007;35(18):6268–79. Pubmed Central PMCID: 2094057.
49. Reis CC, Campbell JL. Contribution of Trf4/5 and the nuclear exosome to genome stability through regulation of histone mRNA levels in Saccharomyces cerevisiae. Genetics. 2007;175(3):993–1010. Pubmed Central PMCID: 1840065.
50. LaCava J, Houseley J, Saveanu C, Petfalski E, Thompson E, Jacquier A, et al. RNA degradation by the exosome is promoted by a nuclear polyadenylation complex. Cell. 2005;121(5):713–24.
51. Vanacova S, Wolf J, Martin G, Blank D, Dettwiler S, Friedlein A, et al. A new yeast poly(A) polymerase complex involved in RNA quality control. PLoS Biol. 2005;3(6), e189. Pubmed Central PMCID: 1079787.
52. Wyers F, Rougemaille M, Badis G, Rousselle JC, Dufour ME, Boulay J, et al. Cryptic pol II transcripts are degraded by a nuclear quality control pathway involving a new poly(A) polymerase. Cell. 2005;121(5):725–37.
53. Arigo JT, Eyler DE, Carroll KL, Corden JL. Termination of cryptic unstable transcripts is directed by yeast RNA-binding proteins Nrd1 and Nab3. Mol Cell. 2006;23(6):841–51.
54. Vasiljeva L, Buratowski S. Nrd1 interacts with the nuclear exosome for 3′ processing of RNA polymerase II transcripts. Mol Cell. 2006;21(2):239–48.
55. Harris ME, Bohni R, Schneiderman MH, Ramamurthy L, Schumperli D, Marzluff WF. Regulation of histone mRNA in the unperturbed cell cycle: evidence suggesting control at two posttranscriptional steps. Mol Cell Biol. 1991;11(5):2416–24. Pubmed Central PMCID: 359999.
56. Schumperli D. Cell-cycle regulation of histone gene expression. Cell. 1986;45(4):471–2.
57. Heintz N. The regulation of histone gene expression during the cell cycle. Biochim Biophys Acta. 1991;1088(3):327–39.
58. Albig W, Doenecke D. The human histone gene cluster at the D6S105 locus. Hum Genet. 1997;101(3):284–94.
59. Pauli U, Chrysogelos S, Stein G, Stein J, Nick H. Protein-DNA interactions in vivo upstream of a cell cycle-regulated human H4 histone gene. Science. 1987;236(4806):1308–11.
60. Birnbaum MJ, Wright KL, van Wijnen AJ, Ramsey-Ewing AL, Bourke MT, Last TJ, et al. Functional role for Sp1 in the transcriptional amplification of a cell cycle regulated histone H4 gene. Biochemistry. 1995;34(23):7648–58.

61. Guo B, Stein JL, van Wijnen AJ, Stein GS. ATF1 and CREB trans-activate a cell cycle regulated histone H4 gene at a distal nuclear matrix associated promoter element. Biochemistry. 1997;36(47):14447–55.
62. Last TJ, van Wijnen AJ, Birnbaum MJ, Stein GS, Stein JL. Multiple interactions of the transcription factor YY1 with human histone H4 gene regulatory elements. J Cell Biochem. 1999;72(4):507–16.
63. van Wijnen AJ, van Gurp MF, de Ridder MC, Tufarelli C, Last TJ, Birnbaum M, et al. CDP/cut is the DNA-binding subunit of histone gene transcription factor HiNF-D: a mechanism for gene regulation at the G1/S phase cell cycle transition point independent of transcription factor E2F. Proc Natl Acad Sci U S A. 1996;93(21):11516–21. Pubmed Central PMCID: 38089.
64. Vaughan PS, van der Meijden CM, Aziz F, Harada H, Taniguchi T, van Wijnen AJ, et al. Cell cycle regulation of histone H4 gene transcription requires the oncogenic factor IRF-2. J Biol Chem. 1998;273(1):194–9.
65. Aziz F, van Wijnen AJ, Stein JL, Stein GS. HiNF-D (CDP-cut/CDC2/cyclin A/pRB-complex) influences the timing of IRF-2-dependent cell cycle activation of human histone H4 gene transcription at the G1/S phase transition. J Cell Physiol. 1998;177(3):453–64.
66. Kelly TJ, Brown GW. Regulation of chromosome replication. Annu Rev Biochem. 2000;69:829–80.
67. Marzluff WF, Pandey NB. Multiple regulatory steps control histone mRNA concentrations. Trends Biochem Sci. 1988;13(2):49–52.
68. Mailand N, Diffley JF. CDKs promote DNA replication origin licensing in human cells by protecting Cdc6 from APC/C-dependent proteolysis. Cell. 2005;122(6):915–26.
69. Duursma A, Agami R. p53-Dependent regulation of Cdc6 protein stability controls cellular proliferation. Mol Cell Biol. 2005;25(16):6937–47. Pubmed Central PMCID: 1190229.
70. Zhao J, Dynlacht B, Imai T, Hori T, Harlow E. Expression of NPAT, a novel substrate of cyclin E-CDK2, promotes S-phase entry. Genes Dev. 1998;12(4):456–61. Pubmed Central PMCID: 316526.
71. Ma T, Van Tine BA, Wei Y, Garrett MD, Nelson D, Adams PD, et al. Cell cycle-regulated phosphorylation of p220(NPAT) by cyclin E/Cdk2 in Cajal bodies promotes histone gene transcription. Genes Dev. 2000;14(18):2298–313. Pubmed Central PMCID: 316935.
72. Zhao J, Kennedy BK, Lawrence BD, Barbie DA, Matera AG, Fletcher JA, et al. NPAT links cyclin E-Cdk2 to the regulation of replication-dependent histone gene transcription. Genes Dev. 2000;14(18):2283–97. Pubmed Central PMCID: 316937.
73. Ghule PN, Dominski Z, Yang XC, Marzluff WF, Becker KA, Harper JW, et al. Staged assembly of histone gene expression machinery at subnuclear foci in the abbreviated cell cycle of human embryonic stem cells. Proc Natl Acad Sci U S A. 2008;105(44):16964–9. Pubmed Central PMCID: 2579361.
74. Liu JL, Buszczak M, Gall JG. Nuclear bodies in the Drosophila germinal vesicle. Chromosome Res. 2006;14(4):465–75.
75. Miele A, Braastad CD, Holmes WF, Mitra P, Medina R, Xie R, et al. HiNF-P directly links the cyclin E/CDK2/p220NPAT pathway to histone H4 gene regulation at the G1/S phase cell cycle transition. Mol Cell Biol. 2005;25(14):6140–53. Pubmed Central PMCID: 1168814.
76. Zheng L, Roeder RG, Luo Y. S phase activation of the histone H2B promoter by OCA-S, a coactivator complex that contains GAPDH as a key component. Cell. 2003;114(2):255–66.
77. Roberts SB, Segil N, Heintz N. Differential phosphorylation of the transcription factor Oct1 during the cell cycle. Science. 1991;253(5023):1022–6.
78. DeRan M, Pulvino M, Greene E, Su C, Zhao J. Transcriptional activation of histone genes requires NPAT-dependent recruitment of TRRAP-Tip60 complex to histone promoters during the G1/S phase transition. Mol Cell Biol. 2008;28(1):435–47. Pubmed Central PMCID: 2223310.
79. Yang XC, Sabath I, Kunduru L, van Wijnen AJ, Marzluff WF, Dominski Z. A conserved interaction that is essential for the biogenesis of histone locus bodies. J Biol Chem. 2014;289(49):33767–82. Pubmed Central PMCID: 4256312.

80. Barcaroli D, Bongiorno-Borbone L, Terrinoni A, Hofmann TG, Rossi M, Knight RA, et al. FLASH is required for histone transcription and S-phase progression. Proc Natl Acad Sci U S A. 2006;103(40):14808–12. Pubmed Central PMCID: 1578501.
81. Burch BD, Godfrey AC, Gasdaska PY, Salzler HR, Duronio RJ, Marzluff WF, et al. Interaction between FLASH and Lsm11 is essential for histone pre-mRNA processing in vivo in Drosophila. RNA. 2011;17(6):1132–47. Pubmed Central PMCID: 3096045.
82. Yang XC, Xu B, Sabath I, Kunduru L, Burch BD, Marzluff WF, et al. FLASH is required for the endonucleolytic cleavage of histone pre-mRNAs but is dispensable for the 5' exonucleolytic degradation of the downstream cleavage product. Mol Cell Biol. 2011;31(7):1492–502. Pubmed Central PMCID: 3135297.
83. Sabath I, Skrajna A, Yang XC, Dadlez M, Marzluff WF, Dominski Z. 3′-End processing of histone pre-mRNAs in Drosophila: U7 snRNP is associated with FLASH and polyadenylation factors. RNA. 2013;19(12):1726–44. Pubmed Central PMCID: 3884669.
84. Eliassen KA, Baldwin A, Sikorski EM, Hurt MM. Role for a YY1-binding element in replication-dependent mouse histone gene expression. Mol Cell Biol. 1998;18(12):7106–18. Pubmed Central PMCID: 109292.
85. Gokhman D, Livyatan I, Sailaja BS, Melcer S, Meshorer E. Multilayered chromatin analysis reveals E2f, Smad and Zfx as transcriptional regulators of histones. Nat Struct Mol Biol. 2013;20(1):119–26.
86. Oswald F, Dobner T, Lipp M. The E2F transcription factor activates a replication-dependent human H2A gene in early S phase of the cell cycle. Mol Cell Biol. 1996;16(5):1889–95. Pubmed Central PMCID: 231176.
87. Gould KL, Nurse P. Tyrosine phosphorylation of the fission yeast cdc2+ protein kinase regulates entry into mitosis. Nature. 1989;342(6245):39–45.
88. Watanabe N, Broome M, Hunter T. Regulation of the human WEE1Hu CDK tyrosine 15-kinase during the cell cycle. EMBO J. 1995;14(9):1878–91. Pubmed Central PMCID: 398287.
89. Mahajan K, Fang B, Koomen JM, Mahajan NP. H2B Tyr37 phosphorylation suppresses expression of replication-dependent core histone genes. Nat Struct Mol Biol. 2012;19(9):930–7.
90. Gao G, Bracken AP, Burkard K, Pasini D, Classon M, Attwooll C, et al. NPAT expression is regulated by E2F and is essential for cell cycle progression. Mol Cell Biol. 2003;23(8):2821–33. Pubmed Central PMCID: 152552.
91. Ye X, Wei Y, Nalepa G, Harper JW. The cyclin E/Cdk2 substrate p220(NPAT) is required for S-phase entry, histone gene expression, and Cajal body maintenance in human somatic cells. Mol Cell Biol. 2003;23(23):8586–600. Pubmed Central PMCID: 262656.
92. Watanabe N, Arai H, Iwasaki J, Shiina M, Ogata K, Hunter T, et al. Cyclin-dependent kinase (CDK) phosphorylation destabilizes somatic Wee1 via multiple pathways. Proc Natl Acad Sci U S A. 2005;102(33):11663–8. Pubmed Central PMCID: 1187955.
93. Lai ZC, Maxson R, Childs G. Both basal and ontogenic promoter elements affect the timing and level of expression of a sea urchin H1 gene during early embryogenesis. Genes Dev. 1988;2(2):173–83.
94. Khochbin S, Wolffe AP. Developmentally regulated expression of linker-histone variants in vertebrates. Eur J Biochem. 1994;225(2):501–10.
95. Duncliffe KN, Rondahl ME, Wells JR. A H1 histone gene-specific AC-box-related element influences transcription from a major chicken H1 promoter. Gene. 1995;163(2):227–32.
96. van Wijnen AJ, Aziz F, Grana X, De Luca A, Desai RK, Jaarsveld K, et al. Transcription of histone H4, H3, and H1 cell cycle genes: promoter factor HiNF-D contains CDC2, cyclin A, and an RB-related protein. Proc Natl Acad Sci U S A. 1994;91(26):12882–6. Pubmed Central PMCID: 45544.
97. Guglielmi B, La Rochelle N, Tjian R. Gene-specific transcriptional mechanisms at the histone gene cluster revealed by single-cell imaging. Mol Cell. 2013;51(4):480–92.

98. Gurley LR, Walters RA, Tobey RA. Sequential phsophorylation of histone subfractions in the Chinese hamster cell cycle. J Biol Chem. 1975;250(10):3936–44.
99. Yasuda H, Matsumoto Y, Mita S, Marunouchi T, Yamada M. A mouse temperature-sensitive mutant defective in H1 histone phosphorylation is defective in deoxyribonucleic acid synthesis and chromosome condensation. Biochemistry. 1981;20(15):4414–9.
100. Halmer L, Gruss C. Effects of cell cycle dependent histone H1 phosphorylation on chromatin structure and chromatin replication. Nucleic Acids Res. 1996;24(8):1420–7. Pubmed Central PMCID: 145815.
101. Alexandrow MG, Hamlin JL. Chromatin decondensation in S-phase involves recruitment of Cdk2 by Cdc45 and histone H1 phosphorylation. J Cell Biol. 2005;168(6):875–86. Pubmed Central PMCID: 2171796.
102. Lu ZH, Sittman DB, Brown DT, Munshi R, Leno GH. Histone H1 modulates DNA replication through multiple pathways in Xenopus egg extract. J Cell Sci. 1997;110(Pt 21):2745–58.
103. Lu ZH, Sittman DB, Romanowski P, Leno GH. Histone H1 reduces the frequency of initiation in Xenopus egg extract by limiting the assembly of prereplication complexes on sperm chromatin. Mol Biol Cell. 1998;9(5):1163–76. Pubmed Central PMCID: 25338.
104. Marzluff WF, Wagner EJ, Duronio RJ. Metabolism and regulation of canonical histone mRNAs: life without a poly(A) tail. Nat Rev Genet. 2008;9(11):843–54. Pubmed Central PMCID: 2715827.
105. Koseoglu MM, Graves LM, Marzluff WF. Phosphorylation of threonine 61 by cyclin a/Cdk1 triggers degradation of stem-loop binding protein at the end of S phase. Mol Cell Biol. 2008;28(14):4469–79. Pubmed Central PMCID: 2447125.
106. Koseoglu MM, Dong J, Marzluff WF. Coordinate regulation of histone mRNA metabolism and DNA replication: cyclin A/cdk1 is involved in inactivation of histone mRNA metabolism and DNA replication at the end of S phase. Cell Cycle. 2010;9(19):3857–63. Pubmed Central PMCID: 3047749.
107. Bracken AP, Ciro M, Cocito A, Helin K. E2F target genes: unraveling the biology. Trends Biochem Sci. 2004;29(8):409–17.
108. Harbour JW, Luo RX, Dei Santi A, Postigo AA, Dean DC. Cdk phosphorylation triggers sequential intramolecular interactions that progressively block Rb functions as cells move through G1. Cell. 1999;98(6):859–69.
109. Truscott M, Raynal L, Premdas P, Goulet B, Leduy L, Berube G, et al. CDP/Cux stimulates transcription from the DNA polymerase alpha gene promoter. Mol Cell Biol. 2003;23(8):3013–28. Pubmed Central PMCID: 152546.
110. Henneke G, Koundrioukoff S, Hubscher U. Phosphorylation of human Fen1 by cyclin-dependent kinase modulates its role in replication fork regulation. Oncogene. 2003;22(28):4301–13.
111. Singh RK, Kabbaj MH, Paik J, Gunjan A. Histone levels are regulated by phosphorylation and ubiquitylation-dependent proteolysis. Nat Cell Biol. 2009;11(8):925–33. Pubmed Central PMCID: 2720428.
112. Singh RK, Liang D, Gajjalaiahvari UR, Kabbaj MH, Paik J, Gunjan A. Excess histone levels mediate cytotoxicity via multiple mechanisms. Cell Cycle. 2010;9(20):4236–44. Pubmed Central PMCID: 3055206.
113. Su C, Gao G, Schneider S, Helt C, Weiss C, O'Reilly MA, et al. DNA damage induces downregulation of histone gene expression through the G1 checkpoint pathway. EMBO J. 2004;23(5):1133–43. Pubmed Central PMCID: 380976.
114. Weinert TA, Kiser GL, Hartwell LH. Mitotic checkpoint genes in budding yeast and the dependence of mitosis on DNA replication and repair. Genes Dev. 1994;8(6):652–65.
115. Bongiorno-Borbone L, De Cola A, Barcaroli D, Knight RA, Di Ilio C, Melino G, et al. FLASH degradation in response to UV-C results in histone locus bodies disruption and cell-cycle arrest. Oncogene. 2010;29(6):802–10.
116. Mullen TE, Marzluff WF. Degradation of histone mRNA requires oligouridylation followed by decapping and simultaneous degradation of the mRNA both 5′ to 3′ and 3′ to 5′. Genes Dev. 2008;22(1):50–65. Pubmed Central PMCID: 2151014.

117. Lyons SM, Ricciardi AS, Guo AY, Kambach C, Marzluff WF. The C-terminal extension of Lsm4 interacts directly with the 3′ end of the histone mRNP and is required for efficient histone mRNA degradation. RNA. 2014;20(1):88–102. Pubmed Central PMCID: 3866647.
118. Azzalin CM, Lingner J. The human RNA surveillance factor UPF1 is required for S phase progression and genome stability. Curr Biol. 2006;16(4):433–9.
119. Herrero AB, Moreno S. Lsm1 promotes genomic stability by controlling histone mRNA decay. EMBO J. 2011;30(10):2008–18. Pubmed Central PMCID: 3098488.
120. Beggs S, James TC, Bond U. The PolyA tail length of yeast histone mRNAs varies during the cell cycle and is influenced by Sen1p and Rrp6p. Nucleic Acids Res. 2012;40(6):2700–11. Pubmed Central PMCID: 3315300.
121. Alabert C, Groth A. Chromatin replication and epigenome maintenance. Nat Rev Mol Cell Biol. 2012;13(3):153–67.
122. Zhao X, McKillop-Smith S, Muller B. The human histone gene expression regulator HBP/SLBP is required for histone and DNA synthesis, cell cycle progression and cell proliferation in mitotic cells. J Cell Sci. 2004;117(Pt 25):6043–51.
123. Mejlvang J, Feng Y, Alabert C, Neelsen KJ, Jasencakova Z, Zhao X, et al. New histone supply regulates replication fork speed and PCNA unloading. J Cell Biol. 2014;204(1):29–43. Pubmed Central PMCID: 3882791.
124. Holland L, Gauthier L, Bell-Rogers P, Yankulov K. Distinct parts of minichromosome maintenance protein 2 associate with histone H3/H4 and RNA polymerase II holoenzyme. Eur J Biochem. 2002;269(21):5192–202.
125. Latreille D, Bluy L, Benkirane M, Kiernan RE. Identification of histone 3 variant 2 interacting factors. Nucleic Acids Res. 2014;42(6):3542–50. Pubmed Central PMCID: 3973350.
126. Richet N, Liu D, Legrand P, Velours C, Corpet A, Gaubert A, et al. Structural insight into how the human helicase subunit MCM2 may act as a histone chaperone together with ASF1 at the replication fork. Nucleic Acids Res. 2015;43(3):1905–17. Pubmed Central PMCID: 4330383.
127. Groth A, Corpet A, Cook AJ, Roche D, Bartek J, Lukas J, et al. Regulation of replication fork progression through histone supply and demand. Science. 2007;318(5858):1928–31.
128. Campisi J. d'Adda di Fagagna F. Cellular senescence: when bad things happen to good cells. Nat Rev Mol Cell Biol. 2007;8(9):729–40.
129. Campisi J. Cellular senescence as a tumor-suppressor mechanism. Trends Cell Biol. 2001;11(11):S27–31.
130. Braig M, Schmitt CA. Oncogene-induced senescence: putting the brakes on tumor development. Cancer Res. 2006;66(6):2881–4.
131. Pang JH, Chen KY. Global change of gene expression at late G1/S boundary may occur in human IMR-90 diploid fibroblasts during senescence. J Cell Physiol. 1994;160(3):531–8.
132. Narita M, Nunez S, Heard E, Narita M, Lin AW, Hearn SA, et al. Rb-mediated heterochromatin formation and silencing of E2F target genes during cellular senescence. Cell. 2003;113(6):703–16.
133. Sherr CJ, McCormick F. The RB and p53 pathways in cancer. Cancer Cell. 2002;2(2):103–12.
134. Geng Y, Lee YM, Welcker M, Swanger J, Zagozdzon A, Winer JD, et al. Kinase-independent function of cyclin E. Mol Cell. 2007;25(1):127–39.
135. O'Sullivan RJ, Kubicek S, Schreiber SL, Karlseder J. Reduced histone biosynthesis and chromatin changes arising from a damage signal at telomeres. Nat Struct Mol Biol. 2010;17(10):1218–25. Pubmed Central PMCID: 2951278.
136. Zhang R, Poustovoitov MV, Ye X, Santos HA, Chen W, Daganzo SM, et al. Formation of MacroH2A-containing senescence-associated heterochromatin foci and senescence driven by ASF1a and HIRA. Dev Cell. 2005;8(1):19–30.
137. Zhang R, Chen W, Adams PD. Molecular dissection of formation of senescence-associated heterochromatin foci. Mol Cell Biol. 2007;27(6):2343–58. Pubmed Central PMCID: 1820509.
138. Funayama R, Saito M, Tanobe H, Ishikawa F. Loss of linker histone H1 in cellular senescence. J Cell Biol. 2006;175(6):869–80. Pubmed Central PMCID: 2064697.

139. Angelov D, Molla A, Perche PY, Hans F, Cote J, Khochbin S, et al. The histone variant macroH2A interferes with transcription factor binding and SWI/SNF nucleosome remodeling. Mol Cell. 2003;11(4):1033–41.
140. Dollard C, Ricupero-Hovasse SL, Natsoulis G, Boeke JD, Winston F. SPT10 and SPT21 are required for transcription of particular histone genes in Saccharomyces cerevisiae. Mol Cell Biol. 1994;14(8):5223–8. Pubmed Central PMCID: 359041.

Chapter 12
The Role of Mcm2–7 in Replication Initiation

Dirk Remus

Abstract The hetero-hexameric Mcm2–7 complex is a multifunctional ATPase that plays essential roles during the initiation of DNA replication in eukaryotic cells. Initially, the Mcm2–7 complex is bound as a catalytically inactive double hexamer around double-stranded DNA, marking potential replication origin sites along the chromosome. Subsequently, upon activation, the Mcm2–7 complex mediates the opening, or "melting," of the parental DNA duplex at the origin, which culminates in the formation of two oppositely oriented DNA replication forks. Eventually, at the fork, the Mcm2–7 complex acts as the catalytic core of the replicative DNA helicase. In addition to unwinding DNA at the fork, the Mcm2–7 helicase complex also serves as the central scaffold around which the replisome is assembled. Due to its varied and essential roles in the initiation of DNA replication, the Mcm2–7 complex is a key target for regulatory mechanisms that govern origin activity in the cell cycle. Activation of the Mcm2–7 helicase entails a large conformational reconfiguration that results in the separation of the Mcm2–7 double hexamer into two individual Mcm2–7 hexamer complexes bound around the single-stranded leading strand template. Recent progress in the structural characterization of the Mcm2–7 complex begins to shed light on the mechanism by which origin unwinding is coupled to Mcm2–7 remodeling.

Keywords Mcm2–7 • Double hexamer • Replication origin • DNA unwinding • DNA helicase • Origin specification • AAA+ • ATPase • Pre-RC

Introduction

The double-stranded structure of eukaryotic chromosomal DNA helps maintain genome integrity by protecting the nucleotide bases from chemical modification and degradation. The base pairing of the complementary DNA strands, however, is incompatible with the DNA copying mechanism of DNA polymerases, which require

D. Remus (✉)
Molecular Biology Program, Memorial Sloan-Kettering Cancer Center,
1275 York Avenue, New York, NY 10065, USA
e-mail: remusd@mskcc.org

D.L. Kaplan (ed.), *The Initiation of DNA Replication in Eukaryotes*,
DOI 10.1007/978-3-319-24696-3_12

single-stranded (ss) DNA as a template. Consequently, to allow access of DNA polymerases during DNA replication the parental DNA duplex needs to be opened at the initiation site, termed replication origin, and needs to be continuously unwound at the replication fork. Initial DNA duplex opening ("melting") and subsequent processive DNA unwinding are critical events during the initiation of DNA replication that are tightly regulated to control origin activity. In eukaryotes, these DNA remodeling reactions are catalyzed by a conserved hetero-hexameric protein complex, Mcm2–7, composed of the *m*ini*c*hromosome *m*aintenance proteins 2–7 [1, 2].

The Mcm2–7 complex performs essential initiator function by marking or "licensing" potential origin sites along each chromosome. At the origin, although not yet formally demonstrated, the Mcm2–7 complex likely catalyzes the melting of the parental DNA duplex during origin activation. And finally, at the replication fork, the Mcm2–7 complex forms the catalytic core of the replicative DNA helicase. Another function of Mcm2–7 is to physically coordinate the numerous factors involved in the establishment and progression of replication forks. For example, in budding yeast the essential factors involved in the early steps of origin activation, including Cdc7-Dbf4 (DDK), Sld3•Sld7, Cdc45, Dpb11, Sld2, GINS, Polε, and Mcm10, associate with origins in an Mcm2–7-dependent manner [3–8]. Moreover, while the replisome of *Escherichia coli* is structurally coordinated by the γ/τ clamp-loader complex, eukaryotic replisomes appear to be organized primarily around the Mcm2–7 helicase [9]. This notion is supported by the analysis of the composition of the replisome progression complex (RPC), a complex of essential replisome components isolated from S phase chromatin of budding yeast cells [10]. The RPC is largely devoid of the clamp-loader, termed RFC in eukaryotes, and the lagging strand synthesis machinery, whereas Polα-primase, the leading strand polymerase, Polε, as well as the replisome components GINS, Cdc45, Ctf4, Mrc1, Dia2, Csm3•Tof1, and the histone chaperone complex FACT are tethered to the RPC directly or indirectly via Mcm2–7 [11–17]. Moreover, observations in vitro, using a reconstituted budding yeast DNA replication system, demonstrate that regulated origin activation and leading strand synthesis can occur in the complete absence of the clamp-loader [8].

The Mcm2–7 complex is composed of six distinct, yet related AAA+ (*A*TPases *a*ssociated with various cellular *a*ctivities) ATPase subunits [18]. Each Mcm2–7 subunit is essential and highly conserved from yeast to humans [2]. MCM proteins are conserved among eukaryotes and archaebacteria, yet archaebacteria often encode only for a single MCM homolog and consequently form homo-hexameric MCM complexes [1]. The heteromeric composition of the Mcm2–7 complex reflects the increased regulation of this complex by eukaryote-specific pathways. The simplified archaeal MCM complexes, however, serve as powerful models to study the catalytic mechanism of the Mcm2–7 complex. Similar to other AAA+ proteins the Mcm2–7 subunits assemble into a ring-shaped oligomer that forms ATPase active sites at specific subunit interfaces. Nonetheless, Mcm2–7 complexes exhibit a remarkable structural flexibility that enables them to perform diverse functions during the DNA replication initiation reaction. This flexibility is for example illustrated by, but not limited to, the distinct structural organization of Mcm2–7 complexes on DNA before and after activation of their helicase activity (Fig. 12.1).

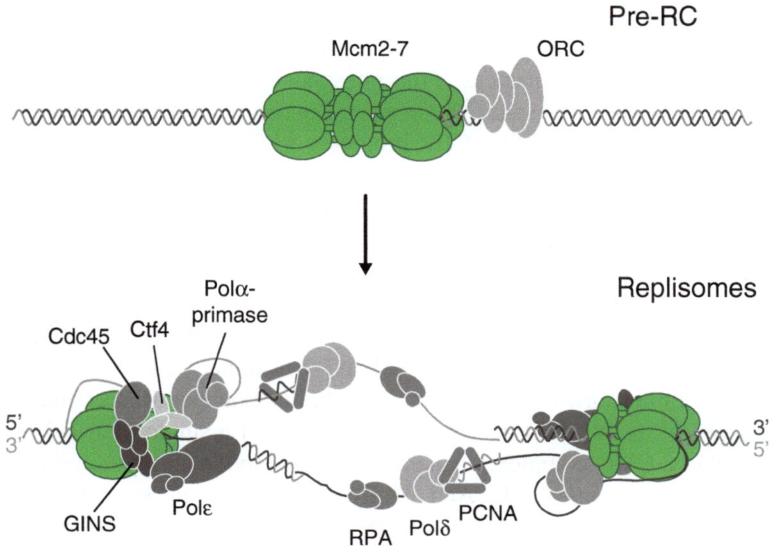

Fig. 12.1 Mcm2–7 complexes mark origins of replication and act as the replicative DNA helicase. Mcm2–7 complexes (*green*) are loaded as inactive double hexamers by the pre-replicative complex (pre-RC) around double-stranded DNA at origins of replication. After origin activation the Mcm2–7 complex forms part of the replisome, acting as the replicative DNA helicase at replication forks. For clarity, only a few select replisome components are depicted. Activation of the Mcm2–7 helicase involves separation of the Mcm2–7 double hexamer and formation of the *C*dc45-*M*cm2–7-*G*INS (CMG) helicase complex, which translocates along single-stranded DNA in 3′ to 5′ direction

To ensure that replication origins fire no more than once per cell cycle eukaryotic cells confine Mcm2–7 loading onto DNA and Mcm2–7 helicase activation during origin firing to distinct cell cycle stages [19]. The Mcm2–7 complex is loaded by the combined action of the origin recognition complex (ORC), Cdc6, and Cdt1, at the end of mitosis and throughout G1 phase as a double hexamer (DH) around double-stranded (ds) DNA [20–22]. In this DH complex two Mcm2–7 hexamers tightly associate via their N-terminal domains in a head-to-head configuration, providing a structural basis for the establishment of bidirectional replication forks at the origin. Importantly, however, the Mcm2–7 DH is inactive for DNA unwinding.

Activation of the Mcm2–7 helicase occurs exclusively in S phase as a pair of replisomes assembles around the Mcm2–7 DH. A key event in this process is the recruitment of two Mcm2–7 helicase accessory factors, Cdc45 and GINS, to form the *C*dc45-*M*cm2–7-*G*INS (CMG) helicase complex, which has greatly increased DNA helicase activity compared to Mcm2–7 alone [10, 23–25]. This event is under the control of two protein kinases, cyclin-dependent kinase (CDK) and Cdc7-Dbf4 (DDK), and involves the activities of numerous initiation factors that either act transiently to mediate Mcm2–7 activation (e.g., Sld3•Sld7, Sld2, and Dpb11), or that are stably incorporated into the replisome at the replication fork (e.g., Cdc45, GINS, and Polε) [3]. DDK exerts its essential function during origin activation by directly phosphorylating the Mcm2–7 complex, specifically the N-termini of Mcm4, 6, and 2 [4]; the molecular consequences of this phosphorylation are not entirely clear, but in budding yeast DDK phosphorylation promotes the recruitment of Sld3•Sld7 and Cdc45 to the Mcm2–7 DH [26]. CDK promotes origin activation via phosphorylation of Sld2 and Sld3, which promotes their association with Dpb11 and mediates the recruitment of Polε and GINS [3]. Thus formation of the CMG depends on the convergence of two separate pathways, controlled by DDK and CDK, respectively, that mediate the incorporation of the two essential helicase cofactors, GINS and Cdc45, into chromatin-bound Mcm2–7 complexes.

Various observations demonstrate that Mcm2–7 activation entails the separation of the Mcm2–7 DH into two individual Mcm2–7 hexamers, each forming the core of a CMG complex unwinding DNA at the fork. For example, RPCs contain only a single copy of Mcm4, consistent with the presence of a single Mcm2–7 hexamer [10]. Similarly, CMG complexes isolated from *Drosophila* embryos contain only a single Mcm2–7 hexamer [24]. And finally, single-molecule studies in *Xenopus* extracts, analyzing the progression of DNA synthesis on linear DNA molecules tethered on both ends to the surface of a flow-cell, demonstrate that the two replication forks emanating from an origin proceed uncoupled from each other in opposite direction [27]. Moreover, while the Mcm2–7 DH encircles dsDNA, the CMG complex encircles ssDNA, translocating with 3′ to 5′ polarity along the leading strand template, and in the process sterically excluding the lagging strand from the central channel [10, 24, 27, 28]. From this follows that Mcm2–7 complexes have to undergo a series of conformational remodeling events during origin activation that involve the opening and closing of the hexamer rings, specific extrusion of the lagging strand template, and separation of the double-hexamer into two uncoupled hexamers. Recent advances in the biochemical and structural characterization of the eukaryotic and archaeal MCM complexes suggest a model wherein Mcm2–7 remodeling is tightly coupled to DNA unwinding at the origin.

Architecture of Mcm2–7 Proteins

Overview

Eukaryotic and archaeal MCM proteins share homology over three core domains: an N-terminal domain (NTD), followed by an AAA+ ATPase domain, and a winged-helix domain (WHD) at the C-terminus (Fig. 12.2a). Several of the six eukaryotic paralogues contain additional, subunit-specific N- or C-terminal extensions (“tails”) that serve important regulatory functions [4, 29–31] or mediate the interaction of Mcm2–7 with the pre-RC [32], replisome components [33], or chromatin [13, 34]. MCM proteins generally assemble into ring-shaped hexameric complexes [22, 35, 36]. The MCM ring features two prominent tiers, one containing the N-terminal domains, and the other comprising the AAA+ domains. Deviations from this

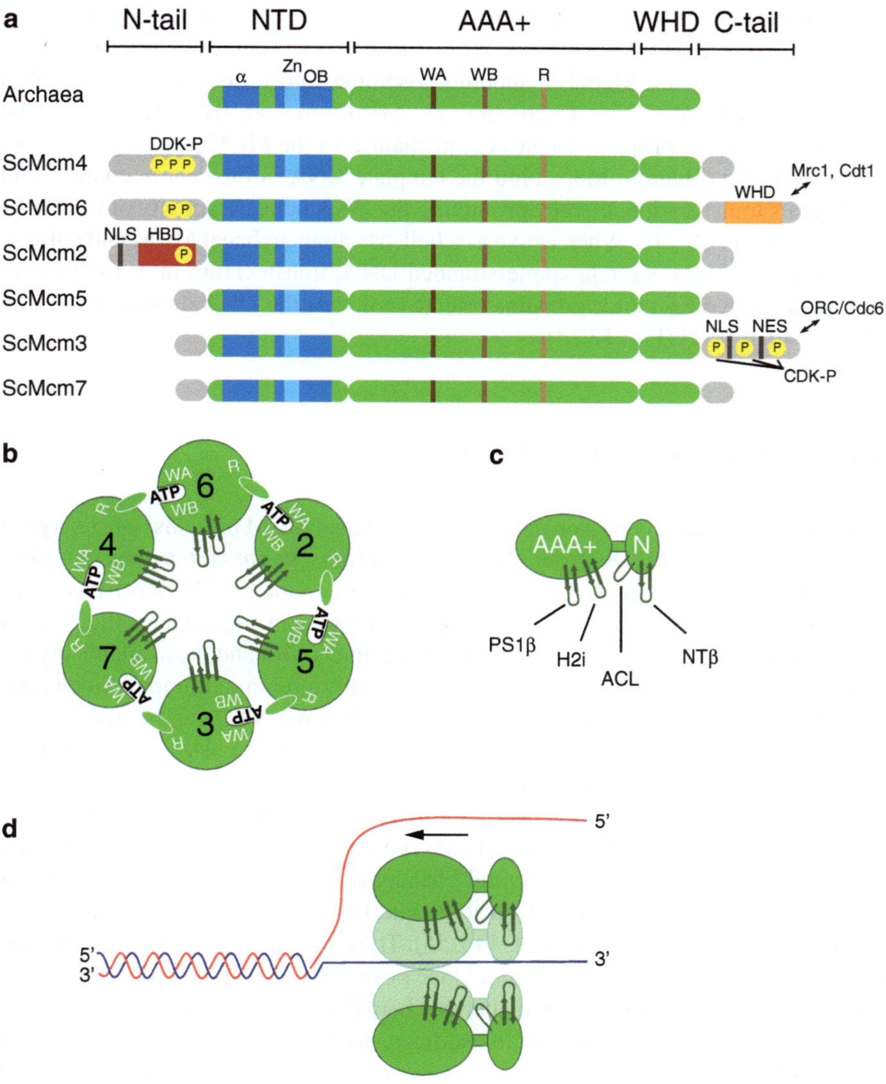

Fig. 12.2 Architecture of Mcm2–7 proteins. (**a**) Domain organization of archaeal (*top*) and eukaryotic MCM proteins. *N-tail* N-terminal tail, *NTD* N-terminal domain, *AAA+* AAA+ ATPase domain, *WHD* winged-helix domain, *C-tail* C-terminal tail, *α* alpha-helical domain (subdomain A), *Zn* zinc-binding domain (subdomain B), *OB* OB-fold (subdomain C), *WA* Walker A motif, *WB* Walker B motif, *R* arginine finger, *NLS* nuclear localization signal, *NES* nuclear export signal, *HBD* histone-binding domain, *DDK-P* DDK phosphorylation sites (*yellow circled* P in N-terminal tails; number of indicated DDK phosphorylation sites does not indicate the actual number of sites, but is intended to reflect the relative extent of phosphorylation); *CDK-P* CDK phosphorylation sites (*yellow circled* P in C-terminal tails; indicated CDK phosphorylation sites indicate relative positions, not the actual number of sites). (**b**) Subunit order in the Mcm2–7 ring as viewed from the C-terminal direction. Catalytic active site elements Walker A (WA), Walker B (WB) and arginine finger (R) motif, and bound ATP at the interface are indicated. β-hairpin loops reach into the central channel. (**c**) Schematic of β-hairpin position in MCM AAA+ ATPase domain and N-terminal domain (N). (**d**) Organization of MCM complexes during DNA unwinding. Two front subunits have been removed in this side-view for clarity. *Arrow* indicates direction of progression

configuration, such as filaments [37, 38], heptameric rings [39], and open ring structures [35, 40–42] are also observed. While the archaeal complexes form homohexamers, the eukaryotic Mcm2–7 subunits assemble into a hexameric ring with a defined subunit order (Fig. 12.2b) [35, 43, 44]. From both the NTD and AAA+ domains β-hairpins project into the central channel of the MCM ring (Fig. 12.2b, c) [45–47]. These β-hairpin loops bind the template DNA in the central channel and mediate its translocation, thereby promoting the DNA helicase activity of the MCM complex [1, 46, 48–50]. When active as a helicase, both archaeal MCM and eukaryotic Mcm2–7/CMG encircle single-stranded DNA, translocating in 3′ to 5′ direction with the AAA+ motor domains oriented towards the DNA duplex at the fork [24, 28, 40, 48, 51] (Fig. 12.2d).

The N-Terminal Domain (NTD)

The NTD is structurally best characterized in archaeal MCM proteins, where crystal structures for the NTDs from *Methanobacterium thermoautotrophicum* [46], *Sulfolobus solfataricus* [45, 47, 52], *Methanopyrus kandleri* [53], and *Pyrococcus furiosus* [54] have been solved. Although the amino acid sequence of the NTD is poorly conserved, structure-based sequence alignment [46] and the available crystal structures demonstrate that the NTD is structurally conserved among eukaryotic and archaeal MCM proteins. Based on these studies the NTD can be divided into three subdomains: An alpha-helical bundle (subdomain A), a zinc (Zn)-binding domain (subdomain B), and an OB-fold (subdomain C).

A striking feature of the NTD is its ability to assemble spontaneously into hexameric rings in isolation of the other MCM core domains [46, 52]. The OB-fold and Zn-binding domains define the central channel of the NTD ring. The Zn-binding domain also mediates the interaction at the interface of a head-to-head double hexamer formed by the MthMCM NTD [46, 55]. The alpha-helical bundle is positioned on the outside of the NTD ring. The relative position of the alpha-helical bundle on the NTD ring appears to be able to alternate between two rigid rotational states [38, 47]. Interestingly, a point mutation in budding yeast Mcm5 (*mcm5-bob1*, P83L) that bypasses the DDK requirement for origin activation [56] is predicted to be located near the point of rotation between the alpha-helical bundle and the OB-fold domains [47]. This may indicate that DDK phosphorylation induces conformational changes in the NTD during Mcm2–7 activation.

The inner channel of the NTD ring exhibits a strongly positive electrostatic surface potential and is wide enough to accommodate either single- or double-stranded DNA, consistent with DNA being threaded through the central channel. While the DNA-free NTD hexamer structures of MthMCM and SsoMCM exhibit clear sixfold symmetry with almost perfectly round central pores [46, 52], the two hexamers in the asymmetric unit of the crystals of ssDNA-bound PfuMCM exhibit more oval-shaped pores and deviate substantially from sixfold symmetry [54], indicating that the NTD ring is inherently flexible.

The OB-fold domain of the NTD has been observed to bind DNA in two distinct modes. The first DNA binding mode involves a β-hairpin (NTβ) insertion that protrudes deeply from the NTD into the central channel. Mutational analysis in archaeal homologs suggests that the NTβ contacts template DNA via a set of conserved basic residues at its tip, thereby contributing to the DNA-binding and -unwinding activities of the active MCM helicase complex [46, 48]. The second DNA binding mode involves the binding of the OB-fold specifically to single-stranded (ss) DNA and is mediated by basic residues of a conserved MCM-ssDNA binding motif (MSSB) at the base of the β-barrel of the OB fold [54]. In this mode ssDNA is not threaded lengthwise through the channel, but instead binds the inner NTD ring perpendicular to the central channel and with a defined polarity. The binding of ssDNA via the MSSB may be relevant for origin melting rather than processive DNA unwinding, as will be discussed further below.

The AAA+ Domain

Mcm2–7 proteins belong to the AAA+ class of ATPases [18], and the ability of MCM proteins to utilize ATP is intimately connected to their biological functions. AAA+ proteins function as multimeric complexes that form composite active sites at subunit interfaces. Each active site is composed of a P-loop ATP-binding domain of one subunit, which comprises the ATP-Mg^{2+}-coordinating Walker A and B motifs, and a conserved arginine finger domain (the AAA+ box VII motif) of an adjacent subunit, which is required for ATP-hydrolysis. This active site configuration promotes the coordination of conformational changes between subunits during the ATP-binding and -hydrolysis cycle and thereby allows AAA+ complexes to convert the chemical energy from ATP-binding and -hydrolysis into mechanical work. In accordance with this general AAA+ configuration, Mcm2–7 subunits form ATPase pairs between specific subunits, which has aided the delineation of the subunit order within the Mcm2–7 ring [44, 57].

Like their archaeal homologs, Mcm2–7 utilize the energy derived from ATP-binding and -hydrolysis to fuel their DNA helicase activity [1, 23, 51]. Moreover, recent evidence suggests that ATP-hydrolysis by Mcm2–7 is also required for Mcm2–7 loading around DNA by the pre-RC [58, 59]. Generally, ring helicases may bind and hydrolyze nucleotides either sequentially or in a concerted manner around the ring [60]. How ATP-binding and -hydrolysis are coordinated around the Mcm2–7 ring is not known. Analysis of the effect of titrating catalytically mutant subunits into the homo-hexameric archaeal SsoMCM complex reveals that its DNA helicase activity can tolerate catalytically inactive subunits, suggesting that SsoMCM hydrolyzes ATP in a semisequential manner [61]. Yet, all six ATP-binding sites present in the eukaryotic Mcm2–7 hexamer are essential for S phase progression and viability of budding yeast cells [58, 62]. Interestingly, equivalent mutations in the ATPase motifs of different Mcm2–7 subunits exhibit slightly different defects in Mcm2–7 loading, origin activation, DNA-unwinding, or S phase progression, respectively [23, 51, 57–59]. This may suggest that the various Mcm2–7 ATPase sites have specialized roles during Mcm2–7 function.

A detailed analysis of the topology of the MCM structure revealed that MCM proteins belong to the presensor 1 (PS1) insert superclade of AAA+ proteins, and within that superclade form part of a specific subgroup, clade 7, of the helix 2-insert (H2i) proteins [18]. The H2i is located between the Walker A and B motifs, while the PS1 insert is located between the Walker B and Sensor 1 motifs of the canonical AAA+ domain. Studies in archaea have shown that both the PS1 and H2 inserts form β-hairpin loops that bind to the DNA backbone in the central channel via basic residues at the tip of the loop, and nucleotide-dependent repositioning of these loops may mediate the translocation of DNA inside the central channel [48, 50, 63] (Fig. 12.2b–d). An analogous DNA translocation mechanism occurs in the distantly related SF3 helicases E1 and large T-antigen, which contain only a single β-hairpin loop that is homologous to the MCM PS1β [64, 65].

During DNA unwinding MCM complexes translocate along the leading strand template, in the process sterically displacing the lagging strand. Various observations with archaeal MCM complexes suggest that the displaced strand is also engaged on the MCM surface, which may stimulate MCM helicase activity. For example, an additional hairpin loop, EXT-hp, is found on the surface of the MCM complexes, which appears to be important for maximal helicase activity of MthMCM [45, 66]. Moreover, FRET analysis indicates that the displaced strand interacts physically with the SsoMCM surface near the EXT-hp and that this interaction stabilizes the MCM hexamer on forked DNA substrates [67, 68].

Curiously, the AAA+ motor domain of SsoMCM exhibits DNA helicase activity even in the absence of the NTD [49]. However, the NTD is required for maximum processivity of the SsoMCM helicase. Due to its ability to form ring shaped hexamers the NTD may stimulate MCM helicase activity by promoting cooperativity between individual AAA+ domains and by acting as a processivity clamp that tethers the AAA+ domains to the template DNA. However, a conserved loop domain, termed the allosteric communication loop (ACL), which is located between the NTD and AAA+ domains, appears to play an important role in coordinating the NTD and AAA+ tiers of the MCM hexamer, suggesting that a specific interplay between the NTD and AAA+ domains mediates optimal DNA translocation and -unwinding by the MCM complex [47, 63].

The Winged-Helix Domain

Immediately C-terminal to the AAA+ domain, downstream of the sensor 2 motif and separated by a short linker, MCM proteins contain a helix-turn-helix domain that is structurally related to winged-helix domains (WHD), although it is degenerate with regard to the two-strand β-sheet wings of the WHD [69]. The function of this domain is unclear. However, it is dispensable for the DNA helicase activity of archaeal MCM homologs, and it may actually negatively regulate both the ATPase and helicase activities of archaeal MCMs [49, 50].

N- and C-Terminal Tails

In addition to the MCM core domains described above, which are conserved among both archaeal and eukaryotic MCM proteins, eukaryotic MCM paralogs contain subunit-specific N- or C-terminal extensions, also referred to as "tails," that play important roles in the regulation of Mcm2–7 function and in mediating the interaction of Mcm2–7 with the pre-RC, the replisome, and histones (Fig. 12.2a). The structure and function of these N- or C-terminal tails are often, but not always, conserved between organisms. For example, while the Mcm6 C-terminal tail is structurally conserved from yeast to humans, *Saccharomyces cerevisiae* Mcm6 contains a significantly longer N-terminal tail than human Mcm6. For illustration purposes the discussion here will primarily focus on the N- and C-terminal tails of budding yeast, *Saccharomyces cerevisiae*, Mcm2–7, unless noted otherwise.

The N-terminal tails of Mcm2, -4, and -6, as well as the C-terminal tail of Mcm3, play important roles in mediating the regulation of budding yeast Mcm2–7 activity by multiple protein kinases, which include DDK, CDK, CK2, and Mec1 [29–31]. For example, budding yeast cells control the subcellular localization of the Mcm2–7 complex to prevent the re-firing of origins within one cell cycle [70]. The Cdt1•Mcm2–7 complex is transported into the nucleus via a bipartite nuclear localization signal (NLS) that is distributed between the N-terminal tail of Mcm2 and the C-terminal tail of Mcm3, while nuclear export is mediated by a nuclear export signal (NES) contained on the C-terminal tail of Mcm3 [29]. CDKs promote the net nuclear export of Cdt1•Mcm2–7 by phosphorylating specific target sites that surround the NLS and NES on the Mcm3 C-terminal tail, thereby restricting Mcm2–7 loading to periods in the cell cycle when CDK activity is low [19].

The C-terminal tails of Mcm6 and Mcm3 also play specialized roles by promoting the interaction of Mcm2–7 with pre-RC components during Mcm2–7 loading around DNA. For example, the conserved C-terminus of the budding yeast Mcm3 C-terminal tail promotes Mcm2–7 loading by bridging the interaction of Cdt1•Mcm2–7 with ORC/Cdc6 [32]. Although it is currently unclear how two Mcm2–7 hexamers are loaded cooperatively and in opposite orientation around DNA to form the Mcm2–7 DH [71], both Mcm2–7 hexamers appear to be recruited to the pre-RC via the Mcm3 C-terminus. The C-terminal tail of human Mcm6, on the other hand, forms a WHD that mediates the interaction of Mcm2–7 with the loading factor Cdt1, which itself is composed of multiple WHDs [72]. The WHD of the Mcm6 C-terminal tail is conserved in budding yeast, and biochemical analysis of the Mcm2–7 loading reaction, as well as the electron-microscopic characterization of an ORC/Cdc6/Cdt1•Mcm2–7 complex obtained in the presence of ATPγS, indicate that the role of the Mcm6–Cdt1 interaction during Mcm2–7 loading is conserved [42, 73].

While CDK negatively regulates Mcm2–7 activity by controlling the subcellular localization of free Cdt1•Mcm2–7, DDK promotes the activation of chromatin-bound Mcm2–7 complexes. The N-terminal tails of budding yeast

Mcm4 and -6, and possibly Mcm2, are the essential phosphorylation targets for DDK during origin activation [8, 30, 74, 75]. DDK intrinsically prefers to phosphorylate Mcm2–7 DHs over free Mcm2–7 complexes, allowing it to target specifically Mcm2–7 loaded at chromosomal origins [5, 30, 74–77]. This specificity may be achieved through the generation of a bipartite DDK binding site on Mcm4 and -2 spanning the N-terminal tiers of the Mcm2–7 DH [76]. The molecular mechanism by which DDK phosphorylation promotes Mcm2–7 activation is not clear, but in budding yeast it induces the association of the essential initiation factors Sld3•Sld7 and Cdc45 with the Mcm2–7 DH at replication origins [5, 6, 8, 26]. The isolation of DDK bypass mutations in Mcm4 and Mcm5 [56, 74] that do not involve phosphomimetic amino-acid substitutions in the Mcm2–7 N-terminal tails suggests that DDK phosphorylation promotes Mcm2–7 activation by inducing a structural change in the Mcm2–7 DH. However, DDK phosphorylation does not grossly affect the structure of the Mcm2–7 DH in vitro [7, 76], leaving open the possibility that the generation of phospho-epitopes for Sld3•Sld7 and Cdc45 on the Mcm2–7 DH also plays a role in DDK-dependent Mcm2–7 activation.

Once activated, the Mcm2–7 complex utilizes specific N- and C-terminal tails to communicate with the replisome and to mediate its interaction with the chromatin template. For example, Mrc1, which is a constitutive replisome component that promotes replication fork progression during normal S phase and that relays signals from the replication fork to the checkpoint during replication stress, integrates with the replication fork via the WHD of the Mcm6 C-terminal tail [33].

During chromosomal DNA replication nucleosomes are disrupted ahead of the replication fork and are reassembled on the daughter strands behind the fork. The molecular mechanisms involved in chromatin replication are little understood. However, Mcm2–7 complexes from yeast and human cells associate with both histones and histone chaperones [13, 78, 79]. This interaction is primarily mediated by the conserved Mcm2 N-terminal tail, which is characterized by an abundance of acidic amino acids and which can interact directly with histones H3 and H4 [34, 80, 81]. Crystallographic analysis of the interaction between the human Mcm2 N-terminal tail and a histone H3-H4 tetramer demonstrates that the Mcm2 N-terminal tail is largely unfolded free in solution, but wraps around H3-H4 at positions that are occupied by DNA in the nucleosome [34]. Intriguingly, the binding of the Mcm2 N-terminal tail to the outside of the H3-H4 dimer permits the formation of a quaternary complex with ASF1, which binds to the opposite face of the H3-H4 dimer and thereby disrupts H3-H4 tetramerization. The N-terminal tail of budding yeast Mcm2 has also been observed to bind simultaneously to both histones and FACT [13]. These observations suggest that the Mcm2 N-terminal tail may cooperate with histone chaperones to reassemble chromatin behind the fork. Such a role for the Mcm2 N-terminal tail would be consistent with the Mcm2–7 N-terminal tier trailing the AAA+ motor domains at the fork [40, 48], and with the observation that nucleosomes are rapidly assembled behind the fork [82].

The Role of Mcm2–7 in Origin Specification

Mcm2–7 Complexes License Replication Origins

DNA can assemble interphase nuclei in *Xenopus* egg extracts, which allows the DNA to undergo exactly one round of replication. Reinitiation of DNA synthesis in this system does not occur unless the nuclear membrane is intermittently permeabilized or degraded during mitosis. These observations led to the hypothesis that a diffusible factor, termed replication licensing factor (RLF), must bind to chromatin to initiate DNA replication, and that this factor is subsequently inactivated or removed from the chromatin as the DNA is being replicated [83]. If RLF were incapable of crossing the nuclear membrane, rebinding of RLF to chromatin, or "licensing," could only occur after nuclear membrane breakdown. Biochemical fractionation of *Xenopus* egg extract found that RLF consists of two components, RLF-M and RLF-B, which were subsequently identified to correspond to Mcm2–7 and Cdt1, respectively [84, 85].

In a separate approach, the isolation of the first eukaryotic replication origin, *Saccharomyces cerevisiae* ARS1 [86], allowed for the identification of proteins that bind to the origin and that are important for its function. Characterization of the protein binding pattern at budding yeast origins by DNase1 footprinting revealed that origins alternate between two states in the cell cycle [87]. Prior to origin firing in S phase, at the end of mitosis and throughout G1 phase, an extended footprint is observed at the origin, corresponding to the so-called pre-replicative complex (pre-RC). Formation of the pre-RC is now known to depend on ORC, Cdc6, Cdt1, and Mcm2–7. After origin firing in S phase, a smaller post-replicative complex is observed at the origin that corresponds to ORC [88, 89], which in budding yeast remains bound to origins throughout the cell cycle.

Replication origin "licensing" and pre-RC formation are now recognized to be equivalent to Mcm2–7 loading. Reconstitution approaches have demonstrated that ORC, Cdc6, Cdt1, and Mcm2–7 are indeed sufficient for Mcm2–7 loading in vitro [20, 22, 90, 91], and that such reconstituted pre-RCs support regulated DNA replication in vitro [6–8, 59, 90]. In the course of the Mcm2–7 loading reaction Cdc6 and Cdt1 only associate transiently with the origin, while ORC remains bound at the origin, but does not maintain contact to the Mcm2–7 DH [22, 76]. Unlike the Mcm2–7 DH, which is topologically bound around the DNA, ORC can be selectively removed from DNA by a high-salt wash after Mcm2–7 loading [22, 92–94]. Using this approach it was found that ORC, Cdc6, and Cdt1 are in fact dispensable for DNA replication after Mcm2–7 loading [6, 8, 94, 95], demonstrating that Mcm2–7 loading is mechanistically the only essential function of the pre-RC. From this follows that the entity that licenses DNA for replication is the Mcm2–7 DH bound to the DNA. This in turn implies that ultimately the position of the Mcm2–7 DH on chromatin determines the site of replication initiation.

Mcm2–7 DH Mobility Permits Origin Plasticity

Mcm2–7 complexes are loaded as inactive double hexamers around DNA by the pre-RC. Nonetheless, comparison of the persistence of Mcm2–7 DHs loaded in vitro on circular and linear DNA revealed that Mcm2–7 DHs are mobile and can slide on DNA [20, 22]. This mobility does not require ATP-hydrolysis, but is dependent on high-salt buffer conditions [22]. By analogy to other ring helicases [60] it may be expected that the interior central channel of the Mcm2–7 DH is positively charged, which is supported by the crystal structure of the archaeal NTD [46]. Considering that each Mcm2–7 DH, at a length of approximately 21 nm, may encircle ~60 bp of dsDNA, extensive electrostatic interactions between the Mcm2–7 central channel interior and the negatively charged DNA backbone might thus limit the extent of Mcm2–7 DH sliding under low salt conditions.

What might be the significance of the Mcm2–7 DH mobility? The topological nature of the Mcm2–7 DH DNA-binding mechanism requires that Mcm2–7 are loaded by specialized factors, comprising the other pre-RC components ORC, Cdc6, and Cdt1, around DNA. This dependency on loading factors for initial DNA binding implies that Mcm2–7 DHs would be unable to spontaneously rebind DNA after accidental dissociation from chromatin. This may be particularly critical at late and dormant origins, which need to be maintained for long periods of time in S phase, when re-replication control mechanisms prevent the reloading of Mcm2–7 by the pre-RC. The intrinsic Mcm2–7 DH mobility may thus be a mechanism that helps maintain Mcm2–7 DH integrity upon collision with other protein complexes on the DNA.

This notion is supported by recent observations (Fig. 12.3). Mcm2–7 DHs remain functional after a high-salt wash, suggesting that salt-induced mobilization of Mcm2–7 DHs on DNA does not interfere with their ability to support replication initiation [6, 8]. Moreover, unpublished results (Gros J, Lynch G, Yadav T, Whitehouse I, and Remus D) demonstrate that RNA polymerases can indeed push Mcm2–7 DHs over several kilo base pairs along naked DNA in vitro and through chromatin in vivo, resulting in Mcm2–7 DHs initiating DNA replication at sites distal to the original loading site. Budding yeast origins, like all eukaryotic origins, are generally located in intergenic regions or within inactive genes, which is expected to limit their collision with RNA polymerases. However, the potential for collisions between pre-RCs and RNA polymerases may persist even in intergenic regions, as pervasive transcription [96], heterogeneity in mRNA transcription [97], and the stochastic nature of transcription termination [98] cause significantly higher levels of intergenic transcription than previously appreciated.

The observation that Mcm2–7 DHs support replication initiation at DNA sequences that are distal to the initial Mcm2–7 loading site demonstrates that Mcm2–7 DH position along the template DNA ultimately determines the position of replication origins. Moreover, the observation that Mcm2–7 DH position is flexible after loading by the pre-RC and can be modulated by collision with other protein complexes suggests a general post-licensing mechanism for eukaryotic origin specification (Fig. 12.4). According to this model, eukaryotic origins are specified globally by the selection of the site of Mcm2–7 loading along each chromosome,

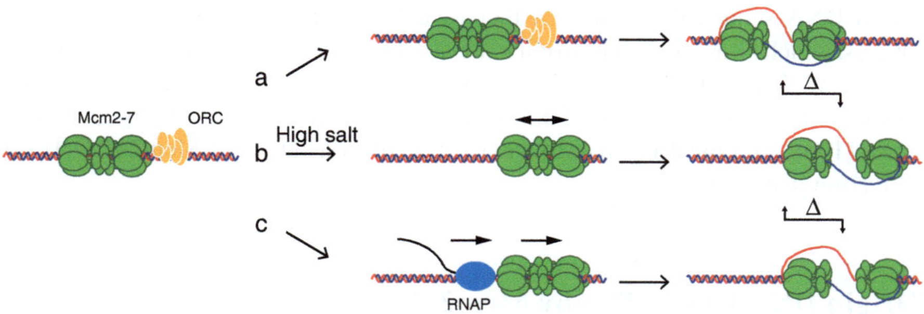

Fig. 12.3 Modulation of origin position by induced Mcm2–7 DH mobility. After Mcm2–7 loading by the pre-RC the Mcm2–7 Dh may remain at its original position (**a**). High-salt treatment of pre-RCs induces sliding of the Mcm2–7 DH, which results in a shift of the origin position (**b**). Mcm2–7 DHs may also get pushed along DNA after collision with RNA polymerase (RNAP), which can also induce a shift in origin position (**c**). Δ origin position shift

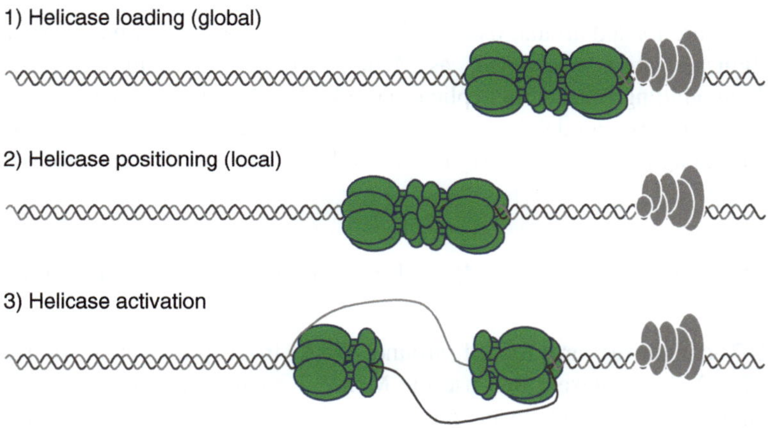

Fig. 12.4 Model for post-licensing origin specification mechanism in eukaryotes. Origin position is specified globally by selection of the Mcm2–7 loading site on the chromosome [1]. Mcm2–7 DH position may subsequently be modulated locally [2]. Origin activation occurs at the site of final Mcm2–7 DH position [3]. Mcm2–7 DH in *green*, ORC is in *grey*

which is primarily mediated through the targeting of ORC to a specific site. In budding yeast ORC recruitment is partly achieved through direct recognition of a specific DNA sequence, while ORC recruitment occurs by alternative means and is modulated throughout development in higher eukaryotes, explaining why budding yeast origins, but not origins from other eukaryotes, contain a conserved sequence element [99]. Outside the ORC binding site, however, all eukaryotic origins appear to lack conserved origin sequences, which is consistent with the observation that specific DNA sequences are mechanistically not required for Mcm2–7 activation in both yeast and higher eukaryotes [6, 7, 99]. This lack of dependency on specific origin DNA sequences confers an increased flexibility on eukaryotic replication origins, which allows origin position to be modulated locally by the repositioning of the Mcm2–7 DH, for example in response to genome traffic around the origin. Finally, since only a fraction of all potential origins is activated in any given cell cycle [100], additional mechanisms exist that select a subset of Mcm2–7 DHs for activation in each S phase.

Origin DNA Unwinding by Mcm2–7

The separation of the origin licensing and origin activation reactions in the cell cycle is essential for the maintenance of genome integrity, as it prevents the re-firing of origins in a single cell cycle [101]. In a normal cell cycle Mcm2–7 activation in S phase may consequently occur several hours after Mcm2–7 are loaded onto DNA in late M/G1 phase. The need to prevent the persistence of unstable ssDNA in the absence of ongoing DNA synthesis in G1 phase may thus explain why Mcm2–7 complexes are loaded around dsDNA. This, however, presents a major obstacle to the origin activation mechanism, as Mcm2–7 complexes encircle ssDNA when actively unwinding DNA at the replication fork [28]. A combination of recent structural data suggests a model for how Mcm2–7 complexes may achieve the transition from encircling dsDNA at the origin to encircling ssDNA at the replication fork.

Mcm2–7 Ring Opening Is Regulated at the Mcm2/5 Interface

Mcm2–7 complexes are loaded around DNA from pre-formed Mcm2–7 or Cdt1•Mcm2–7 complexes. Because the Mcm2–7 subunits are arranged in a ring, transient ring opening and closing must occur during both the Mcm2–7 loading and activation reactions, when dsDNA is inserted into the central channel or a single DNA strand is extruded from it, respectively. Mcm2–7 ring opening appears to be regulated specifically at the interface between the Mcm2 and Mcm5 subunits. Biochemical studies with the Mcm2–7 hexamer from budding yeast indicated that an ATP-dependent discontinuity between Mcm2 and 5 serves as a "gate" that allows the Mcm2–7 ring to bind circular closed ssDNA [51]. The functional importance of this gate was confirmed by fusing Mcm2 and 5 to a pair of inducible FKBP/FRB dimerization domains that allow control of the open and closed states of the gate with rapamycin derivatives both in vitro and in vivo [102]. An intrinsic nucleotide-dependent discontinuity between Mcm2 and 5 was also observed structurally by EM in *Drosophila* Mcm2–7 and CMG [35, 40]. Interestingly, factors that promote Mcm2–7 loading or that promote Mcm2–7 helicase processivity on DNA appear to bind near or across the Mcm2/5 gate, indicating that control of the Mcm2/5 gate partly underlies their mechanism of action. For example, Cdt1, which is essential for Mcm2–7 loading, can be observed to bind immediately adjacent to the Mcm2/5

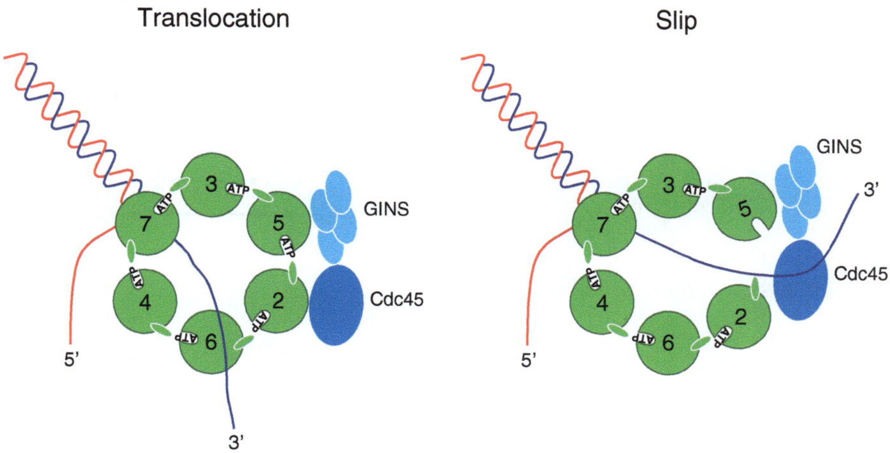

Fig. 12.5 The CMG promotes Mcm2–7 helicase processivity. During normal translocation (*left*) the leading strand template is threaded through the Mcm2–7 ring without making contacts to either Cdc45 or GINS. Upon accidental slippage through the Mcm2/5 gate, Cdc45 captures the leading strand to maintain topological binding of CMG (*right*). Mcm2–7 subunit (*green*) order is as viewed from N-terminus

interface in the EM structure of an ORC/Cdc6/Cdt1·Mcm2–7 loading intermediate assembled in the presence of ATPγS [42]. This may indicate that Cdt1 plays a role in promoting either the opening or closing of the Mcm2–7 ring at the Mcm2/5 gate during Mcm2–7 loading by the pre-RC. EM structures of the *Drosophila* CMG complex, on the other hand, demonstrate that the Mcm2–7 helicase cofactors GINS and Cdc45 bind across the Mcm2/5 discontinuity, suggesting that they stimulate Mcm2–7 helicase activity by promoting the topological linkage of Mcm2–7 to DNA [35, 40]. Consistent with this notion recent protein–DNA cross-linking studies with the *Drosophila* CMG suggest that Cdc45 captures the leading strand in the event of escape from the Mcm2–7 ring through the Mcm2/5 discontinuity (Fig. 12.5) [103].

Mcm2–7 Activity Is Inhibited in the DH Configuration

Mcm2–7 complexes are loaded as head-to-head double hexamers (DHs) around DNA by the pre-RC [20, 22]. The hexamers are held together at their NTD rings, resulting in the AAA+ motor domains facing outwards in opposite directions. Since the AAA+ domains appear to face the parental DNA duplex at the fork during DNA unwinding [40, 48], this configuration provides a mechanistic basis for the establishment of bidirectional replication forks at the origin. However, the two Mcm2–7 hexamers contained within one DH do not form a simple, straight continuous channel. Instead, the two hexamers appear to be stacked slightly off-register, resulting in a kink of the central channel at the hexamer–hexamer interface (Fig. 12.6a) [22, 76].

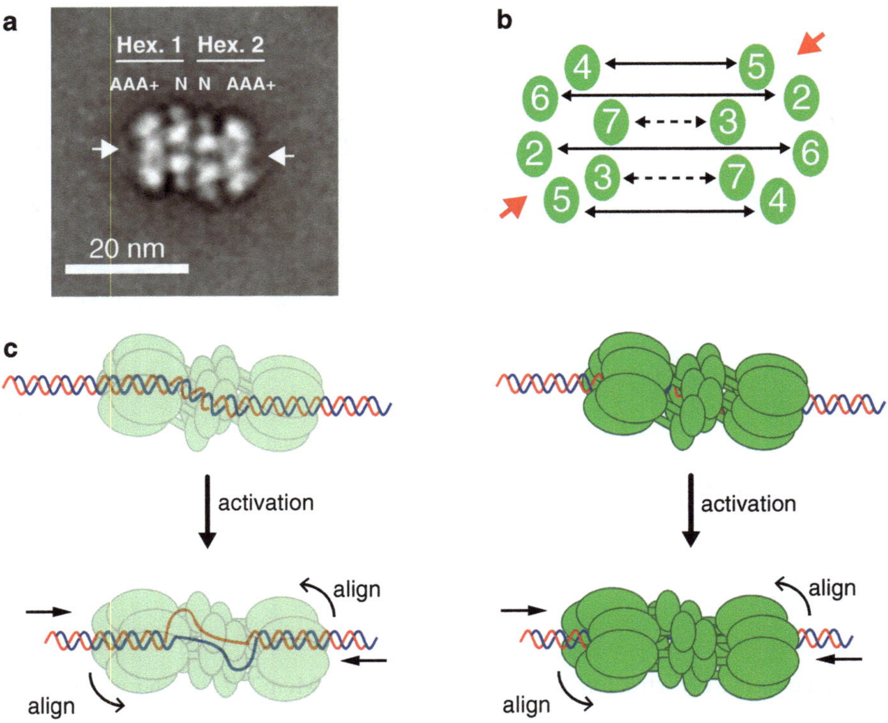

Fig. 12.6 Mcm2–7 DH organization. (**a**) EM analysis of Mcm2–7 DH structure demonstrates off-register stacking of the two Mcm2–7 hexamers in the Mcm2–7 DH. 2D class average of negatively stained Mcm2–7 DH is shown. *Arrows* indicate entry sites of central channel. (**b**) Mcm2–7 subunit order across Mcm2–7 DH interface. *Red arrows* indicate position of the Mcm2/5 gate. (**c**) Model for Mcm2–7 DH activation. In the inactive state, Mcm2–7 subunits are tilted. Alignment of the subunits activates the Mcm2–7 ATPase and DNA translocase activities, resulting in DNA being pumped from both ends into the Mcm2–7 DH, which in turn may induce the melting of the DNA at the Mcm2–7 DH midpoint. Mcm2–7 DH transparency is increased on the left to illustrate changes in DNA structure inside the central channel

Moreover, the recent elucidation of the Mcm2–7 subunit arrangement across the hexamer–hexamer interface revealed that the two component hexamers of the Mcm2–7 DH are rotationally offset from one another in such a way that the two Mcm2/5 gates are almost on opposite sides of the DH barrel, thereby preventing the formation of a continuous Mcm2/5 gate across one DH side (Fig. 12.6b) [76]. Importantly, the opposing orientation of the Mcm2/5 gates in the Mcm2–7 DH would sterically prevent the loss of the Mcm2–7 DH in the event of simultaneous opening of both Mcm2/5 gates, as might occur during Mcm2–7 activation, thus permitting extensive Mcm2–7 remodeling on DNA. The off-register stacking of the Mcm2–7 hexamers and their rotational misalignment may also reflect the opposite polarity of each hexamer during strand separation.

Another interesting feature of the budding yeast Mcm2–7 DH structure that was revealed by the recent higher resolution analysis is the tilted arrangement of the

Mcm2–7 subunits within each hexamer [76]. This tilted arrangement was suggested to result in a potential misalignment of the *cis*- and *trans*-acting catalytic components of the ATPase active sites at the subunit interfaces, which may cause the inhibition of ATP-hydrolysis activity observed with the Mcm2–7 DH. Activation of the Mcm2–7 DH ATPase activity may consequently be induced by realignment of the AAA+ domains. To bring about this conformational change may be one function of the pre-initiation complex (pre-IC), which is required and sufficient for CMG formation [3, 8]. Such a model invokes a certain degree of flexibility of AAA+ domain orientations relative to the NTD, which is also supported by EM structures of the *Drosophila* CMG [40]. Activation of the Mcm2–7 DH ATPase by re-alignment of the AAA+ domains may then induce the translocation of DNA through the central channel (Fig. 12.5c). On the basis of observations that DNA translocates from the C-terminal AAA+ end to the N-terminal NTD end through the central MCM channel [40, 48], activation of the Mcm2–7 ATPase activity prior to Mcm2–7 DH separation would result in the DNA being pumped from both ends into the central, which may thus cause distortion and ultimately melting of the DNA at the hexamer–hexamer interface.

A Model for DNA Unwinding During Mcm2–7 Activation

The data presented so far may be incorporated into a model for how Mcm2–7 activation and DNA unwinding are achieved during origin activation (Fig. 12.7). The model presented here focuses on the role of Mcm2–7 complex, but it needs to be noted that these events depend on the activities of numerous initiation factors, chief among which are the components of the pre-IC, whose individual roles in this process, however, remain to be elucidated.

After being loaded around DNA by the pre-RC, Mcm2–7 double hexamers are inactive, possibly due to the misalignment of the AAA+ domains (Fig. 12.7i) [76]. During origin activation DDK and CDK promote the assembly of the pre-IC around the Mcm2–7 DH [3], which may induce a conformational change in the Mcm2–7 DH that induces the alignment of the AAA+ domains and concomitant activation of Mcm2–7 ATPase (Fig. 12.7ii). Activation of the ATPase activity may subsequently induce the translocation of the dsDNA template into the central channel through the ATP-dependent movement of the H2i and PS1β hairpins. If the two Mcm2–7 hexamers remained associated at this point via their NTDs, DNA would be pumped from both ends into the Mcm2–7 DH channel, causing a steric clash between the increasing number of DNA nucleotides near the NTD rings at the Mcm2–7 DH midpoint, which may distort the B-form of the DNA and eventually force the two parental DNA strands apart. The single DNA strands thus generated may be bound by the MCM single-stranded DNA binding motif, MSSB, of the N-terminal OB-fold domains [54]. The MSSB further prevents reannealing of the unwound DNA strands by contacting the ribose and base moieties of the DNA nucleotides rather than the phospho-backbone [54]. As ssDNA binds with a defined polarity to the MSSBs in the NTD ring, the two

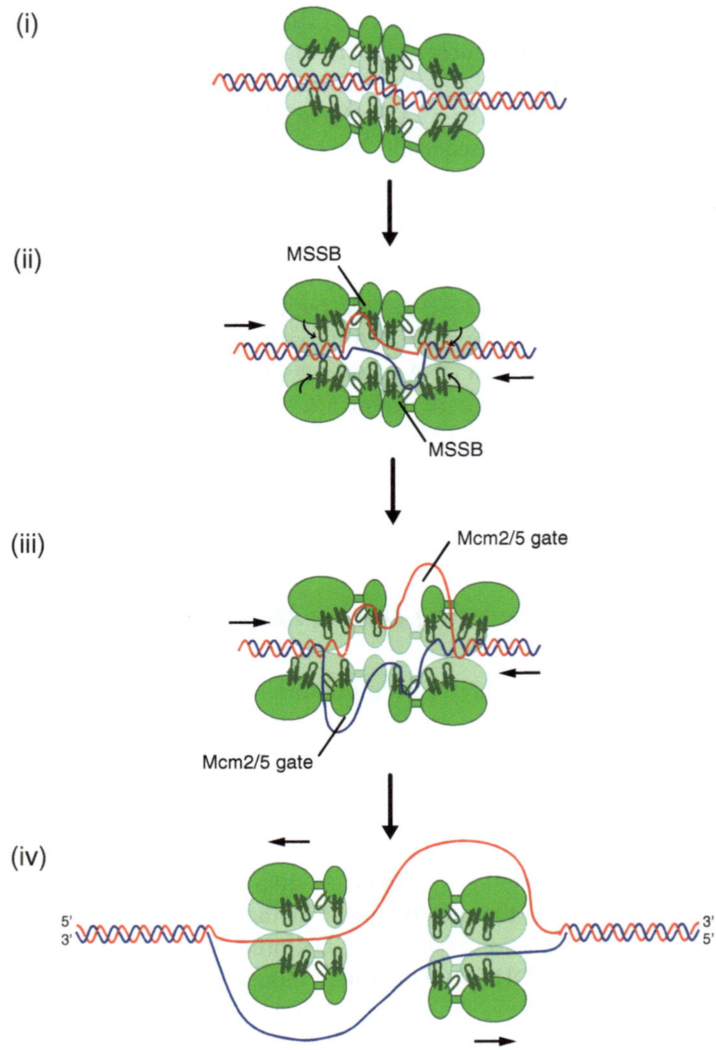

Fig. 12.7 Model for DNA unwinding during Mcm2–7 activation. Mcm2–7 complexes (*green*) are depicted in side-view with two front subunits of each hexamer removed for clarity. See main text for details

Mcm2–7 hexamers would specifically bind to opposite strands, the respective leading-strand template. In the next step opening of the Mcm2/5 gate may be induced (Fig. 12.7iii), allowing extrusion of the unbound lagging-strand template. Extrusion may be promoted by a reduced affinity of Mcm2 and 5 for ssDNA, which contain a less well-conserved MSSB motif [54], and may be further aided by the Sld2/Sld3·Sld7/Dpb11 complex bound to the surface of the Mcm2–7 DH [104]. Opening of the Mcm2/5 gate may coincide with the assumption of a spiral conformation of each Mcm2–7 hexamer [40, 41]. The nonplanar conformation of the Mcm2–7 ring may relate to the mechanism of DNA translocation during DNA unwinding [40], perhaps similar to the one proposed for the SF4 helicase DnaB, which adopts a closed spiral staircase quaternary structure on DNA, and in which NTP hydrolysis may promote the sequential movement of subunits along the helical axis of the staircase [105]. However, given the head-to-head configuration of the Mcm2–7 DH, assumption of a nonplanar conformation may also promote the breaking of the molecular contacts across the NTD interface and thus help dissociate the Mcm2–7 DH into two individual hexamers [10, 24, 27]. Finally, the Mcm2/5 gate is closed again around the leading strand template bound inside the NTD ring, resulting in the formation of two separate Mcm2–7 hexamers being assembled around opposite DNA strands. The roles of Cdc45 and GINS during the remodeling of the Mcm2–7 DH are not known, but it seems likely that both factors are being incorporated into the Mcm2–7 complex during or immediately after lagging strand extrusion to form the CMG. The separation of the Mcm2–7 DH into to individual CMG helicase molecules effectively converts the stationary Mcm2–7 pump into a processive DNA translocase that unwinds duplex DNA at the apex of the replication fork. The generation of a critical amount of ssDNA by the CMG, perhaps aided by the binding of RPA to the ssDNA, may finally promote the recruitment of the Polα-primase to the CMG to induce the priming of DNA synthesis [106].

It is worth noting that the origin melting mechanism described above bears some resemblance to the promoter melting mechanism during open complex (OC) formation by RNA polymerase II (Pol II) [107, 108]. Here, the DNA helicase/translocase subunit of transcription factor TFIIH pumps downstream DNA into the Pol II active site. TFIIH acts against a static complex of promoter-bound Pol II and general transcription factors, such as TFIIB, thereby creating torsional strain that eventually results in the melting of the Pol II-bound promoter DNA. In the case of the Mcm2–7 DH, each hexamer would act as a stator for the opposite hexamer, thereby allowing the generation of torque in the DNA situated between the two oppositely oriented AAA+ motor rings.

The precise sequence of events described in the model above may differ, as the order by which origin DNA unwinding, CMG formation, and DH separation occur is not known. For example, recent evidence suggests that CMG assembly may occur prior to activation of the Mcm2–7 helicase in the absence of Mcm10 [8, 14, 109, 110]. It is not known if the CMG assembled in the absence of Mcm10 is bound around ssDNA or dsDNA, leaving open the order of origin unwinding and CMG formation. However, since extensive DNA unwinding is not observed in this condition, this data may suggest that Mcm2–7 DH separation occurs after CMG formation. Consistent with this notion CMG complexes have been observed to spontaneously associate into head-to-head dimers via their N-terminal tiers in a manner that recapitulates the relative rotational hexamer alignment of the Mcm2–7 DH, indicating that the structure of the CMG is not incompatible with the formation of a stationary Mcm2–7 helicase dimer [40].

References

1. Bell SD, Botchan MR. The minichromosome maintenance replicative helicase. Cold Spring Harb Perspect Biol. 2013;5(11):a012807.
2. Bochman ML, Schwacha A. The Mcm complex: unwinding the mechanism of a replicative helicase. Microbiol Mol Biol Rev. 2009;73(4):652–83.
3. Tanaka S, Araki H. Helicase activation and establishment of replication forks at chromosomal origins of replication. Cold Spring Harb Perspect Biol. 2013;5(12):a010371.
4. Labib K. How do Cdc7 and cyclin-dependent kinases trigger the initiation of chromosome replication in eukaryotic cells? Genes Dev. 2010;24(12):1208–19.
5. Heller RC, Kang S, Lam WM, Chen S, Chan CS, Bell SP. Eukaryotic origin-dependent DNA replication in vitro reveals sequential action of DDK and S-CDK kinases. Cell. 2011;146(1):80–91.
6. Gros J, Devbhandari S, Remus D. Origin plasticity during budding yeast DNA replication in vitro. EMBO J. 2014;33(6):621–36.
7. On KF, Beuron F, Frith D, Snijders AP, Morris EP, Diffley JF. Prereplicative complexes assembled in vitro support origin-dependent and independent DNA replication. EMBO J. 2014;33(6):605–20.
8. Yeeles JT, Deegan TD, Janska A, Early A, Diffley JF. Regulated eukaryotic DNA replication origin firing with purified proteins. Nature. 2015;519(7544):431–5.
9. O'Donnell M, Langston L, Stillman B. Principles and concepts of DNA replication in bacteria, archaea, and eukarya. Cold Spring Harb Perspect Biol. 2013;5(7):pii a010108. Pubmed Central PMCID: 3685895.
10. Gambus A, Jones RC, Sanchez-Diaz A, Kanemaki M, van Deursen F, Edmondson RD, et al. GINS maintains association of Cdc45 with MCM in replisome progression complexes at eukaryotic DNA replication forks. Nat Cell Biol. 2006;8(4):358–66.
11. Simon AC, Zhou JC, Perera RL, van Deursen F, Evrin C, Ivanova ME, et al. A Ctf4 trimer couples the CMG helicase to DNA polymerase alpha in the eukaryotic replisome. Nature. 2014;510(7504):293–7.
12. Sengupta S, van Deursen F, de Piccoli G, Labib K. Dpb2 integrates the leading-strand DNA polymerase into the eukaryotic replisome. Curr Biol. 2013;23(7):543–52.
13. Foltman M, Evrin C, De Piccoli G, Jones RC, Edmondson RD, Katou Y, et al. Eukaryotic replisome components cooperate to process histones during chromosome replication. Cell Rep. 2013;3(3):892–904.
14. van Deursen F, Sengupta S, De Piccoli G, Sanchez-Diaz A, Labib K. Mcm10 associates with the loaded DNA helicase at replication origins and defines a novel step in its activation. EMBO J. 2012;31(9):2195–206.
15. Kilkenny ML, De Piccoli G, Perera RL, Labib K, Pellegrini L. A conserved motif in the C-terminal tail of DNA polymerase alpha tethers primase to the eukaryotic replisome. J Biol Chem. 2012;287(28):23740–7.
16. Morohashi H, Maculins T, Labib K. The amino-terminal TPR domain of Dia2 tethers SCF(Dia2) to the replisome progression complex. Curr Biol. 2009;19(22):1943–9.
17. Gambus A, van Deursen F, Polychronopoulos D, Foltman M, Jones RC, Edmondson RD, et al. A key role for Ctf4 in coupling the MCM2-7 helicase to DNA polymerase alpha within the eukaryotic replisome. EMBO J. 2009;28(19):2992–3004.
18. Erzberger JP, Berger JM. Evolutionary relationships and structural mechanisms of AAA+ proteins. Annu Rev Biophys Biomol Struct. 2006;35:93–114.
19. Siddiqui K, On KF, Diffley JF. Regulating DNA replication in eukarya. Cold Spring Harb Perspect Biol. 2013;5(9):pii a012930.
20. Evrin C, Clarke P, Zech J, Lurz R, Sun J, Uhle S, et al. A double-hexameric MCM2-7 complex is loaded onto origin DNA during licensing of eukaryotic DNA replication. Proc Natl Acad Sci U S A. 2009;106(48):20240–5.

21. Gambus A, Khoudoli GA, Jones RC, Blow JJ. MCM2-7 form double hexamers at licensed origins in Xenopus egg extract. J Biol Chem. 2011;286(13):11855–64.
22. Remus D, Beuron F, Tolun G, Griffith JD, Morris EP, Diffley JF. Concerted loading of Mcm2-7 double hexamers around DNA during DNA replication origin licensing. Cell. 2009;139(4):719–30.
23. Ilves I, Petojevic T, Pesavento JJ, Botchan MR. Activation of the MCM2-7 helicase by association with Cdc45 and GINS proteins. Mol Cell. 2010;37(2):247–58.
24. Moyer SE, Lewis PW, Botchan MR. Isolation of the Cdc45/Mcm2-7/GINS (CMG) complex, a candidate for the eukaryotic DNA replication fork helicase. Proc Natl Acad Sci U S A. 2006;103(27):10236–41.
25. Pacek M, Tutter AV, Kubota Y, Takisawa H, Walter JC. Localization of MCM2-7, Cdc45, and GINS to the site of DNA unwinding during eukaryotic DNA replication. Mol Cell. 2006;21(4):581–7.
26. Tanaka S, Nakato R, Katou Y, Shirahige K, Araki H. Origin association of Sld3, Sld7, and Cdc45 proteins is a key step for determination of origin-firing timing. Curr Biol. 2011;21(24):2055–63.
27. Yardimci H, Loveland AB, Habuchi S, van Oijen AM, Walter JC. Uncoupling of sister replisomes during eukaryotic DNA replication. Mol Cell. 2010;40(5):834–40.
28. Fu YV, Yardimci H, Long DT, Ho TV, Guainazzi A, Bermudez VP, et al. Selective bypass of a lagging strand roadblock by the eukaryotic replicative DNA helicase. Cell. 2011; 146(6):931–41.
29. Liku ME, Nguyen VQ, Rosales AW, Irie K, Li JJ. CDK phosphorylation of a novel NLS-NES module distributed between two subunits of the Mcm2-7 complex prevents chromosomal rereplication. Mol Biol Cell. 2005;16(10):5026–39.
30. Randell JC, Fan A, Chan C, Francis LI, Heller RC, Galani K, et al. Mec1 is one of multiple kinases that prime the Mcm2-7 helicase for phosphorylation by Cdc7. Mol Cell. 2010;40(3):353–63.
31. Sheu YJ, Kinney JB, Lengronne A, Pasero P, Stillman B. Domain within the helicase subunit Mcm4 integrates multiple kinase signals to control DNA replication initiation and fork progression. Proc Natl Acad Sci U S A. 2014;111(18):E1899–908.
32. Frigola J, Remus D, Mehanna A, Diffley JF. ATPase-dependent quality control of DNA replication origin licensing. Nature. 2013;495(7441):339–43.
33. Komata M, Bando M, Araki H, Shirahige K. The direct binding of Mrc1, a checkpoint mediator, to Mcm6, a replication helicase, is essential for the replication checkpoint against methyl methanesulfonate-induced stress. Mol Cell Biol. 2009;29(18):5008–19.
34. Richet N, Liu D, Legrand P, Velours C, Corpet A, Gaubert A, et al. Structural insight into how the human helicase subunit MCM2 may act as a histone chaperone together with ASF1 at the replication fork. Nucleic Acids Res. 2015;43(3):1905–17.
35. Costa A, Ilves I, Tamberg N, Petojevic T, Nogales E, Botchan MR, et al. The structural basis for MCM2-7 helicase activation by GINS and Cdc45. Nat Struct Mol Biol. 2011;18(4):471–7.
36. Costa A, Onesti S. Structural biology of MCM helicases. Crit Rev Biochem Mol Biol. 2009;44(5):326–42.
37. Slaymaker IM, Fu Y, Toso DB, Ranatunga N, Brewster A, Forsburg SL, et al. Mini-chromosome maintenance complexes form a filament to remodel DNA structure and topology. Nucleic Acids Res. 2013;41(5):3446–56.
38. Chen YJ, Yu X, Kasiviswanathan R, Shin JH, Kelman Z, Egelman EH. Structural polymorphism of Methanothermobacter thermautotrophicus MCM. J Mol Biol. 2005;346(2):389–94.
39. Yu X, VanLoock MS, Poplawski A, Kelman Z, Xiang T, Tye BK, et al. The Methanobacterium thermoautotrophicum MCM protein can form heptameric rings. EMBO Rep. 2002; 3(8):792–7.
40. Costa A, Renault L, Swuec P, Petojevic T, Pesavento JJ, Ilves I, et al. DNA binding polarity, dimerization, and ATPase ring remodeling in the CMG helicase of the eukaryotic replisome. eLife. 2014;3:e03273.

41. Lyubimov AY, Costa A, Bleichert F, Botchan MR, Berger JM. ATP-dependent conformational dynamics underlie the functional asymmetry of the replicative helicase from a minimalist eukaryote. Proc Natl Acad Sci U S A. 2012;109(30):11999–2004.
42. Sun J, Evrin C, Samel SA, Fernandez-Cid A, Riera A, Kawakami H, et al. Cryo-EM structure of a helicase loading intermediate containing ORC-Cdc6-Cdt1-MCM2-7 bound to DNA. Nat Struct Mol Biol. 2013;20(8):944–51.
43. Crevel G, Ivetic A, Ohno K, Yamaguchi M, Cotterill S. Nearest neighbour analysis of MCM protein complexes in Drosophila melanogaster. Nucleic Acids Res. 2001;29(23):4834–42.
44. Davey MJ, Indiani C, O'Donnell M. Reconstitution of the Mcm2-7p heterohexamer, subunit arrangement, and ATP site architecture. J Biol Chem. 2003;278(7):4491–9.
45. Brewster AS, Wang G, Yu X, Greenleaf WB, Carazo JM, Tjajadi M, et al. Crystal structure of a near-full-length archaeal MCM: functional insights for an AAA+ hexameric helicase. Proc Natl Acad Sci U S A. 2008;105(51):20191–6.
46. Fletcher RJ, Bishop BE, Leon RP, Sclafani RA, Ogata CM, Chen XS. The structure and function of MCM from archaeal M. thermoautotrophicum. Nat Struct Biol. 2003;10(3):160–7.
47. Miller JM, Arachea BT, Epling LB, Enemark EJ. Analysis of the crystal structure of an active MCM hexamer. eLife. 2014;3:e03433.
48. McGeoch AT, Trakselis MA, Laskey RA, Bell SD. Organization of the archaeal MCM complex on DNA and implications for the helicase mechanism. Nat Struct Mol Biol. 2005;12(9):756–62.
49. Barry ER, McGeoch AT, Kelman Z, Bell SD. Archaeal MCM has separable processivity, substrate choice and helicase domains. Nucleic Acids Res. 2007;35(3):988–98.
50. Jenkinson ER, Chong JP. Minichromosome maintenance helicase activity is controlled by N- and C-terminal motifs and requires the ATPase domain helix-2 insert. Proc Natl Acad Sci U S A. 2006;103(20):7613–8.
51. Bochman ML, Schwacha A. The Mcm2-7 complex has in vitro helicase activity. Mol Cell. 2008;31(2):287–93.
52. Liu W, Pucci B, Rossi M, Pisani FM, Ladenstein R. Structural analysis of the Sulfolobus solfataricus MCM protein N-terminal domain. Nucleic Acids Res. 2008;36(10):3235–43.
53. Bae B, Chen YH, Costa A, Onesti S, Brunzelle JS, Lin Y, et al. Insights into the architecture of the replicative helicase from the structure of an archaeal MCM homolog. Structure. 2009;17(2):211–22.
54. Froelich CA, Kang S, Epling LB, Bell SP, Enemark EJ. A conserved MCM single-stranded DNA binding element is essential for replication initiation. eLife. 2014;3:e01993.
55. Fletcher RJ, Shen J, Gomez-Llorente Y, Martin CS, Carazo JM, Chen XS. Double hexamer disruption and biochemical activities of Methanobacterium thermoautotrophicum MCM. J Biol Chem. 2005;280(51):42405–10.
56. Hardy CF, Dryga O, Seematter S, Pahl PM, Sclafani RA. mcm5/cdc46-bob1 bypasses the requirement for the S phase activator Cdc7p. Proc Natl Acad Sci U S A. 1997;94(7):3151–5.
57. Bochman ML, Bell SP, Schwacha A. Subunit organization of Mcm2-7 and the unequal role of active sites in ATP hydrolysis and viability. Mol Cell Biol. 2008;28(19):5865–73.
58. Coster G, Frigola J, Beuron F, Morris EP, Diffley JF. Origin licensing requires ATP binding and hydrolysis by the MCM replicative helicase. Mol Cell. 2014;55(5):666–77.
59. Kang S, Warner MD, Bell SP. Multiple functions for Mcm2-7 ATPase motifs during replication initiation. Mol Cell. 2014;55(5):655–65.
60. Singleton MR, Dillingham MS, Wigley DB. Structure and mechanism of helicases and nucleic acid translocases. Annu Rev Biochem. 2007;76:23–50.
61. Moreau MJ, McGeoch AT, Lowe AR, Itzhaki LS, Bell SD. ATPase site architecture and helicase mechanism of an archaeal MCM. Mol Cell. 2007;28(2):304–14.
62. Schwacha A, Bell SP. Interactions between two catalytically distinct MCM subgroups are essential for coordinated ATP hydrolysis and DNA replication. Mol Cell. 2001;8(5):1093–104.
63. Barry ER, Lovett JE, Costa A, Lea SM, Bell SD. Intersubunit allosteric communication mediated by a conserved loop in the MCM helicase. Proc Natl Acad Sci U S A. 2009;106(4):1051–6.

64. Enemark EJ, Joshua-Tor L. Mechanism of DNA translocation in a replicative hexameric helicase. Nature. 2006;442(7100):270–5.
65. Gai D, Zhao R, Li D, Finkielstein CV, Chen XS. Mechanisms of conformational change for a replicative hexameric helicase of SV40 large tumor antigen. Cell. 2004;119(1): 47–60.
66. Brewster AS, Slaymaker IM, Afif SA, Chen XS. Mutational analysis of an archaeal minichromosome maintenance protein exterior hairpin reveals critical residues for helicase activity and DNA binding. BMC Mol Biol. 2010;11:62.
67. Graham BW, Schauer GD, Leuba SH, Trakselis MA. Steric exclusion and wrapping of the excluded DNA strand occurs along discrete external binding paths during MCM helicase unwinding. Nucleic Acids Res. 2011;39(15):6585–95.
68. Rothenberg E, Trakselis MA, Bell SD, Ha T. MCM forked substrate specificity involves dynamic interaction with'-tail. J Biol Chem. 2007;282(47):34229–34.
69. Wiedemann C, Szambowska A, Hafner S, Ohlenschlager O, Guhrs KH, Gorlach M. Structure and regulatory role of the C-terminal winged helix domain of the archaeal minichromosome maintenance complex. Nucl Acids Res doi:10.1093/nar/gkv120
70. Nguyen VQ, Co C, Li JJ. Cyclin-dependent kinases prevent DNA re-replication through multiple mechanisms. Nature. 2001;411(6841):1068–73.
71. Yardimci H, Walter JC. Prereplication-complex formation: a molecular double take? Nat Struct Mol Biol. 2014;21(1):20–5.
72. Wei Z, Liu C, Wu X, Xu N, Zhou B, Liang C, et al. Characterization and structure determination of the Cdt1 binding domain of human minichromosome maintenance (Mcm) 6. J Biol Chem. 2010;285(17):12469–73.
73. Fernandez-Cid A, Riera A, Tognetti S, Herrera MC, Samel S, Evrin C, et al. An ORC/Cdc6/MCM2-7 complex is formed in a multistep reaction to serve as a platform for MCM double-hexamer assembly. Mol Cell. 2013;50(4):577–88.
74. Sheu YJ, Stillman B. The Dbf4-Cdc7 kinase promotes S phase by alleviating an inhibitory activity in Mcm4. Nature. 2010;463(7277):113–7.
75. Sheu YJ, Stillman B. Cdc7-Dbf4 phosphorylates MCM proteins via a docking site-mediated mechanism to promote S phase progression. Mol Cell. 2006;24(1):101–13.
76. Sun J, Fernandez-Cid A, Riera A, Tognetti S, Yuan Z, Stillman B, et al. Structural and mechanistic insights into Mcm2-7 double-hexamer assembly and function. Genes Dev. 2014;28(20):2291–303.
77. Francis LI, Randell JC, Takara TJ, Uchima L, Bell SP. Incorporation into the prereplicative complex activates the Mcm2-7 helicase for Cdc7-Dbf4 phosphorylation. Genes Dev. 2009;23(5):643–54.
78. Groth A, Corpet A, Cook AJ, Roche D, Bartek J, Lukas J, et al. Regulation of replication fork progression through histone supply and demand. Science. 2007;318(5858):1928–31.
79. Latreille D, Bluy L, Benkirane M, Kiernan RE. Identification of histone 3 variant 2 interacting factors. Nucleic Acids Res. 2014;42(6):3542–50.
80. Ishimi Y, Ichinose S, Omori A, Sato K, Kimura H. Binding of human minichromosome maintenance proteins with histone H3. J Biol Chem. 1996;271(39):24115–22.
81. Ishimi Y, Komamura Y, You Z, Kimura H. Biochemical function of mouse minichromosome maintenance 2 protein. J Biol Chem. 1998;273(14):8369–75.
82. McKnight SL, Miller Jr OL. Electron microscopic analysis of chromatin replication in the cellular blastoderm Drosophila melanogaster embryo. Cell. 1977;12(3):795–804.
83. Blow JJ, Laskey RA. A role for the nuclear envelope in controlling DNA replication within the cell cycle. Nature. 1988;332(6164):546–8.
84. Tada S, Blow JJ. The replication licensing system. Biol Chem. 1998;379(8-9):941–9.
85. Tada S, Li A, Maiorano D, Mechali M, Blow JJ. Repression of origin assembly in metaphase depends on inhibition of RLF-B/Cdt1 by geminin. Nat Cell Biol. 2001;3(2):107–13.
86. Stinchcomb DT, Struhl K, Davis RW. Isolation and characterisation of a yeast chromosomal replicator. Nature. 1979;282(5734):39–43.

87. Diffley JF, Cocker JH, Dowell SJ, Rowley A. Two steps in the assembly of complexes at yeast replication origins in vivo. Cell. 1994;78(2):303–16.
88. Bell SP, Stillman B. ATP-dependent recognition of eukaryotic origins of DNA replication by a multiprotein complex. Nature. 1992;357(6374):128–34.
89. Diffley JF, Cocker JH. Protein-DNA interactions at a yeast replication origin. Nature. 1992;357(6374):169–72.
90. Gillespie PJ, Li A, Blow JJ. Reconstitution of licensed replication origins on Xenopus sperm nuclei using purified proteins. BMC Biochem. 2001;2:15.
91. Kawasaki Y, Kim HD, Kojima A, Seki T, Sugino A. Reconstitution of Saccharomyces cerevisiae prereplicative complex assembly in vitro. Genes Cells. 2006;11(7):745–56.
92. Bowers JL, Randell JC, Chen S, Bell SP. ATP hydrolysis by ORC catalyzes reiterative Mcm2-7 assembly at a defined origin of replication. Mol Cell. 2004;16(6):967–78.
93. Donovan S, Harwood J, Drury LS, Diffley JF. Cdc6p-dependent loading of Mcm proteins onto pre-replicative chromatin in budding yeast. Proc Natl Acad Sci U S A. 1997;94(11):5611–6.
94. Rowles A, Tada S, Blow JJ. Changes in association of the Xenopus origin recognition complex with chromatin on licensing of replication origins. J Cell Sci. 1999;112(Pt 12):2011–8.
95. Hua XH, Newport J. Identification of a preinitiation step in DNA replication that is independent of origin recognition complex and cdc6, but dependent on cdk2. J Cell Biol. 1998;140(2):271–81.
96. Jensen TH, Jacquier A, Libri D. Dealing with pervasive transcription. Mol Cell. 2013;52(4):473–84.
97. Pelechano V, Wei W, Steinmetz LM. Extensive transcriptional heterogeneity revealed by isoform profiling. Nature. 2013;497(7447):127–31.
98. Mischo HE, Proudfoot NJ. Disengaging polymerase: terminating RNA polymerase II transcription in budding yeast. Biochim Biophys Acta. 2013;1829(1):174–85.
99. Hyrien O. Peaks cloaked in the mist: the landscape of mammalian replication origins. J Cell Biol. 2015;208(2):147–60.
100. Blow JJ, Ge XQ, Jackson DA. How dormant origins promote complete genome replication. Trends Biochem Sci. 2011;36(8):405–14.
101. Diffley JF. Quality control in the initiation of eukaryotic DNA replication. Philos Trans R Soc Lond B Biol Sci. 2011;366(1584):3545–53.
102. Samel SA, Fernandez-Cid A, Sun J, Riera A, Tognetti S, Herrera MC, et al. A unique DNA entry gate serves for regulated loading of the eukaryotic replicative helicase MCM2-7 onto DNA. Genes Dev. 2014;28(15):1653–66.
103. Petojevic T, Pesavento JJ, Costa A, Liang J, Wang Z, Berger JM, et al. Cdc45 (cell division cycle protein 45) guards the gate of the eukaryote replisome helicase stabilizing leading strand engagement. Proc Natl Acad Sci U S A. 2015;112(3):E249–58.
104. Dhingra N, Bruck I, Smith S, Ning B, Kaplan DL. Dpb11 helps control assembly of the Cdc45-Mcm2-7-GINS replication fork helicase. J Biol Chem. 2015;20:7586–601.
105. Itsathitphaisarn O, Wing RA, Eliason WK, Wang J, Steitz TA. The hexameric helicase DnaB adopts a nonplanar conformation during translocation. Cell. 2012;151(2):267–77.
106. Tanaka T, Nasmyth K. Association of RPA with chromosomal replication origins requires an Mcm protein, and is regulated by Rad53, and cyclin- and Dbf4-dependent kinases. EMBO J. 1998;17(17):5182–91.
107. Fishburn J, Tomko E, Galburt E, Hahn S. Double-stranded DNA translocase activity of transcription factor TFIIH and the mechanism of RNA polymerase II open complex formation. Proc Natl Acad Sci U S A. 2015;112(13):3961–6.
108. Kim TK, Ebright RH, Reinberg D. Mechanism of ATP-dependent promoter melting by transcription factor IIH. Science. 2000;288(5470):1418–22.
109. Kanke M, Kodama Y, Takahashi TS, Nakagawa T, Masukata H. Mcm10 plays an essential role in origin DNA unwinding after loading of the CMG components. EMBO J. 2012;31(9):2182–94.
110. Watase G, Takisawa H, Kanemaki MT. Mcm10 plays a role in functioning of the eukaryotic replicative DNA helicase, Cdc45-Mcm-GINS. Curr Biol. 2012;22(4):343–9.

Chapter 13
Role of CDK in Replication Initiation

Seiji Tanaka and Hiroyuki Araki

Abstract Cyclin-dependent kinases (CDKs) are master regulators of cell-cycle progression in eukaryotes. The onset of the synthesis phase (S phase), which is the initiation of chromosomal DNA replication, is a major cell-cycle event that is also regulated by CDKs. Chromosomal DNA replication occurs as a two-step reaction in eukaryotes. In the first reaction, which is termed replication licensing or helicase loading, replicative helicases (the Mcm2–7 complex) are loaded onto replication origins in an inactive form. In the second reaction, which is termed initiation reaction or helicase activation, replicative helicases are activated and double-stranded DNA is unwound to initiate DNA synthesis. The active replicative helicase complex consists of Cdc45, the Mcm2–7 complex, and the GINS complex, and is called the CMG (*C*dc45–*M*cm2–7–*G*INS) complex. In the initiation reaction, many replication proteins assemble onto the Mcm2–7 complex to facilitate the formation of the CMG complex, and this reaction is triggered by two protein kinases, S-phase-specific CDK (S-CDK) and Dbf4-dependent kinase (DDK). Each kinase is responsible for a distinct step of the initiation reaction in yeast: first, Cdc45 is loaded onto licensed origins in a DDK-dependent manner. Subsequently, GINS is loaded onto the origins in an S-CDK-dependent manner. The components of the CMG complex and the requirements of kinases are highly conserved in model eukaryotes. Therefore, the overall rationale of the initiation reaction seems to be highly conserved in eukaryotes, although other replication proteins that are required for the reaction seem to be less conserved, and the set of factors under the control of CDK might be slightly different between species. In this chapter, we focus on the role of S-CDK in the initiation reaction.

Keywords CDK • pre-IC (pre-initiation complex) • Initiation of DNA replication • Dpb11/Cut5/TopBP1 • Sld3/Treslin/Ticrr • Sld2/Drc1/RecQL4

S. Tanaka • H. Araki (✉)
Department of Genetics, SOKENDAI, Mishima, Shizuoka 411-8540, Japan

Division of Microbial Genetics, National Institute of Genetics,
Mishima, Shizuoka 411-8540, Japan
e-mail: hiaraki@lab.nig.ac.jp

D.L. Kaplan (ed.), *The Initiation of DNA Replication in Eukaryotes*,
DOI 10.1007/978-3-319-24696-3_13

Introduction

In the eukaryotic cell cycle, chromosomal DNAs are replicated in the synthesis phase (S phase). Before the synthesis of nascent DNA strands by DNA polymerases, double-stranded DNA must be unwound by the replicative helicase. It has emerged that the activation of this helicase is highly regulated by a two-step reaction, and that this is central to the regulation of the initiation of DNA replication. In the first reaction, which is called "licensing," the core component of the replicative helicase, Mcm2–7, is loaded onto origins as a head-to-head double hexamer via its N-terminal domain [1–3], to assemble the pre-replicative complexes (pre-RCs) in the late M and G1 phases in an ORC-, Cdc6-, and Cdt1-dependent manner (see Chap. 10). At this point, the Mcm2–7 double hexamer in the pre-RC does not exhibit DNA helicase activity. To activate the replicative helicase, other replication factors assemble temporally on licensed origins. In the budding yeast *Saccharomyces cerevisiae*, at least seven additional factors, Cdc45, GINS, Dpb11, Sld2, Sld3, Cdc45, and DNA polymerase ε (Pol ε), are involved in the formation of this complex [4–6]. In this text, we call this complex the preinitiation complex (pre-IC), which was originally defined as a complex formed just before the initiation of DNA replication [7]. The pre-IC seems to be instantly converted to an active replicative helicase, in which two additional factors, Cdc45 and GINS, associate tightly with Mcm2–7 to form the Cdc45–Mcm2–7–GINS (CMG) complex [8, 9].

The assembly of the pre-IC is regulated by two protein kinases, the S-phase cyclin-dependent kinase (S-CDK) and the Dbf4-dependent kinase (DDK) (for reviews: [6, 10–12]). Each kinase is responsible for a distinct step of the assembly of the pre-IC, in the following order in yeast: first, the DDK-dependent phosphorylation of Mcm2–7 leads to the association of Cdc45 with the licensed origins [13, 14]; subsequently, phosphorylation-dependent protein–protein interactions that are caused by S-CDKs promote the loading of GINS onto origins [5]. These associations with origins also require specific initiation proteins: Sld3 for Cdc45 and Dpb11, Sld2, and Pol ε for GINS. Functional homologues of these proteins have been identified in other model eukaryotes. In this chapter, we summarize and discuss the manner in which CDK regulates the initiation of DNA replication by promoting pre-IC assembly. Although there are some discrepancies between species, the process of formation of the pre-IC seems to be fairly well conserved in model eukaryotes.

Overview of the Initiation Reaction in the Budding Yeast *S. cerevisiae*

The initiation reaction is best understood in a unicellular model eukaryote, the budding yeast *Saccharomyces cerevisiae*. Most importantly, only in this organism is it possible to bypass the S-CDK requirement for DNA replication by combining two mutants that bypass the phosphorylation of Sld2 and Sld3, respectively. This indicates that Sld2 and Sld3 constitute the minimal set of essential substrates of S-CDK in the initiation reaction [15, 16] (see below for details).

The outline of the initiation reaction is schematically shown in Fig. 13.1. In the first step, Sld3–Sld7–Cdc45 associates with the licensed origins in a DDK-dependent manner (Fig. 13.1a, step 2-a) [13, 14]. The targets of DDK and the mode of action are described elsewhere (Chap. 14). In the second step, Sld2, Dpb11, Pol ε, and GINS associate with the origins that are bound to Sld3 and Cdc45 in a CDK-dependent manner, and the pre-IC is formed (Fig. 13.1a, step 2-b, 2-c) [5, 13, 15–18]. The details of this reaction will be described in the following section. In budding yeast, the licensing reaction occurs from the end of the M phase to the late G1 phase,

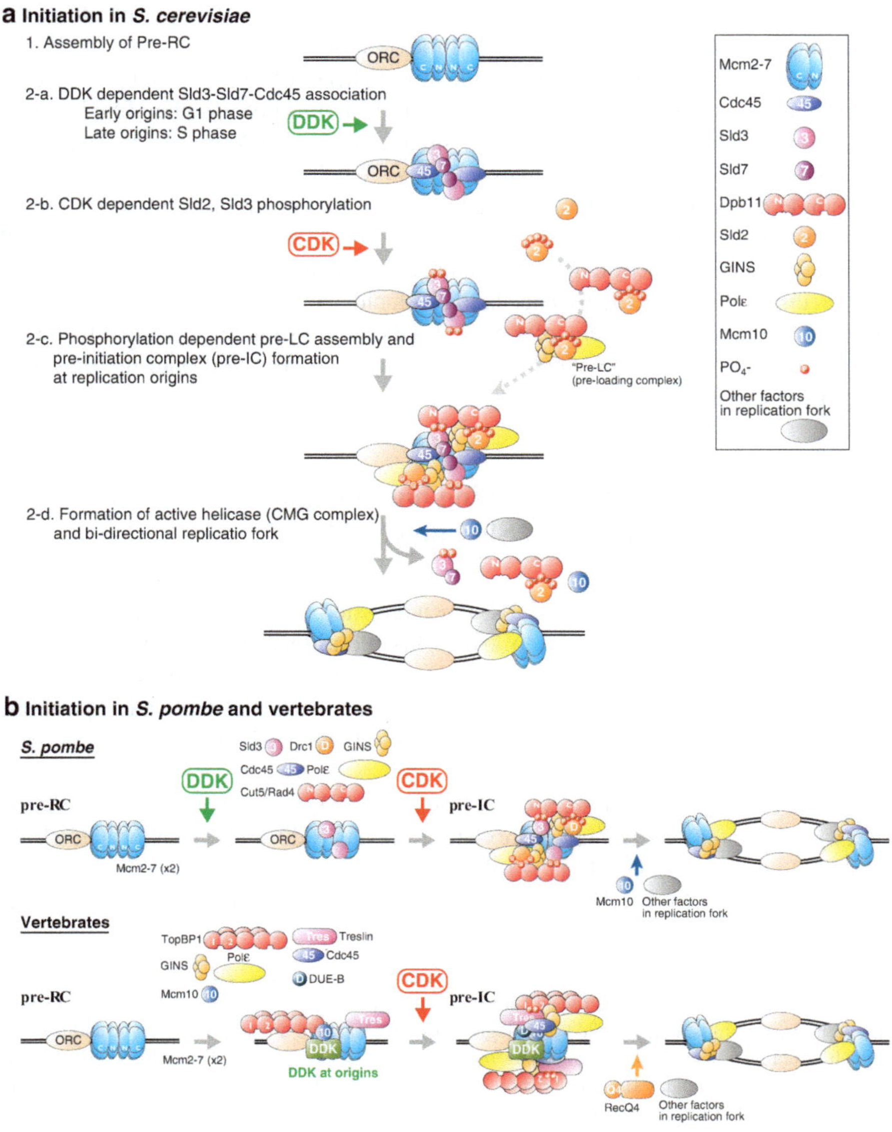

Fig. 13.1 Schematic representation of the formation of the initiation reaction of DNA replication in *S. cerevisiae* (**a**), and *S. pombe* and vertebrates (**b**) See text for details

because any CDK activity inhibits the licensing reaction by downregulating all factors that are required for licensing, such as ORC, Mcm2–7, Cdc6, and Cdt1 (Chap. 10, reviewed by [19, 20]). During this period, it seems that CDK is completely inhibited, because residual CDK activity is genotoxic by diminishing licensing [21, 22]. In contrast, although DDK activity is relatively low in this period because the level of Dbf4, which is the regulatory subunit of DDK, is kept at a low level [23–25], cells have some DDK activity. Based on this DDK activity, the loading of Cdc45 occurs only at early-firing origins, even in G1 (Fig. 13.1a, step 2-a) ([4, 13, 14], also see Chap. 14). This indicates that the order of origin firing is determined by DDK; however, the absolute timing of origin firing is determined by S-CDK. Therefore, the activation of S-CDK defines the onset of the S phase.

After pre-IC assembly, the active replicative helicase (the CMG complex) [8, 9] is immediately formed (Fig. 13.1a, step 2-c and -d). In this process, the Mcm2–7 double hexamer, which embraces the double-stranded origin DNA (Fig. 13.1a, step 1) [1, 2], is converted to the CMG complex with $3' \rightarrow 5'$ helicase activity, which embraces the single-stranded template of the leading strand in replication forks and is highly conserved in eukaryotes (Fig. 13.1a, step 2-d) [26–31]. Mcm10 seems to be required for this conversion reaction in both budding and fission yeasts [13, 32–34], although the precise function of Mcm10 needs to be elucidated in the future (for details, see Chap. 16). Once the CMG complex encircles the single-stranded DNA to unwind the double-stranded DNA, DNA polymerase α (Pol α) is recruited onto the unwound origins, to initiate DNA synthesis (Fig. 13.1a, step 2-d).

S-CDK Promotes GINS Loading onto Origins via the Phosphorylation of Sld2 and Sld3

In the initiation reaction, the S-CDK-dependent step consists in the recruitment of GINS to origins (Fig. 13.1a, step 2-b). This recruitment requires the preceding association of the Sld3–Sld7–Cdc45 complex to the pre-RC, which occurs as a DDK-dependent event. S-CDK phosphorylates two essential replication proteins, Sld2 and Sld3, which promote the formation of the Sld2–Dpb11–Sld3 complex (Fig. 13.1a, step 2-c).

At the onset of the S phase, S-CDK (S-CDK; Clb5– and Clb6–Cdc28 in budding yeast) is activated. S-CDK phosphorylates Sld2 and Sld3, and phosphorylated Sld2 and Sld3 interact with Dpb11 [15–18]. Dpb11 has four BRCA1 C-terminal (BRCT) domains (Fig. 13.2a, BRCT1–4). In general, a tandem pair of BRCT domains con-

Fig. 13.2 (continued) *pale magenta* and are connected by *arrows*. Sld2, Sld3, and GINS interaction domains are also indicated. (**b**) Sld3 orthologues. The Sld3/Treslin domain, a conserved short patch, which contains conserved CDK phosphorylation sites, and the N-terminal region, which is conserved in animal and plant Treslin, are shown in *magenta*, *red*, and *green*, respectively. (**c**) Multiple alignments of the short patches of amino acids that contain conserved CDK phosphorylation sites. The conserved serines and threonines are indicated with *red asterisks*. (**d**) Multiple alignments of the Sld2 orthologues. Full-length amino acids of *S. cerevisiae* Sld2, *S. pombe* Drc1, and *C. elegans* SLD-2, and the N-terminal 54 amino acids of human RecQL4 are shown. The essential threonines T84 in *S. cerevisiae* Sld2 and T111 in *S. pombe* Drc1 are marked with *red asterisk*

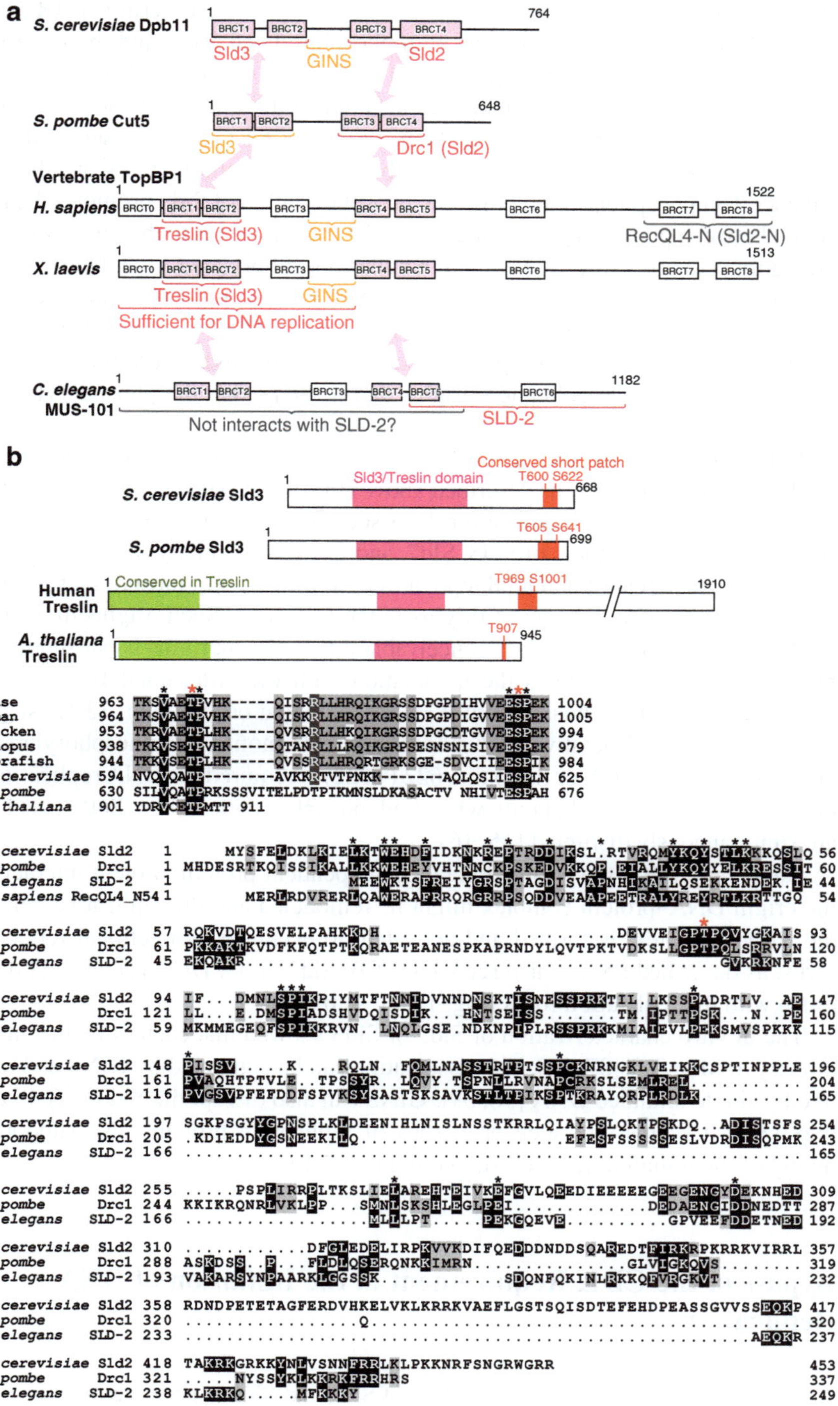

Fig. 13.2 Schematic representation and multiple alignments of the eukaryotic orthologues of Dpb11/TopBP1, Sld3/Treslin, and Sld2/RecQL4. (**a**) Dpb11 orthologues. BRCT domains are shown in boxes. Tandem BRCTs that are conserved in yeasts and animals are shown in

stitutes a phosphopeptide-binding domain [35]; in fact, the N-terminal BRCT pair (BRCT1 and 2) of Dpb11 interacts with phosphorylated Sld3, and the C-terminal BRCT pair interacts with phosphorylated Sld2 [15, 16, 18]. Sld2 has 11 putative CDK phosphorylation motifs that are clustered in the middle of the Sld2 protein (between aa 84 and 257). Among them, threonine 84 (T84) is essential for this phosphor-dependent interaction with Dpb11, and phosphorylations at other sites enhance T84 phosphorylation [18]. The phosphorylation of Sld2 promotes the interaction between phosphorylated Sld2 and Dpb11, which leads to the formation of the preloading complex (pre-LC), which includes phosphorylated Sld2, Dpb11, GINS, and Pol ε (Fig. 13.1a, step 2-c) [5]. The formation of the pre-LC occurs even in the absence of pre-RC, and only requires S-CDK. Once S-CDK is activated, association of pre-LC components with replication origins is observed. Therefore, the pre-LC can be recruited onto replication origins and it seems that the phosphorylation of Sld3 is responsible for this process, as described below.

Sld3, together with Sld7 and Cdc45, associates with the licensed origins in a DDK-dependent manner, as described above (Fig. 13.1a, step 2-a) [13, 14] (Chaps. 14 and 15). This is a prerequisite for the association of the pre-LC with origins [5]. Because the copy numbers of Sld3, Sld7, and Cdc45 (especially Sld3 and Sld7) are much lower than that of potential replication origins [14, 36], most of these proteins might associate with a subset of origins in the G1 phase. Such origins fire early in the S phase [14]. Therefore, it is conceivable that Sld3 on origins is phosphorylated by S-CDK, which promotes the association with the N-terminal BRCT pair of Dpb11 in the pre-LC; thus, the pre-IC is assembled at origins (Fig. 13.1a, step 2-b and -c). Sld3 has 12 putative CDK phosphorylation motifs, and phosphorylation of either serine 600 (S600) or threonine 622 (T622) is required for DNA replication; in other words, cells are lethal only when both phosphorylation sites (S600 and T622) are simultaneously disrupted [15, 16].

Once the pre-IC is assembled via CDK-dependent protein–protein interactions, the origin DNA–protein complex might be remodeled and the replicative helicase, the CMG complex, is formed and activated, to establish bidirectional replication forks, as described above (also see Chap. 19). Sld3–Sld7, Dpb11, and Sld2 do not move with the replication forks (Fig. 13.1a, step 2-d).

The in vitro characterization of Sld2 or Sld3 showed that each of them can bind directly to Mcm2–7 [37, 38]. This association can be competed by GINS and inhibited by single-stranded DNA [38, 39]. GINS can also disrupt the Sld3–Cdc45 complex [39]. It seems that such properties of replication proteins also support the model of the initiation reaction shown in Fig. 13.1a.

Bypass of the CDK Requirement in the Initiation of DNA Replication

As described above, the phosphorylation of Sld2 and Sld3 is necessary for the initiation of DNA replication. This raises the question of whether their phosphorylation is sufficient for initiation. To address this question, phosphorylation-bypass mutations of Sld2 and Sld3 were combined [15, 16]. The phosphorylation of Sld2 was bypassed with phosphomimetic Sld2 (Sld2–T84D or Sld2–11D (aspartic acid substitution for all CDK sites)). Although the phosphomimetic mutant of Sld3 is impossible to isolate, Sld3 phosphorylation can be bypassed in three ways. First, via the Sld3–Dpb11 fusion protein (SD fusion), in which Sld3-3A (Sld3-T600A, T609A, and S622A) and the C-terminal BRCT pair of Dpb11 are artificially fused to achieve a status that mimics the phosphorylated Sld3–Dpb11 complex [15]; second, via a novel mutation termed *JET1-1*, which occurs in *CDC45* and specifically suppresses the lethality of the *sld3-2A* mutant (Sld3-T600A and S622A) [16]; and last, via high-level Dpb11, which can also suppress the lethality of *sld3-2A*, although this suppression is weak [16]. The combining of any of these with phosphomimetic Sld2 can initiate DNA replication, even in the absence of CDK activity. This replication is pre-RC and DDK dependent, indicating that the combination of these mutants bypasses only the CDK requirement. These results show that Sld2 and Sld3 constitute the minimal essential targets of CDK [15, 16]. This is further supported by the in vitro reconstitution of DNA replication; CDK-prephosphorylated Sld2 and Sld3 support DNA replication in the absence of CDK [40].

CDK-Dependent Initiation Reaction in Fission Yeast *Schizosaccharomyces pombe*

The orthologues of Dpb11, Sld2, and Sld3 in the fission yeast *S. pombe* are Cut5 (also called Rad4), Drc1, and Sld3, respectively (the nomenclatures of orthologues are summarized in Table 13.1) [41–44]. The outline of the initiation reaction in *S. pombe*, which is very similar to that of budding yeast, is shown in Fig. 13.1b. The DDK-dependent origin association of Sld3 was first shown in this organism [45], and its association with origins is a prerequisite for the subsequent recruitment of Cut5, GINS, Drc1, and Cdc45 [45]. CDK catalyzes the phosphorylation of Drc1 and Sld3, and these phosphorylations facilitate the interaction between Cut5 and Drc1 and Sld3, respectively, as observed in budding yeast [42, 43, 46]. The conserved threonine (T114) in Sld2 is the target of CDK, and phosphorylation of T114 is essential for the initiation of DNA replication [46]. Because the origin association

Table 13.1 Nomenclature of the Dpb11, Sld2, Sld3, and Sld7 orthologues

Organism	Dpb11 orthologue	Sld2 orthologue	Sld3 orthologue	Sld7 orthologue
S. cerevisiae	Dpb11	Sld2 (Drc1)	Sld3	Sld7
S. pombe	Cut5 (Rad4)	Drc1	Sld3	?
C. elegans	MUS-101	SLD-2	SLD-3	?
D. rerio	TopBP1	RecQL4 (RecQ4)	Ticrr	MTBP?
X. laevis	TopBP1 (Cut5)	RecQ4	Treslin	MTBP?
H. sapiens	TopBP1	RecQL4	Treslin	MTBP

of Cut5 was not observed in an alanine-substitution mutant (T114A) of Sld2, it seems that the phosphorylation-dependent assembly of the pre-IC is also conserved in fission yeast (Fig. 13.1b). Surprisingly, however, phosphorylation of Sld3 is not essential for growth, although an alanine-substitution mutant of *sld3* (*sld3-9A*) shows cold sensitivity, and the association of Cut5 with origins is lost in the sld3-9A mutant at low temperature [43, 46]. Interestingly, in a yeast three-hybrid assay, Sld3, Cut5, and Drc1 form a ternary complex that is dependent on the CDK sites of Sld3 and Drc1, and Drc1–Cut5 binding enhances the Sld3–Cut5 interaction [46]. Therefore, the mechanism of formation of the pre-IC is conserved well in fission and budding yeast, although the manner via which Cdc45 and GINS are recruited onto origins in fission yeast warrants further elucidation.

Each of phosphomimetic mutants of Sld2 and Sld3, Sld2-T111E and Sld3-5D, respectively, can support the cellular growth in fission yeast [46]. This suggests that these phosphomimetic mutants can interact with Cut5 to assemble the pre-IC for the initiation reaction. As described above, phosphorylation of Sld2 by CDK is essential for DNA replication, whereas that of Sld3 is not, in this organism. Therefore, the next question is whether the phosphomimetic mutant of Sld2 alone can bypass the CDK requirement for the initiation. To date, this does not seem very likely. If Sld2 is the sole essential target of CDK in fission yeast, and its phosphomimetic form, Sld2-T111E, bypasses CDK by itself, Sld2-T111E would confer a severe growth defect. Because the CDK-independent initiation reaction causes re-replication in the G1 phase of the cell cycle and is genotoxic, as observed in budding yeast [16, 47]. It seems that this is not occurring in Sld2-T111E. However, it remains possible that Sld2-T111E alone can bypass the CDK requirement. Because virtually no G1 cells exist in the normal cell cycle of *S. pombe*, lethal G1 re-replication by Sld2-111E could be inhibited. Whether Sld2 is the sole essential target of CDK should be elucidated in the future.

CDK-Dependent Initiation Reaction in Model Animals

The factors of the actual replication machinery, such as DNA polymerases and replicative DNA helicase components (Mcm2–7, Cdc45, and GINS), are well conserved in eukaryotes. In contrast, the regulatory factors of the initiation reaction, such as Dpb11, Sld2, and Sld3, have diverged during evolution. This reminds us that the components of the pre-RC are highly conserved, whereas the mechanisms that underlie the inhibition of the licensing reaction, which is regulated by CDK, are divergent [19]. CDK substrates sense CDK activity and promote the initiation of chromosomal DNA replication, as described above. As the cellular environment differs among organisms, the sensing mechanism of CDK activity may have diverged. Nonetheless, the functional homologues of Dpb11, Sld2, and Sld3 (TopBP1 (also denoted as Cut5 and Mus101), RecQ4 (also denoted as RecQL4), and Treslin/Ticrr, respectively) have been identified in other eukaryotes [44, 48–56]. Each of them shows limited sequence similarity to their yeast counterparts, and almost all of them are much larger in size than their yeast counterparts (Fig. 13.2). Although these proteins are required for the initiation of DNA replication, the requirement for, and order of, association with chromatin vary slightly among them.

Vertebrate Treslin/Ticrr Is a Conserved CDK Target in the Initiation of DNA Replication

Treslin/Ticrr was identified by screenings for TopBP1-binding proteins (Treslin) in *Xenopus* [53] and for G2/M checkpoint regulators in zebrafish (Ticrr) [55]. Later, the sequence homology between Sld3 and Treslin/Ticrr was demonstrated and the Sld3–Treslin/Ticrr domain, which is conserved across eukaryotic species, including animals, fungi, and plants, was identified (Fig. 13.2b) [54]. The Sld3–Treslin/Ticrr domain was later shown to be a rhombic structural domain for binding to Cdc45 [57]. In *Xenopus* and humans, Treslin/Ticrr is phosphorylated by CDK and associates with TopBP1 in a phosphorylation-dependent manner. Moreover, Treslin/Ticrr has many putative CDK-phosphorylation sites; notably, two of them (T969 and S1001 in human TopBP1) are located in a conserved short patch of amino acids, which resides at a different position from the above-mentioned Sld3–Treslin/Ticrr domain, and the phosphorylation sites were conserved throughout evolution (Fig. 13.2b, c) [53, 58–60]. The phosphorylation of these sites is required for the interaction with BRCT1-2 of TopBP1, which is homologous to the N-terminal BRCT yeast pair Dpb11/Cut5 and, importantly, is required for DNA replication (Fig. 13.2a) [58, 59]. Therefore, the vertebrate Treslin/Ticrr–TopBP1 interaction and its function in DNA replication seem to mirror those of Sld3-Dpb11 in budding yeast. Because it seems that this is not the case for the Sld2-Dpb11 branch in vertebrates, as described in the next section, whether the phosphorylation of Treslin/Ticrr alone is sufficient for the initiation of DNA replication should be strictly tested in the future, although the overexpression of phosphomimetic Treslin/Ticrr can largely bypass the inhibition of CDK2 [61].

Function of Sld2 Orthologues in the Initiation of DNA Replication Appears to Have Diverged in Animals

In vertebrates, the orthologue of Sld2 is RecQ4 (also called RecQL4) [51, 52]. RECQ4 is a member of the RecQ helicase family in vertebrates, which is known to play various roles in DNA metabolism [62]. RECQ4 has a conserved RecQ helicase domain at its C terminus, and its N-terminal portion shows a weak similarity to Sld2. Mutations in the RecQ helicase domain of human RECQ4 are responsible for the subset of Rothmund–Thomson syndrome, Baller–Gerold syndrome, and RAPADILINO syndrome [63]. However, the N-terminal Sld2-like domain is essential for cell growth and DNA replication, and the RecQ domain is dispensable for

DNA replication [51, 52]. Importantly, during in vitro replication in a *Xenopus* egg extract, the association of RecQL4 with chromatin does not depend on CDK, whereas it depends on TopBP1/Cut5. A recent study showed that RecQ4 is not required for the assembly of the pre-IC; rather, it seems to be required for the conversion of the pre-IC to an active replication fork (Fig. 13.1b) [64]. In both budding and fission yeasts, Sld2 interacts with BRCT3-4 of Dpb11 or Cut5 in a phosphorylation-dependent manner. BRCT4-5 of vertebrate TopBP1 is a counterpart of BRCT3-4 of Dpb11 (Fig. 13.2a); however, BRCT7-8, and not BRCT4-5, has been reported as RecQ4 binding. Importantly, the C-terminal portion located after BRCT4-5 in vertebrate TopBP1 is not required for replication [53, 65]. Moreover, the conserved short N-terminal region (54 aa) of RecQL4 associates with BRCT7-8 (Fig. 13.2a) [66], although it is not known whether phosphorylation is required, and the region corresponding to the essential phosphorylation site in yeast is not included in this short fragment (Fig. 13.2d). Therefore, the phosphorylation-dependent Sld2–Dpb11 interaction pathway for DNA replication is not likely to be conserved in *Xenopus* and mammals, although RecQL4 seems to be required for the assembly of CMG [67].

Recently, the Sld2 orthologue of *Caenorhabditis elegans*, SLD-2, was identified and characterized [56]. Interestingly, *C. elegans* Sld-2 only has a region that is homologous to Sld2 and does not contain the RecQ domain, unlike its vertebrate counterparts (Fig. 13.2d). SLD-2 associates with MUS-101, a Dpb11 orthologue in nematodes, and is required for DNA replication. Importantly, such functions depend on CDK phosphorylation [56]. Although the precise region of SLD-2 that is required for DNA replication and its binding sites in MUS-101 have not been identified, the Sld2–Dpb11 pathway seems to be conserved in this organism. Notably, a Sld3 orthologue, Treslin, also exists in *C. elegans*. However, *C. elegans* Treslin seems to lack a short patch of amino acids containing two conserved phosphorylation sites. Therefore, whether the Sld3–Dpb11 branch is also required for the initiation of DNA replication in this organism is an intriguing question.

Structural Insights on the Sld2– and Sld3–Dpb11/Cut5/TopBP1 Interactions

As described above, a tandem pair of BRCT motifs can associate specifically with phosphopeptides. Recent structural studies revealed that there are variations of the structure of BRCT pairs (reviewed in [68]). Structural analyses of Cut5/Rad4 and human TopBP1 revealed that their N-terminal BRCT pair (BRCT1-2) can accommodate two phosphorylated residues simultaneously, whereas the C-terminal BRCT pair of Cut5/Rad4 can accommodate only one phosphorylated residue [60, 69]. These results perfectly match the observation that the phosphorylation of a single residue of Sld2 (T84 in *S. cerevisiae* and T111 in *S. pombe*) is essential for the interaction with Dpb11/Cut5 [18, 46], and vertebrate Treslins have two conserved amino acid residues that are phosphorylated for the interaction with TopBP1 [58, 59]. Although it is predicted from the amino acid sequence that the BRCT1-2 of budding yeast Dpb11 accommodates only one phosphorylated residue [68], further analysis is required to clarify the association of phosphorylated Sld3 and Dpb11 in budding yeast, because two phosphorylation sites of budding yeast Sld3 participate in the interaction with Dpb11 [15, 16].

Regulatory Aspects of the CDK-Regulated Initiation Process

As described above, although Sld2 and Sld3 are the conserved targets of S-phase CDK for the initiation reaction, the combination of the essentiality of the phosphorylation of these proteins is different in various organisms: the phosphorylation of both proteins is essential in *S. cerevisiae*, the phosphorylation of Sld2 is only essential in *S. pombe*, and the phosphorylation of the Sld3 orthologue is only essential in vertebrates. The mechanism underlying such differences remains a mystery; however, whether the bypass of CDK is possible by single phosphomimetic mutants in *S. pombe* and vertebrates should be tested. Because an untimely initiation reaction in the G1 phase causes re-replication, which is highly genotoxic to cells, the initiation reaction must be tightly regulated. Similar to what is observed in budding yeast, the presence of multiple CDK substrates seems to be an adequate mechanism for the occurrence of this tight switch. Moreover, the multiple phosphorylation events that precede the phosphorylation of the essential T84 of Sld2 and the low abundance of the CDK target protein might help the tight switch for the initiation reaction [18, 47]. In *S. pombe*, the existence of Sld2 enhances the interaction between Sld3 and Cut5/Rad4, as assessed in a yeast two-hybrid assay [46]. Such cooperation might also help make the tight switch for the S phase. The elucidation of whether such regulatory mechanisms are present in other eukaryotes is necessary.

Although the phosphorylation-dependent protein interaction(s) absolutely trigger the assembly of the pre-IC in the initiation reaction, other protein–protein interaction might also be important for the efficient assembly of the pre-IC. The GINS–Dpb11 interaction might be such an example. The inter-BRCT region located between BRCT2 and BRCT3 in Dpb11 can interact with GINS in *S. cerevisiae* (Fig. 13.2a) [65]. Dpb11 mutants that lost this interaction show cold sensitivity and S-phase delay, similar to that observed for the unphosphorylatable Sld3 mutant in *S. pombe* [46]. Therefore, this interaction is required for the efficiency of the initiation reaction. Interestingly, the inter-BRCT region located between BRCT3 and BRCT4 in vertebrate TopBP1 also interacts with GINS, and, importantly, this inter-BRCT region is included in the essential region of TopBP1 (Fig. 13.2a) and is required for efficient DNA replication in *Xenopus* egg extracts [65]. Many physical interactions between components of pre-IC have been reported, some of which might contribute to the stabilization of the pre-IC.

The formation of the pre-IC is also the target of checkpoint kinases in yeast and humans [59, 70, 71]. When the replication fork collapses or is stalled, the checkpoint kinase Rad53 is activated and inhibits further activation of unfired licensed origins in budding yeast. In this checkpoint pathway, the targets of Rad53 are Sld3 and Dbf4, which are inhibited by massive phosphorylation by Rad53. Therefore, both the essential CDK- and DDK-dependent pathways are inhibited [70, 71]. Such a mechanism is conserved in human Treslin/Ticrr [59]. Recent studies also suggest that this inhibitory mechanism is weakly activated even in the normal S phase in both yeast and humans [61, 72]. In addition to the common regulatory mechanism, vertebrate Treslin contains a region that interacts with Chk1, a checkpoint-regulatory kinase, and a corresponding region does not exist in yeast orthologues [73]. Such regulation might be important for the orchestration of the proper initiation of replication and, hence, for proper S-phase control. For details regarding the checkpoint response, see Chap. 24.

Conclusion

Loading of the replicative helicase onto chromatin and activation of the replicative helicase are central for the initiation of DNA replication. CDK promotes the activation of the replicative helicase via the phosphorylation of Sld2 and Sld3. The phosphorylation of Sld2 and Sld3 by CDK promotes the phosphor-dependent association with a phosphor-acceptor protein, Dpb11, and further assembly of the pre-IC, which is a prerequisite for the assembly of the active replicative helicase. Although there are variations in the assembly of the pre-IC between organisms, either or both Sld2 and Sld3 are the absolute gateway to the assembly of the pre-IC by accepting the signal from CDK in eukaryotes. To date, the bypass of the CDK requirement for the initiation reaction is only achieved in budding yeast, and no additional proteins that accept the CDK signal have been identified in other eukaryotes. Therefore, we do not know whether Sld2 and Sld3 are the sole targets of CDK in the initiation of DNA replication in eukaryotes. Future analyses should unveil the outline of the initiation reaction in other eukaryotes and clarify this issue.

Acknowledgements We apologize to the authors whose work could not be cited here because of space limitations. S.T. and H.A. are supported by the Ministry of Education, Culture, Sports, Science, and Technology of Japan.

References

1. Evrin C, Clarke P, Zech J, Lurz R, Sun J, Uhle S, et al. A double-hexameric MCM2-7 complex is loaded onto origin DNA during licensing of eukaryotic DNA replication. Proc Natl Acad Sci U S A. 2009;106(48):20240–5. Pubmed Central PMCID: 2787165 Epub 2009/11/17. eng.
2. Remus D, Beuron F, Tolun G, Griffith JD, Morris EP, Diffley JF. Concerted loading of Mcm2-7 double hexamers around DNA during DNA replication origin licensing. Cell. 2009;139(4):719–30. Pubmed Central PMCID: 2804858 Epub 2009/11/10. eng.

3. Gambus A, Khoudoli GA, Jones RC, Blow JJ. MCM2-7 form double hexamers at licensed origins in Xenopus egg extract. J Biol Chem. 2011;286(13):11855–64. Pubmed Central PMCID: 3064236 Epub 2011/02/02. eng.
4. Kamimura Y, Tak YS, Sugino A, Araki H. Sld3, which interacts with Cdc45 (Sld4), functions for chromosomal DNA replication in Saccharomyces cerevisiae. EMBO J. 2001;20(8):2097–107. Pubmed Central PMCID: 125422 Epub 2001/04/11. eng.
5. Muramatsu S, Hirai K, Tak YS, Kamimura Y, Araki H. CDK-dependent complex formation between replication proteins Dpb11, Sld2, Pol (epsilon}, and GINS in budding yeast. Genes Dev. 2010;24(6):602–12. Pubmed Central PMCID: 2841337 Epub 2010/03/17. eng.
6. Tanaka S, Araki H. Helicase activation and establishment of replication forks at chromosomal origins of replication. Cold Spring Harb Perspect Biol. 2013;5(12):a010371.
7. Zou L, Stillman B. Formation of a preinitiation complex by S-phase cyclin CDK-dependent loading of Cdc45p onto chromatin. Science. 1998;280(5363):593–6. Epub 1998/05/09. eng.
8. Gambus A, Jones RC, Sanchez-Diaz A, Kanemaki M, van Deursen F, Edmondson RD, et al. GINS maintains association of Cdc45 with MCM in replisome progression complexes at eukaryotic DNA replication forks. Nat Cell Biol. 2006;8(4):358–66. Epub 2006/03/15. eng.
9. Moyer SE, Lewis PW, Botchan MR. Isolation of the Cdc45/Mcm2-7/GINS (CMG) complex, a candidate for the eukaryotic DNA replication fork helicase. Proc Natl Acad Sci U S A. 2006;103(27):10236–41. Pubmed Central PMCID: 1482467 Epub 2006/06/27. eng.
10. Labib K. How do Cdc7 and cyclin-dependent kinases trigger the initiation of chromosome replication in eukaryotic cells? Genes Dev. 2010;24(12):1208–19. Pubmed Central PMCID: 2885657 Epub 2010/06/17. eng.
11. Masai H, Matsumoto S, You Z, Yoshizawa-Sugata N, Oda M. Eukaryotic chromosome DNA replication: where, when, and how? Annu Rev Biochem. 2010;79:89–130. Epub 2010/04/09. eng.
12. Tanaka S, Araki H. Regulation of the initiation step of DNA replication by cyclin-dependent kinases. Chromosoma. 2010;119(6):565–74. Epub 2010/08/06. eng.
13. Heller RC, Kang S, Lam WM, Chen S, Chan CS, Bell SP. Eukaryotic origin-dependent DNA replication in vitro reveals sequential action of DDK and S-CDK kinases. Cell. 2011;146(1): 80–91. Epub 2011/07/07. eng.
14. Tanaka S, Nakato R, Katou Y, Shirahige K, Araki H. Origin association of sld3, sld7, and cdc45 proteins is a key step for determination of origin-firing timing. Curr Biol. 2011;21(24):2055–63. Epub 2011/12/16. eng.
15. Zegerman P, Diffley JF. Phosphorylation of Sld2 and Sld3 by cyclin-dependent kinases promotes DNA replication in budding yeast. Nature. 2007;445(7125):281–5. Epub 2006/12/15. eng.
16. Tanaka S, Umemori T, Hirai K, Muramatsu S, Kamimura Y, Araki H. CDK-dependent phosphorylation of Sld2 and Sld3 initiates DNA replication in budding yeast. Nature. 2007;445(7125):328–32. Epub 2006/12/15. eng.
17. Masumoto H, Muramatsu S, Kamimura Y, Araki H. S-Cdk-dependent phosphorylation of Sld2 essential for chromosomal DNA replication in budding yeast. Nature. 2002;415(6872):651–5. Epub 2002/01/25. eng.
18. Tak YS, Tanaka Y, Endo S, Kamimura Y, Araki H. A CDK-catalysed regulatory phosphorylation for formation of the DNA replication complex Sld2-Dpb11. EMBO J. 2006;25(9):1987–96. Pubmed Central PMCID: 1456926 Epub 2006/04/19. eng.
19. Arias EE, Walter JC. Strength in numbers: preventing rereplication via multiple mechanisms in eukaryotic cells. Genes Dev. 2007;21(5):497–518. Epub 2007/03/09. eng.
20. Siddiqui K, On KF, Diffley JF. Regulating DNA replication in eukarya. Cold Spring Harb Perspect Biol. 2013;5(9): a012930
21. Lengronne A, Schwob E. The yeast CDK inhibitor Sic1 prevents genomic instability by promoting replication origin licensing in late G(1). Mol Cell. 2002;9(5):1067–78. Epub 2002/06/07. eng.
22. Tanaka S, Diffley JF. Deregulated G1-cyclin expression induces genomic instability by preventing efficient pre-RC formation. Genes Dev. 2002;16(20):2639–49. Pubmed Central PMCID: 187461 Epub 2002/10/17. eng.

23. Oshiro G, Owens JC, Shellman Y, Sclafani RA, Li JJ. Cell cycle control of Cdc7p kinase activity through regulation of Dbf4p stability. Mol Cell Biol. 1999;19(7):4888–96. Pubmed Central PMCID: 84289. Epub 1999/06/22. eng.
24. Weinreich M, Stillman B. Cdc7p-Dbf4p kinase binds to chromatin during S phase and is regulated by both the APC and the RAD53 checkpoint pathway. EMBO J. 1999;18(19):5334–46. Pubmed Central PMCID: 1171603 Epub 1999/10/03. eng.
25. Ferreira MF, Santocanale C, Drury LS, Diffley JF. Dbf4p, an essential S phase-promoting factor, is targeted for degradation by the anaphase-promoting complex. Mol Cell Biol. 2000;20(1):242–8. Pubmed Central PMCID: 85080. Epub 1999/12/14. eng.
26. Ilves I, Petojevic T, Pesavento JJ, Botchan MR. Activation of the MCM2-7 helicase by association with Cdc45 and GINS proteins. Mol Cell. 2010;37(2):247–58.
27. Kang YH, Galal WC, Farina A, Tappin I, Hurwitz J. Properties of the human Cdc45/Mcm2-7/GINS helicase complex and its action with DNA polymerase epsilon in rolling circle DNA synthesis. Proc Natl Acad Sci U S A. 2012;109(16):6042–7. Pubmed Central PMCID: 3341058.
28. Fu YV, Yardimci H, Long DT, Ho TV, Guainazzi A, Bermudez VP, et al. Selective bypass of a lagging strand roadblock by the eukaryotic replicative DNA helicase. Cell. 2011;146(6):931–41. Pubmed Central PMCID: 3209622 Epub 2011/09/20. eng.
29. Georgescu RE, Langston L, Yao NY, Yurieva O, Zhang D, Finkelstein J, et al. Mechanism of asymmetric polymerase assembly at the eukaryotic replication fork. Nat Struct Mol Biol. 2014;21(8):664–70.
30. Costa A, Renault L, Swuec P, Petojevic T, Pesavento JJ, Ilves I, et al. DNA binding polarity, dimerization, and ATPase ring remodeling in the CMG helicase of the eukaryotic replisome. Elife. 2014;3:e03273.
31. Petojevic T, Pesavento JJ, Costa A, Liang J, Wang Z, Berger JM, et al. Cdc45 (cell division cycle protein 45) guards the gate of the eukaryote replisome helicase stabilizing leading strand engagement. Proc Natl Acad Sci U S A. 2015;112(3):E249–58. Pubmed Central PMCID: 4311868.
32. Watase G, Takisawa H, Kanemaki MT. Mcm10 plays a role in functioning of the eukaryotic replicative DNA helicase, Cdc45-Mcm-GINS. Curr Biol. 2012;22(4):343–9. Epub 2012/01/31. eng.
33. van Deursen F, Sengupta S, De Piccoli G, Sanchez-Diaz A, Labib K. Mcm10 associates with the loaded DNA helicase at replication origins and defines a novel step in its activation. EMBO J. 2012;31(9):2195–206. Pubmed Central PMCID: 3343467.
34. Kanke M, Kodama Y, Takahashi TS, Nakagawa T, Masukata H. Mcm10 plays an essential role in origin DNA unwinding after loading of the CMG components. EMBO J. 2012;31(9):2182–94. Pubmed Central PMCID: 3343466.
35. Glover JN, Williams RS, Lee MS. Interactions between BRCT repeats and phosphoproteins: tangled up in two. Trends Biochem Sci. 2004;29(11):579–85.
36. Mantiero D, Mackenzie A, Donaldson A, Zegerman P. Limiting replication initiation factors execute the temporal programme of origin firing in budding yeast. EMBO J. 2011;30(23):4805–14. Pubmed Central PMCID: 3243606 Epub 2011/11/15. eng.
37. Bruck I, Kaplan DL. Origin single-stranded DNA releases Sld3 protein from the Mcm2-7 complex, allowing the GINS tetramer to bind the Mcm2-7 complex. J Biol Chem. 2011;286(21):18602–13. Pubmed Central PMCID: 3099676.
38. Bruck I, Kanter DM, Kaplan DL. Enabling association of the GINS protein tetramer with the mini chromosome maintenance (Mcm)2-7 protein complex by phosphorylated Sld2 protein and single-stranded origin DNA. J Biol Chem. 2011;286(42):36414–26. Pubmed Central PMCID: 3196077.
39. Bruck I, Kaplan DL. GINS and Sld3 compete with one another for Mcm2-7 and Cdc45 binding. J Biol Chem. 2011;286(16):14157–67. Pubmed Central PMCID: 3077617.
40. Yeeles JT, Deegan TD, Janska A, Early A, Diffley JF. Regulated eukaryotic DNA replication origin firing with purified proteins. Nature. 2015;519:431–5.
41. Saka Y, Fantes P, Sutani T, McInerny C, Creanor J, Yanagida M. Fission yeast cut5 links nuclear chromatin and M phase regulator in the replication checkpoint control. EMBO J. 1994;13(22):5319–29. Pubmed Central PMCID: 395488 Epub 1994/11/15. eng.

42. Noguchi E, Shanahan P, Noguchi C, Russell P. CDK phosphorylation of Drc1 regulates DNA replication in fission yeast. Curr Biol. 2002;12(7):599–605.
43. Nakajima R, Masukata H. SpSld3 is required for loading and maintenance of SpCdc45 on chromatin in DNA replication in fission yeast. Mol Biol Cell. 2002;13(5):1462–72.
44. Garcia V, Furuya K, Carr AM. Identification and functional analysis of TopBP1 and its homologs. DNA Repair. 2005;4(11):1227–39.
45. Yabuuchi H, Yamada Y, Uchida T, Sunathvanichkul T, Nakagawa T, Masukata H. Ordered assembly of Sld3, GINS and Cdc45 is distinctly regulated by DDK and CDK for activation of replication origins. EMBO J. 2006;25(19):4663–74.
46. Fukuura M, Nagao K, Obuse C, Takahashi TS, Nakagawa T, Masukata H. CDK promotes interactions of Sld3 and Drc1 with Cut5 for initiation of DNA replication in fission yeast. Mol Biol Cell. 2011;22(14):2620–33. Pubmed Central PMCID: 3135486 Epub 2011/05/20. eng.
47. Tanaka S, Araki H. Multiple regulatory mechanisms to inhibit untimely initiation of DNA replication are important for stable genome maintenance. PLoS Genet. 2011;7(6):e1002136. Epub 2011/06/24. eng.
48. Makiniemi M, Hillukkala T, Tuusa J, Reini K, Vaara M, Huang D, et al. BRCT domain-containing protein TopBP1 functions in DNA replication and damage response. J Biol Chem. 2001;276(32):30399–406.
49. Van Hatten RA, Tutter AV, Holway AH, Khederian AM, Walter JC, Michael WM. The Xenopus Xmus101 protein is required for the recruitment of Cdc45 to origins of DNA replication. J Cell Biol. 2002;159(4):541–7.
50. Hashimoto Y, Takisawa H. Xenopus Cut5 is essential for a CDK-dependent process in the initiation of DNA replication. EMBO J. 2003;22(10):2526–35.
51. Matsuno K, Kumano M, Kubota Y, Hashimoto Y, Takisawa H. The N-terminal noncatalytic region of Xenopus RecQ4 is required for chromatin binding of DNA polymerase alpha in the initiation of DNA replication. Mol Cell Biol. 2006;26(13):4843–52.
52. Sangrithi MN, Bernal JA, Madine M, Philpott A, Lee J, Dunphy WG, et al. Initiation of DNA replication requires the RECQL4 protein mutated in Rothmund-Thomson syndrome. Cell. 2005;121(6):887–98.
53. Kumagai A, Shevchenko A, Dunphy WG. Treslin collaborates with TopBP1 in triggering the initiation of DNA replication. Cell. 2010;140(3):349–59. Pubmed Central PMCID: 2857569 Epub 2010/02/02. eng.
54. Sanchez-Pulido L, Diffley JF, Ponting CP. Homology explains the functional similarities of Treslin/Ticrr and Sld3. Curr Biol. 2010;20(12):R509–10. Epub 2010/07/14. eng.
55. Sansam CL, Cruz NM, Danielian PS, Amsterdam A, Lau ML, Hopkins N, et al. A vertebrate gene, ticrr, is an essential checkpoint and replication regulator. Genes Dev. 2010;24(2): 183–94. Pubmed Central PMCID: 2807353 Epub 2010/01/19. eng.
56. Gaggioli V, Zeiser E, Rivers D, Bradshaw CR, Ahringer J, Zegerman P. CDK phosphorylation of SLD-2 is required for replication initiation and germline development in C. elegans. J Cell Biol. 2014;204(4):507–22. Pubmed Central PMCID: 3926958.
57. Itou H, Muramatsu S, Shirakihara Y, Araki H. Crystal structure of the homology domain of the eukaryotic DNA replication proteins Sld3/Treslin. Structure. 2014;22(9):1341–7.
58. Kumagai A, Shevchenko A, Dunphy WG. Direct regulation of Treslin by cyclin-dependent kinase is essential for the onset of DNA replication. J Cell Biol. 2011;193(6):995–1007. Pubmed Central PMCID: 3115804 Epub 2011/06/08. eng.
59. Boos D, Sanchez-Pulido L, Rappas M, Pearl LH, Oliver AW, Ponting CP, et al. Regulation of DNA replication through Sld3-Dpb11 interaction is conserved from yeast to humans. Curr Biol. 2011;21(13):1152–7. Epub 2011/06/28. eng.
60. Rappas M, Oliver AW, Pearl LH. Structure and function of the Rad9-binding region of the DNA-damage checkpoint adaptor TopBP1. Nucleic Acids Res. 2011;39(1):313–24. Pubmed Central PMCID: 3017600 Epub 2010/08/21. eng.
61. Sansam CG, Goins D, Siefert JC, Clowdus EA, Sansam CL. Cyclin-dependent kinase regulates the length of S phase through TICRR/TRESLIN phosphorylation. Genes Dev. 2015;29(5): 555–66.

62. Bachrati CZ, Hickson ID. RecQ helicases: guardian angels of the DNA replication fork. Chromosoma. 2008;117(3):219–33. Epub 2008/01/12. eng.
63. Capp C, Wu J, Hsieh TS. RecQ4: the second replicative helicase? Crit Rev Biochem Mol Biol. 2010;45(3):233–42. Pubmed Central PMCID: 2874125 Epub 2010/05/01. eng.
64. Sanuki Y, Kubota Y, Kanemaki MT, Takahashi TS, Mimura S, Takisawa H. RecQ4 promotes the conversion of the pre-initiation complex at a site-specific origin for DNA unwinding in Xenopus egg extracts. Cell Cycle. 2015;14:1010–23.
65. Tanaka S, Komeda Y, Umemori T, Kubota Y, Takisawa H, Araki H. Efficient initiation of DNA replication in eukaryotes requires Dpb11/TopBP1-GINS interaction. Mol Cell Biol. 2013;33(13):2614–22. Pubmed Central PMCID: 3700120.
66. Ohlenschlager O, Kuhnert A, Schneider A, Haumann S, Bellstedt P, Keller H, et al. The N-terminus of the human RecQL4 helicase is a homeodomain-like DNA interaction motif. Nucleic Acids Res. 2012;40(17):8309–24. Pubmed Central PMCID: 3458545.
67. Im JS, Ki SH, Farina A, Jung DS, Hurwitz J, Lee JK. Assembly of the Cdc45-Mcm2-7-GINS complex in human cells requires the Ctf4/And-1, RecQL4, and Mcm10 proteins. Proc Natl Acad Sci U S A. 2009;106(37):15628–32. Pubmed Central PMCID: 2747170 Epub 2009/10/07. eng.
68. Wardlaw CP, Carr AM, Oliver AW. TopBP1: a BRCT-scaffold protein functioning in multiple cellular pathways. DNA Repair. 2014;22:165–74.
69. Qu M, Rappas M, Wardlaw CP, Garcia V, Ren JY, Day M, et al. Phosphorylation-dependent assembly and coordination of the DNA damage checkpoint apparatus by Rad4(TopBP(1)). Mol Cell. 2013;51(6):723–36.
70. Lopez-Mosqueda J, Maas NL, Jonsson ZO, Defazio-Eli LG, Wohlschlegel J, Toczyski DP. Damage-induced phosphorylation of Sld3 is important to block late origin firing. Nature. 2010;467(7314):479–83. Epub 2010/09/25. eng.
71. Zegerman P, Diffley JF. Checkpoint-dependent inhibition of DNA replication initiation by Sld3 and Dbf4 phosphorylation. Nature. 2010;467(7314):474–8. Pubmed Central PMCID: 2948544 Epub 2010/09/14. eng.
72. Magiera MM, Gueydon E, Schwob E. DNA replication and spindle checkpoints cooperate during S phase to delay mitosis and preserve genome integrity. J Cell Biol. 2014;204(2): 165–75. Pubmed Central PMCID: 3897190.
73. Guo C, Kumagai A, Schlacher K, Shevchenko A, Shevchenko A, Dunphy WG. Interaction of Chk1 with Treslin negatively regulates the initiation of chromosomal DNA replication. Mol Cell. 2015;57(3):492–505. Pubmed Central PMCID: 4321788.

Chapter 14
Role of DDK in Replication Initiation

Daniel Rossbach and Robert A. Sclafani

Abstract DDK (Dbf4-dependent kinase) is a serine/threonine protein kinase conserved from yeast to humans. DDK is composed of two subunits, Cdc7 protein kinase and Dbf4 regulatory subunits in 1:1 stoichiometry. Both the *CDC7* and *DBF4* genes were discovered in budding yeast from the analysis of conditional mutants that were defective in the initiation of DNA replication. Cdc7 and Dbf4 homologues have been identified in many eukaryotes and are important in DNA replication, indicating the role of DDK was also conserved. This knowledge has been translated medically as oncogenic DDK overexpression is currently a target of therapeutic inhibitors. DDK activity is cell cycle regulated because it is inactive in G1 phase cells due to the absence of the essential Dbf4 protein as a result of APC-dependent proteolysis. It is clear from both genetic and biochemical studies that several subunits of the hexameric MCM2–7 DNA helicase/ATPase are substrates of DDK. In an allosteric model of DDK function, DDK phosphorylates the Mcm4 protein in the misaligned MCM2–7 double hexamer bound to origins of replication to align the important catalytic residues of the enzyme and to load other proteins to form a CMG (Cdc45-MCM-GINS) holoenzyme. To complete the initiation reaction, the loading of several other replication proteins is also needed, which requires ensuing CDK (cyclin-dependent kinase) phosphorylation. DDK has other substrates important for mutagenesis by TLS (translesion synthesis), meiotic recombination, and chromosome cohesion. Because all these processes are chromatin-based, DDK may have evolved to regulate chromatin-bound proteins in DNA metabolism.

Keywords Replication • Cell cycle • Kinase • Helicase • Cdc7 • Dbf4 • Initiation • Origins • Cancer

---

D. Rossbach • R.A. Sclafani (✉)
Biochemistry and Molecular Genetics, University of Colorado
School of Medicine, MS8101, Aurora, CO 80045, USA
e-mail: Robert.Sclafani@ucdenver.edu

D.L. Kaplan (ed.), *The Initiation of DNA Replication in Eukaryotes*,
DOI 10.1007/978-3-319-24696-3_14

Introduction

In this review, the role of Dbf4-dependent kinase (DDK) in the initiation of DNA replication will be described. DDK, composed of a Cdc7 (cell division cycle 7) subunit and a Dbf4 (dumbbell former 4) subunit, is a critical regulator of the initiation step of DNA replication. *CDC7* and *DBF4* genes were initially discovered using yeast mutants that arrest the cell cycle before the onset of S phase. The two proteins are conserved from yeast through humans and combine to form an active protein kinase with Cdc7 being the catalytic subunit and Dbf4 the regulatory subunit. Protein expression of the Dbf4 subunit regulates kinase activity during the cell cycle. From both genetic and biochemical evidence, DDK was found to regulate the initiation of DNA replication by phosphorylating the Mcm protein complex during S phase. DDK also has functions in many other cellular processes including meiotic recombination and mutagenesis, which are not discussed herein. Due to its role in a wide array of processes, it has become a target of cancer research and kinase inhibitor development.

Identification of *CDC7* and *DBF4*

DNA replication is a highly regulated process that relies on a complex array of proteins to ensure complete duplication of the genome occurs once and only once during the cell cycle. One protein complex required for proper initiation of DNA replication is DDK. DDK is a heterodimer composed of the catalytic subunit, Cdc7, and the regulatory subunit Dbf4. CDC7 was originally discovered in *Saccharomyces cerevisiae* (budding yeast) as part of Hartwell's collection of temperature sensitive mutants that arrest at unique points of the cell cycle [1]. *cdc7* mutants have a dumbbell phenotype, large budded cells with the nucleus stuck at the neck, and cause a cell cycle arrest immediately before the onset of S-phase. Neither Cdc7 protein nor total protein synthesis is required for completion of S phase after Cdc7's function is executed at the G1–S boundary [2]. These two attributes provide evidence that Cdc7 controls a late step immediately at the start of DNA replication initiation. Therefore, it was hypothesized that all proteins necessary for DNA replication are present at the G1–S boundary and Cdc7's role is to activate them [3], which is consistent with Cdc7 being a protein kinase (see below).

DBF4 was isolated in a screen using the collection of cell cycle mutants to only analyze mutants that produced the dumbbell forming phenotype. This specific subset of *cdc* mutants, all having a similar phenotype, were thought to be indicative of a defect in DNA synthesis and thus, closely tied together [4]. *DBF4*, like *CDC7*, was further shown to be an essential gene; and *dbf4* mutants have a defect in DNA replication initiation [5]. A direct interaction between the two genes was first suggested by suppression of *cdc7* by *DBF4* overexpression. Likewise, overexpressed *CDC7* could suppress *dbf4* mutations and double mutants of the two genes are synthetically lethal and not viable [6]. Dbf4 is essential and is likely an activation subunit for Cdc7 kinase activity.

CDC7 Encodes a Protein Kinase

CDC7 was cloned through yeast genomic DNA plasmid complementation of temperature sensitive mutant alleles of *cdc7* revealing a genomic fragment of 1.5 kb located on Chromosome IV. This genomic fragment encodes the approximately 58 kDa Cdc7 protein and based on the predicted amino acid sequence compared to known protein sequences, suggests the protein is a serine/threonine protein kinase [7]. The homology between Cdc7 and other protein kinases was not randomly distributed across the protein, but confined to functionally important kinase domains. Cdc7 contains the conserved ATP binding site as well as phosphorylation acceptor site domains of kinases, but unlike other protein kinases, Cdc7 has longer regions of non-conserved sequences between the conserved kinase domains. *DBF4* was cloned using similar methods to identify a 2.4 kb genomic fragment on Chromosome IV of the yeast genome that encodes a 81 kDa protein [8].

Cdc7 was later confirmed to be a protein kinase using mammalian histone H1 as an exogenous substrate and immunoprecipitates of Cdc7, which also contain Df4 protein [9, 10]. Serine and threonine residues within the histone H1 protein were phosphorylated by Cdc7. An in vivo substrate of Cdc7 has yet to be discovered, but because Cdc7 is found in high concentration within the nucleus, the in vivo substrates are likely nuclear proteins [9, 10]. Mutations made in key residues within the kinase domains abolished all in vivo functions of the Cdc7 protein.

Identification of *CDC7-DBF4* in Other Eukaryotes

Discovery of DDK's role in budding yeast DNA replication initiation allowed for identification of similar proteins and functions in the evolutionary distant yeast, *Schizosaccharomyces pombe*. Degenerate oligonucleotide directed PCR (polymerase chain reaction was used to identify a homolog of Cdc7 in *S. pombe*, Hsk1 (a putative homolog of Cdc7 (seven) kinase 1). Hsk1 was found to share amino acid sequence homology with Cdc7 in the critical conserved kinase domains as well as other regions [11] (Fig. 14.1). Hsk1 shares approximately 60 % identity with Cdc7 from budding yeast, but does not rescue a *cdc7ts* mutant of budding yeast [12]. Furthermore, Hsk1 protein phosphorylated several substrates as well as autophosphorylated; and *hsk1ts* mutants exhibited inhibition of DNA replication. Null mutants of *hsk1* completed mitosis but failed to replicate DNA. Like Cdc7, disruption of the *hsk1* gene is lethal to cell growth and therefore, likely carries out a very similar function. Surprisingly, an *hsk1* deletion is conditional and is viable at 37 °C, indicating DDK is not required at this temperature [13]. Discovery of *hsk1* as a yeast Cdc7 homolog within an evolutionary distant organism paved the way to identify Cdc7 homologs in other eukaryotes, especially vertebrates (Table 14.1). Shortly after Hsk1 was identified in *S. pombe*, Cdc7 homologs were discovered in humans, *Xenopus* and mouse. These proteins do not have exact protein sequence homology, but they all maintain the critical catalytic kinase residues that are required for Cdc7 function [14–16].

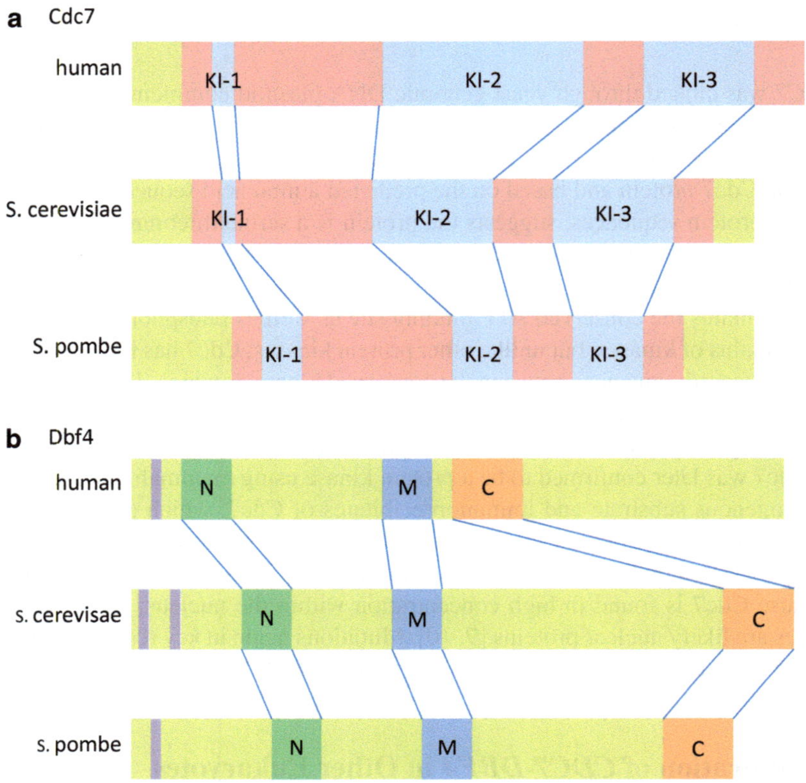

Fig. 14.1 Comparison of DDK subunits from human, *S. cerevisiae*, and *S. pombe*. (**a**) Comparison of Cdc7 homologs. *Red segments* represent conserved kinase domains. *Blue segments* represent less conserved kinase insert sequences. (**b**) Comparison of Dbf4 homologs. *Green*, *Blue*, and *Orange segments* represent conserved Dbf4 motifs (Adapted from Masai et al. 2002)

Table 14.1 CDC7 and DBF4 homologues

Organism	Catalytic subunit	Regulatory subunit	References
S. cerevisiae	CDC7 (507aa)	DBF4 (704aa)	Hartwell et al. 1971 Johnston et al. 1982
S. pombe	HSK1 (507aa)	DFP1/Him1 (545aa)	Masai et al. 1995 Brown et al. 1998
Xenopus laevis	XeCdc7 (483aa)	XeDbf4 (661aa) XeDrf1 (772aa)	Sato et al. 1997 Jares et al. 2004 Takahashi et al. 2005
Mus musculus	muCdc7 (564aa)	muASK (664aa)	Faul et al. 1999 Lepke et al. 1999
Homo sapiens	Hu Cdc7 (574aa)	huASK/huDbf4 (674aa) huDrf1 (615aa)	Sato et al. 1997 Lepke et al. 1999 Montagnoli et al. 2002

Dbf4 homologs have been isolated from other eukaryotes as well providing evidence for the strong evolutionary link between the Cdc7 catalytic subunit and the regulatory subunit (Table 14.1). In *S. pombe*, Hsk1 copurified with a protein, Dfp1. Independent experiments identified the same molecule as Him1, and these were later identified as a homolog of budding yeast Dbf4. Cdc7 association with Dfp1/Him1 was shown to have a similar effect as the budding yeast proteins in stimulating phosphorylation of an exogenous substrate, but does not enhance autophosphorylation of Hsk1 [17]. Dbf4 homologs discovered in human, *Xenopus*, and mouse have all been shown to physically interact with Cdc7 to regulate initiation of DNA replication [18, 19]. The majority of Dbf4 conservation between species is restricted to three different motif regions: M, N, and C. Using the yeast two-hybrid system for protein–protein interactions, it has been shown that Cdc7 and Dbf4 interact directly with each other both in vitro and in vivo to create a functional complex [20–22]. Importantly, both subunits of DDK are conserved and thus preserve the function of the complex from yeast through humans [23]. The essential function of DDK in DNA replication is conserved in evolution as human DDK can complement deletions in yeast *CDC7* and *DBF4* if both human Cdc7 and Dbf4 cDNAs are co-expressed [24].

Regulation of *CDC7-DBF4* Protein Kinase during the Cell Cycle

Cdc7 and Dbf4 proteins are regulated differentially throughout the cell cycle. In budding yeast, the Cdc7 catalytic subunit is stably expressed and subsequently bound to chromatin throughout the majority of the cell cycle [25]. However, Dbf4 protein expression oscillates throughout the cell cycle as a result of protein stability. Dbf4 stability is regulated by the APC (anaphase promoting complex). The protein is rapidly degraded in G1 phase of the cell cycle and mutations in APC subunits or the Dbf4 N-terminus region resembling a destruction box eliminates the APC-dependent degradation [26]. Once Dbf4 is expressed in the cell, it is likely that it immediately associates with chromatin. Dbf4 protein expression oscillation within the cell cycle is equivalent to the role of cyclins in regulating the activity of cyclin-dependent kinases (CDK) [27, 28].

Although the Cdc7 subunit is constitutively expressed during the cell cycle, the kinase activity changes depending on expression of Dbf4 in the cell. Cdc7 kinase activity is low during G1 phase of the cell cycle, increases at the G1–S transition, maintained at high activity through S phase, and then decreases again as S phase in completed [29]. This pattern of kinase activity closely mirrors the expression of the Db4 protein [30]. Similar studies performed using human and *Xenopus* homologs of Cdc7 and Dbf4 also show DDK activity relies on the expression of the Dbf4 subunit within the cell at the G1–S transition [31, 32].

While Dbf4 has long been known to be the main regulatory subunit of the DDK complex, human and *Xenopus* Cdc7 are also regulated by Drf1 (Dbf4-related factor 1) [33]. In *Xenopus* egg extracts, Drf1 is in far more abundance than the Dbf4 protein and removal of the protein by immunodepletion causes the characteristic inhibition of replication. However, as gastrulation completes, Drf1 levels decline and Dbf4 becomes more abundant [34]. Human Drf1 displays similar temporally regulated kinetics as Dbf4, but the function has yet to be established. Drf1 and Dbf4 may bind to different pools of Cdc7 in order to initiate replication from different origin subsets [33]. Another proposed role of human Drf1 is to restart replication forks that have arrested for various reasons throughout S phase [35]. This second regulator of Cdc7 has not been found in the genome of yeast and *Drosophila* indicating another layer of complexity in vertebrate replication initiation that may have developmental or timing implications.

Dbf4 is a limiting factor in DNA replication, along with Dpb11 (DNA polymerase B subunit 11), Sld2 (synthetic lethal with Dpb11), and Sld3. These proteins are found in low abundance in the cell, but overexpression of all four proteins is sufficient to advance the replication timing of late origins and heterochromatic regions [36].

DDK Structure and Function

Crystal structure of human DDK has provided insight into how DDK subunits are held together and lead to activation of the kinase (Fig. 14.2) [37]. Cdc7 has a bilobal structure with a deep cleft between the two domains that houses the active site of the kinase. The N lobe of Cdc7 is comprised of an antiparallel β-sheet as well as α-helices, while the C lobe of Cdc7 is comprised almost entirely of α-helices that encase the catalytic loop of the kinase as well as a Mg^{2+} ion. Dbf4 motif regions M and C are conserved between species and are necessary for binding to Cdc7 in a bipartite manner. Motif C of Dbf4 contains α-helices and a β-strand that forms an essential zinc finger used for binding the N-terminal lobe of Cdc7. Meanwhile, motif M of Dbf4 uses β-strands to pair with β-strands within the kinase insert three domain to create an antiparallel sheet necessary for packing and binding against the C-terminal lobe of Cdc7. Although it is conserved across species, motif N of Dbf4 is not used for binding nor is it required for kinase activation. Motif M and Motif C are necessary to pack Dbf4 against the lobes of Cdc7 and are needed for activation of the kinase. Upon Cdc7 and Dbf4 binding, the kinase subunit maintains a closed nucleotide (Mg-ATP) bound conformation similar to other protein kinases. Mutations in regions of Dbf4 or Cdc7 that disrupt the binding of the two subunits result in extensively reduced activity of the kinase [37].

Cdc7 lacks a threonine residue in the active site present in most other protein kinases. This residue is generally phosphorylated and leads to activation of the serine/threonine kinase. In order to obtain a crystal structure of Cdc7 and the binding interface, the expanded kinase insert region 2 was deleted. Therefore, it cannot be ruled out that this region contains other regulatory elements that make up for the loss of the critical threonine residue found in other kinases [37].

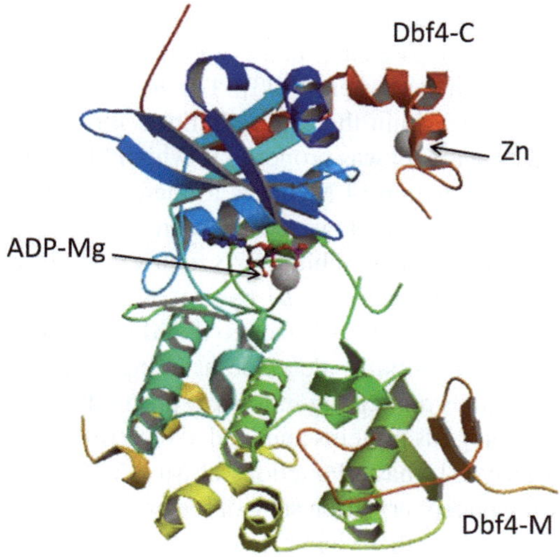

Fig. 14.2 Human DDK atomic crystal structure. Atomic crystal structure of Cdc7-Dbf4 dimer from RCSB PDB (Protein Data Bank): http://www.rcsb.org/pdb/explore/images.do?structureId=4 F99 using coordinates from reference [37]. Note that protein fragments of Cdc7 and Dbf4 were used and not the full-length molecules and the resulting smaller DDK enzyme had about 40–50 % activity. Cdc7 kinase N-terminal Beta-Sheet rich ATP binding domain at top and C-terminal alpha-helical phospho-acceptor domain at bottom. Dbf4-M and Dbf4-C domains are as shown and also depicted in Fig. 14.1. Zn^{2+} and Mg^{2+} ions are shown as *gray spheres*

Budding yeast Cdc7 has a unique C-terminal domain consisting of 55 amino acids that has been shown to be required for binding to Dbf4 [20]. This region is not found in homologous Cdc7 proteins indicating there is redundancy that allows for Cdc7 to bind Dbf4 without this specialized region in other eukaryotes.

The Cdc7 subunit forms dimers as shown both in vivo and with recombinant proteins in vitro [38]. These Cdc7-Cdc7 dimers have low kinase activity which increases substantially when one Dbf4 molecule associates with one Cdc7 molecule to disrupt the Cdc7-Cdc7 dimer and form a Cdc7-Dbf4 heterodimer [39].

Cell Cycle Activity of DDK

All budding yeast replication origins that fire throughout S phase require DDK function [40, 41]. Cdc7 may control the transition from G1 to S as well as activation and initiation from individual origins. Dbf4 has been found to associate directly with origins of replication and thus provides a foundation that Cdc7 functions directly at replication initiation complexes at the origins [21, 42]. However, there is no direct in vivo evidence that Cdc7 also binds to origins. A first indication of the identity of the endogenous substrates of DDK came from studies in which the *mcm5-bob1* (Mcm5 P83L) mutation in the Mcm5 (mini chromosome maintenance) protein was found to be a bypass suppressor of deletions in both *CDC7* and *DBF4*, but not other cell cycle division genes [43]. The *mcm5-bob1* mutation is able to bypass the need for DDK within the cell and allows for the cell to enter S phase earlier. In these studies, a model was proposed in which Mcm5 structure is modified by DDK. An important function of Mcm5 is to therefore inhibit DNA replication and a modification of the protein allows the cell to initiate replication. There is precedent in phage λ for a model where inhibition is removed to allow assembly of DNA replication proteins [44]. The *mcm5-bob1* bypass of DDK indicates that essential functions of the kinase are inherently linked to the Mcm complex.

The Mcm complex was initially identified as being required for stable maintenance of ARS (autonomous replicating sequence) plasmids in budding yeast [45]. It is made up of six homologous proteins and loaded onto origins as a double hexamer in late M phase to early G1 phase. Functions of Mcm2–7 within the cell cycle are likely regulated by the phosphorylation state of the proteins [46, 47]. Comparison of Mcm protein sequences show that Mcm2, 4, and 6 contain serine and threonine residues within long N-terminal domains. Mcm3, 5 and archaeal Mcm proteins do not have this extra region. A mutation within the *MCM2* gene reduces efficiency of DNA replication at origins of DNA replication. To understand how Mcm2 is regulated it is necessary to understand what other genes may interact with the MCM2 gene. It was determined that a *dbf4* mutation was able to suppress the defects seen in *mcm2-1* mutant. Furthermore, Mcm2 and Dbf4 show a strong physical interaction with each other that is compromised by the *mcm2-1* mutation. From these genetic data, a number of investigations focused on the Mcm complex as a substrate of DDK. Recombinant GST-Mcm fusion proteins were made for use in DDK protein kinase assay in vitro [48]. GST-Mcm2, GST-Mcm3, GST-Mcm4, and GST-Mcm6 were all shown to be phosphorylated by DDK in vitro. Surprisingly, GST-Mcm5 was not phosphorylated as was thought from the *mcm5-bob1* mutation. Additionally, a GST-*mcm2-1* fusion was shown to reduce phosphorylation of *mcm2-1* by DDK compared to wild type indicating that the kinase is directly tied to the function of the Mcm complex. Dbf4 is the main subunit required for binding Mcm2, which recruits Cdc7 to phosphorylate amino acids S164 and S170 and leads to DNA replication and cell growth [49].

Mcm2 is phosphorylated during S phase by DDK at two N-terminal serines (Fig. 14.3). Conversion of these two residues to non-phosphorylatable alanines (S164A, S170A) results in a severe growth defect that can be partially rescued by *mcm5-bob1*. Upon mutating these residues, the amount of origin ssDNA is significantly reduced compared to wild type based on the amount of RPA (replication protein A) bound to the origins. These mutations reduced binding of Sld3 and GINS to the Mcm complex. It was also determined that Mcm2 phosphorylation destabilized the interaction with Mcm5 promoting opening of the Mcm2–7 hexameric ring to allow extrusion of ssDNA [50].

Consensus DDK phosphorylation sites are usually S or T followed by an acidic residue (S/T D/E) [51]. These sites are similar to sites preferred by CK2 (casein kinase 2) [52], the most similar kinase to Cdc7 in the human kinome [53]. There

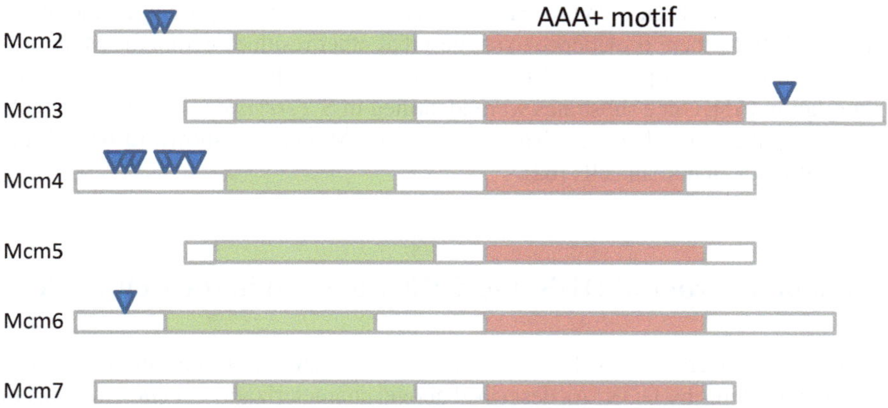

Fig. 14.3 Identification of DDK Target sites on Mcm2–7. Schematic of DDK phosphorylation sites on each Mcm subunit. *Blue triangles* represent DDK phosphorylation sites as determined by mass spectrometry. Extended N-terminal domains of Mcm2, Mcm4, and Mcm6 contain phosphorylated serine and threonine residues (Adapted from Randall et al. 2010)

are also hierarchal DDK sites (S/T-S/T-P) where the second S/T residue is first phosphorylated by CDK creating a priming phosphor-acidic site (S/T-S/T(Phos)-P) and hence DDK phosphorylation will occur at the first S/T residue as seen in human Mcm2 [51, 54] and in yeast Mer2 [55]. Hierarchal sites primed by CDK or by Mec1 (S/T-S/T-Q) are also seen in the N-terminal region of yeast Mcm4 and Mcm6 (Fig. 14.3).

Mcm4 has also been analyzed as a DDK substrate due to the discovery of hierarchal phosphorylation sites. The serine/threonine rich N-terminal domain inhibits DNA replication. Phosphorylation of this region removes the inhibitory effect and cells can transition into S phase. In vitro experiments confirm DDK is the kinase responsible for phosphorylating the N-terminal region of Mcm4 and relieving the inhibitory effect [56, 57]. When the inhibitory region of Mcm4 is deleted, replication can proceed in the absence of DDK similar to the *mcm5-bob1* bypass mutation. Mutating DDK specific serine/threonine residues to alanine within Mcm4 confers lethality since Mcm4 inhibition cannot be removed by DDK. However, if phosphomimetic mutations (S/T to D/E) are made in the same residues, the cell is not only viable but can also bypass the need for DDK. Cell cycle regulation of DNA replication can still occur in this case as CDK (Cdk1-Clb5) function is still required.

If both DDK and CDK steps are made constitutive, lethality occurs [58, 59]. Mcm4 phosphorylation also stimulates association of other replication proteins, such as Cdc45 (cell-division cycle 45), with the chromatin [60]. The lack of Mcm5 phosphorylation may be explained by allosteric effects by Mcm4 phosphorylation such that relief of N-terminal inhibition may result in a conformation change of Mcm5. Mcm5, as evidence from the *mcm5-bob1* mutation, may be an important inhibitor of DNA replication that is removed by phosphorylation of other Mcm subunits such as

Mcm2, Mcm4, or Mcm6. While Mcm2, Mcm4, and Mcm6 appear to be the direct targets of DDK, there is evidence to suggest that other residues within the Mcm proteins need to be phosphorylated first in order to prime the protein for DDK phosphorylation [61]. Once substrates were identified in *S. cerevisiae*, efforts were made to confirm them in other organisms, specifically, Mcm2 was shown to be a direct target of DDK in human cells [62].

Sequential Order of DDK and CDK Function in the Cell Cycle

There is controversy about the sequence of events that lead to phosphorylation of the Mcm proteins by DDK. As described above, characterization of *cdc7* mutants in yeast demonstrate that protein synthesis is no longer required to complete S phase after Cdc7 executes its function indicating that it may be the final factor required for replication initiation. Studies in *S. cerevisiae* also suggest that CDK carries out its phosphorylation function first while DDK phosphorylates the Mcm complex second [63]. However, in vivo studies in *S. pombe*, and in vitro studies in *Xenopus* extracts suggest that DDK acts first to phosphorylate the Mcm complex and then CDK acts second to initiate DNA replication [64, 65]. Both models agree that phosphorylation of the Mcm proteins by DDK is necessary to load Cdc45 onto chromatin at replication origins.

In *Xenopus* and *S. pombe*, DDK loads Cdc45 onto chromatin before CDK carries out its catalytic function and brings in the remaining proteins necessary for DNA replication including the GINS complex. In G1 phase of the cell cycle, CDK activity is low due to the absence of S phase cyclins necessary to transition the cell into S phase [28]. Dbf4 is also low in G1, thus Cdc45 is not loaded onto chromatin [26, 27, 29]. However, in the presence of the *mcm5-bob1* mutation, Cdc45 is loaded onto chromatin in G1 phase of the cell cycle when S phase CDK is inactive. The loading of Cdc45 in the absence of CDK correlates with premature ARS1 origin unwinding as seen with in vivo genomic footprinting in the *mcm5-bob1* mutant arrested in G1 phase [66]. Thus, origin structural changes are directly linked to Cdc45 loading. Furthermore, in an in vitro replication system from *S. cerevisiae* extracts, DDK alone produced Cdc45 loading onto origins in the absence of CDK. In the latter study, there was a clear order of events in that DDK must act before CDK in order to initiate DNA replication as seen in the *Xenopus* in vitro studies.

DDK as a Target of Cancer Therapy

Deregulation of normal cell cycle progression has been a target of previous therapies for cancer treatment. Given that DDK plays such an important role in DNA replication, DNA damage repair and cell cycle progression, it has become a focus of new cancer therapeutics [67–69]. Increased expression of Cdc7 has become a marker for cancer and overexpression of Cdc7 has been shown to inhibit apoptosis and lead to survival of oral squamous cell cancer cells [70]. Cancer cases that show increased Cdc7 expression tend to have poor clinical outcomes [71]. In breast cancer, increased Cdc7 is often correlated with increased Dbf4 that may be linked to the proliferative activity of cancer cells. While increased Cdc7 is not always correlated with proliferative status, it is always correlated with the number of cells that enter S phase. An explanation of this may be that since Cdc7 is involved in the DNA damage repair pathway, it aids in recovery of stalled replication forks to enhance survival of tumor cells. Additionally, p53 mutations or even protein loss are a result of increased Cdc7 and Dbf4 expression [72]. Conversely, loss of Cdc7 in cancer cells induces a p53-independent apoptotic response that leads to cell death without activating the standard checkpoint pathway [73].

The first inhibitors of Cdc7 were developed at Nerviano Medical Sciences as small molecule ATP-competitors in the pyrrolopyridones class [74]. One compound from this screen, PHA-767491, emerged as the leading candidate for Cdc7 inhibition. This compound impairs Mcm2 phosphorylation at DDK-dependent sites and blocks origin firing, but does not prevent progression of replication forks or activate a DNA damage cascade [75]. Antitumor activity of this compound has been seen in AML(acute myeloid leukemia), breast and colon cancer models. Another inhibitor that has been used in cancer models is XL413. However, unlike the robust antiproliferative and apoptotic effects seen in PHA-767491, XL413 only has limited activity in a smaller percentage of tester cancer cell types [76].

Model of DDK Action

A simplified model of DDK action is DDK phosphorylates the Mcm complex, which produces a structural change that activates the helicase and allows for the binding of important replication proteins (Fig. 14.4). The model is bolstered by atomic crystal structures and functional assays of the Mcm complex as deduced from the studies of Archaeal Mcm homomultimeric proteins together with yeast genetic studies [77, 78]. In the Archaeal Mcm protein, the alpha helical bundle (A domain) located in the N-terminus, anchors the proline that is converted to the leucine in the context of the *mcm5-bob1* mutation. Leucine has a large side chain that pushes this A domain out and creates a structural change in the protein. It is possible that this structural change removes an inhibition and allows for other loading proteins that recognize this change. Phosphorylation of the wild type Mcm2 or Mcm4 may cause a structural change in the Mcm complex as a whole and simulate this expansion of the A domain in Mcm5. Thus, the *mcm5-bob1* mutation mimics a conformational change brought about by phosphorylation within the complex. In *S. pombe*, this structural change may occur spontaneously at higher temperatures allowing for DDK bypass [13].

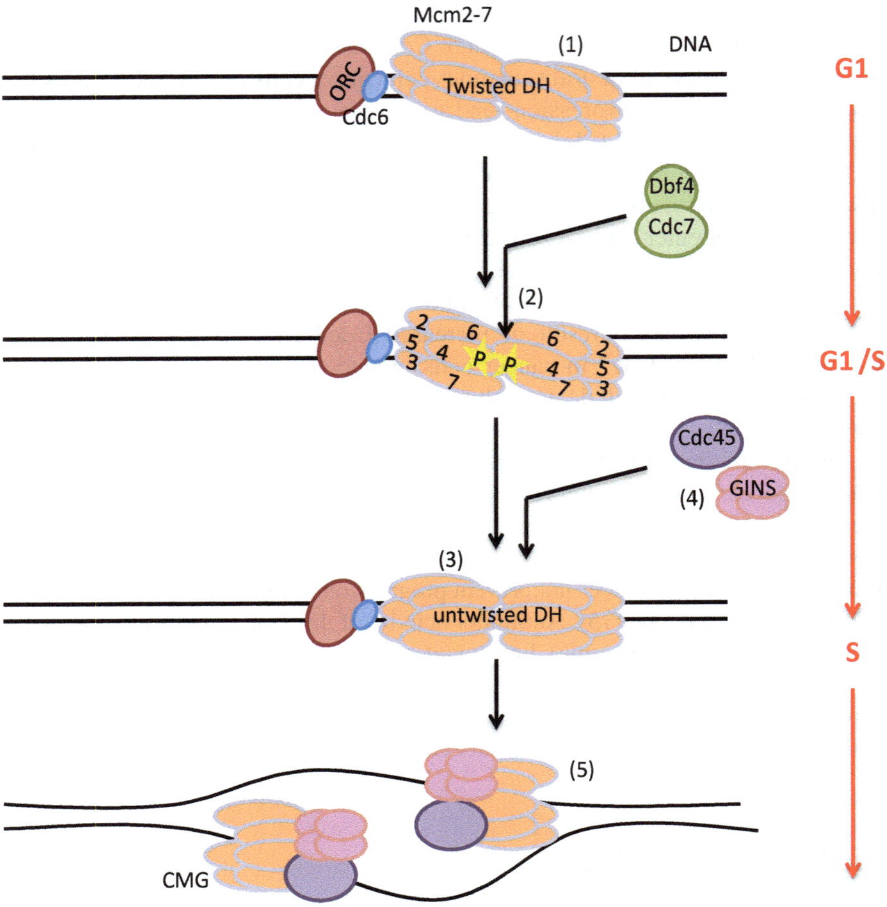

Fig. 14.4 Model of DDK action in replication. (*1*) In late M/early G1 phase of the cell cycle, Mcm2–7 complex is loaded onto origins as a twisted double hexamer. (*2*) At the G1–S transition, DDK phosphorylates the N-terminal serine/threonine residues of Mcm2, Mcm4, and Mcm6. (*3*) The double hexamer untwists to allow additional protein loading. (*4*) Cdc45, GINS and other required proteins bind to activate the helicase. (*5*) A gate between Mcm2 and Mcm5 opens to allow extrusion of ssDNA and form the actively replicating CMG complex

As further evidence of the model, other amino acids were substituted in place of the proline residue within the yeast Mcm5 protein [78, 79]. Only residues with large side chains (K, W) that would give similar expansion of the A domain and not smaller substitutions (A, G) produced a *mcm5-bob1* phenotype and bypassed DDK. Surprisingly, the P to L change in the Archaeal protein only resulted in a 1 Å structural change but the corresponding yeast genetics confirmed that amino acids with large side chains caused a significant effect in the cell. The small structural difference can be explained by the fact the crystallization pushes the domains inward to pack it together tightly, whereas a protein in solution will have more room to expand. Ultracentrifugation of the Archaeal protein with the *mcm5-bob1* mutation showed increased heterogeneity in the sedimentation of the protein [79]. An explanation of this anomaly is that in vivo pushes out the A domain to produce a unique conformation, while the *mcm5-bob1* mutation allows for a greater range in conformations, only a few of which are active. It is possible that the homogeneity due to phosphorylation is preventing the structural studies from seeing the domain push out. Using a construct that contains the N-terminal domain of Mcm from one archaeon fused to the C-terminal of Mcm of another archaeon, the A domain does in fact push out much further than the 1 Å seen previously [80]. Once this A domain of the Mcm is pushed out, the complex can now act as a landing pad for other DNA replication proteins to create the helicase holoenzyme Cdc45-Mcm-GINS complex (CMG). Mcm2 phosphorylation also destabilizes the interaction with Mcm5 promoting opening of the Mcm2–7 hexameric ring to allow extrusion of ssDNA and to allow elongation of DNA replication [50].

Concluding Remarks

Since over 50 years after DDK was initially discovered in yeast, a great deal has been learned about the function and roles DDK plays throughout the cell cycle. Its scope goes well beyond its role in replication initiation to include DNA translesion synthesis and mutagenesis [81, 82], kinetochore function [42], and meiotic recombination [55, 83] (Fig. 14.5). We now know DDK is a serine/threonine protein kinase with the most widely known substrates being the Mcm complex but more substrates will probably be identified in the future. During evolution, DDK may have been recruited for a number of cellular roles that all involve regulating the binding of proteins to chromatin [81]. Clearly, DDK's most well studied and understood role is to initiate DNA replication from individual origins of replication. At the atomic level, crystallography has been used to capture both the binding of the catalytic subunit Cdc7 to its regulatory subunit Dbf4 as well as possible binding to substrates. Furthermore, the substrate specificity of DDK for the Mcm complex has been shown using cryo-EM as the two hexamers in the inactive Mcm complex in the pre-RC are misaligned, which acts as a unique binding site for DDK [84]. Phosphorylation by DDK then aligns the double hexamer correctly to produce the active DNA helicase. The knowledge about DDK has already yielded important translational information for the clinic in that a number of recent studies have identified potential inhibitors of DDK function during cancer development and progression.

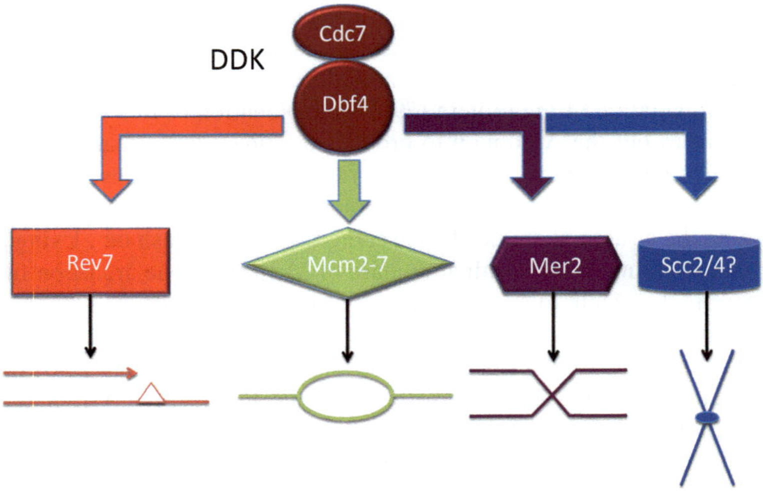

Fig. 14.5 DDK (Cdc7-Dbf4) regulates four different chromatin-bound substrates. DDK (Cdc7-Dbf4) phosphorylates Rev7 to regulate TLS (translesion synthesis) in mutagenesis, MCM complex in the initiation of DNA replication, Mer2 in meiotic recombination, and possibly Scc2-Scc4 in chromosome cohesion (see text for references)

References

1. Hartwell LH. Genetic control of the cell division cycle in yeast II. Genes controlling DNA replication and its initiation. J Mol Biol. 1971;59:183–94.
2. Hartwell LH. Three additional genes required for deoxyribonucleic acid synthesis in Saccharomyces cerevisiae. J Bacteriol. 1973;115:966–74.
3. Hollingsworth RE Jr. Molecular Biology of the CDC7 gene and protein of Saccharomyces cerevisiae [PhD]. University of Colorado; 1990.
4. Johnston LH, Thomas AP. The isolation of new DNA synthesis mutants in the yeast Saccharomyces cerevisiae. Mol Gen Genet. 1982;186(3):439–44.
5. Johnston LH, Thomas AP. A further two mutants defective in initiation of the S phase in the yeast *Saccharomyces cerevisiae*. Mol Gen Genet. 1982;186:445–8.
6. Kitada K, Johnston LH, editors. The Saccharomyces cerevisiae DBF4 gene, required for the G1Æ S phase transition, suppresses temperature-sensitive cdc7 mutations. 1991 Yeast Genetics and Molecular Biology Meeting; 1991; San Francisco, CA: Genetics Society of America.
7. Patterson M, Sclafani RA, Fangman WL, Rosamond J. Molecular characterization of cell cycle gene CDC7 from Saccharomyces cerevisiae. Mol Cell Biol. 1986;6:1590–8.
8. Chapman JW, Johnston LH. The yeast gene, DBF4, essential for entry into the S phase is cell cycle regulated. Exp Cell Res. 1989;180:419–28.
9. Hollingsworth Jr RE, Sclafani RA. DNA metabolism gene CDC7 from yeast encodes a serine (threonine) protein kinase. Proc Natl Acad Sci U S A. 1990;87:6272–6.
10. Yoon H-J, Campbell JL. The CDC7 protein of Saccharomyces cerevisiae is a phosphoprotein that contains protein kinase activity. Proc Natl Acad Sci U S A. 1991;88:3574–8.
11. Masai H, Miyake T, Arai K-I. *hsk1*$^{+}$, a *Schizosaccharomyces pombe* gene related to *Saccharomyces cerevisiae* CDC7, is required for chromosomal replication. EMBO J. 1995;14(13):3094–104.
12. Masai H, Arai K. Cdc7 kinase complex: a key regulator in the initiation of DNA replication. J Cell Physiol. 2002;190(3):287–96.

13. Matsumoto S, Hayano M, Kanoh Y, Masai H. Multiple pathways can bypass the essential role of fission yeast Hsk1 kinase in DNA replication initiation. J Cell Biol. 2011;195(3):387–401.
14. Sato N, Arai K-I, Masai H. Human and *Xenopus* cDNAs encoding budding yeast Cdc7-related kinases: in vitro phosphorylation of MCM subunits by a putative human homolgue of Cdc7. EMBO J. 1997;16:4340–51.
15. Hess GF, Drong RF, Weiland KL, Sligthom JL, Sclafani RA, Hollingsworth Jr RE. A human homolog of the yeast CDC7 gene is overexpressed in some tumors and transformed cell lines. Gene. 1998;211:133–40.
16. Faul T, Staib C, Nanda I, Schmid M, Grummt F. Identification and characterization of mouse homologue to yeast Cdc7 protein and chromosomal localization of the cognate mouse gene Cdc7l. Chromosoma. 1999;108(1):26–31.
17. Brown GW, Kelly TJ. Purification of Hsk1, a minichromosome maintenance protein kinase from fission yeast. J Biol Chem. 1998;273(34):22083–90.
18. Jares P, Luciani MG, Blow JJ. A Xenopus Dbf4 homolog is required for Cdc7 chromatin binding and DNA replication. BMC Mol Biol. 2004;5:5.
19. Lepke M, Putter V, Staib C, Kneissl M, Berger C, Hoehn K, et al. Identification, characterization and chromosomal localization of the cognate human and murine DBF4 genes. Mol Gen Genet. 1999;262(2):220–9.
20. Jackson AL, Pahl PMB, Harrison K, Rosamond J, Sclafani RA. Cell cycle regulation of the yeast Cdc7 protein kinase by association with the Dbf4 protein. Mol Cell Biol. 1993; 13(5):2899–908.
21. Dowell SJ, Romanowski P, Diffley JFX. Interaction of Dbf4, the Cdc7 protein kinase regulatory subunit, with yeast regulatory origins in vivo. Science. 1994;265:1243–6.
22. Harkins V, Gabrielse C, Haste L, Weinreich M. Budding yeast Dbf4 sequences required for Cdc7 kinase activation and identification of a functional relationship between the Dbf4 and Rev1 BRCT domains. Genetics. 2009;183(4):1269–82.
23. Roberts BT, Ying CY, Gautier J, Maller JL. DNA replication in vertebrates requires a homolog of the Cdc7 protein kinase. Proc Natl Acad Sci U S A. 1999;96(6):2800–4.
24. Davey MJ, Andrighetti HJ, Ma X, Brandl CJ. A synthetic human kinase can control cell cycle progression in budding yeast. G3 (Bethesda). 2011;1(4):317–25.
25. Weinreich M, Stillman B. Cdc7p-Dbf4p kinase binds to chromatin during S phase and is regulated by both the APC and the RAD53 checkpoint pathway. EMBO J. 1999;18(19): 5334–46.
26. Cheng L, Collyer T, Hardy CF. Cell cycle regulation of DNA replication initiator factor Dbf4p. Mol Cell Biol. 1999;19(6):4270–8.
27. Ferreira MF, Santocanale C, Drury LS, Diffley JF. Dbf4p, an essential S phase-promoting factor, is targeted for degradation by the anaphase-promoting complex. Mol Cell Biol. 2000;20(1):242–8.
28. Morgan DO (2007) The Cell Cycle. Primers In Biology. New Science Press Ltd., London UK.
29. Oshiro G, Owens JC, Shellman Y, Sclafani RA, Li JJ. Cell cycle control of cdc7p kinase activity through regulation of dbf4p stability. Mol Cell Biol. 1999;19(7):4888–96.
30. Takeda T, Ogino K, Matsui E, Cho MK, Kumagai H, Miyake T, et al. A fission yeast gene, him1(+)/dfp1(+), encoding a regulatory subunit for hsk1 kinase, plays essential roles in S-phase initiation as well as in S-phase checkpoint control and recovery from DNA damage. Mol Cell Biol. 1999;19(8):5535–47.
31. Jiang W, Hunter T. Identification and characterization of a human protein kinase related to budding yeast Cdc7p. Proc Natl Acad Sci U S A. 1997;94:14320–5.
32. Furukohri A, Sato N, Masai H, Arai K, Sugino A, Waga S. Identification and characterization of a Xenopus homolog of Dbf4, a regulatory subunit of the Cdc7 protein kinase required for the initiation of DNA replication. J Biochem. 2003;134(3):447–57.
33. Montagnoli A, Bosotti R, Villa F, Rialland M, Brotherton D, Mercurio C, et al. Drf1, a novel regulatory subunit for human Cdc7 kinase. EMBO J. 2002;21(12):3171–81.
34. Takahashi TS, Walter JC. Cdc7-Drf1 is a developmentally regulated protein kinase required for the initiation of vertebrate DNA replication. Genes Dev. 2005;19(19):2295–300.

35. Yoshizawa-Sugata N, Ishii A, Taniyama C, Matsui E, Arai K, Masai H. A second human Dbf4/ASK-related protein, Drf1/ASKL1, is required for efficient progression of S and M phases. J Biol Chem. 2005;280(13):13062–70.
36. Mantiero D, Mackenzie A, Donaldson A, Zegerman P. Limiting replication initiation factors execute the temporal programme of origin firing in budding yeast. EMBO J. 2011;30(23):4805–14.
37. Hughes S, Elustondo F, Di Fonzo A, Leroux FG, Wong AC, Snijders AP, et al. Crystal structure of human CDC7 kinase in complex with its activator DBF4. Nat Struct Mol Biol. 2012;19(11):1101–7.
38. Shellman YG, Schauer IE, Oshiro G, Dohrmann P, Sclafani RA. Oligomers of yeast Cdc7/Dbf4 protein kinase exist in the cell. Mol Gen Genet. 1998;259:429–36.
39. Kaplan DL, Bruck I. Methods to study kinase regulation of the replication fork helicase. Methods. 2010;51:358–62.
40. Bousset K, Diffley JFX. The Cdc7 protein kinase is required for origin firing during S phase. Genes Dev. 1998;12:480–90.
41. Donaldson AD, Fangman WF, Brewer BJ. Cdc7 is required throughout the yeast S phase to activate replication origins. Genes Dev. 1998;491:491–501.
42. Natsume T, Muller CA, Katou Y, Retkute R, Gierlinski M, Araki H, et al. Kinetochores coordinate pericentromeric cohesion and early DNA replication by cdc7-dbf4 kinase recruitment. Mol Cell. 2013;50(5):661–74.
43. Hardy CFJ, Dryga O, Seematter S, Pahl PMB, Sclafani RA. *mcm5/cdc46-bob1* bypasses the requirement for the S phase activator Cdc7p. Proc Natl Acad Sci U S A. 1997;94:3151–5.
44. Sclafani RA, Holzen TM. Cell cycle regulation of DNA replication. Annu Rev Genet. 2007;41:237–80.
45. Tye B-K. MCM proteins in DNA replication. Annu Rev Biochem. 1999;68:649–86.
46. Todorov IT, Attaran A, Kearsey SE. BM28, a human member of the MCM2-3-5 family, is displaced from chromatin during DNA replication. J Cell Biol. 1995;129:1433–45.
47. Young MR, Tye BK. Mcm2 and Mcm3 are constitutive nuclear proteins that exhibit distinct isoforms and bind chromatin during specific cell cycle stages of *Saccharomyces cerevisiae*. Mol Biol Cell. 1997;8:1587–601.
48. Lei M, Kawasaki Y, Young MR, Kihara M, Sugino A, Tye BK. Mcm2 is a target of regulation by Cdc7-Dbf4 during the initiation of DNA synthesis. Genes Dev. 1997;11:3365–74.
49. Bruck I, Kaplan D. Dbf4-Cdc7 phosphorylation of Mcm2 is required for cell growth. J Biol Chem. 2009;284(42):28823–31.
50. Bruck I, Kaplan DL. The Dbf4-Cdc7 kinase promotes Mcm2-7 ring opening to allow for single-stranded DNA extrusion and helicase assembly. J Biol Chem. 2015;290:1210–21.
51. Cho WH, Lee YJ, Kong SI, Hurwitz J, Lee JK. CDC7 kinase phosphorylates serine residues adjacent to acidic amino acids in the minichromosome maintenance 2 protein. Proc Natl Acad Sci U S A. 2006;103(31):11521–6.
52. Meggio F, Pinna LA. One-thousand-and-one substrates of protein kinase CK2? FASEB J. 2003;17(3):349–68.
53. Manning G, Whyte DB, Martinez R, Hunter T, Sudarsanam S. The protein kinase complement of the human genome. Science. 2002;298(5600):1912–34.
54. Montagnoli A, Valsasina B, Brotherton D, Troiani S, Rainoldi S, Tenca P, et al. Identification of Mcm2 phosphorylation sites by S-phase-regulating kinases. J Biol Chem. 2006;281(15):10281–90.
55. Wan L, Niu H, Futcher B, Zhang C, Shokat KM, Boulton SJ, et al. Cdc28-Clb5 (CDK-S) and Cdc7-Dbf4 (DDK) collaborate to initiate meiotic recombination in yeast. Genes Dev. 2008;22(3):386–97.
56. Sheu YJ, Stillman B. Cdc7-Dbf4 phosphorylates MCM proteins via a docking site-mediated mechanism to promote S phase progression. Mol Cell. 2006;24(1):101–13.
57. Sheu Y-J, Stillman B. The Dbf4-Cdc7 kinase promotes S phase by alleviating an inhibitory activity in Mcm4. Nature. 2010;463(7277):113–7.
58. Tanaka S, Umemori T, Hirai K, Muramatsu S, Kamimura Y, Araki H. CDK-dependent phosphorylation of Sld2 and Sld3 initiates DNA replication in budding yeast. Nature. 2007;445(7125):328–32.

59. Zegerman P, Diffley JF. Phosphorylation of Sld2 and Sld3 by cyclin-dependent kinases promotes DNA replication in budding yeast. Nature. 2007;445(7125):281–5.
60. Masai H, Taniyama C, Ogino K, Matsui E, Kakusho N, Matsumoto S, et al. Phosphorylation of MCM4 by Cdc7 kinase facilitates its interaction with Cdc45 on the chromatin. J Biol Chem. 2006;281(51):39249–61.
61. Randell JC, Fan A, Chan C, Francis LI, Heller RC, Galani K, et al. Mec1 Is one of multiple kinases that prime the Mcm2-7 helicase for phosphorylation by Cdc7. Mol Cell. 2010;40(3):353–63.
62. Jiang W, McDonald D, Hope TJ, Hunter T. Mammalian Cdc7-Dbf4 protein kinase complex is essential for initiation of DNA replication. EMBO J. 1999;18(20):5703–13.
63. Nougarede R, Della Seta F, Zarzov P, Schwob E. Hierarchy of S-phase-promoting factors: yeast Dbf4-cdc7 kinase requires prior S-phase cyclin-dependent kinase activation. Mol Cell Biol. 2000;20(11):3795–806.
64. Walter JC. Evidence for sequential action of cdc7 and cdk2 protein kinases during initiation of DNA replication in Xenopus egg extracts. J Biol Chem. 2000;275(50):39773–8.
65. Heller RC, Kang S, Lam WM, Chen S, Chan CS, Bell SP. Eukaryotic origin-dependent DNA replication in vitro reveals sequential action of DDK and S-CDK kinases. Cell. 2011;146(1):80–91.
66. Geraghty DS, Ding M, Heintz NH, Pederson DS. Premature structural changes at replication origins in a yeast minichromosome maintenance (MCM) mutant. J Biol Chem. 2000;275(24):18011–21.
67. Hollingsworth RE, Ostroff RM, Klein MB, Niswander LA, Sclafani RA. Molecular genetic studies of the Cdc7 protein kinase and induced mutagenesis in yeast. Genetics. 1992;132:53–62.
68. Sclafani RA, Patterson M, Rosamond J, Fangman WL. Differential regulation of the yeast CDC7 gene during mitosis and meiosis. Mol Cell Biol. 1988;8(1):293–300.
69. Pessoa-Brandão L. Genetic and molecular studies of Saccahromyces cerevisiae Cdc7-Dbf4 kinase function in DNA damage-induced mutagenesis [Dissertation/Thesis]. Aurora, Colorado: University of Colorado Health Sciences Center; 2005.
70. Cheng AN, Jiang SS, Fan CC, Lo YK, Kuo CY, Chen CH, et al. Increased Cdc7 expression is a marker of oral squamous cell carcinoma and overexpression of Cdc7 contributes to the resistance to DNA-damaging agents. Cancer Lett. 2013;337(2):218–25.
71. Choschzick M, Lebeau A, Marx AH, Tharun L, Terracciano L, Heilenkotter U, et al. Overexpression of cell division cycle 7 homolog is associated with gene amplification frequency in breast cancer. Hum Pathol. 2010;41(3):358–65.
72. Bonte D, Lindvall C, Liu H, Dykema K, Furge K, Weinreich M. Cdc7-Dbf4 kinase overexpression in multiple cancers and tumor cell lines is correlated with p53 inactivation. Neoplasia. 2008;10(9):920–31.
73. Montagnoli A, Moll J, Colotta F. Targeting cell division cycle 7 kinase: a new approach for cancer therapy. Clin Cancer Res. 2010;16(18):4503–8.
74. Vanotti E, Amici R, Bargiotti A, Berthelsen J, Bosotti R, Ciavolella A, et al. Cdc7 kinase inhibitors: pyrrolopyridinones as potential antitumor agents. 1. Synthesis and structure-activity relationships. J Med Chem. 2008;51(3):487–501.
75. Montagnoli A, Valsasina B, Croci V, Menichincheri M, Rainoldi S, Marchesi V, et al. A Cdc7 kinase inhibitor restricts initiation of DNA replication and has antitumor activity. Nat Chem Biol. 2008;4(6):357–65.
76. Sasi NK, Tiwari K, Soon FF, Bonte D, Wang T, Melcher K, et al. The potent Cdc7-Dbf4 (DDK) kinase inhibitor XL413 has limited activity in many cancer cell lines and discovery of potential new DDK inhibitor scaffolds. PLoS One. 2014;9(11):e113300.
77. Hoang ML, Leon RP, Pessoa-Brandao L, Hunt S, Raghuraman MK, Fangman WL, et al. Structural changes in Mcm5 protein bypass Cdc7-Dbf4 function and reduce replication origin efficiency in Saccharomyces cerevisiae. Mol Cell Biol. 2007;27(21):7594–602.
78. Fletcher RJ, Bishop BE, Leon RP, Sclafani RA, Ogata CM, Chen XS. The structure and function of MCM from archaeal M. Thermoautotrophicum. Nat Struct Biol. 2003;10(3):160–7.
79. Fletcher RJ, Chen XS. Biochemical activities of the BOB1 mutant in Methanobacterium thermoautotrophicum MCM. Biochemistry. 2006;45(2):462–7.

80. Miller JM, Arachea BT, Epling LB, Enemark EJ. Analysis of the crystal structure of an active MCM hexamer. Elife. 2014;3:e03433.
81. Brandao LN, Ferguson R, Santoro I, Jinks-Robertson S, Sclafani RA. The role of Dbf4-dependent protein kinase in DNA polymerase zeta-dependent mutagenesis in Saccharomyces cerevisiae. Genetics. 2014;197:1111–22.
82. Pessoa-Brandao L, Sclafani RA. CDC7/DBF4 functions in the translesion synthesis branch of the RAD6 epistasis group in Saccharomyces cerevisiae. Genetics. 2004;167(4):1597–610.
83. Sasanuma H, Hirota K, Fukuda T, Kakusho N, Kugou K, Kawasaki Y, et al. Cdc7-dependent phosphorylation of Mer2 facilitates initiation of yeast meiotic recombination. Genes Dev. 2008;22(3):398–410.
84. Sun J, Fernandez-Cid A, Riera A, Tognetti S, Yuan Z, Stillman B, et al. Structural and mechanistic insights into Mcm2-7 double-hexamer assembly and function. Genes Dev. 2014;28(20):2291–303.

Chapter 15
Roles of Sld2, Sld3, and Dpb11 in Replication Initiation

Karl-Uwe Reußwig, Dominik Boos, and Boris Pfander

Abstract Replication initiation in eukaryotes is subdivided into two phases—origin licensing and origin firing. These phases have opposite requirements for cyclin-dependent kinase (CDK) phosphorylation and are therefore restricted to different cell cycle phases: origin licensing occurs in the absence of CDK activity (G1 phase) and results in the formation of inactive precursors of the replicative DNA helicase at origins of replication. Origin firing occurs upon CDK activation in S phase and results in the activation of the replicative helicase.

Central to the control of origin firing is a three-partite protein complex, which was first identified in budding yeast, but is apparently conserved among eukaryotes, and consists of Sld3, Dpb11, and Sld2 (SDS complex). Both Sld3 and Sld2 bind to Dpb11 in a CDK-phosphorylation-dependent manner and together form the minimal set of CDK substrates required for origin firing. The SDS complex facilitates helicase activation by promoting the association of accessory helicase subunits (Cdc45 and GINS complex) to form the CMG helicase. Importantly, the SDS complex-mediated CMG formation is a central element to various aspects of replication control, such as the temporal replication program and the regulation of origin firing by the checkpoint.

Keywords DNA replication initiation • Origin firing • Cell cycle regulation • Temporal program of replication • DNA damage checkpoint

K.-U. Reußwig • B. Pfander (✉)
DNA Replication and Genome Integrity, Max Planck Institute of Biochemistry, Martinsried, Munich, Germany
e-mail: bpfander@biochem.mpg.de

D. Boos
Vertebrate DNA Replication, Centre for Medical Biotechnology (ZMB), University of Duisburg-Essen, Essen, Germany

D.L. Kaplan (ed.), *The Initiation of DNA Replication in Eukaryotes*,
DOI 10.1007/978-3-319-24696-3_15

Introduction

To ensure that eukaryotic cells copy their genome precisely once during each cell cycle, the DNA replication process is divided into two phases—origin licensing and origin firing (Fig. 15.1). Both phases are coupled to distinct stages of the cell cycle by the changing activity of cyclin-dependent kinases (CDKs).

Origins of replications—the sites of replication initiation—are determined by binding of the origin recognition complex (ORC). During the origin licensing phase, ORC cooperates with the licensing factors Cdc6 and Cdt1 to facilitate the loading of the Mcm2-7 helicase core onto the DNA [1]. At this stage, these pre-replicative complexes (pre-RCs) contain Mcm2-7 in an inactive double-hexamer conformation.

During the origin firing phase, the missing helicase subunits Cdc45 and GINS are recruited to the pre-RCs to form the active Cdc45/Mcm2-7/GINS (CMG) replicative helicases [2–4]. This activation requires significant remodeling of the pre-RCs. In the process of CMG formation, the Mcm2-7 double hexamer splits into two single hexamers moving into opposite directions [2, 5–7]. Furthermore, Mcm2-7 switches from encircling double-stranded DNA in the pre-RC to encircling single-stranded DNA in the CMG helicase. Following the initial unwinding of DNA at the origin by CMG, additional proteins are recruited to the active CMG to form a replisome and commence DNA synthesis.

A strict separation of origin licensing and origin firing is essential to control the DNA replication process (Fig. 15.1). For this reason, multiple synergizing mechanisms inhibit origin licensing during the origin firing phase. In budding yeast, CDK phosphorylates every single licensing factor [8] and these phosphorylation events trigger the degradation and the nuclear export of licensing factors as well as the

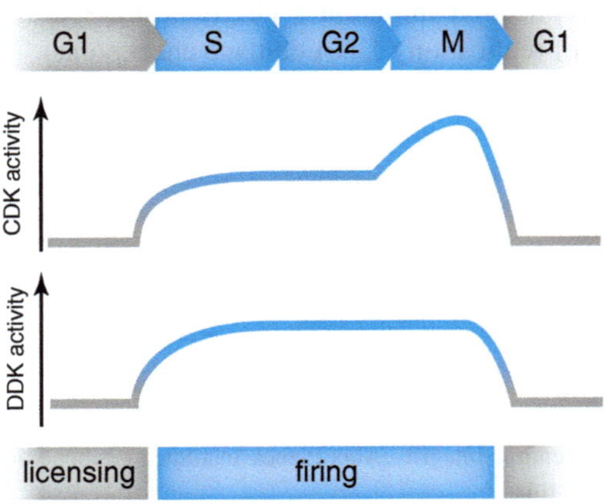

Fig. 15.1 *The cell cycle regulation of replication initiation*. Replication initiation is divided into two temporally separated phases: origin licensing and origin firing. Both phases are tightly linked to the cell cycle. Origin licensing occurs during late M phase and G1 phase when the activity of cyclin-dependent kinase (CDK) is low. Origin firing occurs from S phase to M phase (mitosis) when the activity of both CDK- and Dbf4-dependent kinase (DDK) is high

inhibition of their catalytic activity [9]. In consequence, these regulatory mechanisms lead to a restriction of origin licensing to late mitosis and G1 phase when CDK activity is low.

At the transition from G1 to S phase, the activities of CDK- and Dbf4-dependent kinase (DDK) rise and both are required throughout S phase to facilitate origin firing [10–13]. The major task of DDK is to alleviate an inhibitory interaction within the Mcm2-7 complex [14]. At the same time, CDK drives the conversion of pre-RCs to active CMG helicases. Pivotal to this conversion is the CDK-dependent formation of a protein complex (SDS complex) between the firing factors Sld3 and Sld2 and the scaffold protein Dpb11 [15–18]. The SDS complex is thought to join Cdc45, Mcm2-7, and GINS to initiate CMG formation and hence may be regarded as a key regulator of origin firing.

In this chapter, we provide a comprehensive overview of the latest advances in our knowledge of how the budding yeast SDS complex is regulated and how it mechanistically controls origin firing (sections "The SDS Complex in *Saccharomyces cerevisiae*" and "The SDS Complex is a Hub of Replication Initiation Control"). Furthermore, we discuss to what extent the molecular functions of these proteins are conserved in orthologous proteins in higher eukaryotes (section "Role of the SDS Complex in the CDK-Dependent Regulation of Origin Firing in Higher Eukaryotes").

The SDS Complex in *Saccharomyces cerevisiae*

The firing factors Sld3 and Sld2 were identified as the minimal set of CDK targets required for origin firing in the budding yeast *Saccharomyces cerevisiae*. Both proteins likely form a ternary complex together with the scaffold protein Dpb11 (SDS complex) upon CDK phosphorylation [15, 16] and this complex is thought to drive the conversion of inactive Mcm2-7 helicase precursors to active CMG helicases. Consistent with a role in DNA replication, Sld3, Dpb11, and Sld2 were found to associate with origin DNA by chromatin immunoprecipitation studies. However, none of these proteins was found to be incorporated into the replisome, thereby indicating that these proteins are specific origin firing factors [5, 19, 20].

Dpb11 Connects Sld3 and Sld2

The scaffold protein Dpb11 is an essential firing factor and was discovered as a genetic interactor of the leading strand DNA polymerase ε [21]. Dpb11 harbors four BRCA1 C-terminus (BRCT) domains, arranged as pairs of tandem BRCT domains [22]. Tandem BRCT domains are known to bind phosphorylated proteins [23] and screens with the temperature-sensitive *dpb11-1* allele identified *SLD3* and *SLD2* as genetic interactors and potential binding partners [24, 25].

Indeed, CDK-phosphorylated Sld3 binds to the BRCT 1+2 domain of Dpb11, whereas CDK-phosphorylated Sld2 binds to the BRCT 3+4 domain [15, 16, 26]. Thus, CDK phosphorylation may enable Dpb11 to connect Sld3 and Sld2 and also the protein complexes, in which they reside (Fig. 15.2).

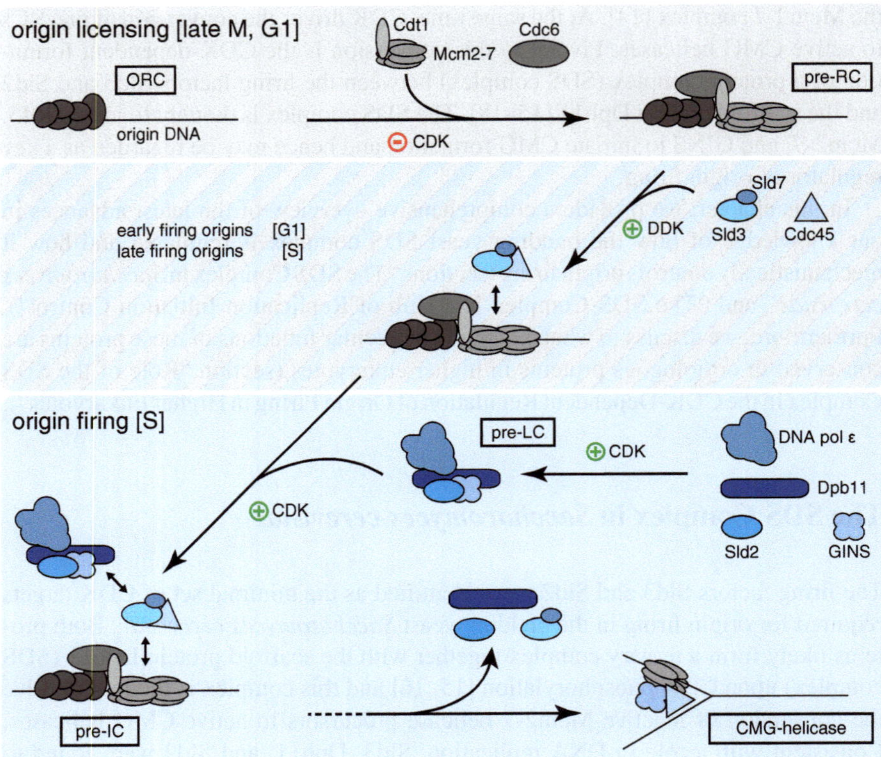

Fig. 15.2 *A mechanistic view of replication initiation*. During the origin licensing phase, the origin recognition complex (ORC) cooperates with the licensing factors Cdt1 and Cdc6 to load the helicase core Mcm2-7 to origin DNA in an inactive double-hexameric conformation and to form a pre-replicative complex at each origin (pre-RC). In a DDK-dependent step, the initiation factors Sld3-Sld7 together with the helicase component Cdc45 are recruited to the pre-RC. At early-firing origins, this recruitment takes place already during G1 phase, while at late-firing origins it does not occur before S phase. At the G1-S transition, the activation of CDK inhibits further origin licensing and promotes origin firing by facilitating the assembly of larger complexes of replication initiation factors. Pre-loading complexes (pre-LC), which contain the helicase component GINS, form in a CDK-dependent manner. Moreover, CDK also promotes the binding of the pre-LC to Cdc45-Sld3-Sld7 to establish pre-initiation complexes (pre-IC) at origins of replication. Critical for the formation of the pre-IC are the CDK-dependent interactions of Sld3 and Sld2 with Dpb11 (SDS complex). This complex comprises the minimal set of CDK targets that is required to trigger origin firing. Once all helicase components are recruited to the origin of replication, the Cdc45-Mcm2-7-GINS helicase (CMG helicase) forms by a yet unknown mechanism and initiates DNA unwinding. At this stage, the subsequent assembly of the replisome (not depicted here) is initiated and the firing factors Sld3, Dpb11, and Sld2 are most likely released

Sld3 Recruits Cdc45 to Pre-replicative Complexes

Similar to Dpb11, Sld3 is required for cell viability. As illustrated in Fig. 15.2, the Sld3 protein forms a complex with the firing factor Sld7 [27] and with the essential helicase component Cdc45 [28]. Importantly, Sld3 and Cdc45 localize to a subset of origins of replication during G1 phase in a mutually dependent manner [28, 29], thereby suggesting that Sld3 may recruit Cdc45 to pre-RCs.

At the transition to S phase, CDK facilitates the binding of Sld3 to Dpb11 by phosphorylating the C-terminus of Sld3. Specifically, phosphorylation of serine 600 and serine 622 appears to generate a binding site for Dpb11 [15, 16]. Mutations of these two phosphorylation sites abolish Sld3 binding to Dpb11 and cells carrying such a phospho-deficient allele of Sld3 are not viable, presumably because they are deficient in origin firing [15, 16]. This interpretation is supported by experiments with a chimeric fusion protein, in which phospho-deficient Sld3 replaces the N-terminal Sld3-binding site of Dpb11. This Sld3-Dpb11ΔN fusion protein rescues cell viability and shows normal replication kinetics [16]. Similarly, the $CDC45^{JET1\text{-}1}$ allele bypasses the requirement for CDK phosphorylation of Sld3. The $\text{Cdc45}^{\text{Jet1-1}}$ mutant is thought to foster the interaction between Sld3 and Dpb11, probably through its interaction with Sld3 [15].

Interestingly, DDK regulation of origin firing also impinges on Sld3. The DDK subunit Dbf4 is required for the recruitment of Sld3 to origins of replication in vivo [29, 30]. Consistently, active DDK promotes the recruitment of Sld3 and Cdc45 to pre-RCs in an in vitro assay for replication initiation [18, 31, 32]. A recent study has suggested that Sld3 itself is a target of DDK phosphorylation [33]. Even though the implications of the DDK phosphorylation of Sld3 are unclear, these findings indicate that Sld3 integrates both CDK and DDK signaling during replication initiation (Figs. 15.2 and 15.3a).

Sld2 Is Part of CDK-Dependent Pre-loading Complexes

At the onset of S phase, CDK phosphorylates not only Sld3 but also Sld2 at multiple sites and thereby stimulates the interaction between Dpb11 and Sld2 [17, 26]. Sld2 is a preferential target of S-phase CDK (Clb5-Cdc28 in budding yeast) [34] and in the current view the first phosphorylation events in the N-terminus of Sld2 lead to conformational changes. These conformational changes in turn allow the phosphorylation of threonine 84, which ultimately promotes the interaction with Dpb11 [26].

CDK phosphorylation does not only modulate the interaction between Dpb11 and Sld2 but also enables the formation of a pre-loading complex (pre-LC). The pre-LC consists of Sld2, Dpb11, GINS, and the leading strand DNA polymerase ε and is a transient, soluble protein complex [35]. In the current model, the pre-LC is proposed to drive CMG formation by delivering GINS to pre-RCs via the interaction of Dpb11 with CDK-phosphorylated, chromatin-localized Sld3-Cdc45 [35, 36].

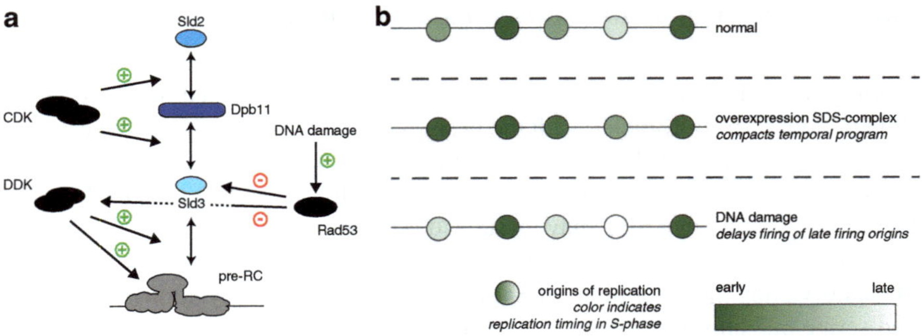

Fig. 15.3 *Regulation of origin firing by the SDS complex.* (**a**) The Sld3-Dpb11-Sld2 complex (SDS complex) plays a pivotal role during origin firing and is tightly regulated. The mechanistically important association of Sld3 with pre-replicative complexes (pre-RCs) depends on DDK. Phosphorylation of Sld2 and Sld3 by CDK triggers binding to Dpb11 and most likely the assembly of the SDS complex. In the presence of DNA damage or replication stress, the checkpoint kinase Rad53 is activated and delays further origin firing. Rad53 phosphorylates and inhibits Sld3, in part by preventing its interaction with Dpb11. Additionally, Rad53 phosphorylates Dbf4 which leads to inhibition of DDK. (**b**) Firing of origins and thereby the temporal program of replication is subject to regulation by the SDS complex. On the one hand, Rad53 blocks the formation of the SDS complex in the presence of DNA damage or replication stress. Consequently, origin firing from late origins is delayed. On the other hand, the temporal replication program can be compacted by overexpression of Sld3, Dpb11, Sld2, and Dbf4. Under these conditions, late-firing origins initiate replication earlier in S phase. This finding supports the idea that formation of the SDS complex constitutes a bottleneck for origin firing, which is important to establish a temporal replication program

The SDS Complex Facilitates Replication Initiation

On a mechanistic level, the SDS complex appears to facilitate the recruitment of several "building blocks" for the replication machinery. The SDS complex recruits Cdc45 (through Sld3) and GINS (through Sld2) to pre-RCs and it is straightforward to speculate that the critical role of the SDS complex in origin firing is to integrate these subunits into the loaded, inactive Mcm2-7 complex to form the active CMG helicase (Fig. 15.2). However, it is currently still unclear how the inactive helicase precursors are converted into active CMG helicases and how the SDS complex is released after replisome formation.

Genetic and biochemical experiments in budding yeast including in vitro reconstitution of replication initiation using cell extracts allowed scrutinizing the role of the SDS complex during the initiation process, and helped to develop a model of origin firing (Fig. 15.2): In the absence of kinase activity, pre-RCs are assembled at origins of replication and serve as precursors for replication [31, 32, 37]. DDK activity enables the recruitment of Sld3 and Cdc45 to the pre-RCs [18, 31, 32]. In a subsequent step that depends on CDK phosphorylation, Sld2 first engages in transient pre-LCs together with Dpb11, GINS, and DNA polymerase ε [35] and these pre-LCs are then recruited to CDK-phosphorylated Sld3 at pre-RCs [18, 31, 32].

This large assembly of proteins at origins of replication is often referred to as pre-initiation complexes (pre-ICs). The molecular architecture of the pre-IC is not well understood in detail, for example the GINS complex binds not only to Sld2, but also to Sld3 and Dpb11 in yeast-two hybrid experiments [38, 39]. Once the pre-IC is formed, all required helicase subunits are recruited to the origin but the complex needs to be remodeled to form the active CMG helicase and initiate DNA unwinding. How the SDS complex may contribute to this remodeling is currently unclear.

A recently established in vitro system of replication initiation using purified proteins provided further insights into the molecular role of the SDS complex [36]. Besides precisely defining the set of proteins required for replication initiation in vitro, this study sheds light on the regulation by CDK and DDK. The study demonstrates that recruitment of Sld3/Cdc45 to the inactive Mcm2-7 double hexamer depends on DDK. Furthermore, this system provides evidence that CDK phosphorylation of Sld3 and Sld2 is necessary and sufficient for replication initiation, thereby emphasizing the pivotal position of the SDS complex in mediating the CDK regulation of this process.

Whereas our knowledge of the regulation of the SDS complexes increases, it cannot be excluded that the SDS complex has additional molecular functions, dependently or independently of its role in integrating Cdc45 and GINS into the Mcm2-7 helicase precursors to form CMG helicases. In line with this, Sld3, Dpb11, and Sld2 may have additional molecular functions although our current view of the SDS complex suggests that it mainly works through mediating protein–protein interactions. To this end, in vitro experiments with Sld3, Dpb11, and Sld2 revealed unexpected roles: These three proteins were demonstrated to bind to single-stranded origin DNA and to modulate the interactions between Cdc45, Mcm2-7, and GINS [40–42]. Interestingly, in vitro studies with the conserved domain of the human Sld2 homolog RecQL4 revealed similar interactions [43, 44]. These data indicate that Sld3, Dpb11, and Sld2 are not just protein scaffolds that facilitate protein–protein interactions, but may also be actively involved in remodeling the protein–DNA intermediates during the activation of the replicative DNA helicase.

The SDS Complex Is a Hub of Replication Initiation Control

The current data suggest that the SDS complex serves as a regulatory platform or hub mediating the activation of the replicative helicase, the committed step during origin firing (Fig. 15.3). In the following, we discuss several mechanisms controlling origin firing during a normal S phase and in the presence of DNA damage or replication impediments that directly impinge on the SDS complex. First, via Sld2 and Sld3 phosphorylation the SDS complex couples replication to the S phase of the cell cycle. Together with CDK-mediated licensing inhibition in S phase this is pivotal to prevent re-replication. Second, protein levels of the components of the SDS complex are low and appear to be rate limiting to origin firing, and this seems to be important to distinguish early- from late-firing origins and thus to enforce the temporal replication program. Third, the SDS complex is a phosphorylation target of kinases of the DNA damage checkpoint, which is critical for the inhibition of origin firing in the presence of DNA damage.

CDK Regulation of the SDS Complex Keeps Origin Firing Separate from Origin Licensing

The principal reason for the separation of origin licensing and origin firing phases is to ensure that each segment of the genome is replicated just once during each cell cycle and to avoid re-replication. Re-replication can arise if licensing and firing are allowed to occur simultaneously, because under these circumstances fired origins can be re-licensed and fire again. Re-replication leads to genome instability, due to over-amplification of parts of the genome and DNA breaks induced by replication fork collapse [45–48]. It is therefore imperative that the licensing and firing phases do not overlap. Critical for this regulation is that both the activation of firing and the inactivation of licensing are under control of the same kinase—CDK, which promotes firing but also turns off licensing. Given that firing eliminates pre-RCs from origins, re-firing of a fired origin cannot occur. A pivotal question is therefore how the cell achieves complete separation of licensing and firing at the G1-S transition (Fig. 15.1).

Two possible answers to this question have been suggested by previous studies: First, it was shown that the formation of the SDS complex is dependent on multi-site phosphorylation. For example, Sld2 needs to be phosphorylated by CDK at several residues in the N-terminus, before threonine 84—the critical residue for Dpb11 binding—becomes accessible for phosphorylation [26]. Multi-site phosphorylation might increase the threshold for CDK activation of firing, or introduce a delay in the response of the initiation machinery between CDK activation and origin firing. Indeed, mathematical modeling has suggested that such a delay of origin firing would be sufficient to achieve the licensing-firing separation [49].

Second, since CDK is controlled by different regulatory cyclin subunits, which are expressed at different phases of the cell cycle and confer substrate selectivity, it is possible that due to substrate specificity of different CDK complexes licensing may be inhibited before firing is activated. Indeed the licensing factor Cdc6 is phosphorylated, and thereby targeted for degradation, by Cln-CDK complexes in late G1 phase [50] before Sld2 and Sld3 are phosphorylated by Clb-CDK complexes in S phase [15–17]. Conversely, origin firing should not respond to Cln-CDK. In line with this, it has been shown that Sld2 comprises interaction motifs, which allow specific binding of S-phase Clb-CDK complexes, explaining why Sld2 is phosphorylated specifically in S phase [34].

No matter how exactly licensing and firing are separated at the level of individual CDK phosphorylation events, it is clear that the SDS complex is at the center of re-replication control. Even partial deregulation of the complex, for example using phosphomimetic mutations on Sld2 or overexpression of either Sld2 or Dpb11, leads to strongly enhanced GCR rates, indicative of genome instability and re-replication [48].

The SDS Complex Creates a Bottleneck to Limit Origin Firing

As all origins are licensed before S phase, one could expect that with the rise of CDK activity at the G1-S transition all origins would fire simultaneously but this is not the case. It has been shown in many eukaryotic model systems that firing of origins occurs throughout S phase [51, 52] and experiments in budding yeast have shown that CDK activity is required continuously during S phase [13]. While firing of an individual origin in an individual cell is largely stochastic, it appears to be determined by a firing probability [53]. The firing probabilities of individual origins determine a temporal program of replication, which can be measured over a cell population and is highly reproducible (discriminating early, late, and passively replicated dormant origins).

Two parameters are important in order to create such a temporal program of replication. First, a timing determinant/probability has to be established at individual origins. While the molecular nature of this timing determinant is still unclear, it has been shown that it is established during origin licensing [54] and that it largely depends on the chromatin context of the origin ([55], see also [52, 56]). Second, in order for the timing determinant to be translated into a temporal order of origin firing, a limiting step or bottleneck must exist during origin firing, which causes an origin with a high firing probability to fire early and an origin with a low firing probability to fire late [53, 57]. In other words, origins need to compete for a limiting factor(s) in order for the timing determinants/probabilities to become apparent. Overall this model may be underlying all eukaryotic replication programs; it can also explain the increase of firing probabilities with S-phase progression and allows mathematical modeling, which fits well with experimental data [53].

Mechanistically, a bottleneck for origin activation could arise from limiting replication proteins, which at a given time can only bind to a fraction of origins and need to be recycled in order to activate more than one origin. In yeast, the components of the SDS complex very likely constitute such limiting factors as they are all expressed at levels (<500 molecules per cell) lower than for example the ORC complex [29, 58]. Moreover, in contrast to other S-phase factors, the SDS complex is released from emerging replication forks and therefore could be easily recycled to be used at other origins [19, 20].

Intriguingly, as shown in Fig. 15.3b, simultaneous overexpression of Sld3, Dpb11, and Sld2 together with the DDK subunit Dbf4 is able to make late origins fire early in S phase [59], suggesting that indeed the SDS complex is part of a bottleneck, which limits origin firing in budding yeast. Under these conditions, the relative order, in which origins fired, appeared to be at least partly maintained suggesting that the bottleneck to origin firing was not completely alleviated.

Importantly however, the time window during which origins fired in these experiments was much shorter and firing too many origins simultaneously gave rise to replication stress, presumably because too many replication forks were active at the same time [59]. Notably, an abrogation or compaction of the temporal program of replication was also observed using overexpression of slightly different combinations of firing factors such as Sld3 + Sld7 + Cdc45, just Sld3 + Sld7, or Cdc7 + Dbf4 [48]. Moreover, consistent with these results, overexpression of Sld3, Dpb11, Sld2, and Cdc45 was key to establish efficient in vitro replication reactions using yeast extracts [18, 31, 32]. Overall, these data suggest that the activation of the replicative helicase, facilitated by the SDS complex and DDK, forms the bottleneck to replication initiation, but that this bottleneck can be overcome in different ways. Future studies will need to address how different steps during the initiation reaction respond to the overexpression of different sets of factors to clarify these discrepancies.

The DNA Damage Checkpoint Targets the SDS Complex

Origin firing is not only regulated during normal S phase, but also in response to DNA damage, which inhibits origin firing. Notably, the SDS complex was found to play a key role also for this regulation. DNA damage or other S-phase problems, such as slow DNA synthesis in the presence of low dNTP levels (e.g., HU-treatment) or after polymerase inhibition (e.g., aphidicolin treatment), lead to replication fork stalling, which not only interferes with the completion of DNA replication but also threatens genome stability. The cellular response to replication fork stalling is controlled by the DNA damage checkpoint, a signal transduction pathway that depends on several checkpoint kinases (Mec1, Tel1, Rad53, and Chk1 in yeast; ATR, ATM, Chk2, and Chk1 in mammalian cells) [60]. One universal checkpoint response is the inhibition of origin firing to prevent excessive replication of damaged DNA templates. Defects in this response were discovered as radio-resistant DNA synthesis (RDS) in mammalian cells or as defects in the block to origin firing in yeast [61–63]. It is thought that the inhibition of origin firing prevents excessive replication of damaged DNA templates.

The discovery of the SDS complex was critical for our understanding of the mechanism by which the checkpoint inhibits origin firing. In particular, it was found that the checkpoint kinase Rad53 extensively phosphorylates Sld3 as well as Dbf4, the regulatory subunit of DDK (Fig. 15.3a and [16, 64, 65]). Each of these phosphorylation events is sufficient to facilitate the block to origin firing, since only a combination of phospho-site mutants of Sld3 and Dbf4, but not mutation of Sld3 or Dbf4 individually, allows firing of late origins in the presence of HU (Fig. 15.3b). Sld3 phosphorylation by Rad53 was shown to inhibit two crucial physical interactions: binding to Dpb11 as well as binding to Cdc45 [58]. How phosphorylation of Dbf4 inhibits DDK is less clear.

Thus, the SDS complex is critical for regulating origin firing also in the presence of DNA damage, but some aspects of this regulation remain poorly understood: first, given that both CDK and DDK are required for origin firing, it is unclear why the checkpoint inhibits both essential replication kinase pathways—CDK and DDK—and if any negative consequences would arise if the checkpoint inhibited only one branch. Is this redundancy a safety mechanism preventing genome instability due to replication failures, highlighting the importance of firing inhibition in DNA damage conditions? In fact, it is unclear what phenotypes arise when cells are unable to block origin firing in response to DNA damage: the published studies disagree on whether mutants that abolish the block are sensitive to HU treatment [58, 64, 65] and so far it has not been investigated whether such mutants show signs of genome instability such as elevated GCR rates.

Despite such open questions, the importance of the SDS complex for the regulation of replication is clear. Interestingly, evidence is emerging that a similar complex is formed during S phase of vertebrate cells, which appears to control replication initiation in a similar way.

Role of the SDS Complex in the CDK-Dependent Regulation of Origin Firing in Higher Eukaryotes

CDK-dependent associations of Sld3 and Sld2 with Dpb11 facilitate origin firing in yeast. This is in part conserved in their higher eukaryotic orthologues Treslin/TICRR, RecQL4, and TopBP1 highlighting that important lessons can be learned from yeast for unraveling principles of replication initiation in humans. However, there seem to be also differences between lower and higher eukaryotic SDS complex-related initiation mechanisms, perhaps accounting for the requirement of more complex regulations in the more complex higher eukaryotic cells (Fig. 15.4). For example, higher eukaryotes appear to have evolved a more sophisticated replication timing program than yeast, the microscopically discernible temporal replication program [70–73]. Since the SDS complex is a known regulation hub for replication initiation in yeast (see above) it is plausible that the appropriate spatial and temporal control of the vertebrate SDS complex function may account for many of these higher eukaryotic regulations of initiation in addition to their partially established roles in conserved core SDS functions in initiation.

The members of the SDS complex are conserved between higher eukaryotes and yeast. Dpb11 and its vertebrate orthologue TopBP1 have conserved phospho-binding BRCT domains [74], but TopBP1 has acquired BRCT repeats that are not present in Dpb11. Sld3 and Sld2 share low but significant levels of sequence conservation with their vertebrate counterparts Treslin and RecQL4 (see below), which is why these proteins long escaped identification.

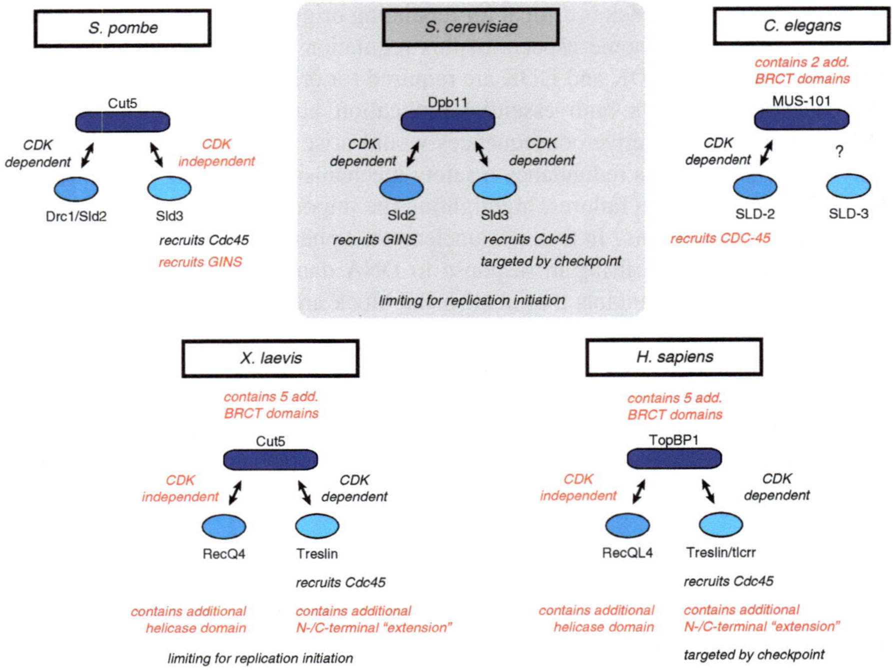

Fig. 15.4 *Conservation of the SDS complex in different organisms*. Orthologues of Sld3, Dpb11, and Sld2 have been discovered in most fungi and metazoans. Despite the low level of sequence conservation, particularly in Sld3 and Sld2 and their orthologues, the SDS complex seems to have retained its principal function in recruiting the missing helicase subunits Cdc45 and GINS to the pre-RCs throughout evolution. However, the regulation of specific protein–protein interactions has diverged. The fission yeast SDS complex seems to form very similarly to the budding yeast equivalent, but the interaction between Sld3 and Cut5 (Dpb11) is independent of CDK even though the CDK phosphorylation sites are conserved [66–68]. Conversely, Sld2 orthologues from frog and human bind to the corresponding Dpb11 orthologues in a CDK-independent manner [69]. Additionally, the metazoan orthologues of all proteins harbor additional domains. For example, RecQL4 comprises a helicase domain, Treslin has large N- and C-terminal extensions, and TopBP1 contains additional BRCT repeats. These additional domains may play important roles in the regulation of the metazoan-specific temporal replication program or have independent functions

Treslin, Vertebrate Sld3, Is Essential for CDK-Dependent Origin Firing

The most recently identified member of the vertebrate SDS complex is Treslin—an essential replication factor [75, 76]. It was shown to be central to mediate CDK dependence of initiation of replication, just like Sld3 in *S. cerevisiae* [77, 78]. After phosphorylation by CDK, Treslin is able to associate with the phosphoprotein-binding BRCT0+1+2 domain of TopBP1 in vitro and in vivo and the CDK phosphorylation sites are essential for replication in cultured human cells. Moreover, thorough protein sequence analysis of Treslin revealed low but significant homology with Sld3. This includes the central Sld3-Treslin domain of Treslin and the region containing the CDK sites essential for TopBP1 binding [77, 79].

In addition to a role in facilitating origin firing, Treslin seems to have also retained through evolution the role of Sld3 as a target of the intra S-phase checkpoint [77, 80]. Although the functional significance of this regulation remains open, this suggests that Treslin might have conserved the function of mediating intra S-phase checkpoint-dependent suppression of origin firing upon DNA damage. Taken together, these studies firmly established that Treslin is the metazoan orthologue of Sld3 and that Treslin and TopBP1 have retained their central role in the control of initiation of DNA replication.

However, Treslin also shows several features that are not conserved with yeast Sld3 making Treslin a good candidate for mediating metazoa-specific regulations of replication initiation. Its C- and N-terminus are specific to metazoans. The C-terminus was recently implicated in mediating Treslin inhibition by the checkpoint kinase Chk1 [80]. Since Chk1 mediates the suppression of origin firing upon DNA damage, this interaction might well reflect the metazoan version of Sld3 inhibition by Rad53 upon DNA damage described in yeast [58]. However, Chk1 was also implicated in regulating origin firing in unperturbed S phases [81–83]. Hence, Treslin inhibition by Chk1 might also play a role in the proper regulation of origin firing during a normal cell cycle and a function in the temporal replication program seems plausible [80, 81].

Interestingly, two additional metazoa-specific replication initiation factors, GEMC1 and Due-B, were found to interact with TopBP1 [84, 85]. The precise function of these factors is currently unclear, but they highlight that changes in the composition and the regulation of the SDS complex may have occurred during evolution.

RecQL4, Vertebrate Sld2, Interacts with TopBP1 Independently of Phosphorylation by CDK

Consistent with RecQL4 being a functional orthologue of yeast Sld2 is the finding that RecQL4 is essential for DNA replication [69, 86]. Rather surprising was therefore that its regulation seems to have diverged from Sld2 during the evolution of vertebrates: the phosphorylation-dependent regulation of the interaction with TopBP1 was probably lost [69], although the tandem BRCT domain of yeast Dpb11 that is responsible for Sld2 binding is conserved in TopBP1 (BRCT 4+5) [74]. In line with this, the BRCT4+5 domain was not required for TopBP1 function in replication initiation in *Xenopus* egg extracts [75] and the CDK sites in Sld2 that mediate Dpb11 interaction [15, 16] are not conserved in RecQL4. Taken together, these findings indicate that RecQL4 regulation has changed during evolution and that vertebrate RecQL4 does not facilitate CDK dependence of origin firing, at least not through TopBP1 binding. However, it remains to be tested if these findings from *Xenopus* oocytes are transferable to the mammalian somatic cell system.

A phosphorylation-independent interaction of RecQL4 with TopBP1 was reported [69]. An attractive yet unexplored hypothesis is that RecQL4 facilitates origin firing through binding to the BRCT3 domain of TopBP1. BRCT3 does not have a counterpart in Dpb11 but is essential for replication [75]. It lacks signature amino acids for phospho-peptide binding [87] and is, thus, likely a protein domain mediating phosphorylation-independent interactions. Although it has not been addressed directly whether the phosphorylation-independent interaction of RecQL4 with TopBP1 is required for initiation, it seems suggestive that the N-terminus of RecQL4, which is the region with homology to Sld2, was sufficient to mediate TopBP1 binding and to support replication in *Xenopus* egg extracts [69].

It needs to be pointed out that phosphorylation-independent RecQL4 binding to TopBP1 might be a particularity of vertebrates since the lower metazoan *C. elegans* was recently shown to require CDK-mediated association of Sld2 with Mus101, *C. elegans* Dpb11, for replication [88].

Thus, although the principle of CDK-dependent initiation of replication through SDS complex proteins is conserved through eukaryotic evolution, the exact molecular mechanisms of CDK regulation seem to be more rapidly evolving. The redundancy of CDK-dependent activation of Sld2 and Sld3 functions in initiation may allow this rapid evolution to occur.

The Vertebrate SDS Complex Is a Regulatory Hub for Origin Firing

Budding yeast uses two CDK targets—Sld2 and Sld3—to control origin firing, but vertebrates seem to have lost CDK regulation of RecQL4. This indicates that regulation by two CDK substrates may not be an essential principle of eukaryotic replication initiation (Fig. 15.4). Thus, it is unclear what advantage the regulation of origin firing by two CDK substrates provides. Theoretically, having two essential CDK substrates gives the cell more opportunities to impinge on CDK control of initiation. One possibility is that having two CDK substrates could be important for the separation of origin licensing and origin firing. Alternatively or additionally, cells may carefully control in space and time CDK-dependent origin firing processes to facilitate such important regulations as the response to DNA damage, the temporal replication program, or as-yet uncharacterized regulations to integrate origin firing into its chromatin environment.

CDK control of origin firing ensures that replication licensing and firing become temporally separated to avoid re-replication The multiple CDK phosphorylation sites on Sld3 and Sld2 may result in a delay of origin firing, creating a time gap between licensing and initiation that is sufficiently wide to inhibit licensing before activating origin firing [49]. Potentially, using two CDK substrates, each a target of multi-site phosphorylation, could be critical to create the specific response kinetics of the initiation machinery to CDK activity levels required to achieve appropriate timing of origin firing.

In light of these pivotal aspects of the regulation of origin firing, it seems counterintuitive that vertebrates have lost CDK regulation of the RecQL4-TopBP1 interaction. Has the recently reported CDK-dependent interaction between RecQL4 and Mcm10-Ctf4/And-1 taken over CDK control of Sld2 [89], or has another initiation factor adopted the role of Sld2 to sense CDK activity levels? It can also be speculated that Treslin, which has about 70 CDK consensus phosphorylation sites, might itself be sufficient to mediate the appropriate response of the initiation machinery to changes in CDK activity.

Alternatively, vertebrate cells might not depend on a finely tuned delay of initiation via multi-site phosphorylation of two CDK substrates to separate origin firing from licensing. Vertebrates may have instead evolved ways of inhibiting licensing sufficiently long before the G1-S transition to create the required time gap. It was for example shown that the licensing factor Cdc6 is inactivated by degradation via APC^{Cdh1}, which is active in G1 phase [90]. Alternatively, CDK-dependent licensing inhibition may have simply become more sensitive to CDK levels than origin firing.

A temporal replication program exists also in metazoans suggesting that, similar to yeast, the SDS complex may also be limiting for replication initiation in metazoans. A particularly interesting case is DNA replication during the early embryonic cell cycles in *Xenopus laevis* eggs. S phase as well as the overall cell cycle are very short at this stage of development and occur in the absence of transcription. Therefore, the amount of replication factors per nucleus constantly decreases during cell divisions. As in yeast, Treslin, Cut5/TopBP1, RecQL4, and Drf1/Dbf4 have been shown to be limiting factors for origin firing in this system [91]. These four firing factors were found to be important for timing the switch from the extremely quick replication in early embryos to slower replication in late embryos (during the mid-blastula transition). Artificially increased concentrations of the four firing factors delayed this transition [91]. These findings strongly suggest that also in metazoans the SDS complex is preserved as a bottleneck to control origin firing and, hence, that the vertebrate SDS complex has remained a central point of the temporal control of origin firing.

Insights into the Molecular Mechanisms of Treslin and RecQL4 in Vertebrate Initiation

Sld3 in yeast facilitates replication initiation most likely by recruiting Cdc45 to pre-RCs in a DDK-dependent manner. Cdc45 is subsequently integrated into the active CMG helicase, a process dependent on CDK-mediated interaction of Sld3 with Dpb11. Whether Treslin also guides Cdc45 to pre-RCs and whether this is controlled by DDK are important issues to be addressed.

Sld3 works in a constitutive complex with Sld7 [27], for which so far no homologue has been identified in higher eukaryotes. Treslin was found to constitutively interact with the MTBP protein (Mdm2-binding protein) [92], which has no homologue in yeast. The metazoa-specific M-region of Treslin, which mediates association with MTBP, proved essential for replication in mammalian cells and MTBP was required for replication initiation. Therefore it appears possible that MTBP fulfills the function of yeast Sld7 in vertebrates.

Yeast Sld2 facilitates CMG formation as part of the pre-LC recruiting GINS to pre-RCs via CDK-dependent interaction of Sld2 with Dpb11. RecQL4 was reported to form a complex with the CMG helicase, Mcm10, Ctf4/And-1, and other replisome proteins in human cells rather than the human equivalents of the pre-RC [89, 93]: As mentioned, the role of RecQL4 interaction with the Dpb11 orthologue TopBP1 for origin firing is unclear. Thus, RecQL4 functions during replication have either changed or an additional function in replication elongation/at replisomes has been acquired during evolution.

Such a role of RecQL4 at progressing replisomes during elongation could explain the observed robust association with replisome components. It is tempting to speculate that the N-terminus of RecQL4, which shows conservation with Sld2, fulfills the essential Sld2-like role during origin firing. Consistent with this is that the N-terminus was sufficient to support replication in immuno-depleted *Xenopus* egg extracts [69] and genetically engineered human cells [94] and was sufficient for viability in RecQL4-knockout chicken cells [95] and in mice [96].

After facilitating initiation, RecQL4 may become integrated into the replisome where it may have functions dependent on its helicase activity, which is located C-terminally of the Sld2-like region. Although the helicase domain is nonessential for replication, vertebrate cells carrying RecQL4-N (lacking the helicase domain) were sensitive to genotoxic agents [94, 95]. It can therefore be speculated that the RecQL4 helicase domain is required to facilitate progression of replication forks encountering specific DNA lesions or sites that are otherwise difficult to replicate. In contrast to vertebrates, the helicase part of RecQL4 is essential for survival in some metazoans, for example in Drosophila [97, 98]. Whether this is due to a function in replication elongation, initiation, or another function remains to be determined.

The molecular functions of RecQL4 for origin firing remain somewhat obscure. It bound to origins of replication in synchronised cells that were approaching the G1-S boundary [89, 93, 99]. It was required for the formation of the CMG helicase in human cells [100] indicating that fewer forks form in the absence of RecQL4. In *Xenopus* egg extracts, immuno-depletion of RecQL4 did not directly interfere with formation of the CMG helicase but affected the recruitment of DNA polymerase α [69, 86]. The link of RecQL4 with DNA polymerase α became corroborated by the finding that the RecQL4 interactors Mcm10 and Ctf4-And1 are required for integrating DNA polymerase α into the replisome [101].

Thus, the present data on RecQL4 in origin firing, integrated in a coherent model, suggest that association of RecQL4 with replication origins might occur before S phase starts and might be required for the subsequent formation of active replisomes. Whether this involves directly assisting CMG formation or recruiting DNA polymerase α to the replisome, or both, remains to be seen.

Concluding Remarks

According to our current knowledge, the SDS complex is the key factor in the control of origin firing in eukaryotes. Not only is formation of the SDS complex an essential intermediate leading to the activation of the replicative helicase, but this step is also targeted by other cellular pathways, which regulate origin firing, such as the DNA damage checkpoint.

It is noteworthy, however, that on sequence level the SDS proteins, particularly Sld3 and Sld2, are less conserved than most other replication proteins. Currently, it is unclear if all eukaryotes possess orthologues of these proteins. This is perhaps not surprising, given that the functions of the SDS complex appear to be mostly regulatory and its components do not have a role as core replication proteins. In line with a relatively fast gene evolution, different eukaryotic organisms may have found partially distinct ways to achieve the appropriate regulation of origin firing. Future research should therefore be directed to reveal which regulatory principles of origin firing have been conserved in eukaryotic organisms and which have changed.

Acknowledgements We are grateful to Markus Räschle, Dirk Remus, and members of the Pfander and Boos labs for helpful discussion and critical reading of the manuscript. KUR is supported by a Boehringer Ingelheim Fonds PhD fellowship. Work in the Pfander lab is supported by the Max Planck Society (MPG) and the German Research Council (DFG). The authors declare that they have no competing financial interest.

References

1. Remus D, Diffley JFX. Eukaryotic DNA replication control: lock and load, then fire. Curr Opin Cell Biol. 2009;21(6):771–7.
2. Moyer SE, Lewis PW, Botchan MR. Isolation of the Cdc45/Mcm2-7/GINS (CMG) complex, a candidate for the eukaryotic DNA replication fork helicase. Proc Natl Acad Sci U S A. 2006;103(27):10236–41.
3. Ilves I, Petojevic T, Pesavento JJ, Botchan MR. Activation of the MCM2-7 helicase by association with Cdc45 and GINS proteins. Mol Cell. 2010;37(2):247–58.
4. Pacek M, Tutter AV, Kubota Y, Takisawa H, Walter JC. Localization of MCM2-7, Cdc45, and GINS to the site of DNA unwinding during eukaryotic DNA replication. Mol Cell. 2006;21(4):581–7.
5. Gambus A, Jones RC, Sanchez-Diaz A, Kanemaki M, van Deursen F, Edmondson RD, et al. GINS maintains association of Cdc45 with MCM in replisome progression complexes at eukaryotic DNA replication forks. Nat Cell Biol. 2006;8(4):358–66.
6. Yardimci H, Loveland AB, Habuchi S, van Oijen AM, Walter JC. Uncoupling of sister replisomes during eukaryotic DNA replication. Mol Cell. 2010;40(5):834–40.
7. Fu YV, Yardimci H, Long DT, Guainazzi A, Bermudez VP, Hurwitz J, et al. Selective bypass of a lagging strand roadblock by the eukaryotic replicative DNA helicase. Cell. 2011; 146(6):931–41.
8. Diffley JFX. Regulation of early events in chromosome replication. Curr Biol. 2004;14(18):R778–86.
9. Arias EE, Walter JC. Strength in numbers: preventing rereplication via multiple mechanisms in eukaryotic cells. Genes Dev. 2007;21(5):497–518.

10. Bousset K, Diffley JF. The Cdc7 protein kinase is required for origin firing during S phase. Genes Dev. 1998;12(4):480–90.
11. Donaldson AD, Fangman WL, Brewer BJ. Cdc7 is required throughout the yeast S phase to activate replication origins. Genes Dev. 1998;12(4):491–501.
12. Schwob E, Nasmyth K. CLB5 and CLB6, a new pair of B cyclins involved in DNA replication in *Saccharomyces cerevisiae*. Genes Dev. 1993;7(7A):1160–75.
13. Donaldson AD, Raghuraman MK, Friedman KL, Cross FR, Brewer BJ, Fangman WL. CLB5-dependent activation of late replication origins in *S. cerevisiae*. Mol Cell. 1998;2(2): 173–82.
14. Sheu Y-J, Stillman B. The Dbf4-Cdc7 kinase promotes S phase by alleviating an inhibitory activity in Mcm4. Nature. 2010;463(7277):113–7.
15. Tanaka S, Umemori T, Hirai K, Muramatsu S, Kamimura Y, Araki H. CDK-dependent phosphorylation of Sld2 and Sld3 initiates DNA replication in budding yeast. Nature. 2007;445(7125):328–32.
16. Zegerman P, Diffley JFX. Phosphorylation of Sld2 and Sld3 by cyclin-dependent kinases promotes DNA replication in budding yeast. Nature. 2007;445(7125):281–5.
17. Masumoto H, Muramatsu S, Kamimura Y, Araki H. S-Cdk-dependent phosphorylation of Sld2 essential for chromosomal DNA replication in budding yeast. Nature. 2002; 415(6872):651–5.
18. Heller RC, Kang S, Lam WM, Chen S, Chan CS, Bell SP. Eukaryotic origin-dependent DNA replication in vitro reveals sequential action of DDK and S-CDK kinases. Cell. 2011;146(1):80–91. Elsevier Inc.
19. Kanemaki M, Labib K. Distinct roles for Sld3 and GINS during establishment and progression of eukaryotic DNA replication forks. EMBO J. 2006;25(8):1753–63.
20. Masumoto H, Sugino A, Araki H. Dpb11 controls the association between DNA polymerases alpha and epsilon and the autonomously replicating sequence region of budding yeast. Mol Cell Biol. 2000;20(8):2809–17.
21. Araki H, Leem SH, Phongdara A, Sugino A. Dpb11, which interacts with DNA polymerase II(epsilon) in *Saccharomyces cerevisiae*, has a dual role in S-phase progression and at a cell cycle checkpoint. Proc Natl Acad Sci U S A. 1995;92(25):11791–5.
22. Garcia V, Furuya K, Carr AM. Identification and functional analysis of TopBP1 and its homologs. DNA Repair (Amst). 2005;4(11):1227–39.
23. Glover JNM, Williams RS, Lee MS. Interactions between BRCT repeats and phosphoproteins: tangled up in two. Trends Biochem Sci. 2004;29(11):579–85.
24. Kamimura Y, Masumoto H, Sugino A, Araki H. Sld2, which interacts with Dpb11 in *Saccharomyces cerevisiae*, is required for chromosomal DNA replication. Mol Cell Biol. 1998;18(10):6102–9.
25. Wang H, Elledge SJ. DRC1, DNA replication and checkpoint protein 1, functions with DPB11 to control DNA replication and the S-phase checkpoint in *Saccharomyces cerevisiae*. Proc Natl Acad Sci U S A. 1999;96(7):3824–9.
26. Tak Y-S, Tanaka Y, Endo S, Kamimura Y, Araki H. A CDK-catalysed regulatory phosphorylation for formation of the DNA replication complex Sld2-Dpb11. EMBO J. 2006;25(9): 1987–96.
27. Tanaka T, Umemori T, Endo S, Muramatsu S, Kanemaki M, Kamimura Y, et al. Sld7, an Sld3-associated protein required for efficient chromosomal DNA replication in budding yeast. EMBO J. 2011;30(10):2019–30.
28. Kamimura Y, Tak YS, Sugino A, Araki H. Sld3, which interacts with Cdc45 (Sld4), functions for chromosomal DNA replication in *Saccharomyces cerevisiae*. EMBO J. 2001;20(8): 2097–107.
29. Tanaka S, Nakato R, Katou Y, Shirahige K, Araki H. Origin association of Sld3, Sld7, and Cdc45 proteins is a key step for determination of origin-firing timing. Curr Biol. 2011;21(24):2055–63. Elsevier Ltd.
30. Natsume T, Müller CA, Katou Y, Retkute R, Gierliński M, Araki H, et al. Kinetochores coordinate pericentromeric cohesion and early DNA replication by Cdc7-Dbf4 kinase recruitment. Mol Cell. 2013;50(5):661–74.

31. On KF, Beuron F, Frith D, Snijders AP, Morris EP, Diffley JFX. Prereplicative complexes assembled in vitro support origin-dependent and independent DNA replication. EMBO J. 2014;33(6):605–20.
32. Gros J, Devbhandari S, Remus D. Origin plasticity during budding yeast DNA replication in vitro. EMBO J. 2014;33(6):621–36.
33. Mattarocci S, Shyian M, Lemmens L, Damay P, Altintas DM, Shi T, et al. Rif1 controls DNA replication timing in yeast through the PP1 phosphatase Glc7. Cell Rep. 2014;7(1): 62–9. The Authors.
34. Loog M, Morgan DO. Cyclin specificity in the phosphorylation of cyclin-dependent kinase substrates. Nature. 2005;434(7029):104–8.
35. Muramatsu S, Hirai K, Tak Y-S, Kamimura Y, Araki H. CDK-dependent complex formation between replication proteins Dpb11, Sld2, Pol (epsilon}, and GINS in budding yeast. Genes Dev. 2010;24(6):602–12.
36. Yeeles JTP, Deegan TD, Janska A, Early A, Diffley JFX. Regulated eukaryotic DNA replication origin firing with purified proteins. Nature. 2015;519:431–5. Nature Publishing Group.
37. Remus D, Beuron F, Tolun G, Griffith JD, Morris EP, Diffley JFX. Concerted loading of Mcm2-7 double hexamers around DNA during DNA replication origin licensing. Cell. 2009;139(4):719–30.
38. Takayama Y, Kamimura Y, Okawa M, Muramatsu S, Sugino A, Araki H. GINS, a novel multiprotein complex required for chromosomal DNA replication in budding yeast. Genes Dev. 2003;17(9):1153–65.
39. Tanaka S, Komeda Y, Umemori T, Kubota Y, Takisawa H, Araki H. Efficient initiation of DNA replication in eukaryotes requires Dpb11/TopBP1--GINS interaction. Mol Cell Biol. 2013;33:2614–22.
40. Bruck I, Kaplan DL. Origin single-stranded DNA releases Sld3 protein from the Mcm2-7 complex, allowing the GINS tetramer to bind the Mcm2-7 complex. J Biol Chem. 2011;286(21):18602–13.
41. Kanter DM, Kaplan DL. Sld2 binds to origin single-stranded DNA and stimulates DNA annealing. Nucleic Acids Res. 2011;39(7):2580–92.
42. Dhingra N, Bruck I, Smith S, Ning B, Kaplan DL. Dpb11 helps control assembly of the Cdc45-Mcm2-7-GINS replication fork helicase. J Biol Chem. 2015. doi:10.1074/jbc.M115.640383.
43. Ohlenschläger O, Kuhnert A, Schneider A, Haumann S, Bellstedt P, Keller H, et al. The N-terminus of the human RecQL4 helicase is a homeodomain-like DNA interaction motif. Nucleic Acids Res. 2012;40(17):8309–24.
44. Keller H, Kiosze K, Sachsenweger J, Haumann S, Ohlenschläger O, Nuutinen T, et al. The intrinsically disordered amino-terminal region of human RecQL4: multiple DNA-binding domains confer annealing, strand exchange and G4 DNA binding. Nucleic Acids Res. 2014;42(20):12614–27.
45. Nguyen VQ, Co C, Li JJ. Cyclin-dependent kinases prevent DNA re-replication through multiple mechanisms. Nature. 2001;411(6841):1068–73.
46. Hansen DV, Loktev AV, Ban KH, Jackson PK. Plk1 regulates activation of the anaphase promoting complex by phosphorylating and triggering SCF-TrCP-dependent destruction of the APC inhibitor Emi1. Mol Biol Cell. 2004;11:5623–34.
47. Green BM, Morreale RJ, Ozaydin B, Derisi JL, Li JJ. Genome-wide mapping of DNA synthesis in *Saccharomyces cerevisiae* reveals that mechanisms preventing reinitiation of DNA replication are not redundant. Mol Biol Cell. 2006;17(5):2401–14.
48. Tanaka S, Araki H. Multiple regulatory mechanisms to inhibit untimely initiation of DNA replication are important for stable genome maintenance. PLoS Genet. 2011;7(6):e1002136.
49. Brümmer A, Salazar C, Zinzalla V, Alberghina L, Höfer T. Mathematical modelling of DNA replication reveals a trade-off between coherence of origin activation and robustness against rereplication. PLoS Comput Biol. 2010;6(5):e1000783.
50. Drury LS, Perkins G, Diffley JF. The cyclin-dependent kinase Cdc28p regulates distinct modes of Cdc6p proteolysis during the budding yeast cell cycle. Curr Biol. 2000;10(5):231–40.

51. Renard-Guillet C, Kanoh Y, Shirahige K, Masai H. Temporal and spatial regulation of eukaryotic DNA replication: from regulated initiation to genome-scale timing program. Semin Cell Dev Biol. 2014;30:110–20.
52. Yoshida K, Poveda A, Pasero P. Time to be versatile: regulation of the replication timing program in budding yeast. J Mol Biol. 2013;425(23):4696–705. Elsevier Ltd.
53. Bechhoefer J, Rhind N. Replication timing and its emergence from stochastic processes. Trends Genet. 2012;28(8):374–81. Elsevier Ltd.
54. Raghuraman MK, Brewer BJ, Fangman WL. Cell cycle-dependent establishment of a late replication program. Science. 1997;276(5313):806–9.
55. Ferguson BM, Fangman WL. A position effect on the time of replication origin activation in yeast. Cell. 1992;68(2):333–9.
56. Aparicio OM. Location, location, location: it's all in the timing for replication origins. Genes Dev. 2013;27(2):117–28.
57. Rhind N. DNA replication timing: random thoughts about origin firing. Nat Cell Biol. 2006;8(12):1313–6.
58. Zegerman P, Diffley JFX. Checkpoint-dependent inhibition of DNA replication initiation by Sld3 and Dbf4 phosphorylation. Nature. 2010;467(7314):474–8.
59. Mantiero D, Mackenzie A, Donaldson A, Zegerman P. Limiting replication initiation factors execute the temporal programme of origin firing in budding yeast. EMBO J. 2011;30(23):4805–14. Nature Publishing Group.
60. Labib K, De Piccoli G. Surviving chromosome replication: the many roles of the S-phase checkpoint pathway. Philos Trans R Soc Lond B Biol Sci. 2011;366(1584):3554–61.
61. Painter RB, Young BR. Radiosensitivity in ataxia-telangiectasia: a new explanation. Proc Natl Acad Sci U S A. 1980;77(12):7315–7.
62. Shirahige K, Hori Y, Shiraishi K, Yamashita M, Takahashi K, Obuse C, et al. Regulation of DNA-replication origins during cell-cycle progression. Nature. 1998;395(6702):618–21.
63. Santocanale C, Diffley JF. A Mec1- and Rad53-dependent checkpoint controls late-firing origins of DNA replication. Nature. 1998;395(6702):615–8.
64. Lopez-Mosqueda J, Maas NL, Jonsson ZO, DeFazio-Eli LG, Wohlschlegel J, Toczyski DP. Damage-induced phosphorylation of Sld3 is important to block late origin firing. Nature. 2010;467(7314):479–83. doi:10.1038/nature09377. Nature Publishing Group.
65. Duch A, Palou G, Jonsson ZO, Palou R, Calvo E, Wohlschlegel J, et al. A Dbf4 mutant contributes to bypassing the Rad53-mediated block of origins of replication in response to genotoxic stress. J Biol Chem. 2011;286(4):2486–91.
66. Noguchi E, Shanahan P, Noguchi C, Russell P. CDK phosphorylation of Drc1 regulates DNA replication in fission yeast. Curr Biol. 2002;12(7):599–605.
67. Yabuuchi H, Yamada Y, Uchida T, Sunathvanichkul T, Nakagawa T, Masukata H. Ordered assembly of Sld3, GINS and Cdc45 is distinctly regulated by DDK and CDK for activation of replication origins. EMBO J. 2006;25(19):4663–74.
68. Fukuura M, Nagao K, Obuse C, Takahashi TS, Nakagawa T, Masukata H. CDK promotes interactions of Sld3 and Drc1 with Cut5 for initiation of DNA replication in fission yeast. Mol Biol Cell. 2011;22(14):2620–33.
69. Matsuno K, Kumano M, Kubota Y, Hashimoto Y, Takisawa H. The N-terminal noncatalytic region of *Xenopus* RecQ4 is required for chromatin binding of DNA polymerase alpha in the initiation of DNA replication. Mol Cell Biol. 2006;26(13):4843–52.
70. Berezney R, Dubey DD, Huberman JA. Heterogeneity of eukaryotic replicons, replicon clusters, and replication foci. Chromosoma. 2000;108(8):471–84.
71. Dimitrova DS, Gilbert DM. Temporally coordinated assembly and disassembly of replication factories in the absence of DNA synthesis. Nat Cell Biol. 2000;2(10):686–94.
72. Jackson DA, Pombo A. Replicon clusters are stable units of chromosome structure: evidence that nuclear organization contributes to the efficient activation and propagation of S phase in human cells. J Cell Biol. 1998;140(6):1285–95.
73. Leonhardt H, Rahn HP, Weinzierl P, Sporbert A, Cremer T, Zink D, et al. Dynamics of DNA replication factories in living cells. J Cell Biol. 2000;149(2):271–80.

74. Mäkiniemi M, Hillukkala T, Tuusa J, Reini K, Vaara M, Huang D, et al. BRCT domain-containing protein TopBP1 functions in DNA replication and damage response. J Biol Chem. 2001;276(32):30399–406.
75. Kumagai A, Shevchenko A, Shevchenko A, Dunphy WG. Treslin collaborates with TopBP1 in triggering the initiation of DNA replication. Cell. 2010;140(3):349–59.
76. Sansam CL, Cruz NM, Danielian PS, Amsterdam A, Lau ML, Hopkins N, et al. A vertebrate gene, ticrr, is an essential checkpoint and replication regulator. Genes Dev. 2010; 24(2):183–94.
77. Boos D, Sanchez-Pulido L, Rappas M, Pearl LH, Oliver AW, Ponting CP, et al. Regulation of DNA replication through Sld3-Dpb11 interaction is conserved from yeast to humans. Curr Biol. 2011;21(13):1152–7. Elsevier Ltd.
78. Kumagai A, Shevchenko A, Shevchenko A, Dunphy WG. Direct regulation of Treslin by cyclin-dependent kinase is essential for the onset of DNA replication. J Cell Biol. 2011;193(6):995–1007.
79. Sanchez-Pulido L, Diffley JFX, Ponting CP. Homology explains the functional similarities of Treslin/Ticrr and Sld3. Curr Biol. 2010;20(12):R509–10.
80. Guo C, Kumagai A, Schlacher K, Shevchenko A, Shevchenko A, Dunphy WG. Interaction of Chk1 with Treslin negatively regulates the initiation of chromosomal DNA replication. Mol Cell. 2014;57:492–505.
81. Maya-Mendoza A, Petermann E, Gillespie DAF, Caldecott KW, Jackson DA. Chk1 regulates the density of active replication origins during the vertebrate S phase. EMBO J. 2007; 26(11):2719–31.
82. Shechter D, Costanzo V, Gautier J. ATR and ATM regulate the timing of DNA replication origin firing. Nat Cell Biol. 2004;6(7):648–55.
83. Syljuåsen RG, Sørensen CS, Hansen LT, Fugger K, Lundin C, Johansson F, et al. Inhibition of human Chk1 causes increased initiation of DNA replication, phosphorylation of ATR targets, and DNA breakage. Mol Cell Biol. 2005;25(9):3553–62.
84. Balestrini A, Cosentino C, Errico A, Garner E, Costanzo V. GEMC1 is a TopBP1-interacting protein required for chromosomal DNA replication. Nat Cell Biol. 2010;12(5):484–91.
85. Chowdhury A, Liu G, Kemp M, Chen X, Katrangi N, Myers S, et al. The DNA unwinding element binding protein DUE-B interacts with Cdc45 in preinitiation complex formation. Mol Cell Biol. 2010;30(6):1495–507.
86. Sangrithi MN, Bernal JA, Madine M, Philpott A, Lee J, Dunphy WG, et al. Initiation of DNA replication requires the RECQL4 protein mutated in Rothmund-Thomson syndrome. Cell. 2005;121(6):887–98.
87. Rappas M, Oliver AW, Pearl LH. Structure and function of the Rad9-binding region of the DNA-damage checkpoint adaptor TopBP1. Nucleic Acids Res. 2011;39(1):313–24.
88. Gaggioli V, Zeiser E, Rivers D, Bradshaw CR, Ahringer J, Zegerman P. CDK phosphorylation of SLD-2 is required for replication initiation and germline development in *C. elegans*. J Cell Biol. 2014;204(4):507–22.
89. Im J-S, Park S-Y, Cho W-H, Bae S-H, Hurwitz J, Lee J-K. RecQL4 is required for the association of Mcm10 and Ctf4 with replication origins in human cells. Cell Cycle. 2015; 14:1001–9.
90. Petersen BO, Wagener C, Marinoni F, Kramer ER, Melixetian M, Lazzerini Denchi E, et al. Cell cycle- and cell growth-regulated proteolysis of mammalian CDC6 is dependent on APC-CDH1. Genes Dev. 2000;14(18):2330–43.
91. Collart C, Allen GE, Bradshaw CR, Smith JC, Zegerman P. Titration of four replication factors is essential for the *Xenopus laevis* midblastula transition. Science. 2013; 341(6148):893–6.
92. Boos D, Yekezare M, Diffley JFX. Identification of a heteromeric complex that promotes DNA replication origin firing in human cells. Science. 2013;340(6135):981–4.
93. Xu X, Rochette PJ, Feyissa EA, Su TV, Liu Y. MCM10 mediates RECQ4 association with MCM2-7 helicase complex during DNA replication. EMBO J. 2009;28(19):3005–14.

94. Kohzaki M, Chiourea M, Versini G, Adachi N, Takeda S, Gagos S, et al. The helicase domain and C-terminus of human RecQL4 facilitate replication elongation on DNA templates damaged by ionizing radiation. Carcinogenesis. 2012;33(6):1203–10.
95. Abe T, Yoshimura A, Hosono Y, Tada S, Seki M, Enomoto T. The N-terminal region of RECQL4 lacking the helicase domain is both essential and sufficient for the viability of vertebrate cells. Role of the N-terminal region of RECQL4 in cells. Biochim Biophys Acta. 2011;1813(3):473–9.
96. Mann MB, Hodges CA, Barnes E, Vogel H, Hassold TJ, Luo G. Defective sister-chromatid cohesion, aneuploidy and cancer predisposition in a mouse model of type II Rothmund-Thomson syndrome. Hum Mol Genet. 2005;14(6):813–25.
97. Capp C, Wu J, Hsieh T-S. Drosophila RecQ4 has a 3′-5′ DNA helicase activity that is essential for viability. J Biol Chem. 2009;284(45):30845–52.
98. Xu Y, Lei Z, Huang H, Dui W, Liang X, Ma J, et al. dRecQ4 is required for DNA synthesis and essential for cell proliferation in Drosophila. PLoS One. 2009;4(7):e6107.
99. Thangavel S, Mendoza-Maldonado R, Tissino E, Sidorova JM, Yin J, Wang W, et al. Human RECQ1 and RECQ4 helicases play distinct roles in DNA replication initiation. Mol Cell Biol. 2010;30(6):1382–96.
100. Im J-S, Ki S-H, Farina A, Jung D-S, Hurwitz J, Lee J-K. Assembly of the Cdc45-Mcm2-7-GINS complex in human cells requires the Ctf4/And-1, RecQL4, and Mcm10 proteins. Proc Natl Acad Sci U S A. 2009;106(37):15628–32.
101. Zhu W, Ukomadu C, Jha S, Senga T, Dhar SK, Wohlschlegel JA, et al. Mcm10 and And-1/CTF4 recruit DNA polymerase alpha to chromatin for initiation of DNA replication. Genes Dev. 2007;21(18):2288–99.

Chapter 16
The Role of Mcm10 in Replication Initiation

Ryan M. Baxley, Yee Mon Thu, and Anja-Katrin Bielinsky

Abstract Minichromosome maintenance protein 10 (Mcm10) is a conserved component of the eukaryotic DNA replication machinery. Mcm10 promotes the initiation of replication by facilitating DNA unwinding and origin firing. Although the molecular details of this action remain unclear, current data support a scaffolding role for Mcm10 via interactions with DNA and other protein partners. Mcm10 binds both single- and double-stranded DNA, as well as components of the CMG helicase complex, DNA polymerase-α, and Ctf4. Upon initiation, Mcm10 becomes part of the replisome, primarily mediating the initiation of Okazaki fragment synthesis, which involves DNA polymerase-α/primase and the replication clamp PCNA. Mcm10 likely contributes to the recruitment of both of these factors. Emerging concepts predict that steady-state levels of Mcm10 are tightly controlled to balance origin firing and fork progression. Investigations into the cellular requirements for Mcm10 have also revealed a key role in maintaining genome stability. Accordingly, it is not surprising that genetic alterations of *MCM10* are associated with cancer. Loss of Mcm10 function is a possible source of DNA damage, whereas overexpression of Mcm10 might serve to facilitate rapid DNA synthesis and proliferation. In this chapter, we provide a comprehensive review of the current literature describing Mcm10's role in replication initiation. Additionally, we consider how contributions to elongation and other potential functions may affect chromosomal integrity.

Keywords Mcm10 • DNA replication • Origin activation • Genome stability • Replication initiation • Replication elongation • CMG helicase

---

R.M. Baxley • Y.M. Thu • A.-K. Bielinsky (✉)
Biochemistry, Molecular Biology and Biophysics, University of Minnesota,
6-155 Jackson Hall, 321 Church Street SE, Minneapolis, MN 55455, USA
e-mail: baxle002@umn.edu; thuxx001@umn.edu; bieli003@umn.edu

D.L. Kaplan (ed.), *The Initiation of DNA Replication in Eukaryotes*,
DOI 10.1007/978-3-319-24696-3_16

Introduction to Mcm10

Discovery of Mcm10 in Eukaryotes

Replication of the eukaryotic genome requires the coordinated action of multiple protein complexes prior to and during DNA synthesis. One essential replication gene, *MCM10* (minichromosome maintenance protein 10), which was originally named *DNA43*, was identified in a *Saccharomyces cerevisiae* screen for temperature-sensitive mutations causing DNA synthesis defects [1]. Subsequent characterization of this allele confirmed the requirement for entry and completion of S phase [2]. Several years later, the same gene was isolated using a strategy previously developed to screen for replication factors based on the maintenance of mitotically stable plasmids, termed minichromosomes [3, 4]. Since then, this gene has been designated *MCM10*. Homologs of *MCM10*, which are unique to eukaryotes, have been described in multiple genomes including those of fission yeast (*S. pombe*), fruit flies (*D. melanogaster*), frogs (*X. laevis*), mice (*M. musculus*), and humans (*H. sapiens*) [5–9].

The comparison of Mcm10 homologs has revealed a protein structure that is conserved both architecturally and functionally. Mcm10 consists of three structural regions that contain discrete functional domains (Fig. 16.1). First, the N-terminal domain (NTD) carries a conserved coiled-coil (CC) motif implicated in protein oligomerization (Fig. 16.1) [11–13]. Second, the well-conserved internal domain (ID) contains multiples motifs that mediate DNA binding and protein–protein interactions (Fig. 16.1) [14–19]. Finally, the C-terminal domain (CTD) is unique to metazoa and functions as an additional platform for DNA and protein binding (Fig. 16.1) [13, 14, 18, 19]. Based on the lack of any catalytic domains, the current literature supports a nonenzymatic scaffolding role for Mcm10 in DNA replication [13, 18].

DNA Binding Properties of Mcm10

Mcm10 carries multiple surfaces, located within the ID and CTD, which facilitate DNA binding. The ID contains two motifs that cooperate to mediate DNA association: a central oligonucleotide/oligosaccharide-binding fold (OB-fold), which forms a canonical DNA-binding cleft, and a single CCCH zinc-finger (ZnF1) (Fig. 16.1) [6, 17, 20–22]. The CTD contains two zinc-chelating structures. The first, a CCCH zinc-finger (ZnF2), mediates DNA binding and the second, a CCCC zinc-ribbon (ZnR), shares homology with the minichromosome maintenance 2-7 (Mcm2-7) ZnR, but lacks a known function (Fig. 16.1) [14, 23, 24]. Although Mcm10 appears to have a modest preference for single-stranded DNA (ssDNA) versus double-stranded DNA (dsDNA), it neither displays any sequence specificity nor a strong preference for more complex substrates such as forks or bubbles [13, 19, 23, 25].

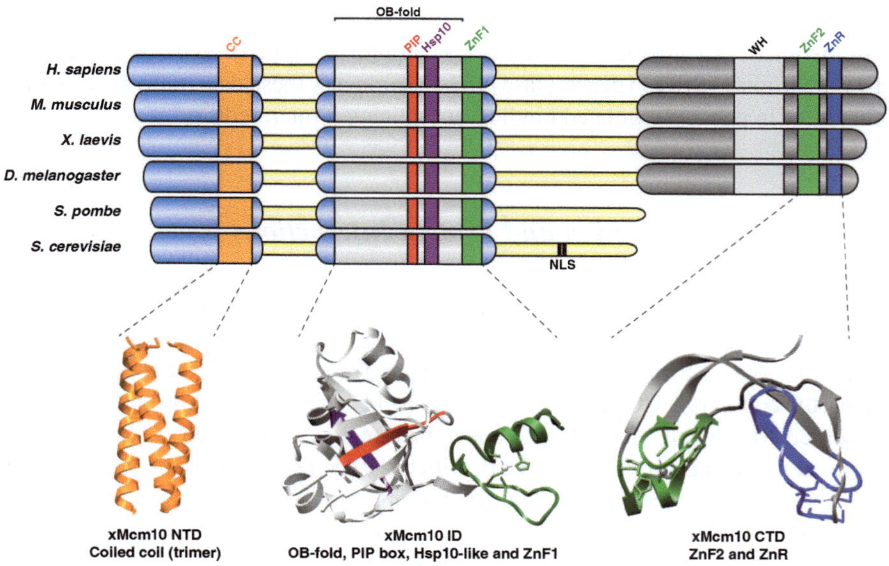

Fig. 16.1 *Structural domains of Mcm10*. Full-length Mcm10 is depicted for *H. sapiens* (874 aa), *M. musculus* (885 aa), *X. laevis* (860 aa), *D. melanogaster* (776 aa), *S. pombe* (593 aa), and *S. cerevisiae* (571 aa). The functional domains of Mcm10 across different species and corresponding crystal structures of *Xenopus* Mcm10 (xMcm10) are shown. The N-terminal domain (NTD) harbors a coiled-coil (CC, *orange*) motif responsible for Mcm10 self-interaction. The evolutionarily conserved internal domain (ID) mediates Mcm10 interactions with PCNA and Pol-α through a PCNA-interacting peptide (PIP) box (*red*) and Hsp10-like domain (*purple*), respectively. Both of these motifs reside in the oligonucleotide/oligosaccharide binding (OB)-fold (*gray*). The OB-fold along with zinc-finger motif 1 (ZnF1, *green*) serves as a DNA-binding domain. The C-terminal domain (CTD) is specific to metazoa and provides additional DNA- and Pol-α-binding regions. The CTD interacts with DNA primarily through ZnF2 (*green*). The metazoan CTD also includes the zinc ribbon (ZnR, *blue*) and winged helix motif (WH, *light gray*); however their functions are currently unknown. A nuclear localization sequence (NLS) has only been identified in *S. cerevisiae*. The crystal structures were generated using the following pdb data files 4JBZ (NTD), 3EBE (ID) and 2KWQ (CTD) and the Chimera program (http://www.cgl.ucsf.edu/chimera) [10]

However, it still remains unclear whether distinct DNA-binding domains confer substrate specificity. One in vitro analysis that disrupted the structure of ZnF1 adding a zinc-chelating competitor showed that ssDNA binding remained unaffected, whereas dsDNA association was significantly reduced [13]. In contrast, when the amino acids in ZnF1 that directly contact DNA were mutated, this resulted in reduced ssDNA association in vitro as well as decreased viability in vivo following replication stress [22]. Taken together, these data suggest that further investigation of the DNA binding properties may yield insight into potential DNA binding preferences of Mcm10, and this knowledge might be crucial to explain Mcm10's function. Regardless of substrate, the presence of both the CTD and ID increases the binding affinity of full-length Mcm10 by at least 100-fold in comparison to each domain alone [13]. Finally, the DNA association of human Mcm10 appears to be regulated by acetylation [14]. This posttranslational modification promotes ID binding to DNA, but disrupts DNA interaction with the CTD. In this system, acetylation is regulated by the Sirtuin 1 (SIRT1) deacetylase [14]. In summary, these data imply that Mcm10 functions in part via regulated binding of replication intermediates.

Mcm10 Oligomerization and Protein Interactions

Direct protein–protein interactions have been defined for several Mcm10 domains. Self-interaction is mediated by the NTD [11, 13]. Biochemical analyses of *Xenopus* Mcm10 revealed that protein dimer- and trimerization are facilitated by an evolutionarily conserved CC domain (Fig. 16.1) [11]. Oligomerization has also been observed for human Mcm10, as biochemical analyses suggested NTD-dependent formation of trimers or hexamers [23], with the latter being consistent with electron microscopy (EM) reconstructions used to propose a hexameric ring structure [21]. If Mcm10 homocomplexes indeed form a ring, then presumably it would require an accessory factor to facilitate loading onto DNA. To date, such a factor has not been identified. In addition, it is not clear that the high-resolution crystal structure of the *Xenopus* Mcm10 ID can be accurately docked into the EM reconstructions of the human Mcm10 hexamer [21–23]. Despite discrepancies regarding the exact nature of oligomerization, there is consensus that the NTD promotes Mcm10 self-interaction and that these interactions occur dynamically. Oligomerization appears to be functionally important, as in vitro characterization of *S. cerevisiae* Mcm10 suggested that ssDNA binding favors trimer formation [25]. In contrast, Mcm10 seems to bind dsDNA as a monomer [25]. Furthermore, a requirement for Mcm10 self-interaction following hydroxyurea-induced replication stress was revealed in the absence of a functional Rad9-Hus1-Rad1 (9-1-1) checkpoint clamp in yeast [26]. Overall, oligomerization of Mcm10 may be needed to increase the binding capacity for DNA and proteins at the replication fork.

Mcm10 is proposed to act as a scaffolding factor, suggesting that protein–protein interactions are central to its function. This notion is consistent with the finding that the Hsp10-like domain interacts with the DNA polymerase required for de novo DNA synthesis, DNA polymerase-α (Pol-α [16, 17, 19, 22]. The Hsp10 motif is located within the OB-fold, implying that DNA and Pol-α compete for a shared interaction surface (Fig. 16.1) [16, 17, 19, 22]. Notably, an additional Pol-α-binding site exists in the CTD [13]. The overlapping roles of the ID and CTD in DNA and Pol-α binding are in agreement with a proposed molecular hand-off mechanism, whereby one domain binds ssDNA while the other recruits Pol-α [19]. Moreover, the ID also mediates interactions with the Mcm2-7 replicative helicase core complex as well as the replication clamp and processivity factor, proliferating cell nuclear antigen (PCNA). Mcm10 binds via residues in the PCNA interacting peptide box (PIP box), which resides within the OB-fold [22, 27]. The location of the PIP box in Mcm10 is unusual, as other PCNA interacting proteins generally carry the domain close to the N- or C-terminus [27–30]. Disruption of the PIP box in *S. cerevisiae* results in lethality, suggesting a critical function for the Mcm10-PCNA interaction during replication [27]. Furthermore, the CTD was recently reported to directly interact with the replisome component cell division cycle 45 (Cdc45), although a specific motif has not been determined [23]. Finally, motifs within the first ~150 amino acids of the NTD of yeast Mcm10 directly bind to the interdomain loop of the mitosis entry checkpoint 3 (Mec3) subunit of the 9-1-1 checkpoint clamp, a key factor in the DNA damage response [26]. Taken together, these data point to important functional roles for Mcm10 protein interactions in unperturbed DNA replication and during replication stress.

The Role of Mcm10 in the Initiation of DNA Replication

General Introduction to Replication Initiation Events

The detailed processes involved in preparation for DNA synthesis have been extensively reviewed in previous chapters, and thus we present a brief description of steps involved in origin licensing and activation of the replicative helicase (Fig. 16.2). The pre-replication (pre-RC) complex assembly during G1 phase of the cell cycle involves the origin recognition complex (ORC), cell division cycle 6 (Cdc6), and Cdc10-dependent transcript 1 (Cdt1), all of which are necessary to subsequently load double hexamers of the Mcm2-7 core helicase onto dsDNA [18, 31–35]. In preparation for the initiation of S phase, two helicase co-activators, Cdc45 and go-ichi-ni-san (GINS), are recruited by the coordinated action of Dbf4-dependent kinase Cdc7 (DDK) and S-phase cyclin-dependent kinase (S-CDK) [32, 36–39]. DDK phosphorylation of the Mcm2-7 complex promotes its association with Cdc45, which is in complex with Sld3 (synthetically lethal with *dpb11* 3) and its binding partner Sld7 [36, 37, 40]. S-CDK phosphorylation of adaptor proteins Sld2 and Sld3 enables the docking of GINS in conjunction with DNA polymerase epsilon (Pol-ε) to assemble the pre-initiation complex [36–39]. The functional replicative helicase—the Cdc45-Mcm2-7-GINS (CMG) complex—is established when Cdc45-Mcm2-7 associates with GINS [41, 42]. The onset of S phase is defined by a conformational change that allows the twin CMG complexes to separate and rearrange to enclose the two leading strand templates (Fig. 16.2) [43–45]. The potential mechanisms underlying Mcm10's function in this process will be discussed in a later section of this chapter.

Evidence Suggesting That Mcm10 Contributes to Replication Initiation

The role of Mcm10 during S phase was first recognized in studies using budding yeast. *mcm10* mutants were unable to maintain minichromosomes and exhibited an S-phase progression defect, characteristics shared by other replication initiation mutants [4]. Subsequent studies have advanced our understanding of Mcm10's contribution to replication initiation. Mcm10 physically interacts with multiple proteins

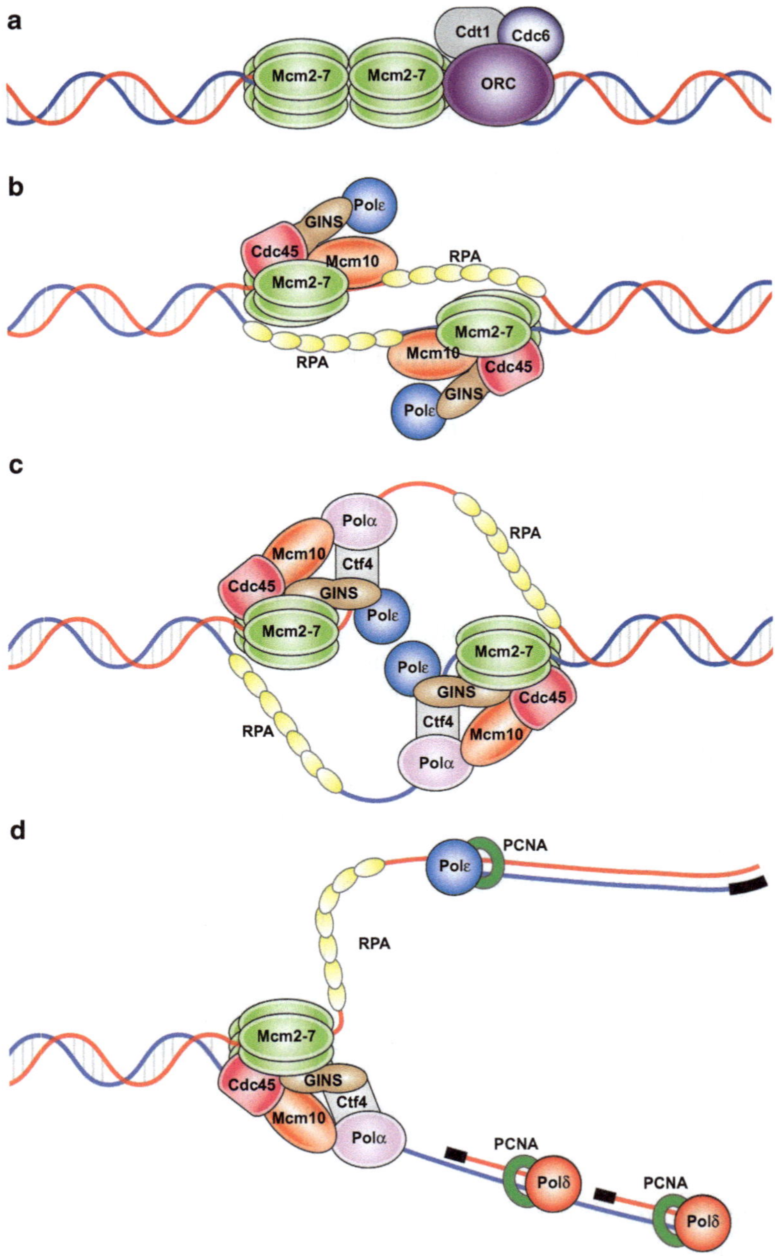

Fig. 16.2 *Roles of Mcm10 in replication initiation and DNA synthesis*. The stepwise assembly of replication complexes is illustrated. (**a**) In G1 phase, ORC, Cdc6, and Cdt1 collaborate to load double-hexameric Mcm2-7 complexes onto origins and this completes the formation of the pre-replication complex. This process is also referred to as origin licensing. (**b**) At the G1/S-phase transition, the functional Cdc45-Mcm2-7-GINS (CMG) helicase is assembled via the activities of DDK and S-CDK. During this step, Pol-ε is delivered in conjunction with GINS. In addition,

participating in initiation, such as subunits of GINS and the Mcm2-7 core helicase [4, 6, 7, 46–48], Cdc45 [6, 23, 49], ORC [6, 7, 46, 50], the Sld2 ortholog RecQ like helicase 4 (RecQl4) [51], as well as the ssDNA-binding complex replication protein A (RPA) [16]. In addition, *mcm10* also exhibits synthetic lethality with mutant alleles encoding various initiation factors such as Cdc45, Orc2, Orc5, Dpb11, Mcm2, 4, 5, 6, and 7, and subunits of Pol-ε and DNA polymerase delta (Pol-δ) [4, 46, 47, 50, 52]. Consistent with a role in replication initiation, oscillation of Mcm10 expression in human cells correlates with the cell cycle. Steady-state protein levels are elevated in G1 and S phase and remain low during mitosis, possibly to prevent re-replication [53–55]. Interestingly, the expression of Mcm10 appears to be co-regulated with other cell cycle proteins by a specific microRNA, miR-215 [56]. Furthermore, Mcm10 degradation has been proposed as a way for cells to inhibit replication in response to certain types of DNA damage [55, 57]. Altogether, these observations imply that Mcm10 participates in S phase, but they do not necessitate a role in origin firing. More direct evidence for Mcm10's contribution to origin activation comes from two-dimensional (2D) gel analyses, which examined replication bubbles indicative of initiation events [4]. These studies documented decreased origin usage in *mcm10* mutants compared to wild-type controls [4]. Similarly, under conditions in which Mcm10 was degraded just before the onset of S phase, a high percentage of cells failed to exit G1 phase and were unable to duplicate the genome [16, 58]. Finally, the timing and regulation of Mcm10 chromatin loading at origins also suggest a clear role in replication initiation. Mcm10 was initially thought to be constitutively associated with chromatin [47]. However, chromatin immunoprecipitation experiments revealed that Mcm10 loading was cell cycle regulated, and that it localized exclusively to replication origins during G1 phase in a Mcm2-7-dependent manner [16]. Independent studies also reported Mcm2-7-mediated chromatin binding of Mcm10 before the initiation of DNA synthesis [9, 59]. Thus, Mcm10 association with replication origins requires pre-RC assembly. However, whether Mcm10 loading occurs *before* or *after* formation of the CMG helicase (Fig. 16.3), and might therefore require DDK and S-CDK activities, remains an open question. The literature to date provides evidence to support both scenarios [9, 16, 49, 58, 60–65]. The timing of Mcm10 loading onto chromatin is crucial to understand whether Mcm10 actively promotes the formation of the CMG helicase [9, 49, 60, 62]. This important aspect will be discussed in the next section of this chapter. What is undisputed is that Mcm10 is necessary for replication initiation and associates with origins after pre-RC assembly.

Fig. 16.2 (continued) Mcm10 is also recruited to chromatin before replication initiation and facilitates origin unwinding by the CMG helicase. RPA stabilizes ssDNA. (**c**) Mcm10 recruits Pol-α to ssDNA together with Ctf4, which is loaded onto chromatin in S phase. Please note that Ctf4 is not strictly required for DNA synthesis in *S. cerevisiae*. (**d**) Mcm10 loading of Pol-α is repeatedly needed to generate RNA/DNA primers (black DNA regions) for Okazaki fragment synthesis. Processive DNA polymerization is executed by Pol-ε (extending the blue leading strand) and Pol-δ (extending the red lagging strand). The steps illustrated in this figure are primarily based on experimental evidence in budding yeast

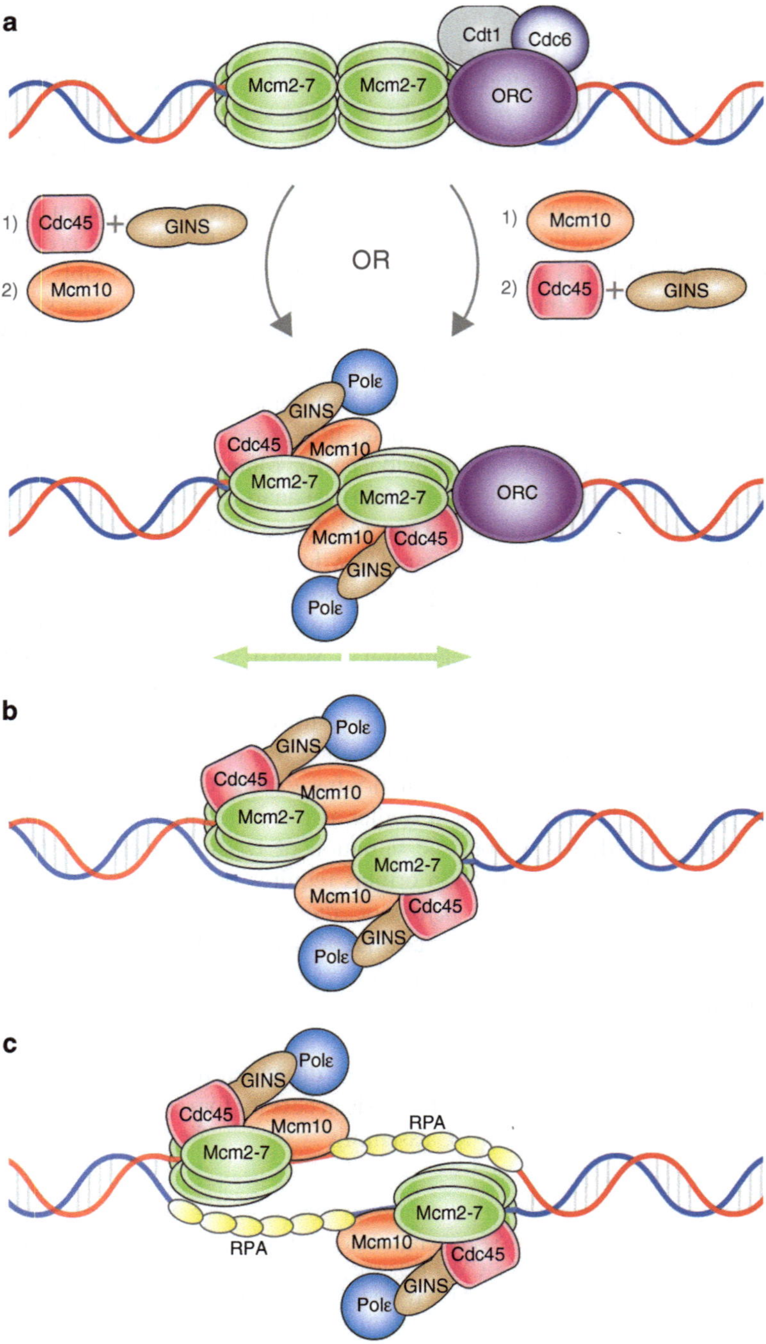

Fig. 16.3 *Proposed mechanism of CMG helicase activation by Mcm10.* (**a**) Pre-replication complex formation precedes the loading of Mcm10 at origins. However, the order of events after this step remains unclear: Mcm10 may be recruited *before* or *after* the helicase co-activators Cdc45 and

Mechanistic Insight into How Mcm10 Facilitates the Initiation of DNA Synthesis

How does Mcm10 stimulate replication initiation as a scaffold protein lacking any enzymatic activity? To answer this question, it is important to determine whether Mcm10 participates in the assembly of the CMG helicase. Indeed, several independent studies proposed that Mcm10 is required to recruit the helicase co-activator Cdc45 to origins of replication [9, 49, 60]. This notion is consistent with evidence that both Mcm10 and RecQl4 were necessary for the formation of the CMG helicase complex in human cells [62]. However, it has been challenging to clearly establish the sequence of events due to their transient nature. Contrary to the results mentioned above, three independent in vivo studies observed that Mcm10 was dispensable for the stable association of Cdc45 with chromatin [16, 61, 62]. This data is in agreement with in vitro experiments, which detected proper assembly of the CMG helicase in the absence of Mcm10 [61, 65]. One caveat is that the above studies that employed depletion methods may not have eliminated Mcm10 completely. It is possible that very little Mcm10 is required to promote Cdc45 loading in the living cell. In fact, this idea has been supported by quantitative measurements of Mcm10 and CMG subunits that stably associate with replication substrates in vitro [64]. Estimated levels of DNA-bound Mcm10 *before* and *after* formation of the CMG helicase were 10–100 times lower than those of several CMG components [64]. Given that CMG complexes are loaded onto DNA in excess [66–69], these observations culminate in a model in which Mcm10 functions as the limiting factor in origin activation. Although this model requires further validation, Mcm10 is clearly needed to assist the CMG helicase in unwinding dsDNA at the onset of S phase.

The first study supporting the notion that Mcm10 stimulates origin firing utilized *Xenopus* egg extract to demonstrate that the ability to unwind circular plasmid DNA was lost upon depletion of Mcm10 [9]. Furthermore, three independent reports showed decreased RPA association at origins when Mcm10 expression was ablated in vivo [58, 70, 71]. Similarly, a recent in vitro study elegantly demonstrated the requirement for Mcm10 to initiate origin unwinding and RPA recruitment [65]. Together, these observations strongly suggest that Mcm10 facilitates helicase activation. Simultaneously, they raise the intriguing question of how Mcm10 accomplishes this task. Two possible scenarios provide insight into this question:

Fig. 16.3 (continued) GINS. Pol-ε is the first polymerase loaded onto DNA via its association with GINS (not shown in the depiction of the two alternate scenarios). Replication initiation begins with the separation of double-hexameric CMG helicases in opposite directions (*green arrows*) and unwinding of dsDNA. (**b**) Subsequently, the CMG helicase encircles the single-stranded template. Mcm10 facilitates helicase activation by stabilizing ssDNA generated after unwinding of origins. At this stage, Mcm10 interacts with both the Mcm2-7 core helicase and ssDNA. (**c**) RPA displaces Mcm10 from DNA, because it has a much higher binding affinity to ssDNA than Mcm10. During this process, Mcm10 remains associated with the Mcm2-7 complex

(1) Mcm10 may play an active role in remodeling the CMG helicase, or (2) Mcm10 may play a more indirect part in DNA unwinding by stabilizing the separated template strands. A direct helicase activator function would predict that Mcm10 induces a conformational change of the CMG helicase to alter its association with DNA at replication origins [43, 44]. However, there is currently very limited experimental evidence to support this notion. One possibility is that Mcm10 promotes the phosphorylation of Mcm subunits by DDK to promote subsequent recruitment of Cdc45 and GINS [48].

An alternative explanation for reduced RPA association following Mcm10 depletion is that Mcm10 stabilizes ssDNA *after* the melting of parental DNA. This model is consistent with current structural information and Mcm10's ability to bind both ss- and dsDNA [12]. Mutations within the ZnF1 domain of Mcm10 reduced association of RPA with replication origins consistent with a defect in DNA unwinding [70]. One attractive model of how Mcm10 facilitates this first step of replication initiation is by engaging with the Mcm2-7 core helicase as it encircles the parental double strand and then transiently stabilizing ssDNA as the CMG double hexamer melts the origin DNA (Fig. 16.3). Mcm10 may be uniquely positioned to serve as a placeholder for RPA on ssDNA until a sufficiently large region is exposed to allow for stable RPA association [16, 65]. Subsequently, RPA may displace Mcm10 from the DNA template given that it has a 40-fold higher affinity for ssDNA than Mcm10 [15]. During this "on-and-off" DNA cycle, Mcm10 likely remains anchored with the Mcm2-7 core helicase (Fig. 16.3) [72].

CMG helicase activation allows for replication to proceed and Mcm10 continues to participate in later steps by facilitating recruitment of the Pol-α/primase complex [16, 61, 73, 74]. Pol-α/primase produces small RNA–DNA primers that are extended into leading and lagging strands by Pol-ε and Pol-δ, respectively [32, 75, 76]. In cooperation with the cohesion factor Ctf4, Mcm10 helps to anchor the Pol-α/primase complex to origin DNA, promoting the initiation of both leading and lagging strand synthesis [16, 59, 73]. In summary, the literature to date clearly suggests that Mcm10 facilitates helicase activation and promotes loading of DNA polymerases at the onset of S phase.

The Role of Mcm10 During DNA Synthesis

Mcm10 Is a Replisome Component at Progressing Replication Forks

The idea that Mcm10 contributes to replication elongation has been strengthened by its co-purification with other known replisome components. In *S. cerevisiae*, Mcm10 interacts with the CMG complex during the G1/S-phase transition [59]. In *Xenopus* egg extracts, Mcm10 has also been identified as a member of functional replisomes [77]. Significantly, independent laboratories have isolated Mcm10 together with other essential replisome components present in nascent chromatin [72, 78].

Interestingly, Mcm10 is slightly enriched on the leading strand of stalled replication forks, consistent with the idea that it is directly anchored to the Mcm2-7 complex, which encircles the leading strand template (Figs. 16.2 and 16.3). These findings are also in agreement with previous reports that showed Mcm10 in origin-flanking regions migrating away from replication origins, similar to the Mcm2-7 complex [16, 79, 80]. Together, these data support the model that Mcm10 travels with replication forks.

The Contribution of Mcm10 to Okazaki Fragment Initiation

Direct interactions between Mcm10 and multiple replication factors provide valuable insight into the molecular role of Mcm10 in lagging strand synthesis. Several groups have identified interactions between Mcm10 and the active CMG complex, including components of the Mcm2-7 complex and the helicase co-activator Cdc45 (Fig. 16.2) [6, 7, 23, 46, 47]. As Mcm10 travels with the progressing helicase, it facilitates loading of Pol-α onto chromatin [16, 61, 74]. In addition to replication initiation, the RNA–DNA primers synthesized by Pol-α/primase are critical for lagging strand synthesis [16, 32, 81]. The coupling of the CMG helicase and Pol-α is proposed to occur via the cooperative actions of Mcm10 and Ctf4 (Figs. 16.2 and 16.3) [59, 62, 73]. Ctf4 is essential in metazoa, but is not required for viability in budding yeast [82]. Consistent with a shared role in replication, Ctf4 knockdown in *Drosophila* phenocopies *Mcm10* mutant flies [83, 84]. Further, these proteins form a common complex in human cells and *Xenopus* egg extracts, with Mcm10–Ctf4 regulating steady-state protein levels of the catalytic subunit of Pol-α [16, 73, 85–89]. Coincidentally, both Mcm10 and Ctf4 have been found to trimerize, suggesting that an architectural similarity coupling Pol-α to the advancing helicase may exist [11, 13, 90]. Unlike Mcm10- and Pol-α-deficient cells, *ctf4Δ* mutants do not accumulate extensive ssDNA regions that trigger replication stress signals, arguing that—at least in budding yeast—Mcm10 can largely compensate for the loss of Ctf4 [91]. Indeed, Ctf4 is not required to initiate DNA synthesis in vitro [65]. Despite the lack of a true homolog in prokaryotes, Mcm10 and Ctf4 may be functionally similar to the tau protein complex, which serves to coordinate the bacterial leading and lagging strand polymerases with the progressing helicase [73, 92–94].

Additional support for a role during elongation comes from investigations demonstrating a direct interaction between Mcm10 and PCNA [27]. The Mcm10 ID carries a PIP box motif that in yeast shows a 3/4 match to the consensus sequence (QxxM/I/LxxF/YF/Y) (Fig. 16.1) [18, 95]. This protein–protein interaction is modulated by ubiquitination of Mcm10 during G1 and S phase of the cell cycle [27]. Significantly, Pol-α is required for generating the substrate for PCNA loading and only unmodified Mcm10 interacts with the polymerase [27]. These observations suggest that following primer synthesis, ubiquitination of Mcm10 may facilitate PCNA loading and simultaneous displacement of Pol-α. Unlike in yeast, the PIP box in other Mcm10 homologs more closely matches the QLsLF consensus site that mediates binding to the prokaryotic β-clamp [96]. Despite its evolutionary conservation, the importance of the PCNA interaction motif in higher eukaryotes has not been determined. Finally, data from yeast suggest that Mcm10 interacts not only with the replication clamp, but also with the 9-1-1 complex that is loaded onto the lagging strand of stalled replication forks [97, 98]. The NTD of Mcm10 binds directly to the 9-1-1 clamp following UV irradiation or nucleotide depletion [26]. These observations are consistent with a role for Mcm10 in stabilizing arrested forks by bridging the 9-1-1 complex with the CMG helicase, thereby possibly preventing nucleolytic degradation of Okazaki fragments [26]. Thus, Mcm10's function is not limited to scaffolding during normal elongation, but is also required to alleviate replication stress.

The Shared Requirement for Mcm10 in Origin Activation and Fork Progression

During S phase, origin firing and replication elongation occur simultaneously at different loci throughout the genome. Mcm10 contributes essential functions to both processes to ensure rapid and high-fidelity duplication of the genome. Interestingly, studies in yeast and human cells suggest that the interplay between origin firing and Okazaki fragment priming may be altered when steady-state levels of Mcm10 are reduced [14, 87, 99, 100]. Experiments in yeast have taken advantage of the temperature-sensitive *mcm10-1* mutant, which exhibits slowed DNA synthesis and growth arrest under restrictive conditions [4, 17, 47]. Importantly, these phenotypes can be suppressed by mutations in *mcm2* [87]. Two models have been proposed to explain these observations. The first model suggests that the *mcm2* suppressor allele alters the Mcm2-7 complex such that helicase activation is no longer dependent on Mcm10 [18, 87]. The second model takes into account that the *mcm2* suppressor allele reduces the activity of the Mcm2-7 core helicase [87]. Presumably, delayed fork progression decreases the requirement for helicase-polymerase coordination, and minimizes the amount of RPA-coated ssDNA that normally triggers a replication stress response and cell cycle arrest. This model would suggest that the critical contribution of Mcm10 might be in elongation. Although seemingly contrary, these models are not mutually exclusive and suppression of the *mcm10-1* allele could result from modulating both Mcm10's role in initiation and elongation.

An independent investigation of *mcm10-1* phenotypes provided further support for a role in elongation. The so-called defective replisome-induced mutagenesis (or DRIM) in *mcm10-1* cells results in elevated levels of PCNA ubiquitination, a readout for the accumulation of ssDNA [99]. A similar phenotype was observed when priming was disrupted in *pol1-1* cells, consistent with previous studies that established a direct connection between Mcm10 and Pol-α [17, 73, 85]. Significantly, increased amounts of ubiquitinated PCNA were not detected in *dbf4-1* mutants, which are defective in DDK regulation [99]. Because DDK activity is required for replication origin firing, these findings suggest that defects in initiation are not the major source of replication stress in *mcm10-1* cells at semi-permissive temperatures [99]. Therefore, partial loss of Mcm10 may become limiting or semi-permissive for replication elongation, but remains fully permissive for origin activation in yeast.

Studies in mammalian cells assessing the roles of Mcm10 in replication initiation and elongation have heavily relied on the DNA fiber technique, which offers a single-molecule approach to analyze dynamics of DNA synthesis [101, 102]. This technique utilizes sequential incorporation of nucleotide analogs and allows for the quantification of origin firing events as well as measurements of fork velocity. Following siRNA-mediated knockdown of Mcm10 in different human cell types, both fork speed and inter-origin distance (IOD) can be measured. The comparison of two independent studies reveals an increased requirement for Mcm10 in faster replicating cells, whereas cells with an intrinsically slower fork speed were not affected by Mcm10 depletion [14, 100]. Further, both studies documented decreased average IODs upon Mcm10 knockdown, arguing that the number of origins initiating replication was increased [14, 100]. Similar to observations in yeast, these data argue that reduction of Mcm10 limits elongation, but does not interfere with increased origin firing in an effort to complete replication of the genome. This increase in origin activation is commonly attributed to the firing of so-called dormant origins that are usually not activated in normal replication cycles [103, 104]. Thus, an alternative explanation for the decreased IODs upon loss of Mcm10 is that dormant origin firing is regulated by a mechanism that foregoes Mcm10 function.

The Contribution of Mcm10 to Genome Stability

The delayed replication program of Mcm10-deficient cells is associated with increased accumulation of DNA damage markers [85, 99, 105, 106]. The disruption of *Mcm10* in mice revealed that homozygous knockouts were embryonic lethal, with null embryos displaying delayed growth, significantly decreased DNA synthesis, and increased DNA damage [8]. In human cells, knockdown of Mcm10 induces DNA damage, G2-phase arrest, and apoptosis [85, 107, 108]. Consistent with these observations, a genome-wide siRNA screen found that Mcm10 knockdown resulted in increased levels of γ-H2AX, an important early marker of DNA double-strand breaks (DSBs) [106]. A complementary but independent siRNA screen reported the increase of a different marker of DSB repair, p53-binding protein 1 (53BP1) foci, in response to Mcm10 knockdown [105]. These data underscore that human Mcm10 is crucial to prevent replication stress and the accumulation of DSBs.

Large-scale genetic interaction studies in yeast have shown that *MCM10* contributes not only to a robust replication program but also to pathways that maintain genomic integrity. Accordingly, *mcm10* mutants display loss of fitness in combination with mutations in checkpoint signaling genes *MEC1* and *RAD53*, the homologs of human ataxia telangiectasia, and Rad3-related kinase (ATR) and checkpoint kinase 2 (Chk2) [87]. Moreover, disruption of *MCM10* negatively interacts with genes involved in resolving stalled replication forks including *DNA2*, *SGS1*, *SRS2*, *MRE11*, and *RAD50* [52, 87, 109]. Finally, several loss-of-function alleles of DSB repair genes including *MRE11*, *RAD50*, *EXO1*, *SGS1*, and *DNA2* have demonstrated deleterious interactions with *mcm10* mutations [52, 87, 109]. Overall, these data argue that Mcm10 contributes to chromosome stability in multiple ways. Whereas the underlying mechanism is not fully understood, it seems likely that increased replication stress in *mcm10* mutants causes increased DSBs due to fork collapse and under-replication of the genome. Consistent with this model, Mcm10 depletion increased common fragile site breakage in human cell culture, a phenotype attributed to decreased replication fork progression [100]. An additional component could be a more direct role in DSB repair. In *Xenopus* egg extract, Mcm10 associates in a complex with DNA replication ATP-dependent helicase/nuclease 2 (Dna2) and Nijmegen breakage syndrome 1 (Nbs1), proteins that perform initial processing steps to facilitate homologous recombination [89]. Thus, future studies investigating a role in DNA repair may be fruitful for understanding the contributions of Mcm10 to genome stability.

Genome instability and defects in DNA repair are associated with a variety of human pathologies, including cancer. Consistent with these observations, there is accumulating evidence supporting a role for Mcm10 in oncogenesis. A bioinformatics analysis of breast and non-small-cell lung cancers ranked *MCM10* in the top-ten cancer-associated genes [110]. Genetic alterations of *MCM10* including mutations (35%), deletions (14%), and amplifications (51%) have been identified in cell lines and tumor samples representing a variety of human tissues (Fig. 16.4) [96, 111–113]. The diverse nature of *MCM10* alterations suggests that these lesions may contribute to cancer progression in several ways that would be consistent with the oncogene-induced DNA damage model of tumorigenesis [114]. Loss of Mcm10 could induce replication stress and DNA damage that may contribute directly to aberrant DNA synthesis and chromosome instability, both of which have a causal role in cancer [85, 107, 108, 115]. Consistent with this postulate, increased levels of the transcriptional repressor ZBTB38 (zinc finger and BTB domain-containing protein 38) lead to reduced *MCM10* expression, slowed replication, and increased damage at common fragile sites [100]. This provides a likely model for the contribution of deletions and loss-of-function mutations to oncogenesis. To date, 82 different point mutations have been mapped to *MCM10* including alterations disrupting splicing, as well as missense and nonsense mutations (Fig. 16.4) [96, 111, 112]. A subset of these point mutations are located in conserved functional domains including the coiled-coil, OB-fold, ZnF1, PIP box, Hsp10-like domain, ZnF2, and ZnR (Fig. 16.4) [96]. Investigations into the consequences of these mutations will be useful in understanding how Mcm10 may impact cancer development. In contrast, amplification of the *MCM10* locus could be a result of "rewiring" the replication program of cancer cells to relieve replication stress and promote rapid cellular proliferation. Interestingly, increased *MCM10* transcription is directly regulated by oncogenes in Ewing's sarcoma (EWS/FLI2) and neuroblastoma (N-MYC) [116, 117]. In addition, a study of cervical tumors and cell lines found that the level of *MCM10* overexpression directly correlated with advanced tumor stage [118]. These data suggest that some cancers become uniquely reliant on

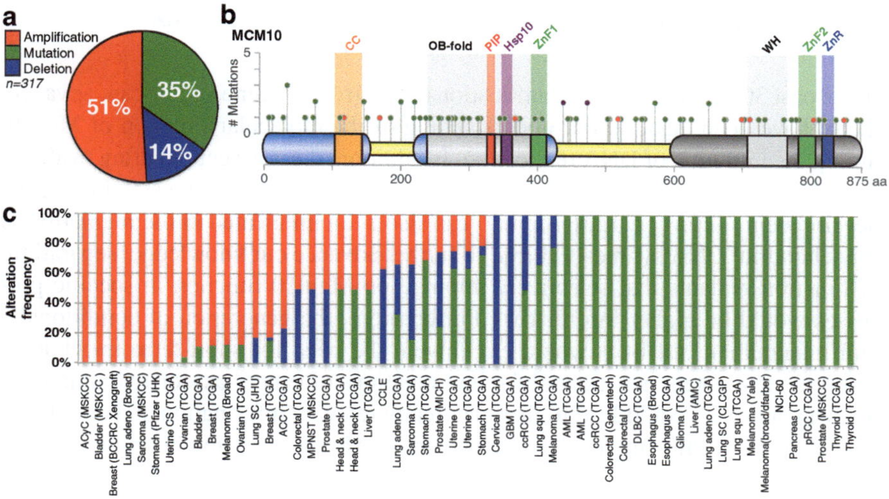

Fig. 16.4 *Genetic alterations in MCM10 identified in human cancer samples*. (**a**) Pie chart depicting the prevalence of *MCM10* amplifications, mutations, and deletions from 317 cancer samples. (**b**) Mapping of cancer-associated point mutations onto a cartoon depiction of the Mcm10 protein structure. Mcm10 domains are indicated, including the coiled-coil (CC, *orange*), OB-fold (*gray*), PIP box (PIP, *red*), Hsp10-like domain (Hsp10, *purple*), zinc-fingers 1 and 2 (ZnF, *green*), winged-helix (WH, *light gray*), and zinc-ribbon (ZnR). The position of each mutation is indicated by *vertical lines topped with green* (missense), *red* (nonsense), or *purple* (multiple types) circles. (**c**) Bar graph showing the alteration frequency of amplifications (*red*), mutations (*green*), and deletions (*blue*) of *MCM10* identified in different cancer types by multiple groups. The data and depictions shown in this figure were accessed via and/or modified from information listed on the cBioPortal for Cancer Genomics (http://www.cbioportal.org/). Cancer-type abbreviations include the following: *ACC* adrenocortical carcinoma, *ACyc* adenoid cystic carcinoma, *AML* acute myeloid leukemia, *CCLE* cancer cell line encyclopedia, *ccRCC* kidney renal clear cell carcinoma, *DLBC* lymphoid neoplasm diffuse large B-cell lymphoma, *GBM* glioblastoma, *Adeno* adenocarcinoma, *SC* small cell, *squ* squamous, *MPNST* malignant peripheral nerve sheath tumor, *NCI* National Cancer Institute, *pRCC* kidney renal papillary cell carcinoma. Source abbreviations include TCGA (The Cancer Genome Atlas), MSKCC (Memorial Sloan Kettering Cancer Center, Pfizer UHK (University of Hong Kong), JHU (Johns Hopkins University), AMC (Asan Medical Center), BCCRC (British Columbia Cancer Research Center)

precise levels of *MCM10* and that targeted depletion offers a therapeutic opportunity to affect cancer cells but not normal tissues. A similar strategy has been suggested for CYCLOPS genes (for *c*opy number alterations *y*ielding *c*ancer *l*iabilities *o*wing to *p*artial los*s*). Following tumor-specific copy number loss of genes involved in key cellular pathways, cancer cells become vulnerable to further depletion [119]. It is notable that tumor samples from specific tissue types tend to carry alterations that group into a single class (Fig. 16.4). For example, breast and bladder cancers primarily display *MCM10* amplifications, whereas lung or colorectal cancers primarily harbor mutations or deletions (Fig. 16.4). These observations suggest that genetic aberrations that alter *MCM10* directly or indirectly may affect oncogenesis in more than one way depending on underlying differences in cell biology within tumors originating from various tissues.

Defining the Molecular Mechanism of Mcm10 Functions

In the past 30 years since the identification of the *MCM10* gene, significant advancements have been made in our understanding of the molecular function of Mcm10 during the initiation of DNA replication. In addition, there is accumulating evidence for a role in maintaining chromatin structure and guarding genome stability during development [120, 121, 122]. Mcm10 has also been implicated in the regulation of mitochondrial DNA synthesis [123]. To further advance our knowledge, separation-of-function mutants that selectively impinge on Mcm10 function in specific processes would be highly informative. Furthermore, the generation of conditionally null mammalian cell lines would circumvent the need for short-term knockdown and provide a potent genetic tool for investigations of cancer-associated mutations. Such an approach could lead to therapeutically relevant information and could provide details regarding molecular mechanisms of Mcm10 function in origin unwinding and elongation.

Abbreviations

2D	Two dimensional
53BP1	p53-binding protein 1
9-1-1	Rad9-Hus1-Rad1
aa	Amino acids
ATR	Ataxia telangiectasia and Rad3-related
CC	Coiled coil
Cdc45	Cell division cycle 45
Cdc6	Cell division cycle 6
Cdt1	Cdc10-dependent transcript 1
Chk2	Checkpoint kinase 2
CMG	Cdc45-Mcm2-7-GINS
CTD	C-terminal domain
Ctf4	Chromosome transmission fidelity 4
CYCLOPs	Copy number alterations yielding cancer liabilities owing to partial loss
DDK	Dbf4-dependent kinase Cdc7
Dna2	ATP-dependent helicase/nuclease 2
DRIM	Defective-replisome-induced-mutagenesis
DSB	Double-strand break
dsDNA	Double-stranded DNA
EM	Electron microscopy
GINS	Go-ichi-ni-san
ID	Internal domain
IOD	Inter origin distance

Mcm10	Minichromosome maintenance protein 10
Mcm2-7	Minichromosome maintenance proteins 2-7
Mec3	Mitosis entry checkpoint 3
Nbs1	Nijmegen breakage syndrome 1
NTD	N-terminal domain
OB-fold	Oligonucleotide/oligosaccharide-binding fold
ORC	Origin recognition complex
PCNA	Proliferating cell nuclear antigen
PIP-box	PCNA interacting peptide box
Pol-α	DNA polymerase alpha
Pol-δ	DNA polymerase delta
Pol-ε	DNA polymerase epsilon
pre-RC	Pre-replication complex
Recql4	RecQ like helicase 4
RPA	Replication protein A
S-CDK	S-phase cyclin-dependent kinase
SIRT1	Sirtuin 1
Sld	Synthetically lethal with *dpb11*
ssDNA	Single-stranded DNA
ZnF	Zinc-finger
ZnR	Zinc-ribbon

Acknowledgements The authors wish to acknowledge funding from the National Institutes of Health R01 GM074917 to A.K.B. R.M.B. was supported by Cancer Biology Training Grant NIH T32 CA009138.

References

1. Dumas LB, Lussky JP, McFarland EJ, Shampay J. New temperature-sensitive mutants of *Saccharomyces cerevisiae* affecting DNA replication. Mol Gen Genet. 1982;187(1):42–6.
2. Solomon NA, Wright MB, Chang S, Buckley AM, Dumas LB, Gaber RF. Genetic and molecular analysis of DNA43 and DNA52: two new cell-cycle genes in *Saccharomyces cerevisiae*. Yeast. 1992;8(4):273–89.
3. Maine GT, Sinha P, Tye BK. Mutants of *S. cerevisiae* defective in the maintenance of minichromosomes. Genetics. 1984;106(3):365–85.
4. Merchant AM, Kawasaki Y, Chen Y, Lei M, Tye BK. A lesion in the DNA replication initiation factor Mcm10 induces pausing of elongation forks through chromosomal replication origins in *Saccharomyces cerevisiae*. Mol Cell Biol. 1997;17(6):3261–71.
5. Aves SJ, Tongue N, Foster AJ, Hart EA. The essential *Schizosaccharomyces pombe* cdc23 DNA replication gene shares structural and functional homology with the *Saccharomyces cerevisiae* DNA43 (MCM10) gene. Curr Genet. 1998;34(3):164–71.
6. Christensen TW, Tye BK. *Drosophila* MCM10 interacts with members of the prereplication complex and is required for proper chromosome condensation. Mol Biol Cell. 2003; 14(6):2206–15.
7. Izumi M, Yanagi K, Mizuno T, Yokoi M, Kawasaki Y, Moon KY, et al. The human homolog of *Saccharomyces cerevisiae* Mcm10 interacts with replication factors and dissociates from nuclease-resistant nuclear structures in G(2) phase. Nucleic Acids Res. 2000;28(23):4769–77.

8. Lim HJ, Jeon Y, Jeon CH, Kim JH, Lee H. Targeted disruption of Mcm10 causes defective embryonic cell proliferation and early embryo lethality. Biochim Biophys Acta. 2011; 1813(10):1777–83.
9. Wohlschlegel JA, Dhar SK, Prokhorova TA, Dutta A, Walter JC. *Xenopus* Mcm10 binds to origins of DNA replication after Mcm2-7 and stimulates origin binding of Cdc45. Mol Cell. 2002;9(2):233–40.
10. Pettersen EF, Goddard TD, Huang CC, Couch GS, Greenblatt DM, Meng EC, et al. UCSF Chimera--a visualization system for exploratory research and analysis. J Comput Chem. 2004;25(13):1605–12.
11. Du W, Josephrajan A, Adhikary S, Bowles T, Bielinsky AK, Eichman BF. Mcm10 self-association is mediated by an N-terminal coiled-coil domain. PLoS One. 2013;8(7):e70518.
12. Du W, Stauffer ME, Eichman BF. Structural biology of replication initiation factor Mcm10. Subcell Biochem. 2012;62:197–216.
13. Robertson PD, Warren EM, Zhang H, Friedman DB, Lary JW, Cole JL, et al. Domain architecture and biochemical characterization of vertebrate Mcm10. J Biol Chem. 2008; 283(6):3338–48.
14. Fatoba ST, Tognetti S, Berto M, Leo E, Mulvey CM, Godovac-Zimmermann J, et al. Human SIRT1 regulates DNA binding and stability of the Mcm10 DNA replication factor via deacetylation. Nucleic Acids Res. 2013;41(7):4065–79.
15. Fien K, Cho YS, Lee JK, Raychaudhuri S, Tappin I, Hurwitz J. Primer utilization by DNA polymerase alpha-primase is influenced by its interaction with Mcm10p. J Biol Chem. 2004;279(16):16144–53.
16. Ricke RM, Bielinsky AK. Mcm10 regulates the stability and chromatin association of DNA polymerase-alpha. Mol Cell. 2004;16(2):173–85.
17. Ricke RM, Bielinsky AK. A conserved Hsp10-like domain in Mcm10 is required to stabilize the catalytic subunit of DNA polymerase-alpha in budding yeast. J Biol Chem. 2006; 281(27):18414–25.
18. Thu YM, Bielinsky AK. Enigmatic roles of Mcm10 in DNA replication. Trends Biochem Sci. 2013;38(4):184–94.
19. Warren EM, Huang H, Fanning E, Chazin WJ, Eichman BF. Physical interactions between Mcm10, DNA, and DNA polymerase alpha. J Biol Chem. 2009;284(36):24662–72.
20. Cook CR, Kung G, Peterson FC, Volkman BF, Lei M. A novel zinc finger is required for Mcm10 homocomplex assembly. J Biol Chem. 2003;278(38):36051–8.
21. Okorokov AL, Waugh A, Hodgkinson J, Murthy A, Hong HK, Leo E, et al. Hexameric ring structure of human MCM10 DNA replication factor. EMBO Rep. 2007;8(10):925–30.
22. Warren EM, Vaithiyalingam S, Haworth J, Greer B, Bielinsky AK, Chazin WJ, et al. Structural basis for DNA binding by replication initiator Mcm10. Structure. 2008;16(12): 1892–901.
23. Di Perna R, Aria V, De Falco M, Sannino V, Okorokov AL, Pisani FM, et al. The physical interaction of Mcm10 with Cdc45 modulates their DNA-binding properties. Biochem J. 2013;454(2):333–43.
24. Robertson PD, Chagot B, Chazin WJ, Eichman BF. Solution NMR structure of the C-terminal DNA binding domain of Mcm10 reveals a conserved MCM motif. J Biol Chem. 2010;285(30):22942–9.
25. Eisenberg S, Korza G, Carson J, Liachko I, Tye BK. Novel DNA binding properties of the Mcm10 protein from *Saccharomyces cerevisiae*. J Biol Chem. 2009;284(37):25412–20.
26. Alver RC, Zhang T, Josephrajan A, Fultz BL, Hendrix CJ, Das-Bradoo S, et al. The N-terminus of Mcm10 is important for interaction with the 9-1-1 clamp and in resistance to DNA damage. Nucleic Acids Res. 2014;42(13):8389–404.
27. Das-Bradoo S, Ricke RM, Bielinsky AK. Interaction between PCNA and diubiquitinated Mcm10 is essential for cell growth in budding yeast. Mol Cell Biol. 2006;26(13):4806–17.
28. Dua R, Levy DL, Li CM, Snow PM, Campbell JL. In vivo reconstitution of *Saccharomyces cerevisiae* DNA polymerase epsilon in insect cells. Purification and characterization. J Biol Chem. 2002;277(10):7889–96.

29. Schmidt KH, Derry KL, Kolodner RD. *Saccharomyces cerevisiae* RRM3, a 5′ to 3′ DNA helicase, physically interacts with proliferating cell nuclear antigen. J Biol Chem. 2002; 277(47):45331–7.
30. Vivona JB, Kelman Z. The diverse spectrum of sliding clamp interacting proteins. FEBS Lett. 2003;546(2–3):167–72.
31. Evrin C, Clarke P, Zech J, Lurz R, Sun J, Uhle S, et al. A double-hexameric MCM2-7 complex is loaded onto origin DNA during licensing of eukaryotic DNA replication. Proc Natl Acad Sci U S A. 2009;106(48):20240–5.
32. Masai H, Matsumoto S, You Z, Yoshizawa-Sugata N, Oda M. Eukaryotic chromosome DNA replication: where, when, and how? Annu Rev Biochem. 2010;79:89–130.
33. Remus D, Beuron F, Tolun G, Griffith JD, Morris EP, Diffley JF. Concerted loading of Mcm2-7 double hexamers around DNA during DNA replication origin licensing. Cell. 2009;139(4):719–30.
34. Bleichert F, Botchan MR, Berger JM. Crystal structure of the eukaryotic origin recognition complex. Nature. 2015;519(7543):321–6.
35. Ticau S, Friedman LJ, Ivica NA, Gelles J, Bell SP. Single-molecule studies of origin licensing reveal mechanisms ensuring bidirectional helicase loading. Cell. 2015;161(3):513–25.
36. Ilves I, Petojevic T, Pesavento JJ, Botchan MR. Activation of the MCM2-7 helicase by association with Cdc45 and GINS proteins. Mol Cell. 2010;37(2):247–58.
37. Sheu YJ, Stillman B. Cdc7-Dbf4 phosphorylates MCM proteins via a docking site-mediated mechanism to promote S phase progression. Mol Cell. 2006;24(1):101–13.
38. Tanaka S, Umemori T, Hirai K, Muramatsu S, Kamimura Y, Araki H. CDK-dependent phosphorylation of Sld2 and Sld3 initiates DNA replication in budding yeast. Nature. 2007;445(7125):328–32.
39. Zegerman P, Diffley JF. Phosphorylation of Sld2 and Sld3 by cyclin-dependent kinases promotes DNA replication in budding yeast. Nature. 2007;445(7125):281–5.
40. Tanaka T, Umemori T, Endo S, Muramatsu S, Kanemaki M, Kamimura Y, et al. Sld7, an Sld3-associated protein required for efficient chromosomal DNA replication in budding yeast. EMBO J. 2011;30(10):2019–30.
41. Bochman ML, Schwacha A. The Mcm2-7 complex has in vitro helicase activity. Mol Cell. 2008;31(2):287–93.
42. Moyer SE, Lewis PW, Botchan MR. Isolation of the Cdc45/Mcm2-7/GINS (CMG) complex, a candidate for the eukaryotic DNA replication fork helicase. Proc Natl Acad Sci U S A. 2006;103(27):10236–41.
43. Fu YV, Yardimci H, Long DT, Ho TV, Guainazzi A, Bermudez VP, et al. Selective bypass of a lagging strand roadblock by the eukaryotic replicative DNA helicase. Cell. 2011; 146(6):931–41.
44. Sun J, Fernandez-Cid A, Riera A, Tognetti S, Yuan Z, Stillman B, et al. Structural and mechanistic insights into Mcm2-7 double-hexamer assembly and function. Genes Dev. 2014; 28(20):2291–303.
45. Bruck I, Kaplan DL. The Dbf4-Cdc7 kinase promotes Mcm2-7 ring opening to allow for single-stranded DNA extrusion and helicase assembly. J Biol Chem. 2015;290(2):1210–21.
46. Hart EA, Bryant JA, Moore K, Aves SJ. Fission yeast Cdc23 interactions with DNA replication initiation proteins. Curr Genet. 2002;41(5):342–8.
47. Homesley L, Lei M, Kawasaki Y, Sawyer S, Christensen T, Tye BK. Mcm10 and the MCM2-7 complex interact to initiate DNA synthesis and to release replication factors from origins. Genes Dev. 2000;14(8):913–26.
48. Lee JK, Seo YS, Hurwitz J. The Cdc23 (Mcm10) protein is required for the phosphorylation of minichromosome maintenance complex by the Dfp1-Hsk1 kinase. Proc Natl Acad Sci U S A. 2003;100(5):2334–9.
49. Sawyer SL, Cheng IH, Chai W, Tye BK. Mcm10 and Cdc45 cooperate in origin activation in *Saccharomyces cerevisiae*. J Mol Biol. 2004;340(2):195–202.
50. Kawasaki Y, Hiraga S, Sugino A. Interactions between Mcm10p and other replication factors are required for proper initiation and elongation of chromosomal DNA replication in *Saccharomyces cerevisiae*. Genes Cells. 2000;5(12):975–89.

51. Xu X, Rochette PJ, Feyissa EA, Su TV, Liu Y. MCM10 mediates RECQ4 association with MCM2-7 helicase complex during DNA replication. EMBO J. 2009;28(19):3005–14.
52. Araki Y, Kawasaki Y, Sasanuma H, Tye BK, Sugino A. Budding yeast mcm10/dna43 mutant requires a novel repair pathway for viability. Genes Cells. 2003;8(5):465–80.
53. Izumi M, Yatagai F, Hanaoka F. Cell cycle-dependent proteolysis and phosphorylation of human Mcm10. J Biol Chem. 2001;276(51):48526–31.
54. Kaur M, Sharma A, Khan M, Kar A, Saxena S. Mcm10 proteolysis initiates before the onset of M-phase. BMC Cell Biol. 2010;11:84.
55. Sharma A, Kaur M, Kar A, Ranade SM, Saxena S. Ultraviolet radiation stress triggers the down-regulation of essential replication factor Mcm10. J Biol Chem. 2010;285(11): 8352–62.
56. Lan W, Chen S, Tong L. MicroRNA-215 regulates fibroblast function: insights from a human fibrotic disease. Cell Cycle. 2015;14(12):1973–84.
57. Yoshida K, Inoue I. Expression of MCM10 and TopBP1 is regulated by cell proliferation and UV irradiation via the E2F transcription factor. Oncogene. 2004;23(37):6250–60.
58. Watase G, Takisawa H, Kanemaki MT. Mcm10 plays a role in functioning of the eukaryotic replicative DNA helicase, Cdc45-Mcm-GINS. Curr Biol. 2012;22(4):343–9.
59. Gambus A, van Deursen F, Polychronopoulos D, Foltman M, Jones RC, Edmondson RD, et al. A key role for Ctf4 in coupling the MCM2-7 helicase to DNA polymerase alpha within the eukaryotic replisome. EMBO J. 2009;28(19):2992–3004.
60. Gregan J, Lindner K, Brimage L, Franklin R, Namdar M, Hart EA, et al. Fission yeast Cdc23/Mcm10 functions after pre-replicative complex formation to promote Cdc45 chromatin binding. Mol Biol Cell. 2003;14(9):3876–87.
61. Heller RC, Kang S, Lam WM, Chen S, Chan CS, Bell SP. Eukaryotic origin-dependent DNA replication in vitro reveals sequential action of DDK and S-CDK kinases. Cell. 2011; 146(1):80–91.
62. Im JS, Ki SH, Farina A, Jung DS, Hurwitz J, Lee JK. Assembly of the Cdc45-Mcm2-7-GINS complex in human cells requires the Ctf4/And-1, RecQL4, and Mcm10 proteins. Proc Natl Acad Sci U S A. 2009;106(37):15628–32.
63. Im JS, Park SY, Cho WH, Bae SH, Hurwitz J, Lee JK. RecQL4 is required for the association of Mcm10 and Ctf4 with replication origins in human cells. Cell Cycle. 2015;14(7):1001–9.
64. On KF, Beuron F, Frith D, Snijders AP, Morris EP, Diffley JF. Prereplicative complexes assembled in vitro support origin-dependent and independent DNA replication. EMBO J. 2014;33(6):605–20.
65. Yeeles JT, Deegan TD, Janska A, Early A, Diffley JF. Regulated eukaryotic DNA replication origin firing with purified proteins. Nature. 2015;519(7544):431–5.
66. Donovan S, Harwood J, Drury LS, Diffley JF. Cdc6p-dependent loading of Mcm proteins onto pre-replicative chromatin in budding yeast. Proc Natl Acad Sci U S A. 1997; 94(11):5611–6.
67. Edwards MC, Tutter AV, Cvetic C, Gilbert CH, Prokhorova TA, Walter JC. MCM2-7 complexes bind chromatin in a distributed pattern surrounding the origin recognition complex in *Xenopus* egg extracts. J Biol Chem. 2002;277(36):33049–57.
68. Lei M, Kawasaki Y, Tye BK. Physical interactions among Mcm proteins and effects of Mcm dosage on DNA replication in *Saccharomyces cerevisiae*. Mol Cell Biol. 1996; 16(9):5081–90.
69. Mahbubani HM, Chong JP, Chevalier S, Thommes P, Blow JJ. Cell cycle regulation of the replication licensing system: involvement of a Cdk-dependent inhibitor. J Cell Biol. 1997;136(1):125–35.
70. Kanke M, Kodama Y, Takahashi TS, Nakagawa T, Masukata H. Mcm10 plays an essential role in origin DNA unwinding after loading of the CMG components. EMBO J. 2012; 31(9):2182–94.
71. van Deursen F, Sengupta S, De Piccoli G, Sanchez-Diaz A, Labib K. Mcm10 associates with the loaded DNA helicase at replication origins and defines a novel step in its activation. EMBO J. 2012;31(9):2195–206.

72. Yu C, Gan H, Han J, Zhou ZX, Jia S, Chabes A, et al. Strand-specific analysis shows protein binding at replication forks and PCNA unloading from lagging strands when forks stall. Mol Cell. 2014;56(4):551–63.
73. Zhu W, Ukomadu C, Jha S, Senga T, Dhar SK, Wohlschlegel JA, et al. Mcm10 and And-1/CTF4 recruit DNA polymerase alpha to chromatin for initiation of DNA replication. Genes Dev. 2007;21(18):2288–99.
74. Yang X, Gregan J, Lindner K, Young H, Kearsey SE. Nuclear distribution and chromatin association of DNA polymerase alpha-primase is affected by TEV protease cleavage of Cdc23 (Mcm10) in fission yeast. BMC Mol Biol. 2005;6:13.
75. Kunkel TA, Burgers PM. Dividing the workload at a eukaryotic replication fork. Trends Cell Biol. 2008;18(11):521–7.
76. Nick McElhinny SA, Gordenin DA, Stith CM, Burgers PM, Kunkel TA. Division of labor at the eukaryotic replication fork. Mol Cell. 2008;30(2):137–44.
77. Pacek M, Tutter AV, Kubota Y, Takisawa H, Walter JC. Localization of MCM2-7, Cdc45, and GINS to the site of DNA unwinding during eukaryotic DNA replication. Mol Cell. 2006;21(4):581–7.
78. Alabert C, Bukowski-Wills JC, Lee SB, Kustatscher G, Nakamura K, de Lima Alves F, et al. Nascent chromatin capture proteomics determines chromatin dynamics during DNA replication and identifies unknown fork components. Nat Cell Biol. 2014;16(3):281–93.
79. Aparicio OM, Weinstein DM, Bell SP. Components and dynamics of DNA replication complexes in *S. cerevisiae*: redistribution of MCM proteins and Cdc45p during S phase. Cell. 1997;91(1):59–69.
80. Taylor M, Moore K, Murray J, Aves SJ, Price C. Mcm10 interacts with Rad4/Cut5(TopBP1) and its association with origins of DNA replication is dependent on Rad4/Cut5(TopBP1). DNA Repair (Amst). 2011;10(11):1154–63.
81. Nasheuer HP, Grosse F. Immunoaffinity-purified DNA polymerase alpha displays novel properties. Biochemistry. 1987;26(25):8458–66.
82. Kouprina N, Kroll E, Bannikov V, Bliskovsky V, Gizatullin R, Kirillov A, et al. CTF4 (CHL15) mutants exhibit defective DNA metabolism in the yeast *Saccharomyces cerevisiae*. Mol Cell Biol. 1992;12(12):5736–47.
83. Apger J, Reubens M, Henderson L, Gouge CA, Ilic N, Zhou HH, et al. Multiple functions for *Drosophila* Mcm10 suggested through analysis of two Mcm10 mutant alleles. Genetics. 2010;185(4):1151–65.
84. Gosnell JA, Christensen TW. *Drosophila* Ctf4 is essential for efficient DNA replication and normal cell cycle progression. BMC Mol Biol. 2011;12:13.
85. Chattopadhyay S, Bielinsky AK. Human Mcm10 regulates the catalytic subunit of DNA polymerase-alpha and prevents DNA damage during replication. Mol Biol Cell. 2007;18(10):4085–95.
86. Haworth J, Alver RC, Anderson M, Bielinsky AK. Ubc4 and Not4 regulate steady-state levels of DNA polymerase-alpha to promote efficient and accurate DNA replication. Mol Biol Cell. 2010;21(18):3205–19.
87. Lee C, Liachko I, Bouten R, Kelman Z, Tye BK. Alternative mechanisms for coordinating polymerase alpha and MCM helicase. Mol Cell Biol. 2010;30(2):423–35.
88. Wang J, Wu R, Lu Y, Liang C. Ctf4p facilitates Mcm10p to promote DNA replication in budding yeast. Biochem Biophys Res Commun. 2010;395(3):336–41.
89. Wawrousek KE, Fortini BK, Polaczek P, Chen L, Liu Q, Dunphy WG, et al. *Xenopus* DNA2 is a helicase/nuclease that is found in complexes with replication proteins And-1/Ctf4 and Mcm10 and DSB response proteins Nbs1 and ATM. Cell Cycle. 2010;9(6):1156–66.
90. Simon AC, Zhou JC, Perera RL, van Deursen F, Evrin C, Ivanova ME, et al. A Ctf4 trimer couples the CMG helicase to DNA polymerase alpha in the eukaryotic replisome. Nature. 2014;510(7504):293–7.
91. Fumasoni M, Zwicky K, Vanoli F, Lopes M, Branzei D. Error-free DNA damage tolerance and sister chromatid proximity during DNA replication rely on the Polalpha/Primase/Ctf4 complex. Mol Cell. 2015;57(5):812–23.

92. Kim S, Dallmann HG, McHenry CS, Marians KJ. tau couples the leading- and lagging-strand polymerases at the *Escherichia coli* DNA replication fork. J Biol Chem. 1996;271(35): 21406–12.
93. Kim S, Dallmann HG, McHenry CS, Marians KJ. Coupling of a replicative polymerase and helicase: a tau-DnaB interaction mediates rapid replication fork movement. Cell. 1996;84(4):643–50.
94. Kurth I, O'Donnell M. New insights into replisome fluidity during chromosome replication. Trends Biochem Sci. 2013;38(4):195–203.
95. Dalrymple BP, Kongsuwan K, Wijffels G, Dixon NE, Jennings PA. A universal protein-protein interaction motif in the eubacterial DNA replication and repair systems. Proc Natl Acad Sci U S A. 2001;98(20):11627–32.
96. Thu YM, Bielinsky AK. MCM10: one tool for all-Integrity, maintenance and damage control. Semin Cell Dev Biol. 2014;30:121–30.
97. Ellison V, Stillman B. Biochemical characterization of DNA damage checkpoint complexes: clamp loader and clamp complexes with specificity for 5′ recessed DNA. PLoS Biol. 2003;1(2):E33.
98. Majka J, Binz SK, Wold MS, Burgers PM. Replication protein A directs loading of the DNA damage checkpoint clamp to 5′-DNA junctions. J Biol Chem. 2006;281(38):27855–61.
99. Becker JR, Nguyen HD, Wang X, Bielinsky AK. Mcm10 deficiency causes defective-replisome-induced mutagenesis and a dependency on error-free postreplicative repair. Cell Cycle. 2014;13(11):1737–48.
100. Miotto B, Chibi M, Xie P, Koundrioukoff S, Moolman-Smook H, Pugh D, et al. The RBBP6/ZBTB38/MCM10 axis regulates DNA replication and common fragile site stability. Cell Rep. 2014;7(2):575–87.
101. Davies SL, North PS, Hickson ID. Role for BLM in replication-fork restart and suppression of origin firing after replicative stress. Nat Struct Mol Biol. 2007;14(7):677–9.
102. Rosenberg C, Florijn RJ, Van de Rijke FM, Blonden LA, Raap TK, Van Ommen GJ, et al. High resolution DNA fiber-fish on yeast artificial chromosomes: direct visualization of DNA replication. Nat Genet. 1995;10(4):477–9.
103. Kawabata T, Luebben SW, Yamaguchi S, Ilves I, Matise I, Buske T, et al. Stalled fork rescue via dormant replication origins in unchallenged S phase promotes proper chromosome segregation and tumor suppression. Mol Cell. 2011;41(5):543–53.
104. Woodward AM, Gohler T, Luciani MG, Oehlmann M, Ge X, Gartner A, et al. Excess Mcm2-7 license dormant origins of replication that can be used under conditions of replicative stress. J Cell Biol. 2006;173(5):673–83.
105. Lukas C, Savic V, Bekker-Jensen S, Doil C, Neumann B, Pedersen RS, et al. 53BP1 nuclear bodies form around DNA lesions generated by mitotic transmission of chromosomes under replication stress. Nat Cell Biol. 2011;13(3):243–53.
106. Paulsen RD, Soni DV, Wollman R, Hahn AT, Yee MC, Guan A, et al. A genome-wide siRNA screen reveals diverse cellular processes and pathways that mediate genome stability. Mol Cell. 2009;35(2):228–39.
107. Park JH, Bang SW, Jeon Y, Kang S, Hwang DS. Knockdown of human MCM10 exhibits delayed and incomplete chromosome replication. Biochem Biophys Res Commun. 2008;365(3):575–82.
108. Park JH, Bang SW, Kim SH, Hwang DS. Knockdown of human MCM10 activates G2 checkpoint pathway. Biochem Biophys Res Commun. 2008;365(3):490–5.
109. Tittel-Elmer M, Alabert C, Pasero P, Cobb JA. The MRX complex stabilizes the replisome independently of the S phase checkpoint during replication stress. EMBO J. 2009; 28(8):1142–56.
110. Wu C, Zhu J, Zhang X. Integrating gene expression and protein-protein interaction network to prioritize cancer-associated genes. BMC Bioinformatics. 2012;13:182.
111. Cerami E, Gao J, Dogrusoz U, Gross BE, Sumer SO, Aksoy BA, et al. The cBio cancer genomics portal: an open platform for exploring multidimensional cancer genomics data. Cancer Discov. 2012;2(5):401–4.

112. Gao J, Aksoy BA, Dogrusoz U, Dresdner G, Gross B, Sumer SO, et al. Integrative analysis of complex cancer genomics and clinical profiles using the cBioPortal. Sci Signal. 2013;6(269):11.
113. Kang G, Hwang WC, Do IG, Wang K, Kang SY, Lee J, et al. Exome sequencing identifies early gastric carcinoma as an early stage of advanced gastric cancer. PLoS One. 2013; 8(12):e82770.
114. Halazonetis TD, Gorgoulis VG, Bartek J. An oncogene-induced DNA damage model for cancer development. Science. 2008;319(5868):1352–5.
115. Burrell RA, McClelland SE, Endesfelder D, Groth P, Weller MC, Shaikh N, et al. Replication stress links structural and numerical cancer chromosomal instability. Nature. 2013;494(7438):492–6.
116. Garcia-Aragoncillo E, Carrillo J, Lalli E, Agra N, Gomez-Lopez G, Pestana A, et al. DAX1, a direct target of EWS/FLI1 oncoprotein, is a principal regulator of cell-cycle progression in Ewing's tumor cells. Oncogene. 2008;27(46):6034–43.
117. Koppen A, Ait-Aissa R, Koster J, van Sluis PG, Ora I, Caron HN, et al. Direct regulation of the minichromosome maintenance complex by MYCN in neuroblastoma. Eur J Cancer. 2007;43(16):2413–22.
118. Das M, Prasad SB, Yadav SS, Govardhan HB, Pandey LK, Singh S, et al. Over expression of minichromosome maintenance genes is clinically correlated to cervical carcinogenesis. PLoS One. 2013;8(7):e69607.
119. Nijhawan D, Zack TI, Ren Y, Strickland MR, Lamothe R, Schumacher SE, et al. Cancer vulnerabilities unveiled by genomic loss. Cell. 2012;150(4):842–54.
120. Liachko I, Tye BK. Mcm10 is required for the maintenance of transcriptional silencing in *Saccharomyces cerevisiae*. Genetics. 2005;171(2):503–15.
121. Liachko I, Tye BK. Mcm10 mediates the interaction between DNA replication and silencing machineries. Genetics. 2009;181(2):379–91.
122. Vo N, Taga A, Inaba Y, Yoshida H, Cotterill S, Yamaguchi M. Drosophila Mcm10 is required for DNA replication and differentiation in the compound eye. PLoS One. 2014; 9(3):e93450.
123. Wang JT, Xu X, Alontaga AY, Chen Y, Liu Y. Impaired p32 regulation caused by the lymphoma-prone RECQ4 mutation drives mitochondrial dysfunction. Cell Rep. 2014; 7(3):848–58.

Chapter 17
Role of Small-Molecule Modifiers in Replication Initiation

Giacomo De Piccoli and Agnieszka Gambus

Abstract Posttranslational modification of proteins through attachment of ubiquitin or ubiquitin-like proteins (UBLs) changes the three-dimensional structure of the modified factors and affects their activity, interactions, and turnover. Ubiquitin and UBLs constitute a very versatile and flexible system of protein modification and regulate almost every aspect of cell biology. In this chapter we focus on the role of these small protein modifiers in regulation of DNA replication initiation. We review the accumulated knowledge showing how ubiquitin-driven proteasomal degradation leads to creation of sequential, non-overlapping stages of the cell cycle allowing DNA replication initiation and how it prevents re-replication during S phase. We also explain the role ubiquitylation plays in the inhibition of DNA replication initiation in response to DNA damage. As the regulation of replication factors is often executed through modifications with both small protein modifiers and phosphorylation we also discuss the important crossovers between these two regulatory mechanisms. Finally, we review our present knowledge of regulation of DNA replication initiation by non-degradative forms of ubiquitylation and modifications with Nedd8 and SUMO.

Keywords Ubiquitin • SUMO • Nedd8 • DNA replication initiation • Proteasomal degradation • Posttranslational modification • Cullins • APC/C

The mitotic cell cycle is an ordered series of events that ensure the generation of two identical daughter cells. Critically, the cell duplication program must be executed in sequence; thus each step can only occur during a specific window of time. An illuminating example of this regulation is provided by the process of DNA replication. The faithful duplication of the chromosomes requires a two-step mechanism involving a first stage in which the inactive form of the replicative DNA helicase is loaded

G. De Piccoli
Warwick Medical School, University of Warwick, Gibbet Hill, Coventry CV4 7AL, UK
e-mail: g.de-piccoli@warwick.ac.uk

A. Gambus (✉)
School of Cancer Sciences, University of Birmingham,
Vincent Drive, Birmingham B15 2TT, UK
e-mail: a.gambus@bham.ac.uk

D.L. Kaplan (ed.), *The Initiation of DNA Replication in Eukaryotes*,
DOI 10.1007/978-3-319-24696-3_17

during late M/G1 phase of the cell cycle [1–3] and a subsequent phase during which origin firing can occur, but further loading of helicase onto origins is blocked. This ensures that DNA replication occurs only once per cell cycle, avoiding multiple rounds of replication and resulting aneuploidy. This two-step mechanism of DNA replication origin loading and activation plays a key role in eukaryotic genome stability maintenance and is evolutionarily conserved. The switch between these two distinct phases of the cell cycle is controlled by positive feedback loops that ensure the quick and irreversible progression from one phase to the other and the regulation of protein synthesis, activity, interactions, and turnover. A key part of this mechanism is executed through regulation of protein stability and activity by posttranslational modifications with small protein modifiers.

In this chapter we focus on the mechanism of regulation of DNA replication initiation by small protein modifiers. We firstly describe the biology of small protein modifiers, before focusing on their role in the regulation of DNA replication initiation. To date, the best-characterized role of small protein modifiers during DNA replication has been described for ubiquitylation, in particular for polyubiquitylation with lysine 48-linked ubiquitin chains, a marker for protein degradation by the proteasome. We discuss how different ubiquitin ligases control the initial steps of origin formation and firing. In addition, we describe how ubiquitylation inhibits DNA replication origin re-establishment and re-replication during S phase and regulates the transition from G1 to S in response to DNA damage. Finally, we present what is known to date for the role of neddylation and sumoylation during DNA replication initiation.

The Biology of Small Protein Modifiers

"Small protein modifiers" or UBLs are terms assigned to small polypeptides, which can be covalently attached to other substrate proteins as posttranslational modifications. The group of small protein modifiers includes ubiquitin, Sumo1, Sumo2/3, Nedd8, ISG15, Urm1, and Ulm1. Although their primary sequences differ, they share important features. One of the most relevant similarities between the UBLs is the presence of two glycine residues at their C-terminus, which are essential for formation of the isopeptide bond between their moiety and the substrate—usually the ε-amino group of a lysine present in the substrate protein. Due to the similar chemistry of the attachment reaction, all small protein modifiers are conjugated to their substrates through a similar cascade of three enzymes. Firstly, the C-terminal glycine of the UBL is activated by an E1-activating enzyme in an ATP-dependent manner. It is then passed on to an E2-conjugating enzyme through the formation of a thioester bond between the glycyl residue and the cysteine found in the active site of E2. Finally, it is attached through an isopeptide bond to the target protein usually with the help of an E3 ligase. Once attached, the small protein modifier changes the three-dimensional structure of the substrate influencing its activity, interactions with other factors, subcellular localization, and/or fate. Having such a profound and variable effect on multitude of substrates, it is not surprising that small protein modifiers are known to regulate almost every aspect of cell biology, including DNA replication.

Importantly, the scope for a variety of modifications outcome is especially wide in the case of UBLs as some of them can be attached not only as single moieties but also in the form of chains. Both, ubiquitin and Sumo2/3 contain lysines within their sequences, which can be used for the attachment of further UBLs and formation of chains. Modified substrates can also be "de-modified" or edited by proteases specialized in removing a particular type of moiety—such as deubiquitylating enzymes (DUBs) or Sumo-specific endopeptidases (SENPs). All these create a flexible and adaptable system of protein regulation. In this chapter we focus on the regulatory role of small protein modifiers on the initiation of DNA replication.

Ubiquitin

Ubiquitin is a highly stable protein that adapts a compact β-grasp fold with a flexible six-residue C-terminal tail [4]. Ubiquitylation is one of the most abundant and versatile posttranslational modifications in cells due to ubiquitin's ability to be attached as monoubiquitin or polyubiquitin chains (Fig. 17.1). Moreover, a protein can be ubiquitylated on a number of different lysine residues (multi-monoubiquitylation or multi-polyubiquitylation), further increasing the signalling complexity of this system. Ubiquitin contains seven lysines within its sequence (K6, K11, K27, K29, K33, K48, and K63) and each of them can be used for further ubiquitin attachment leading to formation of chains with different ubiquitin linkages. Additionally, ubiquitin can also form linear chains linked through N-terminal methionine, or mixed (heterogeneous) and branched chains when different lysine linkages are used throughout the created chain (Fig. 17.1). Importantly, each type of the chain adopts different conformation and can therefore constitute a different outcome for the substrate. Famously, substrates modified with homogenous chains linked through K48 and K11 are most often directed for proteasomal degradation [4, 5]. Interestingly, all chains, apart from K63 chains, accumulate in cells upon the inhibition of proteasomal activity, suggesting that they can also be sometimes selected by the proteasome for degradation [6, 7]. Ubiquitin chains can also be removed or edited by numerous DUBs with different linkage and substrate specificities [4].

The need to specifically target a broad array of substrates accounts for the great diversity among the estimated >600 human ubiquitin ligases (E3s). E3 ligases fall into three classes: the Really Interesting New Gene (RING) family catalyzes direct transfer of ubiquitin from the E2 enzyme to the substrate, while the Homology to E6AP C Terminus (HECT) and the RING-between-RING (RBR) families of E3s ubiquitylate substrates in a two-step reaction (ubiquitin is transferred from the E2 to an active-site cysteine in the E3 and then from the E3 to the substrate) [8]. The largest family of E3 ligases is the RING domain E3s. The RING-domain E3s are subdivided into two more classes: those that are single polypeptide and contain both the substrate recognition domain and the RING domain on the same unit and the multicomponent RING-E3s. The largest family of multicomponent E3s is the cullin-RING ligases (CRLs). CRLs are built around the scaffold cullin subunit. Cullin interacts through

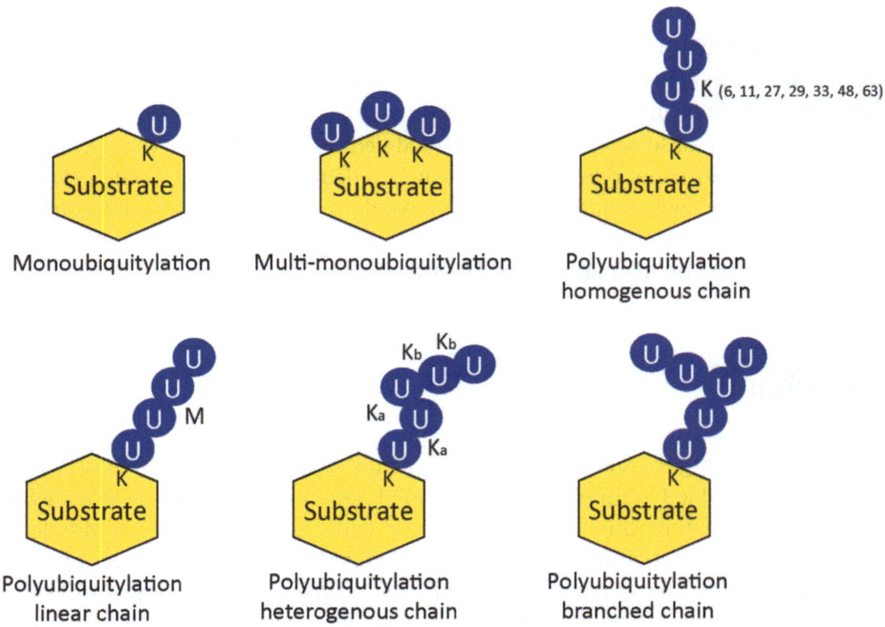

Fig. 17.1 *Types of substrate ubiquitylation*. Monoubiquitylation, multi-monoubiquitylation, and polyubiquitylation are presented. Different types of formed polyubiquitin chains: linked through the same lysine throughout the chain (homogenous chains), linked through N-terminal methionine (linear chains), linked through alternative lysines (heterogenous chains), and chains with one of the chain's ubiquitin modified by two further ubiquitins linked through different lysines (branched chains)

its C-terminus with RING domain proteins Rbx1 and Rbx2, which in turn recruit E2s. On the other hand, the N-terminus of cullin interacts with substrate adaptors, which can bind multiple substrate receptors (Fig. 17.2) [9]. The N-terminal region of the cullins differs between different family members providing them with substrate specificity. On the contrary, cullins C-terminal part is highly conserved [10]. Higher eukaryotes express seven different canonical cullins (Cul1, 2, 3, 4A, 4B, 5, and 7) that each forms a hetero-oligomeric ubiquitin ligase. In addition there are two cullin-like proteins: APC2 in the anaphase-promoting complex (APC) and the p53 cytoplasmic anchor protein PARC [10].

CRLs constitute the major regulators of DNA replication initiation—those members particularly implicated include cullin 1-based complexes (Skp1-Cullin1-F box [SCFs]) and cullin 4-based CRL4s (Fig. 17.1). Additionally, the anaphase-promoting complex or cyclosome (APC/C), which regulates G1-stage entry, also contains a cullin subunit. APC/C is composed of at least 13 subunits, including a RING (Apc11) and a cullin subunit (Apc2) [11]. Together, Apc11 and Apc2 bind to the E2 ubiquitin-conjugating enzymes, which differ depending on the organism: in *S. cerevisiae* these subunits bind Ubc4, which catalyzes the ubiquitylation of lysines of the target protein, and Ubc1, which then extends the chain through modifying K48 of the previously attached ubiquitin [12]. In all organisms, APC/C uses also two adaptor

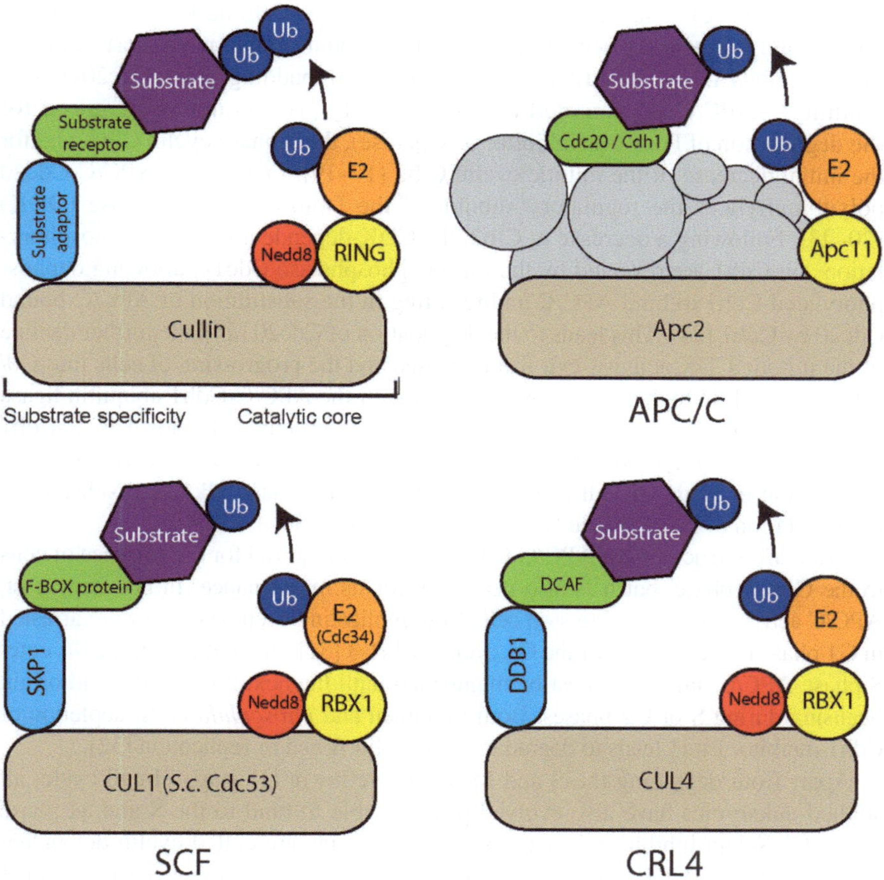

Fig. 17.2 *Model of hetero-oligomeric cullin-RING ubiquitin ligases (CRLs)*. A general model of CRL and models of three CRLs most important for replication initiation regulation are presented: CSF (Cullin1-Skp1-F-box) is built around cullin 1 and uses multiple F-box substrate receptors, CRL4 is built around cullin 4 and uses DCAF substrate receptors, while APC/C is an example of ubiquitin ligase containing cullin-like subunit Apc2 (noncanonical cullin). APC/C forms a large complex containing many other subunits not found in canonical CRLs

proteins, namely Cdc20 and Cdh1, which interact with the APC/C-specific targets for ubiquitylation [13] (Fig. 17.2).

APC/C Regulation of Origin Licensing

Progression through mitosis is driven by high phosphorylation activity of the cyclin-dependent kinase (CDK) associated with mitotic cyclins, while origin licensing in G1 depends upon low CDK activity. The transition from metaphase to anaphase and further to the G1 stage of the cell cycle requires the activation of the APC/C (described above). The substrate adaptor proteins, Cdc20 and Cdh1, bind to the APC/C and promote the polyubiquitylation of proteins containing short sequence motifs, mainly D-boxes and KEN-boxes [14, 15]. In budding yeast, Cdc20, whose binding to APC/C is promoted by CDK activity [16], is mainly required for the degradation of Pds1 (the inhibitor of separase), the S phase cyclin Clb5, and for the initial decrease of the mitotic cyclin Clb2 [17–19]. In addition, APC/C-Cdc20 polyubiquitylates the regulatory subunit of the Dbf4-dependent kinase (DDK) [20–23]. Following a decrease in Clb2, the CDK-dependent inhibitory phosphorylations on Cdh1 are reversed by the mitotic phosphatase Cdc14, allowing dephosphorylated Cdh1 to bind APC/C and resulting in the substitution of APC/C-bound Cdc20 by Cdh1 [24]. This leads to the degradation of Cdc20 [25], the further decline of the mitotic CDK activity, exit from mitosis, and the progression of cells into G0/G1 phase [26]. In higher eukaryotes, activation of the APC/C-Cdh1 ubiquitin ligase also requires the degradation of the inhibitor Emi1, a protein that binds APC/C-Cdh1 and inhibits its polyubiquitin chain-extending activity. This occurs by the phosphorylation of Emi1 by Cdk1-cyclin B and the Polo-kinase Plk1, and subsequent ubiquitylation of Emi1 by the E3 ligase $SCF^{\beta Trcp1}$ [27–30].

Critically, the activity of APC/C-Cdh1 is not only required for the entrance of cells to the G0/G1 phase, but it is also necessary for its maintenance. In budding yeast, APC/C activity is essential to stop cells from replicating even when they are arrested in G1 phase in the absence of the G1 cyclins Cln1-3 [31]. In addition, once cells enter S phase APC/C inhibition must be maintained until the next mitosis to avoid origin licensing during S or G2 phases. Both in human and *Drosophila* cells, depletion of Cdh1 inhibitor Emi1 leads to degradation of cyclin A and re-replication [32].

Apart from degrading the S and M phase cyclins in G1 stage of cell cycle, all studied eukaryotes have also evolved proteins able to bind to the S and M phase cyclin-CDK and inhibit their activity. These proteins are called cyclin-dependent kinase inhibitors (CKIs). Examples of CKIs in fission and budding yeast are Rum1 and Sic1, respectively [33, 34]. During mitotic exit, APC/C-Cdh1 allows the accumulation of CKI proteins by inducing the degradation of cyclins and, in metazoans, of the F-box protein Skp2, a component of the SCF that targets CKI proteins for proteolysis. In addition, in budding yeast, the cullin Cdc53 (cullin 1 homologue) and APC/C-Cdh1 induces, during the M phase, SIC1 transcription by polyubiquitylation and degradation of the transcription repressor Yph1 [35, 36]. This allows a rapid accumulation of Sic1, enforcing a low CDK activity status in the cell and a window of time during which origin licensing can occur. Importantly, premature progression into S phase in the absence of CKI proteins leads to genomic instability caused by an insufficient pool of replication origins [37]. CKIs in higher eukaryotes therefore constitute an important group of tumor-suppressor proteins.

In human cells, APC/C-Cdh1 also promotes the accumulation of the INKa family of proteins (p15, p16, p18, and p19), which control cell cycle progression by inhibiting the binding of the G1 cyclin D1 to Cdk4 and Cdk6 [38]. Finally, APC/C-Cdh1 targets the tyrosine phosphatase Cdc25, an activator of CDK, for ubiquitylation and degradation. An overexpression of a deubiquitylating enzyme Dub3 leads to the deubiquitylation and stabilization of Cdc25 and results in the accumulation of cells in S phase [39, 40].

During the window of time of APC/C-Cdh1 activity and high CKI concentrations, cells are competent for the loading of Mcm2-7 on chromatin and establishment of the pre-replicative complexes (pre-RCs) on replication origins (origin licensing) [41]. In all eukaryotes, this is achieved by ensuring the accumulation in the nucleus of all the components required for the formation of pre-RCs: Mcm2-7 helicase core, the origin recognition complex ORC1-6, and the Mcm2-7 loading factors Cdc6 and Cdt1. All eukaryotic organisms regulate these factors in a way dependent on the APC/C-Cdh1. In higher eukaryotes, Orc1-6 binding to chromatin and pre-RC formation are regulated also by histone and DNA methylation and the ORC-associated protein ORCA/Lrwd1 [42–44].

Mcm2-7 Complex

In budding yeast, low CDK activity (driven through APC/C regulation) is essential for the localization of the Mcm2-7 complex and Cdt1 in the nucleus. The cellular localization of Mcm2-7-Cdt1 complex is regulated by a bipartite nuclear localization signal (NLS) split between Mcm2 and Mcm3, and a nuclear export signal (NES) in Mcm3 [45–47] and depends on the integrity of the Mcm2-7-Cdt1 complex [48]. In addition, it has been reported that Mcm2-7 nuclear import also depends on their de novo synthesis in the M phase, which occurs in an Mcm1-dependent manner following the SCF-induced ubiquitylation and degradation of the Yox1 transcription inhibitor. Indeed, Mcm3 and Mcm4 are unstable during the transition between M and G1, although the mechanism of their degradation is still unknown [49]. The regulation of Mcm2-7 complex in higher eukaryotes is not yet defined.

Cdt1 and Geminin

While Cdt1 is not directly controlled by APC/C in budding yeast, in *S. pombe*, Cdt1 stability is cell cycle regulated and it accumulates only at the end of mitosis following degradation of S and M phase cyclins by APC/C-Cdh1 [50]. In metazoans, similarly to fission yeast, Cdt1 is an unstable protein outside of the late M/early G1 phase and its degradation is controlled by multiple E3 ubiquitin ligases (described below in section "The Prevention of Re-replication"). In addition, metazoan Cdt1 is also inhibited by binding of the small protein geminin. Geminin forms dimers through its coiled-coil domain and binds Cdt1 in a 2:1 ratio. Once bound to geminin, Cdt1 is unable to promote origin licensing. Geminin contains a bipartite NLS and a D-box, both important to regulate its stability during mitosis [51]. Following the metaphase-to-anaphase transition, APC/C-Cdh1 directly targets geminin, inactivating it and degrading it, thereby allowing Cdt1 to engage in origin licensing [52]. Non-degradable geminin mutants block origin licensing and lead to defective DNA replication. Interestingly, it has also been reported that APC/C-Cdh1 directly interacts with Cdt1 and that overexpression of Cdh1 affects Cdt1 protein stability [53], leading to its degradation at the late phase of mitosis, although with a markedly slower kinetic of degradation compared to geminin, thus allowing origin licensing during the late M phase [54].

Cdc6

In all eukaryotes, APC/C-Cdh1 controls the levels and stability of the licensing factor Cdc6. In budding yeast, CDC6 transcription is cell cycle regulated, similarly to the MCM genes, starting during M phase of the cell cycle in an Mcm1- and Swi5-dependent manner [55]. Once expressed, the levels of Cdc6 are finely regulated during the cell cycle. During M phase, Cdc6 turnover depends on the phosphorylation of two CDK consensus sites in the middle of the protein, the kinase Mck1 and the SCF^{Cdc4} ubiquitin ligase [56, 57]. In addition, a fraction of available Cdc6 is phosphorylated on its N-terminal CDK consensus sites and subsequently bound by the mitotic cyclin Clb2, thus allowing the accumulation of an inactivated form of Cdc6 until late mitosis [58]. Interestingly, a phosphorylated peptide of the N-terminal part of Cdc6 can also bind human cyclins E, A, and B1, suggesting that this regulatory mechanism may be conserved in higher eukaryotes. Altogether, APC/C-Cdh1, by inducing the degradation of Clbs, stops the mitotic degradation of Cdc6 and frees it from its interaction with the mitotic cyclins, allowing it to engage in the pre-RC formation. In yeast, the APC/C plays therefore a role in the stabilization of Cdc6 through modulation of CDK activity. Conversely, in human cells ACP/C appears to demonstrate an entirely different activity with respect to Cdc6, resulting in its degradation through its D- and KEN-boxes [59]. This mechanism abolishes origin licensing in cells during the early G1 phase or in a non-proliferative G0 state. Accumulation of Cdc6 requires phosphorylation of the APC/C-Cdh1 targeting boxes by cyclin E, expressed in response to mitogenic signals, stabilizing Cdc6 and allowing origin licensing before the accumulation of S phase cyclin A [59].

Orc1-6

In budding and fission yeast Orc1-6 is constitutively bound to chromatin throughout the cell cycle, while in higher eukaryotes it interacts with chromatin only during the licensing phase. In metazoan, APC/C allows the accumulation of Orc1, the large subunit of Orc1-6, permitting Orc1-6 to bind the DNA. The ability of Orc1-6 to interact with chromatin is also dependent on the methylation state of histones. In human cells, Orc1 binds to methylated histones H3 and H4 [42, 60] and the histone methyltransferase Set8 plays a key role in the regulation of the epigenetic modifications required for origin licensing [60, 61]. Interestingly, similarly to Cdc6 and Cdt1 in human cells, Set8 stability is regulated by APC/C-Cdh1: Set8 is briefly stabilized by Cdk1-cyclin B1 during metaphase, before being dephosphorylated by Cdc14 and targeted for degradation by APC/C-Cdh1 [62].

Role of Protein Ubiquitylation in the G1 to S Phase Transition

During G1 phase, cellular mass and the presence of proliferation signals control the decision of the cell to duplicate. Ultimately, this complex network of stimuli is integrated to allow the accurate control over the activity of CDK, primarily by controlling the levels of G1 cyclins in the cell. Accumulation of G1 cyclins promotes the transcriptional activity that allows cells to move forward with the cell duplication program. For reasons of space, our description of the dynamics of the transition from G1 to S is limited to budding yeast and human cells.

In budding yeast, cell growth controls the levels of the G1 cyclin Cln3. Once active, CDK-Cln3 stimulates both the transcription factor SBF (by promoting the nuclear export of the associated transcriptional repressor Whi5) and the transcription factor MBF (by directing the APC/C-Cdh1-dependent degradation of the MBF repressor Nmr1) [36, 63]. This leads to the expression of about 200 genes required for cell cycle progression and DNA replication, including the late G1 cyclins Cln1/2. Once expressed, CDK-Cln1/2 promote their own transcription and the accumulation of the S-phase cyclins Clb5/6. As they build up, Clb5/6 associate with the CDK but are kept inactive by binding of Sic1 [64, 65]. The levels of Sic1 therefore buffer CDK-Clb activity until it reaches a threshold level. The further progression of the cell cycle into S phase depends on the degradation of Sic1. Sic1 is sequentially phosphorylated by CDK-Cln1/2, followed by CDK-Clb5/6, thus leading to hyperphosphorylation of Sic1 [66]. These phosphorylations create multiple weak phospho-degrons that facilitate the interaction between Sic1 and Cdc4, an F-box protein component of the SCF ubiquitin ligase complex [67]. Sic1 is thus targeted for ubiquitylation and degradation by the SCF^{Cdc4} complex and the E2-conjugating enzyme Cdc34. Sic1 degradation and CDK-Clb5/6 activation promote the phosphorylation of Cdh1, causing its dissociation from the rest of the APC/C, thus allowing the accumulation of APC/C targets, including the regulatory subunit of DDK, Dbf4 [68]. CDK-Clb5/6 activation, therefore, marks the change from a period of high APC/C activity to one of high SCF activity. CDK does not appear to regulate SCF directly but stimulates protein degradation by creating phospho-degrons within their substrates, which are then recognized by the F-box receptors.

In mammalian cells, proliferation requires specific growth signals. When cells are in a quiescent state they exhibit low levels of G1 cyclins. Following growth stimuli, cyclin D1 is induced and interacts with Cdk4 and Cdk6. Similar to budding yeast, the transition from G1 to S phase involves a program of regulated gene transcription that is stimulated by positive feedback. In human cells, Cdk4/6-cyclin D activates transcription by partially phosphorylating the retinoblastoma family of proteins (Rb, p130, p107). This allows the transcription factor E2F1-3 to switch from actively repressing transcription when associated with unphosphorylated Rb to partially inducing genes required for DNA replication and cell cycle progression such as cyclin E (reviewed in [69]). p27 and p21 proteins (CKIs), whose levels are high in G1 phase, bind to Cdk2-cyclin E and inhibit its activity, until they cannot stoichiometrically block the CDK activity anymore. This is accelerated by the activity of the transcription factor Myc (reviewed in [70]). After reaching a critical level, Cdk2-cyclin E then phosphorylates p27 at threonine 187, stimulating the interaction between p27 and the F-box protein Skp2 and leading to p27 polyubiquitylation by the SCF^{Skp2} [71, 72]. This reaction also requires the RING protein Rbx1, and the E2-conjugating enzyme Cdc34. Strikingly, overexpression of cyclin E can suppress a G1 block induced by p27 overexpression as long as p27 is phosphorylated on T187. p21 degradation is also partially regulated by SCF^{Skp2}, and it appears to be stimulated by binding to CDK (reviewed in [73]). Following full activation of Cdk2-cyclin E, the Rb proteins become hyper-phosphorylated. This promotes the synthesis of genes required for S phase progression such as Cdh1 inhibitor Emi1 and the S phase cyclin A, leading in turn to the inhibition of APC/C activity. Importantly, Emi1 accumulation in the cell is favored in late G1 phase by the protein Evi5, which binds Emi1 and blocks its ubiquitylation by $SCF^{\beta TrCP}$ [74]. Furthermore, several other mechanisms co-operate to stop the activity of APC/C-Cdh1. In fact, Cdk2-cyclin E appears to directly phosphorylate Cdh1, thus inactivating it [75]. In addition, APC/C-Cdh1 activity is reduced by a negative feedback loop mechanism, involving APC/C-Cdh1 autoubiquitylation, and degradation of the E2-conjugating enzyme UbcH10 [76]. Moreover, Cdk2-cyclin A and the Polo-kinase Plk1, targets of APC/C-Cdh1 during the late M/G1 phase, phosphorylate Cdh1 during the progression from G1 to S phase, thus promoting its polyubiquitylation by $SCF^{\beta Trcp1}$ [77].

Finally, concomitantly with the progression into S phase, several SCF complexes complete a negative feedback loop that destroys the transcription factors and the cyclins that induced the G1 to S phase transition, thus ensuring the unidirectionality of the cell cycle. In budding yeast, SCF^{Cdc4} degrades Swi5—the transcription factor required for the expression of SIC1 and CDC6 [78] while SCF^{Grr1} degrades Cln1 and Cln2 [79]. Similarly, in human cells, SCF^{Fbx4} promotes the degradation of cyclin D1 [80, 81], SCF^{Skp2} degrades E2F [82], and cyclin E phosphorylation promotes its polyubiquitylation by SCF^{Fbw7}, the human orthologue of yeast SCF^{Cdc4} [83]. In addition, Myc is targeted for degradation by the SCF^{Skp2} and SCF^{Fbw7} [73].

The Prevention of Re-replication

Activation of S phase cyclins/CDK marks the end of the licensing activity in the cell and the beginning of origin firing in S phase. Full activation of CDK and DDK promotes the initiation of DNA replication. At this stage, the cell needs to ensure that no more origins can be licensed to avoid replicating DNA twice (re-replication), which would likely lead to chromosomal instability. At the same time, APC/C activity is substituted by the action of the cullin RING family of ubiquitin ligases (CRLs).

The prevention of DNA re-replication is resolved somehow differently between yeast and metazoans with many overlapping mechanisms working together. In *S. cerevisiae*, the paths that act to block re-replication include the degradation of Cdc6, the phosphorylation of Orc2 and Orc6, and the nuclear export of both Cdt1 and the Mcm2-7 complexes [48, 84–86]. In *S. pombe*, degradation of the Cdc6 homologue Cdc18 and that of Cdt1 prevents re-replication [87], while in vertebrates this is achieved by the inactivation of Cdt1 through its inhibition by geminin and its degradation [88–91], degradation of Orc1 in some cells [92, 93], nuclear export or degradation of Cdc6 [94], and degradation of the histone H4 methyltransferase Set8 [60, 95]. In all cases the time-specific degradation of the licensing factors depends on ubiquitin-mediated proteasomal activity.

Cdt1

Fission yeast and vertebrates degrade Cdt1 through a PCNA-dependent proteolysis pathway. PCNA is a homotrimeric clamp loaded onto DNA to increase the processivity of DNA polymerases. It also acts as a platform to recruit several proteins involved in DNA replication, repair, and sister chromatid cohesion establishment [96]. In addition, PCNA binds proteins required for origin licensing and promotes their degradation by recruiting a specialized E3-ubiquitin ligase, namely the cullin 4-containing ubiquitin ligase $CRL4^{Cdt2}$, where Cdt2 is the substrate-specific adaptor for the ligase [97, 98]. $CRL4^{Cdt2}$ plays a key role in the regulation of cell cycle and prevention of re-replication (reviewed also in [99], and [100]). Most proteins interacting with chromatin-bound PCNA do so through a motif consisting of eight amino acids, called PIP box (PCNA interacting protein box) [96]. However, $CRL4^{Cdt2}$ substrates contain a special high-affinity PIP box (PIP + TD motif) and, four amino acids downstream, a "B+4" basic residue, essential for recruitment of the $CRL4^{Cdt2}$ to the substrate [101, 102]. Additionally, Asp122 in PCNA is essential for the recruitment of $CRL4^{Cdt2}$ to chromatin [99]. Altogether, through these multiple interactions and once replication is initiated, Cdt1 is recruited to chromatin-bound PCNA together with $CRL4^{Cdt2}$, which ubiquitylates Cdt1 leading to its degradation.

In mammalian cells, Cdt1 can also be degraded by an alternative PCNA-independent route, using the cullin 1-based ubiquitin ligase, SCF^{Skp2}. Phosphorylation of Cdt1 by S-phase CDKs is required for the interaction of Cdt1 with Skp2 [103, 104] ensuring S-phase specificity of Cdt1 degradation. Specifically, it is the N-terminally located cyclin-binding consensus motif (Cy-motif), which is essential for the cyclin A and Skp2 association [104]. As both of the above degradation pathways overlap in mammalian cells, mutation of specific pathway motifs does not render Cdt1 non-degradable, but deletion of the N-terminal part of Cdt1, which contains all of the regulatory elements, results in a stable mutant [105–107].

Finally, any Cdt1 remaining until the G2 stage of the cell cycle can be degraded following ubiquitylation by SCF^{FbxO31} ligase in a manner also dependent on the N-terminal end of Cdt1 [108].

Cdc6

Regulation of Cdc6 levels upon entry into S phase is much more variable across the species and even within the genus *Saccharomyces*, this regulation can significantly differ [109]. Nevertheless, in both budding and fission yeast, Cdc6 (Cdc18 in *S.p.*) is degraded upon initiation of replication through the activity of SCF complexes: SCF^{Cdc4} in *S. cerevisiae* and $SCF^{Pop1,2}$ in *S. pombe* [110–112]. This degradation is stimulated by phosphorylation of Cdc6 by CDK activity (both G1- and S-phase CDK complexes) [64, 113]. In human cells, several groups have reported a stable level of Cdc6 throughout most of the cell cycle in proliferating cells [114, 115] and the licensing activity of Cdc6 seems to be regulated through its nuclear export in

S phase, which is also stimulated by CDK phosphorylation [94, 115]. A recent report, however, suggested the involvement of PCNA- and CRL4^{Cdt2}-dependent degradation of Cdc6 upon replication initiation [116]. This degradation mode is possible due to the N-terminal PIP-box within Cdc6 [116].

Finally, CRL4^{Cdt2}- and SCF^{Skp2}-dependent ubiquitylation and degradation of p21 are also important for the inhibition of Cdc6-promoted re-replication. Destruction of CDK inhibitors such as p21 allows CDK phosphorylation of Cdc6 and its nuclear export [94, 117].

Orc1-6

The Orc1-6 complex is differentially regulated in yeast and metazoans. In yeast, the ORC complex remains stably bound to chromatin and at a constant level throughout the cell cycle [41, 118]. Its inactivation occurs on chromatin by direct binding of CDK to the ORC complex through interaction of CDK with Orc2 (in *S. pombe*) or Orc6 (in *S. cerevisiae*) [86, 119]. In metazoans, S-phase regulation of the ORC complex occurs through its large subunit Orc1 and differs not only through evolution but even between different cell lines. In mammals, a number of research groups have reported constant levels of Orc1 throughout the cell cycle, although it was suggested to be selectively released from chromatin at the beginning of S phase [93, 120, 121]. In hamster cells, the released Orc1 is targeted for mono- or di-ubiquitylation, which does not lead to its degradation [120]. On the other hand, in transformed human cells 50–90 % of Orc1 is selectively degraded during S phase by a ubiquitin-dependent mechanism [92, 122] and Orc1 has been shown to be ubiquitylated by SCF^{Skp2} [92].

Dynamic associations of ORCA/Lrwd1 with the licensing factors ORC1-6, Cdt1 and geminin play a role in pre-RC assembly and replication initiation [44, 123]. The levels of ORCA peak in G1 and gradually decrease in S phase [44]. ORCA has been shown to be ubiquitylated at the onset of S phase, most likely by Cul4A(Ddb1) [124]. However, once ubiquitylated, it can still stably bind chromatin and is degraded only upon depletion of Orc2 [124]. The exact nature of this regulation and its importance for replication initiation are still to be determined.

Finally, interaction of the ORC complex with chromatin is further inhibited upon entry into S phase by regulation of epigenetic marks stabilising Orc1-6 binding to replication origins. In fact, Set8 is another protein degraded in S phase by CRL4^{Cdt2} in a PCNA-dependent manner to prevent re-replication. Importantly, while loss of Set8 enzymatic activity causes defects in replication origin firing (as described above) [125, 126], expression of non-degradable (PIP degron mutant) Set8 leads to uncontrolled re-replication [95, 127].

Other Factors Regulated by Ubiquitin upon S Phase Entry

In fission yeast Spd1, an inhibitor of the ribonucleotide reductase (RNR) enzyme that catalyzes the synthesis of dNTPs is also a major substrate of PCNA-dependent CRL4^{Cdt2} [128–130]. Spd1 degradation allows accumulation of dNTPs required for replicative DNA synthesis; thus it regulates DNA replication initiation rather than re-replication. In addition, in budding yeast the S phase checkpoint kinase Rad53, even in the absence of replication stress, finely controls the levels of free histones by binding to them and stimulating their ubiquitylation by the E3 ubiquitin ligase Tom1 [131, 132].

Finally, the activity of $CRL4^{Cdt2}$ is regulated during DNA replication through the modulation of Cdt2 turnover. In human cells, Cdt2 is targeted for degradation both by autoubiquitylation by the CLR4A ubiquitin ligase in a PCNA-independent reaction and by SCF^{FbxO11} [133]. However, Cdk1-cyclin B phosphorylation of Cdt2 at T464 prevents it from interacting with SCF^{FbxO11}, thereby stabilizing Cdt2 in S phase [133, 134]. This protection is executed through 14-3-3 adaptor proteins, which interact with Cdt2 during S phase in a phosphorylation-dependent manner [135]. All these contribute to a very cell-cycle stage-specific degradation of $CRL4^{Cdt2}$ and $CRL4^{Cdt2}$ substrates and prevent re-replication in the cells.

DNA Damage Response and Replication Initiation

In response to DNA damage or replication stress, cells activate checkpoint mechanisms to delay progression of the cell cycle and allow time to fix the lesions and to avoid additional damage. This is a complex response that affects many cellular pathways (for review see [136]). In this section we briefly describe the role of protein ubiquitylation and proteasomal degradation in the regulation of origin firing following the activation of G1/S and intra-S checkpoints. The G1/S checkpoint blocks progression into S phase and the initiation of DNA replication, while the intra-S checkpoint inhibits origin firing in regions of chromatin that have not yet initiated replication and inhibits origin re-licensing [137, 138].

The G1/S checkpoint exploits two complementing strategies, both dependent on protein ubiquitylation, to effectively block origin firing within minutes after DNA damage and establish a longer lasting block in G1, providing sufficient time for repair. The fast response strategy works through the rapid degradation of the Cdc25A phosphatase, while the slow, more prolonged response acts through the stabilization of the transcription factor p53 [137]. The phosphatase activity of Cdc25A, as mentioned above, removes the inhibitory phosphorylation on Cdk2 and is essential for G1/S transition [139]. Upon DNA damage, such as UV and IR exposure, Cdc25A is hyper-phosphorylated by activated Chk1, Chk2, and other kinases, which target it for ubiquitylation and proteasomal degradation [139, 140] and lead to inactivation of Cdk2.

The ubiquitin-dependent slower checkpoint response relies on activity of the p53 tumor-suppressor protein. During unperturbed G1, the amount of p53 protein is maintained at a low level by the E3 ubiquitin ligase Mdm2, which mediates constant p53 ubiquitylation and degradation [141]. Upon DNA damage, p53 is posttranslationally modified, which breaks the interaction with Mdm2 and protects p53 from degradation [142]. Subsequently, the accumulation of p53 activates the transcription of p21, blocking CDK activity and progression into S phase.

DNA damage or replication stress in S phase leads to activation of the checkpoint kinase ATR and CHK1, which stabilize stalled replication forks, delay mitotic entry, and promote lesion repair [143, 144]. In budding yeast, these kinases also inhibit initiation of late origin firing, while in higher eukaryotes they inhibit activation of new replication factories and delay the replication timing program [145]. In budding yeast, the Chk2 homologue Rad53 prevents late origin firing by phosphorylating and inhibiting two key origin firing factors Sld3 and Dbf4 [146, 147]. In metazoans, Chk1 activation also inhibits treslin-TICRR (Sld3 ortholog) from interacting with its origin firing partner Dpb11 [148, 149]. This inhibition may work through direct phosphorylation of these and other factors or indirectly through reducing CDK and/or DDK activity. There is evidence that Chk1 (and other kinases) phosphorylate Cdc25A, targeting it for accelerated degradation by the $SCF^{\beta TrCP}$ ubiquitin ligase and leading to inhibition of Cdk2 activity and further origin firing [150].

In response to DNA damage, many of the substrates of the $CRL4^{Cdt2}$ ubiquitin ligase are also targeted for destruction, including Spd1, Cdt1, p21, and Set8 (reviewed in [100, 151]) in either G1 or S phase. Little is known about the function of degradation of these factors: p21 destruction may be important for the release of PCNA for repair processes, the reduction of Set8 may contribute to the completion of DNA repair, while the role of Cdt1 degradation upon DNA damage is even less clear, as it happens when many origins are already licensed. The importance of this ubiquitin-dependent regulatory mechanism for initiation of DNA replication is therefore still to be determined. It has been suggested also that DNA damage induced by re-replication can stimulate degradation of Cdt1 by CRL4 to reduce the extent of re-replication [152]. Moreover, in response to re-replication and DNA damage (irrespectively of cell cycle stage), Cdc6 can be ubiquitylated by the HECT-family ubiquitin ligase, Huwe1, and degraded to reduce re-replication or to promote checkpoint functions that block further cell cycle progression [152, 153].

NEDD8

Nedd8 (also known as Rub1 in *S. cerevisiae*) was originally identified as a gene that is highly expressed in the embryonic mouse brain [154]. It is however expressed in most eukaryotes and in most tissues and is essential for viability of most model organisms with the exception of budding yeast (reviewed in [155]). Neddylation is the posttranslational modification that is most closely related to ubiquitylation but unlike ubiquitin, only a limited number of Nedd8 attachment substrates have been identified so far. As in ubiquitylation, Nedd8 is attached to substrates by an isopeptide bond between its C-terminal glycine (G76) and lysine of the substrate, involving a classic E1, E2, and E3 enzymatic cascade. Nedd8-activating enzyme E1 NAE (a heterodimer of NAE1 and UBA3) transfers activated Nedd8 to a unique E2 (Ubc12), which passes it to an Nedd8 E3 ligase [156].

The first identified and the major substrate of neddylation are cullins, which function as scaffolds for the assembly of multisubunit cullin E3s: CRLs. Remarkably, all yeast and mammalian canonical cullins (Cul1, Cul2, Cul3, Cul4A, Cul4B, Cul5, and Cul7) and cullin-related PARC are modified with Nedd8 [157], with the exception of non-neddylated, cullin-related Apc2 (a subunit of APC/C) [158]. As described at the beginning of this chapter, the N-terminal part of cullins binds substrate-binding adaptors, while the C-terminal region binds a RING domain adaptor and an E2 enzyme. It is in this C-terminal region that cullins are neddylated (Fig. 17.2). Upon neddylation the C-terminal domain undergoes a substantial conformational change that relieves Rbx1 from an autoinhibitory mechanism and stimulates substrate ubiquitylation [159]. Moreover, unneddylated cullins can interact with the cullin inhibitor CAND1 [160] and thus neddylation disrupts this inhibitory interaction.

Taking into account the above, the main regulatory role of Nedd8 modification upon initiation of DNA replication is therefore stimulation of the activity of multiple CRLs functioning during replication initiation. The roles of Cul1-based CRLs (SCFs) as well as Cul4-based CRL complexes in replication initiation have been described above, but Cul2 has also been reported as essential for replication initiation [161–163]. Moreover, treatment of cells with a small-molecule inhibitor of NAE (MLN4924) results in uncontrolled DNA synthesis through re-replication, leading to DNA damage and induction of apoptosis [164]. This is mainly due to inhibition of $CRL4^{Cdt2}$ E3 ligase-mediated degradation of Cdt1 and other $CRL4^{Cdt2}$ substrates (described above) [165].

Cullins are not the only class of proteins modified by Nedd8, although proteome-wide approaches to identify Nedd8 substrates fail to deliver many non-cullin targets, suggesting that the non-cullin neddylated substrates are only weakly expressed or weakly modified in unsynchronized cell culture [155]. Of importance for replication initiation are the p53 family of proteins and Mdm2 [166, 167]. As described above, p53 is stabilized upon DNA damage and stress and allowed to stimulate expression of replication initiation inhibitors to block entry into S phase. The same ligase that promotes ubiquitylation of p53 can also promote its neddylation [166, 168]. Mdm2-mediated neddylation of p53 inhibits its transcriptional activity. Moreover, as p53 is ubiquitylated and neddylated on overlapping lysines, the two modifications are exclusive and thus regulate each other's abundance [155]. Finally, Mdm2 can auto-neddylate itself, which inhibits its ability to ubiquitylate and degrade p53 [166]. The importance of these regulations for p53 function is still to be determined.

SUMO

The small ubiquitin-like modifier (SUMO) is related to ubiquitin, and its structure shows a similar ubiquitin fold (a globular β-grasp fold), plus an additional flexible N-terminus of about 20 amino acids. In addition to the structural similarities between the two proteins, the SUMO-conjugation pathway shares many features with the ubiquitin-conjugation pathway but shows a much lower degree of complexity and specialization: in human cells, 1 E1, 1 E2, and about 10 E3 ligases exist, compared to the 2 E1, 35 E2, and about 600 E3 components of the ubiquitin-conjugation pathway [169]. Sumoylation usually modifies only a small fraction of the total target protein (less than 1%) and is removed by a small number of SUMO isopeptidases, namely Ulp1-2 in budding yeast and the six SENP proteins in human cells [170]. Mammalian cells have three different SUMO proteins, with SUMO2 sharing 97% of homology with SUMO3 and about 50% homology with SUMO1 (the closest orthologue of Smt3, the only SUMO in yeast). Unbiased mass spectrometry analysis has shown that in about 50% of the sumoylation events, the modification occurs at a consensus site ΨKXE, where Ψ represents a large hydrophobic amino acid and X represents any amino acid.

Protein sumoylation has been associated with many molecular functions. These include the regulation of protein–protein interactions either by recruiting proteins containing a SUMO-interaction motif (SIM), by displacing protein interactors by blocking an interaction surface [96], or by controlling protein stability—either by competing with ubiquitylation, thus stabilizing proteins, or by targeting proteins to degradation via SUMO-dependent ubiquitin ligases [171–173].

One of the proteins involved in replication initiation that we know is modified by SUMO is p53 (in addition to its regulation by ubiquitylation and neddylation). Sumo1 modifies a single lysine of p53—K386—but the consequences of this modification are still unclear due to conflicting reports (reviewed in [174]). Similarly, a number of different Sumo ligases have been identified to mediate Sumo1 conjugation, including Mdm2 and PIAS proteins amongst the others [174]. p53 was also shown to be modified by Sumo2/3 on K386 in response to H_2O_2 or in Kaposi's sarcoma-associated herpes virus infection, both leading to induction of its transcriptional activity [175, 176]. Mdm2 was also shown to promote Sumo2/3 conjugation to p53 resulting in both activation and repression of p53 target genes, including the repression of p21 transcription [177]. The importance of this additional level of p53 regulation on its role in replication initiation is still to be unravelled.

It has been suggested recently that in *Xenopus laevis* egg extract the major Sumo2/3-ylated protein on chromatin during unchallenged DNA replication is cyclin E [178], which is essential in this system for replication initiation. Moreover, blocking sumoylation in the extract led to an increased number of firing origins, thus accelerating DNA synthesis in early S phase [178]. The mechanism behind this phenomenon is, however, yet to be described.

Another replication factor essential for replication initiation and regulated by sumoylation is one of the single-strand binding protein complex RPA subunits, RPA70. In an unperturbed S phase, RPA70 associates with the SUMO-specific protease SENP6/Sentrin that keeps RPA70 in a hypo-sumoylated state. The dissociation of this complex upon fork collapse leads to sumoylation of RPA70 with Sumo2/3 on K449 and K577, which in turn stimulates homologous recombination [179].

Finally, two biochemical screens performed in budding yeast and *Xenopus laevis* egg extract identified a number of replication factors as Sumo targets during unperturbed DNA replication [180, 181], suggesting that Sumo-dependent regulation of DNA replication may soon be revealed to be much more common (Tables 17.1 and 17.2).

Table 17.1 Replication initiation factors regulated by ubiquitin

Substrate	Ligase	Cell cycle stage	Function	Species	Ref.
Cdc25	APC/C-Cdh1	Late M/G1 phase	Blocking CDK activity	*Hs*	[39]
Cdc25A	$SCF^{\beta TrCP}$	S phase	Blocking CDK activity	*Hs*	[182]
Cdc6	APC/C-Cdh1	Late M phase	Block origin licensing	*Hs*	[59]
Cdc6	SCF^{Cdc4}	S phase/M phase	Blocking re-replication	*Sc*	[56, 112]
Cdc6	$CRL4^{Cdt2}$	S phase	Blocking re-replication	*Hs*	[116]
Cdc6	Huwe1	S phase DNA damage	Blocking re-replication?	*Hs*	[152, 153]
Cdc18	$SCF^{Pop1,2}$	Late G1/S phase	Blocking re-replication	*Sp*	[110, 111]
Cdc20	APC/C-Cdh1	Late M phase	Exit from mitosis	*Sc*	[25]
Cdh1	$SCF^{\beta Trcp1}$	Late G1/S phase	Progression in S phase	*Hs*	[77]
Cdt1	APC/C-Cdh1	Late M phase	Block origin licensing	*Hs*	[53]
Cdt1	$CRL4^{Cdt2}$	S phase	Blocking re-replication	*Sp*, *Hs*, *Xl*	[97, 107]
Cdt1	SCF(Skp2)	S phase	Blocking re-replication	*Hs*	[103, 104]
Cdt1	SCF^{FbxO31}	G2 phase	Blocking re-replication	*Hs*	[108]
Cdt2	CRL4A	S phase	Regulation of $CRL4^{Cdt2}$	*Hs*	[133]
Cdt2	SCF^{FbxO11}	S phase	Regulation of $CRL4^{Cdt2}$	*Hs*	[133]
Clb2	APC/C-Cdc20 APC/C-Cdh1	M phase	Exit from mitosis	*Sc*	[19]
Clb5	APC/C-Cdc20	M phase	Progression of M phase	*Sc*	[17, 18]
Cln1-2	SCF^{Grr1}	S phase	Stop late G1 phase transcription	*Sc*	[79]
cyclin A	APC/C-Cdc20	M phase	Exit from mitosis; origin licensing	*Hs*	[183]
cyclin B	APC/C-Cdc20	M phase	Exit from mitosis; origin licensing	*Hs*	[54]
cyclin D1	SCF^{Fbx4}	Late G1/S	Progression into S phase	*Hs*	[80, 81]
cyclin E	SCF^{Fbw7}	Late G1/S	Blocking re-replication	*Hs*	[83]
Dbf4	APC/C-Cdc20	M phase	Blocking origin firing	*Sc*	[20, 21]
E2F	SCF^{Skp2}	Late G1/S	Blocking re-replication	*Hs*	[82]

(continued)

Table 17.1 (continued)

Substrate	Ligase	Cell cycle stage	Function	Species	Ref.
Emi1	$SCF^{\beta Trcp1}$	M/G1 phase	Activation of APC/C	*Hs*	[28, 29]
geminin	APC/C-Cdh1	Late M phase	Origin licensing	*Hs*, *Xl*	[52]
Myc	SCF^{Skp2}	Late G1/S	Blocking re-replication	*Hs*	[184, 185]
Myc	SCF^{Fbw7}	Late G1/S	Blocking re-replication	*Hs*	[186, 187]
Nmr1	APC/C-Cdh1	G1 phase	Allow transcription from MBF	*Sc*	[36]
Orc1	SCF^{Skp2}	S phase	Blocking re-replication	Transformed *Hs* cells	[92]
ORCA	$CRL4A^{Ddb1}$		Blocking re-replication	Hs	[44]
p21	SCF^{Skp2}	S phase	Blocking re-replication	Hs	[107]
p21	$CRL4^{Cdt2}$	S phase	Blocking re-replication	Hs	[94, 117]
p27	SCF^{Skp2}	Late G1-S phase	Progression into S phase	*Hs*	[71, 72]
p53	Mdm2	G1 phase	No delays in the G1-S transition	Hs	[141]
Rum1	$SCF^{Pop1\text{-}Pop2}$	Late G1/S phase	Progression into S phase	*Sp*	[33]
Set8	APC/C-Cdh1	Late M phase	Block origin licensing	*Hs*	[62]
Set8	$CRL4^{Cdt2}$	S phase	Blocking re-replication	Hs	[95]
Sic1	SCF^{Cdc4}	Late G1/S phase	Progression into S phase	*Sc*	[34]
Skp2	APC/C-Cdh1	Late M phase	Origin licensing	*Hs*	[188]
Spd1	$CRL4^{Cdt2}$	S phase	Stimulates nucleotide synthesis	*Sp*	[128–130]
Swi5	SCF^{Cdc4}	Late G1/S phase	Stop late G1 phase transcription	*Sc*	[78]
UbcH10	APC/C-Cdh1	Late G1/S phase	Stop APC/C-Cdh1 activity	*Hs*	[76]
Yox1	Cdc53 (SCF)	Early M phase	Induce transcription of late M/G1 phase	*Sc*	[35]
Yph1	APC/C-Cdh1 Cdc53 (SCF)	M phase, G1 phase	Transcription of Sic1 and other G1 genes	*Sc*	[35, 36]

Sc Saccharomyces cerevisiae, *Hs Homo sapiens*, *Sp Schizosaccharomyces pombe*, *Xl Xenopus laevis*

Table 17.2 Replication initiation factors modified by Nedd8 or SUMO

Substrate	Modification	Function of modification	Reference
Cul1	Nedd8	Activation of SCF	[157]
Cul2	Nedd8	Activation of CRL2	[157]
Cul4	Nedd8	Activation of CRL4	[157]
Cyclin E	Sumo2/3	?	[178]
Mdm2	Nedd8	Inhibition of Mdm2 ubiquitin ligase activity	[166]
p53	Nedd8	Inhibition of p53	[166]
p53	Sumo1	?	[174]
p53	Sumo2/3	?	[177]
RPA	Sumo2/3	Stimulates homologous recombination	[179]

?= unknown function

References

1. Evrin C, Clarke P, Zech J, Lurz R, Sun J, Uhle S, et al. A double-hexameric MCM2-7 complex is loaded onto origin DNA during licensing of eukaryotic DNA replication. Proc Natl Acad Sci U S A. 2009;106(48):20240–5.
2. Gambus A, Khoudoli GA, Jones RC, Blow JJ. MCM2-7 form double hexamers at licensed origins in *Xenopus* egg extract. J Biol Chem. 2011;286(13):11855–64.
3. Remus D, Beuron F, Tolun G, Griffith JD, Morris EP, Diffley JF. Concerted loading of Mcm2-7 double hexamers around DNA during DNA replication origin licensing. Cell. 2009;139(4): 719–30.
4. Komander D, Rape M. The ubiquitin code. Annu Rev Biochem. 2012;81:203–29 [Research Support, N.I.H., Extramural Research Support, Non-U.S. Gov't Review].
5. Behrends C, Harper JW. Constructing and decoding unconventional ubiquitin chains. Nat Struct Mol Biol. 2011;18(5):520–8 [Research Support, N.I.H., Extramural Research Support, Non-U.S. Gov't Review].
6. Peng J, Schwartz D, Elias JE, Thoreen CC, Cheng D, Marsischky G, et al. A proteomics approach to understanding protein ubiquitination. Nat Biotechnol. 2003;21(8):921–6.
7. Kim W, Bennett EJ, Huttlin EL, Guo A, Li J, Possemato A, et al. Systematic and quantitative assessment of the ubiquitin-modified proteome. Mol Cell. 2011;44(2):325–40 [Research Support, N.I.H., Extramural Research Support, Non-U.S. Gov't].
8. Berndsen CE, Wolberger C. New insights into ubiquitin E3 ligase mechanism. Nat Struct Mol Biol. 2014;21(4):301–7 [Research Support, N.I.H., Extramural Research Support, Non-- U.S. Gov't Research Support, U.S. Gov't, Non-P.H.S. Review].
9. Zimmerman ES, Schulman BA, Zheng N. Structural assembly of cullin-RING ubiquitin ligase complexes. Curr Opin Struct Biol. 2010;20(6):714–21 [Research Support, N.I.H., Extramural Research Support, Non-U.S. Gov't Research Support, U.S. Gov't, Non-P.H.S. Review].
10. Sarikas A, Hartmann T, Pan ZQ. The cullin protein family. Genome Biol. 2011;12(4):220 [Research Support, N.I.H., Extramural Research Support, Non-U.S. Gov't Review].
11. Matyskiela ME, Rodrigo-Brenni MC, Morgan DO. Mechanisms of ubiquitin transfer by the anaphase-promoting complex. J Biol. 2009;8(10):92.
12. Rodrigo-Brenni MC, Morgan DO. Sequential E2s drive polyubiquitin chain assembly on APC targets. Cell. 2007;130(1):127–39.
13. Visintin R, Prinz S, Amon A. CDC20 and CDH1: a family of substrate-specific activators of APC-dependent proteolysis. Science. 1997;278(5337):460–3.
14. Fang G, Yu H, Kirschner MW. Direct binding of CDC20 protein family members activates the anaphase-promoting complex in mitosis and G1. Mol Cell. 1998;2(2):163–71.

15. Pfleger CM, Kirschner MW. The KEN box: an APC recognition signal distinct from the D box targeted by Cdh1. Genes Dev. 2000;14(6):655–65.
16. Rahal R, Amon A. Mitotic CDKs control the metaphase-anaphase transition and trigger spindle elongation. Genes Dev. 2008;22(11):1534–48.
17. Shirayama M, Toth A, Galova M, Nasmyth K. APC(Cdc20) promotes exit from mitosis by destroying the anaphase inhibitor Pds1 and cyclin Clb5. Nature. 1999;402(6758):203–7.
18. Thornton BR, Toczyski DP. Securin and B-cyclin/CDK are the only essential targets of the APC. Nat Cell Biol. 2003;5(12):1090–4.
19. Wasch R, Cross FR. APC-dependent proteolysis of the mitotic cyclin Clb2 is essential for mitotic exit. Nature. 2002;418(6897):556–62.
20. Lu D, Hsiao JY, Davey NE, Van Voorhis VA, Foster SA, Tang C, et al. Multiple mechanisms determine the order of APC/C substrate degradation in mitosis. J Cell Biol. 2014;207(1):23–39.
21. Ferreira MF, Santocanale C, Drury LS, Diffley JF. Dbf4p, an essential S phase-promoting factor, is targeted for degradation by the anaphase-promoting complex. Mol Cell Biol. 2000;20(1):242–8.
22. Oshiro G, Owens JC, Shellman Y, Sclafani RA, Li JJ. Cell cycle control of Cdc7p kinase activity through regulation of Dbf4p stability. Mol Cell Biol. 1999;19(7):4888–96.
23. Weinreich M, Stillman B. Cdc7p-Dbf4p kinase binds to chromatin during S phase and is regulated by both the APC and the RAD53 checkpoint pathway. EMBO J. 1999;18(19):5334–46.
24. Jaspersen SL, Charles JF, Morgan DO. Inhibitory phosphorylation of the APC regulator Hct1 is controlled by the kinase Cdc28 and the phosphatase Cdc14. Curr Biol. 1999;9(5):227–36.
25. Robbins JA, Cross FR. Regulated degradation of the APC coactivator Cdc20. Cell Div. 2010; 5:23.
26. Sullivan M, Morgan DO. Finishing mitosis, one step at a time. Nat Rev Mol Cell Biol. 2007;8(11):894–903.
27. Guardavaccaro D, Kudo Y, Boulaire J, Barchi M, Busino L, Donzelli M, et al. Control of meiotic and mitotic progression by the F box protein beta-Trcp1 in vivo. Dev Cell. 2003;4(6):799–812.
28. Moshe Y, Boulaire J, Pagano M, Hershko A. Role of Polo-like kinase in the degradation of early mitotic inhibitor 1, a regulator of the anaphase promoting complex/cyclosome. Proc Natl Acad Sci U S A. 2004;101(21):7937–42.
29. Margottin-Goguet F, Hsu JY, Loktev A, Hsieh HM, Reimann JD, Jackson PK. Prophase destruction of Emi1 by the SCF(betaTrCP/Slimb) ubiquitin ligase activates the anaphase promoting complex to allow progression beyond prometaphase. Dev Cell. 2003;4(6):813–26.
30. Wang W, Kirschner MW. Emi1 preferentially inhibits ubiquitin chain elongation by the anaphase-promoting complex. Nat Cell Biol. 2013;15(7):797–806.
31. Irniger S, Nasmyth K. The anaphase-promoting complex is required in G1 arrested yeast cells to inhibit B-type cyclin accumulation and to prevent uncontrolled entry into S-phase. J Cell Sci. 1997;110(Pt 13):1523–31.
32. Di Fiore B, Pines J. Defining the role of Emi1 in the DNA replication-segregation cycle. Chromosoma. 2008;117(4):333–8.
33. Moreno S, Nurse P. Regulation of progression through the G1 phase of the cell cycle by the rum1+ gene. Nature. 1994;367(6460):236–42.
34. Schwob E, Bohm T, Mendenhall MD, Nasmyth K. The B-type cyclin kinase inhibitor p40SIC1 controls the G1 to S transition in *S. cerevisiae*. Cell. 1994;79(2):233–44.
35. Landry BD, Mapa CE, Arsenault HE, Poti KE, Benanti JA. Regulation of a transcription factor network by Cdk1 coordinates late cell cycle gene expression. EMBO J. 2014;33(9):1044–60.
36. Ostapenko D, Solomon MJ. Anaphase promoting complex-dependent degradation of transcriptional repressors Nrm1 and Yhp1 in *Saccharomyces cerevisiae*. Mol Biol Cell. 2011;22(13): 2175–84.
37. Lengronne A, Schwob E. The yeast CDK inhibitor Sic1 prevents genomic instability by promoting replication origin licensing in late G(1). Mol Cell. 2002;9(5):1067–78.
38. Serrano M, Hannon GJ, Beach D. A new regulatory motif in cell-cycle control causing specific inhibition of cyclin D/CDK4. Nature. 1993;366(6456):704–7.

39. Donzelli M, Squatrito M, Ganoth D, Hershko A, Pagano M, Draetta GF. Dual mode of degradation of Cdc25 A phosphatase. EMBO J. 2002;21(18):4875–84.
40. Pereg Y, Liu BY, O'Rourke KM, Sagolla M, Dey A, Komuves L, et al. Ubiquitin hydrolase Dub3 promotes oncogenic transformation by stabilizing Cdc25A. Nat Cell Biol. 2010;12(4):400–6.
41. Diffley JF, Cocker JH, Dowell SJ, Rowley A. Two steps in the assembly of complexes at yeast replication origins in vivo. Cell. 1994;78(2):303–16.
42. Bartke T, Vermeulen M, Xhemalce B, Robson SC, Mann M, Kouzarides T. Nucleosome-interacting proteins regulated by DNA and histone methylation. Cell. 2010;143(3):470–84.
43. Kuo AJ, Song J, Cheung P, Ishibe-Murakami S, Yamazoe S, Chen JK, et al. The BAH domain of ORC1 links H4K20me2 to DNA replication licensing and Meier-Gorlin syndrome. Nature. 2012;484(7392):115–9.
44. Shen Z, Sathyan KM, Geng Y, Zheng R, Chakraborty A, Freeman B, et al. A WD-repeat protein stabilizes ORC binding to chromatin. Mol Cell. 2010;40(1):99–111.
45. Labib K, Diffley JF, Kearsey SE. G1-phase and B-type cyclins exclude the DNA-replication factor Mcm4 from the nucleus. Nat Cell Biol. 1999;1(7):415–22.
46. Nguyen VQ, Co C, Irie K, Li JJ. Clb/Cdc28 kinases promote nuclear export of the replication initiator proteins Mcm2-7. Curr Biol. 2000;10(4):195–205.
47. Liku ME, Nguyen VQ, Rosales AW, Irie K, Li JJ. CDK phosphorylation of a novel NLS-NES module distributed between two subunits of the Mcm2-7 complex prevents chromosomal rereplication. Mol Biol Cell. 2005;16(10):5026–39.
48. Tanaka S, Diffley JF. Interdependent nuclear accumulation of budding yeast Cdt1 and Mcm2-7 during G1 phase. Nat Cell Biol. 2002;4(3):198–207 [Research Support, Non-U.S. Gov't].
49. Braun KA, Breeden LL. Nascent transcription of MCM2-7 is important for nuclear localization of the minichromosome maintenance complex in G1. Mol Biol Cell. 2007;18(4):1447–56.
50. Nishitani H, Lygerou Z, Nishimoto T, Nurse P. The Cdt1 protein is required to license DNA for replication in fission yeast. Nature. 2000;404(6778):625–8.
51. Benjamin JM, Torke SJ, Demeler B, McGarry TJ. Geminin has dimerization, Cdt1-binding, and destruction domains that are required for biological activity. J Biol Chem. 2004;279(44): 45957–68.
52. McGarry TJ, Kirschner MW. Geminin, an inhibitor of DNA replication, is degraded during mitosis. Cell. 1998;93(6):1043–53.
53. Sugimoto N, Kitabayashi I, Osano S, Tatsumi Y, Yugawa T, Narisawa-Saito M, et al. Identification of novel human Cdt1-binding proteins by a proteomics approach: proteolytic regulation by APC/CCdh1. Mol Biol Cell. 2008;19(3):1007–21.
54. Clijsters L, Ogink J, Wolthuis R. The spindle checkpoint, APC/C(Cdc20), and APC/C(Cdh1) play distinct roles in connecting mitosis to S phase. J Cell Biol. 2013;201(7):1013–26.
55. Piatti S, Lengauer C, Nasmyth K. Cdc6 is an unstable protein whose de novo synthesis in G1 is important for the onset of S phase and for preventing a 'reductional' anaphase in the budding yeast *Saccharomyces cerevisiae*. EMBO J. 1995;14(15):3788–99.
56. Perkins G, Drury LS, Diffley JF. Separate SCF(CDC4) recognition elements target Cdc6 for proteolysis in S phase and mitosis. EMBO J. 2001;20(17):4836–45.
57. Ikui AE, Rossio V, Schroeder L, Yoshida S. A yeast GSK-3 kinase Mck1 promotes Cdc6 degradation to inhibit DNA re-replication. PLoS Genet. 2012;8(12):e1003099.
58. Mimura S, Seki T, Tanaka S, Diffley JF. Phosphorylation-dependent binding of mitotic cyclins to Cdc6 contributes to DNA replication control. Nature. 2004;431(7012):1118–23.
59. Mailand N, Diffley JF. CDKs promote DNA replication origin licensing in human cells by protecting Cdc6 from APC/C-dependent proteolysis. Cell. 2005;122(6):915–26.
60. Tardat M, Brustel J, Kirsh O, Lefevbre C, Callanan M, Sardet C, et al. The histone H4 Lys 20 methyltransferase PR-Set7 regulates replication origins in mammalian cells. Nat Cell Biol. 2010;12(11):1086–93 [Research Support, Non-U.S. Gov't].
61. Beck DB, Burton A, Oda H, Ziegler-Birling C, Torres-Padilla ME, Reinberg D. The role of PR-Set7 in replication licensing depends on Suv4-20h. Genes Dev. 2012;26(23):2580–9.

62. Wu S, Wang W, Kong X, Congdon LM, Yokomori K, Kirschner MW, et al. Dynamic regulation of the PR-Set7 histone methyltransferase is required for normal cell cycle progression. Genes Dev. 2010;24(22):2531–42.
63. de Bruin RA, McDonald WH, Kalashnikova TI, Yates 3rd J, Wittenberg C. Cln3 activates G1-specific transcription via phosphorylation of the SBF bound repressor Whi5. Cell. 2004;117(7):887–98.
64. Drury LS, Perkins G, Diffley JF. The cyclin-dependent kinase Cdc28p regulates distinct modes of Cdc6p proteolysis during the budding yeast cell cycle. Curr Biol. 2000;10(5):231–40.
65. Skotheim JM, Di Talia S, Siggia ED, Cross FR. Positive feedback of G1 cyclins ensures coherent cell cycle entry. Nature. 2008;454(7202):291–6.
66. Koivomagi M, Valk E, Venta R, Iofik A, Lepiku M, Balog ER, et al. Cascades of multisite phosphorylation control Sic1 destruction at the onset of S phase. Nature. 2011;480(7375):128–31.
67. Tang X, Orlicky S, Mittag T, Csizmok V, Pawson T, Forman-Kay JD, et al. Composite low affinity interactions dictate recognition of the cyclin-dependent kinase inhibitor Sic1 by the SCFCdc4 ubiquitin ligase. Proc Natl Acad Sci U S A. 2012;109(9):3287–92.
68. Zachariae W, Schwab M, Nasmyth K, Seufert W. Control of cyclin ubiquitination by CDK-regulated binding of Hct1 to the anaphase promoting complex. Science. 1998;282(5394):1721–4.
69. Bartek J, Lukas J. Pathways governing G1/S transition and their response to DNA damage. FEBS Lett. 2001;490(3):117–22.
70. Bretones G, Delgado MD, Leon J. Myc and cell cycle control. Biochim Biophys Acta. 2014;1849:506–16.
71. Sheaff RJ, Groudine M, Gordon M, Roberts JM, Clurman BE. Cyclin E-CDK2 is a regulator of p27Kip1. Genes Dev. 1997;11(11):1464–78.
72. Sutterluty H, Chatelain E, Marti A, Wirbelauer C, Senften M, Muller U, et al. p45SKP2 promotes p27Kip1 degradation and induces S phase in quiescent cells. Nat Cell Biol. 1999;1(4):207–14.
73. Kitagawa K, Kotake Y, Kitagawa M. Ubiquitin-mediated control of oncogene and tumor suppressor gene products. Cancer Sci. 2009;100(8):1374–81 [Review].
74. Eldridge AG, Loktev AV, Hansen DV, Verschuren EW, Reimann JD, Jackson PK. The evi5 oncogene regulates cyclin accumulation by stabilizing the anaphase-promoting complex inhibitor emi1. Cell. 2006;124(2):367–80.
75. Lau AW, Inuzuka H, Fukushima H, Wan L, Liu P, Gao D, et al. Regulation of APC(Cdh1) E3 ligase activity by the Fbw7/cyclin E signaling axis contributes to the tumor suppressor function of Fbw7. Cell Res. 2013;23(7):947–61.
76. Rape M, Kirschner MW. Autonomous regulation of the anaphase-promoting complex couples mitosis to S-phase entry. Nature. 2004;432(7017):588–95.
77. Fukushima H, Ogura K, Wan L, Lu Y, Li V, Gao D, et al. SCF-mediated Cdh1 degradation defines a negative feedback system that coordinates cell-cycle progression. Cell Rep. 2013;4(4):803–16.
78. Kishi T, Ikeda A, Koyama N, Fukada J, Nagao R. A refined two-hybrid system reveals that SCF(Cdc4)-dependent degradation of Swi5 contributes to the regulatory mechanism of S-phase entry. Proc Natl Acad Sci U S A. 2008;105(38):14497–502.
79. Landry BD, Doyle JP, Toczyski DP, Benanti JA. F-box protein specificity for g1 cyclins is dictated by subcellular localization. PLoS Genet. 2012;8(7):e1002851.
80. Lin DI, Barbash O, Kumar KG, Weber JD, Harper JW, Klein-Szanto AJ, et al. Phosphorylation-dependent ubiquitination of cyclin D1 by the SCF(FBX4-alphaB crystallin) complex. Mol Cell. 2006;24(3):355–66.
81. Vaites LP, Lee EK, Lian Z, Barbash O, Roy D, Wasik M, et al. The Fbx4 tumor suppressor regulates cyclin D1 accumulation and prevents neoplastic transformation. Mol Cell Biol. 2011;31(22):4513–23.
82. Marti A, Wirbelauer C, Scheffner M, Krek W. Interaction between ubiquitin-protein ligase SCFSKP2 and E2F-1 underlies the regulation of E2F-1 degradation. Nat Cell Biol. 1999;1(1):14–9.
83. Koepp DM, Schaefer LK, Ye X, Keyomarsi K, Chu C, Harper JW, et al. Phosphorylation-dependent ubiquitination of cyclin E by the SCFFbw7 ubiquitin ligase. Science. 2001;294(5540):173–7.

84. Nguyen VQ, Co C, Li JJ. Cyclin-dependent kinases prevent DNA re-replication through multiple mechanisms. Nature. 2001;411(6841):1068–73 [Research Support, Non-U.S. Gov't Research Support, U.S. Gov't, P.H.S.].
85. Devault A, Vallen EA, Yuan T, Green S, Bensimon A, Schwob E. Identification of Tah11/Sid2 as the ortholog of the replication licensing factor Cdt1 in *Saccharomyces cerevisiae*. Curr Biol. 2002;12(8):689–94 [Research Support, Non-U.S. Gov't Research Support, U.S. Gov't, P.H.S.].
86. Wilmes GM, Archambault V, Austin RJ, Jacobson MD, Bell SP, Cross FR. Interaction of the S-phase cyclin Clb5 with an "RXL" docking sequence in the initiator protein Orc6 provides an origin-localized replication control switch. Genes Dev. 2004;18(9):981–91.
87. Yanow SK, Lygerou Z, Nurse P. Expression of Cdc18/Cdc6 and Cdt1 during G2 phase induces initiation of DNA replication. EMBO J. 2001;20(17):4648–56 [Research Support, Non-U.S. Gov't].
88. Arias EE, Walter JC. Replication-dependent destruction of Cdt1 limits DNA replication to a single round per cell cycle in *Xenopus* egg extracts. Genes Dev. 2005;19(1):114–26 [Research Support, Non-U.S. Gov't].
89. Li A, Blow JJ. Cdt1 downregulation by proteolysis and geminin inhibition prevents DNA re-replication in *Xenopus*. EMBO J. 2005;24(2):395–404 [Research Support, Non-U.S. Gov't].
90. Yoshida K, Takisawa H, Kubota Y. Intrinsic nuclear import activity of geminin is essential to prevent re-initiation of DNA replication in *Xenopus* eggs. Genes Cells. 2005;10(1):63–73.
91. Zhu W, Chen Y, Dutta A. Rereplication by depletion of geminin is seen regardless of p53 status and activates a G2/M checkpoint. Mol Cell Biol. 2004;24(16):7140–50 [Research Support, U.S. Gov't, P.H.S.].
92. Mendez J, Zou-Yang XH, Kim SY, Hidaka M, Tansey WP, Stillman B. Human origin recognition complex large subunit is degraded by ubiquitin-mediated proteolysis after initiation of DNA replication. Mol Cell. 2002;9(3):481–91 [Research Support, Non-U.S. Gov't Research Support, U.S. Gov't, Non-P.H.S. Research Support, U.S. Gov't, P.H.S.].
93. McNairn AJ, Okuno Y, Misteli T, Gilbert DM. Chinese hamster ORC subunits dynamically associate with chromatin throughout the cell-cycle. Exp Cell Res. 2005;308(2):345–56 [Research Support, N.I.H., Extramural Research Support, Non-U.S. Gov't Research Support, U.S. Gov't, P.H.S.].
94. Kim J, Kipreos ET. Control of the Cdc6 replication licensing factor in metazoa: the role of nuclear export and the CUL4 ubiquitin ligase. Cell Cycle. 2008;7(2):146–50 [Research Support, N.I.H., Extramural].
95. Abbas T, Shibata E, Park J, Jha S, Karnani N, Dutta A. CRL4(Cdt2) regulates cell proliferation and histone gene expression by targeting PR-Set7/Set8 for degradation. Mol Cell. 2010;40(1):9–21 [Research Support, N.I.H., Extramural Research Support, Non-U.S. Gov't].
96. Moldovan GL, Pfander B, Jentsch S. PCNA, the maestro of the replication fork. Cell. 2007;129(4):665–79 [Research Support, Non-U.S. Gov't Review].
97. Jin J, Arias EE, Chen J, Harper JW, Walter JC. A family of diverse Cul4-Ddb1-interacting proteins includes Cdt2, which is required for S phase destruction of the replication factor Cdt1. Mol Cell. 2006;23(5):709–21 [Research Support, N.I.H., Extramural Research Support, Non-U.S. Gov't].
98. Guarino E, Shepherd ME, Salguero I, Hua H, Deegan RS, Kearsey SE. Cdt1 proteolysis is promoted by dual PIP degrons and is modulated by PCNA ubiquitylation. Nucleic Acids Res. 2011;39(14):5978–90 [Research Support, Non-U.S. Gov't].
99. Havens CG, Shobnam N, Guarino E, Centore RC, Zou L, Kearsey SE, et al. Direct role for proliferating cell nuclear antigen in substrate recognition by the E3 ubiquitin ligase CRL4Cdt2. J Biol Chem. 2012;287(14):11410–21 [Research Support, N.I.H., Extramural Research Support, Non-U.S. Gov't].
100. Abbas T, Dutta A. CRL4Cdt2: master coordinator of cell cycle progression and genome stability. Cell Cycle. 2011;10(2):241–9 [Research Support, N.I.H., Extramural Review].
101. Havens CG, Walter JC. Docking of a specialized PIP Box onto chromatin-bound PCNA creates a degron for the ubiquitin ligase CRL4Cdt2. Mol Cell. 2009;35(1):93–104 [Research Support, N.I.H., Extramural].

102. Michishita M, Morimoto A, Ishii T, Komori H, Shiomi Y, Higuchi Y, et al. Positively charged residues located downstream of PIP box, together with TD amino acids within PIP box, are important for CRL4(Cdt2) -mediated proteolysis. Genes Cells. 2011;16(1):12–22 [Research Support, Non-U.S. Gov't].
103. Liu E, Li X, Yan F, Zhao Q, Wu X. Cyclin-dependent kinases phosphorylate human Cdt1 and induce its degradation. J Biol Chem. 2004;279(17):17283–8 [Research Support, Non-U.S. Gov't].
104. Sugimoto N, Tatsumi Y, Tsurumi T, Matsukage A, Kiyono T, Nishitani H, et al. Cdt1 phosphorylation by cyclin A-dependent kinases negatively regulates its function without affecting geminin binding. J Biol Chem. 2004;279(19):19691–7 [Research Support, Non-U.S. Gov't].
105. Nishitani H, Lygerou Z, Nishimoto T. Proteolysis of DNA replication licensing factor Cdt1 in S-phase is performed independently of geminin through its N-terminal region. J Biol Chem. 2004;279(29):30807–16 [Research Support, Non-U.S. Gov't].
106. Takeda DY, Parvin JD, Dutta A. Degradation of Cdt1 during S phase is Skp2-independent and is required for efficient progression of mammalian cells through S phase. J Biol Chem. 2005;280(24):23416–23 [Research Support, N.I.H., Extramural Research Support, U.S. Gov't, P.H.S.].
107. Nishitani H, Sugimoto N, Roukos V, Nakanishi Y, Saijo M, Obuse C, et al. Two E3 ubiquitin ligases, SCF-Skp2 and DDB1-Cul4, target human Cdt1 for proteolysis. EMBO J. 2006;25(5): 1126–36 [Research Support, Non-U.S. Gov't].
108. Johansson P, Jeffery J, Al-Ejeh F, Schulz RB, Callen DF, Kumar R, et al. SCF-FBXO31 E3 ligase targets DNA replication factor Cdt1 for proteolysis in the G2 phase of cell cycle to prevent re-replication. J Biol Chem. 2014;289(26):18514–25 [Research Support, Non-U.S. Gov't].
109. Drury LS, Diffley JF. Factors affecting the diversity of DNA replication licensing control in eukaryotes. Curr Biol. 2009;19(6):530–5 [Research Support, Non-U.S. Gov't].
110. Kominami K, Toda T. Fission yeast WD-repeat protein pop1 regulates genome ploidy through ubiquitin-proteasome-mediated degradation of the CDK inhibitor Rum1 and the S-phase initiator Cdc18. Genes Dev. 1997;11(12):1548–60 [Comparative Study Research Support, Non-U.S. Gov't].
111. Kominami K, Ochotorena I, Toda T. Two F-box/WD-repeat proteins Pop1 and Pop2 form hetero- and homo-complexes together with cullin-1 in the fission yeast SCF (Skp1-Cullin-1-F-box) ubiquitin ligase. Genes Cells. 1998;3(11):721–35 [Research Support, Non-U.S. Gov't].
112. Drury LS, Perkins G, Diffley JF. The Cdc4/34/53 pathway targets Cdc6p for proteolysis in budding yeast. EMBO J. 1997;16(19):5966–76.
113. Elsasser S, Chi Y, Yang P, Campbell JL. Phosphorylation controls timing of Cdc6p destruction: a biochemical analysis. Mol Biol Cell. 1999;10(10):3263–77.
114. Williams RS, Shohet RV, Stillman B. A human protein related to yeast Cdc6p. Proc Natl Acad Sci U S A. 1997;94(1):142–7 [Comparative Study Research Support, U.S. Gov't, P.H.S.].
115. Saha P, Chen J, Thome KC, Lawlis SJ, Hou ZH, Hendricks M, et al. Human CDC6/Cdc18 associates with Orc1 and cyclin-cdk and is selectively eliminated from the nucleus at the onset of S phase. Mol Cell Biol. 1998;18(5):2758–67 [Comparative Study Research Support, Non-U.S. Gov't Research Support, U.S. Gov't, Non-P.H.S. Research Support, U.S. Gov't, P.H.S.].
116. Clijsters L, Wolthuis R. PIP-box-mediated degradation prohibits re-accumulation of Cdc6 during S phase. J Cell Sci. 2014;127(Pt 6):1336–45 [Research Support, Non-U.S. Gov't].
117. Nishitani H, Shiomi Y, Iida H, Michishita M, Takami T, Tsurimoto T. CDK inhibitor p21 is degraded by a proliferating cell nuclear antigen-coupled Cul4-DDB1Cdt2 pathway during S phase and after UV irradiation. J Biol Chem. 2008;283(43):29045–52 [Research Support, Non-U.S. Gov't].
118. Lygerou Z, Nurse P. The fission yeast origin recognition complex is constitutively associated with chromatin and is differentially modified through the cell cycle. J Cell Sci. 1999;112 (Pt 21):3703–12 [Research Support, Non-U.S. Gov't].
119. Wuarin J, Buck V, Nurse P, Millar JB. Stable association of mitotic cyclin B/Cdc2 to replication origins prevents endoreduplication. Cell. 2002;111(3):419–31.

120. Li CJ, DePamphilis ML. Mammalian Orc1 protein is selectively released from chromatin and ubiquitinated during the S-to-M transition in the cell division cycle. Mol Cell Biol. 2002;22(1):105–16.
121. Kreitz S, Ritzi M, Baack M, Knippers R. The human origin recognition complex protein 1 dissociates from chromatin during S phase in HeLa cells. J Biol Chem. 2001;276(9):6337–42 [Research Support, Non-U.S. Gov't].
122. Tatsumi Y, Ohta S, Kimura H, Tsurimoto T, Obuse C. The ORC1 cycle in human cells: I. cell cycle-regulated oscillation of human ORC1. J Biol Chem. 2003;278(42):41528–34 [Research Support, Non-U.S. Gov't].
123. Shen Z, Chakraborty A, Jain A, Giri S, Ha T, Prasanth KV, et al. Dynamic association of ORCA with prereplicative complex components regulates DNA replication initiation. Mol Cell Biol. 2012;32(15):3107–20 [Research Support, N.I.H., Extramural Research Support, U.S. Gov't, Non-P.H.S.].
124. Shen Z, Prasanth SG. Orc2 protects ORCA from ubiquitin-mediated degradation. Cell Cycle. 2012;11(19):3578–89 [Research Support, N.I.H., Extramural Research Support, U.S. Gov't, Non-P.H.S.].
125. Jorgensen S, Elvers I, Trelle MB, Menzel T, Eskildsen M, Jensen ON, et al. The histone methyltransferase SET8 is required for S-phase progression. J Cell Biol. 2007;179(7):1337–45 [Research Support, Non-U.S. Gov't].
126. Tardat M, Murr R, Herceg Z, Sardet C, Julien E. PR-Set7-dependent lysine methylation ensures genome replication and stability through S phase. J Cell Biol. 2007;179(7):1413–26 [Research Support, Non-U.S. Gov't].
127. Centore RC, Havens CG, Manning AL, Li JM, Flynn RL, Tse A, et al. CRL4(Cdt2)-mediated destruction of the histone methyltransferase Set8 prevents premature chromatin compaction in S phase. Mol Cell. 2010;40(1):22–33 [Research Support, N.I.H., Extramural Research Support, Non-U.S. Gov't].
128. Liu C, Poitelea M, Watson A, Yoshida SH, Shimoda C, Holmberg C, et al. Transactivation of *Schizosaccharomyces pombe* cdt2+ stimulates a Pcu4-Ddb1-CSN ubiquitin ligase. EMBO J. 2005;24(22):3940–51 [Research Support, Non-U.S. Gov't].
129. Salguero I, Guarino E, Shepherd ME, Deegan TD, Havens CG, MacNeill SA, et al. Ribonucleotide reductase activity is coupled to DNA synthesis via proliferating cell nuclear antigen. Curr Biol. 2012;22(8):720–6 [Research Support, N.I.H., Extramural Research Support, Non-U.S. Gov't].
130. Moss J, Tinline-Purvis H, Walker CA, Folkes LK, Stratford MR, Hayles J, et al. Break-induced ATR and Ddb1-Cul4(Cdt)(2) ubiquitin ligase-dependent nucleotide synthesis promotes homologous recombination repair in fission yeast. Genes Dev. 2010;24(23):2705–16 [Research Support, Non-U.S. Gov't].
131. Gunjan A, Verreault A. A Rad53 kinase-dependent surveillance mechanism that regulates histone protein levels in *S. cerevisiae*. Cell. 2003;115(5):537–49.
132. Singh RK, Kabbaj MH, Paik J, Gunjan A. Histone levels are regulated by phosphorylation and ubiquitylation-dependent proteolysis. Nat Cell Biol. 2009;11(8):925–33.
133. Abbas T, Mueller AC, Shibata E, Keaton M, Rossi M, Dutta A. CRL1-FBXO11 promotes Cdt2 ubiquitylation and degradation and regulates Pr-Set7/Set8-mediated cellular migration. Mol Cell. 2013;49(6):1147–58 [Research Support, N.I.H., Extramural].
134. Rossi M, Duan S, Jeong YT, Horn M, Saraf A, Florens L, et al. Regulation of the CRL4(Cdt2) ubiquitin ligase and cell-cycle exit by the SCF(Fbxo11) ubiquitin ligase. Mol Cell. 2013;49(6):1159–66 [Research Support, N.I.H., Extramural Research Support, Non-U.S. Gov't].
135. Dar A, Wu D, Lee N, Shibata E, Dutta A. 14-3-3 proteins play a role in the cell cycle by shielding cdt2 from ubiquitin-mediated degradation. Mol Cell Biol. 2014;34(21):4049–61 [Research Support, N.I.H., Extramural].
136. Bartek J, Lukas C, Lukas J. Checking on DNA damage in S phase. Nat Rev Mol Cell Biol. 2004;5(10):792–804 [Research Support, Non-U.S. Gov't Review].
137. Bartek J, Lukas J. Mammalian G1- and S-phase checkpoints in response to DNA damage. Curr Opin Cell Biol. 2001;13(6):738–47 [Research Support, Non-U.S. Gov't Review].

138. Yekezare M, Gomez-Gonzalez B, Diffley JF. Controlling DNA replication origins in response to DNA damage - inhibit globally, activate locally. J Cell Sci. 2013;126(Pt 6):1297–306 [Research Support, Non-U.S. Gov't Review].
139. Busino L, Chiesa M, Draetta GF, Donzelli M. Cdc25A phosphatase: combinatorial phosphorylation, ubiquitylation and proteolysis. Oncogene. 2004;23(11):2050–6 [Research Support, Non-U.S. Gov't Review].
140. Falck J, Mailand N, Syljuasen RG, Bartek J, Lukas J. The ATM-Chk2-Cdc25A checkpoint pathway guards against radioresistant DNA synthesis. Nature. 2001;410(6830):842–7 [Research Support, Non-U.S. Gov't].
141. Haupt Y, Maya R, Kazaz A, Oren M. Mdm2 promotes the rapid degradation of p53. Nature. 1997;387(6630):296–9 [Research Support, Non-U.S. Gov't Research Support, U.S. Gov't, P.H.S.].
142. Midgley CA, Lane DP. p53 protein stability in tumour cells is not determined by mutation but is dependent on Mdm2 binding. Oncogene. 1997;15(10):1179–89 [Research Support, Non-U.S. Gov't].
143. Branzei D, Foiani M. The DNA damage response during DNA replication. Curr Opin Cell Biol. 2005;17(6):568–75 [Research Support, Non-U.S. Gov't Review].
144. Smith J, Tho LM, Xu N, Gillespie DA. The ATM-Chk2 and ATR-Chk1 pathways in DNA damage signaling and cancer. Adv Cancer Res. 2010;108:73–112 [Research Support, Non--U.S. Gov't Review].
145. Ge XQ, Blow JJ. Chk1 inhibits replication factory activation but allows dormant origin firing in existing factories. J Cell Biol. 2010;191(7):1285–97 [Research Support, Non-U.S. Gov't].
146. Lopez-Mosqueda J, Maas NL, Jonsson ZO, Defazio-Eli LG, Wohlschlegel J, Toczyski DP. Damage-induced phosphorylation of Sld3 is important to block late origin firing. Nature. 2010;467(7314):479–83 [Research Support, N.I.H., Extramural Research Support, Non-U.S. Gov't].
147. Zegerman P, Diffley JF. Checkpoint-dependent inhibition of DNA replication initiation by Sld3 and Dbf4 phosphorylation. Nature. 2010;467(7314):474–8 [Research Support, Non-U.S. Gov't].
148. Boos D, Yekezare M, Diffley JF. Identification of a heteromeric complex that promotes DNA replication origin firing in human cells. Science. 2013;340(6135):981–4 [Research Support, Non-U.S. Gov't].
149. Boos D, Sanchez-Pulido L, Rappas M, Pearl LH, Oliver AW, Ponting CP, et al. Regulation of DNA replication through Sld3-Dpb11 interaction is conserved from yeast to humans. Curr Biol. 2011;21(13):1152–7 [Research Support, Non-U.S. Gov't].
150. Honaker Y, Piwnica-Worms H. Casein kinase 1 functions as both penultimate and ultimate kinase in regulating Cdc25A destruction. Oncogene. 2010;29(23):3324–34 [Research Support, N.I.H., Extramural Research Support, Non-U.S. Gov't].
151. Havens CG, Walter JC. Mechanism of CRL4(Cdt2), a PCNA-dependent E3 ubiquitin ligase. Genes Dev. 2011;25(15):1568–82 [Research Support, N.I.H., Extramural Review].
152. Hall JR, Lee HO, Bunker BD, Dorn ES, Rogers GC, Duronio RJ, et al. Cdt1 and Cdc6 are destabilized by rereplication-induced DNA damage. J Biol Chem. 2008;283(37):25356–63 [Research Support, N.I.H., Extramural Research Support, Non-U.S. Gov't].
153. Hall JR, Kow E, Nevis KR, Lu CK, Luce KS, Zhong Q, et al. Cdc6 stability is regulated by the Huwe1 ubiquitin ligase after DNA damage. Mol Biol Cell. 2007;18(9):3340–50 [Research Support, N.I.H., Extramural].
154. Kumar S, Tomooka Y, Noda M. Identification of a set of genes with developmentally down-regulated expression in the mouse brain. Biochem Biophys Res Commun. 1992;185(3):1155–61 [Research Support, Non-U.S. Gov't].
155. Rabut G, Peter M. Function and regulation of protein neddylation. 'Protein modifications: beyond the usual suspects' review series. EMBO Rep. 2008;9(10):969–76 [Review].
156. Parry G, Estelle M. Regulation of cullin-based ubiquitin ligases by the Nedd8/RUB ubiquitin-like proteins. Semin Cell Dev Biol. 2004;15(2):221–9 [Research Support, U.S. Gov't, Non--P.H.S. Research Support, U.S. Gov't, P.H.S. Review].

157. Jones J, Wu K, Yang Y, Guerrero C, Nillegoda N, Pan ZQ, et al. A targeted proteomic analysis of the ubiquitin-like modifier nedd8 and associated proteins. J Proteome Res. 2008;7(3):1274–87 [Research Support, N.I.H., Extramural Research Support, U.S. Gov't, Non-P.H.S.].
158. Pan ZQ, Kentsis A, Dias DC, Yamoah K, Wu K. Nedd8 on cullin: building an expressway to protein destruction. Oncogene. 2004;23(11):1985–97 [Research Support, U.S. Gov't, P.H.S. Review].
159. Duda DM, Borg LA, Scott DC, Hunt HW, Hammel M, Schulman BA. Structural insights into NEDD8 activation of cullin-RING ligases: conformational control of conjugation. Cell. 2008;134(6):995–1006 [Research Support, N.I.H., Extramural Research Support, Non--U.S. Gov't Research Support, U.S. Gov't, Non-P.H.S.].
160. Goldenberg SJ, Cascio TC, Shumway SD, Garbutt KC, Liu J, Xiong Y, et al. Structure of the Cand1-Cul1-Roc1 complex reveals regulatory mechanisms for the assembly of the multisubunit cullin-dependent ubiquitin ligases. Cell. 2004;119(4):517–28 [Research Support, Non--U.S. Gov't Research Support, U.S. Gov't, P.H.S.].
161. Feng H, Zhong W, Punkosdy G, Gu S, Zhou L, Seabolt EK, et al. CUL-2 is required for the G1-to-S-phase transition and mitotic chromosome condensation in *Caenorhabditis elegans*. Nat Cell Biol. 1999;1(8):486–92 [Research Support, Non-U.S. Gov't Research Support, U.S. Gov't, P.H.S.].
162. Craney A, Rape M. Dynamic regulation of ubiquitin-dependent cell cycle control. Curr Opin Cell Biol. 2013;25(6):704–10 [Research Support, N.I.H., Extramural Research Support, Non-U.S. Gov't Research Support, U.S. Gov't, Non-P.H.S. Review].
163. Cukras S, Morffy N, Ohn T, Kee Y. Inactivating UBE2M impacts the DNA damage response and genome integrity involving multiple cullin ligases. PLoS One. 2014;9(7):e101844 [Research Support, Non-U.S. Gov't].
164. Soucy TA, Smith PG, Milhollen MA, Berger AJ, Gavin JM, Adhikari S, et al. An inhibitor of NEDD8-activating enzyme as a new approach to treat cancer. Nature. 2009;458(7239):732–6.
165. Lin JJ, Milhollen MA, Smith PG, Narayanan U, Dutta A. NEDD8-targeting drug MLN4924 elicits DNA rereplication by stabilizing Cdt1 in S phase, triggering checkpoint activation, apoptosis, and senescence in cancer cells. Cancer Res. 2010;70(24):10310–20 [Research Support, N.I.H., Extramural Research Support, U.S. Gov't, Non-P.H.S.].
166. Watson IR, Irwin MS. Ubiquitin and ubiquitin-like modifications of the p53 family. Neoplasia. 2006;8(8):655–66 [Research Support, Non-U.S. Gov't Review].
167. Xirodimas DP, Saville MK, Bourdon JC, Hay RT, Lane DP. Mdm2-mediated NEDD8 conjugation of p53 inhibits its transcriptional activity. Cell. 2004;118(1):83–97 [Research Support, Non-U.S. Gov't].
168. Xirodimas DP. Novel substrates and functions for the ubiquitin-like molecule NEDD8. Biochem Soc Trans. 2008;36(Pt 5):802–6 [Research Support, Non-U.S. Gov't Review].
169. Jackson SP, Durocher D. Regulation of DNA damage responses by ubiquitin and SUMO. Mol Cell. 2013;49(5):795–807.
170. Johnson ES. Protein modification by SUMO. Annu Rev Biochem. 2004;73:355–82.
171. Desterro JM, Rodriguez MS, Hay RT. SUMO-1 modification of IkappaBalpha inhibits NF-kappaB activation. Mol Cell. 1998;2(2):233–9.
172. Prudden J, Pebernard S, Raffa G, Slavin DA, Perry JJ, Tainer JA, et al. SUMO-targeted ubiquitin ligases in genome stability. EMBO J. 2007;26(18):4089–101.
173. Sun H, Leverson JD, Hunter T. Conserved function of RNF4 family proteins in eukaryotes: targeting a ubiquitin ligase to SUMOylated proteins. EMBO J. 2007;26(18):4102–12.
174. Stehmeier P, Muller S. Regulation of p53 family members by the ubiquitin-like SUMO system. DNA Repair (Amst). 2009;8(4):491–8 [Review].
175. Li T, Santockyte R, Shen RF, Tekle E, Wang G, Yang DC, et al. Expression of SUMO-2/3 induced senescence through p53- and pRB-mediated pathways. J Biol Chem. 2006;281(47):36221–7.
176. Chang PC, Izumiya Y, Wu CY, Fitzgerald LD, Campbell M, Ellison TJ, et al. Kaposi's sarcoma-associated herpesvirus (KSHV) encodes a SUMO E3 ligase that is SIM-dependent and SUMO-2/3-specific. J Biol Chem. 2010;285(8):5266–73 [Research Support, N.I.H., Extramural].

177. Stindt MH, Carter S, Vigneron AM, Ryan KM, Vousden KH. MDM2 promotes SUMO-2/3 modification of p53 to modulate transcriptional activity. Cell Cycle. 2011;10(18):3176–88 [Research Support, Non-U.S. Gov't].
178. Bonne-Andrea C, Kahli M, Mechali F, Lemaitre JM, Bossis G, Coux O. SUMO2/3 modification of cyclin E contributes to the control of replication origin firing. Nat Commun. 2013;4:1850 [Research Support, Non-U.S. Gov't].
179. Dou H, Huang C, Singh M, Carpenter PB, Yeh ET. Regulation of DNA repair through deSUMOylation and SUMOylation of replication protein A complex. Mol Cell. 2010;39(3):333–45 [Research Support, N.I.H., Extramural Research Support, Non-U.S. Gov't].
180. Cremona CA, Sarangi P, Yang Y, Hang LE, Rahman S, Zhao X. Extensive DNA damage-induced sumoylation contributes to replication and repair and acts in addition to the mec1 checkpoint. Mol Cell. 2012;45(3):422–32 [Research Support, N.I.H., Extramural].
181. Ma L, Aslanian A, Sun H, Jin M, Shi Y, Yates 3rd JR, et al. Identification of small ubiquitin-like modifier substrates with diverse functions using the *Xenopus* egg extract system. Mol Cell Proteomics. 2014;13(7):1659–75 [Research Support, N.I.H., Extramural Research Support, Non-U.S. Gov't].
182. Jin J, Shirogane T, Xu L, Nalepa G, Qin J, Elledge SJ, et al. SCFbeta-TRCP links Chk1 signaling to degradation of the Cdc25A protein phosphatase. Genes Dev. 2003;17(24):3062–74 [Research Support, Non-U.S. Gov't Research Support, U.S. Gov't, Non-P.H.S.].
183. Wolthuis R, Clay-Farrace L, van Zon W, Yekezare M, Koop L, Ogink J, et al. Cdc20 and Cks direct the spindle checkpoint-independent destruction of cyclin A. Mol Cell. 2008;30(3):290–302.
184. Kim SY, Herbst A, Tworkowski KA, Salghetti SE, Tansey WP. Skp2 regulates Myc protein stability and activity. Mol Cell. 2003;11(5):1177–88.
185. von der Lehr N, Johansson S, Wu S, Bahram F, Castell A, Cetinkaya C, et al. The F-box protein Skp2 participates in c-Myc proteosomal degradation and acts as a cofactor for c-Myc-regulated transcription. Mol Cell. 2003;11(5):1189–200.
186. Welcker M, Orian A, Jin J, Grim JE, Harper JW, Eisenman RN, et al. The Fbw7 tumor suppressor regulates glycogen synthase kinase 3 phosphorylation-dependent c-Myc protein degradation. Proc Natl Acad Sci U S A. 2004;101(24):9085–90.
187. Yada M, Hatakeyama S, Kamura T, Nishiyama M, Tsunematsu R, Imaki H, et al. Phosphorylation-dependent degradation of c-Myc is mediated by the F-box protein Fbw7. EMBO J. 2004;23(10):2116–25.
188. Bashir T, Dorrello NV, Amador V, Guardavaccaro D, Pagano M. Control of the SCF(Skp2-Cks1) ubiquitin ligase by the APC/C(Cdh1) ubiquitin ligase. Nature. 2004;428(6979):190–3.

Chapter 18
Role of Posttranslational Modifications in Replication Initiation

Lei Wei and Xiaolan Zhao

Abstract DNA replication must occur precisely once per cell cycle to maintain a stable genome. An important means to achieve this is through multilayered regulation of replication initiation at both local and global levels. Recent genetic and biochemical studies in several organisms have revealed critical roles of posttranslational modifications (PTMs) in these regulations. While the best-demonstrated class of PTMs is kinase-mediated phosphorylation, additional forms of PTMs including ubiquitylation, methylation, and acetylation also contribute to the control of replication initiation. Here we survey the current understanding of how different types of modifications govern and fine-tune several aspects of replication initiation, including origin licensing, firing, and global replication timing.

Keywords DNA replication initiation • Origin licensing • Origin firing • Replication timing • Posttranslational modifications • Phosphorylation • Ubiquitylation • Methylation • Acetylation

Introduction

Eukaryotic organisms initiate DNA replication at hundreds of origins. Studies in several organisms suggests that the activation of each replication origin requires two temporally separated steps, namely origin licensing and firing

L. Wei • X. Zhao (✉)
Gerstner Sloan-Kettering Graduate School of Biomedical Sciences, Memorial Sloan Kettering Cancer Center, New York, NY 10065, USA

Molecular Biology Program, Memorial Sloan Kettering Cancer Center, New York, NY 10065, USA
e-mail: zhaox1@mskcc.org

D.L. Kaplan (ed.), *The Initiation of DNA Replication in Eukaryotes*,
DOI 10.1007/978-3-319-24696-3_18

(reviewed in [1–4]). During origin licensing, which occurs at late M to G1 phase, the replicative helicase MCM is loaded onto origins, forming the pre-replicative complex (pre-RC). Subsequent origin firing in S phase occurs upon the formation of an active replicative helicase composed of MCM and accessory factors Cdc45 and GINS (together referred to as CMG), leading to DNA unwinding. The overall hierarchy of origin activation is determined by origin nucleotide sequences, chromatin environment, and additional protein factors. Such a global replication timing program establishes the groups of origins that are activated at early or late S phase and determines the efficiencies of their activation (reviewed in [5]).

Origin licensing and firing as well as the replication timing program described above require many layers of regulatory mechanisms. Protein posttranslational modifications (PTMs), including phosphorylation, acetylation, and ubiquitylation, underpin many of these mechanisms. PTMs generally occur rapidly, making them perfectly suited for ensuring the precise deployment of each regulatory measure throughout replication initiation. In addition, most PTMs are reversible, facilitating both positive and negative regulation of a particular reaction. These features make PTMs ideal for the dynamic fine-tuning needed to spatially and temporally organize a balanced replication program. In this chapter, we describe the main functions of different PTMs in regulating each aspect of replication initiation. We discuss how three types of protein kinases and their counteracting phosphatases control origin firing, followed by how other forms of PTMs regulate origin licensing and the global origin firing program. In each section, we provide examples of how these modifications can be used for precise control of key reactions in these processes.

Phosphorylation-Based Regulation of Replication Initiation

Two essential and conserved kinases, DDK (Dbf4-dependent kinase) and CDK (cyclin-dependent kinase), are critical for origin firing mainly through targeting MCM subunits and the Sld2 and Sld3 scaffold proteins that only function during replication initiation. These phosphorylation events trigger the stepwise formation of the active replicative helicase CMG [6–10]. CDK also plays a major role in preventing re-replication by blocking the licensing step at the end of G1. In addition to DDK and CDK, the DNA replication and checkpoint kinases regulate origin firing, particularly under replication stress conditions. Interestingly, the effects of all three types of kinases (DDK, CDK, and checkpoint kinases) can be reversed by different types of protein phosphatases. These balancing effects on protein phosphorylation are important to ensure the order of origin firing. Below we discuss the major phosphorylation events controlled by these kinases and phosphatases. We first describe the findings in budding yeast, which has thus far provided the best understanding of kinase-based regulation of origin firing, and then compare these with findings in other organisms.

DDK and Protein Phosphatase 1-Mediated Regulation of CMG Formation

DDK Is Essential for CMG Formation

A wealth of genetic and biochemical studies have shown that DDK, which is essential for origin firing, supports the assembly of the active replicative helicase CMG [11–15]. This is best illustrated in budding yeast, where it has been shown that DDK-mediated phosphorylation of Mcm4, Mcm6, and perhaps also Mcm2 enables the recruitment of Sld3 and its interacting partner Cdc45 to pre-RCs [12, 16]. DDK associates with the double-hexamer form of MCM on origins, a state referred to as loaded MCM, via the binding to Mcm2 and Mcm4 [14, 17–19]. The three DDK-targeted MCM subunits contain N-terminal serine/threonine-rich domains (NSDs) that harbor the DDK consensus sites S/T-D/E and S/T-S/T-P/Q (where the first S/T is targeted by DDK) [14, 20]. Among these consensus sites, a dozen serines and threonines on Mcm4 are the most critical (Table 18.1) [14, 20, 21]. Expression of Mcm4 phospho-mimetic mutants with these sites changed to acidic amino acids, or removal of the amino acids 74–174 of the Mcm4 NSD, can bypass the DDK requirement for CMG formation, suggesting that DDK-mediated phosphorylation at these sites alleviates an inhibitory effect of Mcm4 NSD on initiation. As these DDK-bypass alleles still show partial defects in Cdc45 recruitment to the pre-RCs and in cell growth, other DDK substrates besides Mcm4 may be relevant [14, 20, 21, 52]. Indeed, phospho-mimetic mutations in the DDK sites of Mcm6 can improve the growth of cells harboring Mcm4 DDK-bypass alleles (Table 18.1) [20]. The significance of Mcm2 phosphorylation is not completely clear, though mutating two putative DDK sites on Mcm2 impairs growth under specific conditions [17, 18, 20, 22].

In other organisms, DDK-mediated phosphorylation of Mcm2, Mcm4, and Mcm6, which also contain NSDs rich in DDK consensus sites, has also been reported [36, 44, 45, 53, 54]. In *S. pombe*, phosphorylated forms of these MCM subunits are enriched in MCM-Cdc45 complex on chromatin, and phosphorylation of individual subunits is redundant to support cell viability [36]. In addition, mutating DDK phosphorylation sites of mammalian MCM2 and MCM4 leads to reduced MCM-Cdc45 association on chromatin and defective replication initiation, respectively [36, 44]. These findings establish that DDK-mediated MCM phosphorylation promotes replication initiation in these organisms as seen in budding yeast. More in-depth studies will flesh out the conserved and different aspects of DDK-mediated phosphorylation in promoting replication initiation.

Mechanisms by Which DDK Supports CMG Formation

Recent biochemical studies using the budding yeast proteins have provided insights into the mechanisms by which DDK phosphorylation of MCM can promote CMG formation. In particular, in vitro studies using reconstituted pre-RCs and DDK-depleted S-phase extracts have shown that the addition of DDK triggers Cdc45-Sld3

Table 18.1 Phosphorylation substrates involved in replication initiation

Species	Substrate	Kinase	Site or region	Functional effects	Reference
S. cerevisiae	Mcm4	DDK	S6, S16, S31, S41, S68 S76, S82, S108, S112, T140, S141, S144, S171, S174	Promote Cdc45-Sld3 recruitment to the pre-RCs	[14, 20, 21]
	Mcm4	Mec1	S87	Prime Mcm4 for DDK phosphorylation	[20]
	Mcm4	–	S2, S7, S17, S32, S42, S52, S56, S69, S77, S87, S145	Prime Mcm4 for DDK phosphorylation	[20]
	Mcm6	DDK	S2, S11, S18, S23, S48, S65, S69, T75, S78, S126, T127	Likely promote Cdc45-Sld3 recruitment to the pre-RCs	[20]
	Mcm6	–	S3, T9, S19, S37, S49, S56, T66, S70	Prime Mcm6 for DDK phosphorylation	[20]
	Mcm2	DDK/casein	S164, S170	Weaken Mcm2-Mcm5 interaction	[17, 18, 20, 22]
	Rif1	DDK/CDK	S106, S110, S119, S124, S125, S133, S134, S138, S140, S159, T180, S181, S211, S228	Weaken Rif1-PP1 interaction	[23, 24]
	Sld2	CDK	T84	Bind Dpb11 C-terminal BRCT	[25–27]
	Sld2	CDK	S100, S128, T168, S172, S208, T241	Prime Sld2 for phosphorylation on T84	[25, 28]
	Sld3	CDK	T600, S622	Bind Dpb11 N-terminal BRCT	[26, 27]
	Cdc6[a]	CDK	T7, T23, T39, S43, T135, S354, T368, S372	Promote degradation	[29]
	Mcm3	CDK	S761, S765, S781, T786, S845	Promote nuclear export	[20, 30]
	Orc2	CDK	S16, T24, T70, T174, S188, S206	Inhibit MCM loading	[31, 32]
	Orc6	CDK	S106, S116, S123, T146	Reduce Cdt1 recruitment and MCM loading	[31, 32]
	Sld3	Rad53	Multiple sites (mainly cluster in the C-terminal region)	Inhibit Sld3-Dpb11 binding	[33, 34]
	Dbf4	Rad53	Multiple sites (not listed)	Inhibit DDK activity	[33–35]

S. pombe	Cdc21 (ScMcm4)	DDK	Likely 1–130aa	May promote Cdc45 recruitment	[36]
	Mis5 (ScMcm6)	DDK	Likely 1–47aa	May promote Cdc45 recruitment	[36]
	Cdc19 (ScMcm2)	DDK	Likely 1–35aa	May promote Cdc45 recruitment	[36]
	Rif1	DDK/CDK	S19, T20, S29, S30, S31, S44, S45, S50, T68, S69, T70, S88	Weaken Rif1-PP1 interaction	[23]
	Drc1 (ScSld2)	CDK	T111	Bind C-terminal BRCT of Cut5	[37]
	Sld3	CDK	T636, T650, S673, T690, S698	Bind N-terminal BRCT of Cut5	[37]
	Cdc18[a] (ScCdc6)	CDK	T10, T46, T60, T104, T134, T374	Promote degradation	[38, 39]
	Orp2 (ScOrc2)	CDK	T44, T57, T62, T65	Suppress re-replication	[40]
X. laevis	Treslin (ScSld3)	CDK	S976	Bind BRCT I-II of TopBP1	[41]
	Cdc6	CDK	S54, S74, S108, S120, S411	Promote nuclear export of excess Cdc6	[42]
C. elegans	SLD2	CDK	8 CDK consensus sites in N-terminus	Bind C-terminus of MUS-101	[43]
M. musculus	Mcm4	DDK	2–88 region	Promote Cdc45 recruitment	[36]
H. sapiens	Mcm2	DDK	S27, S41, S139	Promote MCM ATPase activity	[44]
	Mcm2	DDK	S40, S53, S108	–	[45]
	Treslin (ScSld3)	CDK	T969, S1001	Bind BRCT 0/1/2 of TopBP1	[41, 46]
	Cdt1[a]	CDK	T29	Degradation via SCF^{Skp2}	[47, 48]
	Orc1[a]	CDK	1–387, but not (S/T)PX(K/R) consensus sites	Promote degradation	[49]
	Cdc6	CDK	S45, S54, S74, S106	Promote nuclear export of excess Cdc6	[50]
	Treslin (ScSld3)	Chk1	Depend on motif LTQSPLL	Likely reduce Cdc45 recruitment	[51]

[a]The substrate is subjected to phosphorylation-mediated ubiquitylation and subsequent degradation

recruitment to pre-RCs and concomitant Mcm4 and Mcm6 phosphorylation [12, 16, 55]. Recent electron microscopy studies of yeast MCM found that phosphorylation does not drastically alter the conformation of MCM double hexamers; only a small change to the C-terminal region inside the central MCM channel was observed [55, 56]. Whether this change has any effect on MCM-Cdc45-Sld3 complex formation is unclear. In addition, DDK-mediated Mcm2 phosphorylation also weakens the Mcm2-Mcm5 interaction, an event favoring Mcm2-Mcm5 gate opening and CMG formation [18].

Phosphorylation of DDK sites on budding yeast MCM requires prior phosphorylation events, a process called priming [20, 57, 58]. Such priming events occur on the S/T sites within the S/T-P/Q motifs of Mmc2, Mcm4, and Mcm6 [20]. Mutating all S/T sites within these motifs in Mcm4 and Mcm6 results in a slow growth that can be rescued by phospho-mimetic DDK site mutations of these substrates, suggesting that the key function of priming events is to promote DDK-mediated phosphorylation [20]. The effect of priming at S/T-Q and S/T-P sites appears to be redundant, as mutating one type of sites does not affect replication while simultaneously mutating both types does [20]. Consistent with this, the DNA damage checkpoint kinase Mec1, which targets S/T-Q sites and has a major role under replication stress (see below), is critical for growth when all the S/T-P sites are inactivated in both Mcm4 and Mcm6 [20]. Another priming kinase is CDK, as it enhances phosphorylation by DDK of mouse MCM4 in vitro [36]. Whether additional kinases are involved in the priming events remains to be elucidated.

De-phosphorylation of DDK Substrates via Protein Phosphatase 1

One means of controlling DDK-dependent MCM phosphorylation levels is through protein phosphatase 1 (PP1). In budding and fission yeasts as well as *Xenopus*, PP1 can reverse DDK-mediated phosphorylation of Mcm4 in G1 phase [23, 24, 59, 60]. In both yeasts, PP1 binds to a co-factor, Rif1. Two N-terminal motifs of Rif1, namely SILK and RVXF, mediate interaction with PP1, while the C-terminal tail of Rif1 supports DDK interaction [23, 24, 59]. How the PP1-Rif1 complex is targeted to MCM in G1 phase is not well understood. PP1-mediated inhibition of Mcm4 phosphorylation is overcome in S phase by higher DDK protein levels, achieved by both increased transcription and decreased degradation of the Dbf4 subunit of DDK [61–64]. Higher DDK levels can directly augment Mcm4 phosphorylation [55, 57]. Furthermore, DDK and CDK both phosphorylate regions surrounding the SILK and RVXF motifs in Rif1, weakening the Rif1-PP1 interaction (Table 18.1) [23, 24]. This results in a positive feed-forward mechanism for DDK-dependent phosphorylation. Interestingly, in *rif1*Δ cells, not only Mcm4 phosphorylation levels are higher, as expected, but Sld3 phosphorylation levels also rise [59]. This may suggest that DDK can promote the phosphorylation of Sld3, which is generally regarded as a CDK substrate (see below). Testing this idea may reveal cross talk between the two kinases on this substrate. It is noteworthy that Rif1 has additional roles in regulating genome-wide replication timing in several species including yeasts and mammals, as well as telomere length regulation in budding yeast [65–69]. It will be interesting to see whether and how these functions of Rif1 are interrelated.

CDK-Mediated Regulation of CMG Formation and Prevention of Re-replication

CDK Is Essential for CMG Formation

Both G1- and S-CDKs are critical for the initiation of replication. The G1-CDK triggers S-phase entry via multiple mechanisms, including activation of S-CDK by downregulating its inhibitors (reviewed in [70, 71]). For example, in budding yeast, G1-CDK phosphorylates the S-CDK inhibitor Sic1 leading to Sic1 degradation [72, 73]. Activated S-CDK plays dual roles in replication initiation: it triggers CMG assembly, and also blocks re-replication by preventing pre-RC formation at the end of G1 phase (reviewed in [74, 75]).

In budding yeast, the key S-CDK substrates for CMG formation are the scaffold proteins Sld2 and Sld3 [26, 27]. Phosphorylation levels of Sld2 and Sld3 during the cell cycle mirror S-CDK activity levels [26, 27]. While two main modification sites (T600 and S622) of Sld3 are similarly phosphorylated by CDK, phosphorylation sites of Sld2 show a hierarchy (Table 18.1) [25, 27, 28]. Specifically, phosphorylation of several sites, including S100 and S208 primes phosphorylation of T84 [25]. This conclusion is based on several observations, such as kinetics of T84 phosphorylation are slower than those of S100 and S208 in vitro. This two-step phosphorylation mechanism likely bears biological importance; for example, it can create a threshold requirement for T84 phosphorylation, thus preventing precocious activation of Sld2 [25, 28].

Once phosphorylated by CDK, Sld2 and Sld3 interact with another scaffold Dpb11 [26–28]. Phosphorylated Sld2 and Sld3 bind the C- and N-terminal BRCT domains Dpb11, respectively [26, 27]. The importance of these phosphorylation-mediated interactions is suggested by the ability of a combined phospho-mimetic Sld2-T84D mutation and Sld3-Dpb11 fusion to bypass the S-CDK requirement for replication initiation [27]. Similar bypass can be achieved with Sld2-T84D and a gain-of-function mutation of Cdc45 (*cdc45-JET1*), which is thought to induce CDK-independent Sld3-Dpb11 association via an as-yet unknown mechanism [26]. Though these particular genetic situations enable replication initiation in the absence of CDK, such bypass leads to extensive re-replication, underlining the importance of CDK regulation for blocking re-replication events (see below) [26, 27, 76]. Assembly of the Sld3-Dpb11-Sld2 complex triggers CMG formation through multiple protein-protein interactions [13, 77, 78]. While Sld3 interacts with Cdc45 and MCM, Dpb11-Sld2 interacts with GINS through DNA polymerase epsilon (Pol ε) [78–80]. Sld3-Dpb11 interaction serves as a bridge to bring Sld3-Cdc45-MCM and Dpb11-Sld2-Polε-GINS together [26, 27, 78]. This model is supported by several results using fusion proteins to bypass the need for bridging proteins [27, 81].

The aforementioned findings in budding yeast appear to be largely conserved in other eukaryotes. Both CDK-dependent regulation of the Sld3-Dpb11 interaction and the Sld3 phosphorylation sites are observed in human, fission yeast, and *Xenopus* (Table 18.1) [37, 41, 46, 82]. Phosphorylation of human Sld3 homolog Treslin/Ticrr (referred to as Treslin hereafter) on T969 and S1001 (equivalent to T600 and S622 in Sld3) occurs in vivo and in vitro, and leads to association with TopBP1 (Dpb11 homolog) [41, 46]. These events are required for DNA replication in humans. In *S. pombe*, the interaction of SpSld3 with SpCut5 (Dpb11 homolog) is essential for replication initiation and cell viability [37]. However, phosphorylation of SpSld3 at T636 and S673 (equivalent to T600 and S622 in budding yeast Sld3) is not essential [37]. Mutating additional CDK consensus sites further reduces SpSld3 phosphorylation and impairs SpSld3-SpCut5 association, but does not result in lethality [37]. It is thus likely that there are additional CDK sites or that non-phosphorylation-based mechanisms also contribute to the SpSld3-SpCut5 interaction in fission yeast. CDK phosphorylation of the *S. pombe* and *C. elegans* Sld2 homologs is also essential for binding to Dpb11 homologs and for replication initiation [37, 43]. The phosphorylation sites of the Sld2 homologs are conserved between budding and fission yeast, but not in *C. elegans* [37, 43]. A consensus regarding Sld2 homologs in other metazoans has not been reached, though RecQ4 is thought to be the human homolog (reviewed in [13, 83]). Whether CDK phosphorylation also dictates the interaction between RecQ4 and TopBP1 has not been determined.

CDK Prevents Re-replication

In addition to triggering CMG formation, CDK-mediated phosphorylation is also required for preventing re-replication, which is critical for avoiding aneuploidy and other forms of genome instability [84–90]. Before origin firing begins, G1- and S-CDK phosphorylate several proteins involved in pre-RC formation to inactivate them via multiple mechanisms, such as protein degradation, nuclear export, and inactivation of substrate activities [91, 92].

In budding yeast, CDK-mediated phosphorylation of four substrates plays redundant roles in preventing re-replication (Table 18.1) [32]. One substrate is Cdc6, an AAA^+ ATPase essential for loading MCM onto origins [93–96]. Both G1- and S-CDKs interact with and phosphorylate Cdc6 in late G1 phase and S phase, leading to its recognition by SCF^{Cdc4} ubiquitin ligase and subsequent ubiquitin-mediated degradation [29, 97–102]. The second substrate is Mcm3 [20, 30]. Several results indicate that CDK triggers MCM export to the cytoplasm from late G1 until mitosis through phosphorylating Mcm3 [103, 104]. For example, mutating five CDK consensus sites in the vicinity of yeast Mcm3 nuclear localization and nuclear export signals prevents its nuclear export [30]. Lastly, CDK targets two subunits of the origin recognition complex (ORC), which binds to origin sequences and collaborates with Cdc6 and another loading factor Cdt1 to load MCM [96, 105–112]. S-CDK phosphorylates Orc2 and Orc6 in vitro and in vivo [31, 32, 113]. In vitro phosphorylation of Orc6 disrupts the interaction of its N-terminal region with Cdt1 and reduces MCM loading, while that of Orc2 has minimal effects on MCM loading [31]. The redundancy amongst these four phosphorylation targets is supported by the observation that re-replication only occurs when all phosphorylation events are disrupted [32].

In other organisms, CDKs also prevent re-replication by targeting several pre-RC components (Table 18.1). Only a subset of these is conserved from budding yeast. Even for identical substrates, the effects of CDK phosphorylation vary among organisms. In *S. pombe*, SpCdc18 (Cdc6 homolog) is phosphorylated by CDK and subsequently degraded [38, 39, 114]. Unlike in budding yeast, overexpression of SpCdc18 alone is sufficient to induce re-replication, indicating that *S. pombe* relies more heavily on SpCdc18 to prevent re-replication than budding yeast [115, 116]. In comparison, SpOrc2 phosphorylation by CDK has less dramatic effects as SpCdc18 [40]. Like the yeast proteins, *Xenopus* and human Cdc6 are also targeted by CDKs, but this results in nuclear export rather than degradation [42, 50, 117–121]. Unlike in yeast, human Cdt1 and Orc1 are additional CDK substrates important for preventing re-replication as their phosphorylation leads to SCF^{Skp2}-mediated degradation [47–49, 122–125]. Cdt1 stability regulation plays a key role, as overexpression or Cdt1 or removal of Cdt1 inhibitor geminin is sufficient to cause re-replication. This is achieved through both phosphorylation-dependent, SCF^{Skp2}-mediated and phosphorylation-independent, $PCNA^{Cdt2}$-mediated degradation [126–130]. The conservation and variation among different organisms highlight the importance of blocking re-replication and regulatory flexibility tailored to the special need of each particular organism.

Reversal of CDK-Mediated Phosphorylation

Studies in yeast show that the Cdc14 phosphatase can reverse a myriad of CDK-mediated phosphorylations in mitosis [131–133]. Cdc14 substrates include Sld2 and Sic1, whose functions have been described above [131, 133]. In addition, Cdc14 also targets transcription factor Swi5, which is responsible for G1-specific gene transcription, including that of Sic1 and Cdc6 [133–136]. It is currently unclear which phosphatase reverses Sld3 modification, even though Sld3 phosphorylation oscillates during the cell cycle similarly to Sld2 [26, 28]. Protein degradation and new synthesis thus likely collaborate to reset the stage for CDK function in each cell cycle.

DNA Replication and Damage Checkpoint Kinases in Origin Firing Program

Checkpoint Kinases Regulate Origin Firing under Replication Stress Conditions

Replicative stress poses a challenge for cells as it impedes the progression of the replication machinery. Successful replication under replicative stress requires multiple levels of regulation. One of these entails the inhibition of origin firing by the

DNA replication and damage checkpoint kinases (referred to as checkpoint kinases hereafter) [137–143]. This regulation is thought to both reduce the number of active replication forks, thus the opportunity for them to stall or collapse, and provide extra time for existing forks to complete DNA synthesis. Studies in multiple organisms suggest that checkpoint-mediated regulation of origin firing involves simultaneous targeting of CDK and DDK substrates as well as the kinases themselves, but major differences between organisms exist in the specific details [33, 34, 144].

In budding yeast, the apical checkpoint kinase Mec1 (homolog of human ATR) and the downstream effector kinase Rad53 (homolog of human CHK2) are both required to inhibit late origin firing [141, 143]. Upon replication stress from dNTP depletion or DNA alkylation by MMS (methyl methanesulfonate), Mec1 activates Rad53, which in turn phosphorylates the scaffold protein Sld3 (Table 18.1) [33, 34]. Rad53 phosphorylation of Sld3 is an important means to achieve origin inhibition, since phospho-mimetic mutations in a subset of Rad53 phosphorylation sites on Sld3 impair the Sld3-Dpb11 and Sld3-Cdc45 interactions required for CMG formation [26, 27, 33, 34, 79]. Mec1 and Rad53 also target the DDK subunit Dbf4, and in vitro studies show that such phosphorylation reduces DDK kinase activity (Table 18.1) [33–35, 58]. Checkpoint kinase-mediated phosphorylation of Sld3 and Dbf4 has overlapping functions, since mutating Rad53 phosphorylation sites on both proteins increases firing more strongly than mutating those on only one substrate [33, 34]. We note that checkpoint kinases have additional roles in regulating origin firing. As described above, Mec1 has a role in priming Mcm4 and Mcm6 for DDK phosphorylation [20]. Another example is the requirement of the Tel1 kinase (homolog of human ATM) for origin firing at a subset of short telomeres, which replicate earlier than longer telomeres; this function is presumably due to promoting Cdc45 recruitment to pre-RCs [145].

Interestingly, in human cells under genotoxic stress, unlike in budding yeast, the checkpoint kinases ATM and CHK2 inhibit CDK activity by targeting its activator, CDC25 phosphatase, for degradation [144]. In the mean time, the checkpoint kinase CHK1 (homolog of yeast Chk1) can phosphorylate CDC25, and this modification promotes CDC25 interaction with the phosphoserine binding protein 14-3-3, leading to its inactivation [146–148]. Interestingly, CHK1 can prevent origin firing under unperturbed conditions, by binding to and phosphorylating Treslin [51]. Disrupting the binding leads to increased origin firing and higher levels of chromatin-bound CDC45 [51]. It will be interesting to test whether this mechanism also operates under replicative stress.

Reversal of Checkpoint Kinase-Mediated Phosphorylation

The reversal of checkpoint effects is critical for cells to recover from replicative stress. Four major serine/threonine phosphatases, namely the PP4 phosphatase Pph3-Psy2, two PP2C phosphatases Ptc2 and Ptc3, and the PP1 phosphatase Glc7, contribute to checkpoint inactivation by dephosphorylating Rad53 in budding yeast

[149–153]. Pph3-Psy2 and Ptc2 interact with the kinase domain and FHA1 domain of Rad53, respectively [150, 152]. It is thought that these two phosphatases remove phosphorylation from different sites on Rad53 to engender different outcomes [154]. In addition, Pph3-Psy2 binds Mec1 and dephosphorylates multiple Mec1 substrates [155]. Little is known about whether these phosphatases also dephosphorylate Sld3 and Dbf4, the key factors dictating replication initiation. In human cells, several phosphatases, such as PP2A and PP2C, are required for dephosphorylating and inactivating the checkpoint effector kinases CHK1 and CHK2 [154, 156]. In fission yeast, dephosphorylating SpChk1 appears to be carried out by PP1 [154, 156]. Detailed studies of these dephosphorylation events will clarify their effects on kinase activities and on the reversal of substrate phosphorylation.

Roles of Other PTMs in Origin Licensing and Replication Timing Control

Other PTMs besides phosphorylation also contribute to origin licensing and replication timing control. For example, various types of histone modifications play a major role in altering chromatin accessibility and protein recruitment. Both can influence origin firing efficiency and timing. In addition to histone modifications, modifications of histone chaperones and nucleosome remodeling complexes also affect MCM loading. In this section we describe how writing and erasing of histone methylation and acetylation marks and how the modifications of a histone chaperone and nucleosome remodeling complex contribute to the regulation of replication initiation.

Histone Methylation

Different types of histone methylation are implicated in regulating origin licensing, origin firing, and replication timing through their effects on chromatin structure. As a recent review has summarized many aspects of these regulations, we refer the readers to this review for a thorough understanding on this topic [157]. Here we focus on a well-characterized modification with potential implications in Meier-Gorlin syndrome (MGS), which is characterized primarily by short stature. It has been shown that dimethylated histone H4 at site K20 (H4K20me2) binds to the highly conserved bromo-adjacent homology (BAH) domain of human ORC1 (Table 18.2) [159, 160]. Based on the co-crystal structure of this BAH domain and the K20me2-modified H4 peptide, a few ORC1 residues were identified to be required for this interaction [160]. Mutations of these residues (Y64A and W88A) or deletion of the BAH domain reduce ORC1's interaction with H4K20me2 in vitro [160]. In vivo, these mutations reduce ORC chromatin association, without affecting the ORC complex formation [160, 161]. Interestingly, H4K20me2 binding defects were

Table 18.2 Substrates of PTMs other than phosphorylation that regulate replication initiation

Species	Substrate	PTM	Site	Functional effects	Reference
S. cerevisiae	Spt16	Ubiquitylation	–	Promote FACT recruitment and MCM loading	[158]
	Histone[a]	Acetylation	–	Regulate timing of origin firing	Reviewed in [5]
H. sapiens	Histone H4[b]	Di-methylation	K20	Bind Orc1 BAH domain and promote ORC recruitment	[159, 160]

[a]Also found in *H. sapiens* and *D. melanogaster*
[b]Also found in *D. rerio*

found to be associated with ORC1 mutations underlying MGS [162, 163]. The authors show that the ORC1-BAH-H4K20me2 interaction is conserved in zebrafish and mice, but not in yeasts [160]. In addition, the zebrafish mutants reducing the ORC1-H4K20me2 interaction have an MGS-like dwarfism phenotype [160]. Because other MGS mutations on ORC1 can perturb the role of ORC in centrosome duplication [164], it remains to be determined whether defective origin licensing is the central, evolutionarily conserved link between ORC dysregulation and MGS.

Histone Acetylation

Histone acetylation is a key regulator of chromatin accessibility. Histone hyper-acetylation generally weakens DNA-nucleosome interaction, thus increasing DNA accessibility, while hypo-acetylation has the opposite effect [165, 166]. A direct role for histone acetylation in replication initiation has yet to be demonstrated; however, studies of several histone acetylases (HATs) and deacetylases (HDACs) clearly point to the importance of these enzymes in regulating origin firing both locally and globally (Table 18.2). One example of a locus-specific effect is that tethering the HAT Gcn5 to a late origin results in histone hyper-acetylation and early origin firing in budding yeast [167]. Similarly, tethering the HAT HBO1 to origins in *Drosophila* and human cells also increases local firing [168, 169].

An example of a global effect was recently shown by removal of the Sir2 HDAC in yeast [170]. Sir2 loss reduces silencing at the repetitive ribosomal DNA (rDNA) array that contains about one-third of all yeast origins [171]. Interestingly, *sir2Δ* cells show increased origin firing at rDNA and reduced origin firing at other loci, presumably because the former titrated away limiting replication factors such as Sld3 and Cdc45 from single origins in the rest of the genome [170, 172, 173]. This finding suggests that a global effect on origin firing is tightly linked to silencing at repetitive sequences. The effect of another HDAC called Rpd3 in yeast may be more complex, as its roles involve both a local effect at single origins and a global effect due to changes at repetitive regions [167, 170, 174, 175]. The former scenario is supported by data from *Drosophila*, where tethering of the Rpd3 homolog decreases individual origin firing [168]. The latter is suggested by the finding that Rpd3 removal in yeast leads to decreased origin firing at rDNA and concomitantly increases of firing at other origins [170]. It is worth noting that although histones are the main substrates examined for the aforementioned acetylases and deacetylases, non-histone substrates may also have a role in replication timing control.

Ubiquitylation

Ubiquitylation affects several steps of replication initiation. We have mentioned in earlier sections that ubiquitin-mediated proteasome degradation is essential for G1 entry, enabling origin licensing and preventing re-replication. Details regarding these regulations can be found in several reviews [70, 74, 75, 176, 177]. Here we discuss one example wherein ubiquitylation does not lead to protein degradation. In budding yeast, the cullin ubiquitin ligase Rtt101 promotes ubiquitylation of the Spt16 subunit of histone chaperone complex FACT (FAcilitates Chromatin Transcription) under normal growth conditions (Table 18.2) [158]. Spt16 is modified by K63-linked ubiquitin chains, which do not result in its degradation [158]. Mutating either Rtt101 or FACT subunits partially impairs FACT recruitment and MCM loading at early origins [158]. How such impairments affect replication initiation is not completely understood, since 2D gel electrophoresis data did not uncover noticeable initiation defects [178]. It will be interesting to determine if the defects seen in Rtt101 and FACT mutant cells are caused by lack of Spt16 ubiquitination or other roles that have been proposed for Rtt101, such as histone H3 ubiquitination [179].

Concluding Remarks

Current findings from multiple organisms have established an important role of PTMs in the regulation of replication initiation at multiple levels, from origin licensing to firing, and at both local to global levels. At present, the best understood form of regulation is kinase-based phosphorylation, but the emerging relevance of other PTMs is gradually being revealed. It will take future studies to provide the mechanistic details of their regulatory functions. Besides the forms of PTMs discussed here, additional forms of PTMs should also be considered. For example, several proteins central to the replication machinery have been identified as substrates of sumoylation [180], pointing to a previously unappreciated role for this modification in replication. As most knowledge regarding replication initiation has been gathered from the work in unicellular organisms, an important future direction will be to extrapolate what we already know to more complex eukaryotes, such as humans. Together, these studies will expand our understanding of the conserved and divergent regulatory principles that different organisms have evolved to accommodate the demands of replicating their unique genomes.

References

1. Bell SP, Dutta A. DNA replication in eukaryotic cells. Annu Rev Biochem. 2002;71:333–74.
2. Kelly TJ, Brown GW. Regulation of chromosome replication. Annu Rev Biochem. 2000;69:829–80.
3. Remus D, Diffley JFX. Eukaryotic DNA replication control: lock and load, then fire. Curr Opin Cell Biol. 2009;21(6):771–7.
4. Sclafani RA, Holzen TM. Cell cycle regulation of DNA replication. Annu Rev Genet. 2007;41(1):237–80.
5. Masai H, Matsumoto S, You Z, Yoshizawa-Sugata N, Oda M. Eukaryotic chromosome DNA replication: where, when, and how? Annu Rev Biochem. 2010;79:89–130.
6. Gambus A, Jones RC, Sanchez-Diaz A, Kanemaki M, van Deursen F, Edmondson RD, et al. GINS maintains association of Cdc45 with MCM in replisome progression complexes at eukaryotic DNA replication forks. Nat Cell Biol. 2006;8(4):358–66.
7. Ilves I, Petojevic T, Pesavento JJ, Botchan MR. Activation of the MCM2-7 helicase by association with Cdc45 and GINS proteins. Mol Cell. 2010;37(2):247–58.
8. Kang YH, Galal WC, Farina A, Tappin I, Hurwitz J. Properties of the human Cdc45/Mcm2-7/GINS helicase complex and its action with DNA polymerase epsilon in rolling circle DNA synthesis. Proc Natl Acad Sci U S A. 2012;109(16):6042–7.
9. Moyer SE, Lewis PW, Botchan MR. Isolation of the Cdc45/Mcm2-7/GINS (CMG) complex, a candidate for the eukaryotic DNA replication fork helicase. Proc Natl Acad Sci U S A. 2006;103(27):10236–41.
10. Pacek M, Tutter AV, Kubota Y, Takisawa H, Walter JC. Localization of MCM2-7, Cdc45, and GINS to the site of DNA unwinding during eukaryotic DNA replication. Mol Cell. 2006;21(4):581–7.
11. Hardy CF, Dryga O, Seematter S, Pahl PM, Sclafani RA. mcm5/cdc46-bob1 bypasses the requirement for the S phase activator Cdc7p. Proc Natl Acad Sci U S A. 1997;94(7):3151–5.
12. Heller RC, Kang S, Lam WM, Chen S, Chan CS, Bell SP. Eukaryotic origin-dependent DNA replication in vitro reveals sequential action of DDK and S-CDK kinases. Cell. 2011;146(1):80–91.
13. Labib K. How do Cdc7 and cyclin-dependent kinases trigger the initiation of chromosome replication in eukaryotic cells? Genes Dev. 2010;24(12):1208–19.
14. Sheu Y-J, Stillman B. Cdc7-Dbf4 phosphorylates MCM proteins via a docking site-mediated mechanism to promote S phase progression. Mol Cell. 2006;24(1):101–13.
15. Zou L, Stillman B. Assembly of a complex containing Cdc45p, replication protein A, and Mcm2p at replication origins controlled by S-phase cyclin-dependent kinases and Cdc7p-Dbf4p kinase. Mol Cell Biol. 2000;20(9):3086–96.
16. Gros J, Devbhandari S, Remus D. Origin plasticity during budding yeast DNA replication in vitro. EMBO J. 2014;33(6):621–36.
17. Bruck I, Kaplan D. Dbf4-Cdc7 phosphorylation of Mcm2 is required for cell growth. J Biol Chem. 2009;284(42):28823–31.
18. Bruck I, Kaplan DL. The Dbf4-Cdc7 kinase promotes Mcm2-7 ring opening to allow for single-stranded DNA extrusion and helicase assembly. J Biol Chem. 2014.
19. Ramer MD, Suman ES, Richter H, Stanger K, Spranger M, Bieberstein N, et al. Dbf4 and Cdc7 proteins promote DNA replication through interactions with distinct Mcm2-7 protein subunits. J Biol Chem. 2013;288(21):14926–35.
20. Randell JCW, Fan A, Chan C, Francis LI, Heller RC, Galani K, et al. Mec1 is one of multiple kinases that prime the Mcm2-7 helicase for phosphorylation by Cdc7. Mol Cell. 2010;40(3):353–63.
21. Sheu Y-J, Stillman B. The Dbf4-Cdc7 kinase promotes S phase by alleviating an inhibitory activity in Mcm4. Nature. 2010;463(7277):113–7.

22. Stead BE, Brandl CJ, Davey MJ. Phosphorylation of Mcm2 modulates Mcm2-7 activity and affects the cell's response to DNA damage. Nucleic Acids Res. 2011;39(16):6998–7008.
23. Davé A, Cooley C, Garg M, Bianchi A. Protein phosphatase 1 recruitment by Rif1 regulates DNA replication origin firing by counteracting DDK activity. Cell Rep. 2014;7(1):53–61.
24. Hiraga S-I, Alvino GM, Chang F, Lian H-Y, Sridhar A, Kubota T, et al. Rif1 controls DNA replication by directing protein phosphatase 1 to reverse Cdc7-mediated phosphorylation of the MCM complex. Genes Dev. 2014;28(4):372–83.
25. Tak Y-S, Tanaka Y, Endo S, Kamimura Y, Araki H. A CDK-catalysed regulatory phosphorylation for formation of the DNA replication complex Sld2-Dpb11. EMBO J. 2006;25(9):1987–96.
26. Tanaka S, Umemori T, Hirai K, Muramatsu S, Kamimura Y, Araki H. CDK-dependent phosphorylation of Sld2 and Sld3 initiates DNA replication in budding yeast. Nature. 2007;445(7125):328–32.
27. Zegerman P, Diffley JFX. Phosphorylation of Sld2 and Sld3 by cyclin-dependent kinases promotes DNA replication in budding yeast. Nature. 2007;445(7125):281–5.
28. Masumoto H, Muramatsu S, Kamimura Y, Araki H. S-Cdk-dependent phosphorylation of Sld2 essential for chromosomal DNA replication in budding yeast. Nature. 2002;415(6872):651–5.
29. Calzada A, Sánchez M, Sánchez E, Bueno A. The stability of the Cdc6 protein is regulated by cyclin-dependent kinase/cyclin B complexes in *Saccharomyces cerevisiae*. J Biol Chem. 2000;275(13):9734–41.
30. Liku ME, Nguyen VQ, Rosales AW, Irie K, Li JJ. CDK phosphorylation of a novel NLS-NES module distributed between two subunits of the Mcm2-7 complex prevents chromosomal rereplication. Mol Biol Cell. 2005;16(10):5026–39.
31. Chen S, Bell SP. CDK prevents Mcm2-7 helicase loading by inhibiting Cdt1 interaction with Orc6. Genes Dev. 2011.
32. Nguyen VQ, Co C, Li JJ. Cyclin-dependent kinases prevent DNA re-replication through multiple mechanisms. Nature. 2001;411(6841):1068–73.
33. Lopez-Mosqueda J, Maas NL, Jonsson ZO, Defazio-Eli LG, Wohlschlegel J, Toczyski DP. Damage-induced phosphorylation of Sld3 is important to block late origin firing. Nature. 2010;467(7314):479–83.
34. Zegerman P, Diffley JFX. Checkpoint-dependent inhibition of DNA replication initiation by Sld3 and Dbf4 phosphorylation. Nature. 2010;467(7314):474–8.
35. Weinreich M, Stillman B. Cdc7p-Dbf4p kinase binds to chromatin during S phase and is regulated by both the APC and the RAD53 checkpoint pathway. EMBO J. 1999;18(19):5334–46.
36. Masai H, Taniyama C, Ogino K, Matsui E, Kakusho N, Matsumoto S, et al. Phosphorylation of MCM4 by Cdc7 kinase facilitates its interaction with Cdc45 on the chromatin. J Biol Chem. 2006;281(51):39249–61.
37. Fukuura M, Nagao K, Obuse C, Takahashi TS, Nakagawa T, Masukata H. CDK promotes interactions of Sld3 and Drc1 with Cut5 for initiation of DNA replication in fission yeast. Mol Biol Cell. 2011;22(14):2620–33.
38. Jallepalli PV, Brown GW, MuziFalconi M, Tien D, Kelly TJ. Regulation of the replication initiator protein p65(cdc18) by CDK phosphorylation. Genes Dev. 1997;11(21):2767–79.
39. Lopez-Girona A, Mondesert O, Leatherwood J, Russell P. Negative regulation of Cdc18 DNA replication protein by Cdc2. Mol Biol Cell. 1998;9(1):63–73.
40. Vas A, Mok W, Leatherwood J. Control of DNA rereplication via Cdc2 phosphorylation sites in the origin recognition complex. Mol Cell Biol. 2001;21(17):5767–77.
41. Kumagai A, Shevchenko A, Shevchenko A, Dunphy WG. Direct regulation of Treslin by cyclin-dependent kinase is essential for the onset of DNA replication. J Cell Biol. 2011;193(6):995–1007.
42. Pelizon C, Madine MA, Romanowski P, Laskey RA. Unphosphorylatable mutants of Cdc6 disrupt its nuclear export but still support DNA replication once per cell cycle. Genes Dev. 2000;14(19):2526–33.

43. Gaggioli V, Zeiser E, Rivers D, Bradshaw CR, Ahringer J, Zegerman P. CDK phosphorylation of SLD-2 is required for replication initiation and germline development in C. elegans. J Cell Biol. 2014;204(4):507–22.
44. Tsuji T, Ficarro SB, Jiang W. Essential role of phosphorylation of MCM2 by Cdc7/Dbf4 in the initiation of DNA replication in mammalian cells. Mol Biol Cell. 2006;17(10):4459–72.
45. Montagnoli A, Valsasina B, Brotherton D, Troiani S, Rainoldi S, Tenca P, et al. Identification of Mcm2 phosphorylation sites by S-phase-regulating kinases. J Biol Chem. 2006;281(15):10281–90.
46. Boos D, Sanchez-Pulido L, Rappas M, Pearl LH, Oliver AW, Ponting CP, et al. Regulation of DNA replication through Sld3-Dpb11 interaction is conserved from yeast to humans. Curr Biol. 2011;21(13):1152–7.
47. Li X, Zhao Q, Liao R, Sun P, Wu X. The SCF(Skp2) ubiquitin ligase complex interacts with the human replication licensing factor Cdt1 and regulates Cdt1 degradation. J Biol Chem. 2003;278(33):30854–8.
48. Takeda DY, Parvin JD, Dutta A. Degradation of Cdt1 during S phase is Skp2-independent and is required for efficient progression of mammalian cells through S phase. J Biol Chem. 2005;280(24):23416–23.
49. Méndez J, Zou-Yang XH, Kim S-Y, Hidaka M, Tansey WP, Stillman B. Human origin recognition complex large subunit is degraded by ubiquitin-mediated proteolysis after initiation of DNA replication. Mol Cell. 2002;9(3):481–91.
50. Petersen BO, Lukas J, Sørensen CS, Bartek J, Helin K. Phosphorylation of mammalian CDC6 by cyclin A/CDK2 regulates its subcellular localization. EMBO J. 1999;18(2):396–410.
51. Guo C, Kumagai A, Schlacher K, Shevchenko A, Shevchenko A, Dunphy WG. Interaction of Chk1 with treslin negatively regulates the initiation of chromosomal DNA replication. Mol Cell. 2014;1–14.
52. Lei M, Kawasaki Y, Young MR, Kihara M, Sugino A, Tye BK. Mcm2 is a target of regulation by Cdc7-Dbf4 during the initiation of DNA synthesis. Genes Dev. 1997;11(24):3365–74.
53. Cho W-H, Lee Y-J, Kong S-I, Hurwitz J, Lee J-K. CDC7 kinase phosphorylates serine residues adjacent to acidic amino acids in the minichromosome maintenance 2 protein. Proc Natl Acad Sci U S A. 2006;103(31):11521–6.
54. Lee J-K, Seo Y-S, Hurwitz J. The Cdc23 (Mcm10) protein is required for the phosphorylation of minichromosome maintenance complex by the Dfp1-Hsk1 kinase. Proc Natl Acad Sci U S A. 2003;100(5):2334–9.
55. On KF, Beuron F, Frith D, Snijders AP, Morris EP, Diffley JFX. Prereplicative complexes assembled in vitro support origin-dependent and independent DNA replication. EMBO J. 2014;33(6):605–20.
56. Sun J, Fernández-Cid A, Riera A, Tognetti S, Yuan Z, Stillman B, et al. Structural and mechanistic insights into Mcm2-7 double-hexamer assembly and function. Genes Dev. 2014;28(20):2291–303.
57. Francis LI, Randell JCW, Takara TJ, Uchima L, Bell SP. Incorporation into the prereplicative complex activates the Mcm2-7 helicase for Cdc7-Dbf4 phosphorylation. Genes Dev. 2009;23(5):643–54.
58. Kihara M, Nakai W, Asano S, Suzuki A, Kitada K, Kawasaki Y, et al. Characterization of the yeast Cdc7p/Dbf4p complex purified from insect cells. Its protein kinase activity is regulated by Rad53p. J Biol Chem. 2000;275(45):35051–62.
59. Mattarocci S, Shyian M, Lemmens L, Damay P, Altintas DM, Shi T, et al. Rif1 controls DNA replication timing in yeast through the PP1 phosphatase Glc7. Cell Rep. 2014;7(1):62–9.
60. Poh WT, Chadha GS, Gillespie PJ, Kaldis P, Blow JJ. Xenopus Cdc7 executes its essential function early in S phase and is counteracted by checkpoint-regulated protein phosphatase 1. Open Biol. 2014;4:130138.
61. Chapman JW, Johnston LH. The yeast gene, DBF4, essential for entry into S phase is cell cycle regulated. Exp Cell Res. 1989;180(2):419–28.

62. Cheng LA, Collyer T, Hardy CFJ. Cell cycle regulation of DNA replication initiator factor Dbf4p. Mol Cell Biol. 1999;19(6):4270–8.
63. Ferreira MG, Santocanale C, Drury LS, Diffley JFX. Dbf4p, an essential S phase-promoting factor, is targeted for degradation by the anaphase-promoting complex. Mol Cell Biol. 2000;20(1):242–8.
64. Oshiro G, Owens JC, Shellman Y, Sclafani RA, Li JJ. Cell cycle control of Cdc7p kinase activity through regulation of Dbf4p stability. Mol Cell Biol. 1999;19(7):4888–96.
65. Bianchi A, Shore D. Early replication of short telomeres in budding yeast. Cell. 2007;128(6):1051–62.
66. Cornacchia D, Dileep V, Quivy JP, Foti R, Tili F, Santarella-Mellwig R, et al. Mouse Rif1 is a key regulator of the replication-timing programme in mammalian cells. EMBO J. 2012;31(18):3678–90.
67. Hayano M, Kanoh Y, Matsumoto S, Renard-Guillet C, Shirahige K, Masai H. Rif1 is a global regulator of timing of replication origin firing in fission yeast. Genes Dev. 2012;26(2):137–50.
68. Wotton D, Shore D. A novel Rap1p-interacting factor, Rif2p, cooperates with Rif1p to regulate telomere length in Saccharomyces cerevisiae. Genes Dev. 1997;11(6):748–60.
69. Yamazaki S, Ishii A, Kanoh Y, Oda M, Nishito Y, Masai H. Rif1 regulates the replication timing domains on the human genome. EMBO J. 2012;31(18):3667–77.
70. Diffley JFX. Regulation of early events in chromosome replication. Curr Biol. 2004;14(18):R778–86.
71. Reed SI. Ratchets and clocks: the cell cycle, ubiquitylation and protein turnover. Nat Rev Mol Cell Biol. 2003;4(11):855–64.
72. Nash P, Tang X, Orlicky S, Chen Q, Gertler FB, Mendenhall MD, et al. Multisite phosphorylation of a CDK inhibitor sets a threshold for the onset of DNA replication. Nature. 2001;414(6863):514–21.
73. Verma R, Annan RS, Huddleston MJ, Carr SA, Reynard G, Deshaies RJ. Phosphorylation of Sic1p by G1 Cdk required for its degradation and entry into S phase. Science. 1997;278(5337):455–60.
74. Arias EE, Walter JC. Strength in numbers: preventing rereplication via multiple mechanisms in eukaryotic cells. Genes Dev. 2007;21(5):497–518.
75. Blow JJ, Dutta A. Preventing re-replication of chromosomal DNA. Nat Rev Mol Cell Biol. 2005;6(6):476–86.
76. Tanaka S, Araki H. Multiple regulatory mechanisms to inhibit untimely initiation of DNA replication are important for stable genome maintenance. PLoS Genet. 2011;7(6), e1002136.
77. Araki H. Cyclin-dependent kinase-dependent initiation of chromosomal DNA replication. Curr Opin Cell Biol. 2010;22(6):766–71.
78. Muramatsu S, Hirai K, Tak Y-S, Kamimura Y, Araki H. CDK-dependent complex formation between replication proteins Dpb11, Sld2, Pol (epsilon}, and GINS in budding yeast. Genes Dev. 2010;24(6):602–12.
79. Kamimura Y, Tak Y-S, Sugino A, Araki H. Sld3, which interacts with Cdc45 (Sld4), functions for chromosomal DNA replication in *Saccharomyces cerevisiae*. EMBO J. 2001;20(8):2097–107.
80. Sengupta S, van Deursen F, De Piccoli G, Labib K. Dpb2 integrates the leading-strand DNA polymerase into the eukaryotic replisome. Curr Biol. 2013.
81. Johansson E, Speck C, Chabes A. A top-down view on DNA replication and recombination from 9,000 feet above sea level. Genome Biol. 2011;12(4):304.
82. Kumagai A, Shevchenko A, Shevchenko A, Dunphy WG. Treslin collaborates with TopBP1 in triggering the initiation of DNA replication. Cell. 2010;140(3):349–59.
83. Zegerman P. Evolutionary conservation of the CDK targets in eukaryotic DNA replication initiation. Chromosoma. 2015.
84. Broek D, Bartlett R, Crawford K, Nurse P. Involvement of p34cdc2 in establishing the dependency of S phase on mitosis. Nature. 1991;349(6308):388–93.

85. Correa-bordes J, Nurse P. p25rum1 orders S phase and mitosis by acting as an inhibitor of the p34cdc2 mitotic kinase. Cell. 1995;83(6):1001–9.
86. Dahmann C, Diffley JF, Nasmyth KA. S-Phase-promoting cyclin-dependent kinases prevent re-replication by inhibiting the transition of replication origins to a pre-replicative state. Curr Biol. 1995;5(11):1257–69.
87. Hayles J, Fisher D, Woollard A, Nurse P. Temporal order of S phase and mitosis in fission yeast is determined by the state of the p34cdc2-mitotic B cyclin complex. Cell. 1994;78(5):813–22.
88. Itzhaki JE, Gilbert CS, Porter ACG. Construction by gene targeting in human cells of a 'conditional' CDC2 mutant that rereplicates its DNA. Nat Genet. 1997;15(3):258–65.
89. Jallepalli PV, Kelly TJ. Rum1 and Cdc18 link inhibition of cyclin-dependent kinase to the initiation of DNA replication in *Schizosaccharomyces pombe*. Genes Dev. 1996;10(5):541–52.
90. Moreno S, Nurse P. Regulation of progression through the G1 phase of the cell cycle by the rum1+ gene. Nature. 1994;367(6460):236–42.
91. Detweiler CS, Li JJ. Ectopic induction of Clb2 in early G1 phase is sufficient to block prereplicative complex formation in Saccharomyces cerevisiae. Proc Natl Acad Sci U S A. 1998;95(5):2384–9.
92. Piatti S, Böhm T, Cocker JH, Diffley JF, Nasmyth K. Activation of S-phase-promoting CDKs in late G1 defines a 'point of no return' after which Cdc6 synthesis cannot promote DNA replication in yeast. Genes Dev. 1996;10(12):1516–31.
93. Cocker JH, Piatti S, Santocanale C, Nasmyth K, Diffley JF. An essential role for the Cdc6 protein in forming the pre-replicative complexes of budding yeast. Nature. 1996;379(6561):180–2.
94. Donovan S, Harwood J, Drury LS, Diffley JF. Cdc6p-dependent loading of Mcm proteins onto pre-replicative chromatin in budding yeast. Proc Natl Acad Sci U S A. 1997;94(11):5611–6.
95. Neuwald AF, Aravind L, Spouge JL, Koonin EV. AAA(+): a class of chaperone-like ATPases associated with the assembly, operation, and disassembly of protein complexes. Genome Res. 1999;9(1):27–43.
96. Remus D, Beuron F, Tolun G, Griffith JD, Morris EP, Diffley JFX. Concerted loading of Mcm2-7 double hexamers around DNA during DNA replication origin licensing. Cell. 2009;139(4):719–30.
97. Drury LS, Perkins G, Diffley JF. The Cdc4/34/53 pathway targets Cdc6p for proteolysis in budding yeast. EMBO J. 1997;16(19):5966–76.
98. Drury LS, Perkins G, Diffley JF. The cyclin-dependent kinase Cdc28p regulates distinct modes of Cdc6p proteolysis during the budding yeast cell cycle. Curr Biol. 2000;10(5):231–40.
99. Elsasser S, Chi Y, Yang P, Campbell JL. Phosphorylation controls timing of Cdc6p destruction: a biochemical analysis. Mol Biol Cell. 1999;10(10):3263–77.
100. Elsasser S, Lou F, Wang B, Campbell JL, Jong A. Interaction between yeast Cdc6 protein and B-type cyclin/Cdc28 kinases. Mol Biol Cell. 1996;7(11):1723–35.
101. Perkins G, Drury LS, Diffley JF. Separate SCF(CDC4) recognition elements target Cdc6 for proteolysis in S phase and mitosis. EMBO J. 2001;20(17):4836–45.
102. Sánchez M, Calzada A, Bueno A. The Cdc6 protein is ubiquitinated in vivo for proteolysis in Saccharomyces cerevisiae. J Biol Chem. 1999;274(13):9092–7.
103. Labib K, Diffley JF, Kearsey SE. G1-phase and B-type cyclins exclude the DNA-replication factor Mcm4 from the nucleus. Nat Cell Biol. 1999;1(7):415–22.
104. Nguyen VQ, Co C, Irie K, Li JJ. Clb/Cdc28 kinases promote nuclear export of the replication initiator proteins Mcm2-7. Curr Biol. 2000;10(4):195–205.
105. Bell SP. The origin recognition complex: from simple origins to complex functions. Genes Dev. 2002;16(6):659–72.
106. Duncker BP, Chesnokov IN, McConkey BJ. The origin recognition complex protein family. Genome Biol. 2009;10(3):214.

107. Fernández-Cid A, Riera A, Tognetti S, Herrera MC, Samel S, Evrin C, et al. An ORC/Cdc6/MCM2-7 complex is formed in a multistep reaction to serve as a platform for MCM double-hexamer assembly. Mol Cell. 2013;50(4):577–88.
108. Liang C, Weinreich M, Stillman B. ORC and Cdc6p interact and determine the frequency of initiation of DNA replication in the genome. Cell. 1995;81(5):667–76.
109. Randell JCW, Bowers JL, Rodríguez HK, Bell SP. Sequential ATP hydrolysis by Cdc6 and ORC directs loading of the Mcm2-7 helicase. Mol Cell. 2006;21(1):29–39.
110. Speck C, Chen Z, Li H, Stillman B. ATPase-dependent cooperative binding of ORC and Cdc6 to origin DNA. Nat Struct Mol Biol. 2005;12(11):965–71.
111. Sun J, Kawakami H, Zech J, Speck C, Stillman B, Li H. Cdc6-induced conformational changes in ORC bound to origin DNA revealed by cryo-electron microscopy. Structure. 2012;20(3):534–44.
112. Sun J, Evrin C, Samel SA, Fernández-Cid A, Riera A, Kawakami H, et al. Cryo-EM structure of a helicase loading intermediate containing ORC-Cdc6-Cdt1-MCM2-7 bound to DNA. Nat Struct Mol Biol. 2013;20(8):944–51.
113. Wilmes GM, Archambault V, Austin RJ, Jacobson MD, Bell SP, Cross FR. Interaction of the S-phase cyclin Clb5 with an 'RXL' docking sequence in the initiator protein Orc6 provides an origin-localized replication control switch. Genes Dev. 2004;18(9):981–91.
114. Leatherwood J, Lopez-Girona A, Russell P. Interaction of Cdc2 and Cdc18 with a fission yeast ORC2-like protein. Nature. 1996;379(6563):360–3.
115. Muzi Falconi M, Brown GW, Kelly TJ. cdc18(+) regulates initiation of DNA replication in *Schizosaccharomyces pombe*. Proc Natl Acad Sci U S A. 1996;93(4):1566–70.
116. Nishitani H, Nurse P. p65cdc18 Plays a major role controlling the initiation of DNA replication in fission yeast. Cell. 1995;83(3):397–405.
117. Delmolino LM, Saha P, Dutta A. Multiple mechanisms regulate subcellular localization of human CDC6. J Biol Chem. 2001;276(29):26947–54.
118. Fujita M, Yamada C, Goto H, Yokoyama N, Kuzushima K, Inagaki M, et al. Cell cycle regulation of human CDC6 protein. Intracellular localization, interaction with the human mcm complex, and CDC2 kinase-mediated hyperphosphorylation. J Biol Chem. 1999;274(36):25927–32.
119. Jiang W, Wells NJ, Hunter T. Multistep regulation of DNA replication by Cdk phosphorylation of HsCdc6. Proc Natl Acad Sci U S A. 1999;96(11):6193–8.
120. Méndez J, Stillman B. Chromatin association of human origin recognition complex, cdc6, and minichromosome maintenance proteins during the cell cycle: assembly of prereplication complexes in late mitosis. Mol Cell Biol. 2000;20(22):8602–12.
121. Saha P, Chen J, Thome KC, Lawlis SJ, Hou ZH, Hendricks M, et al. Human CDC6/Cdc18 associates with Orc1 and cyclin-cdk and is selectively eliminated from the nucleus at the onset of S phase. Mol Cell Biol. 1998;18(5):2758–67.
122. Kim Y, Kipreos ET. Cdt1 degradation to prevent DNA re-replication: conserved and non-conserved pathways. Cell Div. 2007;2:18.
123. Liu E, Li X, Yan F, Zhao Q, Wu X. Cyclin-dependent kinases phosphorylate human Cdt1 and induce its degradation. J Biol Chem. 2004;279(17):17283–8.
124. Nishitani H, Lygerou Z, Nishimoto T. Proteolysis of DNA replication licensing factor Cdt1 in S-phase is performed independently of Geminin through its N-terminal region. J Biol Chem. 2004;279(29):30807–16.
125. Sugimoto N, Tatsumi Y, Tsurumi T, Matsukage A, Kiyono T, Nishitani H, et al. Cdt1 phosphorylation by cyclin A-dependent kinases negatively regulates its function without affecting geminin binding. J Biol Chem. 2004;279(19):19691–7.
126. Melixetian M, Ballabeni A, Masiero L, Gasparini P, Zamponi R, Bartek J, et al. Loss of geminin induces rereplication in the presence of functional p53. J Cell Biol. 2004;165(4):473–82.
127. Nishitani H, Sugimoto N, Roukos V, Nakanishi Y, Saijo M, Obuse C, et al. Two E3 ubiquitin ligases, SCF-Skp2 and DDB1-Cul4, target human Cdt1 for proteolysis. EMBO J. 2006;25(5):1126–36.

128. Senga T, Sivaprasad U, Zhu WG, Park JH, Arias EE, Walter JC, et al. PCNA is a cofactor for Cdt1 degradation by CUL4/DDB1-mediated N-terminal ubiquitination. J Biol Chem. 2006;281(10):6246–52.
129. Vaziri C, Saxena S, Jeon Y, Lee C, Murata K, Machida Y, et al. A p53-dependent checkpoint pathway prevents rereplication. Mol Cell. 2003;11(4):997–1008.
130. Zhu WG, Chen YF, Dutta A. Rereplication by depletion of geminin is seen regardless of p53 status and activates a G(2)/M checkpoint. Mol Cell Biol. 2004;24(16):7140–50.
131. Bloom J, Cross FR. Novel role for Cdc14 sequestration: Cdc14 dephosphorylates factors that promote DNA replication. Mol Cell Biol. 2007;27(3):842–53.
132. Bremmer SC, Hall H, Martinez JS, Eissler CL, Hinrichsen TH, Rossie S, et al. Cdc14 phosphatases preferentially dephosphorylate a subset of cyclin-dependent kinase (Cdk) sites containing phosphoserine. J Biol Chem. 2012;287(3):1662–9.
133. Visintin R, Craig K, Hwang ES, Prinz S, Tyers M, Amon A. The phosphatase Cdc14 triggers mitotic exit by reversal of Cdk-dependent phosphorylation. Mol Cell. 1998;2(6):709–18.
134. Knapp D, Bhoite L, Stillman DJ, Nasmyth K. The transcription factor Swi5 regulates expression of the cyclin kinase inhibitor p40SIC1. Mol Cell Biol. 1996;16(10):5701–7.
135. Piatti S, Lengauer C, Nasmyth K. Cdc6 is an unstable protein whose de novo synthesis in G1 is important for the onset of S phase and for preventing a 'reductional' anaphase in the budding yeast *Saccharomyces cerevisiae*. EMBO J. 1995;14(15):3788–99.
136. Toyn JH, Johnson AL, Donovan JD, Toone WM, Johnston LH. The Swi5 transcription factor of Saccharomyces cerevisiae has a role in exit from mitosis through induction of the cdk-inhibitor Sic1 in telophase. Genetics. 1997;145(1):85–96.
137. Harrison JC, Haber JE. Surviving the breakup: the DNA damage checkpoint. Annu Rev Genet. 2006;40:209–35.
138. Larner JM, Lee H, Little RD, Dijkwel PA, Schildkraut CL, Hamlin JL. Radiation down-regulates replication origin activity throughout the S phase in mammalian cells. Nucleic Acids Res. 1999;27(3):803–9.
139. Melo J, Toczyski D. A unified view of the DNA-damage checkpoint. Curr Opin Cell Biol. 2002;14(2):237–45.
140. Recolin B, van der Laan S, Tsanov N, Maiorano D. Molecular mechanisms of DNA replication checkpoint activation. Genes (Basel). 2014;5(1):147–75.
141. Santocanale C, Diffley JF. A Mec1- and Rad53-dependent checkpoint controls late-firing origins of DNA replication. Nature. 1998;395(6702):615–8.
142. Segurado M, Tercero JA. The S-phase checkpoint: targeting the replication fork. Biol Cell. 2009;101(11):617–27.
143. Shirahige K, Hori Y, Shiraishi K, Yamashita M, Takahashi K, Obuse C, et al. Regulation of DNA-replication origins during cell-cycle progression. Nature. 1998;395(6702):618–21.
144. Falck J, Mailand N, Syljuasen RG, Bartek J, Lukas J. The ATM-Chk2-Cdc25A checkpoint pathway guards against radioresistant DNA synthesis. Nature. 2001;410(6830):842–7.
145. Cooley C, Dave A, Garg M, Bianchi A. Tel1ATM dictates the replication timing of short yeast telomeres. EMBO Rep. 2014;15(10):1093–101.
146. Chen MS, Ryan CE, Piwnica-Worms H. Chk1 kinase negatively regulates mitotic function of Cdc25A phosphatase through 14-3-3 binding. Mol Cell Biol. 2003;23(21):7488–97.
147. Peng CY, Graves PR, Thoma RS, Wu ZQ, Shaw AS, PiwnicaWorms H. Mitotic and G(2) checkpoint control: regulation of 14-3-3 protein binding by phosphorylation of Cdc25C on serine-216. Science. 1997;277(5331):1501–5.
148. Sanchez Y, Wong C, Thoma RS, Richman R, Wu RQ, PiwnicaWorms H, et al. Conservation of the Chk1 checkpoint pathway in mammals: Linkage of DNA damage to Cdk regulation through Cdc25. Science. 1997;277(5331):1497–501.
149. Bazzi M, Mantiero D, Trovesi C, Lucchini G, Longhese MP. Dephosphorylation of gamma H2A by Glc7/protein phosphatase 1 promotes recovery from inhibition of DNA replication. Mol Cell Biol. 2010;30(1):131–45.
150. Guillemain G, Ma E, Mauger S, Miron S, Thai R, Guerois R, et al. Mechanisms of checkpoint kinase Rad53 inactivation after a double-strand break in *Saccharomyces cerevisiae*. Mol Cell Biol. 2007;27(9):3378–89.

151. Leroy C, Lee SE, Vaze MB, Ochsenbein F, Guerois R, Haber JE, et al. PP2C phosphatases Ptc2 and Ptc3 are required for DNA checkpoint inactivation after a double-strand break. Mol Cell. 2003;11(3):827–35.
152. O'Neill BM, Szyjka SJ, Lis ET, Bailey AO, Yates JR, Aparicio OM, et al. Pph3-Psy2 is a phosphatase complex required for Rad53 dephosphorylation and replication fork restart during recovery from DNA damage. Proc Natl Acad Sci U S A. 2007;104(22):9290–5.
153. Szyjka SJ, Aparicio JG, Viggiani CJ, Knott S, Xu W, Tavare S, et al. Rad53 regulates replication fork restart after DNA damage in *Saccharomyces cerevisiae*. Genes Dev. 2008;22(14):1906–20.
154. Heideker J, Lis ET, Romesberg FE. Phosphatases, DNA damage checkpoints and checkpoint deactivation. Cell Cycle. 2007;6(24):3058–64.
155. Hustedt N, Seeber A, Sack R, Tsai-Pflugfelder M, Bhullar B, Vlaming H, et al. Yeast PP4 interacts with ATR homolog Ddc2-Mec1 and regulates checkpoint signaling. Mol Cell. 2015;57(2):273–89.
156. Freeman AK, Monteiro AN. Phosphatases in the cellular response to DNA damage. Cell Commun Signal. 2010;8:27.
157. Rivera C, Gurard-Levin ZA, Almouzni G, Loyola A. Histone lysine methylation and chromatin replication. Biochim Biophys Acta. 2014;1839(12):1433–9.
158. Han J, Li Q, McCullough L, Kettelkamp C, Formosa T, Zhang Z. Ubiquitylation of FACT by the Cullin-E3 ligase Rtt101 connects FACT to DNA replication. Genes Dev. 2010;24(14):1485–90.
159. Beck DB, Burton A, Oda H, Ziegler-Birling C, Torres-Padilla M-E, Reinberg D. The role of PR-Set7 in replication licensing depends on Suv4-20 h. Genes Dev. 2012;26(23):2580–9.
160. Kuo AJ, Song J, Cheung P, Ishibe-Murakami S, Yamazoe S, Chen JK, et al. The BAH domain of ORC1 links H4K20me2 to DNA replication licensing and Meier-Gorlin syndrome. Nature. 2012;484(7392):115–9.
161. Noguchi K, Vassilev A, Ghosh S, Yates JL, DePamphilis ML. The BAH domain facilitates the ability of human Orc1 protein to activate replication origins in vivo. EMBO J. 2006;25(22):5372–82.
162. Bicknell LS, Walker S, Klingseisen A, Stiff T, Leitch A, Kerzendorfer C, et al. Mutations in ORC1, encoding the largest subunit of the origin recognition complex, cause microcephalic primordial dwarfism resembling Meier-Gorlin syndrome. Nat Genet. 2011;43(4):350–5.
163. Bicknell LS, Bongers EM, Leitch A, Brown S, Schoots J, Harley ME, et al. Mutations in the pre-replication complex cause Meier-Gorlin syndrome. Nat Genet. 2011;43(4):356–9.
164. Hossain M, Stillman B. Meier-Gorlin syndrome mutations disrupt an Orc1 CDK inhibitory domain and cause centrosome reduplication. Genes Dev. 2012;26(16):1797–810.
165. Eberharter A, Becker PB. Histone acetylation: a switch between repressive and permissive chromatin – second in review series on chromatin dynamics. EMBO Rep. 2002;3(3):224–9.
166. Gorisch SM, Wachsmuth M, Toth KF, Lichter P, Rippe K. Histone acetylation increases chromatin accessibility. J Cell Sci. 2005;118(Pt 24):5825–34.
167. Vogelauer M, Rubbi L, Lucas I, Brewer BJ, Grunstein M. Histone acetylation regulates the time of replication origin firing. Mol Cell. 2002;1223–33.
168. Aggarwal BD, Calvi BR. Chromatin regulates origin activity in Drosophila follicle cells. Nature. 2004;430(6997):372–6.
169. Chen X, Liu G, Leffak M. Activation of a human chromosomal replication origin by protein tethering. Nucleic Acids Res. 2013;41(13):6460–74.
170. Yoshida K, Bacal J, Desmarais D, Padioleau I, Tsaponina O, Chabes A, et al. The histone deacetylases Sir2 and Rpd3 Act on ribosomal DNA to control the replication program in budding yeast. Mol Cell. 2014;54(4):691–7.
171. Smith JS, Boeke JD. An unusual form of transcriptional silencing in yeast ribosomal DNA. Genes Dev. 1997;11(2):241–54.
172. Mantiero D, Mackenzie A, Donaldson A, Zegerman P. Limiting replication initiation factors execute the temporal programme of origin firing in budding yeast. EMBO J. 2011;30(23):4805–14.

173. Tanaka S, Nakato R, Katou Y, Shirahige K, Araki H. Origin association of Sld3, Sld7, and Cdc45 proteins is a key step for determination of origin-firing timing. Curr Biol. 2011;21(24):2055–63.
174. Aparicio JG, Aparicio JG, Viggiani CJ, Viggiani CJ, Gibson DG, Gibson DG, et al. The Rpd3-Sin3 histone deacetylase regulates replication timing and enables intra-S origin control in *Saccharomyces cerevisiae*. Mol Cell Biol. 2004;24(11):4769–80.
175. Knott SRV, Viggiani CJ, Tavare S, Aparicio OM. Genome-wide replication profiles indicate an expansive role for Rpd3L in regulating replication initiation timing or efficiency, and reveal genomic loci of Rpd3 function in *Saccharomyces cerevisiae*. Genes Dev. 2009;23(9):1077–90.
176. Sivakumar S, Gorbsky GJ. Spatiotemporal regulation of the anaphase-promoting complex in mitosis. Nat Rev Mol Cell Biol. 2015;16(2):82–94.
177. Vodermaier HC. APC/C and SCF: controlling each other and the cell cycle. Curr Biol. 2004;14(18):R787–96.
178. Luke B, Versini G, Jaquenoud M, Zaidi IW, Kurz T, Pintard L, et al. The cullin Rtt101p promotes replication fork progression through damaged DNA and natural pause sites. Curr Biol. 2006;16(8):786–92.
179. Han J, Zhang H, Zhang H, Wang Z, Zhou H, Zhang Z. A Cul4 E3 ubiquitin ligase regulates histone hand-off during nucleosome assembly. Cell. 2013;155(4):817–29.
180. Cremona CA, Sarangi P, Yang Y, Hang LE, Rahman S, Zhao X. Extensive DNA damage-induced sumoylation contributes to replication and repair and acts in addition to the Mec1 checkpoint. Mol Cell. 2012;45(3):422–332.

Chapter 19
Assembly of the Cdc45-MCM2-7-GINS Complex, the Replication Helicase

Helmut Pospiech and Anna Szambowska

Abstract In eukaryotes, a crucial step during the initiation of DNA replication is the timely formation and activation of the replicative DNA helicase composed of Cdc45, MCM2-7 and GINS (CMG). The dynamic and spatio-temporal events leading to the ordered and stepwise assembly of the CMG helicase are tightly regulated. Multiple assembly factors ensure in this way that replication occurs only once per cell cycle. The MCM2-7 helicase is loaded in an inactive form onto double-stranded DNA in the G1 phase of the cell cycle, whereas the fully reconstituted CMG complex is assembled and positioned onto single-stranded DNA during the S phase. Thus, DNA plays an important and active role in these events. In this chapter we summarize and discuss our current knowledge about the appropriate recruitment and assembly of the CMG complex into the active eukaryotic replicative DNA helicase, emphasizing the crucial role of DNA in this process. We finally outline how the number of active CMG complexes formed is restricted during unperturbed DNA synthesis.

Keywords DNA replication • DNA replication initiation • Replicative helicase • CMG helicase • Cell division cycle 45 • Cdc45 • Minichromosome maintenance • MCM2-7 • GINS • Pre-replication complex • Pre-initiation complex • Pre-loading complex • Eukaryotes • Dpb11 • TopBP1 • RecQL4 • Guanine quadruplex

---

H. Pospiech (✉)
Research Group Biochemistry, Leibniz Institute on Aging-Fritz Lipmann Institute, Beutenbergstrasse 11, D-07745 Jena, Germany

Faculty of Biochemistry and Molecular Medicine, University of Oulu, 5400, FI-90014 Oulu, Finland
e-mail: helmut.pospiech@leibniz-fli.de; helmut.pospiech@oulu.fi

A. Szambowska
Research Group Biochemistry, Leibniz Institute on Aging-Fritz Lipmann Institute, Beutenbergstrasse 11, D-07745 Jena, Germany
e-mail: anna.szambowska@leibniz-fli.de

D.L. Kaplan (ed.), *The Initiation of DNA Replication in Eukaryotes*,
DOI 10.1007/978-3-319-24696-3_19

Introduction

The eukaryotic Cdc45-MCM2-7-GINS complex (CMG complex) functions as the active replicative helicase that unwinds origins of replication (*origins*) [1, 2], and consists of a single Cdc45 (cell division cycle 45) molecule, the six subunits of the MCM2-7 (minichromosome maintenance 2-7) helicase and the four GINS proteins [3, 4]. All three components of the helicase are essential for its function and conserved across all eukaryotic groups [5, 6]. The MCM complex is composed of six closely related, non-identical proteins that belong to the AAA+ superfamily of ATPases and assemble into a ring-shaped complex, with a defined geometry of MCM 5-3-7-4-6-2 [7–11]. The MCM proteins are composed of three domains. The N-terminal part is responsible for hexamer formation, DNA binding and regulation of the helicase processivity along DNA. The AAA+ domain possesses the catalytic activity and contains several conserved elements that are essential for nucleotide binding and hydrolysis [11]. The C-terminal helix-turn-helix domain mediates allosteric control of the ATPase of the neighbouring MCM subunit, implying a regulatory inter-domain communication within the MCM complex [12, 13]. The assembly of the MCM2-7 proteins onto DNA (also termed pre-replicative complex formation; pre-RC) requires the action of the *origin* recognition complex (ORC), Cdc6, Cdt1 and ATP and is restricted to the G1 phase of the cell cycle [14, 15]. The pre-RC complex formation is initiated by the ORC complex that recruits Cdc6 to the replication sites [16]. Binding of the ORC subunits and Cdc6 to *origins* is an ATP-dependent process [17]. In budding yeast, the ORC-Cdc6 complex mediates recruitment of the MCM2-7 onto chromatin via the actions of Cdt1 that forms a complex with MCM2-7 prior to *origin* binding [18–20]. In eukarya, the MCM2-7 helicase is loaded onto chromatin as a symmetrical head-to-head double hexamer [19, 20], with the N-terminal domains of MCM comprising the dimer interface [20, 21]. The N-terminal tails of MCM2/4/6 undergo phosphorylation in a DDK-dependent manner [22, 23], which leads to alteration of the MCM2-7 configuration [24]. Allosteric changes in the MCM2-7 complex are also induced upon binding to DNA [25, 26]. Reconstruction of the helicase loading reaction in vitro shows that the yeast MCM2-7 double hexamer is salt resistant, in contrast to its loading cofactors, and once encircled onto chromatin can slide on double-stranded DNA, without ATP hydrolysis [27, 28]. The ATPase active sites in the MCM2-7 complex are also involved in the regulation of the helicase recruitment and activation. ATP binding and hydrolysis are required for loading of the helicase onto chromatin and for Cdt1 release [27, 29].

In contrast to bacteria, loading and activation of eukaryotic DNA helicase are separated in time and tightly regulated to ensure that DNA replication occurs only once per cell cycle. Therefore, in the G1 phase, the MCM2-7 helicase is loaded in inactive form onto double-stranded DNA (assembly of the pre-RCs), whereas the active CMG complex encircles single-stranded DNA (ssDNA) during the S phase of the cell cycle, when DNA replication is initiated [30, 31].

Loading of the Cdc45 and GINS Complex

In S phase, upon the action of S-phase kinases (S-CDK and DDK), Cdc45 and GINS proteins are integrated into the MCM2-7 helicase, and together form the fully reconstituted CMG complex [1–4, 32]. The Dbf4-Cdc7 (DDK) activity promotes replication mainly by phosphorylation of MCM2, MCM4 and MCM6 [14, 33–37] (see Chaps. 14 and 18 in this book), thus preparing the pre-RC for helicase activation. Cdc45 and the GINS complex are then recruited to the *origins* and associate with the MCM2-7 helicase in a CDK-dependent manner [33, 35, 38, 39] (see Chaps. 14 and 18 in this book). In higher eukaryotes (*Drosophila melanogaster* and humans), the CMG complex exhibits robust helicase and ATPase activity on Y-shaped DNA in vitro, compared to the MCM2-7 helicase alone, and translocates along the leading strand in 3′–5′ direction [3, 4, 40]. The eukaryotic replicative helicase encircles single-stranded DNA [41], and contains only a single MCM2-7 (see Chap. 20 in this book). Therefore it is likely that transition from the MCM2-7 to the active CMG complex requires not only the association of Cdc45 and GINS but also a number of conformational changes within the MCM2-7 helicase itself. These allosteric rearrangements include splitting of the double hexamer into two single MCM2-7 complexes and opening of the MCM2-7 rings to allow extrusion of one of the DNA strands [41]. The double-MCM2-7 hexamer may be simply separated by turning on the helicase activity, since both hexamers are facing in opposite directions [42]. Another model assumes that two open-gate hexamers of MCM are loaded onto double-strand DNA, and hexamer ring closure occurs with a local DNA melting. The strand capture initiates by association of GINS and Cdc45 [43]. The cryo-EM structure of the entire *Drosophila* CMG complex assumes that the MCM2-7 complex forms an open lock-washer structure and association of Cdc45 and GINS causes a closed-ring formation [4, 26]. The CMG complex exists in the planar form, with Cdc45 and GINS that associate with the MCM2, MCM3 and MCM5 subunits [4]. In this structure, Cdc45 interacts with the N-terminal α-helical domain of Psf2 (a GINS subunit) on one side, and on the opposite side contacts the N-terminus of MCM2, whereas the GINS complex seems to interact with the MCM5 [4]. Between the MCM2 and MCM5 subunits lies a gate through which DNA can access the central channel of the hexamer, whereas the C-terminal ATPase domains of MCM advance on double-strand DNA [26]. It becomes thus clear that interactions with MCM2 and MCM5 are pivotal for CMG assembly (see below). The Cdc45-Psf1 (another GINS subunit) interaction seems to play a critical role in the formation and/or stability of the CMG complex [26]. Structural data provides that Cdc45 and GINS allosterically activate the helicase function of the CMG complex, but so far little is known about how they contribute to this process. Biochemical analyses show that Cdc45 binds to single-stranded DNA [44–46], and exhibit even higher affinity to ss/ds junctions [46]. Cdc45 can slide along ssDNA preferentially in the 3′–5′ direction and at the same time initiate displacement of the other strand. Thus Cdc45 may help in strand separation during movement of the helicase along DNA [46], or it may guide the separated leading strand by tracking it towards the leading strand replicase. The latter would also be in line with studies in yeast postulating that Cdc45-ssDNA interactions are important during the stalling of the helicase, caused by replication stress [45]. Additionally, Cdc45 could capture the leading strand from occasional slippage from the MCM2-7 complex [47]. To date, no enzymatic activity has been shown for GINS, but GINS appears to perform its function as a component of the CMG complex, probably by stabilising the Cdc45-MCM interaction [48]. Cdc45 is required both for the establishment [1, 49, 50] and progression [51–53] of the replication fork in eukaryotic cells. During DNA replication, Cdc45 also interacts with other replication proteins, including the DNA polymerases (Pols) α, δ, ε; MCM10; Orc2, DNA helicase B and TopBP1 [51, 54–57]. Cdc45 represents a distant homologue of the bacterial RecJ protein [58, 59], an exonuclease that degrades single-stranded DNA [60], and plays an important function in bacterial DNA repair and recombination [61]. Homology between Cdc45 and RecJ comprises the N-terminal DHH domain that possesses four sequence motifs containing the conserved residues of aspartic acid (D) and histidine (H), essential for Mn^{2+} or Mg^{2+} binding and catalysis [58]. A structural model of recombinant human Cdc45 revealed that Cdc45 adopts a DHH core-like structure similar to RecJ, with an additional extension between the N- and C-terminal lobes of the DHH core, indicating the presence of a large insertion [44, 46]. In contrast to RecJ, Cdc45 possesses an incomplete phosphoresterase centre, it therefore lost its nuclease activity, but still retains affinity to bind to ssDNA [44, 46]. The eukaryotic GINS complex comprises four related subunits: Sld5; Psf1; Psf2 and Psf3 originally isolated and defined from genetic screens in *Saccharomyces cerevisiae*, where the conditional mutations of Sld5 and Psf1 confer defects in DNA replication under nonpermissive conditions [62]. GINS was isolated in parallel from *Xenopus laevis* [63]. This complex was termed GINS (Go-Ichi-Ni-San, referred as 5-1-2-3 in Japanese), and is arranged in the following order: 2-5-1-3 [62]. All four subunits are highly conserved in eukaryotic cells and are crucial both for the initiation and elongation of chromosome replication [62, 64–68]. The four GINS subunits are all around 25–35 kDa in size, and lack any detectable sequence motifs [62, 67]. Crystal structures of the GINS complex are available at resolutions varying from 2.3 to 3.2 Å [68–70]. Each of the subunits comprises two domains, an alpha–helical A-domain and a B-domain that is rich in β-strands. The overall shape of the GINS complex resembles a trapezium or possesses elliptical shape [69]. The four subunits arrange in two layers, Sld5 and Psf1 form the top layer, whereas Psf2 and Psf3 build the bottom layer [68, 69]. In addition to its contribution to the CMG complex, GINS associates with Polα-primase and Polε and therefore also constitutes an important structural component of the replisome progression complex [71–73].

The Pre-initiation Complex as a Cdc45/GINS Loading Machinery

As already outlined in previous chapters (see Chaps. 13, 14 and 15), the assembly of the CMG complex represents the critical step in the regulation of replication initiation. At the same time, CMG assembly is tightly associated with the establishment of the replication fork, especially the loading of the leading strand replicase Polε. It is a complex protein network regulated by DDK- and S-CDK phosphorylation that brings about CMG assembly at and Polε recruitment to the *origins*. The formation of this so-called pre-initiation complex (pre-IC) has been best characterised in the budding yeast *S. cerevisiae*, where the sequence of events has been recently reconstituted with recombinant and highly purified factors [35]. Upon S-phase onset and DDK activation, Cdc45 is recruited to the *origin*-bound MCM2-7 in a complex with Sld3 and Sld7 [35, 74–76]. Phosphorylation of Mcm2 by DDK has been proposed to be important for opening of the MCM2-7 ring and the expulsion of ssDNA [34, 36]. A second complex comprising phosphorylated Sld2, Dpb11, GINS and Polε (called the pre-loading complex) is recruited to the *origin* once Sld3 has been phosphorylated by S-CDK [35, 38, 39, 77, 78]. Thus, in yeast, GINS loading and subsequent assembly into the CMG complex depend on a prior loading of Sld3, Cdc45 and Dpb11 [35, 65, 66, 79]. In the chain of reactions, Dpb11 represents the central hub of the pre-initiation complex formation, as it mediates not only the association of S-CDK-phosphorylated Sld2 and Sld3 [38, 39, 80] but also direct interactions with Polε [81–83] and GINS [77, 79].

It should be noted that recruitment of Polε is an integral part of the CMG helicase assembly and activation. Paradoxically, the catalytic domain of Polε and thus the polymerase activity are not absolutely essential for viability in yeast [84–86], but other parts of Polε, in particular its second B subunit, play an essential role in initiation complex formation [71, 84, 86–88]. Therefore, the assembly of the CMG helicase should be rather understood as the establishment of the leading strand replisome or "Polε holoenzyme" [40, 72] that is subsequently required for progressive elongation during initiation.

In this context, it seems paradoxical that the leading strand replisome assembly at the *origin* does not require prior loading of the primase-polymerase Polα, as Polε needs a primer to start leading strand DNA synthesis. So how is premature unwinding by CMG prevented? On the one hand, the CMG helicase appears to acquire full processivity only if coupled with Polε DNA synthesis [2, 3, 72, 89]. On the other hand, the first primer laid on the lagging strand could be extended by Polα and/or Polδ into the leading strand of the same initiation bubble until it reaches the opposing fork and is captured by Polε [90]. This hypothetical model is consistent with the variation in strand transition between leading and lagging strand at *origins* [91], and would at the same time explain why there is an occasional use of Polδ on the leading strand near *origins* [92].

As could be expected for such a fundamental cellular pathway, basic principles of CMG assembly and pre-IC formation are conserved among the eukaryotes studied so far [93]. But whereas in fission yeast the factors involved appear to be well conserved in structure and function, in metazoans there is considerable divergence especially in the primary structure of the initiation factors. This is in difference to the factors involved in replication elongation that show a high degree of conservation [6]. Orthologues of yeast Dpb11 have been identified in plants and several metazoans, including human TopBP1, frog XCut5/XMus101 and fruit fly Mus101 (reviewed in [94, 95]). The metazoan homologues generally possess additional BRCT domains, which confer multiple other interactions, mainly to

Cdk-phosphorylated proteins involved in DNA damage response and repair (reviewed in [96]). The metazoan orthologues of Dpb11 have been implicated in CMG and pre-IC assembly in a similar manner as their yeast counterparts [63, 83, 97–100]. One study in human cells did not detect a direct involvement of TopBP1 in CMG assembly [101].

Several studies suggest that RecQL4 represents the functional orthologue of Sld2 in animals. RecQL4 is one of the five RecQ-like helicases identified in humans (reviewed in [102, 103]). Mutations in the RecQL4 gene have been associated with Rothmund-Thomson, RAPADILINO and Baller-Gerold syndromes [104]. Patients with these syndromes exhibit various physical and mental developmental abnormalities, increased risk of osteosarcoma and features of premature aging. Whereas in most metazoans RecQL4 represents a fusion between an Sld2 module and a RecQ helicase, in budding yeast these two parts are coded for by the two separate genes *SLD2* and *HRQ1* [99, 105–110]. Interestingly, separate genes for Sld2 and RecQL4 helicase can also be found in the nematode *Caenorhabditis elegans* [110]. Apart from a conserved N-terminal homeo-like domain, sequence conservation is rather limited between yeast Sld2 and metazoan RecQL4 [99, 100, 109–111]. As would be expected from its central function, the Sld2 homology region (but not the RecQ homology region) of metazoan RecQL4 appears to be essential for viability [99, 100, 112–116]. RecQL4 has been reported to be target of S-CDK phosphorylation in several metazoan species. Although this phosphorylation is essential for its association with the $\text{Dpb11}^{\text{Mus101}}$ in *C. elegans*, this seems not to be the case in the *Xenopus* embryonic system [100].

The initiation factor Treslin/Ticcr represents an orthologue of Sld3 in higher eukaryotes [117–119]. Despite a very limited homology restricted to the Sld3 core region, the CDK-dependent Treslin-TopBP1 interaction mirrors the function of the yeast counterparts during CMG assembly [119–123]. The two key S-CDK phosphorylation sites are conserved from yeast to man, and mediate the interaction with TopBP1, which is required for Cdc45 loading. The situation in higher eukaryotes seems anyway more complicated since two additional factors, Due-B and GEMC1, are also required for the Cdc45 recruitment. Due-B only seems to play a role for initiation of *origins* containing the "DNA unwinding element", which is dependent on a prior phosphorylation by DDK [124, 125], GEMC1, on the other hand, has been described as a Cdc45 loading factor that interacts with $\text{XCut5}^{\text{TopBP1}}$ after phosphorylation by S-CDK [126]. What is more, a direct association of human Cdc45 and TopBP1 has been reported [56]. Although a direct homologue of Sld7 has not been identified in higher eukaryotes yet, the unrelated MDM two binding protein (MTBP) performs an analogous function in humans, as it interacts with treslin throughout the cell cycle. MTBP downregulation by means of RNA interference prevents the CMG assembly [127].

Once the CMG is assembled and the replication fork established, Dpb11, Sld2, Sld3 and Sld7 appear not to be any longer required, and they will not migrate with the progressing fork, as seems to be the case for the human homologues [74, 82, 83, 128, 129].

An Active Role of DNA During the CMG Helicase Assembly

As outlined in the previous section, the CMG assembly is governed by a plethora of molecular interactions that are modified by posttranslational modifications such as phosphorylation. Therefore, to understand the whole dynamic of the process, it may be better to consider helicase assembly in a context of cooperation, competitions and allosteric modifications rather than a sum of bimolecular contacts. It is particularly unfortunate that much of the research has only ascribed a passive role to the DNA during the initiation: Double-stranded DNA represents merely a "landing platform" for the pre-RC and pre-IC. Later during replication initiation DNA is unwound and replication commences. In particular the work of the Kaplan lab points towards a much more active role of the DNA during replication initiation in yeast. S-CDK-phosphorylated Sld2 binds specifically to the thymidine-rich single strand of the ARS sequence marking *origins* in budding yeast [130]. Also Sld3 and Dpb11 bind ssDNA in a phosphorylation-independent manner, and ssDNA binding of Sld3 does not interfere with its interaction to Dpb11 and Cdc45 [128, 131]. Importantly, the interaction of Sld2, Sld3 and Dpb11 with ssDNA all prevent the binding of MCM2-7 by these proteins [128, 131, 132]. Release of these factors from MCM2-7 uncovers the GINS binding site of the MCM hexamer. This suggests that GINS may associate with MCM2-7 once the *origin* DNA has been unwound, thus preventing premature helicase assembly [133]. This model is intriguing as it provides a logical chain of events that lead to CMG assembly and is supported by the fact that expression of mutants of Dpb11 and Sld2 defective in ssDNA binding displays seriously compromised DNA replication. On the other hand, although DDK phosphorylation seems to be required for *origin* unwinding, the molecular mechanism of this unwinding remains still obscure. Moreover, the model does not explain how Cdc45 is integrated into the CMG complex, as ssDNA binding by the Cdc45 loading factor Sld3 does not lead to the release of Cdc45 [128]. It is possible that Sld7 promotes the Cdc45 integration into the CMG complex by destabilising the Sld3-Cdc45 complex [74, 75].

Could this mechanism be generalized for higher eukaryotes? Although many of the CMG loading factors including TopBP1, RecQL4 and Due-B have been shown to display affinity to different DNA structures [109, 124, 134–137], there is no evidence that these DNA binding activities are important for the CMG assembly or replication initiation in general. Instead, clues have been coming up from a different direction: Recent studies demonstrate that the vast majority of human *origins* of replication correlate with G-rich sequences displaying a propensity for guanine quadruplex (G4) formation [138, 139]. G4s are DNA secondary structures formed by guanine-rich sequences, where four guanine bases self-assemble via Hoogsteen hydrogen bonds forming a structure that is further stabilised by the internalisation of potassium. G4s exist in human cells in vivo [140, 141], and their formation is modulated during cell cycle progression [140]. More importantly, the G4 motifs are essential for *origin* function and their orientation determined the position of the replication start site for two model *origins* studied in chicken DT40 cells [142].

Therefore, it is interesting that the Sld2 homology region of human RecQL4 shows high affinity for G4 DNA [143]. In addition, also ssDNA is bound by the same region with a preference of GC-rich over AT-rich sequences [137, 143]. Also the human origin recognition complex ORC has been previously shown to preferentially bind G4s [144]. Therefore, it is tempting to speculate that G4s could serve as a sink or anchor for the CMG loaders in a similar manner as it has been proposed for ARS ssDNA in yeast.

Negative Regulation of Helicase Assembly during Unperturbed DNA Replication

It has become clear that the number of licensed *origins* during a cell cycle by manifold outnumber those that are actually fired [138, 139, 145]. The majority of *origins* represent dormant *origins* that serve as a reserve to guarantee completion of DNA replication. When replication progression is inhibited upon DNA damage or replication stress, these dormant *origins* initiate to ensure that all the chromosomal DNA is replicated [146]. It has become clear that the CMG assembly and activation represent the regulatory target that determines which *origins* are utilised [75, 147]. Therefore one key regulatory mechanism of helicase assembly is the limitation of assembly factors. In yeast, Sld2, Sld3 and Dpb11 are present in extremely low copy number per cell and thus limit replication initiation [75, 148, 149]. Overexpression of these factors brings about firing of *origins* early in S phase that were otherwise fired late. Conversely, four DNA replication factors, Cut5, RecQ4, Treslin and Drf1 (the catalytic subunit of DDK in *Xenopus*), become first limiting for replication initiation after the initial rapid cell divisions in *X. laevis* embryos, and thus determine the midblastula transition when cell cycle slows down [150]. In contrast, the amount of Cdc45 appears to be limiting for replication in human vegetative cells [151–153], and increasing the quantity of this protein leads to increased *origin* firing [152, 154]. Not unexpectedly, Cdc45 has been reported to be overexpressed in many cancers [151, 155]. Agreeing with this view, Cdc45 is recycled to activate nearby dormant *origins* of replication after a replication fork comes to an unscheduled halt [152].

Apparently, there is an additional level of replication limitation during normal growth that directly targets the CMG assembly. ATR and Chk1 regulate replication timing and normal S-phase progression even during unperturbed cell cycle, as ATR or Chk1 inhibition, downregulation as well as inactivation of the *CHEK1* gene all cause slowdown of fork progression and extensive *origin* firing [156–159] (see also Chaps. 18, 22, and 24). A recent study demonstrated Chk1 binding to and phosphorylation of Treslin during normal cell growth [160]. Mutations in Treslin, which abolished Chk1 binding, eliminated Chk1-catalyzed phosphorylation of Treslin. This led to an increased initiation of chromosomal DNA replication during unperturbed cell growth due to an enhanced loading of Cdc45. It will be interesting to see if other components of the CMG assembly machinery are affected in a similar manner, and how the targeted activation of Chk1 for selected, probably dormant or late-firing *origins* is achieved. Considering the prominent role of TopBP1/Dpb11 during ATR/Mec1 activation [96, 161, 162], it may be worth looking for alternative, RPA-independent ways of ATR activation involving the pre-IC.

Outlook

There has been tremendous progress on the understanding of how the replicative CMG helicase complex is assembled during the last 10 years. It has become clear that CMG assembly and activation are closely linked to the establishment of the replication fork, in particular the loading of the leading strand replicase, Polε. Although the molecular networks involved and the post-translational modification regulating this process have been established, current research still has come short to grasp the dynamics by which the interaction network is refashioning during initiation. Moreover, the vivid role that DNA plays during CMG and replication fork assembly should be acknowledged more broadly. In the light of the reconstitution of the initiation process with purified yeast proteins and advances in single-molecule techniques, one can also expect swift progress in the understanding of eukaryotic replication initiation in the near future. Considering the divergences in metazoan compared to yeast initiation, we better be prepared for surprises!

Acknowledgements The Fritz Lipmann Institute (FLI) is member of the Science Association "Gottfried Wilhelm Leibniz" (WGL) and is financially supported by the Federal Government of Germany and the State of Thuringia. The authors are grateful to Frank Grosse for fruitful discussions and critical reading of the manuscript.

References

1. Aparicio T, Ibarra A, Méndez J. Cdc45-MCM-GINS, a new power player for DNA replication. Cell Div. 2006;1:18.
2. Moyer SE, Lewis PW, Botchan MR. Isolation of the Cdc45/Mcm2-7/GINS (CMG) complex, a candidate for the eukaryotic DNA replication fork helicase. Proc Natl Acad Sci U S A. 2006;103:10236–41.
3. Ilves I, Petojevic T, Pesavento JJ, Botchan MR. Activation of the MCM2-7 helicase by association with Cdc45 and GINS proteins. Mol Cell. 2010;37:247–58.
4. Costa A, Ilves I, Tamberg N, Petojevic T, Nogales E, Botchan MR, et al. The structural basis for MCM2-7 helicase activation by GINS and Cdc45. Nat Struct Mol Biol. 2011;18:471–7.
5. Liu Y, Richards TA, Aves SJ. Ancient diversification of eukaryotic MCM DNA replication proteins. BMC Evol Biol. 2009;9:60.
6. Aves SJ, Liu Y, Richards TA. Evolutionary diversification of eukaryotic DNA replication machinery. Subcell Biochem. 2012;62:19–35.
7. Forsburg SL. Eukaryotic MCM proteins: beyond replication initiation. Microbiol Mol Biol Rev. 2004;68:109–31.

8. Bochman ML, Schwacha A. The Mcm complex: unwinding the mechanism of a replicative helicase. Microbiol Mol Biol Rev. 2009;73:652–83.
9. Vijayraghavan S, Schwacha A. The eukaryotic Mcm2-7 replicative helicase. Subcell Biochem. 2012;62:113–34.
10. Crevel G, Ivetic A, Ohno K, Yamaguchi M, Cotterill S. Nearest neighbour analysis of MCM protein complexes in *Drosophila melanogaster*. Nucleic Acids Res. 2001;29:4834–42.
11. Schwacha A, Bell SP. Interactions between two catalytically distinct MCM subgroups are essential for coordinated ATP hydrolysis and DNA replication. Mol Cell. 2001;8:1093–104.
12. Barry ER, McGeoch AT, Kelman Z, Bell SD. Archaeal MCM has separable processivity, substrate choice and helicase domains. Nucleic Acids Res. 2007;35:988–98.
13. Wiedemann C, Szambowska A, Häfner S, Ohlenschläger O, Görlach M. Structure and regulatory role of the C-terminal winged helix domain of the archaeal minichromosome. Nucleic Acids Res. 2015;43:2958–67.
14. Sun J, Fernandez-Cid A, Riera A, Tognetti S, Yuan Z, Stillman B, et al. Structural and mechanistic insights into Mcm2-7 double-hexamer assembly and function. Genes Dev. 2014;28:2291–303.
15. Randell JCW, Bowers JL, Rodríguez HK, Bell SP. Sequential ATP hydrolysis by Cdc6 and ORC directs loading of the Mcm2-7 helicase. Mol Cell. 2006;21:29–39. doi:10.1016/j.molcel.2005.11.023.
16. Weinreich M, Liang C, Chen HH, Stillman B. Binding of cyclin-dependent kinases to ORC and Cdc6p regulates the chromosome replication cycle. Proc Natl Acad Sci U S A. 2001;98:11211–7.
17. Speck C, Chen Z, Li H, Stillman B. ATPase-dependent cooperative binding of ORC and Cdc6 to origin DNA. Nat Struct Mol Biol. 2005;12:965–71.
18. Donovan S, Harwood J, Drury LS, Diffley JF. Cdc6p-dependent loading of Mcm proteins onto pre-replicative chromatin in budding yeast. Proc Natl Acad Sci U S A. 1997;94:5611–6.
19. Evrin C, Clarke P, Zech J, Lurz R, Sun J, Uhle S, et al. A double-hexameric MCM2-7 complex is loaded onto origin DNA during licensing of eukaryotic DNA replication. Proc Natl Acad Sci U S A. 2009;106:20240–5.
20. Remus D, Beuron F, Tolun G, Griffith JD, Morris EP, Diffley JFX. Concerted loading of Mcm2-7 double hexamers around DNA during DNA replication origin licensing. Cell. 2009;139:719–30.
21. Fletcher RJ, Bishop BE, Leon RP, Sclafani RA, Ogata CM, Chen XS. The structure and function of MCM from archaeal M. thermoautotrophicum. Nat Struct Biol. 2003;10:160–7.
22. Labib K. How do Cdc7 and cyclin-dependent kinases trigger the initiation of chromosome replication in eukaryotic cells? Genes Dev. 2010;24:1208–19.
23. Sheu Y-J, Stillman B. The Dbf4-Cdc7 kinase promotes S phase by alleviating an inhibitory activity in Mcm4. Nature. 2010;463:113–7.
24. On KF, Beuron F, Frith D, Snijders AP, Morris EP, Diffley JFX. Prereplicative complexes assembled in vitro support origin-dependent and independent DNA replication. EMBO J. 2014;33:605–20.
25. Hesketh EL, Parker-Manuel RP, Chaban Y, Satti R, Coverley D, Orlova EV, et al. DNA induces conformational changes in a recombinant human minichromosome maintenance complex. J Biol Chem. 2015;290:7973–9.
26. Costa A, Renault L, Swuec P, Petojevic T, Pesavento J, Ilves I, et al. DNA binding polarity, dimerization, and ATPase ring remodeling in the CMG helicase of the eukaryotic replisome. Elife. 2014;3:e03273.
27. Frigola J, Remus D, Mehanna A, Diffley JFX. ATPase-dependent quality control of DNA replication origin licensing. Nature. 2013;495:339–43.
28. Fernández-Cid A, Riera A, Tognetti S, Herrera MC, Samel S, Evrin C, et al. An ORC/Cdc6/MCM2-7 complex is formed in a multistep reaction to serve as a platform for MCM double-hexamer assembly. Mol Cell. 2013;50:577–88.

29. Kang S, Warner MD, Bell SP. Multiple functions for Mcm2-7 ATPase motifs during replication initiation. Mol Cell. 2014;55:655–65.
30. Boos D, Frigola J, Diffley JFX. Activation of the replicative DNA helicase: breaking up is hard to do. Curr Opin Cell Biol. 2012;24:423–30.
31. Diffley JFX. The many faces of redundancy in DNA replication control. Cold Spring Harb Symp Quant Biol. 2010;75:135–42.
32. Sclafani RA, Holzen TM. Cell cycle regulation of DNA replication. Annu Rev Genet. 2007;41:237–80.
33. Sheu Y-J, Stillman B. Cdc7-Dbf4 phosphorylates MCM proteins via a docking site-mediated mechanism to promote S phase progression. Mol Cell. 2006;24:101–13.
34. Ramer MD, Suman ES, Richter H, Stanger K, Spranger M, Bieberstein N, et al. Dbf4 and Cdc7 promote DNA replication through interactions with distinct Mcm2-7 subunits. J Biol Chem. 2013;288:14926–35.
35. Yeeles JTP, Deegan TD, Janska A, Early A, Diffley JFX. Regulated eukaryotic DNA replication origin firing with purified proteins. Nature. 2015;519(7544):431–5.
36. Bruck I, Kaplan DL. The Dbf4-Cdc7 kinase promotes Mcm2-7 ring opening to allow for single-stranded DNA extrusion and helicase assembly. J Biol Chem. 2015;290:1210–21.
37. Bruck I, Kaplan DL. Dbf4-Cdc7 phosphorylation of Mcm2 is required for cell growth. J Biol Chem. 2009;284:28823–31.
38. Zegerman P, Diffley JFX. Phosphorylation of Sld2 and Sld3 by cyclin-dependent kinases promotes DNA replication in budding yeast. Nature. 2007;445:281–5.
39. Tanaka S, Umemori T, Hirai K, Muramatsu S, Kamimura Y, Araki H. CDK-dependent phosphorylation of Sld2 and Sld3 initiates DNA replication in budding yeast. Nature. 2007;445:328–32.
40. Kang Y-H, Galal WC, Farina A, Tappin I, Hurwitz J. Properties of the human Cdc45/Mcm2-7/GINS helicase complex and its action with DNA polymerase epsilon in rolling circle DNA synthesis. Proc Natl Acad Sci U S A. 2012;109:6042–7.
41. Fu YV, Yardimci H, Long DT, Ho TV, Guainazzi A, Bermudez VP, et al. Selective bypass of a lagging strand roadblock by the eukaryotic replicative DNA helicase. Cell. 2011;146:931–41.
42. Tognetti S, Riera A, Speck C. Switch on the engine: how the eukaryotic replicative helicase MCM2-7 becomes activated. Chromosoma. 2015;124:13–26.
43. Bell SD, Botchan MR. The minichromosome maintenance replicative helicase. Cold Spring Harb Perspect Biol. 2013;5:a012807.
44. Krastanova I, Sannino V, Amenitsch H, Gileadi O, Pisani FM, Onesti S. Structural and functional insights into the DNA replication factor Cdc45 reveal an evolutionary relationship to the DHH family of phosphoesterases. J Biol Chem. 2012;287:4121–8.
45. Bruck I, Kaplan DL. Cdc45 protein-single-stranded DNA interaction is important for stalling the helicase during replication stress. J Biol Chem. 2013;288:7550–63.
46. Szambowska A, Tessmer I, Kursula P, Usskilat C, Prus P, Pospiech H, et al. DNA binding properties of human Cdc45 suggest a function as molecular wedge for DNA unwinding. Nucleic Acids Res. 2014;42:2308–19.
47. Petojevic T, Pesavento JJ, Costa A, Liang J, Wang Z, Berger JM, et al. Cdc45 (cell division cycle protein 45) guards the gate of the Eukaryote Replisome helicase stabilizing leading strand engagement. Proc Natl Acad Sci U S A. 2015;112:E249–58.
48. Aparicio T, Guillou E, Coloma J, Montoya G, Méndez J. The human GINS complex associates with Cdc45 and MCM and is essential for DNA replication. Nucleic Acids Res. 2009;37:2087–95.
49. Aparicio OM, Weinstein DM, Bell SP. Components and dynamics of DNA replication complexes in *S. cerevisiae*: redistribution of MCM proteins and Cdc45p during S phase. Cell. 1997;91:59–69.
50. Hardy CF. Identification of Cdc45p, an essential factor required for DNA replication. Gene. 1997;187:239–46.

51. Bauerschmidt C, Pollok S, Kremmer E, Nasheuer H-P, Grosse F. Interactions of human Cdc45 with the Mcm2 – 7 complex, the GINS complex, and DNA polymerases δ and ε during S phase. Genes Cells. 2007;12:745–58.
52. Pacek M, Walter JC. A requirement for MCM7 and Cdc45 in chromosome unwinding during eukaryotic DNA replication. EMBO J. 2004;23:3667–76.
53. Pacek M, Tutter A, Kubota Y, Takisawa H, Walter JC. Localization of MCM2-7, Cdc45, and GINS to the site of DNA unwinding during eukaryotic DNA replication. Mol Cell. 2006;21:581–7.
54. Di Perna R, Aria V, De Falco MM, Sannino V, Okorokov AL, Pisani FM, et al. The physical interaction of Mcm10 with Cdc45 modulates their DNA binding properties. Biochem J. 2013;454:333–43.
55. Zou L, Stillman B. Assembly of a complex containing Cdc45p, replication protein A, and Mcm2p at replication origins controlled by S-phase cyclin-dependent kinases and Cdc7p-Dbf4p kinase. Mol Cell Biol. 2000;20:3086–96.
56. Schmidt U, Wollmann Y, Franke C, Grosse F, Saluz H-P, Hänel F. Characterization of the interaction between the human DNA topoisomerase IIbeta-binding protein 1 (TopBP1) and the cell division cycle 45 (Cdc45) protein. Biochem J. 2008;409:169–77.
57. Gerhardt J, Guler GD, Fanning E. Human DNA helicase B interacts with the replication initiation protein Cdc45 and facilitates Cdc45 binding onto chromatin. Exp Cell Res. 2015;334:283–93.
58. Aravind L, Koonin EV. A novel family of predicted phosphoesterases includes *Drosophila* prune protein and bacterial RecJ exonuclease. Trends Biochem Sci. 1998;23:17–9.
59. Sanchez-Pulido L, Ponting CP. Cdc45: the missing RecJ ortholog in eukaryotes? Bioinformatics. 2011;27:1885–8.
60. Lovett ST, Kolodner RD. Identification and purification of a single-stranded-DNA-specific exonuclease encoded by the recJ gene of Escherichia coli. Proc Natl Acad Sci U S A. 1989;86:2627–31.
61. Courcelle CT, Chow K-H, Casey A, Courcelle J. Nascent DNA processing by RecJ favors lesion repair over translesion synthesis at arrested replication forks in Escherichia coli. Proc Natl Acad Sci U S A. 2006;103:9154–9.
62. Takayama Y, Kamimura Y, Okawa M, Muramatsu S, Sugino A, Araki H. GINS, a novel multiprotein complex required for chromosomal DNA replication in budding yeast. Genes Dev. 2003;17:1153–65.
63. Kubota Y, Takase Y, Komori Y, Hashimoto Y, Arata T, Kamimura Y, et al. A novel ring-like complex of *Xenopus* proteins essential for the initiation of DNA replication. Genes Dev. 2003;17:1141–52.
64. Gómez EB, Angeles VT, Forsburg SL. A screen for Schizosaccharomyces pombe mutants defective in rereplication identifies new alleles of rad4+, cut9+ and psf2+. Genetics. 2005;169:77–89.
65. Yabuuchi H, Yamada Y, Uchida T, Sunathvanichkul T, Nakagawa T, Masukata H. Ordered assembly of Sld3, GINS and Cdc45 is distinctly regulated by DDK and CDK for activation of replication origins. EMBO J. 2006;25:4663–74.
66. Kanemaki M, Labib K. Distinct roles for Sld3 and GINS during establishment and progression of eukaryotic DNA replication forks. EMBO J. 2006;25:1753–63.
67. Gambus A, Jones RC, Sanchez-Diaz A, Kanemaki M, van Deursen F, Edmondson RD, et al. GINS maintains association of Cdc45 with MCM in replisome progression complexes at eukaryotic DNA replication forks. Nat Cell Biol. 2006;8:358–66.
68. Kamada K, Kubota Y, Arata T, Shindo Y, Hanaoka F. Structure of the human GINS complex and its assembly and functional interface in replication initiation. Nat Struct Mol Biol. 2007;14:388–96.
69. Choi JM, Lim HS, Kim JJ, Song O-K, Cho Y. Crystal structure of the human GINS complex. Genes Dev. 2007;21:1316–21.

70. Boskovic J, Coloma J, Aparicio T, Zhou M, Robinson CV, Méndez J, et al. Molecular architecture of the human GINS complex. EMBO Rep. 2007;8:678–84.
71. Sengupta S, van Deursen F, de Piccoli G, Labib K. Dpb2 integrates the leading-strand DNA polymerase into the eukaryotic replisome. Curr Biol. 2013;23:543–52.
72. Langston LD, Zhang D, Yurieva O, Georgescu RE, Finkelstein J, Yao NY, et al. CMG helicase and DNA polymerase ε form a functional 15-subunit holoenzyme for eukaryotic leading-strand DNA replication. Proc Natl Acad Sci U S A. 2014;111:15390–5.
73. De Falco M, Ferrari E, De Felice M, Rossi M, Hübscher U, Pisani FM. The human GINS complex binds to and specifically stimulates human DNA polymerase alpha-primase. EMBO Rep. 2007;8:99–103.
74. Tanaka T, Umemori T, Endo S, Muramatsu S, Kanemaki M, Kamimura Y, et al. Sld7, an Sld3-associated protein required for efficient chromosomal DNA replication in budding yeast. EMBO J. 2011;30:2019–30.
75. Tanaka S, Nakato R, Katou Y, Shirahige K, Araki H. Origin association of Sld3, Sld7, and Cdc45 proteins is a key step for determination of origin-firing timing. Curr Biol. 2011;21:2055–63.
76. Kamimura Y, Tak Y-S, Sugino A, Araki H. Sld3, which interacts with Cdc45 (Sld4), functions for chromosomal DNA replication in *Saccharomyces cerevisiae*. EMBO J. 2001;20:2097–107.
77. Muramatsu S, Hirai K, Tak Y-S, Kamimura Y, Araki H. CDK-dependent complex formation between replication proteins Dpb11, Sld2, Pol ε, and GINS in budding yeast. Genes Dev. 2010;24:602–12.
78. Tak Y-S, Tanaka Y, Endo S, Kamimura Y, Araki H. A CDK-catalysed regulatory phosphorylation for formation of the DNA replication complex Sld2-Dpb11. EMBO J. 2006;25:1987–96.
79. Tanaka S, Komeda Y, Umemori T, Kubota Y, Takisawa H, Araki H. Efficient initiation of DNA replication in eukaryotes requires Dpb11/TopBP1-GINS interaction. Mol Cell Biol. 2013;33:2614–22.
80. Fukuura M, Nagao K, Obuse C, Takahashi TS, Nakagawa T, Masukata H. CDK promotes interactions of Sld3 and Drc1 with Cut5 for initiation of DNA replication in fission yeast. Mol Biol Cell. 2011;22:2620–33.
81. Araki H, Leem SH, Phongdara A, Sugino A. Dpb11, which interacts with DNA polymerase II(ε) in *Saccharomyces cerevisiae*, has a dual role in S-phase progression and at a cell cycle checkpoint. Proc Natl Acad Sci U S A. 1995;92:11791–5.
82. Masumoto H, Sugino A, Araki H. Dpb11 controls the association between DNA polymerases α and ε and the autonomously replicating sequence region of budding yeast. Mol Cell Biol. 2000;20:2809–17.
83. Mäkiniemi M, Hillukkala T, Tuusa J, Reini K, Vaara M, Huang D, et al. BRCT domain-containing protein TopBP1 functions in DNA replication and damage response. J Biol Chem. 2001;276:30399–406.
84. Kesti T, Flick K, Keranen S, Syväoja JE, Wittenberg C. DNA polymerase ε catalytic domains are dispensable for DNA replication, DNA repair, and cell viability. Mol Cell. 1999;3:679–85.
85. Dua R, Levy DL, Campbell JL. Analysis of the essential functions of the C-terminal protein/protein interaction domain of *Saccharomyces cerevisiae* pol epsilon and its unexpected ability to support growth in the absence of the DNA polymerase domain. J Biol Chem. 1999;274:22283–8.
86. Feng W, D'Urso G. Schizosaccharomyces pombe cells lacking the amino-terminal catalytic domains of DNA polymerase epsilon are viable but require the DNA damage checkpoint control. Mol Cell Biol. 2001;21:4495–504.
87. Hiraga S-I, Hagihara-Hayashi A, Ohya T, Sugino A. DNA polymerases α, δ, and ε localize and function together at replication forks in *Saccharomyces cerevisiae*. Genes Cells. 2005;10:297–309.

88. Nuutinen T, Tossavainen H, Fredriksson K, Pirilä P, Permi P, Pospiech H, et al. The solution structure of the amino-terminal domain of human DNA polymerase ε subunit B is homologous to C-domains of AAA+ proteins. Nucleic Acids Res. 2008;36:5102–10.
89. Kang Y, Farina A, Bermudez VP, Tappin I, Du F, Galal WC, et al. Interaction between human Ctf4 and the Cdc45/Mcm2-7/GINS (CMG) replicative helicase. Proc Natl Acad Sci U S A. 2013;110:4–9.
90. Araki H. Cyclin-dependent kinase-dependent initiation of chromosomal DNA replication. Curr Opin Cell Biol. 2010;22:766–71.
91. Clausen AR, Lujan SA, Burkholder AB, Orebaugh CD, Williams JS, Clausen MF, et al. Tracking replication enzymology in vivo by genome-wide mapping of ribonucleotide incorporation. Nat Struct Mol Biol. 2015;2(3):185–91.
92. Daigaku Y, Keszthelyi A, Müller CA, Miyabe I, Brooks T, Retkute R, et al. A global profile of replicative polymerase usage. Nat Struct Mol Biol. 2015;22:192–8.
93. Pospiech H, Grosse F, Pisani FM. The initiation step of eukaryotic DNA replication. Subcell Biochem. 2010;50:79–104.
94. Garcia V, Furuya K, Carr AM. Identification and functional analysis of TopBP1 and its homologs. DNA Repair (Amst). 2005;4:1227–39.
95. Wardlaw CP, Carr AM, Oliver AW. TopBP1: a BRCT-scaffold protein functioning in multiple cellular pathways. DNA Repair (Amst). 2014;22C:165–74.
96. Sokka M, Parkkinen S, Pospiech H, Syväoja JE. Function of TopBP1 in genome stability. Subcell Biochem. 2010;1:119–41.
97. Van Hatten RA, Tutter AV, Holway AH, Khederian AM, Walter JC, Michael WM. The *Xenopus* Xmus101 protein is required for the recruitment of Cdc45 to origins of DNA replication. J Cell Biol. 2002;159:541–7.
98. Hashimoto Y, Takisawa H. *Xenopus* Cut5 is essential for a CDK-dependent process in the initiation of DNA replication. EMBO J. 2003;22:2526–35.
99. Sangrithi MN, Bernal JA, Madine M, Philpott A, Lee J, Dunphy WG, et al. Initiation of DNA replication requires the RECQL4 protein mutated in Rothmund-Thomson syndrome. Cell. 2005;121:887–98.
100. Matsuno K, Kumano M, Kubota Y, Hashimoto Y, Takisawa H. The N-terminal noncatalytic region of *Xenopus* RecQ4 is required for chromatin binding of DNA polymerase α in the initiation of DNA replication. Mol Cell Biol. 2006;26:4843–52.
101. Im J-S, Ki S-H, Farina A, Jung D-S, Hurwitz J, Lee J-K. Assembly of the Cdc45-Mcm2-7-GINS complex in human cells requires the Ctf4/And-1, RecQL4, and Mcm10 proteins. Proc Natl Acad Sci U S A. 2009;106:15628–32.
102. Croteau DL, Popuri V, Opresko PL, Bohr VA. Human RecQ helicases in DNA repair, recombination, and replication. Annu Rev Biochem. 2014;83:519–52.
103. Chu WK, Hickson ID. RecQ helicases: multifunctional genome caretakers. Nat Rev Cancer. 2009;9:644–54.
104. Siitonen HA, Sotkasiira J, Biervliet M, Benmansour A, Capri Y, Cormier-daire V, et al. The mutation spectrum in RECQL4 diseases. Eur J Hum Genet. 2009;17:151–8.
105. Barea F, Tessaro S, Bonatto D. In silico analyses of a new group of fungal and plant RecQ4-homologous proteins. Comput Biol Chem. 2008;32:349–58.
106. Groocock LM, Prudden J, Perry JJ, Boddy MN. The RecQ4 orthologue Hrq1 is critical for DNA interstrand cross-link repair and genome stability in fission yeast. Mol Cell Biol. 2012;32:276–87.
107. Bochman ML, Paeschke K, Chan A, Zakian VA. Hrq1, a homolog of the human RecQ4 helicase, acts catalytically and structurally to promote genome integrity. Cell Rep. 2014;6:346–56.
108. Kamimura Y, Masumoto H, Sugino A, Araki H. Sld2, which interacts with Dpb11 in *Saccharomyces cerevisiae*, is required for chromosomal DNA replication. Mol Cell Biol. 1998;18:6102–9.

109. Ohlenschläger O, Kuhnert A, Schneider A, Haumann S, Bellstedt P, Keller H, et al. The N-terminus of the human RecQL4 helicase is a homeodomain-like DNA interaction motif. Nucleic Acids Res. 2012;40:8309–24.
110. Gaggioli V, Zeiser E, Rivers D, Bradshaw CR, Ahringer J, Zegerman P. CDK phosphorylation of SLD-2 is required for replication initiation and germline development in *C. elegans*. J Cell Biol. 2014;204:507–22.
111. Marino F, Vindigni A, Onesti S. Bioinformatic analysis of RecQ4 helicases reveals the presence of a RQC domain and a Zn knuckle. Biophys Chem. 2013;177–178:34–9.
112. Ichikawa K, Noda T, Furuichi Y. Preparation of the gene targeted knockout mice for human premature aging diseases, Werner syndrome, and Rothmund-Thomson syndrome caused by the mutation of DNA helicases. Nihon Yakurigaku Zasshi. 2002;119:219–26.
113. Mann MB, Hodges CA, Barnes E, Vogel H, Hassold TJ, Luo G. Defective sister-chromatid cohesion, aneuploidy and cancer predisposition in a mouse model of type II Rothmund-Thomson syndrome. Hum Mol Genet. 2005;14:813–25.
114. Capp C, Wu J, Hsieh T. *Drosophila* RecQ4 has a 3′-5′ DNA helicase activity that is essential for viability. J Biol Chem. 2009;284:30845–52.
115. Abe T, Yoshimura A, Hosono Y, Tada S, Seki M, Enomoto T. The N-terminal region of RECQL4 lacking the helicase domain is both essential and sufficient for the viability of vertebrate cells. Role of the N-terminal region of RECQL4 in cells. Biochim Biophys Acta. 2011;1813:473–9.
116. Hoki Y, Araki R, Fujimori A, Ohhata T, Koseki H, Fukumura R, et al. Growth retardation and skin abnormalities of the Recql4-deficient mouse. Hum Mol Genet. 2003;12:2293–9.
117. Kumagai A, Shevchenko A, Shevchenko A, Dunphy WG. Treslin collaborates with TopBP1 in triggering the initiation of DNA replication. Cell. 2010;140:349–59.
118. Sansam CL, Cruz NM, Danielian PS, Amsterdam A, Lau ML, Hopkins N, et al. A vertebrate gene, ticrr, is an essential checkpoint and replication regulator. Genes Dev. 2010;24:183–94.
119. Sanchez-Pulido L, Diffley JFX, Ponting CP. Homology explains the functional similarities of Treslin/Ticrr and Sld3. Curr Biol. 2010;20:R509–10.
120. Kumagai A, Shevchenko A, Shevchenko A, Dunphy WG. Direct regulation of Treslin by cyclin-dependent kinase is essential for the onset of DNA replication. J Cell Biol. 2011;193:995–1007.
121. Boos D, Sanchez-Pulido L, Rappas M, Pearl LH, Oliver AW, Ponting CP, et al. Regulation of DNA replication through Sld3-Dpb11 interaction is conserved from yeast to humans. Curr Biol. 2011;21:1–6.
122. Mueller AC, Keaton MA, Dutta A. DNA replication: mammalian Treslin-TopBP1 interaction mirrors yeast Sld3-Dpb11. Curr Biol. 2011;21:R638–40.
123. Itou H, Muramatsu S, Shirakihara Y, Araki H. Crystal structure of the homology domain of the eukaryotic DNA replication proteins sld3/treslin. Structure. 2014;22:1341–7.
124. Chowdhury A, Liu G, Kemp M, Chen X, Katrangi N, Myers S, et al. The DNA unwinding element binding protein DUE-B interacts with Cdc45 in preinitiation complex formation. Mol Cell Biol. 2010;30:1495–507.
125. Gao Y, Yao J, Poudel S, Romer E, Abu-Niaaj L, Leffak M. Protein phosphatase 2A and Cdc7 kinase regulate the DNA unwinding element-binding protein in replication initiation. J Biol Chem. 2014;289:35987–6000.
126. Balestrini A, Cosentino C, Errico A, Garner E, Costanzo V. GEMC1 is a TopBP1-interacting protein required for chromosomal DNA replication. Nat Cell Biol. 2010;12:484–91.
127. Boos D, Yekezare M, Diffley JF. Identification of a heteromeric complex that promotes DNA replication origin firing in human cells. Science. 2013;340:981–4.
128. Bruck I, Kaplan DL. Origin single-stranded DNA releases Sld3 protein from the Mcm2-7 complex, allowing the GINS tetramer to bind the Mcm2-7 complex. J Biol Chem. 2011;286:18602–13.

129. Thangavel S, Mendoza-maldonado R, Tissino E, Sidorova JM, Yin J, Wang W, et al. Human RECQ1 and RECQ4 helicases play distinct roles in DNA replication initiation. Mol Cell Biol. 2010;30:1382–96.
130. Kanter DM, Kaplan DL. Sld2 binds to origin single-stranded DNA and stimulates DNA annealing. Nucleic Acids Res. 2010;39:2580–92.
131. Dhingra N, Bruck I, Smith S, Ning B, Kaplan DL. Dpb11 helps control assembly of the Cdc45-Mcm2-7-GINS replication fork helicase. J Biol Chem. 2015;290:7586–601.
132. Bruck I, Kanter DM, Kaplan DL. Enabling association of the GINS protein tetramer with the mini chromosome maintenance (Mcm)2-7 protein complex by phosphorylated Sld2 protein and single-stranded origin DNA. J Biol Chem. 2011;286:36414–26.
133. Bruck I, Kaplan DL. The replication initiation protein Sld2 regulates helicase assembly. J Biol Chem. 2014;289:1948–59.
134. Yamane K, Tsuruo T. Conserved BRCT regions of TopBP1 and of the tumor suppressor BRCA1 bind strand breaks and termini of DNA. Oncogene. 1999;18:5194–203.
135. Xu X, Liu Y. Dual DNA unwinding activities of the Rothmund-Thomson syndrome protein, RECQ4. EMBO J. 2009;28:568–77.
136. Macris MA, Krejci L, Bussen W, Shimamoto A, Sung P, Shimamotoc A, et al. Biochemical characterization of the RECQ4 protein, mutated in Rothmund-Thomson syndrome. DNA Repair (Amst). 2006;5:172–80.
137. Sedlackova H, Cechova B, Mlcouskova J, Krejci L. RECQ4 selectively recognizes Holliday junctions. DNA Repair (Amst). 2015;30:80–9.
138. Besnard E, Babled A, Lapasset L, Milhavet O, Parrinello H, Dantec C, et al. Unraveling cell type-specific and reprogrammable human replication origin signatures associated with G-quadruplex consensus motifs. Nat Struct Mol Biol. 2012;19:837–44.
139. Cayrou C, Coulombe P, Puy A, Rialle S, Kaplan N, Segal E, et al. New insights into replication origin characteristics in metazoans. Cell Cycle. 2012;11:658–67.
140. Biffi G, Tannahill D, McCafferty J, Balasubramanian S. Quantitative visualization of DNA G-quadruplex structures in human cells. Nat Chem. 2013;5:182–6.
141. Henderson A, Wu Y, Huang YC, Chavez EA, Platt J, Johnson FB, et al. Detection of G-quadruplex DNA in mammalian cells. Nucleic Acids Res. 2014;42:860–9.
142. Valton A-L, Hassan-Zadeh V, Lema I, Boggetto N, Alberti P, Saintomé C, et al. G4 motifs affect origin positioning and efficiency in two vertebrate replicators. EMBO J. 2014;33:732–46.
143. Keller H, Kiosze K, Sachsenweger J, Haumann S, Ohlenschläger O, Nuutinen T, et al. The intrinsically disordered amino-terminal region of human RecQL4 : multiple DNA-binding domains confer annealing, strand exchange and G4 DNA binding. Nucleic Acids Res. 2014;42:12614–27.
144. Hoshina S, Yura K, Teranishi H, Kiyasu N, Tominaga A, Kadoma H, et al. Human origin recognition complex binds preferentially to G-quadruplex-preferable RNA and single-stranded DNA. J Biol Chem. 2013;288:30161–71.
145. Ge XQ, Jackson DA, Blow JJ. Dormant origins licensed by excess Mcm2-7 are required for human cells to survive replicative stress. Genes Dev. 2007;21:3331–41.
146. McIntosh D, Blow JJ. Dormant origins, the licensing checkpoint, and the response to replicative stresses. Cold Spring Harb Perspect Biol. 2012;4:a012955.
147. Li Y, Araki H. Loading and activation of DNA replicative helicases: the key step of initiation of DNA replication. Genes Cells. 2013;18:266–77.
148. Mantiero D, Mackenzie A, Donaldson A, Zegerman P. Limiting replication initiation factors execute the temporal programme of origin firing in budding yeast. EMBO J. 2011;30:4805–14.
149. Tanaka S, Araki H. Multiple regulatory mechanisms to inhibit untimely initiation of DNA replication are important for stable genome maintenance. PLoS Genet. 2011;7:e1002136.
150. Collart C, Allen GE, Bradshaw CR, Smith JC, Zegerman P. Titration of four replication factors is essential for the *Xenopus laevis* midblastula transition. Science. 2013;341:893–6.

151. Pollok S, Bauerschmidt C, Sänger J, Nasheuer H-P, Grosse F. Human Cdc45 is a proliferation-associated antigen. FEBS J. 2007;274:3669–84.
152. Wong PG, Winter SL, Zaika E, Cao TV, Oguz U, Koomen JM, et al. Cdc45 limits replicon usage from a low density of preRCs in mammalian cells. PLoS One. 2011;6:e17533.
153. Broderick R, Ramadurai S, Tóth K, Togashi DM, Ryder AG, Langowski J, et al. Cell cycle-dependent mobility of Cdc45 determined in vivo by fluorescence correlation spectroscopy. PLoS One. 2012;7:e35537.
154. Srinivasan SV, Dominguez-Sola D, Wang LC, Hyrien O, Gautier J. Cdc45 is a critical effector of myc-dependent DNA replication stress. Cell Rep. 2013;3:1629–39.
155. Tomita Y, Imai K, Senju S, Irie A, Inoue M, Hayashida Y, et al. A novel tumor-associated antigen, cell division cycle 45-like can induce cytotoxic T-lymphocytes reactive to tumor cells. Cancer Sci. 2011;102:697–705.
156. Maya-Mendoza A, Petermann E, Gillespie DA, Caldecott KW, Jackson DA. Chk1 regulates the density of active replication origins during the vertebrate S phase. EMBO J. 2007;26:2719–31.
157. Petermann E, Caldecott KW. Evidence that the ATR/Chk1 pathway maintains normal replication fork progression during unperturbed S phase. Cell Cycle. 2006;5:2203–9.
158. Petermann E, Woodcock M, Helleday T. Chk1 promotes replication fork progression by controlling replication initiation. Proc Natl Acad Sci U S A. 2010;107:16090–5.
159. Syljuåsen RG, Sørensen CS, Hansen LT, Fugger K, Lundin C, Johansson F, et al. Inhibition of human Chk1 causes increased initiation of DNA replication, phosphorylation of ATR targets, and DNA breakage. Mol Cell Biol. 2005;25:3553–62.
160. Guo C, Kumagai A, Schlacher K, Shevchenko A, Shevchenko A, Dunphy WG. Interaction of Chk1 with treslin negatively regulates the initiation of chromosomal DNA replication. Mol Cell. 2015;57:492–505.
161. Friedel AM, Pike BL, Gasser SM. ATR/Mec1: coordinating fork stability and repair. Curr Opin Cell Biol. 2009;21:237–44.
162. Cimprich KA, Cortez D. ATR: an essential regulator of genome integrity. Nat Rev Mol Cell Biol. 2008;9:616–27.

Chapter 20
Structure and Activity of the Cdc45-Mcm2–7-GINS (CMG) Complex, the Replication Helicase

Barbara Medagli, Patrizia Di Crescenzio, Matteo De March, and Silvia Onesti

Abstract Highly efficient unwinding of double-stranded DNA is an essential prerequisite for duplication of the genetic material. In eukaryotes this process is catalysed by the CMG (Cdc45-Mcm2–7-GINS) complex, which is composed by 11 proteins: the Cdc45 replication factor, the hetero-hexameric Mcm2–7 and the tetrameric GINS complex. Although the Mcm2–7 is the helicase motor, in the absence of the other components the purified Mcm complex displays only a weak and labile helicase activity; in contrast, a stable complex comprising Cdc45, Mcm2–7 and GINS can be co-expressed and/or reconstituted, and shows a robust helicase activity.

Despite its importance, many questions regarding the function of the CMG are still open. Although we do not yet have all the answers, over the last few years a number of studies have been published, including detailed biochemical characterisation of the CMG, the atomic structures of active archaeal proteins and the electron microscopy models of larger macromolecular complexes. All of this information is contributing to build a coherent picture of the structure and function of this complex machinery. In this chapter the current knowledge on the structure and function of the single components of the CMG and the whole complex is summarised, and the putative conformational changes that are necessary to trigger the helicase activity are discussed.

As all the components of the CMG complex are hallmarks of cell proliferation and therefore are putative tumour biomarkers and potential drug targets, a full understanding of the structure and activity of this complex is therefore a prerequisite for the full exploitation of its potential in cancer diagnostics and therapy.

B. Medagli • M. De March • S. Onesti (✉)
Structural biology laboratory, Elettra-Sincrotrone Trieste S.C.p.A., Area Science Park, Trieste, Italy
e-mail: silvia.onesti@elettra.eu

P. Di Crescenzio
Structural biology laboratory, Elettra-Sincrotrone Trieste S.C.p.A., Area Science Park, Trieste, Italy

Dipartimento di Scienze della Vita, Università degli Studi di Trieste, Trieste, Italy

D.L. Kaplan (ed.), *The Initiation of DNA Replication in Eukaryotes*,
DOI 10.1007/978-3-319-24696-3_20

Keywords Mcm2–7 helicase • Cdc45 • GINS complex • Sld5 • Psf1 • Psf2 • Psf3 • Eukaryotic DNA replication • Archaeal DNA replication • Cancer diagnostic and therapy

Introduction

In eukaryotic cells, the unwinding of double-stranded DNA before replication is catalysed by the CMG complex (Fig. 20.1), a macromolecular assembly of three distinct replication factors: the **C**dc45 protein, the **M**cm2–7 complex and the **G**INS complex [1]. Each of these factors is essential for CMG function and each is conserved across all eukaryotes [2].

The Mcm complex is the helicase motor and is conserved in evolution, with most archaea possessing a single copy, while eukaryotes have at least six different paralogues (Mcm2–7) [3]. Mcm proteins can be divided into three domains: an N-terminal domain, a AAA+ motor domain and a C-terminal WH domain (Fig. 20.1). Whereas the archaeal proteins are constitutively active, the eukaryotic Mcm complex alone has no or little helicase activity: the activity of the Mcm2–7 complex requires the presence of additional factors (such as Cdc45 and GINS), though the exact mechanisms of activation are not completely understood. GINS is a tetramer composed of four subunits (Sld5, Psf1, Psf2, Psf3, Fig. 20.1), likely to derive from a single protein through gene duplication [4, 5]. Cdc45 is an essential factor required for the establishment and progression of DNA replication in eukaryotic cells. Recent reports have revealed an unsuspected similarity with the DHH superfamily of phosphoesterases [6–9], shedding light on the evolution of this factor (Fig. 20.1).

Many replication proteins are abundant in transforming or cancer cell lines when compared to normal cells, making them, in principle, potential biomarkers for cancer detection and prognosis. Mcm proteins have been found to be overexpressed in a variety of tumours; most results show that they are more sensitive and specific markers than the conventional proliferative markers Ki-67 and PCNA [10]. More recently, many reports have shown that also Cdc45 and GINS are very promising candidates for novel proliferation markers and potential drug targets [11–15]. Moreover, defects in MCM and GINS subunit themselves can cause cancer [16–18]. A detailed knowledge of the structure and activity of the CMG complex and its components is therefore a prerequisite for the full exploitation of their potential in cancer diagnostic and therapy [19].

Despite its importance, many questions regarding the CMG remain unanswered. Which is the exact mechanism of action of Mcm proteins? How is the helicase activated? How does it load onto DNA? How are GINS and Ccd45 assembled? Precisely what functions are performed by the GINS and Cdc45 proteins in initiation and elongation? Although we do not have all the answers, over the last few years a large body of data has been published, ranging form detailed characterisation of the CMG biochemical properties [20–22], to crystallographic structure of active archaeal proteins [23–25] to electron microscopy structures of larger macromolecular complexes

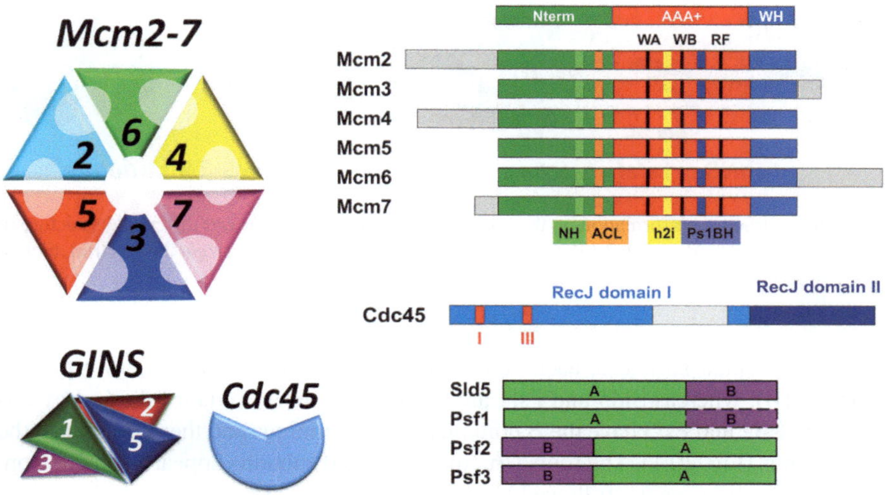

Fig. 20.1 A simplified representation of the CMG components. The **Mcm2, 3, 4, 5, 6, 7** proteins form a hexameric arrangement with the ATP binding sites (*white ovals*) at the interface between subunits and a central channel that accommodates the DNA. Schematic diagrams of the sequences of the human Mcm proteins are shown on the *right*. Each subunit includes three conserved domains: the N-terminal domain (*green*), AAA+ catalytic domain (*red*) and a C-terminal WH fold (*blue*); additional domains are present in some subunits, either at the N- or C-terminus. Beside the Walker A (WA), Walker B (WB) and Arginine finger (RF), three conserved loops have been identified: the allosteric communication look (ACL, *orange*), the helix 2 insertion (h2i, *yellow*) and the pre-sensor I b-hairpin (PS1BH, *blue*). **GINS** is a tetramer formed by four subunits (Sld5, Psf1, Psf2, Psf3) which are likely to be derived from a single gene through gene duplication and domain swapping. Each subunit is made up of an α-helical (A, in *green*) and a β-rich domain (B, in *magenta*). Whereas Sld5 and Psf1 have the same architecture (A–B), Psf2 and Psf3 show an inversion in the order of the domains (B–A). The B domain of Psf1 is connected by a flexible link and is not visible in the GINS crystal structure. **Cdc45** shows some sequence homology with domains I (*light blue*) and II (*dark blue*) of the bacterial RecJ exonuclease, but only retains two of the seven conserved motifs (motifs I and III, in *red*). An insertion domain that is unique to the archaeal and eukaryotic Cdc45 homologues is shown in *grey*

[6, 26–28]. All of this information is contributing to build a coherent picture of the structure and function of the CMG complex.

Structure and Function of the GINS Complex

The only high resolution structure of a eukaryotic CMG component is the human GINS complex, which comprises four subunits: Sld5, Psf1, Psf2 and Psf3. Three independent crystal structures of the tetrameric human GINS complex and a structure of an archaeal homologue have been determined [28–31]: they all show the same trapezoid shape with the four subunits forming a tight bundle, held by extensive inter-subunit contacts, mediated by hydrophobic interactions (Fig. 20.2).

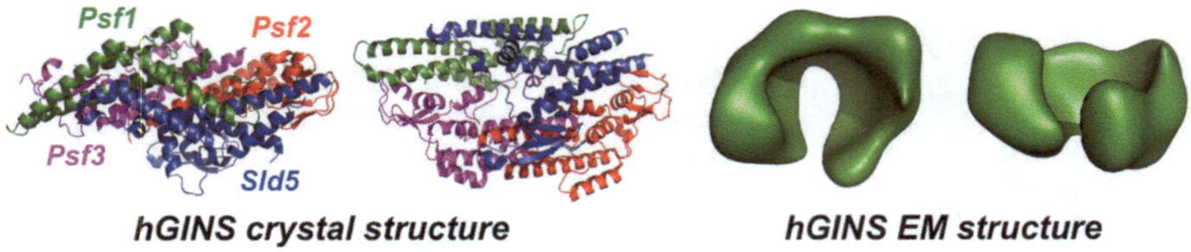

Fig. 20.2 Structural information on the human GINS complex. *On the left* the crystal structure of human GINS ([31], PDB ID: 2E9X) forming a compact trapezoid; *on the right* the electron microscopy structure showing an open ring [33]

In the human structure, each subunit comprises a α-helical (A) and a β-sheet domain (B): whereas Sld5 and Psf1 are made up by an A domain followed by a B domain, Psf2 and Psf3 have the B domain at the N-terminus and the A domain at the C-terminus (Fig. 20.1). The fours structurally related subunits appear to derive from gene duplication events, followed by domain swapping. The C-terminal domain of the Psf1 subunit (i.e. the B domain) is flexible and prone to proteolysis: indeed in two of the crystal structures the domain has been deleted to favour the formation of a tight lattice [30, 31], whereas in the other, although it is present, it is not visible in the electron density, and an atomic model could not be built [29].

The evolutionary origin of GINS from one or more gene duplication events is confirmed by an analysis of archaeal genomes: in the majority of Euryarchaeota a single subunit can be identified, similar to Sld5 and Psf1 (Gins51), whereas a subset of Archaea possesses two subunits with the topology of Sld5/Psf1 (Gins51) and Psf2/Psf3 (Gins23), respectively. The archaeal structure available (from *Thermococcus kodakaraensis*) is a dimer of dimers (Fig. 20.2), with two copies of Gins51 and two copies of Gins23 arranged in a similar way to the human GINS complex [32].

A puzzling discrepancy exists among the GINS crystal structures and a low resolution structure obtained by single particle electron microscopy (EM) showing a C-shaped molecule (Fig. 20.2) [33]; a range of possible explanations have been proposed [29, 31, 33], but none can convincingly explain the remarkable difference.

As a component of the CMG, GINS has an essential function in the regulation of the Mcm2–7 helicase activity and the progression of the replication fork: the role and position of GINS within the CMG is discussed in "Structure and Function of the CMG Complex". Beside its role in the CMG, a wealth of data connects GINS with most of the players in DNA replication, giving it a central position in the protein network which is assembled at replication origins [34]. For example GINS has a critical role in origin licencing and the assembling of the pre-initiation complex: although the details may be diverse in different organisms, it is a key factor in the establishment of the replication fork, via its interaction with the replication factors Sld2, Sld3 and Mcm10.

Moreover, the unstructured region of the N-terminal domain of Sld5 interacts with the C-terminal region of Ctf4; as Ctf4 is a trimeric molecule that binds the polymerase α/primase, GINS contributes to the physical coupling between helicase and polymerase on the lagging strand [35]. Biochemical data also suggest that the B domain interacts with the Dpb2 subunit of polymerase ε [36] and the interaction is critical to link the helicase to the leading strand polymerase during the progression of the replication fork. Thus, during fork progression GINS appears to have a key role not only in the activation of the helicase activity but also in stabilising the replisome by contributing to the physical connection between the Mcm2–7 helicase and the polymerases on both the leading and lagging strand, to avoid the dangers of uncoupling DNA unwinding and DNA synthesis.

Structure and Function of Cdc45

Cdc45 is an essential factor conserved in all eukaryotes and is required for the establishment and progression of the DNA replication fork in eukaryotic cells. As many other DNA replication factors, Cdc45 is more abundant in proliferating cells, whereas it is almost absent from long-term quiescent, terminally differentiated and senescent cells [37]. Several genetic studies, two-hybrid screens and co-immunoprecipitation analyses revealed that Cdc45 interacts with a number of other replication factors, including the Mcm2–7 complex, GINS, Mcm10, replication protein A (RPA), DNA polymerase α, δ and ε, the origin recognition complex subunit 2 (Orc2) and TopBP1 [38].

Whereas homologues of both GINS and Mcm proteins in Archaea have been long detected, no counterpart for Cdc45 was known. Recently a weak but significant relationship was described between eukaryotic Cdc45 proteins and a large family of phosphoesterases (DHH family), which includes inorganic pyrophosphatases and RecJ ssDNA exonucleases [6–9, 39]. These proteins catalyse the hydrolysis of phosphodiester bonds via a mechanism involving two Mn^{2+} ions and are characterised by a number of conserved motifs involved in metal binding and catalysis. Indeed Cdc45-like nucleases have been found associated with GINS and MCM proteins in Archaea [34, 39].

Whereas some of the archaeal Cdc45 homologues retain the full set of catalytic residues and can display exonuclease activity [39, 40], the eukaryotic proteins have lost most of the conserved motifs, with the exception of motifs I and III (Fig. 20.2). No metal was found associated with the protein, nor a nuclease or pyrophosphatase activity could be detected [8]. An insertion in the RecJ scaffolding is unique to the archaeal and eukaryotic Cdc45 homologues. The protein retains the ability to bind a single stranded DNA (ssDNA) in a sequence-independent manner, and has a preference for longer ssDNA substrates [41]. The DNA binding affinity appears to be enhanced by interactions with the replication factor Mcm10 [42].

No atomic structure has been determined for Cdc45 and only low resolution models are available: a SAXS envelope has been obtained by small-angle X-ray scattering [8] and an electron microscopy structure can be extracted by the structure of the CMG complex [6, 27]. Within the CMG Cdc45 is located at “gate” between the Mcm2/Mcm5 subunit of the hexameric ring; the role of Cdc45 within the CMG is discussed in “Structure and Function of the CMG Complex”.

Cdc45 binds to yeast Sld3 and its metazoan homologue Treslin both in vivo and in vitro [43–45]. Cdc45 also has been shown to interact with human DNA helicase B: the exact role of this metazoan helicase is still under investigation, although it has been suggested that it may be involved in the initiation of mammalian replication [46].

Structure and Function of the Mcm2–7 Complex

The Mcm2–7 complex is the actual motor of the replicative helicase and comprises six homologous proteins (Mcm2, 3, 4, 5, 6, 7; reviewed in Ref. [3]). Mcm proteins are members of the AAA+ family of ATPases (ATPases Associated with various cellular Activities) that use energy derived from cycles of ATP hydrolysis to do mechanical work. As most AAA+ members, the Mcm2–7 complex assembles in a heterohexameric ring, with the ATPase active sites located between two adjacent subunits (Fig. 20.1). Archaeal homologues have been used for long time as model systems to study the structure and function of the MCM helicases, as in most Archaea a single Mcm subunit forms a homo-hexamer or double-hexamer. Thus, until very recently, most of the biochemical and structural work has been carried out on the archaeal proteins.

Mcm proteins can be divided into three domains: the N-terminal, AAA+ core and C-terminal domains. Both the N- and the C-terminal domains are less conserved and may present additional extensions that may have regulatory roles and/or mediate further protein–protein interactions.

The N-Terminal Domain

The N-terminal domain binds ssDNA and dsDNA, is essential for hexamerisation and influences the processivity and the polarity of the helicase [47–51].

A few crystal structures are available for the N-terminal domain from different archaeal organism. The crystal structure of the *Methanothermobacter thermautotrophicus* MCM (*Mth*MCM) N-terminal domain, revealed a double hexamer in a head-to-head configuration, with a central channel that can accommodate dsDNA (Fig. 20.3) [52]. The N-terminal domain of *Sulfolobus solfataricus* MCM (*Sso*MCM) is a single planar hexamer with a smaller channel that can only accommodate ssDNA [51]. In contrast, the crystal structure of the N-terminal domain of *Thermoplasma acidophilum* MCM (*Tap*MCM) shows a right handed helix [23]; although it is not unusual for hexameric proteins to crystallise as a 6_1 helix, it may indicate a certain degree of flexibility that may be functionally important. The structure of the N-terminal domain of *Pyrococcus furiosus* MCM (N-*Pfu*MCM) showed for the first time the hexameric ring interacting with a short ssDNA [24]. Upon this interaction the N-terminal ring becomes asymmetric, generating an elliptical pore that accommodates two short ssDNA stretches in an unexpected configuration: each

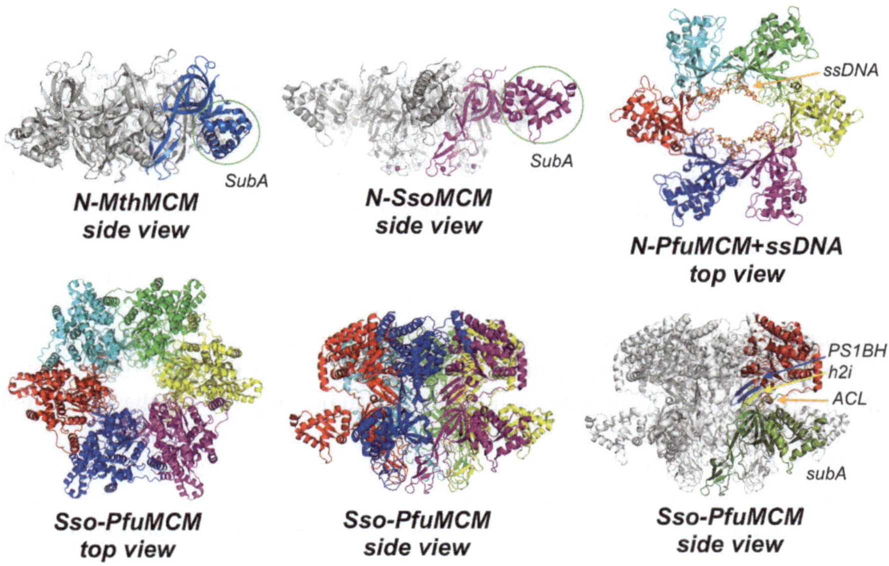

Fig. 20.3 A selection of crystallographic structure of the archaeal Mcm proteins. A side view of a single hexamer of the **N-terminal domain of *Mth*MCM** ([52], PDB ID: 1LTL) is shown in *grey*, with one subunit highlighted in *blue*: a *green circle* shows the position of subdomain A in the closed conformation. The **N-terminal hexamer of SsoMCM**, taken from the structure of the *Sso-PfuMCM* fusion construct ([25], PDB ID: 4R7Y) is shown in *grey* with one subunit highlighted in *magenta*: a *green circle* shows the position of subdomain A in the open conformation. A top view of the structure of ***Pfu*MCM crystallised in the presence of ssDNA** ([24], PDB ID: 4POG) shows an unusual mode of binding, with the DNA forming an arc perpendicular to the channel axis. Each subunit is shown in a different colour, with the nucleic acid in *orange*. The crystal structure of an **active MCM hexamer (*Sso-Pfu*MCM**, [25], PDB ID: 4R7Y) is shown from the top and the side. The position of one monomer is highlighted in colours within a *grey* hexamer: the N-terminal domain is in green and the AAA+ domain in *red*. The interaction of the ACL (*orange*), h2i (*yellow*), and PS1BH (*blue*) elements is visible at the interface between the N-terminal and AAA+ domains

short DNA segment (4–11 nucleotides) forms an arch which is perpendicular to the ring axis, rather than extending across the pore, as expected (Fig. 20.3). The protein–DNA contacts involve sites that are protected in the DNA double helix possibly suggesting that the structure may be a snapshot of the origin melting process. The amino-acid residues involved in the interactions are conserved in a subset of eukaryotic Mcm proteins that are adjacent and may provide a single extended single-stranded DNA binding surface. The central channel is significantly larger than that observed in most hexameric helicases [53].

Despite the differences in the hexameric arrangements each N-terminal domain folds into three different subdomains. Subdomain A is located in the peripheral belt of the MCM ring and gives the assembly its characteristic "dumbbell" shape. A number of hints had suggested that it is flexible and can undergo large conformational changes, from a close to and open configuration [25, 52, 54–57].

Remarkably, whereas the crystal structure of the N-terminal domain of *Sso*MCM showed subdomain A in the closed conformation [51], when it was fused to the AAA+ domain of *Pyrococcus furiosus* [25], its conformation was strikingly different, with a large swing of subdomain A (Fig. 20.3). Subdomain B is found at the double-ring interface and contains a conserved Zn motif that provides the interface for „the formation of a double hexamer [52, 57]. Subdomain C comprises an OB fold and is the main determinant for hexamer formation. It contains a β-hairpin element (NBH) that projects toward the central channel. Biochemical studies have identified a number of positively charged residues important for the nucleic acid interactions, both within the NBH and in the channel [52, 58, 59]. The structure of N-*Pfu*MCM bound to ssDNA confirms the role of some of these residues in interaction with ssDNA; mutations of these residues in yeast MCM2–7 are associated to defects in helicase loading and activation [24], suggesting a role of the N-terminal domain in origin melting.

The N-terminal domain is also involved in inter-subunit communications, as its deletion abolishes the cooperativity between AAA+ domains. This role relies on a conserved loop (allosteric communication loop, ACL) situated in subdomain C at the interface with the AAA+ domain [3, 60–62].

The AAA+ Domain

The motor of the MCM helicase is the central AAA+ domain, which is responsible for the ATP-dependent DNA unwinding. Beside many of the canonical motifs that are shared by other AAA+ proteins (Walker A, Walker B, Arg finger, sensor I, sensor II, Glu switch) MCM proteins have two unique features that have been implicated in DNA binding and catalysis: a helix 2 insertion (h2i) and an insertion before the sensor 1 motif (PS1BH) [3].

Three different crystal structures are available for the AAA+ domain. Two of these structures show a single monomer and, as the ATP binding site is at the interface of two adjacent subunits, are not very useful to understand the mechanism of action [61, 63]. Recently the crystallographic structure of a chimeric MCM helicase (in which the N-terminal domain from *S. solfataricus* was fused to the AAA+ domain from *P. furiosus: Sso-Pfu*MCM) bound to ADP showed the hexameric conformation of the catalytic core and the arrangement of the ATP binding site [25]. As expected, both the PS1BH and h2i are projected into the channel, providing a number of positively charged and aromatic residues that can interact with the DNA backbone and bases. The ACL loop is positioned near the PS1BH and h2i loops, as predicted based on biochemical data and electron density maps [3, 61, 62]. Interestingly, subdomain A is found in an "open" conformation, so far inferred by biochemical data and clearly visualised only in the EM structure of a mutant [57], confirming the previous predictions (Fig. 20.3).

The C-Terminal Domain

The less conserved C-terminal domain folds as winged-helix (WH) domain. A bioinformatic prediction was confirmed by the NMR structure of the C-terminal domain of human Mcm6 [64]. This domain is important for the interaction with Cdt1 and single point mutations disrupt this interaction. Models derived from electron microscopy locate this domain on top of the AAA+ core, but the weak electron density in the EM as well as the lack of order in the X-ray structures suggests high flexibility [63, 65, 66].

The archaeal homolog of this domain was shown to have a regulatory effect on the helicase activity. Deletion of the C-terminal domain showed an increase in helicase and ATPase activity, suggesting that it may act as a brake [48, 49]. Despite a poorly conserved sequence, the presence and fold of this domain seem to be conserved in all MCM homologues.

Electron Microscopy Structures of Archaeal MCM Proteins

Due to the difficulties encountered in crystallising full-length MCM proteins, a number of studies have been carried out by electron microscopy, revealing a plethora of stoichiometric arrangements and variable conformations. The complex can form single and double hexamer, single and double heptamer as well as helical filaments, depending on the species, temperature, salt and protein concentration, presence of nucleotides or nucleic acid substrate interactions [57, 61, 65–70]. This variability raised a number of questions as to the in vivo relevance and the physiological function of the single and the double hexamer, or the hexameric and heptameric forms (reviewed in Ref. [3]).

One of these structures revealed an unexpected interaction with dsDNA, with the nucleic acid wrapped on the outside of a single hexameric ring, rather than threading through the central hole, as expected from an active helicase. The structure also suggested the swinging of subdomain A to accommodate the dsDNA, and could provide a structural model for the initial interaction between MCM and DNA before loading occurs [56].

Electron Microscopy Structures of Eukaryotic MCM Complexes

A number of EM models have been generated over the last few years for eukaryotic complexes, including the Mcm2–7 complex on its own, bound to nucleotides and nucleic acids, or in the presence of additional factors such as Cdt1, ORC, ORC-Cdc6, as well as GINS and Cdc45, producing a series of 3D models that elucidated some of the steps necessary for the loading and activation of the MCM complex.

In vitro studies with *S. cerevisiae* MCM2–7 revealed that pre-incubation of the complex with ATP, or a poorly hydrolysable analogue, impaired the binding to circular ssDNA but had negligible effects on the ability of the complex to bind linear DNA, suggesting the existence of a "gate" in the ring that is closed in the presence of ATP, thus preventing the DNA to enter in the central channel; mutagenesis studies suggested that the gate could be localised at the interface between these two subunits [71]. The EM reconstructions of *Drosophila* MCM2–7 provided structural confirmation for the presence of a gap between Mcm2 and Mcm5 [6]. Two distinct conformations of MCM2–7 were seen. One form resembles an almost symmetric ring, but possesses a notch in the AAA+ tier. The second and more abundant form had an asymmetric opened configuration (a left-handed "lock washer") with a gap extending the full length of the complex between Mcm2 and Mcm5. The structure of a "minimalist" Mcm2–7 from the eukaryotic parasite *Encephalitozoon cuniculi* (*Ecu*Mcm2–7) also shows a left-handed open-ring structure with a gap [28]. Using a rapamycin-inducible linkage the essential functional role of the Mcm2/Mcm5 gate in helicase loading has been confirmed [72].

Reconstitution of the *S. cerevisiae* Mcm2–7 loading reaction in vitro has been achieved with purified budding yeast protein. Two hexameric Mcm2–7 assemble on dsDNA in a head-to-head configuration, suggesting a physiological role for a double hexamer in loading onto origin DNA [73, 74]. A 30 Å reconstruction of the Mcm2–7 double hexamer shows a structure similar to that seen for the *Mh*MCM bound to dsDNA [66], but with the two hexamers slightly off-register (Fig. 20.4).

A higher resolution structure of the double hexamer by EM also shows the presence of two rings slightly offset from the central axis, with a significant twist [26]. Remarkably, mapping of the subunits through the visualisation of the location of maltose binding proteins fused either to the N-terminus or the C-terminus seems to suggest a significant mismatch between the N-terminal ring and the AAA+ ring, so

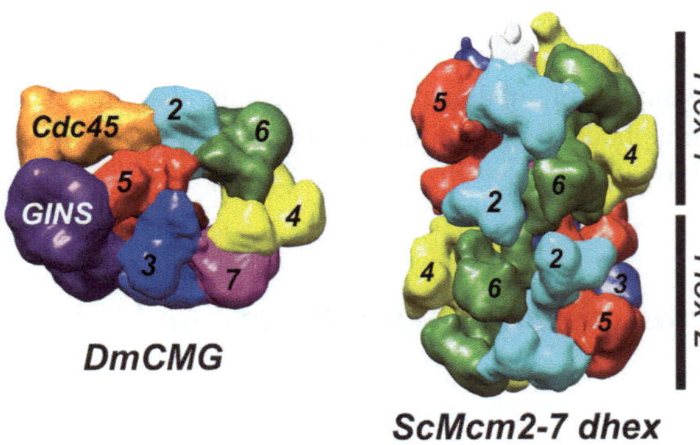

Fig. 20.4 *On the left* EM model of the CMG complex from *D. melanogaster* in the presence of a 3′ tailed nucleic acid substrate and the non-hydrolysable nucleotide analogue ATPγS ([27] EMDB ID: 2772). *On the right* a side view of the EM model for the *S. cerevisiae* Mcm2–7 double hexamer loaded onto dsDNA ([26] EMBD ID: 5857)

that each subunit is stretched by 30 °C (Fig. 20.4). The two hexamers are arranged in such a way that the two potential Mcm2/Mcm5 gates are not colinear; this may ensure that the gate can open without the protein falling off from the dsDNA. It has been suggested that the tilted structure may provide a mechanism for preventing ATP hydrolysis and stabilising the double hexamer on licensed origins thought G1.

Thus, consistent with the polymorphism observed for the archaeal MCM proteins [3], even the eukaryotic proteins show a very dynamic structure that alternated between planar rings, gapped rings and lock-washer arrangements. The Mcm2–7 complex is generally a hexamer, but it is loaded on origin or dsDNA as a double hexamers, with the two hexamers slightly staggered and twisted.

Structure and Function of the CMG Complex

The CMG complex translocates along the DNA with a 3′ → 5′ polarity, acting as the replicative helicase at the core of the replication fork. Whereas the Mcm2–7 hexamer alone displays a very weak and labile ATPase and helicase activity, the CMG complex is an efficient and processive helicase. While the Mcm2–7 complex can bind both ssDNA and dsDNA (however with a preference for the former), the CMG complex only associates with ssDNA. Active recombinant CMG complexes from *Drosophila* and human proteins have been obtained from baculovirus-infected cells and biochemically characterised [20, 21].

Structural insights on the overall organisation of the CMG complex came from the EM structure of the *Drosophila* recombinant CMG (*Dm*CMG, [6]). GINS and Cdc45 were seen to bind across the Mcm2/Mcm5 gate, thus bridging the gate and generating a very large central channel: GINS and Cdc45 preferentially interact with the Mcm N-terminal ring. Upon binding an ATP analogue (ADP·BeF_3), the CMG undergoes a conformational change that seals the Mcm2/Mcm5 gate, closing the Mcm2–7 ring and creating two distinct channels; in this conformation GINS further interacts with the AAA+ domains of Mcm3 and Mcm5, possibly explaining the enhanced ATPase activity compared to Mcm2–7.

A higher resolution EM map (18 Å) obtained for the *Dm*CMG in the presence of a 3′ tailed DNA substrate and the non-hydrolysable ATP analogue ATPγS provides a glimpse of how the various components of the CMG engage with nucleic acid strands (Fig. 20.4) [27]. The CMG complex binds preferentially the single-stranded region, with the 3′ end towards the N-terminal domain and the 5′ end towards the C-terminal side; this is consistent with the orientation predicted by FRET experiments [75, 76] but is different from that seen in other SF3 helicases such as E1 [77]. A comparison with the apo CMG shows that in the presence of the DNA the AAA+ domains move from a planar ring to a right-handed spiral, whereas the N-terminal domains remain almost planar. GINS and Cdc45 play a role in stabilising this spiral configuration, which could have important implications for DNA unwinding. The width of the central channel (35 Å) is significantly larger than that of other helicases such as E1 or even DNAB. Within the CMG the orientation of RecJ suggest that it may bind the leading strand escaping from the helicase channel either during stalling or the inadvertent opening of the gate [22].

A small but significant population of particles can be interpreted as a dimer of CMG, assembling head-to head via the Mcm2–7 N-terminal domains [27]; identification of the subunits suggests that the two Mcm2/Mcm5 gates are staggered between the two hexamers; this configuration and the predicted arrangement of the subunits resemble that observed in the loaded Mcm2–7 double hexamer [26], suggesting that the CMG complex may initially assemble on the double hexamer, prior to the separation upon origin melting and fork progression.

Acknowledgments Funding from the Associazione Italiana per la Ricerca sul Cancro (IG10646 & IG14718) and under the Cross-Border Cooperation Programme Italy-Slovenia 2007–2013 by the European Regional Development Fund and national funds (PROTEO) are gratefully acknowledged.

References

1. Onesti S, MacNeill SA. Structure and evolutionary origins of the CMG complex. Chromosoma. 2013;122:47–53.
2. Aves SJ, Liu Y, Richards TA. Evolutionary diversification of eukaryotic DNA replication machinery. Subcell Biochem. 2012;62:19–35.
3. Costa A, Onesti S. Structural biology of MCM helicases. Crit Rev Biochem Mol Biol. 2009;44:326–42.
4. MacNeill SA. Structure and function of the GINS complex, a key component of the eukaryotic replisome. Biochem J. 2010;425:489–500.
5. Kamada K. The GINS complex: structure and function. Subcell Biochem. 2012;62:135–56.
6. Costa A, Ilves I, Tamberg N, Petojevic T, Nogales E, Botchan MR. The structural basis for MCM2-7 helicase activation by GINS and Cdc45. Nat Struct Mol Biol. 2011;18:471–7.
7. Sanchez-Pulido L, Ponting CP. Cdc45: the missing recj ortholog in eukaryotes? Bioinformatics. 2011;27:1885–8.
8. Krastanova I, Sannino V, Amenitsch H, Gileadi O, Pisani FM, Onesti S. Structural and functional insights into the DNA replication factor Cdc45 reveal an evolutionary relationship to the DHH family of phosphoesterases. J Biol Chem. 2012;287:4121–8.
9. Makarova KS, Koonin EV, Kelman Z. The CMG (CDC45/RecJ, MCM, GINS) complex is a conserved component of the DNA replication system in all archaea and eukaryotes. Biol Direct. 2012;7:7.
10. Giaginis C, Vgenopoulou S, Vielh P, Theocharis S. MCM proteins as diagnostic and prognostic tumor markers in the clinical setting. Histol Histopathol. 2010;25:351–70.
11. Li J, Feng C, Lu Y, Li H, Tu Z, Liao C, Liang C. mRNA expression of the DNA replication-initiation proteins in epithelial dysplasia and squamous cell carcinoma of the tongue. BMC Cancer. 2008;8:395.
12. Tomita Y, Imai K, Senju S, Irie A, Inoue M, Hayashida Y, Shiraishi K, Mori T, Daigo Y, Tsunoda T, Ito T, Nomori H, Nakamura Y, Kohrogi H, Nishimura Y. A novel tumor-associated antigen, cell division cycle 45-like can induce cytotoxic T-lymphocytes reactive to tumor cells. Cancer Sci. 2011;102:697–705.
13. Hokka D, Maniwa Y, Tane S, Nishio W, Yoshimura M, Okita Y, et al. Psf3 is a prognostic biomarker in lung adenocarcinoma. Lung Cancer. 2013;79:77–82.
14. Zheng M, Zhou Y, Yang X, Tang J, Wei D, Zhang Y, Jiang JL, Chen ZN, Zhu P. High GINS2 transcript level predicts poor prognosis and correlates with high histological grade and endo-

crine therapy resistance through mammary cancer stem cells in breast cancer patients. Breast Cancer Res Treat. 2014;148:423–36.
15. Haider S, Wang J, Nagano A, Desai A, Arumugam P, Dumartin L, et al. A multi-gene signature predicts outcome in patients with pancreatic ductal adenocarcinoma. Genome Med. 2014;6:1–11.
16. Shima N, Alcaraz A, Liachko I, Buske TR, Andrews CA, Munroe RJ, Hartford SA, Tye BK, Schimenti JC. A viable allele of Mcm4 causes chromosome instability and mammary adenocarcinomas in mice. Nat Genet. 2007;39:93–8.
17. Dutta A. Chaotic license for genetic instability and cancer. Nat Genet. 2007;39:11–2.
18. Barkley LR, Ihn YS, Zou Y, Vaziri C. Reduced expression of GINS complex members induces hallmarks of pre-malignancy in primary untransformed human cells. Cell Cycle. 2009;8: 1577–88.
19. Simon NE, Schwacha A. The Mcm2-7 replicative helicase: a promising chemotherapeutic target. Biomed Res Int. 2014;2014:549719.
20. Ilves I, Petojevic T, Pesavento JJ, Botchan MR. Activation of the MCM2-7 helicase by association with Cdc45 and GINS proteins. Mol Cell. 2010;37:247–58.
21. Kang Y-H, Galal WC, Farina A, Tappin I, Hurwitz J. Properties of the human Cdc45/Mcm2-7/GINS helicase complex and its action with DNA polymerase epsilon in rolling circle DNA synthesis. Proc Natl Acad Sci U S A. 2012;109:6042–7.
22. Petojevic T, Pesavento JJ, Costa A, Liang J, Wang Z, Berger JM, Botchan MR. Cdc45 (cell division cycle protein 45) guards the gate of the eukaryote replisome helicase stabilizing leading strand engagement. Proc Natl Acad Sci U S A. 2015;112:E249–58.
23. Fu Y, Slaymaker IM, Wang J, Wang G, Chen XS. The 1.8-Å crystal structure of the N-terminal domain of an archaeal MCM as a right-handed filament. J Mol Biol. 2014;426:1512–23.
24. Froelich CA, Kang S, Epling LB, Bell SP, Enemark EJ. A conserved MCM single-stranded DNA binding element is essential for replication initiation. Elife. 2014;3:e01993.
25. Miller JM, Arachea BT, Epling LB, Enemark EJ. Analysis of the crystal structure of an active MCM hexamer. Elife. 2014;3:1–13.
26. Sun J, Fernandez-cid A, Riera A, Tognetti S, Yuan Z, Stillman B, et al. Structural and mechanistic insights into Mcm2 – 7 double-hexamer assembly and function. Genes Dev. 2014;28:2291–303.
27. Costa A, Renault L, Swuec P, Petojevic T, Pesavento J, Ilves I, MacLellan-Gibson K, Fleck RA, Botchan MR, Berger JM. DNA binding polarity, dimerization, and ATPase ring remodeling in the CMG helicase of the eukaryotic replisome. Elife. 2014;3:e03273.
28. Lyubimov AY, Costa A, Bleichert F, Botchan MR, Berger JM. ATP-dependent conformational dynamics underlie the functional asymmetry of the replicative helicase from a minimalist eukaryote. Proc Natl Acad Sci U S A. 2012;109:11999–2004.
29. Chang YP, Wang G, Bermudez VP, Hurwitz J, Chen XS. Crystal structure of the GINS complex and functional insights into its role in DNA replication. Proc Natl Acad Sci U S A. 2007;104:12685–90.
30. Choi JM, Lim HS, Kim JJ, Song O, Cho Y. Crystal structure of the human GINS complex service. Genes Dev. 2007;21:1316–21.
31. Kamada K, Kubota Y, Arata T, Shindo Y, Hanaoka F. Structure of the human GINS complex and its assembly and functional interface in replication initiation. Nat Struct Mol Biol. 2007;14:388–96.
32. Oyama T, Ishino S, Fujino S, Ogino H, Shirai T, Mayanagi K, et al. Architectures of archaeal GINS complexes, essential DNA replication initiation factors. BMC Biol. 2011;9:28.
33. Boskovic J, Coloma J, Aparicio T, Zhou M, Robinson CV, Méndez J, Montoya G. Molecular architecture of the human GINS complex. EMBO Rep. 2007;8:678–84.
34. Marinsek N, Barry ER, Makarova KS, Dionne I, Koonin EV, Bell SD. GINS, a central nexus in the archaeal DNA replication fork. EMBO Rep. 2006;7:539–45.
35. Simon AC, Zhou JC, Perera RL, van Deursen F, Evrin C, Ivanova ME, Kilkenny ML, Renault L, Kjaer S, Matak-Vinković D, Labib K, Costa A, Pellegrini L. A Ctf4 trimer couples the CMG helicase to DNA polymerase α in the eukaryotic replisome. Nature. 2014;510:293–7.

36. Sengupta S, Van Deursen F, De Piccoli G, Labib K. Dpb2 Integrates the leading-strand DNA polymerase into the eukaryotic replisome. Curr Biol. 2013;23:543–52.
37. Pollok S, Bauerschmidt C, Sänger J, Nasheuer HP, Grosse F. Human Cdc45 is a proliferation-associated antigen. FEBS J. 2007;274:3669–84.
38. Pospiech H, Grosse F, Pisani MF. The initiation step of eukaryotic DNA replication. Subcell Biochem. 2010;50:17–42.
39. Li Z, Pan M, Santangelo TJ, Chemnitz W, Yuan W, Edwards JL, Hurwitz J, Reeve JN, Kelman Z. A novel DNA nuclease is stimulated by association with the GINS complex. Nucleic Acids Res. 2011;39:6114–23.
40. Yuan H, Liu XP, Han Z, Allers T, Hou JL, Liu JH. RecJ-like protein from Pyrococcus furiosus has 3′-5′ exonuclease activity on RNA: implications for proofreading of 3′-mismatched RNA primers in DNA replication. Nucleic Acids Res. 2013;41:5817–26.
41. Bruck I, Kaplan DL. Cdc45 protein-single-stranded DNA interaction is important for stalling the helicase during replication stress. J Biol Chem. 2013;288:7550–63.
42. Di Perna R, Aria V, De Falco M, Sannino V, Okorokov AL, Pisani FM, De Felice M. The physical interaction of Mcm10 with Cdc45 modulates their DNA-binding properties. Biochem J. 2013;454:333–43.
43. Tanaka T, Umemori T, Endo S, Muramatsu S, Kanemaki M, Kamimura Y, Obuse C, Araki H. Sld7, an Sld3-associated protein required for efficient chromosomal DNA replication in budding yeast. EMBO J. 2011;30:2019–30.
44. Bruck I, Kaplan DL. GINS and Sld3 compete with one another for Mcm2-7 and Cdc45 binding. J Biol Chem. 2011;286:14157–67.
45. Itou H, Muramatsu S, Shirakihara Y, Araki H. Crystal structure of the homology domain of the eukaryotic DNA replication proteins Sld3/Treslin. Structure. 2014;22:1341–7.
46. Gerhardt J, Guler GD, Fanning E. Human DNA helicase B interacts with the replication initiation protein Cdc45 and facilitates Cdc45 binding onto chromatin. Exp Cell Res. 2015;334:283–93.
47. Kasiviswanathan R, Shin JH, Melamud E, Kelman Z. Biochemical characterization of the Methanothermobacter thermautotrophicus minichromosome maintenance (MCM) helicase N-terminal domains. J Biol Chem. 2004;279:28358–66.
48. Jenkinson ER, Chong JPJ. Minichromosome maintenance helicase activity is controlled by N- and C-terminal motifs and requires the ATPase domain helix-2 insert. Proc Natl Acad Sci U S A. 2006;103:7613–8.
49. Barry ER, McGeoch AT, Kelman Z, Bell SD. Archaeal MCM has separable processivity, substrate choice and helicase domains. Nucleic Acids Res. 2007;35:988–98.
50. Pucci B, De Felice M, Rocco M, Esposito F, De Falco M, Esposito L, et al. Modular organization of the Sulfolobus solfataricus mini-chromosome maintenance protein. J Biol Chem. 2007;282:12574–82.
51. Liu W, Pucci B, Rossi M, Pisani FM, Ladenstein R. Structural analysis of the Sulfolobus solfataricus MCM protein N-terminal domain. Nucleic Acids Res. 2008;36:3235–43.
52. Fletcher RJ, Bishop BE, Leon RP, Sclafani RA, Ogata CM, Chen XS. The structure and function of MCM from archaeal M. Thermoautotrophicum. Nat Struct Biol. 2003;10:160–7.
53. Medagli B, Onesti S. Structure and mechanism of hexameric helicases. Adv Exp Med Biol. 2013;973:75–95.
54. Fletcher RJ, Chen XS. Biochemical activities of the BOB1 mutant in Methanobacterium thermoautotrophicum MCM. Biochemistry. 2006;45:462–7.
55. Hoang ML, Leon RP, Pessoa-Brandao L, Hunt S, Raghuraman MK, Fangman WL, Brewer BJ, Sclafani RA. Structural changes in Mcm5 protein bypass Cdc7-Dbf4 function and reduce replication origin efficiency in Saccharomyces cerevisiae. Mol Cell Biol. 2007;27:7594–602.
56. Costa A, van Duinen G, Medagli B, Chong J, Sakakibara N, Kelman Z, Nair SK, Patwardhan A, Onesti S. Cryo-electron microscopy reveals a novel DNA-binding site on the MCM helicase. EMBO J. 2008;27:2250–8.
57. Jenkinson ER, Costa A, Leech AP, Patwardhan A, Onesti S, Chong JP. Mutations in subdomain B of the minichromosome maintenance (MCM) helicase affect DNA binding and modulate conformational transitions. J Biol Chem. 2009;284:5654–61.

58. Pucci B, De Felice M, Rossi M, Onesti S, Pisani FM. Amino acids of the Sulfolobus solfataricus mini-chromosome maintenance-like DNA helicase involved in DNA binding/remodeling. J Biol Chem. 2004;279:49222–8.
59. Sakakibara N, Kasiviswanathan R, Kelman Z. Different residues on the surface of the Methanothermobacter thermautotrophicus MCM helicase interact with single- and double-stranded DNA. Archaea. 2010;2010:505693.
60. Sakakibara N, Kasiviswanathan R, Melamud E, Han M, Schwarz FP, Kelman Z. Coupling of DNA binding and helicase activity is mediated by a conserved loop in the MCM protein. Nucleic Acids Res. 2008;36:1309–20.
61. Bae B, Chen YH, Costa A, Onesti S, Brunzelle JS, Lin Y, Cann IK, Nair SK. Insights into the architecture of the replicative helicase from the structure of an archaeal MCM homolog. Structure. 2009;17:211–22.
62. Barry ER, Lovett JE, Costa A, Lea SM, Bell SD. Intersubunit allosteric communication mediated by a conserved loop in the MCM helicase. Proc Natl Acad Sci U S A. 2009;106:1051–6.
63. Brewster AS, Wang G, Yu X, Greenleaf WB, Carazo JM, Tjajadia M, Klein MG, Chen XS. Crystal structure of a near-full-length archaeal MCM: functional insights for an AAA+ hexameric helicase. Proc Natl Acad Sci U S A. 2008;105:20191–6.
64. Wei Z, Liu C, Wu X, Xu N, Zhou B, Liang C, Zhu G. Characterization and structure determination of the cdt1 binding domain of human minichromosome maintenance (Mcm) 6. J Biol Chem. 2010;285:12469–73.
65. Costa A, Pape T, van Heel M, Brick P, Patwardhan A, Onesti S. Structural studies of the archaeal MCM complex in different functional states. J Struct Biol. 2006;156:210–9.
66. Costa A, Pape T, van Heel M, Brick P, Patwardhan A, Onesti S. Structural basis of the Methanothermobacter thermautotrophicus MCM helicase activity. Nucleic Acids Res. 2006;34:5829–38.
67. Yu X, VanLoock MS, Poplawski A, Kelman Z, Xiang T, Tye BK, Egelman EH. The Methanobacterium thermoautotrophicum MCM protein can form heptameric rings. EMBO Rep. 2002;3:792–7.
68. Pape T, Meka H, Chen S, Vicentini G, van Heel M, Onesti S. Hexameric ring structure of the full-length archaeal MCM protein complex. EMBO Rep. 2003;4:1079–83.
69. Chen YJ, Yu X, Kasiviswanathan R, Shin JH, Kelman Z, Egelman EH. Structural polymorphism of Methanothermobacter thermautotrophicus MCM. J Mol Biol. 2005;346:389–94.
70. Gómez-Llorente Y, Fletcher RJ, Chen XS, Carazo JM, Martin CS. Polymorphism and double hexamer structure in the archaeal minichromosome maintenance (MCM) helicase from Methanobacterium thermoautotrophicum. J Biol Chem. 2005;280:40909–15.
71. Bochman ML, Schwacha A. The Mcm2-7 complex has in vitro helicase activity. Mol Cell. 2008;31:287–93.
72. Samel SA, Fernández-Cid A, Sun J, Riera A, Tognetti S, Carmen Herrera M, Li H, Speck C. A unique DNA entry gate serves for regulated loading of the eukaryotic replicative helicase MCM2-7 onto DNA. Genes Dev. 2014;28:1653–66.
73. Remus D, Beuron F, Tolun G, Griffith JD, Morris EP, Diffley JFX. Concerted loading of Mcm2-7 double hexamers around DNA during DNA replication origin licensing. Cell. 2009;139: 719–30.
74. Evrin C, Clarke P, Zech J, Lurz R, Sun J, Uhle S, Li H, Stillman B, Speck C. A double-hexameric MCM2-7 complex is loaded onto origin DNA during licensing of eukaryotic DNA replication. Proc Natl Acad Sci U S A. 2009;106:20240–5.
75. McGeoch AT, Trakselis MA, Laskey RA, Bell SD. Organization of the archaeal MCM complex on DNA and implications for the helicase mechanism. Nat Struct Mol Biol. 2005;12: 756–62.
76. Graham BW, Schauer GD, Leuba SH, Trakselis MA. Steric exclusion and wrapping of the excluded DNA strand occurs along discrete external binding paths during MCM helicase unwinding. Nucleic Acids Res. 2011;39:6585–95.
77. Enemark EJ, Joshua-Tor L. Mechanism of DNA translocation in a replicative hexameric helicase. Nature. 2006;442:270–5.

Chapter 21
Structure and Function Studies of Replication Initiation Factors

Jingchuan Sun, Zuanning Yuan, Bruce Stillman, Christian Speck, and Huilin Li

Abstract We have used negative stain EM and cryo-EM to visualize step by step the replication initiation events in *S. cerevisiae*, as the process is driven forward by the interplay of a dozen or so macromolecular initiation factors, leading to the establishment of pre-replication complexes (pre-RC) at each origin of DNA replication. This work took advantage of our ability to reconstitute the Mcm2-7 loading reaction with purified proteins. We determined the architecture of several previously known replication initiation complexes such as ORC, ORC-Cdc6 on DNA, and the Mcm2-7 double hexamer. We also captured by EM reaction intermediates such as the ORC-Cdc6-Cdt1-Mcm2-7 (OCCM) and the ORC-Cdc6-Mcm2-7-Mcm2-7 (OCMM) that had evaded previous biochemical identification. In this chapter, we describe what we have learnt about the structure and interaction with origin DNA of the replication initiators. We further discuss what may be expected in the coming years as cryo-EM is becoming a near-atomic resolution structural tool, thanks to the recent advent of the direct electron detector.

Keywords Origin replication complex (ORC) • Pre-replication complex (pre-RC) • Mcm2-7 • Replicative helicase • Replication initiator • cryo-EM • Structural biology

J. Sun
Biosciences Department, Brookhaven National Laboratory, Upton, NY 11973, USA

Z. Yuan • H. Li (✉)
Biosciences Department, Brookhaven National Laboratory, Upton, NY 11973, USA

Department of Biochemistry and Cell Biology, Stony Brook University, Stony Brook, NY 11794, USA
e-mail: hli@bnl.gov

B. Stillman (✉)
Cold Spring Harbor Laboratory, Cold Spring Harbor, NY 11724, USA
e-mail: stillman@cshl.edu

C. Speck (✉)
DNA Replication Group, Imperial College London, Faculty of Medicine, Institute of Clinical Sciences, Hammersmith Hospital Campus, Du Cane Road, London W12 0NN, UK
e-mail: christian.speck@imperial.ac.uk

D.L. Kaplan (ed.), *The Initiation of DNA Replication in Eukaryotes*,
DOI 10.1007/978-3-319-24696-3_21

Introduction

In eukaryotes, initiation of DNA replication is a key regulation point in cell division cycle. As such an elaborate control system has evolved to specify replication timing during S phase and to ensure that initiation events only occur once at each replication origin [1–3]. The central player in eukaryotic replication initiation is the highly conserved Origin Recognition Complex (ORC) [1, 4]. *S. cerevisiae* ORC constitutively binds to the replication origins in an ATP-dependent manner throughout the cell cycle [5, 6]. ORC becomes activated when another initiation factor Cdc6 is recruited to the origin at the end of mitosis [6–10]. The origin-bound ORC-Cdc6 then functions as a loading machine to recruit the Cdt1-bound Mcm2-7 and then load two copies of the Mcm2-7 hexamer ring onto DNA [10, 11]. Recently, using the ARS1-containing DNA and purified yeast proteins, an in vitro Mcm2-7 loading reaction was developed that recapitulated the molecular events of in vivo origin licensing in the G1 phase [12, 13]. The in vitro system led to the discovery that Mcm2-7 helicase core—although it is hexameric in solution—assembled into a head-to-head double hexamer on the DNA [14]. The double hexamer and ORC, which in yeast remains bound to the DNA, most likely represents the pre-replication complex (pre-RC) that was defined in vivo [15] and each Mcm2-7 hexamer will eventually be transformed into the active Cdc45-Mcm2-7-GINS (CMG) helicase at the G1 to S transition [16–20].

This chapter focuses on the recent EM structural analyses of the protein complexes assembled on the origin DNA during helicase loading [11, 18, 21–25]. We have used negative stain EM as well as cryo-EM when feasible. In negative stain EM, the biological structures are embedded in a thin layer of heavy metal such as uranyl acetate. In cryo-EM, purified protein–nucleic acid complexes are embedded in a thin layer of vitreous ice. Negative staining produces stronger contrast and requires less concentrated samples, but limits the structure to a lower resolution [26]. Cryo-EM requires higher concentration samples and produces less image contrast due to the lack of a staining agent but has the potential to reach higher resolution [27].

Structures of ORC and ORC-Cdc6

EM images showed that the *S. cerevisiae* ORC is a slightly twisted, nascent-like open ring structure that is 16 nm long and 12 nm wide [11, 21, 24] (Fig. 21.1a). Consistent with the conserved nature of eukaryotic ORC subunit proteins, DmORC was found to be of very similar size and shape [25]. In order to delineate the individual proteins in ORC, we implemented a systematic maltose binding protein (MBP) fusion strategy [21]. We fused the 36-kDa MBP to each ORC subunit, one protein and one peptide end (amino or carboxyl) at a time, expressed and then purified the MBP-fused ORC complexes, and tested the origin DNA binding ability for

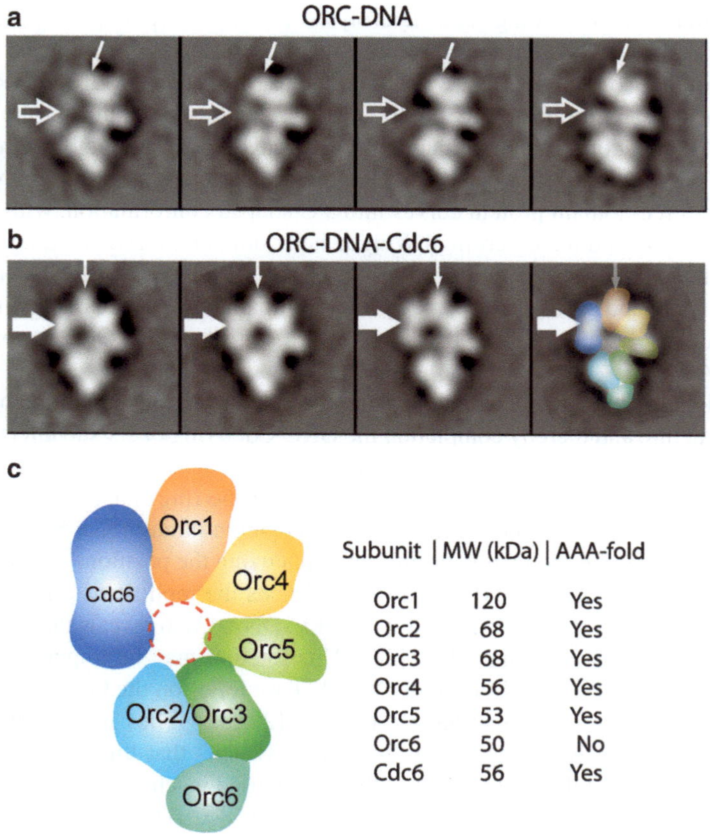

Fig. 21.1 Overall molecular architecture of the *S. cerevisiae* helicase loader ORC-Cdc6. (**a**) 2D class averages of the purified ORC in the presence of the 66 bp ARS1-containing dsDNA. (**b**) Class averages of ORC-Cdc6-DNA. The *thicker horizontal arrows* mark the absence (A) or presence (B) of the Cdc6 density. *Thin arrows* mark the conformational change of Orc1 upon Cdc6 binding. (**c**) A sketch of the subunit arrangement of ORC-Cdc6. Six of the seven proteins contain an ATPase associated with diverse cellular activities (AAA+) fold. The top three subunits, Orc1, Orc5, and Cdc6, have demonstrated ATPase activity

every one of these fusion complexes. Only the complexes whose functions were intact were used for EM structural characterization. 2D image classification and averaging was used to localize the flexibly fused MBP. The smallest Orc6 was mapped by comparing structures of ORC and the Orc1-5 subcomplex. The six subunits Orc1-6 were arranged into two lobes with an upper lobe containing Orc1, Orc4, and Orc5, the ATPase lobe, and a lower lobe containing Orc2, Orc3, and Orc6. The six proteins were arranged sequentially as Orc1:Orc4:Orc5:Orc2/Orc3, with the smallest Orc6 binding to Orc3 [21]. Although Orc1-5 each is predicted to contain an AAA+-like domain, only budding yeast Orc1 and Orc5 have demonstrated ATP-binding and hydrolysis activity [11, 28] (Fig. 21.1a–c). Therefore, the ORC ATPase resides in the top lobe.

On comparison of ORC that was not bound to DNA (apo ORC) and ORC-DNA structures, we found that ORC binding to DNA causes a rotation in the ATPase lobe [24]. This conformation change in ORC may explain why DNA inhibits the ORC ATPase activity [5]. In the crystal structure of an archaeal Orc1-DNA complex, the three-domain protein curves into a C-shaped conformation, with both the AAA+ domain and the C-terminal Winged helix domain (WHD) binding to DNA, much like two claws of a lobster [29, 30]. ORC is comprised of five sequentially stacked C-shaped proteins (Orc1, 4, 5, 2/3). Comparison of the DNA binding mode of the Archaeal Orc1 with the ORC architecture, it is likely that ORC wraps around the DNA.

EM images show that Cdc6 binds from the side, filling the gap in the bio-lobed ORC structure and thereby completing the ORC-Cdc6 ring of six subunits that have structural similarity to AAA+ ATPases (Fig. 21.1b). The newly formed Cdc6-ORC ring structure, but not ORC itself, is the loader of the ring-shaped MCM2-7 hexamer, suggesting that in the absence of the replication initiator Cdc6, ORC is dormant, binding to the replication origins and organizing the positions of surrounding nucleosomes [31]. Cdc6 binding activates ORC as the replicative helicase loader. Further, a comparison of ORC-DNA with ORC-DNA-Cdc6 reveals that upon Cdc6 binding, Orc1 rotates against Orc4 [24]. This rotation may be important because biochemical studies have shown that ORC ATPase relies on the Orc4 Arginine finger that probes the ATP binding site in Orc1 [28, 32]. Thus it is likely that the observed Orc1 rotation is part of a "molecular switch" that flips on the helicase-loading activity of ORC. Consistent with the proposed mechanism of Orc1 rotation [24], a recent crystal structure of *D. melanogaster* ORC (DmORC) in the absence of DNA revealed that Orc1 AAA+ domain was rotated >90° out-of-plane, disrupting the interactions with the catalytic residues in Orc4 [33].

ORC-Cdc6 Loads the First Cdt1-Mcm2-7 onto DNA: The OCCM Complex

The initiation factor Cdt1 was previously shown to interact with Orc6 [34–36]. Following the ORC-Cdc6 EM structure, we wondered if we could capture by EM a ternary complex of ORC-Cdc6-Cdt1. Despite the fact that Cdt1 can be pulled down by ORC-Cdc6, we were unable to observe Cdt1 in the ORC-Cdc6 and Cdt1 mixture solution. Cdt1 is also known to bind the Mcm2-7 hexamer directly [37], however, we also failed to observe a well-defined Cdt1 density in the Cdt1-Mcm2-7 preparation. Perhaps Cdt1 binds Mcm2-7 only weakly in the absence of the loader ORC-Cdc6.

We then asked if we could capture more complex intermediates in the assembly of pre-RCs on DNA, i.e., whether we could visualize ORC-Cdc6 in the process of

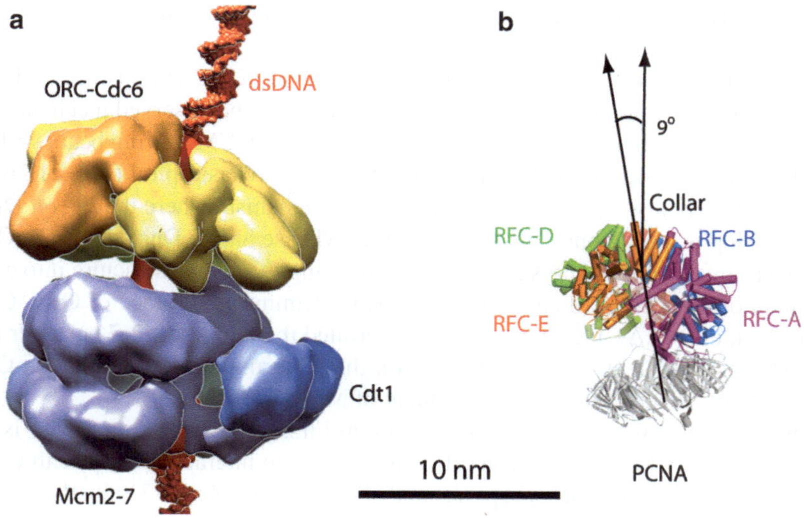

Fig. 21.2 The loading of Cdt1-Mcm2-7 onto dsDNA by ORC-Cdc6 ATPase spiral resembles the loading mechanism of PCNA ring by the RF-C ATPase spiral. (**a**) Cryo-EM 3D map of the OCCM complex. ORC is in *yellow*, Cdc6 *orange*, Mcm2-7 *purple*, Cdt1 *blue*, and DNA in *red*. ORC-Cdc6 forms a spiral structure with helical pitch that matches the dsDNA. Atomic structures of short DNA segments are modeled on the top and bottom of the structure. The first Mcm2-7 hexamer is already loaded on the dsDNA. (**b**) Crystal structure of the yeast RFC-PCNA complex (PDB 1SXJ). There is a 9° mismatch between the central axis of the PCNA ring and the spiral axis of the RFC-A-RFC-E pentamer and the spiral arrangement of the AAA+ motor domains in RFC

loading the Cdt1-bound Mcm2-7 hexamer on the ARS1-containing DNA. Because the helicase loading reaction is a dynamic process, we needed to slow down the loading reaction by using the poorly hydrolyzable ATPγS [22]. The reaction intermediate we captured by EM is a three-tiered structure, with a slanted top tier and two nearly parallel tiers below (Fig. 21.2a). The top tier has a dome-like shape, is the 473-kDa ORC-Cdc6 complex. The lower two tiers are Mcm2-7 hexamer with a total mass of 603 kDa. A distinct density at the lower left side of the structure is the 66-kDa Cdt1. Through an MBP insertion strategy, we confirmed the molecular identities in the complex, and found that the face of the Mcm2-7 hexamer that contains the C-termini of each Mcm subunit (the CT face) engages ORC-Cdc6, leaving the face containing the N-termini of the Mcm subunits (the NT face) exposed and available for binding the next Mcm2-7 hexamer. This orientation is conceptually consistent with the final loading product of a head-to-head double hexamer of Mcm2-7 [12, 13]. Therefore, cryo-EM caught ORC-Cdc6 in the act of loading the first MCM2-7 hexamer. The OCCM loading intermediate consisted of a 1.1-MDa 14-protein complex comprised of ORC-Cdc6-Cdt1-Mcm2-7 formed in the presence of ATP-γS. There is a long rod-like density passing through the middle of Mcm2-7 ring. This rod passes nearly continuously from the top center of ORC-Cdc6 through the central channel of Mcm2-7 ring, and is shown to be dsDNA [22, 38]. This means that the first Mcm2-7 hexamer had already been loaded onto DNA in the absence of

ATP hydrolysis. Therefore, OCCM represents a late stage loading intermediate of the first Cdt1-Mcm2-7.

Interestingly, upon recruiting the Mcm2-7 hexamer, ORC is transformed into a spiral with a 34-Å pitch that matches the dsDNA. Thus, the observed mechanism in which the ORC-Cdc6 ATPase spiral on DNA loads the MCM2-7 ring bears striking similarity to the loading of PCNA ring by the RF-C ATPase spiral (Fig. 21.2b). A recent EM work also revealed a similar loading mechanism where the *E. coli* DnaB helicase ring is loaded onto DNA by the spiral ATPase loader DnaC [39]. DmORC alone was quasi-spiral in crystal lattice, and docking the crystal structure into cryo-EM map of the ScOCCM suggested that the C-terminal WHD face of ORC-Cdc6 rather than the AAA+ face contacted and recruited the first MCM2-7 hexamer [22, 33]. This is consistent with the observation that C-terminal MBP fusions to Orc1 and Orc4 specifically block the recruitment of MCM2-7 [22].

Two questions require further investigation. First, the exact role of Cdt1 is not clear. In the OCCM structure, Cdt1 does not appear to interact directly with ORC-Cdc6. The elongated Cdt1 wraps around the waist of Mcm2-7, mostly outside Mcm2, with one end interacts with Mcm6, the other end touches Mcm5. Interestingly, Cdt1 was shown to overcome the Mcm6 C-terminal autoinhibitory domain, leading to Orc1/Cdc6 ATP hydrolysis, release of Cdt1 from OCCM, and formation of OCM [40]. Therefore, instead of functioning as a recruiter that bridges ORC-Cdc6 and Mcm2-7, Cdt1 may facilitate DNA gate opening between Mcm2 and Mcm5. It is also possible that Cdt1 plays a structural role to stabilize the Mcm2-7 hexamer. Second, because OCCM is a late stage intermediate of loading reaction for the first Mcm2-7 hexamer, we do not yet know at a structural level how the initial encounter complex looks like, although it is known that a conserved C-terminal domain of Mcm3 first approaches and stimulates the ATPase activity of ORC-Cdc6 on DNA [41].

The OCM Complex Loads the Second Cdt1-Mcm2-7: The OCMM Complex

The OCCM described above captures loading of the first Mcm2-7 hexamer by using ATPγS to prevent ATP hydrolysis (Fig. 21.3a); ATP hydrolysis is required for loading and assembly of the Mcm2-7 double hexamer [12, 13, 32, 42, 43]. We therefore examined complexes by EM in the presence of ATP hydrolysis, although these complexes had to be isolated very early in the Mcm2-7 loading reaction. In the presence of 1 mM ATP, many particles of different sizes and shapes were observed. 2D classification and averaging showed three complexes: the smallest complex was similar to OCCM since it also had a three-tiered structure, but had lost the density belonging to Cdt1. This was the ORC-Cdc6-Mcm2-7 complex (OCM) [40] (Fig. 21.3b). The largest complex was a five-tiered ORC-Cdc6-Mcm2-7-Mcm2-7 complex (OCMM) [23] (Fig. 21.3c). This complex could be interpreted either as ORC-Cdc6

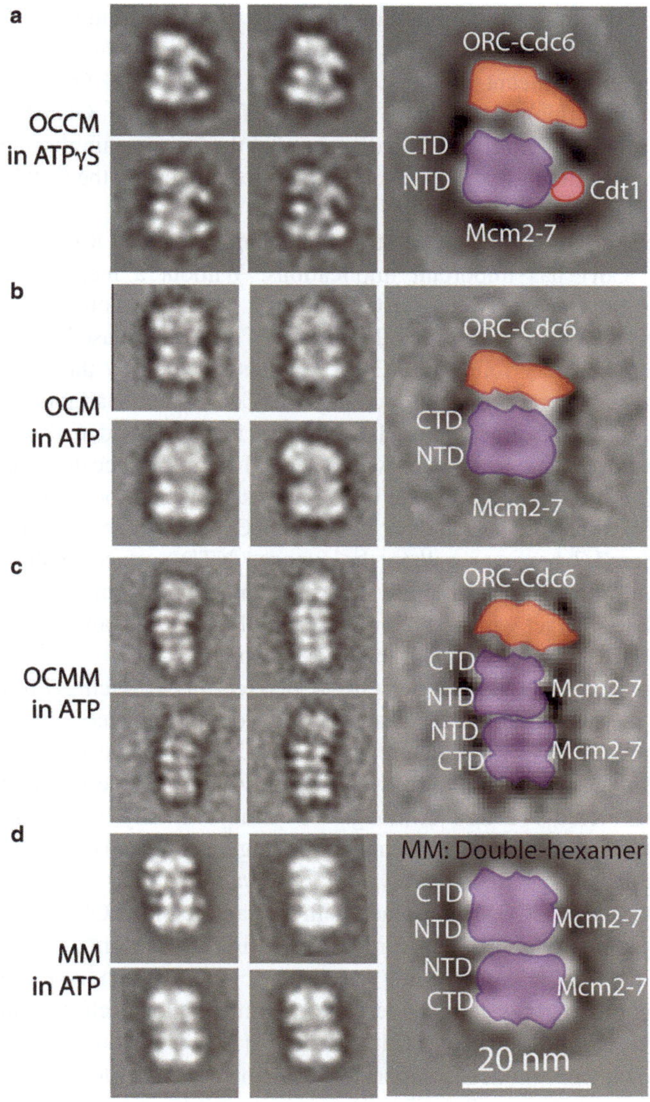

Fig. 21.3 Loading intermediates captured by EM in the presence of ATPγS or ATP. (**a**) Cryo-EM 2D class averages of OCCM in 1 mM ATPγS. (**b–d**) Four class averages of three loading intermediates observed by negative stain EM in 1 mM ATP: OCM (**b**), OCMM (**c**), and MM the Double hexamer (**d**). Note that panel (**c**) for OCMM is on a smaller scale

binding on the assembled double hexamer, or an OCM recruiting a second Mcm2-7 hexamer. The third complex was a four-tiered structure with an apparent twofold symmetry that is distinctively the Mcm2-7 double hexamer (Fig. 21.3d). In a time course study, we found in the 2-min reaction sample the OCM particles dominated (96 %), with a small percentage of OCMM particles (4 %) and without observable

DH. In the 30-min sample the double hexamer particles dominated (99 %), with occasional OCMM particles (<1 %). In the 7-min sample, all three types of particles were present in significant numbers: ~15 % were OCM, ~10 % OCMM, and ~75 % fully assembled double hexamers [23]. The fact that the OCMM population increased from 2 to 7 min, and then decreased in the 30 min sample suggested to us that OCMM was an on-pathway reaction intermediate during the ORC-Cdc6 mediated double hexamer assembly.

Our observation that the OCM structure is similar to the OCCM except for the absence of Cdt1 has important implications. Without a major conformational change, the interface between ORC-Cdc6 and Mcm2-7 should be the same in OCCM and OCM. This means that the surface of ORC-Cdc6 used for recruiting the first Mcm2-7 hexamer found in OCCM remains shielded by the first hexamer in OCM, therefore the recruiting surface of ORC-Cdc6 is not available for binding to the next Mcm2-7 hexamer. It is also unlikely that ORC-Cdc6 could utilize the opposing exposed face to interact with a second Mcm2-7, because we have never observed in the EM images of one ORC-Cdc6 sandwiched between two Mcm2-7 hexamers. We therefore suggest that it is not ORC-Cdc6 per se that directly recruits the next Mcm2-7 hexamer, rather, the exposed NTD surface of the first loaded Mcm2-7 within the OCM structure may have undergone conformational changes that allow it to interact with and recruit a second Mcm2-7 hexamer. This possibility is supported by our observation of the OCMM complex.

It was generally thought that two ORC-Cdc6 recruit two Mcm2-7 hexamers onto DNA, then the two hexamers associate to form a double hexamer [44]. Observation of a six-tiered OCM-OCM structure would have been indicative of this possibility. However, among tens of thousands of assembly intermediate particles, we never found the six-tiered structure. Although we cannot formally rule out the existence of the intermediate as the system is under ongoing reaction and one of the ORC may be less stable and get lost rapidly, our observation of the structural similarity between OCM with OCCM, combined with the capture of the OCMM structure, led us to propose that one ORC-Cdc6 recruits the first Cdt1-Mcm2-7, and upon ATP hydrolysis and Cdt1 release, the first recruited Mcm2-7, still bound to ORC-Cdc6 in the OCM complex, may have undergone a conformational change and becomes competent to recruit the second Cdt1-Mcm2-7 [23]. This "one ORC loading two Mcm2-7 hexamers" model is supported by two recent single molecule analyses [45]. In a single molecule fluorescence assay, it was demonstrated that in 80 % of the time, only a single ORC was present on DNA when the second Mcm2-7 hexamer was loaded [46]. In a DNA curtain assay, it was shown that Mcm2-7 double hexamer still assembled on DNA even when free ORC in solution had been washed away [47]. This observation rules out any pre-RC formation model that requires a second ORC.

The Architecture of the Mcm2-7 Double Hexamer

In eukaryotic cells, helicase loading at origins of DNA replication is temporarily separate from activation of the helicase [48]. Loading occurs in G1 phase, and after loading, the Mcm2-7 double hexamer has to be stable on chromatin for a long time before the cell is ready to enter the S (DNA synthesis) phase when the helicase is activated [49]. Mcm2-7 is not just passively loaded by ORC-Cdc6; it participates in its own loading by contributing the ATPase activity [43, 50]. There is a wealth of structural information on archaeal Mcm proteins [51]. An important difference between the Archaeal and eukaryotic Mcm proteins is the additional regulatory domains in the eukaryotic Mcm N-termini. Eukaryotic Mcm2, Mcm4, and Mcm6 possess N-terminal serine and threonine rich domains (NSD) and a proximal Dbf4-Cdc7 kinase-docking domain (DDD, in Mcm2 and Mcm4) [52–55].

Two earlier studies have revealed that the Mcm2-7 hexamer is loaded on DNA as a head-to-head double hexamer [12, 13], but exactly which subunit interacts with which in the double hexamer was unknown. The six Mcm protein subunits in the hexamer are similar to each other. In order to distinguish the different subunits in negative stain EM map, we needed to make MBP fusion proteins more rigid so we could localize the MBP in 3D map. We inserted MBP into the N-terminal region or C-terminal domains of Mcm proteins, rather than by simply appending MBP to N-terminus or C-terminus of Mcm proteins. Design of the MBP insertion site was facilitated by several available archaeal Mcm protein crystal structures such that the yeast Mcm sequences can be aligned and inter-domain region identified. In this manner, MBP insertion in six locations in five of the Mcm proteins was achieved and each of these modified proteins was able to form double hexamer structure on dsDNA. We then determined 3D maps for each of the six purified double hexamers, and resolved the MBP density in 3D maps for all six MBP-inserted double hexamers [23]. On the basis of these MBP locations, and the previously determined subunit arrangement order of Mcm5-3-7-4-6-2 [18, 19, 22, 56, 57], we were able to unambiguously identity all six Mcm proteins, and more importantly, specify their respective NTD and CTD positions in the double hexamer. Two Mcm hexamers were linked by a horizontal twofold axis that ran approximately from the Mcm2 NTD in the front view to Mcm3 NTD in the back view (Fig. 21.4).

The architecture of Mcm2-7 double hexamer provided three important biological insights. (1) DDK was known to bind both Mcm2 and Mcm4 NTDs in vivo [53, 55]. This was a little mysterious because these two domains are far apart in the Mcm2-7 hexamer. They come together in our double hexamer structure, explaining why DDK acts only on the double but not the single hexamer. In vitro phosphorylation assay with purified DDK, single hexamer and double hexamer further confirmed this observation [23]. (2) Mcm2/5 is the known gate for DNA loading [38, 58]. Our double hexamer structure revealed that the two hexamers are staggered, forming an inter-locked structure, particularly at Mcm2/5 region that was blocked by Mcm4 NTD. So the architecture potentially explains why the double hexamer could be so stable on DNA, and why the two hexamers have to be loaded one at a time. (3)

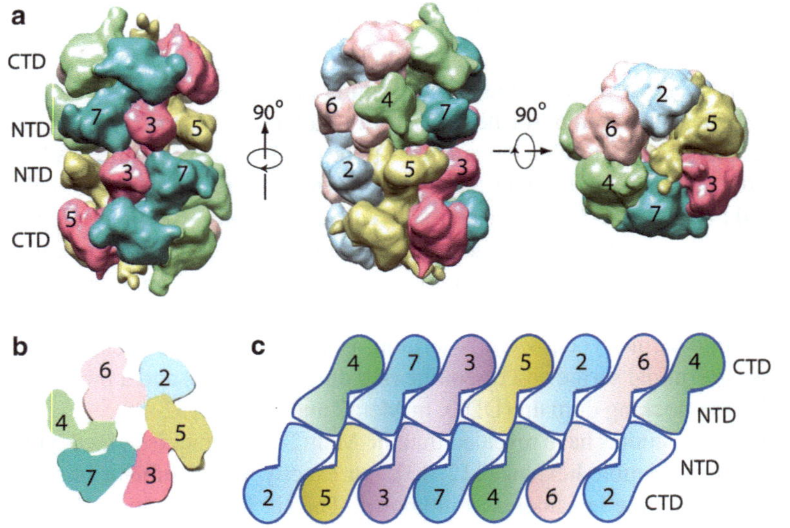

Fig. 21.4 The architecture of the Mcm2-7 double hexamer. (**a**) Front (*left*), side (*middle*), and top (*right*) views of the surfaced-rendered and segmented 3D map of the double hexamer. (**b**) A horizontal section of the double hexamer through the NTD ring of the top Mcm2-7 hexamer, showing the central channel for the dsDNA. (**c**) Illustration of Mcm subunit arrangement as the cylindrical double hexamer is cut-open and flattened. The number refers to Mcm subunit identity; for example, 2 refers to Mcm2

Unique to the double hexamer, Mcm proteins are tilted towards the right by nearly 30°. It is known that ATP is bound at the interface between the Mcm subunits and that ATPase activity relies on accurate positioning of the catalytic residues in the Walker A and B motifs and the arginine finger of the adjacent AAA+ subunit [51, 56, 59]. The large tilt would uncouple the ATP-hydrolysis motifs and inhibit the ATPase activity. Indeed, the purified Mcm2-7 double hexamer had virtually no measurable the ATP hydrolysis activity [23].

Summary and Future Perspectives

Combining an extensive literature on genetic and biochemical studies of the eukaryotic DNA replication initiation [3] and the EM structural characterizations of assembly intermediates [11, 21–25], and recent single molecule analysis of loading events [46], we propose the following series of events that likely occur during the pre-RC assembly in vivo [23] (Fig. 21.5). Step 1 refers to the fact that ScORC resides on origin DNA throughout the cell cycle, perhaps in a dormant state. The initiation event commences at step 2 when Cdc6 binds ORC on origin DNA. Cdc6 may be considered as the activator of ORC. However, because Cdc6 forms an integral part of the loading platform it is be more accurate to regard the ORC-Cdc6 binary complex as the true helicase loader, and ORC is an incomplete loader. Formation of

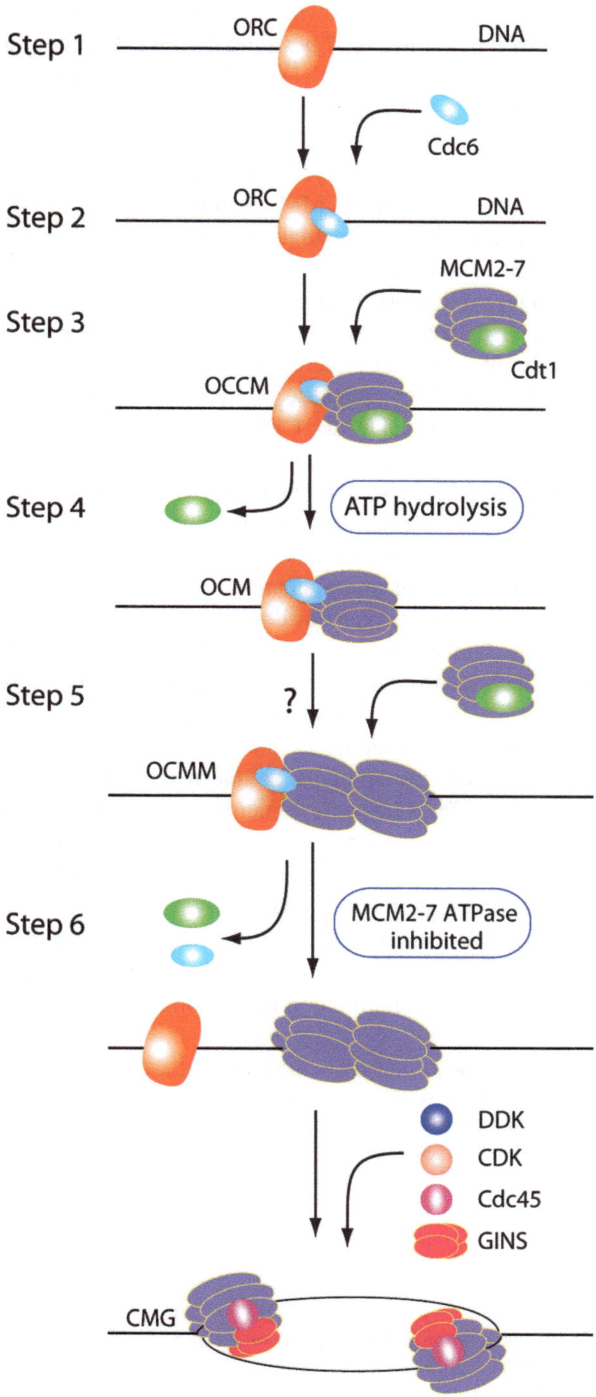

Fig. 21.5 Key molecular events likely occur during pre-RC assembly in vivo. Each of the illustrated intermediates has been captured and visualized by EM. See text for details

OCCM in the presence of ATP as shown in step 3 represents the recruitment and loading onto DNA of the first Mcm2-7 hexamer by ORC-Cdc6. A subsequent ATP hydrolysis event in step 4 leads to the release of Cdt1 and formation of OCM, which we consider the loader of the second Cdt1-Mcm2-7. In step 5, the OCMM captures the OCM in the middle of recruiting and loading of the second Mcm2-7. This step likely requires ATP binding and hydrolysis. Cdt1 has been released from the second Mcm2-7, and the two hexamers in OCMM resembles the fully loaded double hexamer. So OCMM is a late stage intermediate in recruitment of the second MCM2-7, and the hexamer appears to have already been loaded on DNA. The question mark in step 5 indicates a possible short-lived intermediate preceding OCMM that has yet to be captured and visualized. In step 6, ATP hydrolysis by ORC-Cdc6 leads to maturation of the double hexamer structure on DNA and its separation from ORC that remains bound to the DNA. Because the double hexamer is inactive, further modifications by the actions of DDK and CDK and binding of Cdc45 and GINS will eventually lead to the separation of the double hexamer and formation of two CMG complexes—two active helicases each encircling ssDNA. Because the Mcm proteins within the double hexamer is tilted, we hypothesized that an untwisting event accompany the DDK/CDK modifications may cause the initial dsDNA melting and extrusion of one strand from the central channel.

The work described in this chapter was done at low resolution as the focus of our research had been on identifying reaction intermediates and establishing their basic architectures. It is clear that even at very low resolution and with negative stain EM, important biological insights can be obtained. But it is also true that a deeper and detailed chemical level understanding of the molecular functions requires much higher resolution structures, preferably at atomic resolution. For many years such high-resolution information were only obtainable from protein crystallography, which requires relatively stable structures and the ability to produce large amount of samples for crystallization. The direct electron detector has enabled near-atomic resolution structural determination by cryo-EM, and is changing the landscape of structural biology [60–62]. We anticipate that cryo-EM coupled with the direct detector will help to elucidate some of the complexes discussed here at higher resolution and perhaps identify more replication initiation intermediates in the coming years.

Acknowledgements Many people in the labs of Huilin Li, Christian Speck, and Bruce Stillman have helped in this work. The work was supported by National Institutes of Health grant numbers GM45436 (to B.S.) and GM74985 (to H.L.) and the United Kingdom Biotechnology and Biological Sciences Research Council (to C.S.).

References

1. Stillman B. Origin recognition and the chromosome cycle. FEBS Lett. 2005;579(4):877–84.
2. Bell SP, Dutta A. DNA replication in eukaryotic cells. Annu Rev Biochem. 2002;71:333–74.
3. O'Donnell M, Langston L, Stillman B. Principles and concepts of DNA replication in bacteria, archaea, and eukarya. Cold Spring Harb Perspect Biol. 2013;5(7):a010108.

4. Li H, Stillman B. The origin recognition complex: a biochemical and structural view. Subcell Biochem. 2012;62:37–58.
5. Bell SP, Stillman B. ATP-dependent recognition of eukaryotic origins of DNA replication by a multiprotein complex. Nature. 1992;357(6374):128–34.
6. Liang C, Stillman B. Persistent initiation of DNA replication and chromatin-bound MCM proteins during the cell cycle in cdc6 mutants. Genes Dev. 1997;11(24):3375–86.
7. Tanaka T, Knapp D, Nasmyth K. Loading of an Mcm protein onto DNA replication origins is regulated by Cdc6p and CDKs. Cell. 1997;90(4):649–60.
8. Weinreich M, Liang C, Stillman B. The Cdc6p nucleotide-binding motif is required for loading mcm proteins onto chromatin. Proc Natl Acad Sci U S A. 1999;96(2):441–6.
9. Perkins G, Diffley JF. Nucleotide-dependent prereplicative complex assembly by Cdc6p, a homolog of eukaryotic and prokaryotic clamp-loaders. Mol Cell. 1998;2(1):23–32.
10. Speck C, Stillman B. Cdc6 ATPase activity regulates ORC x Cdc6 stability and the selection of specific DNA sequences as origins of DNA replication. J Biol Chem. 2007;282(16): 11705–14.
11. Speck C, Chen Z, Li H, Stillman B. ATPase-dependent cooperative binding of ORC and Cdc6 to origin DNA. Nat Struct Mol Biol. 2005;12(11):965–71.
12. Remus D, Beuron F, Tolun G, Griffith JD, Morris EP, Diffley JF. Concerted loading of Mcm2-7 double hexamers around DNA during DNA replication origin licensing. Cell. 2009;139(4):719–30.
13. Evrin C, Clarke P, Zech J, Lurz R, Sun J, Uhle S, et al. A double-hexameric MCM2-7 complex is loaded onto origin DNA during licensing of eukaryotic DNA replication. Proc Natl Acad Sci U S A. 2009;106(48):20240–5.
14. Botchan M, Berger J. DNA replication: making two forks from one prereplication complex. Mol Cell. 2010;40(6):860–1.
15. Diffley JF, Cocker JH, Dowell SJ, Rowley A. Two steps in the assembly of complexes at yeast replication origins in vivo. Cell. 1994;78(2):303–16.
16. Ilves I, Petojevic T, Pesavento JJ, Botchan MR. Activation of the MCM2-7 helicase by association with Cdc45 and GINS proteins. Mol Cell. 2010;37(2):247–58.
17. Moyer SE, Lewis PW, Botchan MR. Isolation of the Cdc45/Mcm2-7/GINS (CMG) complex, a candidate for the eukaryotic DNA replication fork helicase. Proc Natl Acad Sci U S A. 2006;103(27):10236–41.
18. Costa A, Ilves I, Tamberg N, Petojevic T, Nogales E, Botchan MR, et al. The structural basis for MCM2-7 helicase activation by GINS and Cdc45. Nat Struct Mol Biol. 2011;18(4):471–7.
19. Costa A, Renault L, Swuec P, Petojevic T, Pesavento J, Ilves I, et al. DNA binding polarity, dimerization, and ATPase ring remodeling in the CMG helicase of the eukaryotic replisome. Elife. 2014;3:e03273.
20. Petojevic T, Pesavento JJ, Costa A, Liang J, Wang Z, Berger JM, et al. Cdc45 (cell division cycle protein 45) guards the gate of the eukaryote replisome helicase stabilizing leading strand engagement. Proc Natl Acad Sci U S A. 2015;112(3):E249–58.
21. Chen Z, Speck C, Wendel P, Tang C, Stillman B, Li H. The architecture of the DNA replication origin recognition complex in Saccharomyces cerevisiae. Proc Natl Acad Sci U S A. 2008;105(30):10326–31.
22. Sun J, Evrin C, Samel SA, Fernandez-Cid A, Riera A, Kawakami H, et al. Cryo-EM structure of a helicase loading intermediate containing ORC-Cdc6-Cdt1-MCM2-7 bound to DNA. Nat Struct Mol Biol. 2013;20(8):944–51.
23. Sun J, Fernandez-Cid A, Riera A, Tognetti S, Yuan Z, Stillman B, et al. Structural and mechanistic insights into Mcm2-7 double-hexamer assembly and function. Genes Dev. 2014;28(20):2291–303.
24. Sun J, Kawakami H, Zech J, Speck C, Stillman B, Li H. Cdc6-induced conformational changes in ORC bound to origin DNA revealed by cryo-electron microscopy. Structure. 2012; 20(3):534–44.

25. Clarey MG, Erzberger JP, Grob P, Leschziner AE, Berger JM, Nogales E, et al. Nucleotide-dependent conformational changes in the DnaA-like core of the origin recognition complex. Nat Struct Mol Biol. 2006;13(8):684–90.
26. Ohi M, Li Y, Cheng Y, Walz T. Negative staining and image classification - powerful tools in modern electron microscopy. Biol Proced Online. 2004;6:23–34.
27. Baker ML, Zhang J, Ludtke SJ, Chiu W. Cryo-EM of macromolecular assemblies at near-atomic resolution. Nat Protoc. 2010;5(10):1697–708.
28. Klemm RD, Austin RJ, Bell SP. Coordinate binding of ATP and origin DNA regulates the ATPase activity of the origin recognition complex. Cell. 1997;88(4):493–502.
29. Dueber EL, Corn JE, Bell SD, Berger JM. Replication origin recognition and deformation by a heterodimeric archaeal Orc1 complex. Science. 2007;317(5842):1210–3.
30. Gaudier M, Schuwirth BS, Westcott SL, Wigley DB. Structural basis of DNA replication origin recognition by an ORC protein. Science. 2007;317(5842):1213–6.
31. Lipford JR, Bell SP. Nucleosomes positioned by ORC facilitate the initiation of DNA replication. Mol Cell. 2001;7(1):21–30.
32. Bowers JL, Randell JC, Chen S, Bell SP. ATP hydrolysis by ORC catalyzes reiterative Mcm2-7 assembly at a defined origin of replication. Mol Cell. 2004;16(6):967–78.
33. Bleichert F, Botchan MR, Berger JM. Crystal structure of the eukaryotic origin recognition complex. Nature. 2015;519:321–6.
34. Chen S, de Vries MA, Bell SP. Orc6 is required for dynamic recruitment of Cdt1 during repeated Mcm2-7 loading. Genes Dev. 2007;21(22):2897–907.
35. Takara TJ, Bell SP. Multiple Cdt1 molecules act at each origin to load replication-competent Mcm2-7 helicases. EMBO J. 2011;30(24):4885–96.
36. Liu C, Wu R, Zhou B, Wang J, Wei Z, Tye BK, et al. Structural insights into the Cdt1-mediated MCM2-7 chromatin loading. Nucleic Acids Res. 2012;40(7):3208–17.
37. You Z, Masai H. Cdt1 forms a complex with the minichromosome maintenance protein (MCM) and activates its helicase activity. J Biol Chem. 2008;283(36):24469–77.
38. Samel SA, Fernandez-Cid A, Sun J, Riera A, Tognetti S, Herrera MC, et al. A unique DNA entry gate serves for regulated loading of the eukaryotic replicative helicase MCM2-7 onto DNA. Genes Dev. 2014;28(15):1653–66.
39. Arias-Palomo E, O'Shea VL, Hood IV, Berger JM. The bacterial DnaC helicase loader is a DnaB ring breaker. Cell. 2013;153(2):438–48.
40. Fernandez-Cid A, Riera A, Tognetti S, Herrera MC, Samel S, Evrin C, et al. An ORC/Cdc6/MCM2-7 complex is formed in a multistep reaction to serve as a platform for MCM double-hexamer assembly. Mol Cell. 2013;50(4):577–88.
41. Frigola J, Remus D, Mehanna A, Diffley JF. ATPase-dependent quality control of DNA replication origin licensing. Nature. 2013;495(7441):339–43.
42. Randell JC, Bowers JL, Rodriguez HK, Bell SP. Sequential ATP hydrolysis by Cdc6 and ORC directs loading of the Mcm2-7 helicase. Mol Cell. 2006;21(1):29–39.
43. Kang S, Warner MD, Bell SP. Multiple functions for Mcm2-7 ATPase motifs during replication initiation. Mol Cell. 2014;55:655–65.
44. Yardimci H, Walter JC. Prereplication-complex formation: a molecular double take? Nat Struct Mol Biol. 2014;21(1):20–5.
45. Chistol G, Walter JC. Single-molecule visualization of MCM2-7 DNA loading: seeing is believing. Cell. 2015;161(3):429–30.
46. Ticau S, Friedman LJ, Ivica NA, Gelles J, Bell SP. Single-molecule studies of origin licensing reveal mechanisms ensuring bidirectional helicase loading. Cell. 2015;161(3):513–25.
47. Duzdevich D, Warner MD, Ticau S, Ivica NA, Bell SP, Greene EC. The dynamics of eukaryotic replication initiation: origin specificity, licensing, and firing at the single-molecule level. Mol Cell. 2015;58:483–94.
48. Remus D, Diffley JF. Eukaryotic DNA replication control: lock and load, then fire. Curr Opin Cell Biol. 2009;21(6):771–7.

49. Yardimci H, Loveland AB, Habuchi S, van Oijen AM, Walter JC. Uncoupling of sister replisomes during eukaryotic DNA replication. Mol Cell. 2010;40(5):834–40.
50. Coster G, Frigola J, Beuron F, Morris EP, Diffley JF. Origin licensing requires ATP binding and hydrolysis by the MCM replicative helicase. Mol Cell. 2014;55:666–77.
51. Slaymaker IM, Chen XS. MCM structure and mechanics: what we have learned from archaeal MCM. Subcell Biochem. 2012;62:89–111.
52. Sheu YJ, Stillman B. Cdc7-Dbf4 phosphorylates MCM proteins via a docking site-mediated mechanism to promote S phase progression. Mol Cell. 2006;24(1):101–13.
53. Sheu YJ, Stillman B. The Dbf4-Cdc7 kinase promotes S phase by alleviating an inhibitory activity in Mcm4. Nature. 2010;463(7277):113–7.
54. Sheu YJ, Kinney JB, Lengronne A, Pasero P, Stillman B. Domain within the helicase subunit Mcm4 integrates multiple kinase signals to control DNA replication initiation and fork progression. Proc Natl Acad Sci U S A. 2014;111(18):E1899–908.
55. Ramer MD, Suman ES, Richter H, Stanger K, Spranger M, Bieberstein N, et al. Dbf4 and Cdc7 proteins promote DNA replication through interactions with distinct Mcm2-7 protein subunits. J Biol Chem. 2013;288(21):14926–35.
56. Davey MJ, Indiani C, O'Donnell M. Reconstitution of the Mcm2-7p heterohexamer, subunit arrangement, and ATP site architecture. J Biol Chem. 2003;278(7):4491–9.
57. Bochman ML, Bell SP, Schwacha A. Subunit organization of Mcm2-7 and the unequal role of active sites in ATP hydrolysis and viability. Mol Cell Biol. 2008;28(19):5865–73.
58. Bochman ML, Schwacha A. The Saccharomyces cerevisiae Mcm6/2 and Mcm5/3 ATPase active sites contribute to the function of the putative Mcm2-7 'gate'. Nucleic Acids Res. 2010;38(18):6078–88.
59. Bochman ML, Schwacha A. The Mcm complex: unwinding the mechanism of a replicative helicase. Microbiol Mol Biol Rev. 2009;73(4):652–83.
60. Henderson R. Structural biology: ion channel seen by electron microscopy. Nature. 2013;504(7478):93–4.
61. Kuhlbrandt W. Biochemistry. The resolution revolution. Science. 2014;343(6178):1443–4.
62. Li X, Mooney P, Zheng S, Booth CR, Braunfeld MB, Gubbens S, et al. Electron counting and beam-induced motion correction enable near-atomic-resolution single-particle cryo-EM. Nat Methods. 2013;10(6):584–90.

Chapter 22
Regulation of the Initiation of DNA Replication upon DNA Damage in Eukaryotes

Kerstin Köhler*, Pedro Ferreira*, Boris Pfander, and Dominik Boos

Abstract Cycling cells must ensure homeostasis of the genetic information during repeated chromosome replication-segregation cycles. To guarantee genome stability in normal and DNA damage conditions the initiation of DNA replication in eukaryotes is regulated by the cell cycle machinery and the intra S-phase checkpoint (ISC). The cell cycle kinases CDK and DDK induce initiation specifically in S phase, and the ISC inhibits both kinase pathways, suppressing initiation upon DNA damage and replisome stalling to prevent the replication machinery from having to copy damaged DNA templates. Despite this ISC-mediated inhibition, dormant origins are allowed to fire in genomic regions that are actively engaged in replication when the DNA damage occurs. Forks from dormant origins can rescue replisomes that have stalled at DNA lesions, helping to ensure that no part of these replicating regions is left unreplicated in DNA damage conditions. This replisome rescue also helps prevent stalled and collapsed forks from causing genome rearrangements. In higher eukaryotes, these principles of regulating initiation upon DNA damage must be implemented into a particularly complex temporal regulation programme of genome replication. Molecular details of how the ISC, which poses an important barrier against tumour formation, achieves the regulation of initiation upon DNA damage is only beginning to emerge.

Keywords Intra S-phase checkpoint (ISC) • Initiation of DNA replication • DNA damage • Replisome stalling • Radio-resistant DNA synthesis (RDS) • CDK • DDK • ATR • CHK1 • Sld3 • Treslin/TICRR • Sld2 • Dpb11 • TopBP1

---

*Both are contributed equally

K. Köhler • P. Ferreira • D. Boos (✉)
Molecular Genetics II, Vertebrate DNA Replication, Centre for Medical Biotechnology (ZMB), University of Duisburg-Essen, Essen 45117, Germany
e-mail: dominik.boos@uni-due.de

B. Pfander
DNA Replication and Genome Integrity, Max-Planck Institute of Biochemistry, Martinsried 82152, Germany

D.L. Kaplan (ed.), *The Initiation of DNA Replication in Eukaryotes*,
DOI 10.1007/978-3-319-24696-3_22

Introduction

DNA damage changes the structure of the genetic material, preventing the replication machinery to make a complete and accurate copy of the genome. Hence, DNA damage in the S phase of the cell cycle has to be dealt with by special mechanisms to keep the genetic information stable throughout the generations of cells. Eukaryotes have evolved checkpoint pathways to properly respond to DNA damage. Here we discuss how the checkpoint operating in S phase, the intra S checkpoint (ISC), controls the initiation of DNA replication.

If DNA damage occurs during G1 phase the G1-S checkpoint blocks or delays cell cycle progression before entry into S phase. This avoids copying structurally compromised DNA, and, thus, prevents genetic alterations. Cells that reside in S phase when the damage occurs activate the ISC. The ISC is also induced under certain types of damage that escape detection in G1 phase, but which hamper the progression of replisomes, leading to detection of these lesions in S phase. The ISC coordinates various processes at replication forks encountering DNA damage sites to prevent genetic alterations. Among these processes is the protection of replication forks from gaining damage, called `replisome collapse [1, 2]. The ISC also activates DNA repair and lesion bypass processes to make progression of stalled replisome possible, and it elevates nucleotide levels by inducing ribonucleotide reductase. Moreover, it blocks mitotic entry to avoid erroneous attempts to segregate incompletely replicated chromosomes (Fig. 22.1). We focus here on how the ISC regulates the initiation of replication. It inhibits initiation in most genome regions in order to limit the extent of DNA replication under fork stalling conditions, thereby preventing mutations. However the ISC allows initiation from usually dormant replication origins specifically in regions where replication is ongoing at the time the damage occurs [3–5]. Dormant origin firing is thought to compensate for stalled replication forks, and, thus, to help ensure that genome duplication is complete under DNA damage. This differential regulation of initiation is thought to help prevent genetic instability and, thus, to contribute to protecting humans from developing genetic diseases like cancer.

Eukaryotic initiation of DNA replication is separated into two phases, origin licensing and origin firing. Licensing in G1 phase is the loading of an inactive double hexamer of the replicative helicase Mcm2-7 onto origins of replication [6]. The loaded Mcm2-7 helicase is called pre-RC (pre-replicative complex). Pre-RC formation is followed by origin firing in S phase, during which the pre-RC is extensively remodelled to become activated [7], and to assemble the full replisome. Origin firing is dependent on several firing factors (see below) and the two cell cycle kinases cyclin-dependent kinase (CDK) and Dbf4-dependent kinase (DDK), which are both activated at the G1-S phase transition. Importantly, firing is the step at which initiation becomes inhibited by the ISC, as discussed later.

CDK is central for cell cycle control of replication. Not only does it mediate S-phase specificity of origin firing, but it also inhibits licensing in all cell cycle phases except G1 when CDK activity is low [8]. This CDK-mediated licensing

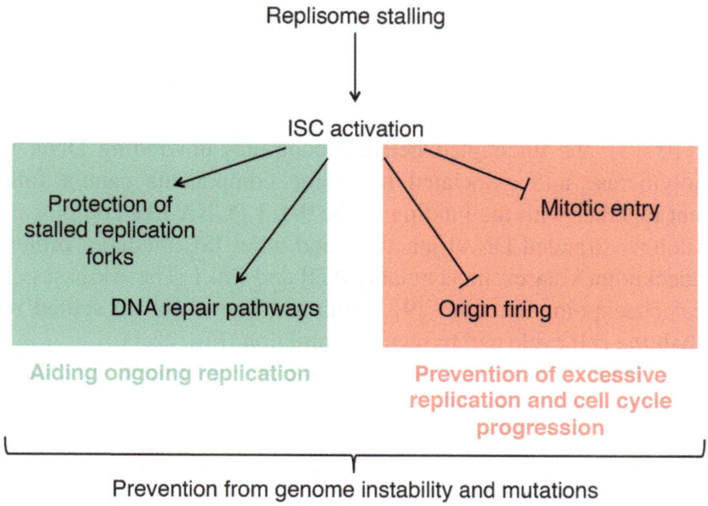

Fig. 22.1 The ISC helps maintain genome integrity. DNA damage and other obstacles on chromatin stall replisomes, which induces the ISC. The ISC helps maintain the genetic information by aiding ongoing replication, for which it must protect stalled replisomes from collapse and control DNA repair and lesion bypass processes in order to remove replisome-blocking lesions and/or help the replisomes bypass the DNA lesion. The ISC also prevents excessive replication by inhibiting the initiation of replication in order to limit having to copy damaged DNA templates, and it halts cell cycle progression by blocking mitotic entry to avoid errors in the distribution of the genetic information due to damaged chromosomes

inhibition facilitates that origins that have fired in a given S phase cannot be licensed again until the CDK activity drops in the following G1 phase. Thus, no origin can fire more than once in each S phase, ensuring that the genome is not copied more than once per cell cycle.

The Intra S-Phase Checkpoint Inhibits Origin Firing to Attenuate Replication under DNA Damage Conditions

Damaged DNA cannot serve as a template for faithful DNA replication. The lesions block the progression of replisomes, often by interfering with the replicative DNA polymerases. This ultimately leads to induction of the ISC [9]. Replisomes also stall when replication faces other kinds of challenging conditions. For example, when cells enter S phase prematurely due to expression of certain oncogenes they are not metabolically prepared for DNA synthesis, causing replicative stress. Experimental induction of the ISC is often achieved by chemically stalling polymerases, for example by reduction of dNTP levels using hydroxyurea (HU), or due to inhibition of DNA polymerase activity by aphidicolin.

Replication fork stalling generates excessive amounts of single-stranded DNA (ssDNA), leading to the recruitment of the RPA protein, the eukaryotic ssDNA-binding protein. In aphidicolin-treated *Xenopus* egg extracts excessive ssDNA generation has been attributed to an uncoupling of DNA polymerase from the replicative helicase [10–13]: the uncoupled helicase continues unwinding DNA whilst the stalled polymerase and associated replisome components cannot follow. RPA recruitment together with the binding of the 9-1-1 PCNA-like complex to ssDNA-dsDNA (double-stranded DNA) junctions and other ISC mediator proteins finally activate checkpoint kinases, most notably ATR and Chk1. These kinases orchestrate the critical checkpoint functions [9] mentioned above: to aid stalled replication forks, to halt the cell cycle and to regulate initiation (Fig. 22.1).

That the DNA damage checkpoint inhibits replication first became evident in cells with a compromised checkpoint—from patients with the Ataxia telangiectasia-mutated (ATM) disease, who carry mutations in the ATM kinase. Both ATM and ATR belong to the PI3-kinase family and have partially overlapping functions. ATM cells showed radio-resistant DNA synthesis (RDS) after treatment with ionising radiation, just like Hela cells upon treatment with caffeine, a PI3 kinase inhibitor. RDS was proposed to result from new origin firing events [14–16]. It was clarified later that RDS in checkpoint-compromised cells partially results from both firing of origins and failure to prevent progression from G1 into S phase [17–19].

The ISC has been suggested to be a major barrier that needs to be overcome for tumours to form [20–23]: expression of oncogenes deregulating the G1-S transition leads to premature entry into S phase and replication stress, which becomes apparent by induction of DNA damage. This DNA damage, in turn, was proposed to result in ISC activation and oncogene-induced senescence. Experimental inactivation of the ISC in human cells in these conditions suppressed oncogene-induced senescence and increased cell transformation [20], highlighting that the ISC is a cancer barrier. How much the inhibition of replication initiation, as opposed to other ISC functions, contributes to oncogene-induced senescence and how it attenuates transformation is unknown. Clarification will require selective inhibition of the individual branches of the ISC in mammalian tissue culture and mouse transformation models.

Origin Firing Inhibition upon DNA Damage in Yeast

After the initial observations of RDS in vertebrate cells, important insight into the underlying cellular and molecular mechanisms was gained using the budding yeast model. In 1995 Paulovich and Hartwell observed that DNA damage slowed down replication in budding yeast. Similar to the situation higher eukaryotes, this slowing down was found to be an active process, which depends on the checkpoint kinase Mec1, the yeast homolog of ATR [24].

Important for understanding ISC function is that replication origins in budding yeast are well defined and each origin fires with a characteristic reproducible timing during S phase [25]. Therefore, origins are categorised as early or late firing. Importantly, the ISC inhibits DNA synthesis by specifically blocking the firing of late origins [2, 4, 5, 26]. The following sequence of events upon DNA damage in S phase was deduced: (1) early origins fire and give rise to replisomes, (2) these replisomes stall upon encountering DNA lesions and thereby create the checkpoint signal and (3) this checkpoint signal inhibits firing of late origins (Fig. 22.3a). Studies using different fork stalling agents (such as HU and MMS) showed that the blocking of late origins is a universal response that occurs independently of how replisomes are stalled and checkpoint activation is induced [2, 4]. Moreover, the checkpoint pathway responsible for blocking late origin firing has also been elucidated. It involves induction of the Rad53 checkpoint kinase that takes over functions of vertebrate Chk1. Rad53 becomes activated downstream of the damage-induced Mec1 [4, 5, 24].

It turned out that Rad53 inhibits origin firing upon DNA damage by counteracting CDK and DDK functions that are required for origin firing. The essential CDK substrates for origin firing are the initiation factors Sld3 and Sld2. After CDK phosphorylation, Sld3 and Sld2 bind to the Dpb11 protein, most likely forming a ternary complex [27, 28]. The Sld3-Dpb11-Sld2 complex facilitates the activation of the replicative DNA helicase (Mcm2-7) and the assembly of the replisome by mechanisms that are not fully understood. They involve the interaction of Sld3 with the essential initiation factor Cdc45, recruiting Cdc45 to Mcm2-7 to activate the helicase. Essential phosphorylation sites of DDK are in the Mcm2-7 helicase complex [29], but how exactly their phosphorylations contribute to replication initiation is unclear.

Because CDK and DDK are each essential for initiation, inactivation of either CDK or DDK function would in principle be sufficient to inhibit origin firing. However, yeast cells inhibit both kinase pathways (Fig. 22.2): the Rad53 check-

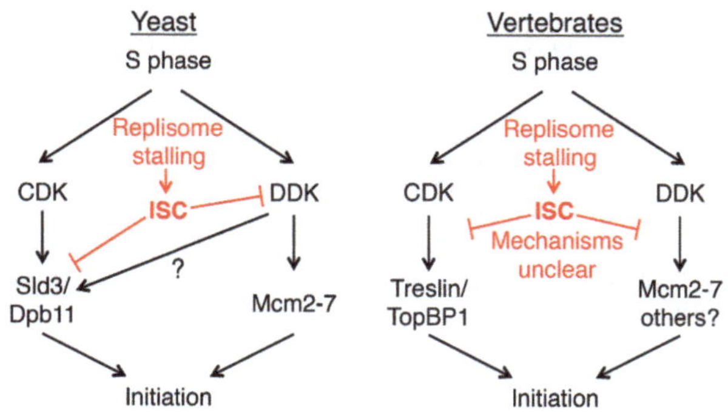

Fig. 22.2 The ISC prevents the initiation of DNA replication by inhibiting CDK and DDK processes during origin firing. In yeast and vertebrates CDK and DDK are essential for replication initiation, and for making initiation S phase specific, as their kinase activities increase when cells progress from G1 into S phase. In yeast, the ISC inhibits both pathways. The CDK pathway is blocked by inhibiting the initiation function of its substrate Sld3, whereas DDK is inhibited by the ISC at the level of the kinase, by phosphorylation of the DDK subunit Dbf4. In vertebrates, the principle of CDK and DDK pathway inhibition to achieve suppression of origin firing is probably conserved, although the underlying molecular mechanisms that are poorly characterized seem to be different

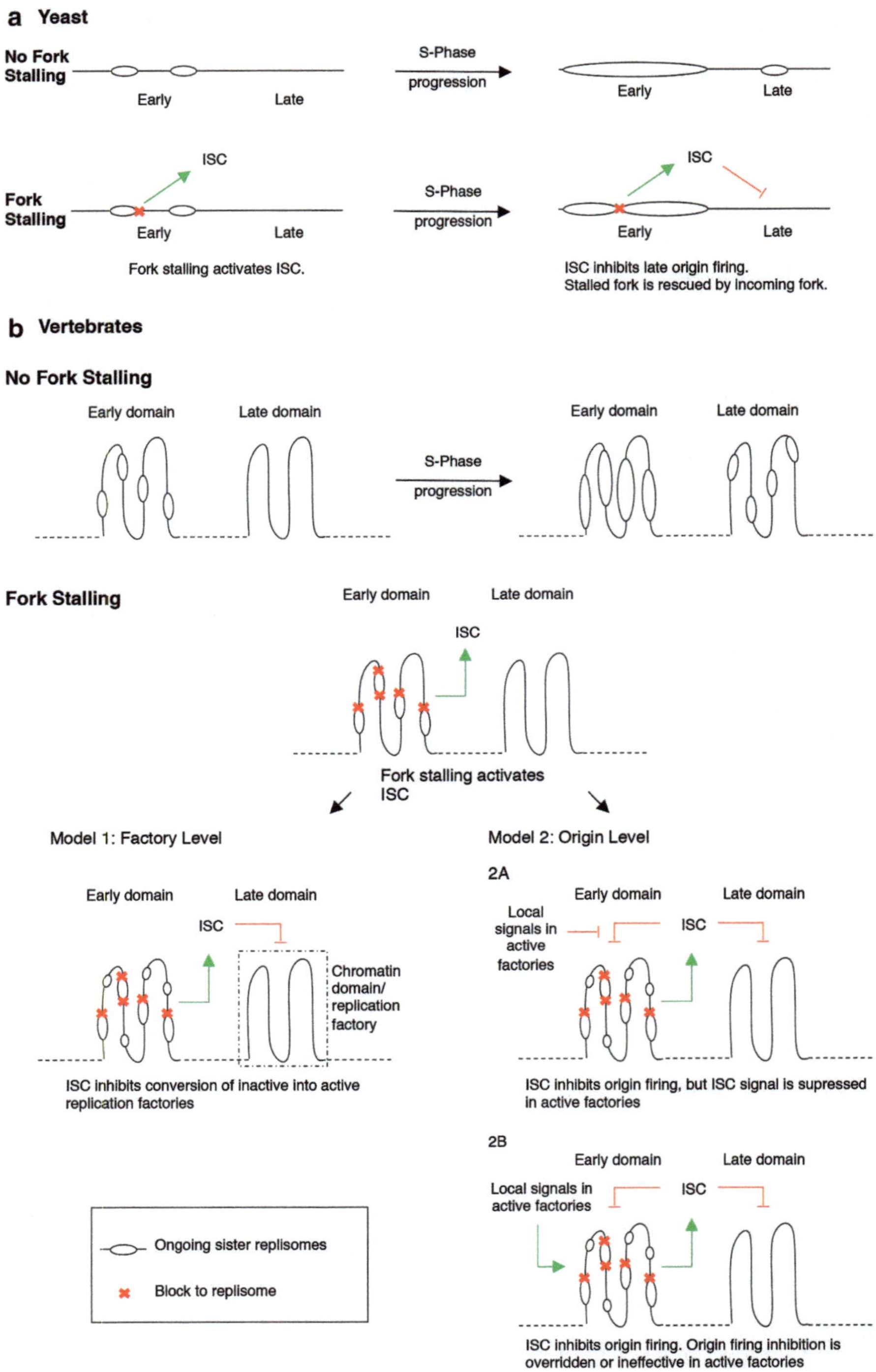

Fig. 22.3 Hypothetical models for ISC-mediated control of origin firing in yeast and vertebrates. (**a**) In yeast, early replication origins fire before late origins in unperturbed S phases. DNA damage or other obstacles lead to replication fork stalling, which activates the ISC, subsequently suppressing

point kinase phosphorylates the CDK target Sld3 as well as the DDK subunit Dbf4 at multiple sites, and only the simultaneous mutation of the phosphorylation sites on both proteins relieves the block to late origin firing [30, 31]. How Sld3 and Dbf4 are inhibited by Rad53 phosphorylation is only beginning to be unravelled: so far it has been shown that Rad53 phosphorylation of Sld3 interferes with Sld3 binding to Dpb11 and Cdc45, suggesting that two essential Sld3 functions are blocked [30].

The importance of origin firing inhibition for yeast cells is currently unclear. Two studies found a mild sensitivity of mutant cells, in which the block was bypassed, to genotoxic agents, whereas another study could not detect sensitivity using a slightly different mutant set-up [30–32]. Future research should address this more closely, and also why redundant mechanisms—inhibition of CDK and DDK—exist to block origin firing. It is tempting to speculate that this redundancy makes the block to origin firing more robust, which might present an evolutionary advantage by preventing genome instability.

Finally, it is interesting to note that whereas the checkpoint in yeast acts directly on DDK in order to inhibit its function in replication, it does not inhibit CDK directly, but rather at the level of its substrate Sld3 (Fig. 22.2). Perhaps such a mechanism has evolved because full inactivation of CDK upon checkpoint induction must be avoided, given that S-phase CDK is pivotal to inhibit re-licensing of origins that have already fired. Full inactivation of CDK during S phase by the ISC would therefore risk re-replication, which has to be avoided at any cost because even mild re-replication causes genome instability.

Fig. 22.3 (continued) the firing of late origins. Stalled forks are often rescued by an incoming fork generated from a neighbouring dormant origin or from another early origin that fired before ISC activation. This model is simplified and more complicated ISC mechanisms similar to the ones described for higher eukaryotes in (**b**) might apply also in yeast. (**b**) In higher eukaryotes, the temporal replication programme becomes visible as replication factories—actively replicating chromatin domains, whose replication occurs at reproducible times during S phase (*upper panel*, "No Fork Stalling"). Upon fork stalling, the ISC is induced and inhibits replication in inactive factories while allowing dormant origins to fire in active factories to rescue stalled replisomes. To explain this seeming paradox, two non-exclusive models are proposed. *Model 1*: The ISC acts at the level of chromatin domains. It inhibits the conversion of inactive into active replication domains, perhaps by preventing initiation factors from accessing inactive replication factories. Therefore, initiation factors, of which some are limiting, become available for the firing of dormant origins in active factories. *Model 2*: The ISC inhibits initiation at the origin level rather than at the factory level. Specifically in active factories, the ISC is not allowed to be active in order to relieve the inhibition of firing. This could be achieved by inhibiting the ISC signal locally in active but not in inactive factories (*Model 2A*). The relieving signal could be dependent on or independent of ongoing replication and the ISC itself. Alternatively, the signal relieving ISC-mediated firing suppression could work not by inhibiting the ISC itself but by overriding its effects at replication origins (model 2B). Local signals in active factories could, for example, remove the firing-inhibiting phosphorylations from ISC kinase substrates at origins. Alternatively, the relieving of ISC-mediated firing suppression could also be achieved passively if dormant origins in active factories had already passed the stage of origin firing where they can be inhibited by the ISC

Molecular Mechanisms of Suppression of Origin Firing in Vertebrates

Whether the inhibition of CDK and DDK to suppress origin firing is conserved in vertebrates is a matter of debate. Like in yeast, CDK is essential for replication initiation by phosphorylation of the Sld3 orthologue Treslin/TICRR, promoting its association with vertebrate Dpb11, TopBP1 [33, 34]. However, whereas in *C. elegans* CDK must phosphorylate Sld2 for the binding to the TopBP1 orthologue Mus101 to initiate replication [35], this CDK dependency does not appear to be conserved in the embryonic *Xenopus* egg extract system [36, 37]. This indicates that the molecular mechanisms of CDK-dependent initiation have partially diverged in higher eukaryotes but the principle of CDK dependency of replication initiation is conserved. Hence, inhibition of CDK function could be a means to inhibit origin firing upon DNA damage also in metazoa (Fig. 22.2).

This has hardly been addressed at a mechanistic level, partly because key players such as Treslin/TICRR have only recently been discovered [37, 38]. Moreover, observing molecular events leading to inhibition of origin firing remains difficult due to origin positions in the genome being poorly defined in vertebrates. Binding experiments using bacterially expressed TopBP1 fragments indicated that the interaction with Treslin/TICRR in cell lysates is reduced upon ISC activation [33]. This correlated with lower overall phosphorylation of Treslin/TICRR, which might well reflect lower phosphorylation of Treslin/TICRR by CDK as a result of ISC induction. Decreased Treslin/TICRR phosphorylation by CDK would lead to diminished binding of the protein to TopBP1 and inhibition of replication initiation.

Although the described experiments clearly need confirmation by methods analysing CDK phosphorylation of Treslin/TICRR specifically at replication origins, a decrease in CDK phosphorylation would be consistent with the finding that CDK activity becomes partially inhibited upon activation of the ISC in cultured cells [39, 40]. CDK inhibition occurs by degradation of the CDK-activating Cdc25A phosphatase. Cdc25A activates CDK2 by removing phosphorylations from CDK2-T14 and Y15. Phosphorylation of the CDK2-inhibitory T14 and Y15 was in turn suggested to be required for proper execution of the ISC. To test this, genomic knock-in into cultured human cells of a mutant CDK2 allele was used, which codes for CDK2 that is resistant to inhibition by phosphorylation [41]. The mutant carries point mutations of T14 and Y15 to A and F (CDK2-AF) that cannot be phosphorylated. CDK2-AF cells failed to recover from stalling of replisomes by replication inhibitors, and showed elevated levels of DNA damage upon exposure to the same agents. Partial chemical inhibition of CDK2-AF kinase activity rescued this phenotype [41]. This set of phenotypes can be explained by a scenario in which CDK-AF cells fail to suppress origin firing upon ISC activation by replication fork stalling, subsequently causing DNA damage. In line with this, over-initiation does occur if CDK inhibition is compromised, as modified cells in which Cdc25A cannot be degraded in response to ISC induction showed radio-resistant DNA synthesis [42, 43].

However, cellular mechanisms other than a lack of inhibition of initiation could also contribute to the observed phenotypes in CDK2-AF cells. Inhibition of CDK by the ISC could have functions independent of the suppression of origin firing. For example, CDK2-AF cells enter S phase pre-maturely, which could induce replication stress, fork stalling and DNA damage independently of firing inhibition. In addition, CDK inhibition could have a role in the ISC-mediated protection of replisomes from collapse, which is known to be important for the checkpoint to prevent excessive DNA damage upon fork stalling [1, 2]. Final clarification will be possible when the molecular details of the block of initiation have been unravelled, and mutants can be generated that bypass the inhibition of origin firing, but not other functions of CDK inhibition by the ISC.

Like in budding yeast, full downregulation of CDK activity upon checkpoint induction might provoke re-replication in vertebrate cells because CDK is central for the inhibition of origin licensing. However, in contrast to yeast, higher eukaryotes have also CDK-independent mechanisms to inhibit licensing [44–46], which might allow a degree of CDK downregulation without triggering re-licensing.

However, pathways that do not involve direct CDK inhibition to suppress origin firing might also play a role in vertebrates: Treslin/TICRR was recently proposed to become inhibited through phosphorylation by the ISC kinase Chk1 [47]. The molecular details of this inhibition remain unknown. Although this study tested DNA damage-independent Chk1 functions, it is a distinct possibility that Chk1-mediated Treslin/TICRR inhibition contributes to the inhibition of origin firing also under ISC-inducing conditions. This would be highly reminiscent of the direct phosphorylation of Sld3 by Rad53 in yeast, which blocks Sld3 interaction with Dpb11 and Cdc45, inhibiting initiation upon replisome stalling [30].

Like CDK, DDK is essential for replication initiation in vertebrates although the molecular mechanisms have not been elucidated [48, 49]. This means that, potentially, ISC-mediated inhibition of DDK to suppress initiation, which was described in yeast, could be conserved in metazoa (Fig. 22.2).

It has been controversially discussed whether DDK is inhibited by the ISC in higher eukaryotes. Initial reports suggested that DDK is inhibited upon DNA damage induced by exposing replicating *Xenopus* egg extracts to the topoisomerase II inhibitor etoposide [50, 51]. Dissociation of DDK from chromatin was reported to be responsible for its suppression. For some time these findings could neither be confirmed in *Xenopus* nor in human tissue culture using various ISC-inducing conditions [51–57]. As it is difficult to investigate events at specific origins in vertebrates these studies left open the possibility that DDK activity might be suppressed only locally at origins of replication, potentially explaining why DDK inhibition might have often escaped detection.

More recent work using *Xenopus* egg extracts provides mechanistic insight, arguing in favour of an ISC-mediated inhibition of DDK to inhibit initiation [51]: etoposide and other replication inhibitors lead to the recruitment of the PP1 phosphatase to chromatin. This contributes to the dephosphorylation of the DDK substrate Mcm4. Thus, it seems that although the molecular mechanism of DDK inhibition by the checkpoint changed during evolution of higher eukaryotes, counteracting the DDK pathway has been conserved as a means to exert ISC function (Fig. 22.2): In yeast, direct inhibition of the DDK kinase activity by its phosphorylation by checkpoint kinases downregulates DDK activity. In vertebrates, reversing phosphorylations on DDK substrates is the current proposed mechanism.

Preventing Mcm4 dephosphorylation upon etoposide treatment using chemical inhibition of PP1 was not sufficient to relieve the block of replication imposed by the ISC [51]. This suggests that inhibition of CDK, which presumably still occurs under these conditions, is sufficient to achieve the block of origin firing. This is consistent with the fact that CDK is essential for origin firing. Taken together, and assuming these results can be transferred from the embryonic *Xenopus* system to somatic metazoan cells, ISC-dependent inhibition of the two cell cycle kinases essential for initiation of DNA replication, CDK and DDK, appears to be conserved among eukaryotes to suppress replication initiation upon DNA damage (Fig. 22.2). Ultimate clarification will be possible by unravelling the molecular details of how CDK and DDK work and are regulated to control replication initiation.

Replication Initiation Control Is Complex in Higher Eukaryotes and Implicates Global and Local Regulations

In budding yeast, a subgroup of origins reproducibly fire early in S phase, whilst others fire late. If yeast cells enter S phase under conditions that block replication, early origin firing generates forks that stall and activate the ISC. CDK and DDK inhibition then prevents the firing of late origins (Fig. 22.3a).

Radio-resistant DNA synthesis in checkpoint-compromised cells indicates that origin firing inhibition upon DNA damage is conserved between yeast and vertebrates. But various reports demonstrate that the cellular organisation of replication in vertebrates is dynamic and more complex [58–60]. Importantly, ISC-mediated regulation of initiation must account for this complexity. Thus, regulation of origin firing in metazoans cannot be discussed without considering the cellular organisation of replication.

Visualising active replication by pulse labelling of sites of DNA synthesis using the nucleotide analogue BrdU, or by GFP-PCNA, reveals patterns of replication foci, also called replication factories, in the nucleus of cells. These patterns are distinctive for early, mid and late S phase [58, 60]. They represent the reproducible order of replication of different genomic regions, called the temporal replication program.

Corroborating these results, whole-genome analyses of nascent replication products showed that mammalian cells possess megabase-size chromosome domains, within which DNA is replicated at the same time [61]. These timing domains probably represent the microscopically defined replication factories. Replication domains seem to constitute actual structural units, as analysis of proximity between different chromatin regions indicated [62]. It turns out that euchromatic gene-rich domains are typically replicated in early S phase, whereas heterochromatic regions replicate late, although exceptions exist.

Activation of the ISC in vertebrate cells has been shown to differentially regulate initiation in different replication domains [3, 63–66]. As mentioned, initiation of replication becomes inhibited upon ISC induction. This reflects the inhibition of late origins in budding yeast, and is consistent with the idea that replication must be limited under replication-challenging conditions. In contrast to this general suppression of initiation, firing of origins in domains actively engaged in replication at the time of ISC induction is allowed. Even more, usually inactive replication origins, so-called dormant origins, are allowed to fire in actively replicating domains of human cells [63], and in *Xenopus* egg extracts [64]. This is thought to rescue replication of genomic regions close to replisomes that have stalled due to the challenge to replication present: while in some cases stalled replisomes may resume replication after removal of the block, or bypass the block, in other cases stalled replisomes may have collapsed and be unable to resume replication [1, 67]. In order to fully replicate the genomic region in front of a collapsed fork another replication fork generated by a dormant origin may be necessary.

Proper Control of Replication Initiation Is Important for Genome Stability

Dormant origins appear to be important for genome stability in higher eukaryotes, particularly under ISC-inducing conditions. This is supported by experiments using mammalian cells with lowered function of Mcm proteins. Mcm2-7 form pre-RCs in G1 phases of unperturbed cell cycles (licensing) that are converted into replisomes in S phase (firing). Cells with low Mcm levels displayed sensitivity to drug-induced replication stress [63], and mice carrying a hypomorphic allele of Mcm4 showed an increased risk for cancer [68]. This can be explained by a scenario, in which low Mcm levels cause fewer than normal origins to be licensed. Licensing is high enough, however, to allow the generation of a sufficient number of replisomes for replication to proceed with relatively normal speed in the absence of exogenous challenges to replication [63]. However, insufficient dormant origin firing results in a higher incidence of stalled replisomes that cannot be rescued by generation of new replisomes, causing genetic instability. Since replisome stalling occurs in normal cell cycles but is elevated under conditions that challenge replication forks, this phenotype becomes aggravated after exposure to replication inhibitors in cells and mice [63, 68].

Specific loci in the genome where the described mechanism of replisome stalling and rescue by dormant origins might become particularly evident are common fragile sites (CFS). CFSs have a propensity to break when replication is perturbed, for example by treatment with a low dose of aphidicolin [69], or various conditions inhibiting the ISC [70]. Although the nature of CFSs is not fully understood they are thought to arise from stalled replication forks that cannot be rescued because of a lack of dormant origins, leaving behindpartially unreplicated DNA. Supporting this, some of the most prominent CFSs show a paucity of initiation events [71, 72]. As a result of origin paucity the chance for unrescued fork breakdown is increased because single replisomes have to travel long distances. Although origin paucity likely contributes to CFS formation, other characteristics of CFS chromatin that pose challenges to replisomes might also play into this.

Emerging Molecular Mechanisms Origin Firing Control

As mentioned, origin firing becomes inhibited by the ISC in replication factories that are inactive at the time of ISC induction, but dormant origins initiate in actively replicating factories. How cells differentiate between active and inactive replication domains to appropriately adapt the checkpoint response is unknown. Theoretically, the checkpoint signal could be generated locally in inactive replication factories, inhibiting initiation selectively in these regions. However, this option can be excluded because we know that the ISC signal is generated in active factories: ISC induction depends on replication—on exposure of ssDNA at stalled replisomes. The ISC signal must then diffuse out of these domains, and inhibit initiation in inactive factories. This means that the initiation-inhibiting activity of the ISC must be ineffective at the site of its generation, in active factories. The following models based on those proposed by Yekezare et al. [19] integrate these theoretical considerations with experimental data (Fig. 22.3b).

Model 1: The ISC Acts at the Level of Replication Factories

Differential regulation of origin firing in active and inactive replication domains upon ISC induction could be accomplished if the checkpoint worked at the level of replication domains rather than individual origins (Fig. 22.3b). Supporting this idea, it was suggested that ISC-mediated CDK suppression inhibits the conversion of inactive into active replication factories [73, 74]. This potentially involves the regulation of chromatin architecture, which is thought to control the accessibility of origins for initiation factors. Suppression of factory activation would be sufficient to explain the differential effects of the ISC on initiation in replicating vs. non-replicating regions, if the ISC had no capability to inhibit firing at the level of individual origins. This regulation would leave initiation untouched in active factories but would inhibit initiation in inactive domains.

Regulation at the level of chromatin architecture for factory activation implies functions of CDK and/or DDK in the organisation of chromosome domains that remain largely unknown. If these functions require higher levels of CDK and DDK activity than the process of origin firing, moderate inhibition of CDK and DDK activity upon induction of the checkpoint will result in a block of factory activation whilst allowing origin firing in already active factories. It appears that ISC-mediated downregulation of CDK activity could indeed inhibit initiation in inactive factories via preventing factory activation: Although mechanistically uncharacterised, CDK activity was recently proposed to induce the activation of new replication factories, at least in unperturbed S phases [75].

Dormant origin firing selectively in active factories could be achieved passively in this model. These origins could constitute inefficient replication start sites that usually do not fire. Upon stalling or slowing of replisomes inefficient origins would have more time to initiate because, in contrast to normal S phases, they are not inactivated by passive replication. This mechanism could be aided by increased availability of initiation factors in active factories, given that several initiation factors have been shown to be limiting [76–78]. These proteins could become available for dormant origin firing in active factories because the generation of new active factories is suppressed in ISC-inducing conditions.

Model 2: The ISC Acts at the Level of Individual Origins

Differential control of initiation in active vs. inactive replication factories upon ISC induction could also be achieved if the ISC regulated firing at the level of individual origins rather than at the level of replication domains. In this scenario, the ISC inhibits CDK and DDK functions sufficiently well to suppress the conversion of pre-RC to replisomes at replication start sites. However, residual CDK activity in addition to CDK-independent inhibition of re-licensing must be high enough to prevent re-replication.

In order to restrict firing inhibition to factories that are inactive at the time the ISC is induced, the ISC could become blocked in active factories so that the ISC signal itself is switched off in these domains, e.g. by inhibition of ATR or Chk1 (model 2A, Fig. 22.3b). In model 2B, the ISC signal itself could be left untouched, but inhibited further downstream. For example, the inhibition of initiation factor activity by the ISC could be bypassed.

A checkpoint-inhibitory function that would be required for model 2 was proposed for several proteins, e.g. DDK [52] and the orthologues of the Polo kinase (Plx1) in *Xenopus laevis* [79, 80] and budding yeast [81]. Also protein phosphatase 2A (PP2A) was suggested to possess the capability to suppress the inhibition of replication in response to a double-strand break-induced ISC [82]. Thus, the distinct regulations of initiation in active vs. inactive replication factories could involve tight spatial control of DDK, Polo like kinase 1 and PP2A activities, directing these activities exclusively to active replication factories.

Very recently, the inter strand cross-link repair factor FANCI was implicated in the control of origin firing upon DNA damage [83]. Thus, its regulation could also be important for the differential control of firing by the ISC.

The processes leading to the relieving of the ISC-mediated firing inhibition selectively in active factories could depend on ongoing replication in these domains, as opposed to inactive factories where no replication is taking place. These processes might be dependent on or independent of the ISC itself, which we know becomes induced by replisome stalling in active factories.

There is some experimental evidence supporting an involvement of Polo kinases in this model. The vertebrate Polo like kinase 1 might become specifically activated at some origins of replication. *Xenopus* Plx1 is recruited to origins dependently on checkpoint-mediated phosphorylation of Mcm2 on serine 92 via Plx1's phosphoprotein-binding polo-box domain [79, 84]. This recruitment could position Plx1 appropriately for efficient phosphorylation of substrates localised at origins. For example, human Polo, Plk1, phosphorylates Orc2 on serine 188, and this was induced upon exposure of cells to replicative stress [85].

Plk1 was also reported to phosphorylate and downregulate the important ISC mediator claspin [80, 86–88], attenuating the checkpoint under certain conditions. Plk1 phosphorylation turns claspin into a β-TRCP-binding protein. β-TRCP mediates ubiquitylation of claspin by the SCF ubiquitin ligase, followed by proteasome-dependent degradation.

It remains to be investigated whether and how checkpoint-dependent Polo recruitment to origins and subsequent ORC2 and/or claspin phosphorylation contribute to the regulatory mechanisms to allow origin firing selectively in active but not in inactive replication domains. If they do contribute specific molecular mechanisms will be required to target these pathways to active replication factories upon ISC induction.

How DDK exerts its proposed checkpoint-overriding activity [52] is also poorly characterised. The question of DDK involvement in relieving origin firing is complicated by the fact that DDK is essential for initiation, and that replication is required for induction of the ISC. Final clarification will require genetic separation of DDK functions in initiation and checkpoint override.

A distinct possibility is that DDK overrides inhibition of origin firing via regulating the initiation inhibitor Rif1 [89]. Rif1 was suggested to control the timing of origin firing in unperturbed cell cycles [89–91] in budding and fission yeast [92], as well as in mammalian cells [93, 94]. In budding yeast, Rif1-mediated suppression of late origins involves the recruitment of the PP1 phosphatase by a motif in Rif1 that is conserved in higher eukaryotes. PP1 might subsequently inhibit origin firing by dephosphorylating Mcm proteins, whose phosphorylation by DDK is important for initiation in yeast [29, 95]. DDK was proposed to counteract PP1 association with Rif1 by binding to and subsequently phosphorylating Rif1 [89], potentially relieving the block to origin firing. If true, Rif1 would be an integrator of positive and negative signals to initiation. However, it needs to be mentioned that it has not been addressed directly whether budding yeast Rif1 is required for ISC-mediated inhibition of initiation.

Regulation of origin firing might occur by similar molecular mechanisms in vertebrates. A recent report using *Xenopus* egg extracts indicates that vertebrate PP1 mediates Mcm4 dephosphorylation upon ISC induction [51]. However, Rif1 involvement was not addressed directly in this study.

Thus, although progress is being made as to how initiation of DNA replication is controlled by the ISC on the molecular level, much more research is required to finally clarify this question that is central to understand genome instability.

Abbreviations

A	Alanine
CFS	Common fragile site
dsDNA	Double-stranded DNA
F	Phenylalanine

HU	Hydroxyurea
ISC	Intra S-phase checkpoint
MMS	Methyl methanesulfonate
Pre-RC	Pre-replicative complex
RDS	Radio-resistant DNA synthesis
ssDNA	Single-stranded DNA
Y	Tyrosine

Acknowledgements We are grateful to Belen Gomez Gonzales, Juliane Pfrötzschner, Jenny Bormann, Melisa Merdanovic, Gerben Vader and the members of the Pfander lab for helpful discussions and critical reading of the manuscript. The lab of D. Boos is supported by an NRW Rückkehrerförderprogramm fellowship from the Ministry for Innovation, Science and Research of North Rhine-Westphalia, Germany. The Pfander lab is supported by the Max-Planck Society and the German Research Council (DFG). The authors declare that they have no competing financial interest.

References

1. Lopes M, et al. The DNA replication checkpoint response stabilizes stalled replication forks. Nature. 2001;412(6846):557–61.
2. Tercero JA, Diffley JFX. Regulation of DNA replication fork progression through damaged DNA by the Mec1/Rad53 checkpoint. Nature. 2001;412(6846):553–7.
3. McIntosh D, Blow JJ. Dormant origins, the licensing checkpoint, and the response to replicative stresses. Cold Spring Harb Perspect Biol. 2012;4(10):012955.
4. Santocanale C, Diffley JFX. A Mec1- and Rad53-dependent checkpoint controls late-firing origins of DNA replication. Nature. 1998;395(6702):615–8.
5. Shirahige K, et al. Regulation of DNA-replication origins during cell-cycle progression. Nature. 1998;395(6702):618–21.
6. Bell SP, Dutta A. DNA replication in eukaryotic cells. Annu Rev Biochem. 2002;71:333–74.
7. Boos D, Frigola J, Diffley JFX. Activation of the replicative DNA helicase: breaking up is hard to do. Curr Opin Cell Biol. 2012;24(3):423–30.
8. Diffley JFX. Regulation of early events in chromosome replication. Curr Biol. 2004;14(18): R778–86.
9. Flynn RL, Zou L. ATR: a master conductor of cellular responses to DNA replication stress. Trends Biochem Sci. 2011;36(3):133–40.
10. Walter J, Newport J. Initiation of eukaryotic DNA replication: origin unwinding and sequential chromatin association of Cdc45, RPA, and DNA polymerase alpha. Mol Cell. 2000;5(4):617–27.
11. Pacek M, Walter JC. A requirement for MCM7 and Cdc45 in chromosome unwinding during eukaryotic DNA replication. EMBO J. 2004;23(18):3667–76.
12. Byun TS, et al. Functional uncoupling of MCM helicase and DNA polymerase activities activates the ATR-dependent checkpoint. Genes Dev. 2005;19(9):1040–52.
13. Pacek M, et al. Localization of MCM2-7, Cdc45, and GINS to the site of DNA unwinding during eukaryotic DNA replication. Mol Cell. 2006;21(4):581–7.
14. Painter RB. Inhibition of initiation of HeLa cell replicons by methyl methanesulfonate. Mutat Res. 1977;42(2):299–303.
15. Painter RB. Inhibition and recovery of DNA synthesis in human cells after exposure to ultraviolet light. Mutat Res. 1985;145(1-2):63–9.
16. Painter RB, Young BR. Radiosensitivity in ataxia-telangiectasia: a new explanation. Proc Natl Acad Sci U S A. 1980;77(12):7315–7.

17. Larner JM, et al. Radiation down-regulates replication origin activity throughout the S phase in mammalian cells. Nucleic Acids Res. 1999;27(3):803–9.
18. Lee H, Larner JM, Hamlin JL. A p53-independent damage-sensing mechanism that functions as a checkpoint at the G1/S transition in Chinese hamster ovary cells. Proc Natl Acad Sci U S A. 1997;94(2):526–31.
19. Yekezare M, Gomez-Gonzalez B, Diffley JF. Controlling DNA replication origins in response to DNA damage – inhibit globally, activate locally. J Cell Sci. 2013;126(Pt 6):1297–306.
20. Bartkova J, et al. DNA damage response as a candidate anti-cancer barrier in early human tumorigenesis. Nature. 2005;434(7035):864–70.
21. Bartkova J, et al. Oncogene-induced senescence is part of the tumorigenesis barrier imposed by DNA damage checkpoints. Nature. 2006;444(7119):633–7.
22. Di Micco R, et al. Oncogene-induced senescence is a DNA damage response triggered by DNA hyper-replication. Nature. 2006;444(7119):638–42.
23. Gorgoulis VG, et al. Activation of the DNA damage checkpoint and genomic instability in human precancerous lesions. Nature. 2005;434(7035):907–13.
24. Paulovich AG, Hartwell LH. A checkpoint regulates the rate of progression through S phase in S. cerevisiae in response to DNA damage. Cell. 1995;82(5):841–7.
25. Yoshida K, Poveda A, Pasero P. Time to be versatile: regulation of the replication timing program in budding yeast. J Mol Biol. 2013;425(23):4696–705.
26. Santocanale C, Sharma K, Diffley JFX. Activation of dormant origins of DNA replication in budding yeast. Genes Dev. 1999;13(18):2360–4.
27. Zegerman P, Diffley JFX. Phosphorylation of Sld2 and Sld3 by cyclin-dependent kinases promotes DNA replication in budding yeast. Nature. 2007;445(7125):281–5.
28. Tanaka S, et al. CDK-dependent phosphorylation of Sld2 and Sld3 initiates DNA replication in budding yeast. Nature. 2007;445(7125):328–32.
29. Sheu YJ, Stillman B. The Dbf4-Cdc7 kinase promotes S phase by alleviating an inhibitory activity in Mcm4. Nature. 2010;463(7277):113–7.
30. Zegerman P, Diffley JFX. Checkpoint-dependent inhibition of DNA replication initiation by Sld3 and Dbf4 phosphorylation. Nature. 2010;467(7314):474–8.
31. Lopez-Mosqueda J, et al. Damage-induced phosphorylation of Sld3 is important to block late origin firing. Nature. 2010;467(7314):479–83.
32. Duch A, et al. A Dbf4 mutant contributes to bypassing the Rad53-mediated block of origins of replication in response to genotoxic stress. J Biol Chem. 2011;286(4):2486–91.
33. Boos D, et al. Regulation of DNA replication through Sld3-Dpb11 interaction is conserved from yeast to humans. Curr Biol. 2011;21(13):1152–7.
34. Kumagai A, Shevchenko A, Dunphy WG. Direct regulation of Treslin by cyclin-dependent kinase is essential for the onset of DNA replication. J Cell Biol. 2011;193(6):995–1007.
35. Gaggioli V, et al. CDK phosphorylation of SLD-2 is required for replication initiation and germline development in C. elegans. J Cell Biol. 2014;204(4):507–22.
36. Matsuno K, et al. The N-terminal noncatalytic region of Xenopus RecQ4 is required for chromatin binding of DNA polymerase alpha in the initiation of DNA replication. Mol Cell Biol. 2006;26(13):4843–52.
37. Kumagai A, Shevchenko A, Dunphy WG. Treslin collaborates with TopBP1 in triggering the initiation of DNA replication. Cell. 2010;140(3):349–59.
38. Sansam CL, et al. A vertebrate gene, ticrr, is an essential checkpoint and replication regulator. Genes Dev. 2010;24(2):183–94.
39. Mailand N, et al. Rapid destruction of human Cdc25A in response to DNA damage. Science. 2000;288(5470):1425–9.
40. Molinari M, et al. Human Cdc25 A inactivation in response to S phase inhibition and its role in preventing premature mitosis. EMBO Rep. 2000;1(1):71–9.
41. Hughes BT, et al. Essential role for Cdk2 inhibitory phosphorylation during replication stress revealed by a human Cdk2 knockin mutation. Proc Natl Acad Sci U S A. 2013;110(22):8954–9.
42. Falck J, et al. The ATM-Chk2-Cdc25A checkpoint pathway guards against radioresistant DNA synthesis. Nature. 2001;410(6830):842–7.

43. Busino L, et al. Degradation of Cdc25A by beta-TrCP during S phase and in response to DNA damage. Nature. 2003;426(6962):87–91.
44. McGarry TJ, Kirschner MW. Geminin, an inhibitor of DNA replication, is degraded during mitosis. Cell. 1998;93(6):1043–53.
45. Arias EE, Walter JC. Replication-dependent destruction of Cdt1 limits DNA replication to a single round per cell cycle in Xenopus egg extracts. Genes Dev. 2005;19(1):114–26.
46. Arias EE, Walter JC. PCNA functions as a molecular platform to trigger Cdt1 destruction and prevent re-replication. Nat Cell Biol. 2006;8(1):84–90.
47. Guo C, et al. Interaction of Chk1 with Treslin negatively regulates the initiation of chromosomal DNA replication. Mol Cell. 2014;57(3):492–505.
48. Roberts BT, et al. DNA replication in vertebrates requires a homolog of the Cdc7 protein kinase. Proc Natl Acad Sci U S A. 1999;96(6):2800–4.
49. Walter JC. Evidence for sequential action of cdc7 and cdk2 protein kinases during initiation of DNA replication in Xenopus egg extracts. J Biol Chem. 2000;275(50):39773–8.
50. Costanzo V, et al. An ATR- and Cdc7-dependent DNA damage checkpoint that inhibits initiation of DNA replication. Mol Cell. 2003;11(1):203–13.
51. Poh WT, et al. Xenopus Cdc7 executes its essential function early in S phase and is counteracted by checkpoint-regulated protein phosphatase 1. Open Biol. 2014;4:130138.
52. Tsuji T, et al. The role of Dbf4/Drf1-dependent kinase Cdc7 in DNA-damage checkpoint control. Mol Cell. 2008;32(6):862–9.
53. Heffernan TP, et al. Cdc7-Dbf4 and the human S checkpoint response to UVC. J Biol Chem. 2007;282(13):9458–68.
54. Tenca P, et al. Cdc7 is an active kinase in human cancer cells undergoing replication stress. J Biol Chem. 2007;282(1):208–15.
55. Silva T, et al. Xenopus CDC7/DRF1 complex is required for the initiation of DNA replication. J Biol Chem. 2006;281(17):11569–76.
56. Petersen P, et al. Protein phosphatase 2A antagonizes ATM and ATR in a Cdk2- and Cdc7-independent DNA damage checkpoint. Mol Cell Biol. 2006;26(5):1997–2011.
57. Liu P, et al. The Chk1-mediated S-phase checkpoint targets initiation factor Cdc45 via a Cdc25A/Cdk2-independent mechanism. J Biol Chem. 2006;281(41):30631–44.
58. Dimitrova DS, Gilbert DM. The spatial position and replication timing of chromosomal domains are both established in early G1 phase. Mol Cell. 1999;4(6):983–93.
59. Berezney R, Dubey DD, Huberman JA. Heterogeneity of eukaryotic replicons, replicon clusters, and replication foci. Chromosoma. 2000;108(8):471–84.
60. Leonhardt H, et al. Dynamics of DNA replication factories in living cells. J Cell Biol. 2000;149(2):271–80.
61. Hiratani I, et al. Global reorganization of replication domains during embryonic stem cell differentiation. PLoS Biol. 2008;6(10), e245.
62. Ryba T, et al. Evolutionarily conserved replication timing profiles predict long-range chromatin interactions and distinguish closely related cell types. Genome Res. 2010;20(6):761–70.
63. Ge XQ, Jackson DA, Blow JJ. Dormant origins licensed by excess Mcm2-7 are required for human cells to survive replicative stress. Genes Dev. 2007;21(24):3331–41.
64. Woodward AM, et al. Excess Mcm2-7 license dormant origins of replication that can be used under conditions of replicative stress. J Cell Biol. 2006;173(5):673–83.
65. Anglana M, et al. Dynamics of DNA replication in mammalian somatic cells: nucleotide pool modulates origin choice and interorigin spacing. Cell. 2003;114(3):385–94.
66. Courbet S, et al. Replication fork movement sets chromatin loop size and origin choice in mammalian cells. Nature. 2008;455(7212):557–60.
67. Desany BA, et al. Recovery from DNA replicational stress is the essential function of the S-phase checkpoint pathway. Genes Dev. 1998;12(18):2956–70.
68. Shima N, et al. A viable allele of Mcm4 causes chromosome instability and mammary adenocarcinomas in mice. Nat Genet. 2007;39(1):93–8.
69. Glover TW, et al. DNA polymerase alpha inhibition by aphidicolin induces gaps and breaks at common fragile sites in human chromosomes. Hum Genet. 1984;67(2):136–42.

70. Casper AM, et al. ATR regulates fragile site stability. Cell. 2002;111(6):779–89.
71. Letessier A, et al. Cell-type-specific replication initiation programs set fragility of the FRA3B fragile site. Nature. 2011;470(7332):120–3.
72. Ozeri-Galai E, et al. Failure of origin activation in response to fork stalling leads to chromosomal instability at fragile sites. Mol Cell. 2011;43(1):122–31.
73. Dimitrova DS, Gilbert DM. Temporally coordinated assembly and disassembly of replication factories in the absence of DNA synthesis. Nat Cell Biol. 2000;2(10):686–94.
74. Ge XQ, Blow JJ. Chk1 inhibits replication factory activation but allows dormant origin firing in existing factories. J Cell Biol. 2010;191(7):1285–97.
75. Thomson AM, Gillespie PJ, Blow JJ. Replication factory activation can be decoupled from the replication timing program by modulating Cdk levels. J Cell Biol. 2010;188(2):209–21.
76. Mantiero D, et al. Limiting replication initiation factors execute the temporal programme of origin firing in budding yeast. EMBO J. 2011;30(23):4805–14.
77. Tanaka S, et al. Origin association of Sld3, Sld7, and Cdc45 proteins is a key step for determination of origin-firing timing. Curr Biol. 2011;21(24):2055–63.
78. Collart C, et al. Titration of four replication factors is essential for the Xenopus laevis midblastula transition. Science. 2013;341(6148):893–6.
79. Trenz K, Errico A, Costanzo V. Plx1 is required for chromosomal DNA replication under stressful conditions. EMBO J. 2008;27(6):876–85.
80. Yoo HY, et al. Adaptation of a DNA replication checkpoint response depends upon inactivation of Claspin by the Polo-like kinase. Cell. 2004;117(5):575–88.
81. Donnianni RA, et al. Elevated levels of the polo kinase Cdc5 override the Mec1/ATR checkpoint in budding yeast by acting at different steps of the signaling pathway. PLoS Genet. 2010;6(1), e1000763.
82. Chou DM, et al. Protein phosphatase 2A regulates binding of Cdc45 to the prereplication complex. J Biol Chem. 2002;277(43):40520–7.
83. Chen YH, et al. ATR-mediated phosphorylation of FANCI regulates dormant origin firing in response to replication stress. Mol Cell. 2015;58(2):323–38.
84. Yoo HY, et al. Mcm2 is a direct substrate of ATM and ATR during DNA damage and DNA replication checkpoint responses. J Biol Chem. 2004;279(51):53353–64.
85. Song B, et al. Plk1 phosphorylation of Orc2 promotes DNA replication under conditions of stress. Mol Cell Biol. 2011;31(23):4844–56.
86. Peschiaroli A, et al. SCFbetaTrCP-mediated degradation of Claspin regulates recovery from the DNA replication checkpoint response. Mol Cell. 2006;23(3):319–29.
87. Mamely I, et al. Polo-like kinase-1 controls proteasome-dependent degradation of Claspin during checkpoint recovery. Curr Biol. 2006;16(19):1950–5.
88. Mailand N, et al. Destruction of Claspin by SCFbetaTrCP restrains Chk1 activation and facilitates recovery from genotoxic stress. Mol Cell. 2006;23(3):307–18.
89. Hiraga S, et al. Rif1 controls DNA replication by directing protein phosphatase 1 to reverse Cdc7-mediated phosphorylation of the MCM complex. Genes Dev. 2014;28(4):372–83.
90. Mattarocci S, et al. Rif1 controls DNA replication timing in yeast through the PP1 phosphatase Glc7. Cell Rep. 2014;7(1):62–9.
91. Dave A, et al. Protein phosphatase 1 recruitment by Rif1 regulates DNA replication origin firing by counteracting DDK activity. Cell Rep. 2014;7(1):53–61.
92. Tazumi A, et al. Telomere-binding protein Taz1 controls global replication timing through its localization near late replication origins in fission yeast. Genes Dev. 2012;26(18):2050–62.
93. Yamazaki S, et al. Rif1 regulates the replication timing domains on the human genome. EMBO J. 2012;31(18):3667–77.
94. Cornacchia D, et al. Mouse Rif1 is a key regulator of the replication-timing programme in mammalian cells. EMBO J. 2012;31(18):3678–90.
95. Sheu YJ, Stillman B. Cdc7-Dbf4 phosphorylates MCM proteins via a docking site-mediated mechanism to promote S phase progression. Mol Cell. 2006;24(1):101–13.

Chapter 23
Protein Phosphatases and DNA Replication Initiation

Michael J.R. Stark, Shin-ichiro Hiraga, and Anne D. Donaldson

Abstract Eukaryotic DNA replication is controlled by regulated cycles of protein phosphorylation. While the controls over these cycles of kinase activity have been the subject of intense investigation, controls over the removal of phosphorylation, carried out by protein phosphatases, are potentially of equal importance for regulating DNA replication but have in comparison been largely neglected. In this chapter we will first present a brief overview of the families of phosphatases occurring in eukaryotic cells, with emphasis on the PP1 and PP2A subtypes that have been implicated in direct control of replication origin initiation. We will then review our current knowledge of how these phosphatases interact with established control pathways to impact on replication initiation, outlining how PP1 activity is required to prevent premature origin initiation, and its potential involvement in dephosphorylating ORC to enable pre-replication complex formation. Possible pathways for the involvement of PP2A in promoting replication initiation will also be introduced, highlighting the gaps in our understanding and areas of ongoing investigation.

Keywords Phosphatase • PP1 • PP2A • Chromatin • Rif1 • ORC • DNA replication • Replication • Origin • Cell cycle

Introduction

DNA replication is controlled during the cell cycle by regulated protein kinase activity, in particular the activity of Cyclin-dependent and Dbf4-dependent kinases. In any process where phosphorylation is central to the regulation, regulated dephosphorylation has equal potential to act as a critical control. Nonetheless, understanding of how

M.J.R. Stark
Centre for Gene Regulation & Expression, College of Life Sciences, University of Dundee, Dundee DD1 5EH, UK

S.-i. Hiraga • A.D. Donaldson (✉)
Institute of Medical Sciences, University of Aberdeen, Foresterhill, Aberdeen AB25 2ZD, UK
e-mail: a.d.donaldson@abdn.ac.uk

D.L. Kaplan (ed.), *The Initiation of DNA Replication in Eukaryotes*,
DOI 10.1007/978-3-319-24696-3_23

phosphatase activity regulates mechanisms of DNA replication lags well behind our insights into how different kinases control replication. Here we outline our knowledge of the effects of protein phosphatases on DNA replication initiation. We begin by reviewing the molecular biology of protein phosphatases, with an emphasis on PP1 and PP2A as the two phosphatases known to impact on replication control. We then outline studies that have investigated roles of phosphatases in controlling the initiation of replication—outlining advances made and progress towards identification of dephosphorylation targets relevant to replication control.

The human genome encodes over 500 protein kinases and around 150 protein phosphatases (www.genenames.org [1]). Although more than 98 % of protein phosphorylation occurs on serine or threonine residues, around 90 of the 150 protein kinases are tyrosine-specific and are matched by a similar number of protein tyrosine phosphatases [2]. In terms of catalytic subunits there are only around 40 dedicated protein serine/threonine phosphatases to oppose the function of over 400 kinases that phosphorylate these two residues. Coupling this information with initial biochemical studies indicating low specificity of isolated phosphatase catalytic subunits, the idea emerged that phosphatases provide largely constitutive and promiscuous activities, against which changes in kinase activity are played to effect the regulation of biological processes. However, it has now become clear that this idea is not at all true, and that the regulation of phosphatase activity does indeed play critical roles. The complexity required however lies not in the number of phosphatase catalytic subunits but rather in the way they are regulated through interactions with other proteins, which control their localization, substrate specificity, and activity. Thus the relatively few phosphatase catalytic subunits combine with a diverse multiplicity of regulators to generate protein phosphatase families, some of which contain many distinct members. PP1 and PP2A, the most abundant and best studied phosphatases that account for the majority of phosphatase activity within cells [3], probably therefore collectively encompass hundreds of distinct phosphatase specificities.

Unlike kinases, which frequently target for phosphorylation specific amino acid motifs within their substrates, phosphatases generally show little specificity for dephosphorylating a particular consensus sequence. Instead they select their substrates through more complex interactions involving both the catalytic subunit and its regulators. As well as proteins that determine substrate recognition, phosphatase regulators include inhibitory polypeptides that suppress phosphatase activity by occupying the catalytic site.

Protein Phosphatase Families

Protein phosphatase catalytic subunits fall into a small number of distinct families [4] (see Table 23.1). Most *protein tyrosine phosphatases (PTPs)* are related in sequence and contain a conserved cysteine and arginine at their active site. In addition to transmembrane receptor and intracellular non-receptor PTPs, this group includes dual-specificity phosphatases that can dephosphorylate phosphoserine and phosphothreonine, as well as phosphotyrosine. The *S. cerevisiae* Cdc14

Table 23.1 Protein phosphatase classification

Family	Class	Human gene number[a]	Regulatory subunits
PTP superfamily (-CX_5R-)[b]			
Class I	Classic receptor PTP	21	
	Classic non-receptor PTP	18	
	Dual-specificity	58	
Class II	Low molecular weight PTP	1	
Class III	Cdc25s (dual specificity)	3	
Ser/Thr phosphatases			
PPP family	PP1	3	>90
	PP2A	2	A (2 genes), B (13 genes)[b]
	PP3 (PP2B)	3	B (2 genes), calmodulin
	PP4	1	R1, R2, R4
	PP5	1	none
	PP6	1	SAPs (3 genes), ankyrin repeat proteins (3 genes)[b]
	PP7	2	unknown
PPM family	PP2C	17	none
Asp-based (-DXDX[TV]-)[c]	FCP/SCP family	8[d]	TFIIF, unknown
	HAD family	4[d]	

[a]Gene number data taken from www.genenames.org (accessed 24 Mar 2015)
[b]Only the major regulators are indicated
[c]Denotes active site motif
[d]Additional members of these families may also have phosphatase activity

dual-specificity phosphatase falls in this category. Another set of dual-specificity PTP enzymes comprises the Cdc25 PTP subgroup, which includes the enzymes that mediate dephosphorylation of CDK1 on Thr-14 and Tyr-15 upon mitotic entry. Dedicated *protein serine/threonine phosphatases* fall into three main groups: (1) the *PPP* or PhosphoProtein Phosphatase family (13 human genes, including three encoding PP1 and two encoding PP2A catalytic subunits), (2) the *PPM* or Protein Phosphatase Mg^{2+} or Mn^{2+}-dependent family (17 human genes), and (3) a small group of *Aspartate-based* phosphatases (eight human genes) of which most are involved in targeting the C-terminal domain repeats of the largest subunit of RNA polymerase II (see www.genenames.org).

The PPP and PPM phosphatases are unrelated in sequence, but through presumed convergent evolution do adopt related overall structures, in which two divalent metal ions play a key role at the active site [5, 6]. Unlike PPP family members, PPM phosphatases are not characterized by interaction with multiple regulatory proteins [7] but instead are frequently embedded within additional domains that confer specificity [4]. In this chapter we will focus primarily on the PP1 and PP2A members of the PPP family, which are the phosphatase types with known roles in regulating DNA replication initiation.

PP1

Three human genes encode PP1 catalytic subunits, expressing a total of five different isoforms (namely PP1α1, α2, β, γ1, and γ2 [3, 8–11]). In budding yeast in contrast one gene encodes the single PP1 isoform present. However, a variety of approaches in a number of organisms have revealed that PP1 catalytic subunits may have literally hundreds of interaction partners [12], so that PP1 enzymes in effect comprise the largest family of PPP Phosphatases. While in certain organisms the small number of PP1 catalytic subunit isoforms may impart some degree of specificity, it is the wide variety of PP1-interacting proteins that lead to its great diversity of specific roles. These interacting proteins include inhibitory proteins such as Inhibitor-2, in addition to 'substrate-targeting' regulators that help to determine substrate specificity and localization of PP1 isoforms [3]. A striking example where PP1 localization is regulated by a targeting subunit is provided by Repo-Man, which dramatically regulates chromatin association of PP1γ during anaphase in human cells [13, 14]. Binding to the myosin phosphatase targeting subunit MYPT1 in contrast targets PP1 to dephosphorylate myosin and polo-like kinase 1 amongst other substrates. Interaction of the PP1 catalytic subunit with such regulators is mediated by multiple, short amino acid motifs, often acting in a combinatorial fashion, which interact with the surface of the catalytic subunit some distance from the phosphatase active site (Fig. 23.1a). The active site itself is shallow, and association of PP1 with regulatory subunits may provide the additional points of contact for substrates needed to confer a high degree of selectivity and specificity [15]. There are around ten defined motifs known to mediate interaction of regulatory subunits with PP1, with the so-called RVxF, SILK, and MyPhoNE motifs the best known [16]. The RVxF motif (also termed RxVxF and more properly represented as [KR][KR][VI]{FIMYDP}[FW], where {} indicate excluded residues) is present in the vast majority of PP1-interacting proteins, functioning as a primary PP1-binding motif and binding in an extended conformation to a hydrophobic groove away from the phosphatase active site [17]. The SILK motif ([GS]IL[RK]) is present in a smaller proportion of PP1 interactors and is generally located N-terminal to the RVxF motif [3]. Inhibitor-2 for example contains variants of both these motifs, together with a third region that contacts the PP1 active site [18]. MyPhoNE (myosin phosphatase N-terminal element) is a third consensus motif indicative of PP1 binding and having the consensus RxxQ[VIL][KR]x[YW]. These characteristic sequences are specific for binding to PP1, and the requirement for the PP1 surface to interact with a wide range of different, short sequences during interactions with numerous other proteins may in part be the driving force for the high level of conservation shown by PP1 [12]. Mutation of key residues in these PP1-binding motifs frequently abolishes interaction with PP1 and is therefore a good way to test the functional significance of PP1 interaction. This theme of interacting motifs carries through to other PPP family phosphatases—for example, PPP3 (also called PP2B or calcineurin) is a trimeric phosphatase whose catalytic subunit interacts with other proteins through short PxIxIT and LxVP motifs [19].

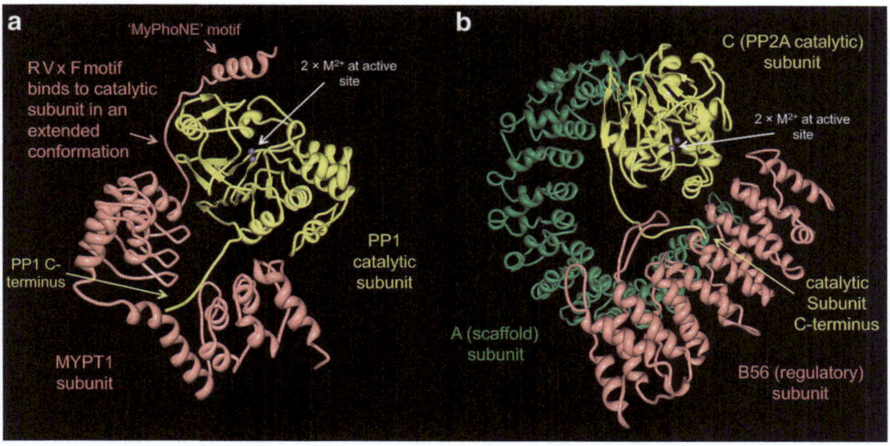

Fig. 23.1 Structures of PP1 and PP2A. (**a**) PP1β bound to the N-terminal domain of the myosin regulatory subunit MYPT1 [PDB entry 1S70] [86] showing the catalytic subunit in *yellow* and the MYPT1 regulatory subunit in *red*. The RVxF motif binds to PP1 in an extended conformation some distance away from the catalytic site, but secondary structure elements on either side of it contribute to re-shaping the catalytic cleft. Another short sequence element in the extreme N-terminal region of MYPT1 ('MyPhoNE') confers specificity for the PP1β isoform [87]. Other interactions between the two proteins involve the C-terminal region of PP1β (**b**) PP2AB56γ1 [PDB entry 2IAE] [88], with the catalytic 'C' subunit in *yellow*, scaffold 'A' subunit in *green*, and the B56γ1 (B′) subunit in *red*. Note the interaction between methylated C-terminus of PP2A Cα and B56γ1. Structures were displayed from the PDB entries using Protein Workshop [89]

As monomers many PP1-interacting proteins show intrinsically disordered structures, a feature that may help them to bind to PP1 in an extended conformation permitting larger interaction surfaces and facilitating interaction with multiple binding determinants [3]. Furthermore, different interacting proteins may compete with each other for binding to PP1, such that overexpression of one interacting protein may lead to effects caused by displacement of others [20]. Although present throughout the cell, PP1 is enriched in the nuclear compartment [2].

PP2A

Like PP1, the PP2A catalytic subunit functions within the context of multimeric protein complexes. PP2A holoenzymes are normally assembled on a U-shaped scaffold (the 'A' subunit) composed of multiple HEAT repeats and contain one of a range of regulatory or 'B' subunits in addition to the catalytic ('C') subunit (Fig. 23.1b). The B subunit is a major determinant of substrate specificity and localization of the assembled holoenzyme. Budding yeast contains at least three different B subunits, while in mammals multiple isoforms and splice variants derived from four B subunit gene families may lead to a diversity of PP2A holoenzymes rivalling the diversity of PP1 with its regulatory subunits [15]. As with PP1, the particular combination of B and C subunits appears to confer a high level of specificity on particular forms of PP2A. PP2A holoenzymes containing different B subunits can show exquisite selectivity, for example targeting distinct phosphorylated residues in the same phosphoprotein substrate ([15], see Fig. 23.1b). Moreover PP2A inhibitory proteins such as ENSA and Bod1 can suppress the activity of specific isoforms (PP2AB55 and PP2AB56 respectively). PP4 is related to PP2A and shows a similar organization, with specific scaffolding (PP4R2) and regulatory (PP4R3β) subunits involved in targeting PP4 to dephosphorylate γ-H2AX during S phase, while a small number of other PP4 regulators specify additional functions. PP2A is methylated on its carboxy-terminal leucine residue and a conserved, DYFL sequence at the extreme C-terminus of PP2A (shared by PP4 and PP6) is probably related to this modification. Carboxymethylation is reversible and has been proposed to regulate assembly of the multimeric PP2A complexes [21], while association with the carboxylmethylesterase may also serve to inhibit PP2A that has not yet correctly assembled into an 'ABC' trimer [22].

A Phosphatase Network Controlling Cell Cycle Progression

Contrary to initial suggestions that phosphatase activity was relatively unspecific and constitutive, it has become clear that phosphatases can form important elements of complex biological phosphoregulatory networks, sometimes involving the multiplicity of specific phosphatase activities available (as outlined above). A good example is provided by a recently elucidated network that controls mitotic entry and exit. Mitotic entry requires activation of the cyclin-dependent kinase Cdk1, through dephosphorylation by Cdc25 phosphatase of Cdk1 Thr-14 and Tyr-15. This activation involves feed-forward regulation in which Cdk1 either directly or indirectly promotes activatory dephosphorylation of Cdc25 by PP1, and overcomes inhibitory dephosphorylation of Cdc25 by PP2A, at least in part through active inhibition of PP2A [23, 24]. As cells enter mitosis, PP1 is inhibited by Cdk1 phosphorylation [25] and PP2A activity is suppressed through interaction with specific inhibitory proteins following their phosphorylation by Cdk1 or Greatwall, a kinase that acts downstream of Cdk1 [24, 26–28]. Mitotic exit involves sequential reactivation of three different kinase activities: first PP1 which is then required to reactivate PP2A-B55 and lastly PP2A-B56, the latter being PP2A holoenzymes with different B subunits [25]. Reactivation of these phosphatases enables cells to reverse Cdk-dependent mitotic phosphorylation events, so that active regulation of these phosphatase functions is critical for proper progression through mitosis. The elucidation of this sophisticated network of phosphatase regulatory events illustrates how phosphatases can no longer be treated as the 'poor relations' of protein kinases. Instead they must be considered as potentially active elements of any process that is regulated by reversible protein phosphorylation.

Protein Phosphatase 1 in Replication Control

The two different kinase activities essential for replication initiation are DDK (Dbf4-Dependent Kinase) consisting of Cdc7 kinase and its activating subunit Dbf4, and the better known Cyclin-Dependent Kinase (CDK). The substrates of these kinases in replication have been identified and studied primarily in the budding yeast *Saccharomyces cerevisiae* [29, 30]. The critical target of DDK for replication initiation is the MCM complex and particularly Mcm4, which has an N-terminal region that is multiply phosphorylated by DDK to activate MCM helicase function. DDK phosphorylation is also thought to promote recruitment of the initiation factor Sld3 to the MCM complex [31, 32]. The critical CDK targets are also best understood in yeast: phosphorylation of Sld3 by CDK promotes its interaction with Dpb11. Dbp11 in turn recruits Sld2 (also phosphorylated by CDK) and the GINS complex to mediate polymerase recruitment. Sld3 also binds Cdc45, which is essential for recruitment of Cdc45 to the MCM complex and formation of the active CMG helicase complex. Homologous pathways are now believed to operate in mammalian cells [30], where TopBP11 and Treslin correspond to yeast Dbp11 and Sld3, respectively, and appear to operate similarly to promote replication initiation.

Yeast Rif1 Acts as a Protein Phosphatase 1-Targeting Subunit to Regulate DNA Replication

Several investigations have implicated PP1 as important for restraining replication origin initiation in both budding and fission yeasts [33–35]. This role for PP1 is directed by a protein called Rif1, identified by these studies as a new PP1 substrate-targeting subunit. Yeast Rif1 contains a series of PP1 interaction (RVxF and SILK) motifs, and interacts with PP1 as shown by co-immunoprecipitation and 2-hybrid analysis. As outlined above, a critical step in replication initiation is activation of the DNA unwinding function of MCM helicase (Fig. 23.2, pink downward arrows) through phosphorylation of Mcm4 and other MCM subunits by DDK. It was discovered that yeast Rif1 affects replication by directing PP1 to dephosphorylate Mcm4 (Fig. 23.2, blue upward arrow) [33, 34, 36]. The Rif1-PP1 control module therefore counteracts DDK, restricting MCM helicase function to the correct cell cycle phase and limiting origin initiation. One consequence of this function of Rif1-PP1 in antagonizing DDK is that cells lacking Rif1 have a diminished requirement for DDK activity, and a *rif1Δ* mutation partially relieves the effects of *cdc7*^{ts} mutations in either *S. pombe* or *S. cerevisiae* [33, 34, 36, 37]. Rif1-PP1 therefore controls S phase entry by determining the DDK level at which the first initiation events can occur.

Interestingly, replication origins vary in their sensitivity to the removal of Rif1 activity: while some origins are highly prone to initiate prematurely in the absence of Rif1, others appear to be largely immune [33, 37]. Still others may actually be

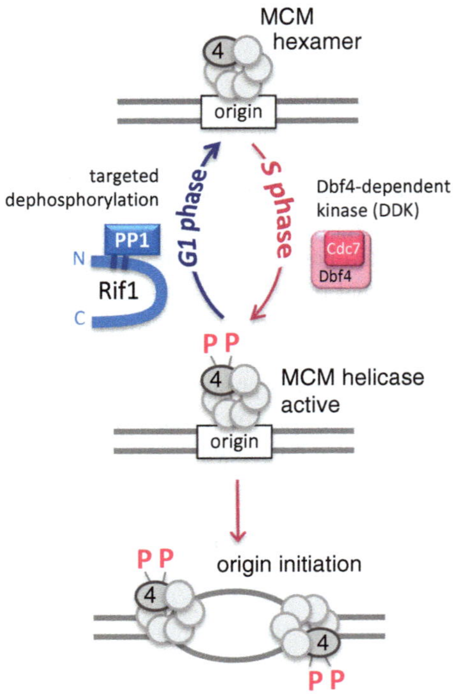

Fig. 23.2 Role of yeast Rif1-PP1 in restricting DNA replication initiation. Origin initiation during S phase requires MCM helicase activation (*pink downward arrows*), through phosphorylation of Mcm4 (and other MCM subunits) by DDK (composed of Cdc7 and Dbf4, *pink oblongs*). In G1 phase (*blue arrow*), Rif1 binds PP1 (*blue shapes*) through its PP1 interaction motifs (*dark blue bars*), directing it to dephosphorylate Mcm4 at origins. During G1 phase Dbf4 levels are low and dephosphorylation by Rif1-PP1 predominates, preventing premature origin initiation. As S phase begins, Dbf4 levels rise and DDK phosphorylates Mcm4, favouring replication initiation. Simultaneously, the Rif1-PP1 inhibitory effect is attenuated by DDK/CDK phosphorylation of Rif1 close to its PP1 interaction motifs, preventing PP1 interaction with Rif1 and targeting to the MCM complex [33–35]—so that in S phase, DDK-mediated origin initiation predominates

delayed in initiation, although this latter effect could be an indirect result of altered availability of initiation-limiting factors [38]. One likely determinant of susceptibility of an origin to effects of Rif1 removal is its proximity to chromosomal Rif1 binding sites [37]. Consistently, Rif1 is strongly associated with yeast telomeres, and telomere-proximal origins initiate replication aberrantly early in a *rif1Δ* mutant [34, 36, 39]. This effect on telomere replication highlights the fact that Rif1-PP1 not only controls S phase entry, but equally is important for specifying the replication timing programme once S phase has begun [37, 40], affecting origins throughout the genome to differing degrees. This dual role of Rif1 in S phase entry and the timing programme reflects the close intertwinement of replication timing control with initiation mechanisms. It is perhaps not surprising that a protein which opposes DDK activity determines both the S phase entry threshold and the replication timing programme, given that the levels of Dbf4 expression are critical both for regulating S phase entry and as one of the factors whose levels limit progression through the timing programme [38].

An intriguing twist to the effect of Rif1 on replication is that the interaction of Rif1 with PP1 is itself controlled by DDK-mediated phosphorylation. Dbf4 directly interacts with the C-terminal domain of Rif1 to mediate phosphorylation of residues surrounding its PP1 interaction motifs, preventing PP1 binding [33–35]. This control presumably serves to switch off the repression of replication by Rif1-PP1 as cells enter S phase. Execution of the temporal programme of origin initiation might be associated with progressive attenuation of the Rif1-PP1 control module through this mechanism.

Phosphorylation of Mcm4 is strongly downregulated by Rif1-PP1; other subunits of the yeast Mcm complex have not yet been tested and might also be affected. In fact it remains unclear exactly how many other substrates may be targeted by Rif1-PP1 to regulate replication. Phosphorylation of the critical CDK substrate Sld3 is also increased when *S. cerevisiae RIF1* is deleted [36], hinting that the Rif1-PP1 module may act more broadly to downregulate other phosphorylation events important for replication initiation, including CDK-mediated phosphorylation events. Phosphorylation of the second critical CDK substrate Sld2 is however not affected by Rif1 removal [36].

Is the Rif1-PP1 Replication Control Module Conserved?

Rif1 is present in all eukaryotic organisms so far examined, and Rif1 also affects S phase progression in vertebrate cells. Two different studies [41, 42] identified Rif1 as a critical determinant of the replication timing programme that stimulates origin initiation in early S phase. PP1 interaction RVxF and SILK motifs exist in all Rif1 proteins so far analysed (Fig. 23.3) [43] despite evolutionary divergence in Rif1 primary structure. MCM

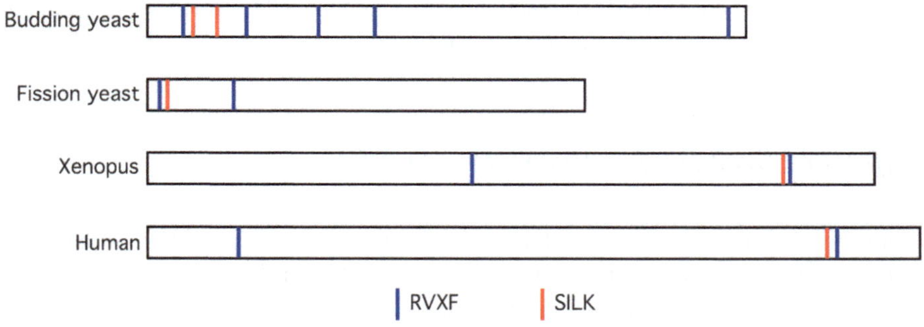

Fig. 23.3 PP1 interaction motifs in eukaryotic Rif1 proteins. Location of matches to PP1 interaction motifs RVxF (*blue*) and SILK (*red*) within the Rif1 protein sequences of budding yeast *S. cerevisiae*, fission yeast *S. pombe*, *X. laevis*, and *H. sapiens*. For *S. cerevisiae* and *S. pombe*, mutational analysis has shown that the N-terminal RVxF and SILK motifs are the most functionally significant of the motifs present for controlling replication

subunit phosphorylation is increased upon depletion of Rif1 from HeLa cells [42], mirroring the findings in yeast. Together these facts suggest the Rif1-PP1 replication control module is probably conserved and counteracts DDK function throughout eukaryotes. Although this suggestion has yet to be directly confirmed, a study in *Xenopus* provides data consistent with this idea, by demonstrating that PP1 is recruited to chromatin in vitro and acts continuously to reverse DDK-mediated phosphorylation of MCM subunits, thereby preventing MCM hyperphosphorylation and regulating replication progression [44]. The relevant *Xenopus* PP1 substrate-targeting subunit has yet to be identified, but could be Rif1. A number of high-throughput studies have shown that mammalian Rif1 interacts with PP1α and γ isoforms [13, 45], consistent with the notion that vertebrate Rif1 is a PP1 substrate-targeting subunit.

Are Other Effects of Rif1 Mediated by PP1?

Rif1 was initially identified for its repressive effect on telomere length in *S. cerevisiae* [46–48]; and for many years it was assumed the primary function of Rif1 was control of telomere length in yeast. However Rif1 is now known to carry out various functions in addition to its role in controlling DNA replication: in particular, Rif1 appears to regulate checkpoint recovery and repair pathway choice throughout eukaryotes [49–56], while in mammalian cells Rif1 has been shown to affect chromatin loop size [42]. Rif1 was identified by two independent studies as interacting with a HP1α- and SetDB1-containing H3K9 methylation complex [57, 58], a role which might relate to the control of replication timing in heterochromatic domains, or to a different Rif1 function.

The discovery that Rif1 directs PP1 activity to control replication raises the question of whether Rif1 also acts as a PP1 targeting subunit in controlling DNA repair, checkpoint recovery, chromatin loop size, and yeast telomere length. Whether any or all of these Rif1 chromosome stability functions are mediated through PP1 has yet to be directly addressed, but seems likely to be the case for at least some functions, given the conservation of PP1 interaction sites within the Rif1 sequence [43], and the demonstrated interaction of mammalian Rif1 with PP1α and γ isoforms [13, 45].

Effects of Orc2 Dephosphorylation by PP1

Phosphorylation events are central not only for replication initiation [29, 30] but also for preventing re-replication in a single cell cycle [59, 60]. It is not well understood whether and how phosphatases are needed to reverse replication-associated phosphorylation in re-establishing competence for pre-replication complex formation, but a series of interesting studies from the Hwang laboratory suggest a role for PP1 in such control. These investigations analysed effects of phosphorylation of the Origin Recognition Complex subunit Orc2 at CDK sites Thr-116 and Tyr-226, which occurs during S phase. In mammalian cells, ORC dissociates from chromatin (and presumably replication origins) during S phase, only re-associating as cells enter the next cell cycle. This cycle of ORC chromatin association presumably contributes to preventing re-replication within a single cell cycle [61, 62], since origins lacking ORC will be unable to re-form a pre-replication complex. Lee et al. [63] found that phosphorylation at Thr-116 and Tyr-226 is cyclically regulated, with phosphorylation at these sites increasing as cells pass through S phase and then removed following mitosis at entry into the next G1 phase. Various lines of evidence implicate all three major isoforms of PP1 (α, β, and γ) as involved in dephosphorylation of Orc2 at mitotic exit. Inhibition of these PP1 isoforms (by siRNA or using tautomycin) caused increased phosphorylation at these residues, while overexpression of any PP1 isoform reduces phosphorylation at the Orc2 Thr-116 and Tyr-226 residues [64]. Orc2 and PP1 can be co-precipitated, and Orc2 moreover contains a match to an RVxF PP1 interaction motif whose deletion prevents PP1 interaction and leads to increased Orc2 phosphorylation [65]. Nonphosphorylatable or phosphomimetic mutations at Thr-116 and Tyr-226 lead to shifts in ORC chromatin association. Overall, the effects are consistent with a model in which CDK-mediated Thr-116/Tyr-226 phosphorylation must be removed by PP1 upon mitotic exit to promote ORC binding to replication origins, necessary for pre-RC formation in preparation for entry into the next S phase. However, direct effects of PP1-mediated dephosphorylation of Orc2 on pre-RC formation or replication itself remain to be described. Interestingly however, the RVxF interaction motif is conserved in other vertebrate Orc2 proteins, and budding and fission yeast Orc2 proteins also contain a similar PP1 interaction motif [65].

Direct impact of Orc2 dephosphorylation by PP1 on replication has yet to be demonstrated. The papers described above however raise the intriguing suggestion that the action of PP1 early in the cell cycle may be stimulatory for replication (by re-establishing the capability of ORC for pre-RC formation upon mitotic exit [64, 65]), while at later cell cycle stages PP1 limits replication (by controlling the buildup of phosphorylated, activated MCM helicase [33, 44]). One emerging theme is that PP1 regulation may be viewed as a direct contributor to the bi-stable cycle of replication control where G1 phase corresponds to a 'low phosphorylation' state permissive for pre-RC formation, and S/G2 phase a 'high phosphorylation' state enabling origin activation.

Other Phosphatase Functions in Replication Control

A Role for PP2A in Stimulating DNA Replication

One of the first implications of a role for phosphatase activity in stimulating DNA replication came in 1998 from the Walter laboratory, with the finding that depletion of PP2A from a *Xenopus* extract cell-free system interfered with DNA replication [66]. Binding of ORC and the Mcm complex were unaffected by PP2A depletion, leading the authors to suggest that PP2A is required to stimulate an event necessary for the origin initiation step. Subsequent analysis confirmed this suggestion, with the demonstration that PP2A activity is required for loading of the initiation factor Cdc45 onto chromatin [67], although Cdc45 itself did not appear to be the target for PP2A in stimulating replication. A further study [68] designed to identify relevant target(s) of PP2A implicated Chk1 kinase as important for limiting DNA replication (particularly in response to damage) and as one possible target for PP2A dephosphorylation. However there is still no conclusive evidence for the relevant target(s) for PP2A in promoting replication in the *Xenopus* system, and another interesting possibility is raised by a recent investigation [69] that dissected roles for the *Xenopus* Chk1 kinase in controlling DNA replication. This study found that Chk1 negatively regulates the Treslin-mediated loading of Cdc45 onto chromatin, antagonizing replication initiation even during an unperturbed cell cycle. Chk1 interacts directly with a seven amino acid docking site at the C-terminus of Treslin, and the investigators proposed that phosphorylation of Treslin by Chk1 interferes with Cdc45 loading and so its stimulation of replication. If so, reversal of this inhibitory phosphorylation would be one possible mode through which PP2A may promote replication initiation. Indeed, Petersen et al. had suggested [68] that the then unidentified *Xenopus* homolog of Sld3 (now known to be Treslin) presents one likely dephosphorylation target through which PP2A might stimulate DNA replication. Petersen et al. [68] had implicated PP2A as involved in controlling replication through a number of routes, so the suggestion that PP2A stimulates replication by dephosphorylating Treslin does not preclude the possibility that Chk1 may also be directly targeted and inactivated by PP2A, particularly in mediating recovery from checkpoint-inducing conditions. Experiments in these studies generally did not distinguish between different PP2A complexes, so it remains an open question which PP2A complexes might be involved in replication control.

A further protein called DUE-B, identified as interacting with the mammalian c-myc replication origin sequence, has also been suggested as required to load Cdc45 at replication origins [70], with PP2A potentially contributing to control of its activity during the cell cycle [71]. Another possible role for PP2A in stimulating replication was raised by the identification of a regulatory B subunit of PP2A, called PR48, as interacting with the Cdc6 replication licensing factor, suggesting that PP2A-mediated dephosphorylation of Cdc6 could be important for replication control [72].

Cdc14 Acts Indirectly to Promote DNA Replication Initiation

Plasmid loss analyses and genetic interactions observed in *cdc14* mutants led to the suggestion that the *S. cerevisiae* dual-specificity PTP Cdc14 is required for replication origin initiation, e.g. [73, 74]. However it has gradually emerged these effects on replication are largely indirect, occurring as a consequence of the requirement for Cdc14 phosphatase in cell cycle 'resetting' to low CDK activity upon mitotic exit [75]. In particular, Cdc14 dephosphorylates the CDK inhibitor Sic1 upon mitotic exit, stabilizing Sic1 to help ensure repression of CDK activity during G1 phase, important for effective formation of pre-replication complexes in G1 phase [76–78]. Compromised CDK repression during late mitosis and G1 phase in *cdc14* mutants impacts on the effectiveness of pre-RC formation. Cdc14 therefore affects execution of S phase by impacting on control of Sic1 and various other cell cycle proteins [79], leading to the effects on replication that originally implicated Cdc14 in replication control

Concluding Remarks

The cellular role of phosphatases, originally regarded as providing a fairly constitutive, unregulated activity, has progressed to be viewed more correctly as encompassing a multifaceted, highly regulated set of enzymes with specificities of critical importance for cell cycle control. While now clear that the mechanisms governing DNA replication must be included in the network of cell cycle events controlled by regulated protein dephosphorylation, many important questions remain to be answered. In particular, as described above there are instances where one or more phosphatase activities have been implicated at a specific step in the DNA replication process, but the precise phosphatase complex involved remains unclear—or indeed whether that step is regulated by a single phosphatase or a series of distinct phosphatase enzymes. A good example is the role of phosphatase activities in recovery from replication and DNA damage checkpoint arrests, where in yeast PP1, PP4, and other phosphatases have all been implicated in various aspects of the process including H2AX dephosphorylation [80–83]. In vertebrate cells PP2A and PP4 appear to be involved in analogous H2AX deactivation [84, 85]. In such cases it will be a priority to understand the exact nature of phosphatase complexes involved, whether they act under distinct circumstances or else perhaps in a relay akin to that controlling mitotic exit [25], and to what extent the dephosphorylation circuitry is conserved from yeast to human cells. Further investigations promise to yield important insights into the significance of regulated protein dephosphorylation as a counterpoint to the kinase functions so important for controlling DNA replication.

Acknowledgments Work in the Donaldson/Hiraga lab is funded by Cancer Research UK grants C1445/A13356 and C1445/A19059, and BBSRC grant K006304/1.

References

1. Gray KA, Yates B, Seal RL, Wright MW, Bruford EA. Genenames.org: the HGNC resources in 2015. Nucleic Acids Res. 2015;43(Database issue):D1079–85.
2. Moorhead GB, Trinkle-Mulcahy L, Ulke-Lemee A. Emerging roles of nuclear protein phosphatases. Nat Rev Mol Cell Biol. 2007;8(3):234–44.
3. Bollen M, Peti W, Ragusa MJ, Beullens M. The extended PP1 toolkit: designed to create specificity. Trends Biochem Sci. 2010;35(8):450–8.
4. Moorhead GBG, De Wever V, Templeton G, Kerk D. Evolution of protein phosphatases in plants and animals. Biochem J. 2009;417:401–9.

5. Das AK, Helps NR, Cohen PT, Barford D. Crystal structure of the protein serine/threonine phosphatase 2C at 2.0 A resolution. EMBO J. 1996;15(24):6798–809.
6. Egloff MP, Cohen PT, Reinemer P, Barford D. Crystal structure of the catalytic subunit of human protein phosphatase 1 and its complex with tungstate. J Mol Biol. 1995;254(5):942–59.
7. Brautigan DL. Protein Ser/Thr phosphatases—the ugly ducklings of cell signalling. FEBS J. 2013;280(2):324–45.
8. Barker HM, Craig SP, Spurr NK, Cohen PT. Sequence of human protein serine/threonine phosphatase 1 gamma and localization of the gene (PPP1CC) encoding it to chromosome bands 12q24.1-q24.2. Biochim Biophys Acta. 1993;1178(2):228–33.
9. Durfee T, Becherer K, Chen PL, Yeh SH, Yang Y, Kilburn AE, et al. The retinoblastoma protein associates with the protein phosphatase type 1 catalytic subunit. Genes Dev. 1993;7(4):555–69.
10. Yoshida K, Watanabe M, Kato H, Dutta A, Sugano S. BH-protocadherin-c, a member of the cadherin superfamily, interacts with protein phosphatase 1 alpha through its intracellular domain. FEBS Lett. 1999;460(1):93–8.
11. Tran HT, Ulke A, Morrice N, Johannes CJ, Moorhead GB. Proteomic characterization of protein phosphatase complexes of the mammalian nucleus. Mol Cell Proteomics. 2004;3(3):257–65.
12. Heroes E, Lesage B, Gornemann J, Beullens M, Van Meervelt L, Bollen M. The PP1 binding code: a molecular-lego strategy that governs specificity. FEBS J. 2013;280(2):584–95.
13. Trinkle-Mulcahy L, Andersen J, Lam YW, Moorhead G, Mann M, Lamond AI. Repo-Man recruits PP1 gamma to chromatin and is essential for cell viability. J Cell Biol. 2006;172(5):679–92.
14. Trinkle-Mulcahy L, Andrews PD, Wickramasinghe S, Sleeman J, Prescott A, Lam YW, et al. Time-lapse imaging reveals dynamic relocalization of PP1gamma throughout the mammalian cell cycle. Mol Biol Cell. 2003;14(1):107–17.
15. Virshup DM, Shenolikar S. From promiscuity to precision: protein phosphatases get a makeover. Mol Cell. 2009;33(5):537–45.
16. Hendrickx A, Beullens M, Ceulemans H, Den Abt T, Van Eynde A, Nicolaescu E, et al. Docking motif-guided mapping of the interactome of protein phosphatase-1. Chem Biol. 2009;16(4):365–71.
17. Egloff MP, Johnson DF, Moorhead G, Cohen PT, Cohen P, Barford D. Structural basis for the recognition of regulatory subunits by the catalytic subunit of protein phosphatase 1. EMBO J. 1997;16(8):1876–87.
18. Hurley TD, Yang J, Zhang L, Goodwin KD, Zou Q, Cortese M, et al. Structural basis for regulation of protein phosphatase 1 by inhibitor-2. J Biol Chem. 2007;282(39):28874–83.
19. Roy J, Cyert MS. Cracking the phosphatase code: docking interactions determine substrate specificity. Sci Signal. 2009;2(100):9.
20. Pinsky BA, Kotwaliwale CV, Tatsutani SY, Breed CA, Biggins S. Glc7/protein phosphatase 1 regulatory subunits can oppose the Ipl1/aurora protein kinase by redistributing Glc7. Mol Cell Biol. 2006;26(7):2648–60.
21. Janssens V, Longin S, Goris J. PP2A holoenzyme assembly: in cauda venenum (the sting is in the tail). Trends Biochem Sci. 2008;33(3):113–21.
22. Xing Y, Li Z, Chen Y, Stock JB, Jeffrey PD, Shi Y. Structural mechanism of demethylation and inactivation of protein phosphatase 2A. Cell. 2008;133(1):154–63.
23. Margolis SS, Perry JA, Forester CM, Nutt LK, Guo Y, Jardim MJ, et al. Role for the PP2A/B56delta phosphatase in regulating 14-3-3 release from Cdc25 to control mitosis. Cell. 2006;127(4):759–73.
24. Hunt T. On the regulation of protein phosphatase 2A and its role in controlling entry into and exit from mitosis. Adv Biol Regul. 2013;53(2):173–8.
25. Grallert A, Boke E, Hagting A, Hodgson B, Connolly Y, Griffiths JR, et al. A PP1-PP2A phosphatase relay controls mitotic progression. Nature. 2015;517(7532):94–8.
26. Yu J, Zhao Y, Li Z, Galas S, Goldberg ML. Greatwall kinase participates in the Cdc2 autoregulatory loop in Xenopus egg extracts. Mol Cell. 2006;22(1):83–91.

27. Mochida S, Maslen SL, Skehel M, Hunt T. Greatwall phosphorylates an inhibitor of protein phosphatase 2A that is essential for mitosis. Science (New York, NY). 2010;330(6011): 1670–3.
28. Porter IM, Schleicher K, Porter M, Swedlow JR. Bod1 regulates protein phosphatase 2A at mitotic kinetochores. Nat Commun. 2013;4:2677.
29. Labib K. How do Cdc7 and cyclin-dependent kinases trigger the initiation of chromosome replication in eukaryotic cells? Genes Dev. 2010;24(12):1208–19.
30. Zegerman P. Evolutionary conservation of the CDK targets in eukaryotic DNA replication initiation. Chromosoma. 2015.
31. Heller RC, Kang S, Lam WM, Chen S, Chan CS, Bell SP. Eukaryotic origin-dependent DNA replication in vitro reveals sequential action of DDK and S-CDK kinases. Cell. 2011;146(1):80–91.
32. Yabuuchi H, Yamada Y, Uchida T, Sunathvanichkul T, Nakagawa T, Masukata H. Ordered assembly of Sld3, GINS and Cdc45 is distinctly regulated by DDK and CDK for activation of replication origins. EMBO J. 2006;25(19):4663–74.
33. Hiraga S, Alvino GM, Chang F, Lian HY, Sridhar A, Kubota T, et al. Rif1 controls DNA replication by directing protein phosphatase 1 to reverse Cdc7-mediated phosphorylation of the MCM complex. Genes Dev. 2014;28(4):372–83.
34. Dave A, Cooley C, Garg M, Bianchi A. Protein phosphatase 1 recruitment by Rif1 regulates DNA replication origin firing by counteracting DDK activity. Cell Rep. 2014;7(1):53–61.
35. Sridhar A, Kedziora S, Donaldson AD. At short telomeres Tel1 directs early replication and phosphorylates Rif1. PLoS Genet. 2014;10(10), e1004691.
36. Mattarocci S, Shyian M, Lemmens L, Damay P, Altintas DM, Shi T, et al. Rif1 controls DNA replication timing in yeast through the PP1 phosphatase Glc7. Cell Rep. 2014;7(1):62–9.
37. Hayano M, Kanoh Y, Matsumoto S, Renard-Guillet C, Shirahige K, Masai H. Rif1 is a global regulator of timing of replication origin firing in fission yeast. Genes Dev. 2012;26(2):137–50.
38. Mantiero D, Mackenzie A, Donaldson A, Zegerman P. Limiting replication initiation factors execute the temporal programme of origin firing in budding yeast. EMBO J. 2011;30(23): 4805–14.
39. Lian HY, Robertson ED, Hiraga S, Alvino GM, Collingwood D, McCune HJ, et al. The effect of Ku on telomere replication time is mediated by telomere length but is independent of histone tail acetylation. Mol Biol Cell. 2011;22(10):1753–65.
40. Peace JM, Ter-Zakarian A, Aparicio OM. Rif1 regulates initiation timing of late replication origins throughout the S. cerevisiae genome. PLoS One. 2014;9(5), e98501.
41. Cornacchia D, Dileep V, Quivy JP, Foti R, Tili F, Santarella-Mellwig R, et al. Mouse Rif1 is a key regulator of the replication-timing programme in mammalian cells. EMBO J. 2012; 31(18):3678–90.
42. Yamazaki S, Ishii A, Kanoh Y, Oda M, Nishito Y, Masai H. Rif1 regulates the replication timing domains on the human genome. EMBO J. 2012;31(18):3667–77.
43. Sreesankar E, Senthilkumar R, Bharathi V, Mishra RK, Mishra K. Functional diversification of yeast telomere associated protein, Rif1, in higher eukaryotes. BMC Genomics. 2012;13:255.
44. Poh WT, Chadha GS, Gillespie PJ, Kaldis P, Blow JJ. Xenopus Cdc7 executes its essential function early in S phase and is counteracted by checkpoint-regulated protein phosphatase 1. Open Biol. 2014;4(1):130138.
45. Esteves SL, Domingues SC, da Cruz e Silva SC, Fardilha M, da Cruz e Silva EF. Protein phosphatase 1alpha interacting proteins in the human brain. J Integr Biol. 2012;16(1-2):3–17.
46. Marcand S, Gilson E, Shore D. A protein-counting mechanism for telomere length regulation in yeast. Science (New York, NY). 1997;275(5302):986–90.
47. Lue NF. Closing the feedback loop: how cells "count" telomere-bound proteins. Mol Cell. 2009;33(4):413–4.
48. Hardy CF, Sussel L, Shore D. A RAP1-interacting protein involved in transcriptional silencing and telomere length regulation. Genes Dev. 1992;6(5):801–14.
49. Xue Y, Rushton MD, Maringele L. A novel checkpoint and RPA inhibitory pathway regulated by Rif1. PLoS Genet. 2011;7(12), e1002417.

50. Ribeyre C, Shore D. Anticheckpoint pathways at telomeres in yeast. Nat Struct Mol Biol. 2012;19(3):307–13.
51. Chapman JR, Barral P, Vannier JB, Borel V, Steger M, Tomas-Loba A, et al. RIF1 is essential for 53BP1-dependent nonhomologous end joining and suppression of DNA double-strand break resection. Mol Cell. 2013;49(5):858–71.
52. Escribano-Diaz C, Orthwein A, Fradet-Turcotte A, Xing M, Young JT, Tkac J, et al. A cell cycle-dependent regulatory circuit composed of 53BP1-RIF1 and BRCA1-CtIP controls DNA repair pathway choice. Mol Cell. 2013;49(5):872–83.
53. Zimmermann M, Lottersberger F, Buonomo SB, Sfeir A, de Lange T. 53BP1 regulates DSB repair using Rif1 to control 5' end resection. Science (New York, NY). 2013;339(6120):700–4.
54. Di Virgilio M, Callen E, Yamane A, Zhang W, Jankovic M, Gitlin AD, et al. Rif1 prevents resection of DNA breaks and promotes immunoglobulin class switching. Science (New York, NY). 2013;339(6120):711–5.
55. Kumar S, Yoo HY, Kumagai A, Shevchenko A, Shevchenko A, Dunphy WG. Role for Rif1 in the checkpoint response to damaged DNA in Xenopus egg extracts. Cell Cycle. 2012;11(6):1183–94.
56. Martina M, Bonetti D, Villa M, Lucchini G, Longhese MP. Saccharomyces cerevisiae Rif1 cooperates with MRX-Sae2 in promoting DNA-end resection. EMBO Rep. 2014.
57. Loyola A, Tagami H, Bonaldi T, Roche D, Quivy JP, Imhof A, et al. The HP1alpha-CAF1-SetDB1-containing complex provides H3K9me1 for Suv39-mediated K9me3 in pericentric heterochromatin. EMBO Rep. 2009;10(7):769–75.
58. Dan J, Liu Y, Liu N, Chiourea M, Okuka M, Wu T, et al. Rif1 maintains telomere length homeostasis of ESCs by mediating heterochromatin silencing. Dev Cell. 2014;29(1):7–19.
59. Diffley JF. Regulation of early events in chromosome replication. Curr Biol. 2004;14(18): R778–86.
60. Nguyen VQ, Co C, Li JJ. Cyclin-dependent kinases prevent DNA re-replication through multiple mechanisms. Nature. 2001;411(6841):1068–73.
61. Kreitz S, Ritzi M, Baack M, Knippers R. The human origin recognition complex protein 1 dissociates from chromatin during S phase in HeLa cells. J Biol Chem. 2001;276(9): 6337–42.
62. Siddiqui K, Stillman B. ATP-dependent assembly of the human origin recognition complex. J Biol Chem. 2007;282(44):32370–83.
63. Lee KY, Bang SW, Yoon SW, Lee SH, Yoon JB, Hwang DS. Phosphorylation of ORC2 protein dissociates origin recognition complex from chromatin and replication origins. J Biol Chem. 2012;287(15):11891–8.
64. Lee KY, Bae JS, Kim GS, Hwang DS. Protein phosphatase 1 dephosphorylates Orc2. Biochem Biophys Res Commun. 2014;447(3):437–40.
65. Lee KY, Bae JS, Yoon S, Hwang DS. Dephosphorylation of Orc2 by protein phosphatase 1 promotes the binding of the origin recognition complex to chromatin. Biochem Biophys Res Commun. 2014;448(4):385–9.
66. Lin XH, Walter J, Scheidtmann K, Ohst K, Newport J, Walter G. Protein phosphatase 2A is required for the initiation of chromosomal DNA replication. Proc Natl Acad Sci U S A. 1998;95(25):14693–8.
67. Chou DM, Petersen P, Walter JC, Walter G. Protein phosphatase 2A regulates binding of Cdc45 to the prereplication complex. J Biol Chem. 2002;277(43):40520–7.
68. Petersen P, Chou DM, You Z, Hunter T, Walter JC, Walter G. Protein phosphatase 2A antagonizes ATM and ATR in a Cdk2- and Cdc7-independent DNA damage checkpoint. Mol Cell Biol. 2006;26(5):1997–2011.
69. Guo C, Kumagai A, Schlacher K, Shevchenko A, Shevchenko A, Dunphy WG. Interaction of Chk1 with Treslin negatively regulates the initiation of chromosomal DNA replication. Mol Cell. 2015;57(3):492–505.
70. Chowdhury A, Liu G, Kemp M, Chen X, Katrangi N, Myers S, et al. The DNA unwinding element binding protein DUE-B interacts with Cdc45 in preinitiation complex formation. Mol Cell Biol. 2010;30(6):1495–507.

71. Gao Y, Yao J, Poudel S, Romer E, Abu-Niaaj L, Leffak M. Protein phosphatase 2A and Cdc7 kinase regulate the DNA unwinding element-binding protein in replication initiation. J Biol Chem. 2014;289(52):35987–6000.
72. Yan Z, Fedorov SA, Mumby MC, Williams RS. PR48, a novel regulatory subunit of protein phosphatase 2A, interacts with Cdc6 and modulates DNA replication in human cells. Mol Cell Biol. 2000;20(3):1021–9.
73. Hartwell LH, Smith D. Altered fidelity of mitotic chromosome transmission in cell cycle mutants of S. cerevisiae. Genetics. 1985;110(3):381–95.
74. Hogan E, Koshland D. Addition of extra origins of replication to a minichromosome suppresses its mitotic loss in cdc6 and cdc14 mutants of Saccharomyces cerevisiae. Proc Natl Acad Sci U S A. 1992;89(7):3098–102.
75. Fitzpatrick PJ, Toyn JH, Millar JB, Johnston LH. DNA replication is completed in Saccharomyces cerevisiae cells that lack functional Cdc14, a dual-specificity protein phosphatase. Mol Gen Genet. 1998;258(4):437–41.
76. Visintin R, Craig K, Hwang ES, Prinz S, Tyers M, Amon A. The phosphatase Cdc14 triggers mitotic exit by reversal of Cdk-dependent phosphorylation. Mol Cell. 1998;2(6):709–18.
77. Noton E, Diffley JF. CDK inactivation is the only essential function of the APC/C and the mitotic exit network proteins for origin resetting during mitosis. Mol Cell. 2000;5(1):85–95.
78. Lengronne A, Schwob E. The yeast CDK inhibitor Sic1 prevents genomic instability by promoting replication origin licensing in late G(1). Mol Cell. 2002;9(5):1067–78.
79. Dulev S, de Renty C, Mehta R, Minkov I, Schwob E, Strunnikov A. Essential global role of CDC14 in DNA synthesis revealed by chromosome underreplication unrecognized by checkpoints in cdc14 mutants. Proc Natl Acad Sci U S A. 2009;106(34):14466–71.
80. Keogh MC, Kim JA, Downey M, Fillingham J, Chowdhury D, Harrison JC, et al. A phosphatase complex that dephosphorylates gammaH2AX regulates DNA damage checkpoint recovery. Nature. 2006;439(7075):497–501.
81. Bazzi M, Mantiero D, Trovesi C, Lucchini G, Longhese MP. Dephosphorylation of gamma H2A by Glc7/protein phosphatase 1 promotes recovery from inhibition of DNA replication. Mol Cell Biol. 2010;30(1):131–45.
82. Travesa A, Duch A, Quintana DG. Distinct phosphatases mediate the deactivation of the DNA damage checkpoint kinase Rad53. J Biol Chem. 2008;283(25):17123–30.
83. Hustedt N, Seeber A, Sack R, Tsai-Pflugfelder M, Bhullar B, Vlaming H, et al. Yeast PP4 interacts with ATR homolog Ddc2-Mec1 and regulates checkpoint signaling. Mol Cell. 2015;57(2):273–89.
84. Chowdhury D, Keogh MC, Ishii H, Peterson CL, Buratowski S, Lieberman J. Gamma-H2AX dephosphorylation by protein phosphatase 2A facilitates DNA double-strand break repair. Mol Cell. 2005;20(5):801–9.
85. Nakada S, Chen GI, Gingras AC, Durocher D. PP4 is a gamma H2AX phosphatase required for recovery from the DNA damage checkpoint. EMBO Rep. 2008;9(10):1019–26.
86. Terrak M, Kerff F, Langsetmo K, Tao T, Dominguez R. Structural basis of protein phosphatase 1 regulation. Nature. 2004;429(6993):780–4.
87. Scotto-Lavino E, Garcia-Diaz M, Du G, Frohman MA. Basis for the isoform-specific interaction of myosin phosphatase subunits protein phosphatase 1c beta and myosin phosphatase targeting subunit 1. J Biol Chem. 2010;285(9):6419–24.
88. Cho US, Xu W. Crystal structure of a protein phosphatase 2A heterotrimeric holoenzyme. Nature. 2007;445(7123):53–7.
89. Moreland JL, Gramada A, Buzko OV, Zhang Q, Bourne PE. The molecular biology toolkit (MBT): a modular platform for developing molecular visualization applications. BMC Bioinformatics. 2005;6:21.

Chapter 24
DNA Replication Checkpoint Signaling

Nanda Kumar Sasi and Michael Weinreich

Abstract Eukaryotic cells respond to perturbations in DNA replication by activating checkpoint signaling pathways to maintain genome integrity. This ensures complete and accurate replication of DNA before chromosomes segregate during mitosis. Checkpoints can be activated upon replication fork stalling or DNA damage, both from extrinsic or intrinsic sources. Studies from several different eukaryotic model organisms have provided a clear picture of the broadly conserved replication checkpoint. In particular, the ATR and Chk1 kinases emerge as key regulators that mediate cellular responses to replication stress. The replication checkpoint stabilizes replication forks, coordinates repair, and induces a G2/M cell cycle arrest. It also promotes activation of nearby dormant origins while inhibiting late origin activation throughout the genome, which lengthens the S phase. In this chapter, we focus on the activation of replication checkpoint signaling, emphasizing its effect on origin activation.

Keywords Replication initiation • DNA replication checkpoint • Fork arrest • Origin • Chk1 • Chk2 • ATR • ATM • Rad53 • MCM helicase • Dormant origins

Introduction

Eukaryotes have large genomes that are tightly packed into chromatin and dispersed among multiple chromosomes and therefore utilize multiple origins of replication to replicate their genomes. Tight regulation of the initiation of DNA replication is required since re-initiation from any single origin during S phase would result in

N.K. Sasi
Laboratory of Genome Integrity and Tumorigenesis, Van Andel Research Institute, Grand Rapids, MI, USA

Graduate Program in Genetics, Michigan State University, East Lansing, MI, USA

M. Weinreich (✉)
Laboratory of Genome Integrity and Tumorigenesis, Van Andel Research Institute, Grand Rapids, MI, USA
e-mail: Michael.Weinreich@vai.org

D.L. Kaplan (ed.), *The Initiation of DNA Replication in Eukaryotes*,
DOI 10.1007/978-3-319-24696-3_24

over-replicated regions that could cause chromosome breakage during chromosome segregation. When DNA replication is perturbed or DNA is damaged during S phase, checkpoint mechanisms inhibit initiation events at late origins, which conserves limiting initiation proteins, and stabilize replication forks. Checkpoints also inhibit mitotic entry to allow time for the repair of DNA damage and replisome reactivation.

The initiation of eukaryotic DNA replication is separated into two mutually exclusive steps during the cell cycle. Origins are "licensed" in late M to early G1 phase by the loading of an inactive form of the replicative MCM (mini-chromosome maintenance) helicase. The MCM helicase is subsequently remodeled by the recruitment of Cdc45 and the four-subunit GINS (Go-Ichi-Ni-San) complex to form the active CMG (Cdc45-MCM-GINS) helicase, which is required to unwind origin DNA and moves with each replication fork. Helicase activation signals the beginning of S phase but occurs continuously at each individual origin that is utilized (or fires) during S phase.

Origin Licensing

Origin licensing requires binding of the hetero-hexameric origin recognition complex (ORC) to DNA (Fig. 24.1). Budding yeast origins are comprised of specific DNA sequences that bind ORC, but fission yeast and more complex eukaryotes specify ORC-binding sites with little or no DNA sequence specificity. Instead secondary DNA structures and chromatin features have been proposed to be important in origin determination in mammalian cells [1]. ORC is an ATP-binding protein comprised of Orc1 through Orc6 subunits, five of which (Orc1 to Orc5) have AAA$^+$ (or AAA$^+$-like) ATPase domains [2, 3]. To initiate DNA replication ORC (likely the Orc1 and/or Orc2 subunits) first recruits Cdc6, another AAA+ ATPase protein. Cdc6 is homologous to the Orc1 subunit and its ATPase activity is also important for replication initiation [2, 3]. The next step in origin licensing is recruitment of the eukaryotic replicative MCM helicase, which is a hetero-hexamer of Mcm2-7 subunits [2, 4]. Each of the six subunits of the MCM complex also contains AAA+ ATPase domains. The Mcm2-7 complex interacts with a mediator protein Cdt1. In budding yeast Cdt1 aids in the nuclear import of MCM complex, which allows docking of Cdt1-Mcm2-7 complex onto the DNA-ORC-Cdc6 complex [2]. The recruitment of Cdt1-Mcm2-7 to ORC-Cdc6 is thought to be mediated by Mcm3 interactions with ORC-Cdc6 and a Cdt1 interaction with Orc6 protein [5, 6]. These reactions lead to the formation of an ORC-Cdc6-Cdt1-Mcm2-7 complex (OCCM), which upon ATP hydrolysis leads to loss of Cdt1 forming an ORC-Cdc6-Mcm2-7 (OCM) complex [2, 7]. Subsequent recruitment of another Cdt1-Mcm2-7 complex forms a stable head-to-head double hexamer (dhMCM) followed by the dissociation of Cdt1 and Cdc6 [7]. The dhMCM bound to origin DNA forms what is known as the "pre-replicative complex" (pre-RC) and these origins are considered licensed for initiation. Although the precise biochemical roles of ATP hydrolysis by Cdc6, ORC, and the MCM subunits in pre-RC formation are not fully understood, several models have been proposed [2].

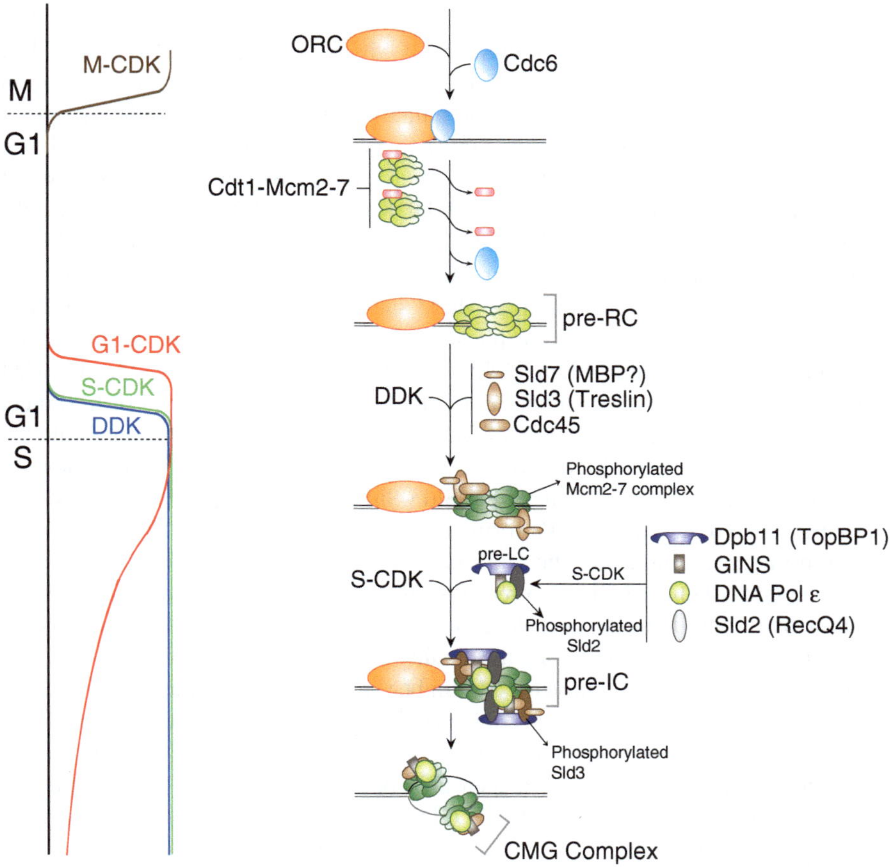

Fig. 24.1 Initiation of eukaryotic DNA replication. In G1-phase ORC-Cdc6 recruits Cdt1-Mcm2-7 to form a double-hexameric form of the MCM helicase (dhMCM) encircling dsDNA. This is also called the pre-replicative complex (pre-RC). Upon entering S phase the dhMCM helicase is activated by two protein kinases, DDK and CDK. These kinases facilitate formation of a pre-initiation complex (pre-IC) that ultimately forms the active Cdc45-MCM-GINS (CMG) helicase. The *S. cerevisiae* pathway shown here is broadly conserved in higher eukaryotes with some notable differences in pre-IC formation. See text for details. Also shown are the regulated levels of cell cycle-specific CDKs, and DDK

Origin Firing

The dhMCM complex can slide freely on dsDNA but it has no helicase activity [4, 7]. As mentioned above, the Cdc45 and GINS proteins form a complex with MCM to generate the active Cdc45-MCM-GINS (CMG) helicase [3, 7]. The transition of dhMCM to two CMG complexes involves multiple loading factors and another intermediate known as the pre-initiation complex (pre-IC, see Fig. 24.1). MCM helicase loading and activation are thus highly regulated and later sections will justify the need for these regulatory networks.

Two Ser/Thr protein kinases, DDK (Dbf4-dependent kinase) and CDK (cyclin-dependent kinase), are critical regulators of MCM loading and activation. CDK activity is lowest in G1 phase but increases at the G1- to S-phase transition (Fig. 24.1). G1-CDK inhibits the ubiquitin ligase APC/C, thereby stabilizing targets of this degradation pathway (e.g., Dbf4) to promote S-phase entry [8]. Hence G1-CDK indirectly promotes DDK activity. At the G1-S-phase transition, distinct S-CDKs are activated and cooperate with DDK to initiate DNA replication at individual origins. S-CDK and additional kinases like ATR phosphorylate the MCM complex in the pre-RC. This priming phosphorylation can facilitate phosphorylation by DDK, which targets multiple MCM subunits [9, 10]. DDK phosphorylates S/T residues and prefers acidic or phospho-S/T in the +1 position. Purified DDK phosphorylates individual Mcm2,3,4,6,7 subunits in vitro [11] and the dhMCM complex [12], but DDK phosphorylation does not cause dissociation of hexamers [13]. Instead one essential function of DDK in budding yeast is to phosphorylate the N-terminii of Mcm4 and Mcm6 to relieve an inhibitory effect on DNA replication [9, 14] and promote recruitment of the Sld3-Sld7-Cdc45 complex to dhMCM [3]. S-CDK is also required for CMG formation by phosphorylating Sld3 and Sld2 [15, 16] to prime interaction with the scaffolding protein Dpb11. Dpb11 loads GINS to form the CMG helicase and also recruits DNA polymerase ε to the origin [3, 7]. Each active CMG helicase complex, in association with the DNA polymerase ε, encircles ssDNA and moves along the leading strand in the 3′–5′ direction [17, 18].

CMG formation in fission yeast and higher eukaryotes is similar but with some notable differences [3, 7] since phosphorylation of Drc1^{Sld2} and Sld3 by S-CDK is less important in fission yeast than in budding yeast. In metazoans S-CDK-mediated phosphorylation of RecQ4^{Sld2} is dispensable but that of Treslin^{Sld3} is essential for recruitment of TopBP1^{Dpb11}, Cdc45, and GINS [19–21]. See Table 24.1 for comparison of protein names.

The precise role(s) of Cdc45 and GINS in promoting the helicase activity of the Mcm2-7 complex is under active investigation. In recent years, a number of additional proteins, e.g., Mcm10 [22], Ctf4 [23], DUE-B [24], and GEMC1 [25], have also been shown to be important for origin unwinding (Mcm10), Cdc45 recruitment (GEMC1), and coupling of polymerases and CMG helicase at the replication fork (Ctf4). Finally, the CMG complex recruits DNA polymerase α-primase, which is the only polymerase capable of initiating DNA synthesis de novo [7, 23].

DNA Replication Checkpoint

Eukaryotic DNA replication occurs efficiently and accurately due to the high number of replication origins and the fidelity of replicative polymerases. Coupling DNA repair with replication also increases overall accuracy. Nevertheless, DNA replication faces many hurdles even in an unperturbed cell cycle. Tight coordination of replication with other DNA-specific processes like transcription and chromatin remodeling poses major challenges since conflict between these processes can

Table 24.1 Conserved replication initiation and checkpoint proteins in yeast and metazoans

S. cerevisiae	*S. pombe*	Metazoans
Replication initiation proteins		
ORC1-6	ORC1-6	ORC1-6
Cdc6	Cdc6	Cdc6
Cdt1	Cdt1	Cdt1
Mcm2-7	Mcm2-7	Mcm2-7
DDK	DDK	DDK
Cdc7	Hsk1	Cdc7
Dbf4	Dfp1/Him1	Dbf4
–	–	Drf1
S-CDK	S-CDK	S-CDK
Cdc28	Cdc2	Cdk2
Clb5/Clb6	Cig1/Cig2	CyclinA/CyclinE
Sld7	–	MBP (?)
Sld3	Sld3	Treslin/ticrr
Cdc45	Cdc45/Sna41	Cdc45
Dpb11	Cut5/Rad4	TopBP1/Cut5/Rad4
GINS	GINS	GINS
Pol ε	Pol ε	Pol ε
Sld2	Drc1	RecQ4/RecQL4
Mcm10	Mcm10	Mcm10
Replication checkpoint proteins		
RPA	RPA	RPA
Ddc2	Rad26	ATRIP
Mec1	Rad3	ATR
Rad24	Rad17	Rad17
RFC2-5	RFC2-5	RFC2-5
Ddc1	Rad9	Rad9
Mec3	Hus1	Hus1
Rad17	Rad1	Rad1
Chk1	Chk1/Rad27	Chk1
Rad53	Cds1	Chk2
Mrc1	Mrc1	Claspin
Csm3	Swi3	Tipin
Tof1	Swi1	Timeless (Tim1)
Ctf4	Mcl1	And1
Swe1	Mik1	Wee1, Myt1
Mih1	Cdc25	Cdc25A-C
Bmh1, Bmh2	Rad24, Rad25	14-3-3
M-CDK	M-CDK	M-CDK
Cdc28	Cdc2	Cdk1
Clb1/Clb2/ Clb3/Clb4	Cig2/Cdc13	CyclinA/CyclinB

result in genomic instability. Oncogene-driven tumor cells are more susceptible to such conflicts since they have increased replication initiation events [26]. Other challenges arise due to the complex nature of eukaryotic genomes, which contain repetitive elements and heterochromatin. For example, chromosomal fragile sites often occur in late-replicating or heterochromatic regions where replication is more prone to stall [27]. Genotoxic agents including reactive oxygen species, heavy metals, by-products of metabolic processes, and exposure to harmful radiation from sunlight are all sources of replication stress. The DNA replication checkpoint is activated in response to stalled or damaged forks to help ensure genome integrity.

Checkpoint Activation by Compounds

The replication checkpoint is activated by DNA structures that are formed in the vicinity of stalled replication forks. Stalling or pausing occurs either when the fork encounters a DNA lesion or when there is a paucity of essential replication factors or nucleotides. Reagents that are normally used to study checkpoint pathways induce replication stress through different means [28]. One group of reagents inhibits the replicative polymerases but not the helicases. This can result in large stretches of ssDNA formed due to uncoupling of polymerase and helicase activities [29]. This kind of replication stress includes dNTP depletion caused by hydroxyurea (HU), an inhibitor of ribonucleotide reductase, and inhibition of Pol α-primase by aphidicolin [28]. Gemcitabine, a deoxycytidine analog, induces checkpoint activation by two mechanisms. Gemcitabine triphosphate competes with endogenous dCTP for incorporation into elongating DNA strands leading to inhibition of replication fork progression [30]. Gemcitabine triphosphate also inhibits ribonucleotide reductase, which depletes dNTP pools enhancing the replication fork arrest [28, 30].

Another group of reagents, like the platinum-based drugs cisplatin and carboplatin, form interstrand cross-links that restrict progression of both helicases and polymerases. These lesions activate various DNA repair pathways or form dsDNA breaks, both of which eventually lead to the generation of ssDNA [28]. DNA-alkylating agents form both kinds of lesions described above. Monofunctional alkylating agents like methylmethanesulfonate (MMS) form DNA adducts like N^7-methylguanine, N^3-methyladenine (3meA), N^1-methyladenine, and N^3-methylcytosine. The 3meA adduct can actively inhibit DNA polymerase and result in helicase-polymerase decoupling, generating ssDNA and checkpoint activation [31]. Other adducts can induce DNA repair pathways and indirectly activate checkpoint mechanisms. Bifunctional alkylating agents like mitomycin-C and cyclophosphamide form interstrand cross-links and block progression of both helicases and polymerases [31]. Another major source of replication stress is topoisomerase inhibition. Topoisomerases are critical enzymes that introduce an ssDNA nick (Top1) or dsDNA break (Top2) to relieve torsional stress and prevent collisions between replication and transcription machineries. Inhibitors of Top1 (e.g., camptothecin) and Top2 (e.g., etoposide) severely impair replication and transcription. These inhibitors trap the topoisomerase on the

DNA and subsequent collision of replication fork with Top1/2-DNA complexes can lead to dsDNA breaks, which is then followed by checkpoint activation [32]. Cells that do not have Top1 exhibit severe defects in replication fork progression [33] mainly due to the formation of aberrant DNA:RNA hybrids called R-loops. R-loops are structures formed when nascent mRNA transcripts hybridize with DNA exposing an ssDNA strand. Accumulation of torsional stress upon inhibition of topoisomerases promotes stable R-loop formation, which triggers DNA damage and checkpoint responses.

Mechanism of Checkpoint Activation

The general mechanism of replication checkpoint activation is shown in Fig. 24.2. Long stretches of ssDNA formed at stalled replication forks are stabilized and protected from nucleolytic degradation by association with the single-stranded binding protein, replication protein A (RPA). RPA-bound ssDNA is at least partially responsible for initiating the replication checkpoint response [34]. Reduced levels of RPA result in attenuated checkpoint activation in response to replication stress [35]. RPA-bound ssDNA recruits ATR through interactions with an essential ATR cofactor, ATRIP [28, 36, 37]. ATR in turn phosphorylates the 32 kDa subunit of the RPA complex and also mediates the recruitment of an ubiquitin ligase PRP19 that preferentially ubiquitylates hyper-phosphorylated RPA [38]. Polyubiquitylated RPA induces the recruitment of additional ATR-ATRIP complexes onto RPA-coated ssDNA, forming a feed-forward loop that is important for amplification of the replication checkpoint response [38]. Some forms of replication stress, however, do not result in large stretches of RPA-bound ssDNA. In such cases, DNA resection by repair mechanisms or collapse of stalled forks results in replication-associated double-strand breaks (DSBs). DSB-binding proteins amplify ATR checkpoint response at such DNA structures [39]. The DSBs are bound by the Ku70/Ku80 heterodimer, which then recruits the DNA repair protein DNA-dependent protein kinase catalytic subunit (DNA-PKcs). ATR kinase, initially activated by the small stretch of RPA-ssDNA, phosphorylates DNA-PKcs, which in turn phosphorylates RPA and other proteins downstream of ATR, thereby amplifying the ATR signal [39].

In the ATR checkpoint pathway a donut-shaped clamp composed of Rad9, Rad1, and Hus1 (the 9-1-1-complex) is loaded onto dsDNA adjacent to RPA-coated ssDNA by the Rad17-Rfc2-5 clamp loader complex [40, 41]. Independent recruitment of 9-1-1 and ATR-ATRIP complexes to stalled replication forks promotes autophosphorylation of ATR and kinase activation [42]. ATR phosphorylates Rad9 protein in the 9-1-1-complex [40], which then recruits TopBP1, another mediator of the checkpoint response. The Rad9, Rad1, Hus1 interacting nuclear orphan (RHINO) protein promotes the stable association of TopBP1 with 9-1-1-complex [43]. TopBP1, the homolog of budding yeast Dpb11, further stimulates ATR kinase activity and also acts as a platform to bring several other targets of ATR to the vicinity of stalled replication forks [40, 41].

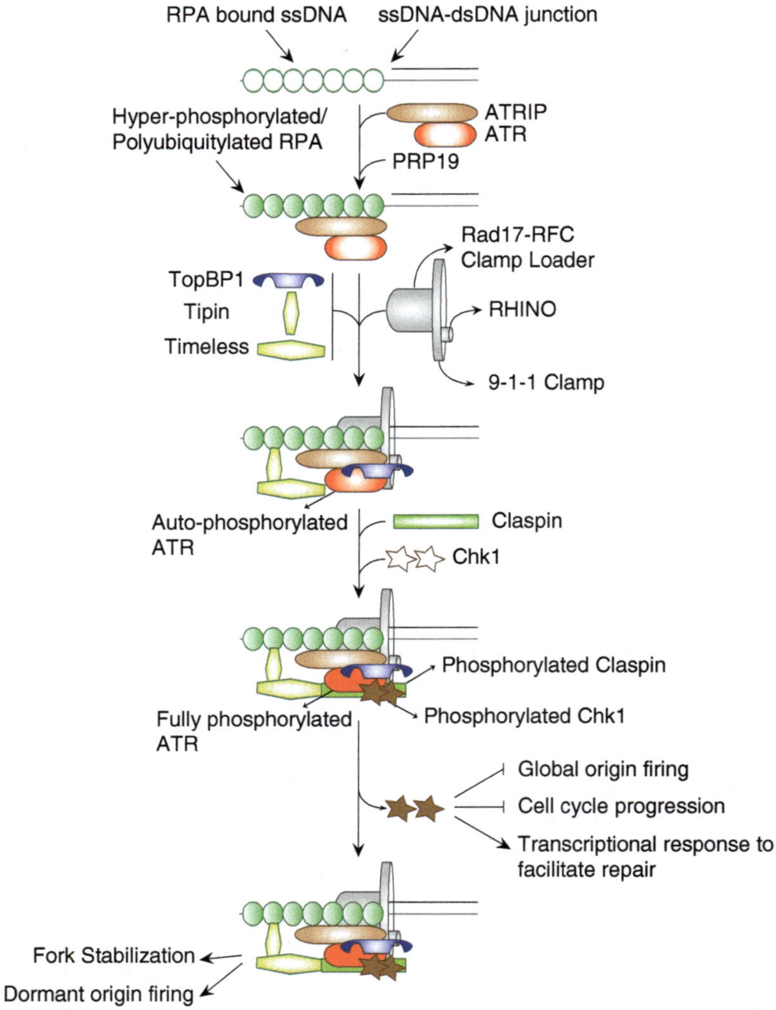

Fig. 24.2 Activation and transduction of the replication checkpoint. Replication stress generates long stretches of RPA-coated ssDNA, which recruits ATR-ATRIP. ATR phosphorylates RPA and also mediates the recruitment of the PRP19 ubiquitin ligase. Hyper-phosphorylation and polyubiquitylation of RPA form a feed-forward loop that recruits multiple ATR-ATRIP complexes. Subsequent interaction with 9-1-1-RHINO complex and TopBP1 promotes auto-phosphorylation of ATR kinase. Recruitment of Claspin, aided by Tipin and Timeless proteins, is essential for full activation of ATR kinase. Claspin also stabilizes and activates the downstream effector kinase Chk1, which is released from the chromatin to execute global checkpoint responses while ATR executes the local response. The core pathway is well conserved from yeast to humans with several additional proteins being involved in higher eukaryotes

In recent years several other proteins have been shown to be required for ATR activation. The NIMA (never in mitosis gene A)-related kinase 1 (NEK1) interacts with ATR-ATRIP, stabilizes the ATR-ATRIP interaction, and is required for efficient ATR activation [44]. The human homolog of *C. elegans* protein Rad5/Clk2 (HCLK2) stabilizes ATR kinase, and promotes autophosphorylation of ATR kinase and also binding of TopBP1 to ATR-ATRIP complex [45]. HCLK2 also interacts with the DNA processing enzymes FANCM and FAAP24, components of Fanconi anemia DNA repair pathway [45], which too are required for ATR signaling. FANCM and FAAP24 might be important for the proper processing of stalled forks upstream of ATR activation. Another HCLK2-interacting protein LARG [46], leukemia-associated Rho guanine nucleotide exchange factor, has recently been shown to promote efficient ATR-dependent checkpoint signaling [46] but its mechanism of action is unclear. Finally, a pro-apoptotic protein BH3-interacting death domain agonist (BID) has been shown to promote replication checkpoint activation. BID interacts with RPA-bound ssDNA and is thought to mediate the initial stages of ATR-ATRIP recruitment and stabilization on chromatin [47].

Major downstream ATR targets are the checkpoint kinase Chk1, and replisome components like RPA and Claspin [40]. ATR directly phosphorylates Chk1, which then transduces the replication checkpoint signals to downstream effector proteins [40]. However, complete activation of Chk1 depends on its interactions with several other checkpoint proteins. Nek9 interacts with Chk1 and is critical for its autophosphorylation, and full Chk1 kinase activity [48]. Claspin, Timeless, Tipin, and And1, called the fork-protection complex, are components of the replication fork and are important for normal fork progression rates in unperturbed cells [49]. At stalled forks, Tipin binds to RPA-coated ssDNA facilitating a stable Timeless-Tipin complex [50], which is required for an efficient replication checkpoint response [51]. Timeless-Tipin complex then recruits Claspin, which in turn facilitates the hyper-phosphorylation and stabilization of Chk1 [52]. The Claspin-Chk1 interaction is central to the replication checkpoint response and hence is heavily regulated. This complex regulation enables quick amplification of the checkpoint response and also turns off the replication checkpoint after replication stress has been resolved.

Claspin-Chk1 regulation is complex and occurs at three levels: phosphorylation, transcriptional regulation, and ubiquitylation. Initially, Chk1 phosphorylates Claspin at T916 to prevent its proteasomal degradation [53], thereby promoting the Claspin-Chk1 interaction and enhancing Chk1 activation. The DNA repair protein DNA-PKcs promotes the Claspin-Chk1 interaction, at least partly through transcriptional regulation of Claspin gene expression [54]. Multiple ubiquitylation and deubiquitylation enzymes also regulate the stability of Claspin and Chk1. ATR-mediated phosphorylation on Chk1-S345 not only activates its kinase activity but also marks Chk1 for ubiquitylation and degradation [55]. The ubiquitin ligases βTrCP1/2-SCF and Cdh1-APC/C ubiquitylate Claspin and Chk1 leading to proteasome-mediated degradation. The ubiquitin-specific protease 7 (USP7), USP20, USP28, and USP29 remove ubiquitin and stabilize Claspin-Chk1 in a checkpoint-dependent manner. USP7 has been implicated in the stability of both Claspin and Chk1 [56, 57] while

USP20, USP28, and USP29 stabilize Claspin [58–60]. Under unstressed conditions USP20 activity is inhibited by its interaction with the ubiquitin ligase HERC2, which promotes USP20 degradation [59]. Upon replication stress, ATR phosphorylates USP20 and disrupts the USP20-HERC2 interaction, thereby stabilizing USP20, which in turn deubiquitylates and stabilizes Claspin [59]. Claspin is essential for hyper-phosphorylation of multiple Chk1 proteins at stalled forks and for sustained maintenance of the checkpoint response [40, 49].

Activated Chk1 kinase is subsequently released from the chromatin to target various downstream effector proteins. The human homolog of *C. elegans* sex determination fem1 protein (FEM1B) is thought to be important for the release of active Chk1 from chromatin [61]. FEM1B directly interacts with Chk1 kinase and with Rad9, a component of the 9-1-1 complex, which could facilitate the recruitment of Chk1 kinase to stalled forks. Upon phosphorylation of Chk1 by upstream proteins like ATR, FEM1B-Chk1 interaction is disrupted suggesting a mechanism by which active Chk1 could be released from chromatin [61]. The multiple effects of activated DNA replication checkpoint signaling influence both local (replication fork stabilization, DNA repair, dormant origin firing, fork restart) and global processes (cell cycle arrest, inhibition of origin firing, transcriptional regulation) that preserve genome integrity (Fig. 24.2).

Inhibition of Late Origin Firing

Early insights into mammalian checkpoint signaling came from studies on an autosomal recessive disorder known as ataxia telangiectasia (AT). Cells derived from AT patients display higher sensitivity to radiation-induced DNA damage and an increased frequency of chromosomal aberrations [62]. The heightened response to radiation is not due to defect in a particular DNA repair mechanism but due to an inability of AT cells to stop DNA replication and cell cycle progression upon DNA damage. Upon exposure to X-rays, AT cells exhibit increased rates of DNA replication and faster progression into mitosis compared to normal cells, causing increased chromosomal aberrations and cell death [63]. Phenotypes similar to those in AT cells were observed by treating irradiated HeLa and Chinese hamster cells with caffeine [64]. These studies revealed that cells actively inhibit DNA replication and delay cell cycle progression in the face of DNA damage. We now know that AT cells are defective in the ataxia telangiectasia-mutated (ATM) protein kinase and that caffeine is an inhibitor of ATM and ATR (ATM and Rad3-related) protein kinases indicating that these kinases are important in DNA damage checkpoint signaling.

In normal cells, X-ray and UV-induced DNA damage results in inhibition of DNA synthesis mainly by preventing further initiation events and, to a lesser extent, by slowing replication fork elongation [65, 66]. Consequently, radio-resistant DNA replication observed in irradiated AT or caffeine-treated cells was also largely due to increased origin firing [63, 64]. This suggested that inhibition of origin firing is an important mechanism for increasing the length of S phase upon DNA damage.

The mechanism for inhibiting origin firing is well studied in budding yeast. Mutants in *MEC1*, the budding yeast ATR ortholog, and *RAD53* (the CHK2 ortholog) inappropriately activate late-replicating origins in response to the alkylating agent MMS or to nucleotide depletion by HU [67–69]. *mec1* and *rad53* mutants exhibited no effect on early origin firing and also did not disrupt the temporal nature of origin initiation [70]. Replication fork progression in HU-treated cells was greatly reduced (caused by decreased dNTP pools) but fork rates were similar in MMS-treated *mec1* and *MEC1* cells [71]. Late origin firing was inhibited by blocking an early step in initiation, presumably MCM helicase activation [68, 72, 73]. As described above, CDK and DDK are the main kinases involved in helicase activation, and in budding yeast the replication checkpoint blocks origin firing by modulating the functional activity of these kinases.

CDK phosphorylates Sld2 and Sld3 [15, 16] to promote interaction with Dpb11, which results in the recruitment of a Dpb11-Sld2-GINS-Pol ε complex to the chromatin-bound MCM-Cdc45-Sld3 complex (Fig. 24.1). When the replication checkpoint is activated, Rad53 phosphorylates Sld3 to disrupt its interaction with Dpb11 [72, 73] and Cdc45 [73], thus inhibiting pre-IC formation and late origin firing (Fig. 24.3a) [72, 73]. Similar mechanisms to regulate activities of downstream CDK targets exist in other eukaryotic organisms. The vertebrate homolog of Dpb11, TopBP1, is a large protein with roles in both DNA replication and checkpoint activation [74]. The functional homologs of Sld2 and Sld3 in vertebrates are RecQ4 and Treslin, respectively. Like in budding yeast, both RecQ4 and Treslin interact with TopBP1 and are essential for DNA replication [20, 21]. Furthermore, replication checkpoint-mediated Chk1 activation disrupts the TopBP1-Treslin interaction (Fig. 24.3b) similar to the effect of Rad53 on the Sld3-Dpb11 interaction in budding yeast. The mechanism of action in metazoans, however, seems different [75]. HU treatment in budding yeast activates Rad53, which hyper-phosphorylates Sld3 resulting in disruption of Sld3-Dpb11 and Sld3-Cdc45 interactions, which inhibits late origin firing (Fig. 24.3a). In metazoans, Chk1 activation upon HU treatment instead causes *hypo*-phosphorylation of Treslin; this phenotype is lost upon pretreatment with a Chk1 inhibitor. Both modifications, however, result in similar outcomes. The different modes of action could be due to differential regulation of S-CDK activities in yeast and metazoans. In yeast, CDK activity is not inhibited directly by checkpoint activation because S-CDK is essential to prevent relicensing of origins during S phase [8]. In metazoans, relicensing can be prevented by mechanisms independent of S-CDK like binding of the Cdt1 inhibitor, Geminin, and Cul4-mediated Cdt1 degradation [8]. Therefore, direct inhibition of CDK1 upon checkpoint activation is possible in metazoans, which could explain the hypo-phosphorylation of its target proteins like Treslin (Fig. 24.3b). On the other hand, Chk1 could recruit phosphatases to actively dephosphorylate Treslin in metazoans. One recent study has also shown that the C terminus of Treslin binds and is phosphorylated by Chk1 in Xenopus egg extracts and human cells [76]. This phosphorylation leads to inhibition of DNA replication in an unperturbed cell suggesting a role for basal Chk1 activity in maintaining the temporal order of origin firing during a normal S phase. However, it is also possible that Chk1-Treslin controls yet another mechanism by which the replication checkpoint regulates late

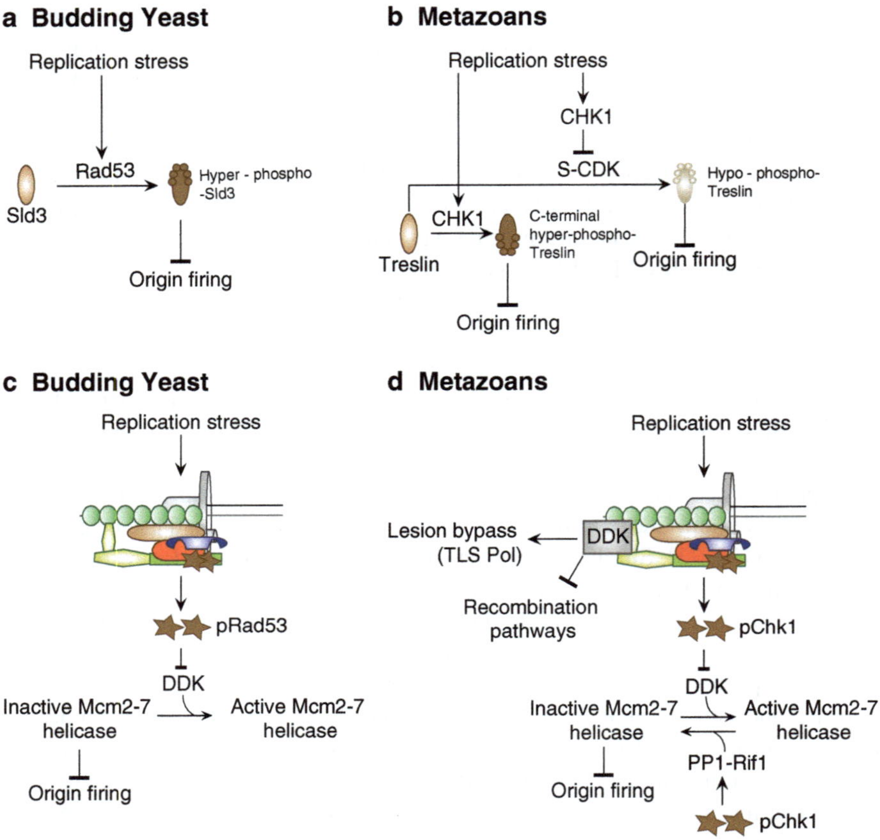

Fig. 24.3 Replication checkpoint-mediated inhibition of late origins. The replication checkpoint inhibits late origin firing primarily by targeting CDK and Sld3 (Treslin) (*panels* **a** and **b**) and the Dbf4-dependent kinase (DDK) (*panels* **c** and **d**). The differences in the mechanism of action between budding yeast and metazoans have been highlighted

origin firing (Fig. 24.3b). Additionally, the dual role of TopBP1 in replication initiation and checkpoint activation could also be exploited in response to replication stress. TopBP1 uses a single BRCT protein domain to bind Treslin as well as the 9-1-1 clamp and the Mre11-Rad50-Nbs1 (MRN) complex, which processes double-stranded DNA (dsDNA) breaks and initiates dsDNA repair. A phosphopeptide derived from the Rad9 subunit of 9-1-1 complex has been shown to compete with Treslin for interaction with TopBP1 [75]. Hence, activation of the 9-1-1 complex in the presence of replication stress could disrupt the Treslin-TopBP1 interaction (required for MCM helicase activation) resulting in inhibition of global origin firing.

DDK, or Dbf4-dependent Cdc7 kinase, is an essential S-phase kinase that regulates replication initiation. While Cdc7 kinase levels remain constant, the levels of its regulatory subunit Dbf4 (and/or Drf1 in metazoans) are cell cycle regulated. Drf1 is a Dbf4 homolog expressed during embryonic cell cycles in *Xenopus* and perhaps other organisms. Dbf4 expression peaks in S phase and remains high through early M phase followed by APC/C-mediated degradation [8]. Among the many target proteins of DDK are the Mcm2-7 helicase subunits. Phosphorylation of the helicase is required for its activation and thereby for initiation of replication. Moreover, Dbf4 is among the limiting factors that determine replication timing in budding yeast [77]. Therefore, upon exposure to replication stress DDK activity at origins is blocked to inhibit global origin firing. Studies from multiple organisms support this idea.

In budding yeast, Dbf4 is a direct target of Rad53 phosphorylation in response to HU [78, 79] and the hyper-phosphorylated version of DDK has modestly reduced kinase activity [78]. Rad53-phosphorylated Dbf4 inhibits late origin firing by an unknown mechanism [73] (Fig. 24.3c). In fission yeast, HU treatment leads to Cds1 (Rad53)-dependent hyper-phosphorylation of Dfp1 (Dbf4) [80]. An early study using *Xenopus* egg extracts also showed that Cdc7-Dbf4 complex dissociates upon treatment with etoposide, a topoisomerase inhibitor [37]. Studies using human cell lines also supported that DDK is a target of replication checkpoint. In a BCR-ABL tumor cell line, etoposide treatment resulted in dissociation of Cdc7-Dbf4 complex [81]. Multiple cancer cell lines exposed to sublethal doses of UV light showed similar phenotypes [82]. Chk1 was shown to interact with Dbf4 in vivo and phosphorylate it in vitro. Increased expression of Dbf4 also abrogated the ATR-Chk1-mediated intra-S-phase checkpoint induced by UV.

Although these studies indicate that the replication checkpoint targets DDK to inhibit origin firing, several studies point towards a more complex regulation. A non-essential N-terminal region of Dbf4 that interacts with Cdc5 and Rad53 has been found to be critical for survival in *rad53* but not *mec1* mutants; Mec1 is upstream of Rad53 [83]. In Xenopus egg extracts, Cdc7-Drf1 complex, the dominant form of DDK, was unaffected by aphidicolin treatment and the overall kinase activity of DDK was also unperturbed [84]. Finally, Mcm2 was hyper-phosphorylated at Cdc7-dependent sites in human cells arrested in S phase with HU [85], and Cdc7-Dbf4/Cdc7-Drf1 complexes were stable upon etoposide and HU treatment [86, 87]. The more recent study has shown that chromatin-bound Cdc7-Dbf4 complex is stabilized upon replication stress in an ATR-Chk1-dependent pathway. Chk1 phosphorylates and inhibits Cdh1, a component of the APC/C ubiquitin ligase. Chk1 also phosphorylates other components of the APC/C complex. Upon inhibition by Chk1, APC/C is unable to degrade Dbf4, thereby resulting in DDK stabilization on chromatin. The stable form of DDK at stalled replication forks recruits trans-lesion synthesis polymerase to replicate through the DNA lesion (Fig. 24.3d). This could act as a switch in determining the type of DNA repair pathway that is recruited to the stalled fork. Only in the absence of DDK does cisplatin treatment lead to recruitment of Rad51 (mediates homologous recombination) and mono-ubiquitylated forms of FancD2 (mediates Fanconi anemia DNA repair pathway) at stalled forks. Therefore, DDK might stabilize stalled forks by preventing aberrant recombination-mediated repair and thereby avoid genomic instability (Fig. 24.3d).

DDK's role in checkpoint signaling is complex. Perhaps a soluble fraction of DDK is phosphorylated to block firing of late origins upon replication stress but a chromatin-bound fraction is altered to enable lesion bypass through TLS. Replication checkpoint signaling could also regulate the role of DDK through phosphatases that are specific to DDK targets. Protein phosphatase 1 (PP1) is recruited to chromatin in a checkpoint-dependent manner and dephosphorylates DDK target sites, which could inhibit origin firing [88]. PP1 interacts with Rif1, an important determinant of origin firing timing [89], and Rif1 targets PP1 to DDK-phosphorylated proteins [90]. Through this mechanism DDK function could be regulated by the replication checkpoint without directly inhibiting DDK kinase activity (Fig. 24.3c), similar to the counteraction of CDK activity seen in budding yeast by Rad53 phosphorylation of Sld3.

Cell Cycle Arrest

Another global effect of the replication checkpoint is to arrest the cell cycle. Cell cycle progression requires the ordered activation of multiple CDKs at each stage of cell cycle. Inhibitory kinases Wee1 and Myt1 prevent cell cycle progression by inactivating mitotic CDKs. They phosphorylate two key residues in the ATP-binding domain: T14 and Y15. These phosphates are removed by the dual-specificity phosphatase Cdc25. While yeasts have a single Cdc25 phosphatase, mammalian cells have three isoforms: Cdc25A, B, and C. All three isoforms have been shown to promote G1-S and G2-M cell cycle progression with Cdc25A being more important for G1-S while Cdc25B and Cdc25C being primarily responsible for G2-M transition [91]. The inactivating kinases and activating phosphatases described above are important downstream targets of the ATR-Chk1 signaling induced by replication stress [91]. Chk1 directly phosphorylates Cdc25A resulting in its ubiquitin-mediated degradation [92]. Other kinases like Nek11, casein kinase 1α, and glycogen synthase kinase 3β work together with Chk1 kinase in mediating Cdc25A degradation [93]. Chk1-mediated phosphorylation of Cdc25B and Cdc25C causes increased binding with 14-3-3 proteins and subsequent sequestration in the cytoplasm [94, 95]. In the absence of Cdc25 phosphatase activity CDK complexes remain inactive and the cell cycle is arrested. Chk1 can also phosphorylate and activate Wee1 to enhance cell cycle arrest [96]. While these mechanisms are broadly conserved in fission yeast and mammalian cells, budding yeast rely on distinct set of pathways to induce cell cycle arrest. In the presence of replication stress, budding yeast activate Chk1 and Rad53 kinases, which then inhibit sister chromatid separation, spindle elongation, and mitotic exit, thereby inducing a cell cycle arrest. Sister chromatids are held together by a protein complex called cohesin, which can be cleaved by separase. Separase is kept inactive through its association with securin, but at the metaphase-to-anaphase transition securin is targeted for degradation by the APC^{Cdc20} ubiquitin ligase [97]. Chk1 phosphorylates securin and makes it resistant to APC^{Cdc20}-mediated degradation, thereby preventing separase activation and sister chromatid separation [97, 98]. Rad53 induces cell cycle arrest by phosphorylating Cdc5 (Polo kinase) and Bfa1, which are integral components of the mitotic exit network [97, 99]. Cdc5-mediated phosphorylation is also essential for the inactivation of APC^{Cdh1} ubiquitin ligase. Hence Rad53-mediated Cdc5 inhibition can lead to stabilization of APC^{Cdh1} ubiquitin ligase. Rad53 also phosphorylates the microtubule-associated proteins Cin8 and Stu2, which prevents spindle elongation and further restricts sister chromatid segregation [100].

Dormant Origin Firing

Although checkpoint activation inhibits global initiation events, one way to replicate DNA in the vicinity of a stalled replication fork is through activation of nearby dormant origins. Eukaryotic cells initiate DNA replication from origins spaced from 40 to ~200 kb apart depending on the organism; however more origins are licensed than are actually used [101]. Current estimates suggest that only ~10 % of licensed origins are used in each S phase of metazoan cells [101]. As stated above, under normal growth conditions Chk1 kinase has an inhibitory effect on origin activation likely through interaction of Chk1 with the replication initiation factor, Treslin [76]. This suggests a possible mechanism by which many origins are kept dormant. However, when replication forks stall these dormant origins are activated to complete replication in the stressed regions of the genome, although the mechanism for this is unclear.

Several models have been proposed for dormant origin activation following replication stress [101, 102]. One model predicts a passive activation of dormant origins. In unstressed cells, replication forks originating from active origins passively replicate the nearby dormant origin DNA and render it incapable of firing during that S phase. In the presence of replication stress, however, forks stall and the possibility of dormant origins being passively replicated is reduced. These dormant origins might then fire and replicate through the stalled replication fork, thereby completing replication. Of course, to do this they must somehow bypass the inhibition to replication origin firing induced by replication stress. Even if dormant origins in the vicinity of stalled forks are passively activated, it is necessary for the cell to inhibit global initiation events. Activation of dormant origins throughout the genome in response to fork stalling would exhaust limiting replication factors, which would be deleterious for the cell [77, 103, 104]. Upon exposure to replication stress, global origin firing is inhibited by the ATR-Chk1-mediated replication checkpoint (discussed above). Hence, while dormant origins in the vicinity of stalled replicated fork are activated, most distal origins are inhibited. Evidence for such a mechanism has come from studies in human cells. Human Chk1 activated by low doses of HU can selectively inhibit origin firing in new replication factories but not in existing factories [105]. How dormant origins in active replication factories escape the inhibitory effect of Chk1 is not known. Presence of an active mechanism that protects these factories from Chk1-mediated inhibition has been suggested [102].

Conversely, dormant origin activation could be mediated by checkpoint proteins directly recruited to stalled replication forks (Fig. 24.4). ATR in particular is known

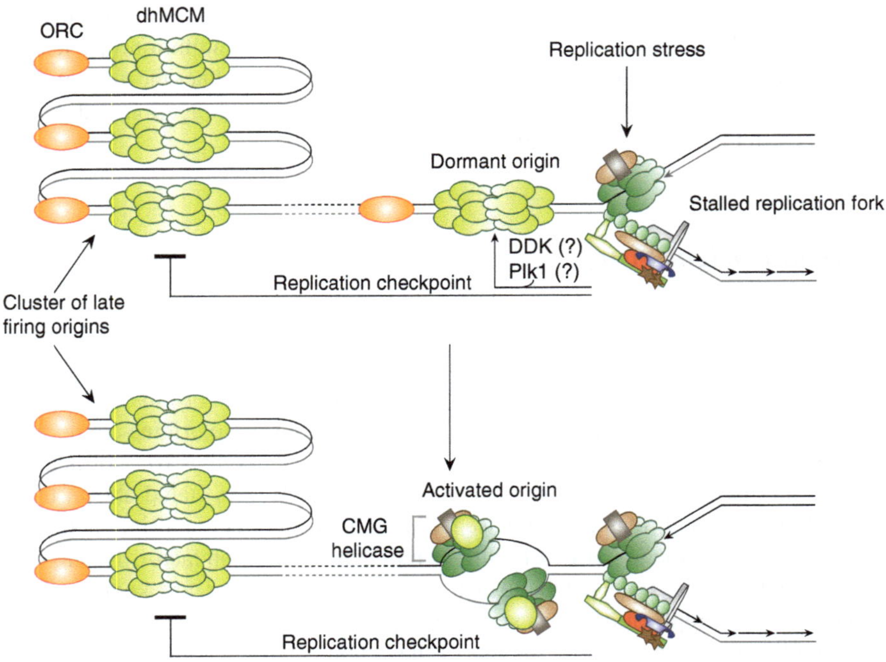

Fig. 24.4 Replication checkpoint-mediated activation of dormant origins. Stalled forks can be rescued by initiating replication from an adjacent dormant origin. The mechanism by which such dormant origins escape global inhibition of origin firing is not known. One possibility is the checkpoint-mediated recruitment of proteins like Plk1 or DDK could phosphorylate and activate a nearby inactive replicative helicase and promote dormant origin firing

to phosphorylate several components of the replication machinery including the Mcm2-7 complex, ORC, DNA polymerase ε, RF-C, RPA1, and RPA2 [40, 106]. ATR-mediated phosphorylation of Mcm2-7 is needed for recruitment of Polo-like kinase1 (Plk1) to chromatin and for DNA replication under replication stress [107]. Plk1 also phosphorylates Orc2 and promotes replication in the presence of several replication inhibitors [108]. Therefore, checkpoint proteins that are recruited and activated locally at stalled forks (like ATR) could promote dormant origin firing whereas activated effector checkpoint kinases (like Chk1) would ensure global inhibition of initiation, transcriptional control of replication factors, and cell cycle arrest.

Dormant origin firing is also important for completion of DNA replication in normal cells [109]. Mice with hypomorphic alleles of Mcm4 exhibit greatly reduced binding of Mcm2-7 complex to chromatin, which results in fewer licensed origins and therefore fewer dormant origins. These mice are highly prone to tumorigenesis and exhibit increased genomic instability [109, 110]. Dormant origins are thought to be important for complete replication of late-replicating regions of the genome. These are also important for rescuing replication when two converging forks stall in response to replication stress [111] because the intervening DNA cannot be passively replicated and would therefore remain unduplicated causing genomic instability.

In such a scenario, activation of dormant origins could complete replication. Based on modeling studies, such double-fork stalling events have been estimated to occur every 10 Mb of the genome [111]. This is larger than any yeast chromosome, so yeast might be much less dependent on dormant origins than eukaryotes with larger genome sizes. Hence, activation of dormant origins is a localized method of rescuing DNA replication under stressful conditions.

Stabilization of Stalled Forks

Stalled forks are prone to aberrant recombination events and collision with active transcriptional and co-transcriptional machineries [112]. The tethering of transcriptionally active genes to the nuclear pore complex (NPC) also increases torsional stress in the DNA increasing the chance of nearby fork reversal and collapse [113]. Collapsed forks pose a challenge to DNA replication restart and promote genomic instability. Stabilization of forks and the resumption of DNA replication are especially important when a particular genomic region is devoid of extra origins, such as fragile site loci and telomeres. Hence one of the key roles played by the DNA replication checkpoint is to maintain the integrity of stalled replication forks [112].

Stability of replication forks has to be maintained not only during periods of replication stress but also during normal DNA replication. This function is primarily performed by components of the "fork protection complex," composed of Tim1, Tipin, Claspin, and And1 proteins [49]. They maintain fork stability mainly by coupling the CMG helicase with DNA polymerases at replication forks. These proteins are also important for complete activation of the replication checkpoint response in the presence of replication stress [49]. While homologous recombination is a DNA repair pathway that can be used by cells to repair collapsed or reversed forks [112], aberrant recombination at stalled forks would be deleterious to cells. Therefore, both positive and negative regulation of recombination by the replication checkpoint has been reported but the mechanism of differential regulation is not clearly understood [28, 112]. Several nucleases and helicases that affect fork stability are also targets of checkpoint kinases [28, 112]. Some nucleases like Exo1 and Mus81-Eme1 initiate deleterious fork cleavage and DNA resection at stalled forks. Their activities, therefore, are inhibited upon phosphorylation by replication checkpoint proteins [28, 112]. Helicases like Dna2, which also possess nuclease activity, and Sgs1/WRN/BLM, however, are required to maintain fork stability and are recruited to stalled forks upon phosphorylation [28, 112]. Other targets of the replication checkpoint include replisome components like DNA polymerases and helicases [40]. Although dissociation of the replisome at stalled forks was thought to accompany fork collapse [114], one recent study has shown that replisome components remain intact even in checkpoint mutant cells exposed to replication stress [115]. This study, however, did show that the replisome components were lost from a small set of early origins. It is possible that analysis of such early origins in previous ChIP studies led to the conclusion that the replisome is destabilized in checkpoint mutants upon replication stress. The emerging model currently is that the replication checkpoint regulates the function of the replisome rather than its stability. Another group also seems to have come to a similar conclusion using the iPOND (isolation of proteins on nascent DNA) technique in human cells [116]. The checkpoint-mediated stability of replisome components and its role in maintaining fork stability warrant further study.

Finally, stalled forks can be stabilized by preventing collision with transcriptional machinery and by reducing the topological force induced by transcription. Actively transcribing genes in budding yeast are tethered to the nuclear periphery [117], which can impose topological constraints on progression of replication forks especially in the presence of replication stress [118]. Factors involved in this tethering phenomenon are targets of checkpoint kinases [119], and activation of checkpoint kinases leads to the release of genes tethered to nuclear periphery [113]. The replication stress checkpoint pathway also actively regulates transcription of tRNA genes. Because the pre-initiation complex formed by RNA polymerase III interferes with the activity of DNA polymerases, transcribing tRNA genes act as barriers to replication fork progression. A Mec1-Rad53-Mrc1 pathway represses tRNA gene expression [120]. The master repressor Maf1 carries out the repression. Rad53 could activate Maf1 to repress tRNA gene repression, thereby preventing fork collapse by reducing the chances of fork arrest at tRNA genes [120]. Other aspects of transcription like splicing and chromatin remodeling also affect fork stability, both of which are targeted by active replication checkpoint kinases [28, 106, 121]. These studies indicate that regulation of transcriptional processes is an important way by which checkpoint pathways promote fork stability.

Summary

Preserving genome integrity is essential for accurate transmission of the genetic material and to prevent tumorigenesis. Replication stress forms a major type of DNA damage that can compromise genome integrity and cells respond to such damage by activating replication checkpoint pathways. The checkpoint mounts a multipronged response that tackles DNA damage at a local level while also regulating cellular wide processes like origin firing and cell cycle progression. Tumor cells experience high levels of replication stress owing to their high replicative potential, which in turn is attained through inactivating mutations in tumor-suppressor proteins. This leaves tumor cells at the mercy of few functioning checkpoint proteins. This overdependence can be exploited to develop strategies to specifically target tumor cells as has been shown with chemical inhibition of checkpoint proteins like RPA, ATR, Chk1, and Wee1. A better understanding of the replication checkpoint would present more targets for therapeutic intervention and might allow synergistic enhancement of existing drugs.

References

1. Leonard AC, Méchali M. DNA replication origins. Cold Spring Harb Perspect Biol. 2013;5(10):010116.
2. Yardimci H, Walter JC. Prereplication-complex formation: a molecular double take? Nat Struct Mol Biol. 2013;21(1):20–5.
3. Yan L, Hiroyuki A. Loading and activation of DNA replicative helicases: the key step of initiation of DNA replication. Genes Cells. 2013;18(4):266–77.
4. Remus D, Beuron F, Tolun G, Griffith JD, Morris EP, Diffley JF. Concerted loading of Mcm2-7 double hexamers around DNA during DNA replication origin licensing. Cell. 2009;139(4): 719–30.
5. Sun J, Evrin C, Samel SA, Fernández-Cid A, Riera A, Kawakami H, et al. Cryo-EM structure of a helicase loading intermediate containing ORC-Cdc6-Cdt1-MCM2-7 bound to DNA. Nat Struct Mol Biol. 2013;20(8):944–51.
6. Frigola J, Remus D, Mehanna A, Diffley JF. ATPase-dependent quality control of DNA replication origin licensing. Nature. 2013;495(7441):339–43.
7. Silvia T, Alberto R, Christian S. Switch on the engine: how the eukaryotic replicative helicase MCM2–7 becomes activated. Chromosoma. 2014.
8. Khalid S, Kin Fan O, John FXD. Regulating DNA replication in eukarya. Cold Spring Harb Perspect Biol. 2013;5(9):a012930.
9. Randell JC, Fan A, Chan C, Francis LI, Heller RC, Galani K, et al. Mec1 is one of multiple kinases that prime the Mcm2-7 helicase for phosphorylation by Cdc7. Mol Cell. 2010;40(3):353–63.
10. Sheu Y-JJ, Stillman B. Cdc7-Dbf4 phosphorylates MCM proteins via a docking site-mediated mechanism to promote S phase progression. Mol Cell. 2006;24(1):101–13.
11. Karim L. How do Cdc7 and cyclin-dependent kinases trigger the initiation of chromosome replication in eukaryotic cells? Genes Dev. 2010;24(12):1208–19.
12. Sun J, Fernandez-Cid A, Riera A, Tognetti S, Yuan Z, Stillman B, et al. Structural and mechanistic insights into Mcm2-7 double-hexamer assembly and function. Genes Dev. 2014;28(20):2291–303.
13. On KF, Beuron F, Frith D, Snijders AP, Morris EP, Diffley JF. Prereplicative complexes assembled in vitro support origin-dependent and independent DNA replication. EMBO J. 2014;33(6): 605–20.
14. Sheu Y-JJ, Stillman B. The Dbf4-Cdc7 kinase promotes S phase by alleviating an inhibitory activity in Mcm4. Nature. 2010;463(7277):113–7.
15. Tanaka S, Umemori T, Hirai K, Muramatsu S, Kamimura Y, Araki H. CDK-dependent phosphorylation of Sld2 and Sld3 initiates DNA replication in budding yeast. Nature. 2007;445(7125): 328–32.
16. Zegerman P, Diffley JF. Phosphorylation of Sld2 and Sld3 by cyclin-dependent kinases promotes DNA replication in budding yeast. Nature. 2007;445(7125):281–5.
17. Fu YV, Yardimci H, Long DT, Ho TV, Guainazzi A, Bermudez VP, et al. Selective bypass of a lagging strand roadblock by the eukaryotic replicative DNA helicase. Cell. 2011;146(6):931–41.
18. Kang Y-HH, Galal WC, Farina A, Tappin I, Hurwitz J. Properties of the human Cdc45/ Mcm2-7/GINS helicase complex and its action with DNA polymerase epsilon in rolling circle DNA synthesis. Proc Natl Acad Sci U S A. 2012;109(16):6042–7.
19. Itou H, Muramatsu S, Shirakihara Y, Araki H. Crystal structure of the homology domain of the eukaryotic DNA replication proteins Sld3/Treslin. Structure. 2014;22(9):1341–7.
20. Kumagai A, Shevchenko A, Shevchenko A, Dunphy WG. Treslin collaborates with TopBP1 in triggering the initiation of DNA replication. Cell. 2010;140(3):349–59.
21. Sangrithi MN, Bernal JA, Madine M, Philpott A, Lee J, Dunphy WG, et al. Initiation of DNA replication requires the RECQL4 protein mutated in Rothmund-Thomson syndrome. Cell. 2005;121(6):887–98.
22. Watase G, Takisawa H, Kanemaki MT. Mcm10 plays a role in functioning of the eukaryotic replicative DNA helicase, Cdc45-Mcm-GINS. Curr Biol. 2012;22(4):343–9.

23. Simon AC, Zhou JC, Perera RL, van Deursen F, Evrin C, Ivanova ME, et al. A Ctf4 trimer couples the CMG helicase to DNA polymerase α in the eukaryotic replisome. Nature. 2014;510(7504):293–7.
24. Chowdhury A, Liu G, Kemp M, Chen X, Katrangi N, Myers S, et al. The DNA unwinding element binding protein DUE-B interacts with Cdc45 in preinitiation complex formation. Mol Cell Biol. 2010;30(6):1495–507.
25. Balestrini A, Cosentino C, Errico A, Garner E, Costanzo V. GEMC1 is a TopBP1-interacting protein required for chromosomal DNA replication. Nat Cell Biol. 2010;12(5):484–91.
26. Jones RM, Mortusewicz O, Afzal I, Lorvellec M, García P, Helleday T, et al. Increased replication initiation and conflicts with transcription underlie Cyclin E-induced replication stress. Oncogene. 2013;32(32):3744–53.
27. Debatisse M, Le Tallec B, Letessier A, Dutrillaux B, Brison O. Common fragile sites: mechanisms of instability revisited. Trends Genet. 2011;28(1):22–32.
28. Hustedt N, Gasser SM, Shimada K. Replication checkpoint: tuning and coordination of replication forks in s phase. Genes. 2012;4(3):388–434.
29. Byun TS, Pacek M, Yee M-cC, Walter JC, Cimprich KA. Functional uncoupling of MCM helicase and DNA polymerase activities activates the ATR-dependent checkpoint. Genes Dev. 2005; 19(9):1040–52.
30. Ewald B, Sampath D, Plunkett W. Nucleoside analogs: molecular mechanisms signaling cell death. Oncogene. 2008;27(50):6522–37.
31. Dragony F, Jennifer AC, Leona DS. Balancing repair and tolerance of DNA damage caused by alkylating agents. Nat Rev Cancer. 2012;12(2):104–20.
32. Mohamed EA, Reham A, Sherif FE-K. Topoisomerase-mediated chromosomal break repair: an emerging player in many games. Nat Rev Cancer. 2015;15(3):137–51.
33. Sandie T, Laure C, Chiara C, Hélène T, Heidi H-G, Anna J, et al. Topoisomerase I suppresses genomic instability by preventing interference between replication and transcription. Nat Cell Biol. 2009;11(11):1315–24.
34. Alexandre M, Lee Z. RPA-coated single-stranded DNA as a platform for post-translational modifications in the DNA damage response. Cell Res. 2015;25(1):9–23.
35. Zou L, Liu D, Elledge SJ. Replication protein A-mediated recruitment and activation of Rad17 complexes. Proc Natl Acad Sci U S A. 2003;100(24):13827–32.
36. Zou L, Elledge SJ. Sensing DNA damage through ATRIP recognition of RPA-ssDNA complexes. Science. 2003;300(5625):1542–8.
37. Costanzo V, Shechter D, Lupardus PJ, Cimprich KA, Gottesman M, Gautier J. An ATR- and Cdc7-dependent DNA damage checkpoint that inhibits initiation of DNA replication. Mol Cell. 2002;11(1):203–13.
38. Alexandre M, Ju-Mei L, Xiao Ye J, Ching-Shyi W, Stephanie AY, Hai Dang N, et al. PRP19 transforms into a sensor of RPA-ssDNA after DNA damage and drives ATR activation via a ubiquitin-mediated circuitry. Mol Cell. 2014;53(2):235–46.
39. Sophie V-E, Chantal D, Jihane B, Angelos C. DNA structure-specific priming of ATR activation by DNA-PKcs. J Cell Biol. 2013;202(3):421–9.
40. Cimprich KA, Cortez D. ATR: an essential regulator of genome integrity. Nat Rev Mol Cell Biol. 2008;9(8):616–27.
41. Edward AN, David C. ATR signalling: more than meeting at the fork. Biochem J. 2011; 436(3):527–36.
42. Liu S, Shiotani B, Lahiri M, Maréchal A, Tse A, Leung CC, et al. ATR autophosphorylation as a molecular switch for checkpoint activation. Mol Cell. 2011;43(2):192–202.
43. Cecilia C-R, McDonald ER, Kristen H, Mathew ES, Harper JW, Stephen JE. A DNA damage response screen identifies RHINO, a 9-1-1 and TopBP1 interacting protein required for ATR signaling. Science. 2011;332(6035):1313–7.
44. Shizhou L, Chu Kwen H, Jian O, Lee Z. Nek1 kinase associates with ATR–ATRIP and primes ATR for efficient DNA damage signaling. Proc Natl Acad Sci. 2013.
45. Zuzana H, Spencer JC, Simon JB. FANCM-FAAP24 and HCLK2: roles in ATR signalling and the Fanconi anemia pathway. Cell Cycle. 2009;8(8):1133–7.

46. Ryan DB, Christopher JS, Abhijit AP, Katie NM, Sarah M, Skehel JM, et al. The leukemia-associated Rho guanine nucleotide exchange factor LARG is required for efficient replication stress signaling. Cell Cycle. 2014;13(21):3450–9.
47. Yang L, Sivaraja V, Qiong S, Walter JC, Sandra SZ. BID binds to replication protein A and stimulates ATR function following replicative stress. Mol Cell Biol. 2011;31(21):4298–309.
48. Scott CS, Aleksandra VP, Matthew ZM, Hongyan W, Yunfeng P, Matthew DW, et al. A gemcitabine sensitivity screen identifies a role for NEK9 in the replication stress response. Nucleic Acids Res. 2014;42(18):11517–27.
49. Errico A, Costanzo V. Mechanisms of replication fork protection: a safeguard for genome stability. Crit Rev Biochem Mol Biol. 2011;47(3):222–35.
50. Justine W, Eva W, Naoko M. Architecture and ssDNA interaction of the timeless-Tipin-RPA complex. Nucleic Acids Res. 2014;42(20):12912–27.
51. Michael GK, Zafer A, Seçil Y, Mary G, Stephanie LS-R, Tae-Hong K, et al. Tipin-replication protein A interaction mediates Chk1 phosphorylation by ATR in response to genotoxic stress. J Biol Chem. 2010;285(22):16562–71.
52. Yoshizawa-Sugata N, Masai H. Human tim/timeless-interacting protein, tipin, is required for efficient progression of S phase and DNA replication checkpoint. J Biol Chem. 2007;282(4):2729–40.
53. Chini CCS, Wood J, Chen J. Chk1 is required to maintain claspin stability. Oncogene. 2006;25(30):4165–71.
54. Lin Y-FF, Shih H-YY, Shang Z, Matsunaga S, Chen BP. DNA-PKcs is required to maintain stability of Chk1 and claspin for optimal replication stress response. Nucleic Acids Res. 2014;42(7):4463–73.
55. Zhang Y-WW, Otterness DM, Chiang GG, Xie W, Liu Y-CC, Mercurio F, et al. Genotoxic stress targets human Chk1 for degradation by the ubiquitin-proteasome pathway. Mol Cell. 2005;19(5):607–18.
56. Faustrup H, Bekker-Jensen S, Bartek J, Lukas J, Mailand N. USP7 counteracts SCFbetaTrCP- but not APCCdh1-mediated proteolysis of claspin. J Cell Biol. 2009;184(1):13–9.
57. Alonso-de Vega I, Martín Y, Smits VA. USP7 controls Chk1 protein stability by direct deubiquitination. Cell Cycle. 2013;13(24):3921–6.
58. Bassermann F, Frescas D, Guardavaccaro D, Busino L, Peschiaroli A, Pagano M. The Cdc14B-Cdh1-Plk1 axis controls the G2 DNA-damage-response checkpoint. Cell. 2008;134(2):256–67.
59. Jian Y, Kuntian L, Min D, Yunhui L, Ping Y, Bowen G, et al. HERC2-USP20 axis regulates DNA damage checkpoint through claspin. Nucleic Acids Res. 2014;42(21):13110–21.
60. Martín Y, Cabrera E, Amoedo H, Hernández-Pérez S, Domínguez-Kelly R, Freire R. USP29 controls the stability of checkpoint adaptor claspin by deubiquitination. Oncogene. 2014.
61. Sun TP, Shieh SY. Human FEM1B is required for Rad9 recruitment and CHK1 activation in response to replication stress. Oncogene. 2009;28(18):1971–81.
62. Taylor AM, Harnden DG, Arlett CF, Harcourt SA, Lehmann AR, Stevens S, et al. Ataxia telangiectasia: a human mutation with abnormal radiation sensitivity. Nature. 1975;258(5534):427–9.
63. Painter RB, Young BR. Radiosensitivity in ataxia-telangiectasia: a new explanation. Proc Natl Acad Sci U S A. 1980;77(12):7315–7.
64. Painter RB. Effect of caffeine on DNA synthesis in irradiated and unirradiated mammalian cells. J Mol Biol. 1980;143(3):289–301.
65. Walters RA, Hildebrand CE. Evidence that x-irradiation inhibits DNA replicon initiation in Chinese hamster cells. Biochem Biophys Res Commun. 1975;65(1):265–71.
66. Kaufmann WK, Cleaver JE. Mechanisms of inhibition of DNA replication by ultraviolet light in normal human and xeroderma pigmentosum fibroblasts. J Mol Biol. 1981;149(2):171–87.
67. Shirahige K, Hori Y, Shiraishi K, Yamashita M, Takahashi K, Obuse C, et al. Regulation of DNA-replication origins during cell-cycle progression. Nature. 1998;395(6702):618–21.
68. Santocanale C, Diffley JF. A Mec1- and Rad53-dependent checkpoint controls late-firing origins of DNA replication. Nature. 1998;395(6702):615–8.
69. José Antonio T, John FXD. Regulation of DNA replication fork progression through damaged DNA by the Mec1/Rad53 checkpoint. Nature. 2001;412(6846):553–7.

70. Wenyi F, David C, Max EB, Lindsay AF, Gina MA, Walton LF, et al. Genomic mapping of single-stranded DNA in hydroxyurea-challenged yeasts identifies origins of replication. Nat Cell Biol. 2006;8(2):148–55.
71. Paulovich AG, Hartwell LH. A checkpoint regulates the rate of progression through S phase in S. cerevisiae in response to DNA damage. Cell. 1995;82(5):841–7.
72. Jaime L-M, Nancy LM, Zophonias OJ, Lisa GD-E, James W, David PT. Damage-induced phosphorylation of Sld3 is important to block late origin firing. Nature. 2010;467(7314):479–83.
73. Zegerman P, Diffley JF. Checkpoint-dependent inhibition of DNA replication initiation by Sld3 and Dbf4 phosphorylation. Nature. 2010;467(7314):474–8.
74. Hashimoto Y, Tsujimura T, Sugino A, Takisawa H. The phosphorylated C-terminal domain of Xenopus Cut5 directly mediates ATR-dependent activation of Chk1. Genes Cells. 2006;11(9): 993–1007.
75. Boos D, Sanchez-Pulido L, Rappas M, Pearl LH, Oliver AW, Ponting CP, et al. Regulation of DNA replication through Sld3-Dpb11 interaction is conserved from yeast to humans. Curr Biol. 2011;21(13):1152–7.
76. Guo C, Kumagai A, Schlacher K, Shevchenko A, Shevchenko A, Dunphy WG. Interaction of Chk1 with Treslin negatively regulates the initiation of chromosomal DNA replication. Mol Cell. 2014.
77. Mantiero D, Mackenzie A, Donaldson A, Zegerman P. Limiting replication initiation factors execute the temporal programme of origin firing in budding yeast. EMBO J. 2011;30(23):4805–14.
78. Weinreich M, Stillman B. Cdc7p-Dbf4p kinase binds to chromatin during S phase and is regulated by both the APC and the RAD53 checkpoint pathway. EMBO J. 1999;18(19): 5334–46.
79. Duncker BP, Shimada K, Tsai-Pflugfelder M, Pasero P, Gasser SM. An N-terminal domain of Dbf4p mediates interaction with both origin recognition complex (ORC) and Rad53p and can deregulate late origin firing. Proc Natl Acad Sci U S A. 2002;99(25):16087–92.
80. Brown GW, Kelly TJ. Cell cycle regulation of Dfp1, an activator of the Hsk1 protein kinase. Proc Natl Acad Sci U S A. 1999;96(15):8443–8.
81. Dierov J, Dierova R, Carroll M. BCR/ABL translocates to the nucleus and disrupts an ATR-dependent intra-S phase checkpoint. Cancer Cell. 2004;5(3):275–85.
82. Heffernan TP, Unsal-Kaçmaz K, Heinloth AN, Simpson DA, Paules RS, Sancar A, et al. Cdc7-Dbf4 and the human S checkpoint response to UVC. J Biol Chem. 2007;282(13): 9458–68.
83. Gabrielse C, Miller CT, McConnell KH, DeWard A, Fox CA, Weinreich M. A Dbf4p BRCA1 C-terminal-like domain required for the response to replication fork arrest in budding yeast. Genetics. 2006;173(2):541–55.
84. Yanow SK, Gold DA, Yoo HY, Dunphy WG. Xenopus Drf1, a regulator of Cdc7, displays checkpoint-dependent accumulation on chromatin during an S-phase arrest. J Biol Chem. 2003;278(42):41083–92.
85. Montagnoli A, Valsasina B, Brotherton D, Troiani S, Rainoldi S, Tenca P, et al. Identification of Mcm2 phosphorylation sites by S-phase-regulating kinases. J Biol Chem. 2006;281(15): 10281–90.
86. Tenca P, Brotherton D, Montagnoli A, Rainoldi S, Albanese C, Santocanale C. Cdc7 is an active kinase in human cancer cells undergoing replication stress. J Biol Chem. 2007;282(1): 208–15.
87. Yamada M, Watanabe K, Mistrik M, Vesela E, Protivankova I, Mailand N, et al. ATR-Chk1-APC/CCdh1-dependent stabilization of Cdc7-ASK (Dbf4) kinase is required for DNA lesion bypass under replication stress. Genes Dev. 2013;27(22):2459–72.
88. Poh WT, Chadha GS, Gillespie PJ, Kaldis P, Blow JJ. Xenopus Cdc7 executes its essential function early in S phase and is counteracted by checkpoint-regulated protein phosphatase 1. Open Biol. 2013;4:130138.
89. Satoshi Y, Motoshi H, Hisao M. Replication timing regulation of eukaryotic replicons: Rif1 as a global regulator of replication timing. Trends Genet. 2013;29(8):449–60.

90. Hiraga S-I, Alvino GM, Chang F, Lian H-YY, Sridhar A, Kubota T, et al. Rif1 controls DNA replication by directing protein phosphatase 1 to reverse Cdc7-mediated phosphorylation of the MCM complex. Genes Dev. 2014;28(4):372–83.
91. Christina K-R, Jonathan M. Cdc25: mechanisms of checkpoint inhibition and recovery. Trends Cell Biol. 2006.
92. Mailand N. Rapid destruction of human Cdc25A in response to DNA damage. Science. 2000.
93. Young LM, Pagano M. Cdc25 phosphatases: differential regulation by ubiquitin-mediated proteolysis. Cell Cycle. 2010;9(23):4613–4.
94. Lindqvist A, Källström H, Karlsson Rosenthal C. Characterisation of Cdc25B localisation and nuclear export during the cell cycle and in response to stress. J Cell Sci. 2004;117(Pt 21):4979–90.
95. Peng CY, Graves PR, Thoma RS, Wu Z, Shaw AS, Piwnica-Worms H. Mitotic and G2 checkpoint control: regulation of 14-3-3 protein binding by phosphorylation of Cdc25C on serine-216. Science. 1997;277(5331):1501–5.
96. Raleigh JM, O'Connell MJ. The G(2) DNA damage checkpoint targets both Wee1 and Cdc25. J Cell Sci. 2000;113(Pt 10):1727–36.
97. Yolanda S, Jeff B, Hong W, Fenghua H, Dou L, Michael T, et al. Control of the DNA damage checkpoint by Chk1 and Rad53 protein kinases through distinct mechanisms. Science. 1999;286(5442):1166–71.
98. Hong W, Dou L, Yanchang W, Jun Q, Stephen JE. Pds1 phosphorylation in response to DNA damage is essential for its DNA damage checkpoint function. Genes Dev. 2001.
99. Hu F, Wang Y, Liu D, Li Y, Qin J, Elledge SJ. Regulation of the Bub2/Bfa1 GAP complex by Cdc5 and cell cycle checkpoints. Cell. 2001;107(5):655–65. Epub 2001/12/06.
100. Tao Z, Saurabh N, Hong Hwa L, Indrajit S, Uttam S. DNA damage checkpoint maintains Cdh1 in an active state to inhibit anaphase progression. Dev Cell. 2009;17(4):541–51.
101. Blow JJ, Ge XQ, Jackson DA. How dormant origins promote complete genome replication. Trends Biochem Sci. 2011;36(8):405–14.
102. Yekezare M, Gómez-González B, Diffley JF. Controlling DNA replication origins in response to DNA damage – inhibit globally, activate locally. J Cell Sci. 2013;126(Pt 6):1297–306.
103. Toledo LI, Altmeyer M, Rask M-BB, Lukas C, Larsen DH, Povlsen LK, et al. ATR prohibits replication catastrophe by preventing global exhaustion of RPA. Cell. 2013;155(5):1088–103.
104. Poli J, Tsaponina O, Crabbé L, Keszthelyi A, Pantesco V, Chabes A, et al. dNTP pools determine fork progression and origin usage under replication stress. EMBO J. 2012;31(4):883–94.
105. Ge XQ, Blow JJ. Chk1 inhibits replication factory activation but allows dormant origin firing in existing factories. J Cell Biol. 2010;191(7):1285–97.
106. Matsuoka S, Ballif BA, Smogorzewska A, McDonald ER, Hurov KE, Luo J, et al. ATM and ATR substrate analysis reveals extensive protein networks responsive to DNA damage. Science. 2007;316(5828):1160–6.
107. Trenz K, Errico A, Costanzo V. Plx1 is required for chromosomal DNA replication under stressful conditions. EMBO J. 2008;27(6):876–85.
108. Song B, Liu XS, Davis K, Liu X. Plk1 phosphorylation of Orc2 promotes DNA replication under conditions of stress. Mol Cell Biol. 2011;31(23):4844–56.
109. Kawabata T, Luebben SW, Yamaguchi S, Ilves I, Matise I, Buske T, et al. Stalled fork rescue via dormant replication origins in unchallenged S phase promotes proper chromosome segregation and tumor suppression. Mol Cell. 2011;41(5):543–53.
110. Shima N, Alcaraz A, Liachko I, Buske TR, Andrews CA, Munroe RJ, et al. A viable allele of Mcm4 causes chromosome instability and mammary adenocarcinomas in mice. Nat Genet. 2006;39(1):93–8.
111. Newman TJ, Mamun MA, Nieduszynski CA, Blow JJ. Replisome stall events have shaped the distribution of replication origins in the genomes of yeasts. Nucleic Acids Res. 2013;41(21):9705–18.
112. Branzei D, Foiani M. Maintaining genome stability at the replication fork. Nat Rev Mol Cell Biol. 2010;11(3):208–19.
113. Bermejo R, Capra T, Jossen R, Colosio A, Frattini C, Carotenuto W, et al. The replication checkpoint protects fork stability by releasing transcribed genes from nuclear pores. Cell. 2011;146(2):233–46.

114. Lucca C, Vanoli F, Cotta-Ramusino C, Pellicioli A, Liberi G, Haber J, et al. Checkpoint-mediated control of replisome-fork association and signalling in response to replication pausing. Oncogene. 2004;23(6):1206–13.
115. De Piccoli G, Katou Y, Itoh T, Nakato R, Shirahige K, Labib K. Replisome stability at defective DNA replication forks is independent of S phase checkpoint kinases. Mol Cell. 2012;45(5): 696–704.
116. David C. Preventing replication fork collapse to maintain genome integrity. DNA Repair. 2015.
117. Ptak C, Aitchison JD, Wozniak RW. The multifunctional nuclear pore complex: a platform for controlling gene expression. Curr Opin Cell Biol. 2014;28:46–53.
118. Helmrich A, Ballarino M, Nudler E, Tora L. Transcription-replication encounters, consequences and genomic instability. Nat Struct Mol Biol. 2013;20(4):412–8.
119. Smolka MB, Albuquerque CP, Chen S-hH, Zhou H. Proteome-wide identification of in vivo targets of DNA damage checkpoint kinases. Proc Natl Acad Sci U S A. 2007;104(25):10364–9.
120. Vesna CN, Brett WC, Darren JH, Sonya LK-C, Holly EM, Michael CS. Replication stress checkpoint signaling controls tRNA gene transcription. Nat Struct Mol Biol. 2010;17(8):976–81.
121. Paulsen RD, Soni DV, Wollman R, Hahn AT, Yee M-CC, Guan A, et al. A genome-wide siRNA screen reveals diverse cellular processes and pathways that mediate genome stability. Mol Cell. 2009;35(2):228–39.

Chapter 25
Meier-Gorlin Syndrome

Manzar Hossain and Bruce Stillman

Abstract Proteins required for the earliest stages of the initiation of DNA replication, including the origin recognition complex, Cdc6, Cdt1, and the Mcm2-7 proteins, cooperate to assemble pre-replicative complexes at all origins of DNA replication prior to the actual start of DNA synthesis from each origin during S phase of the cell division cycle. These initiation proteins are also involved in processes at centrosomes and centromeres during mitosis that ensure the correct segregation of the duplicated sister chromatids after DNA replication. Rare, recessive mutations in genes encoding some of these proteins result in Meier-Gorlin syndrome (MGS), characterized by microcephaly and primordial dwarfism, but normal intelligence. Biochemical and cell biology studies show that MGS mutations affect DNA replication, but some mutations affect both DNA replication and chromosome segregation. The observations have implications related to control of tissue and body size.

Keywords Origin recognition complex • Cdc6 • Cdt1 • Centriole • Mitosis • Centrosome

Introduction

On an evolutionary scale, the curious case of microcephalic human dwarfism came to the fore with the discovery of a small-bodied hominin named Liang Bua 1 (LB1) on the island of Flores, Indonesia [1]. Based on the cranial and other traits, the LB1 skeleton has been attributed to a new species of *Homo* called *Homo floresiensis*. Because LB1 is short in stature and has small cranial capacity, it was speculated that LB1 is a microcephalic version of modern human [2, 3]. Since then there has been intense debate whether to include LB1 as new species or whether LB1 just represents a pathological, microcephalic dwarf homo sapiens [4]. In most cases of human dwarfism the growth of the individuals is halted because of various endocrine conditions and those occurring due to abnormal growth hormone level can be treated [5]. There are other kinds of dwarfism that do not involve hormonal imbalance, many

M. Hossain • B. Stillman (✉)
Cold Spring Harbor Laboratory, 1 Bungtown Road, Cold Spring Harbor, NY 11724, USA
e-mail: hossain@cshl.edu; stillman@cshl.edu

D.L. Kaplan (ed.), *The Initiation of DNA Replication in Eukaryotes*,
DOI 10.1007/978-3-319-24696-3_25

caused by genetic mutations or chromosome aberrations. Primordial dwarfism is one such example in which the affected individuals display proportionate dwarfism and inherit the recessive gene mutations from their carrier parents. In this chapter, we discuss in detail a type of primordial dwarfism called Meier-Gorlin syndrome (MGS), including its pathogenic genetic mutations and the implications for tissue growth.

History of Meier-Gorlin Syndrome

In 1959, Meier et al. and, later on in 1975, Gorlin et al. described a new syndrome where the patients presented with microtia (external ear is underdeveloped), micrognathia (jaw is undersized), arthrogryposis of the joints of the lower limbs (joints have contracture), absent patellae (kneecap), cryptorchidism (absence of one or both testes from the scrotum), and short stature [6, 7]. After these two initial reports more cases unfolded in subsequent years with similar symptoms, as diagnosed and described by Meier and Gorlin [8–16]. In 1991 Cohen et al., based on the diagnostic symptoms or clinical features, named this condition in patients as ear-patella-short stature syndrome [8]. Subsequently, Boles et al. in 1994 coined the term MGS (Online Mendelian Inheritance in man, OMIM 224690) named after its discoverers.

Genetics of Meier-Gorlin Syndrome

MGS is a rare autosomal recessive disorder based on its occurrence in siblings with equal sex ratio [10]. Lacombe et al. suggested that the symptoms indicated in MGS might be human equivalent of short ear (se) mutation found in mouse [12]. The *se* mutation in mouse is caused from inactivation of the gene for a bone morphogenetic protein (BMP5), a member of the transforming growth factor β superfamily. Bongers et al. [17] studied in great detail eight more cases of MGS and found that all the patients display characteristic features of bilateral microtia, aplasia/hypoplasia of the patellae, and pre- and postnatal growth retardation. They further suggested that the genes involved in the BMP5 signaling pathway, such as homeobox genes, might be the underlying genetic defects in patients with MGS. Later on, Cohen et al. [18] analyzed the BMP5 gene in MGS patients and found no mutations occurring in the gene. Apart from BMP5, they also studied LMX1B and SHOX genes, but these proved normal in those MGS patients studied [18]. Recently, a series of genetic mapping studies of individuals with MGS from various parts of world reported the surprising observations that genetic mutations in DNA replication proteins with related functions in initiation of DNA replication may contribute to the clinical features of MGS [19–22].

Meier-Gorlin Syndrome in Relation to Microcephalic Primordial Syndrome

The vast majority of microcephalic (small head size) primordial dwarfism syndromes include disorders where the affected individuals show severe growth failure, with onset before birth (prenatal) and continuing into postnatal life. Based on the phenotype, microcephalic primordial dwarfism is broadly classified into three basic syndromic subtypes: Seckel syndrome [23, 24], microcephalic osteodysplastic primordial dwarfism (MOPD type I and type II) [25, 26], and MGS [7, 17]. Short stature and small head size are the prominent features shared among these different groups. Primordial dwarfism is rare and caused by autosomal recessive mutations. Frequently the affected individuals in this group also have mild-to-severe mental retardation, but this is usually not the case in MGS.

The molecular defects responsible for large number of disorders in microcephalic primordial syndrome (MPD) have been identified and studies of the molecular consequences of these mutations have revealed pathways disrupted in the affected individuals. These disorders affect important cellular processes as diverse as mitosis, DNA damage response, RNA splicing machinery, DNA replication licensing, and centrosome duplication (Table 25.1).

Pre-replicative Complex Proteins in Meier-Gorlin Syndrome

DNA replication is a basic functional process that ensures the propagation of all cells, from single-cell bacteria and archaea to cells present in eukaryotic, multicellular organisms. In eukaryotes, the DNA replication process is precisely controlled to allow the duplication of the genome only once per cell division cycle and is coupled to the processes of chromosome segregation during mitosis or meiosis. Initiation of

Table 25.1 Diversity of proteins affected and its function involved in different subtypes of Microcephalic Primordial Dwarfism (MPD)

Syndromes	Protein	Protein functions	References
Seckel syndrome	ATR	DNA damage response	[27]
	CPAP (CENPJ)	Centrosome duplication	[28, 29]
	CEP152	Centrosome duplication	[30]
	CtIP	DNA repair	[31]
MOPD I	U4atac	mRNA splicing	[32]
MOPD II	PCNT	Centrosome duplication	[33, 34]
Meier-Gorlin syndrome	ORC1	DNA replication	[19–22, 35–39]
	ORC4	Centrosome duplication	
	ORC6	Cytokinesis	
	CDC6		
	CDT1		

DNA duplication occurs at a specific time during the cell cycle, involving well-ordered assembly of protein complexes on DNA in a stepwise manner [40–47]. DNA replication starts with the origin recognition complex (ORC) binding to or assembly on DNA or chromatin. ORC was initially identified in budding yeast *Saccharomyces cerevisiae* [48]; however all six subunits are conserved in all eukaryotes [49] and even archaea species have at least one origin-binding protein that is highly related to the largest subunit of ORC, Orc1, and its sequence-related protein Cdc6 [50–52]. Orc1 and Cdc6 are structurally related to each other and are members of the AAA+ family of ATPases, as are Orc2, Orc3, Orc4, and Orc5 subunits of ORC (Fig. 25.1) [51, 53, 54]. In yeast, ORC is stable and binds to origin sequences throughout the cell division cycle, but in cells from mammalian species and particularly in human cells, ORC is assembled and disassembled during the cell division cycle and thus is a very dynamic complex [41, 42, 47, 55–62]. As cells enter and progress through mitosis and then enter G1 phase, ORC binds to chromatin and is part of a molecular machine that forms pre-replicative complex (pre-RC) in cooperation with Cdc6, Cdt1, and MCM (minichromosome maintenance) proteins [40, 42, 43, 63]. Formation of pre-RCs at every potential origin in the genome licenses the chromosomes for a new round of DNA replication, but

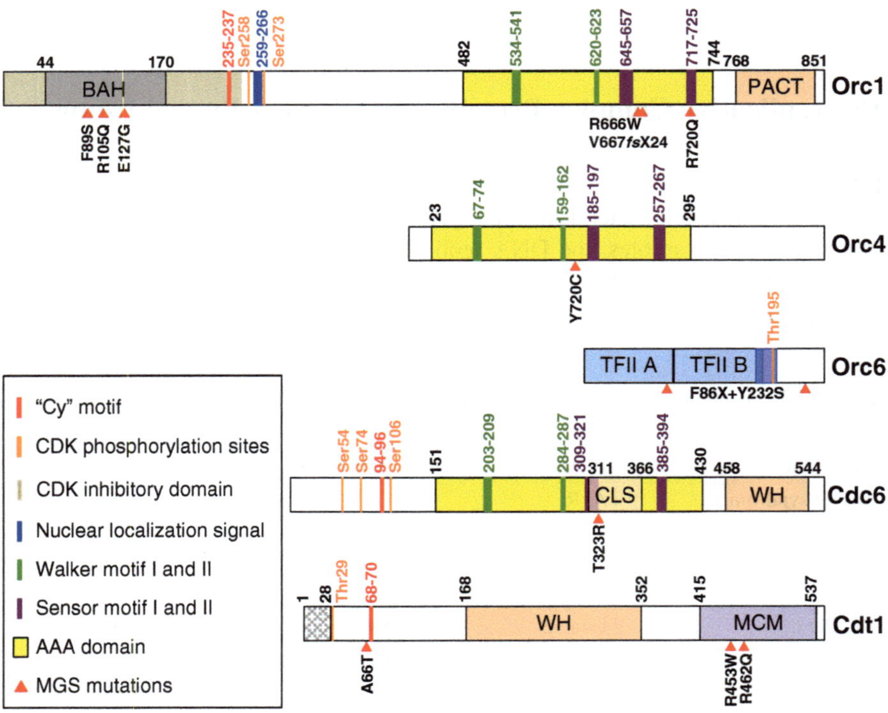

Fig. 25.1 Schematics of pre-RC proteins involved in Meier-Gorlin syndrome. The different functional and structural domains of human Orc1, Orc4, Orc6, Cdc6, and Cdt1 are indicated using different *colored boxes* and *bars*. The mutations within pre-RC proteins occurring in MGS patients are marked with *inverted red color arrowheads*

actual initiation of DNA synthesis is restricted to a temporally regulated program during S phase [41, 42, 46, 47, 64, 65]. Once the pre-RC is used at each origin, it is destroyed and thus reinitiation cannot occur until cells enter into mitosis, starting the process over again. The dependence of the initiation of DNA replication on progressing through mitosis ensures that DNA duplicates only once per cell cycle.

Recently, studies of the genetics of MGS identified a number of mutations in genes encoding DNA replication proteins that are involved in pre-RC formation [19, 20, 22]. We discuss these and other gene mutations and their biochemical implications.

Orc1

Orc1 is the largest subunit of ORC and initial studies showed that it was multifunctional, having a role in DNA replication as well as transcriptional gene silencing in the yeast *S. cerevisiae*, and the amino acid sequence of Orc1 protein in budding yeast shows similarity not only to its sister initiation protein Cdc6, but also to the transcriptional gene silencing protein Sir3 [66–68]. Subsequently, many studies confirmed that Orc1 in mammalian cells is an essential protein for the formation of pre-RCs leading to DNA replication but there is increasing evidence for a role in transcriptional repression and maintenance of the heterochromatin protein 1 (HP1) protein at both facultative and constitutive heterochromatin [50, 51, 56–62, 67, 69–84]. ORC in human cells is very dynamic in its assembly and disassembly during the cell division cycle, with Orc1 binding to mitotic chromosomes during mitosis and assembling ORC during G1 phase, but then being degraded at the G1- to S-phase transition as cells have committed to a new round of DNA replication [55–57, 59, 74, 85]. It should be noted that assembly, stability, and disassembly of ORC during the mammalian cell cycle vary depending on the species, since the Chinese hamster Orc1 behaves differently than human Orc1 in some aspects of cell cycle regulation [55, 58, 61]. Thus, for the discussion of the MGS, we will focus on human ORC subunits.

Orc1 in mammalian cells binds to proteins other than ORC subunits, including cyclin-dependent protein kinases [38, 86], histone-modifying enzymes [87–89], the retinoblastoma protein [90], heterochromatin protein 1 [51, 83, 91, 92], and the ORCA/LRWD1 heterochromatin-organizing protein [87, 93–96]. The Orc1 subunit from different organisms shows multiple well-conserved domains: an N-terminal bromo adjacent homology (BAH) domain, a domain containing the conserved AAA+ fold that suggests an ATPase function, and a predicted winged helix (WH) motif near the C-terminus that overlaps with a centrosome-targeting domain (PACT) (Fig. 25.1). The structures of the AAA+ and WH domains have recently been confirmed in the crystal structure of ORC from *Drosophila* [97]. The BAH domain of both yeast and mouse Orc1 has been crystallized and the mammalian Orc1 BAH domain forms an aromatic cage that binds the histone H4K20 dimethylation mark (the lysine at position 4 in histone H4 is modified with two methyl groups), implicating the interaction with modified histones in epigenetic regulation of gene expression or origin licensing [72, 98]. The mammalian Orc1 BAH domain also contributes to chromosome binding [99], as well as functioning as a cyclin-CDK2 kinase binding and kinase inhibitory domain (CID) that can inhibit the activity of both cyclin E-CDK2 and cyclin A-CDK2 protein kinases [38, 86]. Nevertheless, the two kinase inhibitory activities are functionally separate since binding and inhibition of the cyclin A-CDK2 kinase require an "RxL" or "Cy" motif that is not required for binding to or inhibition of the cyclin E-CDK2 kinase [38, 59, 86]. The predicted winged helix motif of human Orc1 at its C-terminal harbors an overlapping centrosome localization signal similar to centrosome-targeting domains (PACT domains) in other centrosome-localized proteins and, when fused to the CID domain, is involved in control of centriole and centrosome copy number in human cells [38].

The connection between ORC and MGS emerged from human genetic studies. Bicknell et al. screened 206 individuals from different parts of the world who had microcephalic primordial dwarfism and found five individuals from four families (Saudi Arabia, USA, and UK) where the affected patients harbored homozygous missense mutations in the Orc1 gene [20]. The affected patients in the cohort changed amino acids in the Orc1 gene at Phe89Ser, Arg105Gln, Glu127Gly, and Arg720Gln with the Arg105Gln mutation in Orc1 gene occurring more frequently in patients. Furthermore in an accompanying study Bicknell et al. identified one more patient with severe growth retardation and microcephaly and his sibling was found to contain compound heterozygous mutations (Arg105Gln and Val667*fs*X24) in the Orc1 gene [19]. Overall ten individuals have now been reported with MGS having biallelic compound heterozygous or homozygous mutations in Orc1 with one individual bearing a monoallelic missense mutation (Fig. 25.3) [21].

Orc4

The Orc4 protein identified in human, mouse, and Xenopus are similar to its counterpart in *S. cerevisiae* Orc4 (ScOrc4) and sequence comparison shows that it is also sequence related to Orc1p and Cdc6p proteins [100, 101]. Among different Orc4 proteins from various organisms, the fission yeast *S. pombe* Orc4 (SpOrc4) is unique in that it contains nine AT hooks that specifically bind to AT-rich sequences and specifying genomic loci as DNA replication origins [102–105]. Although the mammalian Orc4 protein does not have AT hooks, some studies predict the involvement of human Orc4 in binding to nonclassical DNA structures [106–108]. It is noteworthy that except for SpOrc4, the AAA+ domain of Orc4 in all other organisms mentioned above contains well-conserved sequence that is related to the walker A and walker B motifs that is involved in nucleotide binding (Figs. 25.1 and 25.2) [101,

Fig. 25.3 (continued) individual with Orc6 mutation; *4c*: one individual with Orc4 mutation, one individual with Orc6 mutation, and two individuals with Cdt1 mutation; *29d*: nine individuals with Orc1 mutations, six individuals with Orc4 mutations, five individuals with Orc6 mutations, eight individuals with Cdt1 mutations, and one individual with Cdc6 mutation

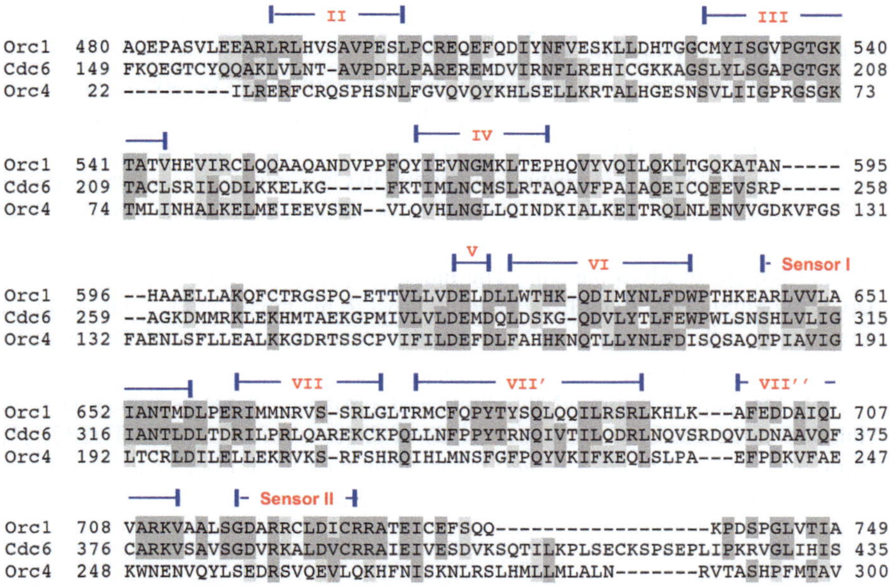

Fig. 25.2 Alignment of closely related pre-RC proteins implicated in MGS. The alignment of human Orc1, Orc4, and Cdc6 proteins was done using online ClustalW software. The different RFC *boxes* (II–VII) and sensor motifs of the pre-RC proteins are indicated with *closed lines* above the alignment

35 MGS patients with mutations in Orc1, Orc4, Orc6, Cdc6 and Cdt1 proteins.

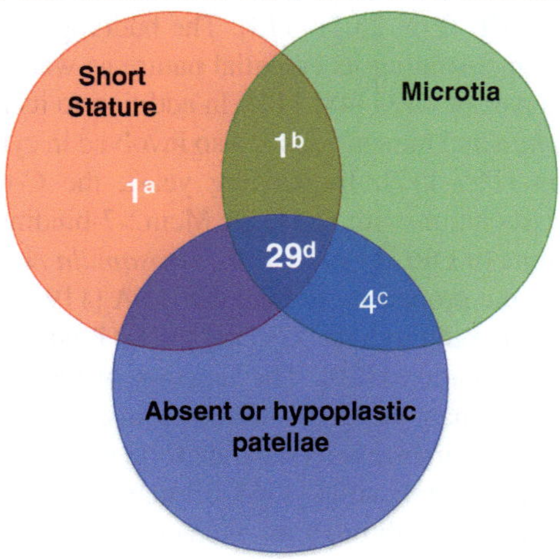

1[a] (1 individual with Orc1 mutation)
1[b] (1 individual with Orc6 mutation)
4[c] (1 individual with Orc4 mutation, 1 individual with Orc6 mutation and 2 individuals with Cdt1 mutation)
29[d] (9 individuals with Orc1 mutations, 6 individuals with Orc4 mutations, 5 individuals with Orc6 mutations 8 individuals with Cdt1 mutations, and 1 individual with Cdc6 mutation)

Fig. 25.3 Venn diagram showing the prominent triad phenotypic features shared among pre-RC mutations in MGS patients. The *numbers* in the diagram indicate the individuals sharing the features or where mutually exclusive and the statistics of occurrence in MGS patients with pre-RC mutations are described below the Venn diagram. *1a*: one individual with Orc1 mutation; *1b*: one

109, 110], but more importantly, Orc4 has an arginine finger that activates the ATPase of the Orc1 subunit [111].

Screening of patients with MGS led to the identification of seven individuals with mutations in Orc4 gene [19, 21, 22]. The Orc4 gene in MGS individuals mostly contains homozygous missense mutations causing pTyr174Cys substitution in the protein. The compound heterozygote mutation in Orc4 mainly arises due to frame-shift or gene deletion events in combination with the pTyr174Cys mutation. The frame-shift mutation also leads to deletion of 125 amino acids from the C-terminal of Orc4 protein in MGS patients. The mutated tyrosine residue of Orc4 in MGS patients is highly conserved from budding yeast to mammals and is found to lie in between the walker B and sensor I motif of the AAA+ domain (Fig. 25.2). Although the human Orc4 protein shows only 29 % sequence identity to budding yeast ScOrc4, the sequence around the conserved tyrosine residue is highly similar. Mutation of the conserved tyrosine residue of *S. cerevisiae* ScOrc4 (p.Tyr232Cys) results in a substantial growth retardation phenotype with a defect in the G1- to S-phase transition based on bud size of the emerging daughter cell [22].

Orc6

Among the components of ORC, the Orc6 is the smallest in size and the amino acid sequence is divergent from the other ORC subunits [112]. The structure of N-terminal half of human Orc6 protein shows similarity to the cyclin fold of the transcription factor TFIIB (Fig. 25.1) [113]. The budding yeast mutant that lacks Orc6 is not viable, demonstrating its essential nature as well as its involvement in origin recognition as part of ORC [48, 114]. In addition to its role in DNA replication, Orc6 in *Drosophila* and human cells is also involved in cytokinesis and heterochromatin formation [115–117]. In budding yeast, the C-terminus of ScOrc6 establishes a tight association with Cdt1, an Mcm2-7-binding protein and to the Orc3 and Orc5 subunits of ORC [118]. Unlike *Drosophila melanogaster* DmOrc6, ScOrc6 does not have the propensity to bind the DNA [119]. Furthermore, ScOrc6 binds S-phase cyclin Clb5 through its “Cy” or “cyclin binding” motif and prevents reinitiation of DNA replication [120]. The function of Orc6 is quite divergent in different organisms. In human, the Orc6 is involved in cytokinesis through its localization at the midbody as daughter cells separate form each other, the separate event apart from its role in DNA replication [116]. Similar to human Orc6, the DmOrc6 is also involved in cytokinesis through its interaction with Pnut, a member of the septin protein family [115, 121, 122]. A recent report further extends the role of DmOrc6 in both DNA replication and cytokinesis and suggests functional conservation of Orc6 activities in metazoans [119, 121].

Bicknell et al. [19] found three individuals from one family bearing compound heterozygous mutations in the Orc6 gene in MGS patients, leading to a change of residue p.Tyr232Ser in combination with a loss-of-function mutation resulting from a 2-bp deletion. The residue mutated in MGS patients is well conserved from lower eukaryotes to higher eukaryotes and the domain encompassing the residue is known to involve interaction with the Pnut protein, a member of septin family of proteins important for cytokinesis [115], although a recent analysis of the functional consequences of the Orc6 mutation shows that the Meier-Gorlin mutation impairs formation of the ORC hexamer [37]. Munnik et al. identified four more MGS patients from three families having compound heterozygous missense and splice site mutations in Orc6 [21].

Cdc6

Cdc6 protein is a member of the AAA+ superfamily, containing walker A and B motifs, sensor motifs I and II, and well-conserved RFC boxes II–VII (Figs. 25.1 and 25.2) [123, 124]. Cdc6 is loaded onto origins of DNA during pre-RC assembly through its interactions to ORC subunits [71, 125–127]. Cdc6 protein in turn binds the Mcm2-7 complex via an interaction with the essential domain of Mcm3, consistent with pre-RC assembly occurring by the sequential loading of ORC, Cdc6, and the Mcm2-7 proteins [54, 75, 76, 128–132]. The primary sequence of Cdc6 shows that the protein is highly related to Orc1, Orc4, and Orc5 [101, 133]. The Cdc6 protein has a critical function in ensuring that the DNA replication occurs only once during the cell cycle, thus preventing re-replication or genome instability [134–138].

Only one case of a mutation in Cdc6 in MGS patients has been reported where the affected individual carries a homozygous missense mutation leading to a change of a Thr323 to arginine residue in the protein [19]. The mutated Cdc6 residue in MGS patients is well conserved across different species, ranging from lower to higher eukaryotes, and lies within the AAA+ domain of Cdc6 protein between the sensor motif I and RFC box VII (Fig. 25.2). Recently, Kim et al. [139] reported that Cdc6 contains centrosome localization signal from residues 311 to 366 amino acids and the conserved Thr323 falls within this region, although the functional consequences of this mutation have not been studied.

Cdt1

Like Cdc6, the Cdt1 protein is also dependent on its prior association with ORC subunits for binding to chromosomal replication origins [140, 141]. It has been suggested that the loading of Cdt1 to ORC might be occurring after the initial binding of Cdc6 to ORC proteins [142]. The primary role of Cdt1 is to interact with the Mcm2-7 hexamer and help assembly the Mcm2-7 onto DNA to form a double-Mcm2-7 hexamer [128, 132, 143–146]. Thus, Cdt1 protein acts as bridge between Mcm2-7 and the ORC-Cdc6 complex. The licensing activity of Cdt1 protein is inhibited through its interaction with Geminin protein to prevent re-replication of the genome [147–152]. The Cdt1 protein contains two functionally conserved regions: an N-terminal regulatory region and a domain required for replication licensing activity located in the middle and C-terminal domains [143, 144]. The middle and C-terminal domains of Cdt1 protein are required for its binding to Geminin and the Mcm2-7 complex, respectively [144, 153]. The regulatory N-terminal region of Cdt1 contributes to weak binding to Geminin protein and contains a PCNA-binding motif (PIP box), a cyclin-binding consensus motif (Cy-motif), and CDK phosphorylation sites [151, 154–157]. Cdt1 is unstable and is degraded at the G1- to S-phase transition or in response to DNA damage, in a process important for maintaining once per cell cycle replication [140, 147, 149–151, 154–156, 158–163].

Bicknell and Guernsey reported eight individuals from the cohort of MGS patients bearing mutations in the Cdt1 protein [20, 22]. Out of eight patients, seven individuals from five families had heterozygous mutations while only one of the MGS patients had biallelic, missense homozygous mutations. The Cdt1 mutations in two individuals having MGS phenotypes bear monoallelic missense mutation (Fig. 25.1). Most of the Cdt1 mutations reported in MGS patients fall within the C-terminal of the protein and this region is critical for its interaction to Mcm2-7 complex. It has been demonstrated that mutation of p.Arg474Gln in the mouse Cdt1 protein, in combination with other mutations, is required for binding activity to the Mcm4/6/7 hexamer [164], although the functional significance of the Mcm4/6/7 hexamer is questionable.

Phenotype of Patients with Meier-Gorlin Syndrome

The MGS was initially diagnosed based on the presence of three prominent phenotypic features called the triad, consisting of bilateral small ears (microtia), aplasia/hypoplasia of the patellae, and short stature. The 45 patients described by de Munnik et al. share all the three clinical features consistent with the diagnosis of MGS [35], but 6 individuals out of the 35 MGS patients with mutations in pre-RC proteins (Orc1, Orc4, Orc6, Cdc6, and Cdt1) did not show all the features (Fig. 25.3). The phenotypic effect of Orc1 and Orc4 mutations in MGS patients is more pronounced based on height and head circumference of individuals [21, 35]. The short stature and head circumference of MGS patients with Orc1 mutations are most severe compared to the phenotype in patients with other pre-RC proteins, suggesting either that Orc1 mutations have more penetrance or that Orc1 mutation affect multiple processes. A more severe, lethal phenotype has been mostly correlated with the patient's genotype bearing compound heterozygous missense and loss-of-function mutations. In their phenotype-genotype correlation studies, they have included additional clinical features such as deformity in facial appearance, respiratory and gastrointestinal problems, skeletal and limb anomalies, as well as genitourinary problems [21]. The intelligence of MGS patients varies with normal intellect in most of the cases to moderately retarded in one of the individuals harboring a mutation in the gene encoding Orc1. Recently, a very severe homozygous mutation in the gene encoding Orc6 was identified as a lethal mutation with fetuses dying at 20–36 weeks, and in addition to the known phenotypes, clubfeet, micro-retrognathia (small, receding jaw), gracile bones (slender and thinner bines), and abnormal genitalia [36]. It will be important to create mouse models with each of the MGS mutations in order to understand the exact cause of the phenotypes in more detail, starting from prenatal development during embryogenesis to the postnatal growth in adult animals. To date, no such mouse models have been described, but studies with cells in culture and other animal models have been informative (Fig. 25.4).

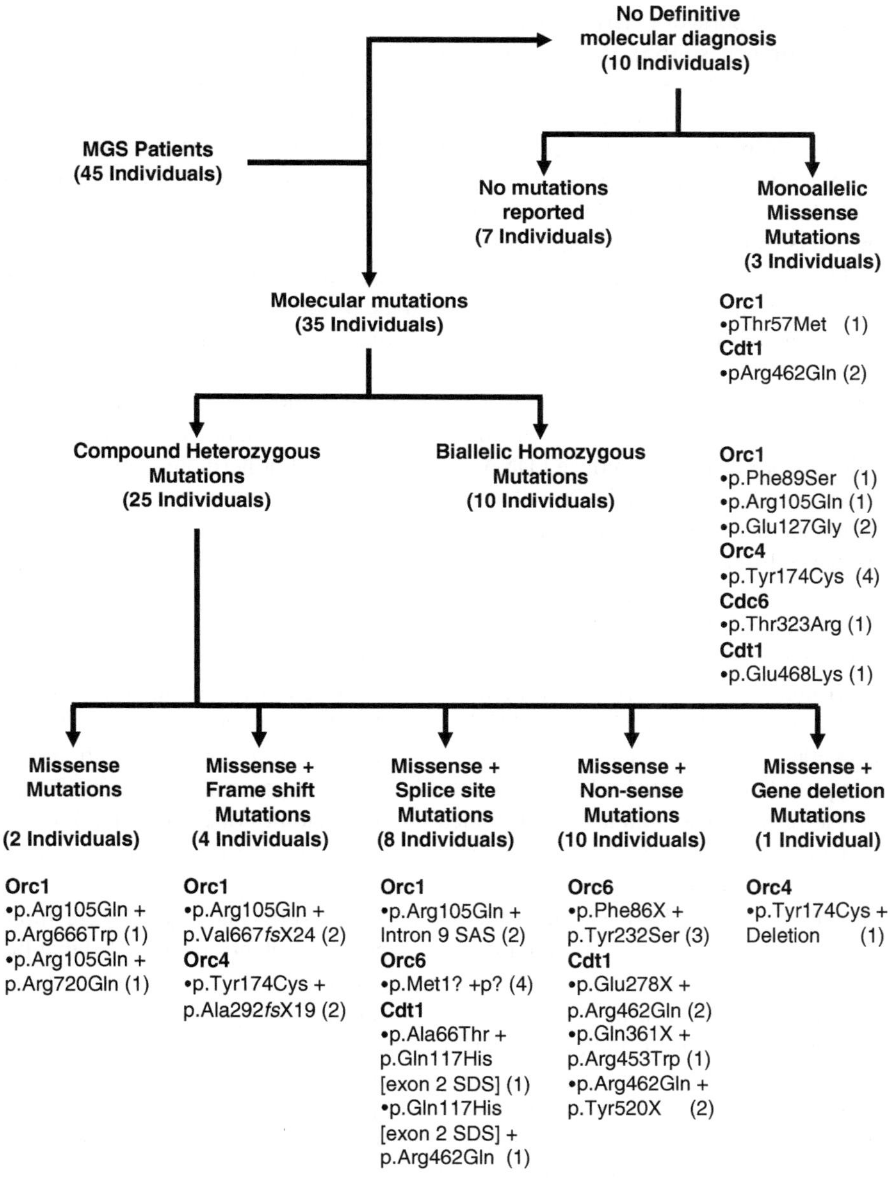

Fig. 25.4 The flowchart shows genetic heterogeneity among MGS patients with mutations in pre-RC proteins. The number of MGS patients with different types of genetic mutations for each pre-RC protein is indicated in *closed brackets*. Data from de Munnik et al. [21, 35]

Biological Consequences of Pre-RC Protein Mutations in Meier-Gorlin Syndrome

DNA replication is one of the fundamental, essential processes required by living organisms for cell proliferation and any kind of perturbations to the process due to stress or mutations is deleterious to an organism. In the literature, there have been a number of reports of inherited genetic disorders occurring due to mutations in DNA replication genes [165]. Recently, the genetic mutations were found in proteins of pre-RC and are linked to growth retardation in certain kind of disorders. The homozygous recessive mutation of MCM4 in six Irish patients leads to an autosomal genetic disorder due to a natural killer (NK) cell deficiency and DNA breakage. The genetic disorder due to partial MCM4 deficiency in humans also results in patients with short stature, adrenal insufficiency and prone to viral infections [166–168]. But the MGS mutations are distinct enough to from a syndrome.

In order to understand the functional implications of mutations in the patients, Bicknell et al. established two cell lines from the MGS patients bearing Orc1 mutations. One of them was a lymphoblastoid cell line (LBL) made from patient having the Orc1 pGlu127Gly mutation, and another was a skin fibroblast line made from patient harboring Orc1 pArg105Gln. The results from studies of these cell lines show that the Orc1 mutations caused insufficient loading of components of the pre-RC, subsequently leading to reduced DNA replication [20]. Furthermore, they generated zebrafish with mutations in Orc1 that generated embryos with marked growth retardation, perhaps related to the primordial dwarfism phenotype in humans. The studies further indicated that the mutations in Orc1 cause reduced cellular proliferation due to slow S-phase progression [20]. Most of the mutations within the Orc1 protein (Phe89Ser, Arg105Gln, and Glu127Gly) in MGS patients are situated within the overlapping, N-terminal BAH or CID domains. Recently, a crystal structure of the N-terminal region (9–170 aa) of the mouse Orc1 protein revealed that the region is involved in binding to histone H4K20 dimethylated peptide (H4K20me2). The H4K20me2-binding mutant of Orc1 was not able to rescue a growth retardation phenotype of Orc1-depleted zebrafish compared to its wild-type counterpart [20]. Based on the results, it was naturally concluded that the inability of H4K20me2-binding mutant of Orc1 to recognize the replication origins causes slow cell cycle progression and reduced cellular proliferation, which might lead to primordial dwarfism in humans [72, 98]. However, since the Orc1 Meier-Gorlin phenotypes were more severe than the phenotype observed with other pre-RC mutants, it suggested that the Orc1 mutations might alter a non-DNA replication function.

In addition to its role in DNA replication, the human Orc1 protein is also involved in controlling centriole and hence centrosome copy number in human cells [86]. In the absence of Orc1, centrioles and centrosomes re-duplicate in a single cell cycle, a condition that can lead to aneuploidy or chromosome miss-segregation. The N-terminal domain of Orc1, wherein most of the MGS mutations lie, functions as an inhibitor of cyclin-CDK2 kinases while the C-terminus of Orc1 harbors homology to the centrosome localization of PACT domain [38]. The two domains in Orc1, when coupled together in a single recombinant protein that lacks DNA replication activity, are necessary and sufficient for controlling the amplification of centrosomes or centrioles. Most interestingly, the MGS mutation Arg105Gln in Orc1 ablates its kinase inhibition activity towards cyclin E-CDK2, but not cyclin A-CDK2, and thus loses its control in maintaining normal centrosome or centriole copy number, similar to cells that over-express cyclin E [38]. The loss-of-function mutation in human Orc1 also leads to a marked reduction in cellular proliferation. In a similar context, Orc1-deficient primary fibroblast cells impair ciliogenesis, leading to markedly reduced cilia formation, suggesting that defects in centrioles that form the basis of cilia in cells might be affected in MGS patients, similar to mutations in the percentrin (PCNT) protein that is also involved in ciliogenesis, which is mutated in the primordial dwarfism syndrome MOPDII [39].

The domain architecture of human Orc6 protein shows that the MGS mutation lies within its unstructured C-terminal region. The several reports show that the C-terminal domain of Orc6 protein is involved in a non-replicative function such as cytokinesis and also interacts with Pnut (a septin protein) in *Drosophila* [115, 116, 169]. Interestingly, Bleichert et al. have recently found that the C-terminus of the Orc6 protein is also involved in its interaction with Orc3 protein in *Drosophila* and in human [37]. The corresponding MGS mutation in *Drosophila* Orc6 (Tyr225Ser) reduced its affinity for association with ORC via its interaction Orc3 protein. The MGS mutation in Orc6 that decreased its association to ORC resulted in a reduction of MCM chromatin association, suggesting a defect in pre-RC assembly and thus reduced DNA replication licensing. Thus it is possible that all the MGS mutations in Orc1 and Orc6 cause a defect in DNA replication, but the Orc1 mutations have an additional defect in centriole copy number control.

The cellular pathways affected in MGS patients with mutations in Orc4, Cdc6, and Cdt1 genes have not been characterized, although the equivalent mutation in yeast Orc4 reduced cell growth and delayed cell cycle progression, but the precise molecular mechanism involved is yet to be discovered. The non-redundant functional consequences of MGS mutations of pre-RC proteins not only affect DNA replication but also might involve non-replicative functions such as centrosome/ centriole duplication, ciliogenesis, and cytokinesis.

A major issue that remains to be resolved is how do defects in DNA replication and centriole copy number control lead to a phenotype with a small, but proportional body size? Is cell proliferation compromised so that the number of cell divisions in a particular tissue is less than would normally occur in normal people or do cells accumulate defects at some increased rate and induced apoptosis removes the cells during development, yielding smaller tissue size? Studies in cell culture have not resolved these issues, making studies of organisms with precise MGS mutations of great future interest. Another major issue is whether the ancient humans on the island of Flores in Indonesia, possibly and new species called *Homo floresiensis*, are in reality humans that are primordial microcephalic dwarfs with normal intelligence that harbored mutations in genes pre-RC proteins, like modern MGS patients. Unfortunately the tropical conditions in which the bones were found preclude isolation of DNA and sequencing, so we may never know.

Acknowledgements Research in the author's laboratory is supported by grants from the National Institutes of Health, CA13106 and GM45436.

References

1. Brown P, Sutikna T, Morwood MJ, Soejono RP, et al. A new small-bodied hominin from the Late Pleistocene of Flores, Indonesia. Nature. 2004;431:1055–61. doi:10.1038/nature02999.
2. An P. An early Minoan microcephale. Anthropos. 1975;2:40–7.
3. Henneberg M, Thorne A. Flores human may be pathological Homo sapiens. Before Farming. 2004;4:2–4.
4. Vannucci RC, Barron TF, Holloway RL. Craniometric ratios of microcephaly and LB1, Homo floresiensis, using MRI and endocasts. Proc Natl Acad Sci U S A. 2011;108:14043–8. doi:10.1073/pnas.1105585108.
5. Siklar Z, Berberoglu M. Syndromic disorders with short stature. J Clin Res Pediatr Endocrinol. 2014;6:1–8. doi:10.4274/Jcrpe.1149.
6. Meier Z, Poschiavo, Rothschild M. [Case of arthrogryposis multiplex congenita with mandibulofacial dysostosis (Franceschetti syndrome)]. Helv Paediatr Acta. 1959;14:213–6.
7. Gorlin RJ, Cervenka J, Moller K, Horrobin M, Witkop Jr CJ. Malformation syndromes. A selected miscellany. Birth Defects Orig Artic Ser. 1975;11:39–50.
8. Cohen B, Temple IK, Symons JC, Hall CM, Shaw DG, Bhamra M, et al. Microtia and short stature: a new syndrome. J Med Genet. 1991;28:786–90.
9. Hurst JA, Winter RM, Baraitser M. Distinctive syndrome of short stature, craniosynostosis, skeletal changes, and malformed ears. Am J Med Genet. 1988;29:107–15. doi:10.1002/ajmg.1320290113.
10. Boles RG, Teebi AS, Schwartz D, Harper JF. Further delineation of the ear, patella, short stature syndrome (Meier-Gorlin syndrome). Clin Dysmorphol. 1994;3:207–14.
11. Buebel MS, Salinas CF, Pai GS, Macpherson RI, Greer MK, PerezComas A. A new Seckel-like syndrome of primordial dwarfism. Am J Med Genet. 1996;64:447–52.
12. Lacombe D, Toutain A, Gorlin RJ, Oley CA, Battin J. Clinical identification of a human equivalent to the short ear (se) murine phenotype. Ann Genet. 1994;37:184–91.
13. Fryns JP. Meier-Gorlin syndrome: the adult phenotype. Clin Dysmorphol. 1998;7:231–2.
14. Loeys BL, Lemmerling MM, Van Mol CE, Leroy JG. The Meier-Gorlin syndrome, or ear patella short stature syndrome, in sibs. Am J Med Genet. 1999;84:61–7. doi:10.1002/(Sici)1096-8628(19990507)84:1<61::Aid-Ajmg12>3.0.Co;2-6.
15. Teebi AS, Gorlin RJ. Not a new Seckel-like syndrome but ear-patella short stature syndrome. Am J Med Genet. 1997;70:454. doi:10.1002/(Sici)1096-8628(19970627)70:4<454::Aid-Ajmg23>3.0.Co;2-F.
16. Verhallen JTCM vdLN, Kant SG. Het syndroom van Meier-Gorlin. Tijdschr Kindergeneesk 1999;67:32–5.
17. Bongers EMHF, Opitz JM, Fryer A, Sarda P, Hennekam RCM, Hall BD, et al. Meier-Gorlin syndrome: report of eight additional cases and review. Am J Med Genet. 2001;102:115–24. doi:10.1002/Ajmg.1452.
18. Cohen A, Mulas R, Seri M, Gaiero A, Fichera G, Marini M, et al. Meier-Gorlin syndrome (ear-patella-short stature syndrome) in an Italian patient: Clinical evaluation and analysis of possible candidate genes. Am J Med Genet. 2002;107:48–51. doi:10.1002/Ajmg.10083.
19. Bicknell LS, Bongers EM, Leitch A, Brown S, Schoots J, Harley ME, et al. Mutations in the pre-replication complex cause Meier-Gorlin syndrome. Nat Genet. 2011;43:356–9. doi:10.1038/ng.775.
20. Bicknell LS, Walker S, Klingseisen A, Stiff T, Leitch A, Kerzendorfer C, et al. Mutations in ORC1, encoding the largest subunit of the origin recognition complex, cause microcephalic primordial dwarfism resembling Meier-Gorlin syndrome. Nat Genet. 2011;43:350–5. doi:10.1038/ng.776.

21. de Munnik SA, Bicknell LS, Aftimos S, Al-Aama JY, van Bever Y, Bober MB, et al. Meier-Gorlin syndrome genotype-phenotype studies: 35 individuals with pre-replication complex gene mutations and 10 without molecular diagnosis. Eur J Hum Genet. 2012;20:598–606. doi:10.1038/Ejhg.2011.269.
22. Guernsey DL, Matsuoka M, Jiang H, Evans S, Macgillivray C, Nightingale M, et al. Mutations in origin recognition complex gene ORC4 cause Meier-Gorlin syndrome. Nat Genet. 2011;43:360–4. doi:10.1038/ng.777.
23. Majewski F, Goecke T. Studies of micro-cephalic primordial dwarfism. 1. Approach to a delineation of the Seckel syndrome. Am J Med Genet. 1982;12:7–21. doi:10.1002/Ajmg.1320120103.
24. Seckel HPG. Bird-headed dwarfs: studies in developmental anthropology including human proportions. Springfield, IL: Charles C. Thomas; 1960.
25. Hall JG, Flora C, Scott CI, Pauli RM, Tanaka KI. Majewski osteodysplastic primordial dwarfism type II (MOPD II): natural history and clinical findings. Am J Med Genet A. 2004;130A:55–72. doi:10.1002/Ajmg.A.30203.
26. Majewski F, Ranke M, Schinzel A. Studies of micro-cephalic primordial dwarfism. 2. The osteodysplastic type-II of primordial dwarfism. Am J Med Genet. 1982;12:23–35. doi:10.1002/Ajmg.1320120104.
27. O'Driscoll M, Ruiz-Perez VL, Woods CG, Jeggo PA, Goodship JA. A splicing mutation affecting expression of ataxia-telangiectasia and Rad3-related protein (ATR) results in Seckel syndrome. Nat Genet. 2003;33:497–501. doi:10.1038/ng1129.
28. Al-Dosari MS, Shaheen R, Colak D, Alkuraya FS. Novel CENPJ mutation causes Seckel syndrome. J Med Genet. 2010;47:411–4. doi:10.1136/jmg.2009.076646.
29. McIntyre RE, Lakshminarasimhan Chavali P, Ismail O, Carragher DM, Sanchez-Andrade G, Forment JV, et al. Disruption of mouse Cenpj, a regulator of centriole biogenesis, phenocopies Seckel syndrome. PLoS Genet. 2012;8, e1003022. doi:10.1371/journal.pgen.1003022.
30. Kalay E, Yigit G, Aslan Y, Brown KE, Pohl E, Bicknell LS, et al. CEP152 is a genome maintenance protein disrupted in Seckel syndrome. Nat Genet. 2011;43:23–6. doi:10.1038/Ng.725.
31. Qvist P, Huertas P, Jimeno S, Nyegaard M, Hassan MJ, Jackson SP, et al. CtIP mutations cause Seckel and Jawad syndromes. PLoS Genet. 2011;7, e1002310. doi:10.1371/journal.pgen.1002310.
32. Nagy R, Wang H, Albrecht B, Wieczorek D, Gillessen-Kaesbach G, Haan E, et al. Microcephalic osteodysplastic primordial dwarfism type I with biallelic mutations in the RNU4ATAC gene. Clin Genet. 2012;82:140–6. doi:10.1111/j.1399-0004.2011.01756.x.
33. Griffith E, Walker S, Martin CA, Vagnarelli P, Stiff T, Vernay B, et al. Mutations in pericentrin cause Seckel syndrome with defective ATR-dependent DNA damage signaling. Nat Genet. 2008;40:232–6. doi:10.1038/ng.2007.80.
34. Rauch A, Thiel CT, Schindler D, Wick U, Crow YJ, Ekici AB, et al. Mutations in the pericentrin (PCNT) gene cause primordial dwarfism. Science. 2008;319:816–9. doi:10.1126/science.1151174.
35. de Munnik SA, Otten BJ, Schoots J, Bicknell LS, Aftimos S, Al-Aama JY, et al. Meier-Gorlin syndrome: growth and secondary sexual development of a microcephalic primordial dwarfism disorder. Am J Med Genet A. 2012;158A:2733–42. doi:10.1002/ajmg.a.35681.
36. Shalev SA, Khayat M, Etty DS, Elpeleg O. Further insight into the phenotype associated with a mutation in the ORC6 gene, causing Meier-Gorlin syndrome 3. Am J Med Genet A. 2015;167A:607–11. doi:10.1002/ajmg.a.36906.
37. Bleichert F, Balasov M, Chesnokov I, Nogales E, Botchan MR, Berger JM. A Meier-Gorlin syndrome mutation in a conserved C-terminal helix of Orc6 impedes origin recognition complex formation. eLife. 2013;2, e00882. doi:10.7554/eLife.00882.
38. Hossain M, Stillman B. Meier-Gorlin syndrome mutations disrupt an Orc1 CDK inhibitory domain and cause centrosome reduplication. Genes Dev. 2012;26:1797–810. doi:10.1101/gad.197178.112.
39. Stiff T, Alagoz M, Alcantara D, Outwin E, Brunner HG, Bongers EM, et al. Deficiency in origin licensing proteins impairs cilia formation: implications for the aetiology of Meier-Gorlin syndrome. PLoS Genet. 2013;9, e1003360. doi:10.1371/journal.pgen.1003360.

40. Chistol G, Walter JC. Single-molecule visualization of MCM2-7 DNA loading: seeing is believing. Cell. 2015;161:429–30. doi:10.1016/j.cell.2015.04.006.
41. Diffley JF. Regulation of early events in chromosome replication. Curr Biol. 2004;14:R778–86. doi:10.1016/j.cub.2004.09.019.
42. Diffley JF. Quality control in the initiation of eukaryotic DNA replication. Philos Trans R Soc Lond B Biol Sci. 2011;366:3545–53. doi:10.1098/rstb.2011.0073.
43. Gerbi SA, Strezoska Z, Waggener JM. Initiation of DNA replication in multicellular eukaryotes. J Struct Biol. 2002;140:17–30.
44. Lei M, Tye BK. Initiating DNA synthesis: from recruiting to activating the MCM complex. J Cell Sci. 2001;114:1447–54.
45. Riera A, Tognetti S, Speck C. Helicase loading: how to build a MCM2-7 double-hexamer. Semin Cell Dev Biol. 2014;30:104–9. doi:10.1016/j.semcdb.2014.03.008.
46. Sclafani RA, Holzen TM. Cell cycle regulation of DNA replication. Annu Rev Genet. 2007;41:237–80. doi:10.1146/annurev.genet.41.110306.130308.
47. Stillman B. Cell cycle control of DNA replication. Science. 1996;274:1659–64.
48. Bell SP, Stillman B. ATP-dependent recognition of eukaryotic origins of DNA replication by a multiprotein complex. Nature. 1992;357:128–34. doi:10.1038/357128a0.
49. Bell SP. The origin recognition complex: from simple origins to complex functions. Genes Dev. 2002;16:659–72. doi:10.1101/gad.969602.
50. Aves SJ, Liu Y, Richards TA. Evolutionary diversification of eukaryotic DNA replication machinery. Subcell Biochem. 2012;62:19–35. doi:10.1007/978-94-007-4572-8_2.
51. Duncker BP, Chesnokov IN, McConkey BJ. The origin recognition complex protein family. Genome Biol. 2009;10:214. doi:10.1186/gb-2009-10-3-214.
52. Samson RY, Xu Y, Gadelha C, Stone TA, Faqiri JN, Li D, et al. Specificity and function of archaeal DNA replication initiator proteins. Cell Rep. 2013;3:485–96. doi:10.1016/j.celrep.2013.01.002.
53. Iyer LM, Leipe DD, Koonin EV, Aravind L. Evolutionary history and higher order classification of AAA+ ATPases. J Struct Biol. 2004;146:11–31. doi:10.1016/j.jsb.2003.10.010.
54. Speck C, Chen Z, Li H, Stillman B. ATPase-dependent cooperative binding of ORC and Cdc6 to origin DNA. Nat Struct Mol Biol. 2005;12:965–71. doi:10.1038/nsmb1002.
55. DePamphilis ML. Cell cycle dependent regulation of the origin recognition complex. Cell Cycle. 2005;4:70–9.
56. Kara N, Hossain M, Prasanth SG, Stillman B. Orc1 binding to mitotic chromosomes precedes spatial patterning during G1 phase and assembly of the origin recognition complex in human cells. J Biol Chem. 2015. doi:10.1074/jbc.M114.625012.
57. Kreitz S, Ritzi M, Baack M, Knippers R. The human origin recognition complex protein 1 dissociates from chromatin during S phase in HeLa cells. J Biol Chem. 2001;276:6337–42. doi:10.1074/jbc.M009473200.
58. McNairn AJ, Okuno Y, Misteli T, Gilbert DM. Chinese hamster ORC subunits dynamically associate with chromatin throughout the cell-cycle. Exp Cell Res. 2005;308:345–56. doi:10.1016/j.yexcr.2005.05.009.
59. Mendez J, Zou-Yang XH, Kim SY, Hidaka M, Tansey WP, Stillman B. Human origin recognition complex large subunit is degraded by ubiquitin-mediated proteolysis after initiation of DNA replication. Mol Cell. 2002;9:481–91.
60. Siddiqui K, Stillman B. ATP-dependent assembly of the human origin recognition complex. J Biol Chem. 2007;282:32370–83. doi:10.1074/jbc.M705905200.
61. Okuno Y, McNairn AJ, den Elzen N, Pines J, Gilbert DM. Stability, chromatin association and functional activity of mammalian pre-replication complex proteins during the cell cycle. EMBO J. 2001;20:4263–77. doi:10.1093/emboj/20.15.4263.
62. Tatsumi Y, Ohta S, Kimura H, Tsurimoto T, Obuse C. The ORC1 cycle in human cells: I. cell cycle-regulated oscillation of human ORC1. J Biol Chem. 2003;278:41528–34. doi:10.1074/jbc.M307534200.
63. Bell SP, Kaguni JM. Helicase loading at chromosomal origins of replication. Cold Spring Harb Perspect Biol. 2013;5:a010124. doi:10.1101/cshperspect.a010124.

64. Donaldson AD, Blow JJ. The regulation of replication origin activation. Curr Opin Genet Dev. 1999;9:62–8.
65. Renard-Guillet C, Kanoh Y, Shirahige K, Masai H. Temporal and spatial regulation of eukaryotic DNA replication: from regulated initiation to genome-scale timing program. Semin Cell Dev Biol. 2014;30:110–20. doi:10.1016/j.semcdb.2014.04.014.
66. Bell SP, Mitchell J, Leber J, Kobayashi R, Stillman B. The multidomain structure of Orc1p reveals similarity to regulators of DNA replication and transcriptional silencing. Cell. 1995;83:563–8.
67. Loo S, Fox CA, Rine J, Kobayashi R, Stillman B, Bell S. The origin recognition complex in silencing, cell cycle progression, and DNA replication. Mol Biol Cell. 1995;6:741–56.
68. Hickman MA, Rusche LN. Transcriptional silencing functions of the yeast protein Orc1/Sir3 subfunctionalized after gene duplication. Proc Natl Acad Sci U S A. 2010;107:19384–9. doi:10.1073/pnas.1006436107.
69. Abdurashidova G, Danailov MB, Ochem A, Triolo G, Djeliova V, Radulescu S, et al. Localization of proteins bound to a replication origin of human DNA along the cell cycle. EMBO J. 2003;22:4294–303. doi:10.1093/emboj/cdg404.
70. Giordano-Coltart J, Ying CY, Gautier J, Hurwitz J. Studies of the properties of human origin recognition complex and its Walker A motif mutants. Proc Natl Acad Sci U S A. 2005;102:69–74. doi:10.1073/pnas.0408690102.
71. Grallert B, Nurse P. The ORC1 homolog orp1 in fission yeast plays a key role in regulating onset of S phase. Genes Dev. 1996;10:2644–54. doi:10.1101/Gad.10.20.2644.
72. Kuo AJ, Song J, Cheung P, Ishibe-Murakami S, Yamazoe S, Chen JK, et al. The BAH domain of ORC1 links H4K20me2 to DNA replication licensing and Meier-Gorlin syndrome. Nature. 2012;484:115–9. doi:10.1038/nature10956.
73. Ladenburger EM, Keller C, Knippers R. Identification of a binding region for human origin recognition complex proteins 1 and 2 that coincides with an origin of DNA replication. Mol Cell Biol. 2002;22:1036–48.
74. Ohta S, Tatsumi Y, Fujita M, Tsurimoto T, Obuse C. The ORC1 cycle in human cells: II. Dynamic changes in the human ORC complex during the cell cycle. J Biol Chem. 2003;278:41535–40. doi:10.1074/jbc.M307535200.
75. Romanowski P, Madine MA, Rowles A, Blow JJ, Laskey RA. The Xenopus origin recognition complex is essential for DNA replication and MCM binding to chromatin. Curr Biol. 1996;6:1416–25. doi:10.1016/S0960-9822(96)00746-4.
76. Rowles A, Chong JPJ, Brown L, Howell M, Evan GI, Blow JJ. Interaction between the origin recognition complex and the replication licensing system in Xenopus. Cell. 1996;87:287–96. doi:10.1016/S0092-8674(00)81346-X.
77. Vujcic M, Miller CA, Kowalski D. Activation of silent replication origins at autonomously replicating sequence elements near the HML locus in budding yeast. Mol Cell Biol. 1999;19:6098–109.
78. Wu PY, Nurse P. Establishing the program of origin firing during S phase in fission yeast. Cell. 2009;136:852–64. doi:10.1016/j.cell.2009.01.017.
79. Muzi-Falconi M, Kelly TJ. Orp1, a member of the Cdc18/Cdc6 family of S-phase regulators, is homologous to a component of the origin recognition complex. Proc Natl Acad Sci U S A. 1995;92:12475–9.
80. Vashee S, Cvetic C, Lu W, Simancek P, Kelly TJ, Walter JC. Sequence-independent DNA binding and replication initiation by the human origin recognition complex. Genes Dev. 2003;17:1894–908. doi:10.1101/gad.1084203.
81. Vashee S, Simancek P, Challberg MD, Kelly TJ. Assembly of the human origin recognition complex. J Biol Chem. 2001;276:26666–73. doi:10.1074/jbc.M102493200.
82. Wu M, Lu W, Santos RE, Frattini MG, Kelly TJ. Geminin inhibits a late step in the formation of human pre-replicative complexes. J Biol Chem. 2014;289:30810–21. doi:10.1074/jbc.M114.552935.
83. Prasanth SG, Shen Z, Prasanth KV, Stillman B. Human origin recognition complex is essential for HP1 binding to chromatin and heterochromatin organization. Proc Natl Acad Sci U S A. 2010;107:15093–8. doi:10.1073/pnas.1009945107.

84. Sasaki T, Gilbert DM. The many faces of the origin recognition complex. Curr Opin Cell Biol. 2007;19:337–43. doi:10.1016/j.ceb.2007.04.007.
85. McNairn AJ, Gilbert DM. Overexpression of ORC subunits and increased ORC-chromatin association in transformed mammalian cells. J Cell Biochem. 2005;96:879–87. doi:10.1002/jcb.20609.
86. Hemerly AS, Prasanth SG, Siddiqui K, Stillman B. Orc1 controls centriole and centrosome copy number in human cells. Science. 2009;323:789–93. doi:10.1126/science.1166745.
87. Beck DB, Burton A, Oda H, Ziegler-Birling C, Torres-Padilla ME, Reinberg D. The role of PR-Set7 in replication licensing depends on Suv4-20 h. Genes Dev. 2012;26:2580–9. doi:10.1101/gad.195636.112.
88. Burke TW, Cook JG, Asano M, Nevins JR. Replication factors MCM2 and ORC1 interact with the histone acetyltransferase HBO1. J Biol Chem. 2001;276:15397–408. doi:10.1074/jbc.M011556200.
89. Iizuka M, Stillman B. Histone acetyltransferase HBO1 interacts with the ORC1 subunit of the human initiator protein. J Biol Chem. 1999;274:23027–34.
90. Mendoza-Maldonado R, Paolinelli R, Galbiati L, Giadrossi S, Giacca M. Interaction of the retinoblastoma protein with Orc1 and its recruitment to human origins of DNA replication. PLoS One. 2010;5, e13720. doi:10.1371/journal.pone.0013720.
91. Auth T, Kunkel E, Grummt F. Interaction between HP1alpha and replication proteins in mammalian cells. Exp Cell Res. 2006;312:3349–59. doi:10.1016/j.yexcr.2006.07.014.
92. Lidonnici MR, Rossi R, Paixao S, Mendoza-Maldonado R, Paolinelli R, Arcangeli C, et al. Subnuclear distribution of the largest subunit of the human origin recognition complex during the cell cycle. J Cell Sci. 2004;117:5221–31. doi:10.1242/jcs.01405.
93. Bartke T, Vermeulen M, Xhemalce B, Robson SC, Mann M, Kouzarides T. Nucleosome-interacting proteins regulated by DNA and histone methylation. Cell. 2010;143:470–84. doi:10.1016/j.cell.2010.10.012.
94. Giri S, Aggarwal V, Pontis J, Shen Z, Chakraborty A, Khan A, et al. The preRC protein ORCA organizes heterochromatin by assembling histone H3 lysine 9 methyltransferases on chromatin. eLife. 2015;4:PMCID:PMC4442312. doi:10.7554/eLife.06496.
95. Shen Z, Chakraborty A, Jain A, Giri S, Ha T, Prasanth KV, et al. Dynamic association of ORCA with prereplicative complex components regulates DNA replication initiation. Mol Cell Biol. 2012;32:3107–20. doi:10.1128/MCB.00362-12.
96. Shen Z, Sathyan KM, Geng Y, Zheng R, Chakraborty A, Freeman B, et al. A WD-repeat protein stabilizes ORC binding to chromatin. Mol Cell. 2010;40:99–111. doi:10.1016/j.molcel.2010.09.021.
97. Bleichert F, Botchan MR, Berger JM. Crystal structure of the eukaryotic origin recognition complex. Nature. 2015;519:321–6. doi:10.1038/nature14239.
98. Zhang W, Sankaran S, Gozani O, Song J. A Meier-Gorlin syndrome mutation impairs the ORC1-nucleosome association. ACS Chem Biol. 2015. doi:10.1021/cb5009684.
99. Noguchi K, Vassilev A, Ghosh S, Yates JL, DePamphilis ML. The BAH domain facilitates the ability of human Orc1 protein to activate replication origins in vivo. EMBO J. 2006;25:5372–82. doi:10.1038/sj.emboj.7601396.
100. Quintana DG, Hou Z, Thome KC, Hendricks M, Saha P, Dutta A. Identification of HsORC4, a member of the human origin of replication recognition complex. J Biol Chem. 1997;272:28247–51.
101. Tugal T, Zou-Yang XH, Gavin K, Pappin D, Canas B, Kobayashi R, et al. The Orc4p and Orc5p subunits of the Xenopus and human origin recognition complex are related to Orc1p and Cdc6p. J Biol Chem. 1998;273:32421–9. doi:10.1074/Jbc.273.49.32421.
102. Chuang RY, Kelly TJ. The fission yeast homologue of Orc4p binds to replication origin DNA via multiple AT-hooks. Proc Natl Acad Sci U S A. 1999;96:2656–61.
103. Lee JK, Moon KY, Jiang Y, Hurwitz J. The Schizosaccharomyces pombe origin recognition complex interacts with multiple AT-rich regions of the replication origin DNA by means of the AT-hook domains of the spOrc4 protein. Proc Natl Acad Sci U S A. 2001;98:13589–94. doi:10.1073/pnas.251530398.

104. Kong D, DePamphilis ML. Site-specific DNA binding of the Schizosaccharomyces pombe origin recognition complex is determined by the Orc4 subunit. Mol Cell Biol. 2001;21:8095–103. doi:10.1128/MCB.21.23.8095-8103.2001.
105. Kong D, DePamphilis ML. Site-specific ORC binding, pre-replication complex assembly and DNA synthesis at Schizosaccharomyces pombe replication origins. EMBO J. 2002;21:5567–76.
106. Kusic J, Tomic B, Divac A, Kojic S. Human initiation protein Orc4 prefers triple stranded DNA. Mol Biol Rep. 2010;37:2317–22. doi:10.1007/s11033-009-9735-8.
107. Stefanovic D, Kusic J, Divac A, Tomic B. Formation of noncanonical DNA structures mediated by human ORC4, a protein component of the origin recognition complex. Biochemistry. 2008;47:8760–7. doi:10.1021/bi800684f.
108. Stefanovic D, Stanojcic S, Vindigni A, Ochem A, Falaschi A. In vitro protein-DNA interactions at the human lamin B2 replication origin. J Biol Chem. 2003;278:42737–43. doi:10.1074/jbc.M307058200.
109. Klemm RD, Austin RJ, Bell SP. Coordinate binding of ATP and origin DNA regulates the ATPase activity of the origin recognition complex. Cell. 1997;88:493–502.
110. Chesnokov I, Remus D, Botchan M. Functional analysis of mutant and wild-type Drosophila origin recognition complex. Proc Natl Acad Sci U S A. 2001;98:11997–2002. doi:10.1073/pnas.211342798.
111. Bowers JL, Randell JC, Chen S, Bell SP. ATP hydrolysis by ORC catalyzes reiterative Mcm2-7 assembly at a defined origin of replication. Mol Cell. 2004;16:967–78. doi:10.1016/j.molcel.2004.11.038.
112. Dhar SK, Dutta A. Identification and characterization of the human ORC6 homolog. J Biol Chem. 2000;275:34983–8. doi:10.1074/jbc.M006069200.
113. Liu S, Balasov M, Wang H, Wu L, Chesnokov IN, Liu Y. Structural analysis of human Orc6 protein reveals a homology with transcription factor TFIIB. Proc Natl Acad Sci U S A. 2011;108:7373–8. doi:10.1073/pnas.1013676108.
114. Li JJ, Herskowitz I. Isolation of ORC6, a component of the yeast origin recognition complex by a one-hybrid system. Science. 1993;262:1870–4.
115. Chesnokov IN, Chesnokova ON, Botchan M. A cytokinetic function of Drosophila ORC6 protein resides in a domain distinct from its replication activity. Proc Natl Acad Sci U S A. 2003;100:9150–5. doi:10.1073/pnas.1633580100.
116. Prasanth SG, Prasanth KV, Stillman B. Orc6 involved in DNA replication, chromosome segregation, and cytokinesis. Science. 2002;297:1026–31. doi:10.1126/science.1072802.
117. Semple JW, Da-Silva LF, Jervis EJ, Ah-Kee J, Al-Attar H, Kummer L, et al. An essential role for Orc6 in DNA replication through maintenance of pre-replicative complexes. EMBO J. 2006;25:5150–8. doi:10.1038/sj.emboj.7601391.
118. Chen S, de Vries MA, Bell SP. Orc6 is required for dynamic recruitment of Cdt1 during repeated Mcm2-7 loading. Genes Dev. 2007;21:2897–907. doi:10.1101/gad.1596807.
119. Balasov M, Huijbregts RP, Chesnokov I. Role of the Orc6 protein in origin recognition complex-dependent DNA binding and replication in Drosophila melanogaster. Mol Cell Biol. 2007;27:3143–53. doi:10.1128/MCB.02382-06.
120. Wilmes GM, Archambault V, Austin RJ, Jacobson MD, Bell SP, Cross FR. Interaction of the S-phase cyclin Clb5 with an "RXL" docking sequence in the initiator protein Orc6 provides an origin-localized replication control switch. Genes Dev. 2004;18:981–91. doi:10.1101/gad.1202304.
121. Balasov M, Huijbregts RP, Chesnokov I. Functional analysis of an Orc6 mutant in Drosophila. Proc Natl Acad Sci U S A. 2009;106:10672–7. doi:10.1073/pnas.0902670106.
122. Huijbregts RP, Svitin A, Stinnett MW, Renfrow MB, Chesnokov I. Drosophila Orc6 facilitates GTPase activity and filament formation of the septin complex. Mol Biol Cell. 2009;20:270–81. doi:10.1091/mbc.E08-07-0754.
123. Patel S, Latterich M. The AAA team: related ATPases with diverse functions. Trends Cell Biol. 1998;8:65–71. doi:10.1016/S0962-8924(97)01212-9.
124. Neuwald AF, Aravind L, Spouge JL, Koonin EV. AAA(+): a class of chaperone-like ATPases associated with the assembly, operation, and disassembly of protein complexes. Genome Res. 1999;9:27–43.

125. Liang C, Weinreich M, Stillman B. ORC and Cdc6p interact and determine the frequency of initiation of DNA replication in the genome. Cell. 1995;81:667–76.
126. Leatherwood J, LopezGirona A, Russell P. Interaction of Cdc2 and Cdc18 with a fission yeast ORC2-like protein. Nature. 1996;379:360–3. doi:10.1038/379360a0.
127. Coleman TR, Carpenter PB, Dunphy WG. The Xenopus Cdc6 protein is essential for the initiation of a single round of DNA replication in cell-free extracts. Cell. 1996;87:53–63. doi:10.1016/S0092-8674(00)81322-7.
128. Evrin C, Clarke P, Zech J, Lurz R, Sun J, Uhle S, et al. A double-hexameric MCM2-7 complex is loaded onto origin DNA during licensing of eukaryotic DNA replication. Proc Natl Acad Sci U S A. 2009;106:20240–5. doi:10.1073/pnas.0911500106.
129. Frigola J, Remus D, Mehanna A, Diffley JF. ATPase-dependent quality control of DNA replication origin licensing. Nature. 2013;495:339–43. doi:10.1038/nature11920.
130. Sun J, Fernandez-Cid A, Riera A, Tognetti S, Yuan Z, Stillman B, et al. Structural and mechanistic insights into Mcm2-7 double-hexamer assembly and function. Genes Dev. 2014;28:2291–303. doi:10.1101/gad.242313.114.
131. Sun J, Kawakami H, Zech J, Speck C, Stillman B, Li H. Cdc6-induced conformational changes in ORC bound to origin DNA revealed by cryo-electron microscopy. Structure. 2012;20:534–44. doi:10.1016/j.str.2012.01.011.
132. Remus D, Beuron F, Tolun G, Griffith JD, Morris EP, Diffley JF. Concerted loading of Mcm2-7 double hexamers around DNA during DNA replication origin licensing. Cell. 2009;139:719–30. doi:10.1016/j.cell.2009.10.015.
133. Perkins G, Diffley JFX. Nucleotide-dependent prereplicative complex assembly by Cdc6p, a homolog of eukaryotic and prokaryotic clamp-loaders. Mol Cell. 1998;2:23–32. doi:10.1016/S1097-2765(00)80110-0.
134. Liang C, Stillman B. Persistent initiation of DNA replication and chromatin-bound MCM proteins during the cell cycle in cdc6 mutants. Genes Dev. 1997;11:3375–86. doi:10.1101/Gad.11.24.3375.
135. Jallepalli PV, Brown GW, MuziFalconi M, Tien D, Kelly TJ. Regulation of the replication initiator protein p65(cdc18) by CDK phosphorylation. Genes Dev. 1997;11:2767–79. doi:10.1101/Gad.11.21.2767.
136. Nishitani H, Nurse P. P65(Cdc18) plays a major role controlling the initiation of DNA-replication in fission yeast. Cell. 1995;83:397–405. doi:10.1016/0092-8674(95)90117-5.
137. Tanaka TU, Knapp D, Nasmyth K. Loading of an Mcm protein onto DNA replication origins is regulated by Cdc6p and CDKs. Cell. 1997;90:649–60. doi:10.1016/S0092-8674(00)80526-7.
138. Nguyen VQ, Co C, Li JJ. Cyclin-dependent kinases prevent DNA re-replication through multiple mechanisms. Nature. 2001;411:1068–73. doi:10.1038/35082600.
139. Kim GS, Kang J, Bang SW, Hwang DS. Cdc6 localizes to S- and G2-phase centrosomes in a cell cycle-dependent manner. Biochem Biophys Res Commun. 2015;456:763–7. doi:10.1016/j.bbrc.2014.12.018.
140. Nishitani H, Lygerou Z, Nishimoto T. Proteolysis of DNA replication licensing factor Cdt1 in S-phase is performed independently of Geminin through its N-terminal region. J Biol Chem. 2004;279:30807–16. doi:10.1074/Jbc.M312644200.
141. Maiorano D, Rul W, Mechali M. Cell cycle regulation of the licensing activity of Cdt1 in Xenopus laevis. Exp Cell Res. 2004;295:138–49. doi:10.1016/J.Yexcr.2003.11.018.
142. Tsuyama T, Tada S, Watanabe S, Seki M, Enomoto T. Licensing for DNA replication requires a strict sequential assembly of Cdc6 and Cdt1 onto chromatin in Xenopus egg extracts. Nucleic Acids Res. 2005;33:765–75. doi:10.1093/Nar/Gki226.
143. Yanagi K, Mizuno T, You ZY, Hanaoka F. Mouse geminin inhibits not only Cdt1-MCM6 interactions but also a novel intrinsic Cdt1 DNA binding activity. J Biol Chem. 2002;277:40871–80. doi:10.1074/Jbc.M206202200.
144. Ferenbach A, Li A, Brito-Martins M, Blow JJ. Functional domains of the Xenopus replication licensing factor Cdt1. Nucleic Acids Res. 2005;33:316–24. doi:10.1093/Nar/Gki176.
145. Tanaka S, Diffley JFX. Interdependent nuclear accumulation of budding yeast Cdt1 and Mcm2-7 during G1 phase. Nat Cell Biol. 2002;4:198–207. doi:10.1038/Ncb757.

146. Cook JG, Chasse DAD, Nevins JR. The regulated association of Cdt1 with minichromosome maintenance proteins and Cdc6 in mammalian cells. J Biol Chem. 2004;279:9625–33. doi:10.1074/Jbc.M311933200.
147. Mihaylov IS, Kondo T, Jones L, Ryzhikov S, Tanaka J, Zheng JY, et al. Control of DNA replication and chromosome ploidy by geminin and cyclin A. Mol Cell Biol. 2002;22:1868–80. doi:10.1128/Mcb.22.6.1868-1880.2002.
148. Vaziri C, Saxena S, Jeon Y, Lee C, Murata K, Machida Y, et al. A p53-dependent checkpoint pathway prevents rereplication (vol 11, pg 997, 2003). Mol Cell. 2003;11:1415. doi:10.1016/S1097-2765(03)00202-8.
149. Zhong W, Feng H, Santiago FE, Kipreos ET. CUL-4 ubiquitin ligase maintains genome stability by restraining DNA-replication licensing. Nature. 2003;423:885–9. doi:10.1038/nature01747.
150. Melixetian M, Ballabeni A, Masiero L, Gasparini P, Zamponi R, Bartek J, et al. Loss of Geminin induces rereplication in the presence of functional p53. J Cell Biol. 2004;165:473–82. doi:10.1083/Jcb.200403106.
151. Liu EB, Li XH, Yan F, Zhao QP, Wu XH. Cyclin-dependent kinases phosphorylate human Cdt1 and induce its degradation. J Biol Chem. 2004;279:17283–8. doi:10.1074/Jbc.C300549200.
152. McGarry TJ, Kirschner MW. Geminin, an inhibitor of DNA replication, is degraded during mitosis. Cell. 1998;93:1043–53. doi:10.1016/S0092-8674(00)81209-X.
153. Ballabeni A, Melixetian M, Zamponi R, Masiero L, Marinoni F, Helin K. Human Geminin promotes pre-RC formation and DNA replication by stabilizing CDT1 in mitosis. EMBO J. 2004;23:3122–32. doi:10.1038/Sj.Emboj.7600314.
154. Arias EE, Walter JC. PCNA functions as a molecular platform to trigger Cdt1 destruction and prevent re-replication. Nat Cell Biol. 2006;8:84-U33. doi:10.1038/Ncb1346.
155. Senga T, Sivaprasad U, Zhu WG, Park JH, Arias EE, Walter JC, et al. PCNA is a cofactor for Cdt1 degradation by CUL4/DDB1-mediated N-terminal ubiquitination. J Biol Chem. 2006;281:6246–52. doi:10.1074/Jbc.M512705200.
156. Sugimoto N, Tatsumi Y, Tsurumi T, Matsukage A, Kiyono T, Nishitani H, et al. Cdt1 phosphorylation by cyclin A-dependent kinases negatively regulates its function without affecting geminin binding. J Biol Chem. 2004;279:19691–7. doi:10.1074/Jbc.M313175200.
157. Saxena S, Yuan P, Dhar SK, Senga T, Takeda D, Robinson H, et al. A dimerized coiled-coil domain and an adjoining part of geminin interact with two sites on Cdt1 for replication inhibition. Mol Cell. 2004;15:245–58. doi:10.1016/J.Molcel.2004.06.045.
158. Kondo T, Kobayashi M, Tanaka J, Yokoyama A, Suzuki S, Kato N, et al. Rapid degradation of Cdt1 upon UV-induced DNA damage is mediated by SCFSkp2 complex. J Biol Chem. 2004;279:27315–9. doi:10.1074/jbc.M314023200.
159. Thomer M, May NR, Aggarwal BD, Kwok G, Calvi BR. Drosophila double-parked is sufficient to induce re-replication during development and is regulated by cyclin E/CDK2. Development. 2004;131:4807–18. doi:10.1242/dev.01348.
160. Xouri G, Lygerou Z, Nishitani H, Pachnis V, Nurse P, Taraviras S. Cdt1 and geminin are down-regulated upon cell cycle exit and are over-expressed in cancer-derived cell lines. Eur J Biochem. 2004;271:3368–78. doi:10.1111/J.1432-1033.2004.04271.X.
161. Takeda DY, Parvin JD, Dutta A. Degradation of Cdt1 during S phase is Skp2-independent and is required for efficient progression of mammalian cells through S phase. J Biol Chem. 2005;280:23416–23. doi:10.1074/jbc.M501208200.
162. Jin JP, Arias EE, Chen J, Harper JW, Walter JC. A family of diverse Cul4-Ddb1-interacting proteins includes Cdt2, which is required for S phase destruction of the replication factor Cdt1. Mol Cell. 2006;23:709–21. doi:10.1016/J.Molcel.2006.08.010.
163. Kim Y, Kipreos ET. Cdt1 degradation to prevent DNA re-replication: conserved and non-conserved pathways. Cell Div. 2007;2:18. doi:10.1186/1747-1028-2-18.
164. Jee J, Mizuno T, Kamada K, Tochio H, Chiba Y, Yanagi K, et al. Structure and mutagenesis studies of the C-terminal region of licensing factor Cdt1 enable the identification of key residues for binding to replicative helicase Mcm proteins. J Biol Chem. 2010;285:15931–40. doi:10.1074/jbc.M109.075333.

165. Jackson AP, Laskey RA, Coleman N. Replication proteins and human disease. Cold Spring Harb Perspect Biol. 2014;6:pii:a013060. doi:10.1101/cshperspect.a013060.
166. Gineau L, Cognet C, Kara N, Lach FP, Dunne J, Veturi U, et al. Partial MCM4 deficiency in patients with growth retardation, adrenal insufficiency, and natural killer cell deficiency. J Clin Invest. 2012;122:821–32. doi:10.1172/JCI61014.
167. Hughes CR, Guasti L, Meimaridou E, Chuang CH, Schimenti JC, King PJ, et al. MCM4 mutation causes adrenal failure, short stature, and natural killer cell deficiency in humans. J Clin Invest. 2012;122:814–20. doi:10.1172/JCI60224.
168. Casey JP, Nobbs M, McGettigan P, Lynch S, Ennis S. Recessive mutations in MCM4/PRKDC cause a novel syndrome involving a primary immunodeficiency and a disorder of DNA repair. J Med Genet. 2012;49:242–5. doi:10.1136/jmedgenet-2012-100803.
169. Bernal JA, Venkitaraman AR. A vertebrate N-end rule degron reveals that Orc6 is required in mitosis for daughter cell abscission. J Cell Biol. 2011;192:969–78. doi:10.1083/jcb.201008125.

Chapter 26
Mechanisms and Consequences of Break-Induced Replication

Roberto A. Donnianni and Lorraine S. Symington

Abstract Homology-dependent repair of chromosomal double-strand breaks (DSBs) usually involves short tracts of DNA synthesis. However, repair of DSBs that have only one end with homology to a donor chromosome can occur by extensive DNA synthesis from the site of strand invasion to the telomere, a process referred to as break-induced replication (BIR). Recent studies of BIR initiated at DSBs have shown that DNA synthesis occurs by a conservative mechanism involving a migrating D-loop intermediate and is associated with a much higher rate of mutagenesis than normal S-phase synthesis. Furthermore, the invading strand is unstable and can switch to different templates increasing the risk of chromosome rearrangements. The mutagenic potential of BIR suggests it may play an important role in genome evolution as well as cancer etiology; however, it has raised the question of whether similar processes occur at collapsed replication forks.

Keywords DNA replication • DNA repair • Homologous recombination • Mutagenesis • Rad51 • Pol32 • Pif1

Introduction

Chromosomal double-strand breaks (DSBs) are cytotoxic lesions that occur spontaneously during normal cell metabolism or following treatment of cells with DNA-damaging agents, such as ionizing radiation or certain chemotherapeutic drugs. The two main mechanisms used to repair DSBs in eukaryotic cells are non-homologous end joining (NHEJ) and homologous recombination (HR) [1, 2]. NHEJ requires DNA end protection the Ku heterodimer and direct ligation by DNA ligase IV (Dnl4/Lig4) with the help of accessory proteins [1]. Although NHEJ occurs with high fidelity at cohesive ends, such as those produced by restriction endonucleases, it is generally considered to be an error-prone mechanism because of the potential for loss or gain of nucleotides at the junction, particularly at blunt or mismatched ends.

R.A. Donnianni • L.S. Symington (✉)
Department of Microbiology & Immunology, Columbia University Medical Center,
701 W. 168th St., New York, NY 10032, USA
e-mail: lss5@cumc.columbia.edu

D.L. Kaplan (ed.), *The Initiation of DNA Replication in Eukaryotes*,
DOI 10.1007/978-3-319-24696-3_26

In addition, *trans* joining of different DSB ends can lead to chromosome translocations or other types of rearrangements. NHEJ is favored in the G1 phase of the cell cycle when resection of DNA ends is repressed [3–5]. HR occurs in S and G2 phases when end resection is activated and a homologous sister chromatid is available to template repair of a broken chromosome [4]. HR is considered to be an error-free mechanism when the donor is the preferred sister chromatid. However, if repair occurs from a non-sister chromatid, such as the homolog in diploid cells or repeated sequence elsewhere in the genome, HR can lead to loss of heterozygosity (LOH) or chromosome rearrangements.

DSBs created by endonucleases generally have two ends with homology to a donor sequence (two-ended DSBs) and are repaired by direct ligation or by HR associated with a short tract of DNA synthesis. By contrast, collapsed replication forks, eroded telomeres, or mis-segregated fragmented chromosomes present only one end with homology to a donor sequence, the so called one-ended or single-end DSBs [6, 7]. The invading end of a one-ended DSB can prime extensive DNA synthesis in a process referred to as break-induced replication (BIR) [8, 9]. This process has been modeled in yeast by creating an endonuclease-induced DSB between sequence homologous to a donor duplex and heterologous downstream sequence forcing extensive DNA synthesis from the site of strand invasion to the telomere [10–15]. Recent studies have shown that the properties of the replication machinery during BIR are grossly different from those at a normal replication fork [12, 16–18], raising the issue of whether collapsed replication forks repair by a BIR-like mechanism. Here, we discuss models for BIR, the mutagenic potential of this process, and compare BIR as observed at DSBs with recombination-dependent replication fork restart.

Mechanisms for Homology-Directed DSB Repair

Gene Conversion and Associated Crossovers

Much of our understanding of how DSBs are repaired has come from physical analysis of intermediates formed following induction of site-specific DSBs in yeast, the phenotype of HR mutants in genetic and physical assays that measure recombination proficiency, and biochemical analysis of purified proteins [19]. In general, mechanisms for homology-directed repair of chromosomal DSBs initiate by 5′–3′ degradation of the broken DNA ends to create 3′ single-stranded DNA (ssDNA) tails, a process referred to as end resection (Fig. 26.1) [20]. The Mre11-Rad50-Xrs2 complex (MRX/N, Xrs2 is replaced by Nbs1 in vertebrates), activated by Sae2 (CtIP in vertebrates), initiates end resection by nicking the 5′ strand internal to the DSB end followed by 3′–5′ resection by the Mre11 nuclease to create a short 3′ ssDNA tail. More extensive processing of the partially resected DNA ends requires the 5′–3′ exonuclease Exo1 or Dna2 endonuclease in collaboration with the Sgs1 helicase (BLM or WRN in vertebrates) [21–24]. The Fun30 chromatin remodeler facilitates

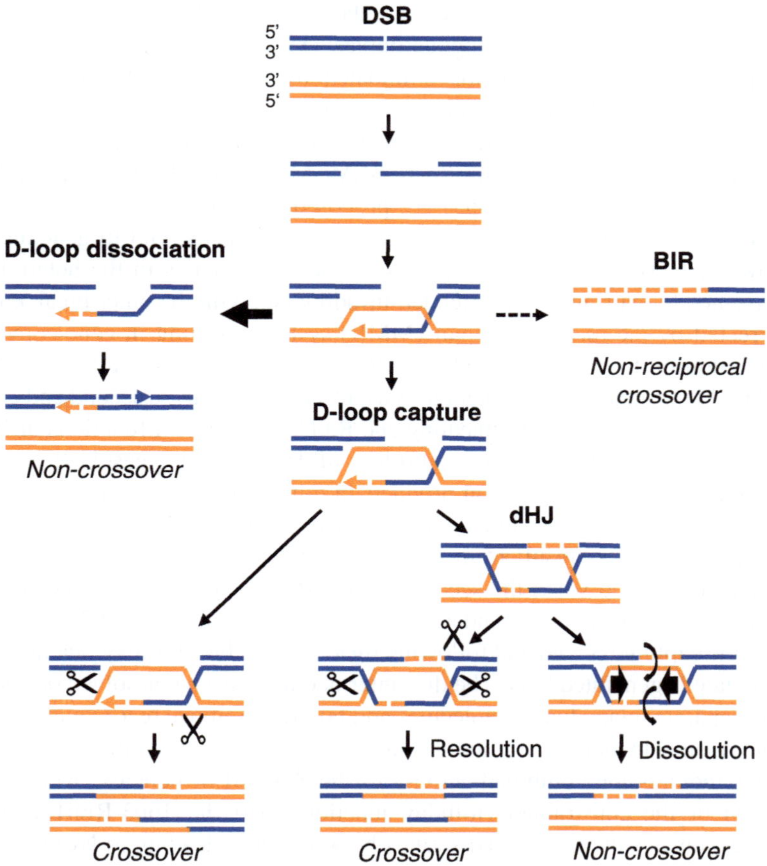

Fig. 26.1 Models for homologous recombination. DSB ends are resected to yield 3′ ssDNA tails used for Rad51-dependent strand invasion to generate a D-loop. After extension of the 3′ end, the invading end can be displaced to anneal to the other DSB end yielding NCO products. Alternatively, the D-loop is captured by the other DSB end and can be directly cleaved by Mus81-Mms4 to generate CO products. DNA synthesis primed from the 3′ end of the D-loop captures intermediate and ligation results in formation of a dHJ intermediate that can be dissolved by Sgs1-Top3-Rmi1 (STR) or resolved by endonucleases (only the CO product of resolution is shown). If the other end of the break is lost, or homology is restricted to one side of the DSB, replication proceeds to the chromosome terminus (BIR)

extensive resection [25–27], which occurs at a rate of ~4 kb/h and can proceed for up to 50 kb in the absence of a homologous sequence to template repair [24, 28].

The ssDNA formed by end resection is initially bound by replication protein A (RPA), which is then replaced by the Rad51 recombinase in a reaction requiring Rad52 in yeast or BRCA2 in many other organisms [29]. RPA facilitates loading of Rad51 by removing secondary structure from the ssDNA and shields ssDNA from degradation [30–32]. The Rad51 paralogs (Rad55 and Rad57 in yeast) stabilize the Rad51-ssDNA complex and prevent dissociation of Rad51 by the Srs2 helicase [33, 34].

The Rad51 nucleoprotein complex catalyzes the key step of HR—homology searching and strand invasion with homologous duplex DNA to form a displacement-loop (D-loop) intermediate—a reaction stimulated by the Rad54 dsDNA-dependent ATPase (Fig. 26.1) [29]. The invading 3′ end from the broken chromosome primes DNA synthesis by DNA polymerase δ (Polδ), templated by the donor duplex [35, 36]. Several DNA helicases, including Sgs1/BLM, Mph1/Fml1/FANCM and RTEL1, and DNA Topoisomerase 3 (Top3) are proposed to unwind the D-loop to allow annealing of the extended 3′ strand with the other resected end to generate non-crossover (NCO) products [37–42]. Mismatched nucleotides in the heteroduplex DNA formed by pairing of the extended invading end with the other break end can be corrected by the mismatch repair (MMR) system resulting in gene conversion [43]. In addition to its role in mediating Rad51 assembly on RPA-coated ssDNA, Rad52 is able to anneal complementary ssDNA and can catalyze second end capture in vitro [44–46]. The ssDNA annealing and Rad51 recruitment functions of Rad52 are thought to account for its essential role in all HR events in yeast [19].

The majority of recombination products in mitotically dividing cells are NCOs while around 1–20 %, depending on the organism studied, arise by a crossover (CO) mechanism [19]. COs result from endonucleolytic processing of branched DNA intermediates formed when the second break end captures the displaced strand of the D-loop instead of the extended 3′ invading end (Fig. 26.1). DNA synthesis extends the captured 3′ end and following ligation a double Holliday junction (dHJ) intermediate is generated [47]. The dHJ intermediate must be dissolved or resolved to yield separate intact duplex molecules and failure to remove recombination intermediates results in meiotic or mitotic catastrophe [48, 49]. Dissolution of dHJ intermediates requires the combined activity of the Sgs1/BLM helicase, which drives migration of the constrained Holliday junctions, and the Top3-Rmi1 complex, which decatenates the interlinked strands between them, leading to NCO products [50, 51]. Resolution of dHJ intermediates by endonucleolytic cleavage can yield CO or NCO products. Several structure-selective nucleases, including Mus81-Mms4/EME1, SLX1-SLX4, and Yen1/GEN1, able to cut HJs in vitro have been identified as candidate resolvases in yeast and mammals [49]. Genetic studies have shown that Mus81-Mms4 is the main nuclease responsible for DSB-induced mitotic COs in budding yeast with Yen1 serving a back-up function [52]. The preferred substrates for Mus81-Mms4 cleavage in vitro are 3′ flaps and nicked HJs raising the possibility that Mus81-Mms4 directly cleaves the intermediate formed by second end capture rather than the dHJ intermediate (Fig. 26.1) [53, 54].

BIR Initiated from DSBs

If coordination of the two ends of a DSB is lost, or if homology is present at only one of the two ends, the invading 3′ end primes extensive DNA synthesis from the site of strand invasion to the end of the chromosome. BIR is a rare outcome of repair (<1 % of repaired products) at a two-ended DSB [52, 55]; however, the frequency of BIR is much higher when homology is limited to one side of the DSB suggesting a second homologous break end plays an important role in preventing BIR [11, 12, 14]. Elimination of Mph1 and Mus81 increases the frequency of BIR at two-ended DSBs indicating that D-loop dissociation and cleavage, respectively, facilitate second end capture to prevent BIR [54]. Gene conversion between non-sister chromatids (chromosome homologs in diploid cells) results in minimal LOH, though more extensive LOH can occur if there is an associated CO and the recombinant chromatids segregate to opposite poles during mitosis. In contrast, BIR between non-sister chromatids results in extensive LOH, and if it occurs between dispersed repeats then nonreciprocal translocations or more complex rearrangements can occur.

BIR initiated from endonuclease-induced DSBs has mostly been studied in budding yeast. Recombination products with some characteristics of BIR have been detected at a DSB generated within one copy of a direct repeat recombination reporter in mammalian cells and are generally referred to as long-tract gene conversion [56, 57]. These events are thought to arise by strand invasion between misaligned repeats of sister chromatids and after copying both repeats and extending beyond them the invading end displaces and joins to the other break end by NHEJ. A similar type of event initiated from a broken plasmid was reported in yeast [58], but was very rare, presumably due to the low efficiency of NHEJ at noncohesive ends; indeed, the greater capacity of mammalian cells for NHEJ could prevent BIR at DSBs.

In Vivo Assays to Study BIR

Chromosomal BIR Systems

Chromosomal BIR reporter systems have been developed in yeast by inserting the HO endonuclease cut site between a sequence homologous to a donor site elsewhere in the genome followed by a heterologous sequence; the constructs are designed so that sequences centromere distal to the DSB are not essential for cell viability and are lost during BIR. HO is regulated using an inducible promoter. Although erroneous end joining can ligate the two ends of the DSB, this mechanism is much less efficient in yeast than BIR. The reporter developed in our laboratory consists of overlapping nonfunctional fragments of the *lys2* gene inserted on different chromosomes of a haploid strain with an HO cut site downstream of the 3′ truncated *lys2* fragment [12]. The location of the donor 5′truncated fragment can be varied to monitor BIR efficiency at differing distances from the telomere. BIR generates a functional copy of the *LYS2* gene enabling selection on medium lacking lysine (Fig. 26.2a). Physical methods (PCR or Southern blot hybridization of intact chromosomes or digested genomic DNA) can be used to monitor the kinetics of BIR. Similar assays using fragments of the *CAN1* and *URA3* genes have also been reported [13, 59]. One potential limitation with the ectopic BIR assays is that the region of shared homology between the recipient and donor is short and could be lost by end resection. Although resection initiates on the 5′ strand, 3′ end loss is also detected 2 h after HO induction [30, 60]. Indeed, slowing extensive resection increases the efficiency of BIR between ectopic repeats suggesting BIR is slow and resection can remove the short homologies needed to drive BIR in these assays [26, 61, 62].

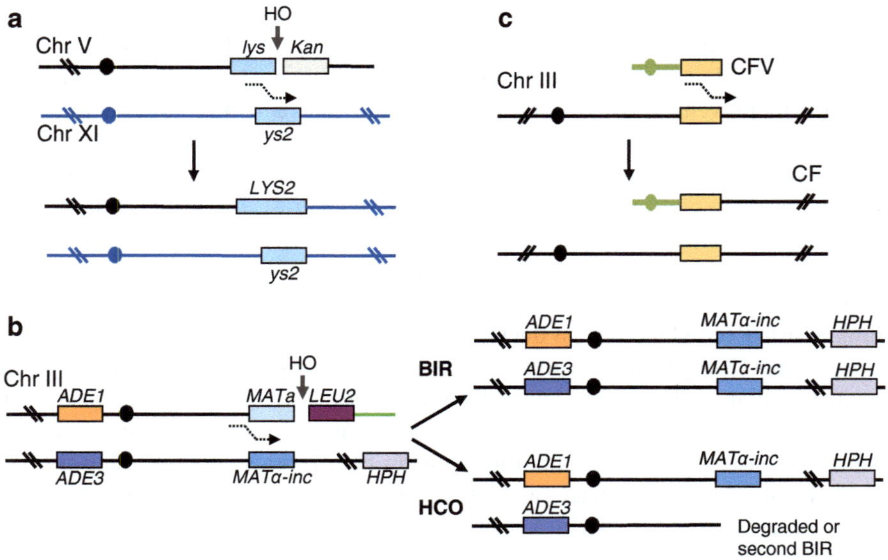

Fig. 26.2 Assays for BIR. (**a**) Ectopic assay with homology restricted to only one side of the HO cut site to force repair by BIR forming a Chr V-XI translocation. The nonessential sequence to the left of the DSB, including the Kan marker, is degraded. (**b**) Repair of an HO-induced DSB at the *MAT* locus of a haploid strain disomic for Chr III. The recipient chromosome is truncated to restrict homology to the centromere proximal side of the DSB. After degradation of *MATa* sequence, invasion of the homolog occurs copying to the end of the donor chromosome (BIR). Some events are resolved as a half crossover (HCO); the fragmented chromosome is degraded or invades the HCO product to repair by BIR. (**c**) Transformation of yeast with the linearized chromosome fragmentation vector (CFV) generates a stable chromosome fragment (CF) following strand invasion and replication to the telomere and telomere addition at the telomere seed sequence present at the other end of the linear CFV. These events are selected by the *URA3* marker carried on the CFV and are stably maintained by the centromere on the CF. Centromeres are shown by solid circles

An alternative BIR reporter makes use of the natural HO cut site at the *MAT* locus in a haploid strain modified to be disomic for chromosome (Chr) III. Sequence centromere distal to the *MAT***a** HO cut site of one Chr III homolog was replaced with a *LEU2* cassette and telomere, while the *MATα-inc* allele, which is resistant to cutting by HO, was introduced on the other homolog (Fig. 26.2b). After DSB induction, centromere proximal sequence invades the homolog and copies to the end of the donor chromosome (~100 kb). Failure to repair the DSB results in chromosome loss and can be distinguished from BIR using heterozygous markers on the left arm of Chr III. In some mutant backgrounds where strand invasion occurs but DNA synthesis is limited, the recombination intermediate is resolved as a "half crossover" (HCO); these events are characterized by retention of the left arm of the recipient chromosome and acquisition of sequences from the donor *MAT* locus to the telomere, detected using the heterozygous markers present on each chromosome arm [63]. The more extensive homology provided by the homolog might be the reason for the higher efficiency of BIR using the disome system compared with the ectopic reporters.

Plasmid Transformation Assay

A chromosome fragmentation vector (CFV) has also been used to study BIR [10, 15]. The CFV contains the *URA3* selectable marker, *CEN4*, a $(G_{1-3}T)_n$ tract to provide a site for telomere addition and a 2–5 kb sequence homologous to a yeast chromosomal sequence. The CFV is linearized between the Chr homology region and telomere seeding sequence with a restriction endonuclease and used to transform yeast, selecting for Ura^+ colonies (Fig. 26.2c). Most transformants arise by *de novo* telomere addition to heal one end of the CFV and strand invasion at the other end into the endogenous yeast locus to copy the entire chromosome arm yielding a stable chromosome fragment (CF). This assay can be used in haploid or diploid cells, but is not amenable to physical analysis of BIR intermediates due to low transformation efficiency.

Recombination-Dependent Replication Fork Restart

BIR in all of the above assays involves repair of the DSB from a non-sister chromatid, a situation that may not be directly relevant to repair of a collapsed replication fork by sister-chromatid recombination. With the modification of I-SceI and CRSPR/Cas9 nucleases to generate site-specific nicks instead of DSBs, it should be possible to establish systems where a nick is converted to a single-ended DSB by passage of the replication fork requiring strand invasion into the sister chromatid to restore the collapsed fork [64, 65]. A mutant of the FLP recombinase, which generates a long-lived single-strand break by covalently attaching to the 3′ end at the FRT site, has been developed to study replication-dependent DSBR in yeast and could be used to study BIR [66]. Aguilera and colleagues have shown that a shortened HO cut site undergoes single- instead of double-stranded DNA cleavage and can be used to monitor repair at collapsed replication forks [67]; however, it has only been used on plasmids to study sister chromatid or intramolecular repair between inverted repeats and has not been tested in the chromosomal context.

Proposed Mechanisms for BIR

As with other HR processes, BIR initiates by Rad51-dependent strand invasion [10], but what distinguishes BIR from other HR mechanisms is the vast length of newly synthesized DNA: hundreds of kb of DNA can be synthesized during BIR even when cells are arrested in the G2/M phase [14, 68]. Physical methods to detect BIR intermediates and products have yielded insight into the mechanism and have been used to test the requirement for essential replication proteins using conditional alleles of the relevant genes. Two models have been proposed for the mechanism of DNA synthesis (Fig. 26.3).

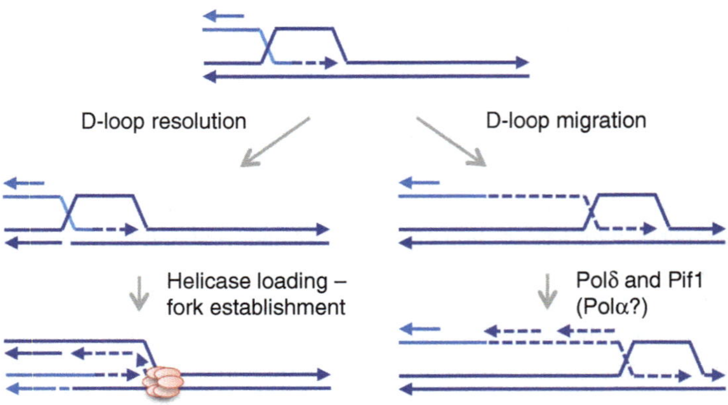

Fig. 26.3 Mechanisms proposed for BIR. Cleavage of the D-loop intermediate by a structure-selective nuclease can convert it to an intermediate resembling a replication fork to synthesize DNA by a semiconservative mechanism. By the migrating D-loop model both newly synthesized strands segregate with the recipient chromosome

Semiconservative DNA Synthesis

In this model, the D-loop is cleaved by a structure-selective endonuclease to form a replication fork where, after completion of DNA synthesis, one newly synthesized strand is associated with the broken chromosome and the other to the donor chromosome, resulting in semiconservative inheritance of DNA as in normal S-phase synthesis. This model has support from studies showing BIR requires the MCM helicase complex, Cdt1, Cdc45, Dpb11, Cdc7, Mcm10, the GINS complex, and all three replicative DNA polymerases [13, 69]. As expected for an origin-independent mechanism for DNA synthesis, *cdc6-1* and *orc6-td* conditional mutants were reported to be proficient for BIR at the restrictive temperature; however, it is not known how the MCM helicase is recruited to the strand invasion intermediate. Conversion of a D-loop intermediate to a replication fork is predicted to require the activity of structure-selective nucleases. Conversely Mus81 and Yen1, the two main resolvases in yeast, are dispensable for BIR at endonuclease-induced DSBs [18, 70].

Conservative DNA Synthesis

D-loop migration to the end of the chromosome results in the newly synthesized "leading strand" extruded from the trailing end of the D-loop acting as a template for synthesis of the complementary strand, similar to a classic mechanism envisioned for bacteriophage T4 recombination-initiated DNA synthesis [71]. Both newly synthesized strands segregate with the repaired chromosome, while the donor chromosome remains unchanged. To distinguish between conservative and semiconservative modes of DNA synthesis, two groups followed bromodeoxyuridine (BrdU) incorporation into chromosomes during BIR by pulsed field gel electrophoresis of intact chromosomes or DNA combing and found BrdU associated only with the recipient chromosome, consistent with conservative DNA synthesis [12, 17]. Moreover, by native two-dimensional gel electrophoresis, Saini et al. [17] detected a long-lived ssDNA intermediate associated with D-loop migration indicating that synthesis of the two strands is uncoupled. Gene conversion during yeast mating-type switching is also by a conservative mechanism suggesting BIR has similar characteristics to long-tract gene conversion [72, 73].

Unlike normal S-phase DNA synthesis, BIR is highly dependent on Pol32, a nonessential subunit of DNA Polδ [13, 63, 74]. Products due to half crossovers were recovered from *pol32* mutants using the Chr III disome and plasmid-based assays, consistent with proficient strand invasion and a defect in extension of the invading strand [63, 74]. A reduced frequency of BIR and increased half crossovers were also found in mutants expressing conditional mutations in *POL3*, which encodes the catalytic subunit of DNA Polδ, suggesting that DNA Polδ catalyzes "leading strand" synthesis during BIR [74, 75]. Long-tract gene conversion is also reduced in *pol32* and *pol3-ct* mutants [36, 52, 76].

The Pif1 helicase has been shown to play an important role in BIR in genetic assays [17, 18, 70], and an in vitro study demonstrated that Pif1, DNA Polδ, and PCNA promote transition of a Rad51-catalyzed strand invasion intermediate into a migrating D-loop that is capable of extensive DNA synthesis [18]. Wilson et al. [18] reported no BIR defect in the *mcm4-td* mutant, in contrast to an earlier study [69], and instead proposed that Pif1 acts as the replicative helicase for BIR. It is puzzling why two groups came to different conclusions using *mcm* mutants. One possible explanation is that BIR is slower and less efficient using the *CAN1* ectopic assay [69] than the Chr III disome system [18]. To assess the requirement for essential replication proteins, cells with the ectopic BIR assay expressing temperature-sensitive alleles were grown at permissive temperature, arrested in G2/M phase by addition of nocodazole, and switched to the nonpermissive temperature, and then HO was induced [69]. Even for wild type cells the BIR product was barely detected 6 h after HO induction and the BIR defect observed for most of the replication mutants tested was most pronounced at 24 h. It is possible that cells override the nocodazole arrest during the long time course and then repair the break in the next cell cycle. Since HR is restricted to the S-G2 phase, strand invasion would be defective if replication-defective cells had arrested in G1 of the next cell cycle.

Mutagenesis Associated with BIR

Elevated Mutagenesis During DNA Synthesis

Pioneering studies by Malkova and colleagues showed that DNA synthesis during BIR is accompanied by a highly increased frequency of mutagenesis [16, 77]. To study mutagenesis associated with BIR, a *lys2-Ins* frameshift reporter cassette was inserted at different positions downstream of the *MATα-inc* locus of the donor chromosome in the Chr III disome system (Fig. 26.2b) [16]. The frequency of reversion to Lys^+ increased by up to 2800-fold during BIR, as compared with normal S-phase DNA synthesis (spontaneous errors). The increased mutagenesis was partially due to activation of the DNA damage checkpoint and elevation of dNTP levels [16]. Elimination of mismatch repair (MMR) resulted in increased mutagenesis, but the differential between BIR and S-phase associated mutations was lower in the *msh2* mutant than wild type suggesting MMR is much less efficient at correcting BIR-induced errors than spontaneous errors. The mutagenesis data are consistent with the migrating D-loop model for BIR, which postulates the newly synthesized strand is extruded behind the D-loop and would therefore not be a substrate for MMR; the errors would then be fixed by second strand synthesis. Mutation of *RAD30*, which encodes the Polη translesion synthesis DNA polymerase, did not significantly reduce the frequency of mutagenesis and only a modest decrease at some locations was found in the absence of Polζ suggesting that most errors are caused by the replicative DNA polymerases. A defect in the proofreading function of DNA Polδ, but not of DNA Polε, caused a higher frequency of mutagenesis consistent with DNA Polδ catalyzing DNA synthesis during BIR [16].

Uncoupling of first and second strand synthesis during BIR results in the accumulation of a long tract of ssDNA that is expected to be susceptible to base damage [17, 18]. Indeed, when BIR was completed in the presence of the DNA alkylating agent, methyl methanesulfonate (MMS), around 50 % of the products contained mutation clusters of 4–115 kb in length in the area of BIR on the right arm of Chr III [77]. The mutations were generally restricted to the recipient chromosome at low MMS doses, but at a higher MMS dose more complex clusters were recovered due to mutations in the recipient and donor chromosomes. Because an increase in half crossovers was also found at the higher MMS dose, it is likely that the broken chromosome resulting from a half crossover invaded the newly generated BIR product setting off another round of BIR.

Template Switching During BIR

Smith et al. [78] showed that when a linearized CFV was targeted to invade Chr III of a diploid with polymorphic Chr III homologs around 15 % of the CFs recovered had sequences from both homologs. These data suggest an unstable strand invasion intermediate that can invade, dissociate, and then invade another template, a process referred to as template switching (TS). In addition to switching between homologs, if the invading end of the CFV dissociated from a repeated sequence (Ty transposon or long terminal repeat (LTRs) derived from Tys) downstream of the site of strand invasion, then the second invasion could occur at a Ty or LTR elsewhere in the genome resulting in a nonreciprocal translocation. D-loop dissociation is required for TS and the *mph1* mutant, while showing a higher frequency of BIR due to stabilization of the D-loop intermediate, exhibited a defect in TS [70, 79]. In a chromosomal assay to monitor TS between a fragment of the *MAT* locus inserted on

Chr VII and the native *MAT*, *HML*, and *HMR* loci on Chr III, a high frequency of complex rearrangements dependent on the Mus81 and Yen1 resolvases was observed, suggesting cleavage of recombination intermediates is required for TS [80]. The Rdh54 translocase is also required for TS during BIR, but how it promotes this process is not understood [59].

Half Crossover-Induced Genome Instability

As noted above, the strand invasion intermediate can be cleaved linking the recipient chromosome to the donor chromosome and leaving a truncated fragment of the donor chromosome (Fig. 26.2b). This fragment could be degraded or engage in a second BIR event targeting the HCO, or the homolog in diploid cells. Consistent with this suggestion a significant increase in recombination between chromosome homologs was found in the presence of a linear CFV [70]. The DNA damage checkpoint is activated by BIR, presumably because BIR is slow and involves a long-lived ssDNA intermediate, and the checkpoint is required to maintain the integrity of chromosomes during BIR. Cycles of BIR, often terminating with chromosome rearrangements formed when the truncated Chr III was degraded to expose Ty elements located 30 kb upstream of the initiating DSB, were observed in cells defective for the DNA damage checkpoint [75].

Microhomology-Mediated BIR

Most of the studies of BIR in yeast have utilized pairs of sequence with >1 kb of homology. Even when homology was reduced to 62 bp, or if repeats were used that have 29 % sequence divergence, Rad51-dependent BIR was observed, albeit at a low frequency [59, 81]. In a study designed to detect de novo gene duplication in yeast, most of events recovered had intrachromosomal segmental duplications (SD) of ~115 kb flanked by LTRs derived from Ty3 transposons, or intra- or inter-chromosomal large duplications (>115 kb) with LTRs or microhomologies at the junctions. The frequency of SD was unaffected by a *dnl4* mutation suggesting they are not due to classical NHEJ, though a role for microhomology-mediated end joining (MMEJ), a Ku and Ligase IV independent mechanism, could not be excluded. Interestingly, all classes of events were dependent on Pol32 and interpreted to result from BIR. While Rad52 is required for BIR in all other assays, it was only required for the SDs resulting from LTR recombination; moreover, there was no requirement for Rad51. Large SDs have been detected in human cells undergoing replication stress and were partially dependent on *POL3D* [82]. The frequency of SDs was significantly increased by replication stress consistent with broken forks engaging in BIR [82, 83]; alternatively, broken forks at each end of a replicon might join via microhomologies (or the greater homology provided by LTRs) creating an SD [84]. Pol32 is required for MMEJ in yeast, in addition to BIR, making it difficult to distinguish between these hypotheses by genetic means [85].

Complex genomic rearrangements are associated with a number of human genomic disorders and are frequently observed in cancer cells [86–88]. Most of the rearrangements observed have microhomologies at the junctions and are likely formed by classical NHEJ or MMEJ. However, some of the rearrangements are comprised of duplications and deletions interspersed with nonduplicated or triplicated segments, most easily explained by a replicative mechanism. These types of events could occur by template switching at stalled replication forks or possibly by BIR initiated at microhomologies [86].

Physiological Relevance of BIR

Restart of Collapsed Replication Forks

Replication fork collapse is thought to result from the replisome running into a nick on one of the template strands to produce a one-ended DSB (Fig. 26.4) [19]. Restoration of the replication fork requires Rad51-dependent strand invasion of the intact sister chromatid by the broken chromatid [6], and how the resulting strand invasion intermediate is converted to a structure competent for extensive replication is currently under investigation. The mutagenic nature of BIR at endonuclease-induced DSBs (see above) has raised the question of whether the migrating D-loop mechanism operates at collapsed replication forks. To date, no functions equivalent to the *E. coli* Pri proteins, which are required to load the DnaB helicase at collapsed replication forks [8], have been identified in eukaryotes raising the question of how the MCM helicase is recruited to collapsed replication forks or if another helicase is able to promote processive DNA synthesis after replication restart. Eukaryotes have multiple origins of replication and it is possible that a one-ended DSB is converted to a two-ended DSB when the fork from the adjacent replicon encounters the collapsed fork (Fig. 26.4). Alternatively, a fork converging with the D-loop intermediate would result in a single HJ requiring resolution by a structure-selective nuclease such as Mus81-Mms4. Yeast *mus81* mutants show high sensitivity to camptothecin, a Topoisomerase I inhibitor that causes replication fork collapse, consistent with a role for Mus81 in replication fork restart [89–91]. Furthermore, human cells deficient for MUS81 or EME2 (human cells have two MUS81-containing endonucleases, MUS81-EME1 and MUS81-EME2) are defective for replication restart after hydroxyurea treatment [92]. A recent study found that human cells subjected to replication stress by overexpression of cyclin E are highly dependent on *POL3D* and *POL4D* (encoding the two smallest subunits of the DNA Polδ complex) for survival, interpreted as evidence for a BIR-like mechanism in mammals [82].

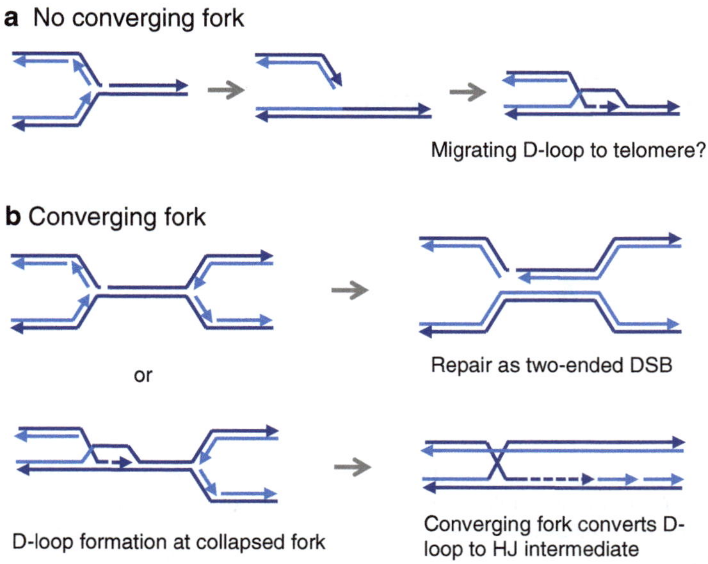

Fig. 26.4 Replication restart at a collapsed fork. (**a**) The replication fork collapses when it encounters a nick on the template strand. The leading strand ligates to the nick creating an intact duplex for invasion by the resected broken arm to form a D-loop intermediate. If there is no converging fork, the D-loop might migrate to the telomere as shown for endonuclease-induced DSBs or be resolved to form a replication fork (not shown, see Fig. 26.3). (**b**) If a converging fork approaches the collapsed fork before strand invasion, the one-ended DSB is converted to a two-ended DSB and could then repair by gene conversion. On the other hand, if a converging fork approaches the D-loop intermediate a single HJ is formed that would require Mus81-Mms4 or Yen1 for cleavage to separate the replicated sister chromatids

Telomere Maintenance by BIR

Perhaps the most physiologically relevant function of BIR is to elongate telomeres that are lost when telomerase is absent or when telomeres are uncapped [93]. Early studies of a yeast mutant defective for telomerase showed that the cells started to senesce as the telomere repeats were eroded, but then recovered by a *RAD52*-dependent mechanism [94]. Two *RAD52* and *POL32*-dependent pathways of "survivors," referred to as type I and type II survivors, have been identified based on the arrangement of telomeric DNA (Fig. 26.5) [13, 95, 96]. Type I survivors arise by amplification of the subtelomeric Y′repeats but the ends still have very short tracts of duplex telomeric repeat DNA [94, 95]. Type I survivors contain extrachromosomal circular Y′elements that are thought to be by-products of recombination or to serve as templates for recombination [97]. Formation of type I survivors requires *RAD51*, *RAD54*, *RAD57* (*RAD55* has not been tested, but is assumed to be required based on its requirement for other *RAD57*-dependent recombination events) and *PIF1* helicase, in addition to *RAD52* and *POL32* [98–100]. Type II survivors show extensive elongation of the telomeric repeats, but the telomeres are highly

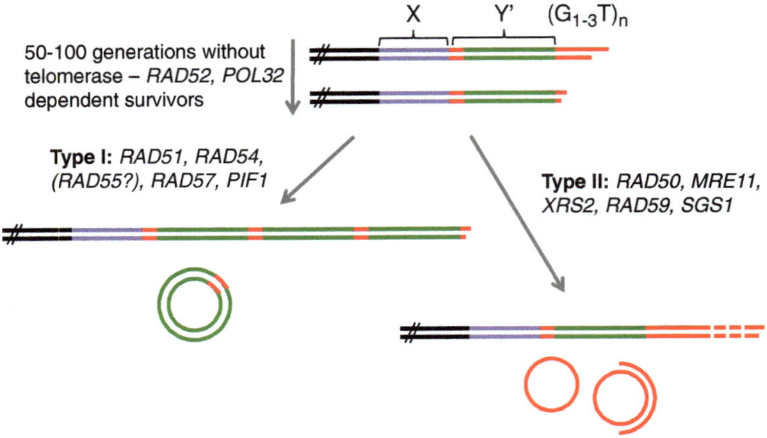

Fig. 26.5 Telomere maintenance by BIR. Yeast telomeres are comprised of ~300 bp of telomere repeats, $(G_{1\text{-}3}T)_n$, 0–4 copies of the Y′ element plus ~100 bp of telomere repeats and X elements. Telomeres become progressively shorter in the absence of telomerase and enter crisis. Rad52 and Pol32-dependent survivors arise by amplification of the Y′ elements (and associated telomere repeats) or by formation of long heterogeneous telomeric tracts. Adapted from [96]

heterogeneous with some very short telomeres and others exceeding 12 kb in length [95]. The telomeres of type II survivors are unstable and undergo progressive shrinkage and rapid lengthening [95]. Extrachromosomal circles (double-stranded or partially single-stranded) consisting of telomere repeats are found in type II survivors and the dramatic lengthening of telomeres is proposed to occur by a rolling circle mechanism [96, 97]. Annealing of the ssDNA at telomeres with a partially ssDNA circular template would explain the requirement for *RAD52* to generate type II survivors. The generation of type II survivors requires *MRE11*, *RAD50*, *XRS2*, *RAD59*, and *SGS1* as well as *RAD52* and *POL32* [96]. Type I survivors grow slowly and easily convert to type II survivors due to their faster growth rate indicating that the two survival mechanisms are not mutually exclusive [95]. Since survival of telomerase-negative cells requires BIR, this phenotype can be used to identify genes involved in BIR [100].

Similar recombination-dependent mechanisms have been proposed for at least some of the telomere elongation events in ALT (alternative lengthening of telomeres) positive tumors [101]. The telomeric tracts of ALT cells are heterogeneous in length and associated with extrachromosomal circles containing telomeric DNA. Evidence for a recombination-based telomere maintenance mechanism in human ALT cell lines derives from the high frequency of telomere sister chromatid exchange and spreading of a marker inserted within the telomere of one chromosome to other chromosomes [102–104]. Telomere clustering and elongation in ALT cells requires the MRN complex, consistent with yeast studies. RAD51 localizes to damaged telomeres and is required for telomere clustering and long-range movement of telomeres in ALT cells [105].

Concluding Remarks

Yeast cells have the capacity for extensive DNA synthesis initiated from a one-ended DSB by a mechanism that is quite different from normal S-phase synthesis. The mutagenic nature of BIR suggests it could play an important role in remodeling genomes during evolution, and contribute to the elevated mutagenesis and chromosome rearrangements observed in tumor cells. A challenge for the future will be to determine whether collapsed fork repair occurs by the mutagenic migrating D-loop mechanism or whether systems exist to limit mutagenesis during single-end invasion. One attractive hypothesis is that a converging fork from an adjacent replicon limits D-loop extension; indeed, the activation of dormant origins by DSBs may be a mechanism to prevent extensive D-loop migration [106]. Understanding how a D-loop formed at one-ended DSBs is able to migrate over long distances (>50 kb) and how the second strand is synthesized are provocative questions that are sure to lead to new and exciting experimentation in the field.

Abbreviations

ALT	Alternative lengthening of telomere
BIR	Break-induced replication
CEN	Centromere
CF	Chromosome fragment
CFV	Chromosome fragmentation vector
Chr	Chromosome
CO	Crossover
dHJ	Double Holliday junction
D-Loop	Displacement loop
DSB	Double-strand break
DSBR	Double-strand break repair
dsDNA	Double-stranded DNA
HCO	Half crossover
HR	Homologous recombination
LOH	Loss of heterozygosity
LTR	300 bp long terminal repeat of Ty transposable element
MMEJ	Microhomology-mediated end joining
MMS	Methyl methanesulfonate
MRX	Mre11-Rad50-Xrs2
NCO	Non-crossover
NHEJ	Non-homologous end joining
RPA	Replication protein A
SD	Segmental duplication
sHJ	Single Holliday junction
ssDNA	Single-stranded DNA
TS	Template switching

Acknowledgements We thank P. Ruff and W.K. Holloman for comments on the manuscript. Research in our laboratory cited here was supported by grants from the National Institutes of Health (GM041784 and GM094386). R.A.D. was supported by an International Fellowship in Cancer Research co-funded by AIRC and Marie Curie Actions.

References

1. Chiruvella KK, Liang Z, Wilson TE. Repair of double-strand breaks by end joining. Cold Spring Harb Perspect Biol. 2013;5(5):012757. PubMed Epub 2013/05/03.eng.
2. Jasin M, Rothstein R. Repair of strand breaks by homologous recombination. Cold Spring Harb Perspect Biol. 2013;5(11):012740. PubMed Epub 2013/10/08. eng.
3. Aylon Y, Liefshitz B, Kupiec M. The CDK regulates repair of double-strand breaks by homologous recombination during the cell cycle. EMBO J. 2004;23(24):4868–75. PubMed Epub 2004/11/19. eng.
4. Ira G, Pellicioli A, Balijja A, Wang X, Fiorani S, Carotenuto W, et al. DNA end resection, homologous recombination and DNA damage checkpoint activation require CDK1. Nature. 2004;431(7011):1011–7. PubMed Epub 2004/10/22. eng.
5. Zhang Y, Shim EY, Davis M, Lee SE. Regulation of repair choice: Cdk1 suppresses recruitment of end joining factors at DNA breaks. DNA Repair. 2009;8(10):1235–41. Pubmed Central PMCID: 2748135. Epub 2009/08/25. eng.
6. Hashimoto Y, Puddu F, Costanzo V. RAD51- and MRE11-dependent reassembly of uncoupled CMG helicase complex at collapsed replication forks. Nat Struct Mol Biol. 2012;19:17–24.
7. Prado F, Aguilera A. Impairment of replication fork progression mediates RNA polII transcription-associated recombination. EMBO J. 2005;24(6):1267–76. Pubmed Central PMCID: 556405.
8. Anand RP, Lovett ST, Haber JE. Break-induced DNA replication. Cold Spring Harb Perspect Biol. 2013;5:a010397.
9. Llorente B, Smith CE, Symington LS. Break-induced replication: what is it and what is it for? Cell Cycle. 2008;7(7):859–64. PubMed Epub 2008/04/17. eng.
10. Davis AP, Symington LS. RAD51-dependent break-induced replication in yeast. Mol Cell Biol. 2004;24:2344.
11. Bosco G, Haber JE. Chromosome break-induced DNA replication leads to nonreciprocal translocations and telomere capture. Genetics. 1998;150:1037–47.
12. Donnianni RA, Symington LS. Break-induced replication occurs by conservative DNA synthesis. Proc Natl Acad Sci U S A. 2013;110:13475–80.
13. Lydeard JR, Jain S, Yamaguchi M, Haber JE. Break-induced replication and telomerase-independent telomere maintenance require Pol32. Nature. 2007;448:1–5.
14. Malkova A, Naylor ML, Yamaguchi M, Ira G, Haber JE. RAD51-dependent break-induced replication differs in kinetics and checkpoint responses from RAD51-mediated gene conversion. Society. 2005;25:933–44.
15. Morrow DM, Connelly C, Hieter P. "Break copy" duplication: a model for chromosome fragment formation in Saccharomyces cerevisiae. Genetics. 1997;147(2):371–82. Pubmed Central PMCID: 1208164. Epub 1997/10/23. eng.
16. Deem A, Keszthelyi A, Blackgrove T, Vayl A, Coffey B, Mathur R, et al. Break-induced replication is highly inaccurate. PLoS Biol. 2011;9, e1000594.
17. Saini N, Ramakrishnan S, Elango R, Ayyar S, Zhang Y, Deem A, et al. Migrating bubble during break-induced replication drives conservative DNA synthesis. Nature. 2013; 502(7471):389–92.
18. Ma W, Kwon Y, Xu Y, Chung W-H, Chi P, Niu H. Pif1 helicase and Polδ promote recombination-coupled DNA synthesis via bubble migration. Nature. 2013;502(7471):393–6.

19. Symington LS, Rothstein R, Lisby M. Mechanisms and regulation of mitotic recombination in Saccharomyces cerevisiae. Genetics. 2014;198(3):795–835. Pubmed Central PMCID: 4224172.
20. Symington LS, Gautier J. Double-strand break end resection and repair pathway choice. Annu Rev Genet. 2011;45:247–71.
21. Cannavo E, Cejka P. Sae2 promotes dsDNA endonuclease activity within Mre11-Rad50-Xrs2 to resect DNA breaks. Nature. 2014;514(7520):122–5. PubMed Epub 2014/09/19. eng.
22. Gravel S, Chapman JR, Magill C, Jackson SP. DNA helicases Sgs1 and BLM promote DNA double-strand break resection. Genes Dev. 2008;22(20):2767–72. PubMed Epub 2008/10/17. eng.
23. Mimitou EP, Symington LS. Sae2, Exo1 and Sgs1 collaborate in DNA double-strand break processing. Nature. 2008;455(7214):770–4. Pubmed Central PMCID: 3818707.
24. Zhu Z, Chung WH, Shim EY, Lee SE, Ira G. Sgs1 helicase and two nucleases Dna2 and Exo1 resect DNA double-strand break ends. Cell. 2008;134(6):981–94. Pubmed Central PMCID: 2662516.
25. Chen X, Cui D, Papusha A, Zhang X, Chu CD, Tang J, et al. The Fun30 nucleosome remodeller promotes resection of DNA double-strand break ends. Nature. 2012;489(7417):576–80. Pubmed Central PMCID: 3640768. Epub 2012/09/11. eng.
26. Costelloe T, Louge R, Tomimatsu N, Mukherjee B, Martini E, Khadaroo B, et al. The yeast Fun30 and human SMARCAD1 chromatin remodellers promote DNA end resection. Nature. 2012;489(7417):581–4. Pubmed Central PMCID: 3493121. Epub 2012/09/11. eng.
27. Eapen VV, Sugawara N, Tsabar M, Wu WH, Haber JE. The Saccharomyces cerevisiae chromatin remodeler Fun30 regulates DNA end resection and checkpoint deactivation. Mol Cell Biol. 2012;32(22):4727–40. Pubmed Central PMCID: 3486187. Epub 2012/09/26. eng.
28. Chung W-H, Zhu Z, Papusha A, Malkova A, Ira G. Defective resection at DNA double-strand breaks leads to de novo telomere formation and enhances gene targeting. PLoS Genet. 2010;6:e1000948.
29. San Filippo J, Sung P, Klein H. Mechanism of eukaryotic homologous recombination. Annu Rev Biochem. 2008;77:229–57.
30. Chen H, Lisby M, Symington LS. RPA coordinates DNA end resection and prevents formation of DNA hairpins. Mol Cell. 2013;50(4):589–600. Pubmed Central PMCID: 3855855.
31. Sugiyama T, Zaitseva EM, Kowalczykowski SC. A single-stranded DNA-binding protein is needed for efficient presynaptic complex formation by the Saccharomyces cerevisiae Rad51 protein. J Biol Chem. 1997;272(12):7940–5. PubMed Epub 1997/03/21. eng.
32. Sung P. Catalysis of ATP-dependent homologous DNA pairing and strand exchange by yeast RAD51 protein. Science. 1994;265(5176):1241–3. PubMed Epub 1994/08/26. eng.
33. Fung CW, Mozlin AM, Symington LS. Suppression of the double-strand-break-repair defect of the Saccharomyces cerevisiae rad57 mutant. Genetics. 2009;181(4):1195–206. Pubmed Central PMCID: 2666491.
34. Liu J, Renault L, Veaute X, Fabre F, Stahlberg H, Heyer WD. Rad51 paralogues Rad55-Rad57 balance the antirecombinase Srs2 in Rad51 filament formation. Nature. 2011; 479(7372):245–8. Pubmed Central PMCID: 3213327.
35. Li X, Stith CM, Burgers PM, Heyer WD. PCNA is required for initiation of recombination-associated DNA synthesis by DNA polymerase delta. Mol Cell. 2009;36(4):704–13. Pubmed Central PMCID: 2784891.
36. Maloisel L, Fabre F, Gangloff S. DNA polymerase delta is preferentially recruited during homologous recombination to promote heteroduplex DNA extension. Mol Cell Biol. 2008;28(4):1373–82. Pubmed Central PMCID: 2258756.
37. Adams MD, McVey M, Sekelsky JJ. Drosophila BLM in double-strand break repair by synthesis-dependent strand annealing. Science. 2003;299(5604):265–7. PubMed Epub 2003/01/11. eng.
38. Barber LJ, Youds JL, Ward JD, McIlwraith MJ, O'Neil NJ, Petalcorin MI, et al. RTEL1 maintains genomic stability by suppressing homologous recombination. Cell. 2008;135(2): 261–71. PubMed Epub 2008/10/30. eng.

39. Fasching CL, Cejka P, Kowalczykowski SC, Heyer WD. Top3-rmi1 dissolve rad51-mediated d loops by a topoisomerase-based mechanism. Mol Cell. 2015;57(4):595–606. Pubmed Central PMCID: 4338411. Epub 2015/02/24. eng.
40. Ira G, Malkova A, Liberi G, Foiani M, Haber JE. Srs2 and Sgs1-Top3 suppress crossovers during double-strand break repair in yeast. Cell. 2003;115(4):401–11.
41. Prakash R, Satory D, Dray E, Papusha A, Scheller J, Kramer W, et al. Yeast Mph1 helicase dissociates Rad51-made D-loops: implications for crossover control in mitotic recombination. Genes Dev. 2009;23(1):67–79. Pubmed Central PMCID: 2632165.
42. Whitby MC. The FANCM family of DNA helicases/translocases. DNA Repair. 2010; 9(3):224–36. PubMed Epub 2010/02/02. eng.
43. Boiteux S, Jinks-Robertson S. DNA repair mechanisms and the bypass of DNA damage in Saccharomyces cerevisiae. Genetics. 2013;193(4):1025–64. Pubmed Central PMCID: 3606085.
44. Lao JP, Oh SD, Shinohara M, Shinohara A, Hunter N. Rad52 promotes postinvasion steps of meiotic double-strand-break repair. Mol Cell. 2008;29(4):517–24. PubMed Epub 2008/03/04. eng.
45. Mortensen UH, Bendixen C, Sunjevaric I, Rothstein R. DNA strand annealing is promoted by the yeast Rad52 protein. Proc Natl Acad Sci U S A. 1996;93(20):10729–34.
46. Sugiyama T, Kantake N, Wu Y, Kowalczykowski SC. Rad52-mediated DNA annealing after Rad51-mediated DNA strand exchange promotes second ssDNA capture. EMBO J. 2006;25(23):5539–48. Pubmed Central PMCID: 1679760. Epub 2006/11/10. eng.
47. Szostak JW, Orr-Weaver TL, Rothstein RJ, Stahl FW. The double-strand-break repair model for recombination. Cell. 1983;33(1):25–35.
48. Bizard AH, Hickson ID. The dissolution of double Holliday junctions. Cold Spring Harb Perspect Biol. 2014;6(7):a016477.
49. Wyatt HD, West SC. Holliday junction resolvases. Cold Spring Harb Perspect Biol. 2014;6(9):023192. PubMed Epub 2014/09/04. eng.
50. Cejka P, Plank JL, Bachrati CZ, Hickson ID, Kowalczykowski SC. Rmi1 stimulates decatenation of double Holliday junctions during dissolution by Sgs1-Top3. Nat Struct Mol Biol. 2010;17(11):1377–82. Pubmed Central PMCID: 2988882. Epub 2010/10/12. eng.
51. Wu L, Hickson ID. The Bloom's syndrome helicase suppresses crossing over during homologous recombination. Nature. 2003;426(6968):870–4.
52. Ho CK, Mazon G, Lam AF, Symington LS. Mus81 and Yen1 promote reciprocal exchange during mitotic recombination to maintain genome integrity in budding yeast. Mol Cell. 2010;40(6):988–1000. Pubmed Central PMCID: 3021384.
53. Osman F, Dixon J, Doe CL, Whitby MC. Generating crossovers by resolution of nicked Holliday junctions: a role for Mus81-Eme1 in meiosis. Mol Cell. 2003;12(3):761–74.
54. Mazon G, Symington LS. Mph1 and Mus81-Mms4 prevent aberrant processing of mitotic recombination intermediates. Mol Cell. 2013;52(1):63–74. Pubmed Central PMCID: 3818723.
55. Malkova A, Ivanov EL, Haber JE. Double-strand break repair in the absence of RAD51 in yeast: a possible role for break-induced DNA replication. Proc Natl Acad Sci. 1996;93:7131.
56. Chandramouly G, Kwok A, Huang B, Willis NA, Xie A, Scully R. BRCA1 and CtIP suppress long-tract gene conversion between sister chromatids. Nat Commun. 2013;4:2404. Pubmed Central PMCID: 3838905. Epub 2013/09/03. eng.
57. Johnson RD, Jasin M. Sister chromatid gene conversion is a prominent double-strand break repair pathway in mammalian cells. EMBO J. 2000;19(13):3398–407.
58. Kraus E, Leung WY, Haber JE. Break-induced replication: a review and an example in budding yeast. Proc Natl Acad Sci U S A. 2001;98(15):8255–62. Pubmed Central PMCID: 37429. Epub 2001/07/19. eng.
59. Anand RP, Tsaponina O, Greenwell PW, Lee CS, Du W, Petes TD, et al. Chromosome rearrangements via template switching between diverged repeated sequences. Genes Dev. 2014;28(21):2394–406. Pubmed Central PMCID: 4215184. Epub 2014/11/05. eng.
60. Zierhut C, Diffley JF. Break dosage, cell cycle stage and DNA replication influence DNA double strand break response. EMBO J. 2008;27(13):1875–85. Pubmed Central PMCID: 2413190. Epub 2008/05/31. eng.

61. Lydeard JR, Lipkin-Moore Z, Jain S, Eapen VV, Haber JE. Sgs1 and exo1 redundantly inhibit break-induced replication and de novo telomere addition at broken chromosome ends. PLoS Genet. 2010;6, e1000973.
62. Marrero VA, Symington LS. Extensive DNA end processing by exo1 and sgs1 inhibits break-induced replication. PLoS Genet. 2010;6(7), e1001007. Pubmed Central PMCID: 2900301. Epub 2010/07/16. eng.
63. Deem A, Barker K, Vanhulle K, Downing B, Vayl A, Malkova A. Defective break-induced replication leads to half-crossovers in Saccharomyces cerevisiae. Genetics. 2008;179: 1845–60.
64. Katz SS, Gimble FS, Storici F. To nick or not to nick: comparison of I-SceI single- and double-strand break-induced recombination in yeast and human cells. PLoS One. 2014;9(2), e88840. Pubmed Central PMCID: 3928301. Epub 2014/02/22. eng.
65. Vriend LE, Jasin M, Krawczyk PM. Assaying break and nick-induced homologous recombination in mammalian cells using the DR-GFP reporter and Cas9 nucleases. Methods Enzymol. 2014;546:175–91. PubMed Epub 2014/11/16. eng.
66. Nielsen I, Bentsen IB, Lisby M, Hansen S, Mundbjerg K, Andersen AH, et al. A Flp-nick system to study repair of a single protein-bound nick in vivo. Nat Methods. 2009;6(10):753–7. PubMed Epub 2009/09/15. eng.
67. Cortes-Ledesma F, Aguilera A. Double-strand breaks arising by replication through a nick are repaired by cohesin-dependent sister-chromatid exchange. EMBO Rep. 2006;7(9):919–26. Pubmed Central PMCID: 1559660. Epub 2006/08/05. eng.
68. Llorente B, Smith CE, Symington LS. BIR, what is and what is for? 2008:859–64.
69. Lydeard JR, Lipkin-Moore Z, Sheu Y-J, Stillman B, Burgers PM, Haber JE. Break-induced replication requires all essential DNA replication factors except those specific for pre-RC assembly. Genes Dev. 2010;24:1133–44.
70. Stafa A, Donnianni RA, Timashev LA, Lam AF, Symington LS. Template switching during break-induced replication is promoted by the Mph1 helicase in Saccharomyces cerevisiae. Genetics. 2014;196:1017–28.
71. Formosa T, Alberts BM. DNA synthesis dependent on genetic recombination: characterization of a reaction catalyzed by purified bacteriophage T4 proteins. Cell. 1986;47(5):793–806.
72. Arcangioli B, de Lahondes R. Fission yeast switches mating type by a replication-recombination coupled process. EMBO J. 2000;19(6):1389–96. Pubmed Central PMCID: 305679.
73. Ira G, Satory D, Haber JE. Conservative inheritance of newly synthesized DNA in double-strand break-induced gene conversion. Mol Cell Biol. 2006;26(24):9424–9. Pubmed Central PMCID: 1698534.
74. Smith CE, Lam AF, Symington LS. Aberrant double-strand break repair resulting in half crossovers in mutants defective for Rad51 or the DNA polymerase {delta} complex. Mol Cell Biol. 2009;29(6):1432–41. PubMed Epub 2009/01/14. Eng.
75. Vasan S, Deem A, Ramakrishnan S, Argueso JL, Malkova A. Cascades of genetic instability resulting from compromised break-induced replication. PLoS Genet. 2014;10, e1004119.
76. Jain S, Sugawara N, Lydeard J, Vaze M, Tanguy N, Gac L, et al. A recombination execution checkpoint regulates the choice of homologous recombination pathway during DNA double-strand break repair. Genes Dev. 2009;291–303.
77. Sakofsky CJ, Roberts SA, Malc E, Mieczkowski PA, Resnick MA, Gordenin DA, et al. Break-induced replication is a source of mutation clusters underlying kataegis. Cell Rep. 2014;7(5):1640–8. Pubmed Central PMCID: 4274036. Epub 2014/06/03. eng.
78. Smith CE, Llorente B, Symington LS. Template switching during break-induced replication. Nature. 2007;447:102–5.
79. Luke-Glaser S, Luke B. The Mph1 helicase can promote telomere uncapping and premature senescence in budding yeast. PLoS One. 2012;7(7), e42028. Pubmed Central PMCID: 3407055. Epub 2012/08/01. eng.
80. Pardo B, Aguilera A. Complex chromosomal rearrangements mediated by break-induced replication involve structure-selective endonucleases. PLoS Genet. 2012;8(9), e1002979. Pubmed Central PMCID: 3459980.

81. Ruiz JF, Gomez-Gonzalez B, Aguilera A. Chromosomal translocations caused by either pol32-dependent or pol32-independent triparental break-induced replication. Mol Cell Biol. 2009;29(20):5441–54. Pubmed Central PMCID: 2756893. Epub 2009/08/05. eng.
82. Costantino L, Sotiriou SK, Rantala JK, Magin S, Mladenov E, Helleday T, et al. Break-induced replication repair of damaged forks induces genomic duplications in human cells. Science. 2014;343:88–91.
83. Payen C, Koszul R, Dujon B, Fischer G. Segmental duplications arise from Pol32-dependent repair of broken forks through two alternative replication-based mechanisms. PLoS Genet. 2008;4(9), e1000175. Pubmed Central PMCID: 2518615. Epub 2008/09/06. eng.
84. Finn KJ, Li JJ. Single-stranded annealing induced by re-initiation of replication origins provides a novel and efficient mechanism for generating copy number expansion via non-allelic homologous recombination. PLoS Genet. 2013;9(1), e1003192. Pubmed Central PMCID: 3536649.
85. Villarreal DD, Lee K, Deem A, Shim EY, Malkova A, Lee SE. Microhomology directs diverse DNA break repair pathways and chromosomal translocations. PLoS Genet. 2012;8(11), e1003026. Pubmed Central PMCID: 3493447. Epub 2012/11/13. eng.
86. Hastings PJ, Ira G, Lupski JR. A microhomology-mediated break-induced replication model for the origin of human copy number variation. PLoS Genet. 2009;5, e1000327.
87. Lee JA, Carvalho CM, Lupski JR. A DNA replication mechanism for generating nonrecurrent rearrangements associated with genomic disorders. Cell. 2007;131(7):1235–47. PubMed Epub 2007/12/28. eng.
88. Stephens PJ, Greenman CD, Fu B, Yang F, Bignell GR, Mudie LJ, et al. Massive genomic rearrangement acquired in a single catastrophic event during cancer development. Cell. 2011;144(1):27–40. Pubmed Central PMCID: 3065307.
89. Deng C, Brown JA, You D, Brown JM. Multiple endonucleases function to repair covalent topoisomerase I complexes in Saccharomyces cerevisiae. Genetics. 2005;170(2):591–600. Pubmed Central PMCID: 1450417. Epub 2005/04/19. eng.
90. Liu C, Pouliot JJ, Nash HA. Repair of topoisomerase I covalent complexes in the absence of the tyrosyl-DNA phosphodiesterase Tdp1. Proc Natl Acad Sci U S A. 2002;99(23):14970–5. Pubmed Central PMCID: 137529. Epub 2002/10/25. eng.
91. Regairaz M, Zhang YW, Fu H, Agama KK, Tata N, Agrawal S, et al. Mus81-mediated DNA cleavage resolves replication forks stalled by topoisomerase I-DNA complexes. J Cell Biol. 2011;195(5):739–49. Pubmed Central PMCID: 3257568. Epub 2011/11/30. eng.
92. Pepe A, West SC. MUS81-EME2 promotes replication fork restart. Cell Rep. 2014; 7(4):1048–55. PubMed Epub 2014/05/13. eng.
93. McEachern MJ, Haber JE. Break-induced replication and recombinational telomere elongation in yeast. Annu Rev Biochem. 2006;75:111–35.
94. Lundblad V, Blackburn EH. An alternative pathway for yeast telomere maintenance rescues est1- senescence. Cell. 1993;73(2):347–60. Epub 1993/04/23. eng.
95. Teng SC, Zakian VA. Telomere-telomere recombination is an efficient bypass pathway for telomere maintenance in Saccharomyces cerevisiae. Mol Cell Biol. 1999;19(12):8083–93. Pubmed Central PMCID: 84893. Epub 1999/11/24. eng.
96. Wellinger RJ, Zakian VA. Everything you ever wanted to know about Saccharomyces cerevisiae telomeres: beginning to end. Genetics. 2012;191(4):1073–105. Pubmed Central PMCID: 3415994.
97. Larrivee M, Wellinger RJ. Telomerase- and capping-independent yeast survivors with alternate telomere states. Nat Cell Biol. 2006;8(7):741–7. PubMed Epub 2006/06/13. eng.
98. Chen Q, Ijpma A, Greider CW. Two survivor pathways that allow growth in the absence of telomerase are generated by distinct telomere recombination events. Mol Cell Biol. 2001;21(5):1819–27. Pubmed Central PMCID: 86745. Epub 2001/03/10. eng.
99. Le S, Moore JK, Haber JE, Greider CW. RAD50 and RAD51 define two pathways that collaborate to maintain telomeres in the absence of telomerase. Genetics. 1999;152(1):143–52. Pubmed Central PMCID: 1460580. Epub 1999/05/04. eng.

100. Hu Y, Tang HB, Liu NN, Tong XJ, Dang W, Duan YM, et al. Telomerase-null survivor screening identifies novel telomere recombination regulators. PLoS Genet. 2013;9(1), e1003208. Pubmed Central PMCID: 3547846.
101. Cesare AJ, Reddel RR. Alternative lengthening of telomeres: models, mechanisms and implications. Nat Rev Genet. 2010;11(5):319–30.
102. Londono-Vallejo JA, Der-Sarkissian H, Cazes L, Bacchetti S, Reddel RR. Alternative lengthening of telomeres is characterized by high rates of telomeric exchange. Cancer Res. 2004;64(7):2324–7.
103. Bechter OE, Zou Y, Walker W, Wright WE, Shay JW. Telomeric recombination in mismatch repair deficient human colon cancer cells after telomerase inhibition. Cancer Res. 2004;64(10):3444–51.
104. Dunham MA, Neumann AA, Fasching CL, Reddel RR. Telomere maintenance by recombination in human cells. Nat Genet. 2000;26(4):447–50.
105. Cho NW, Dilley RL, Lampson MA, Greenberg RA. Interchromosomal homology searches drive directional ALT telomere movement and synapsis. Cell. 2014;159(1):108–21. Pubmed Central PMCID: 4177039.
106. Doksani Y, Bermejo R, Fiorani S, Haber JE, Foiani M. Replicon dynamics, dormant origin firing, and terminal fork integrity after double-strand break formation. Cell. 2009;137(2): 247–58. Pubmed Central PMCID: 2789576.

Index

D.L. Kaplan (ed.), *The Initiation of DNA Replication in Eukaryotes*,
DOI 10.1007/978-3-319-24696-3

I

L

M

N

O

W

Y

MIX
Papier aus verantwortungsvollen Quellen
Paper from responsible sources
FSC® C105338

If you have any concerns about our products,
you can contact us on
ProductSafety@springernature.com

In case Publisher is established outside the EU,
the EU authorized representative is:
Springer Nature Customer Service Center GmbH
Europaplatz 3, 69115 Heidelberg, Germany

Printed by Libri Plureos GmbH
in Hamburg, Germany